Differential Geometrical Theory of Statistics

Special Issue Editors

Frédéric Barbaresco
Frank Nielsen

MDPI

Special Issue Editors
Frédéric Barbaresco
Thales Air Systems
France

Frank Nielsen
École Polytechnique
France
Sony Computer Science Laboratories Inc
Japan

Editorial Office
MDPI AG
St. Alban-Anlage 66
Basel, Switzerland

This edition is a reprint of the Special Issue published online in the open access journal *Entropy* (ISSN 1099-4300) from 2016–2017 (available at: http://www.mdpi.com/journal/entropy/special_issues/entropy-statistics).

For citation purposes, cite each article independently as indicated on the article page online and as indicated below:

Author 1; Author 2; Author 3 etc. Article title. *Journal Name*. **Year**. Article number/page range.

ISBN 978-3-03842-424-6 (Pbk)
ISBN 978-3-03842-425-3 (PDF)

Table of Contents

About the Guest Editors

Frédéric Barbaresco received his State Engineering degree from the French Grand Ecole CENTRALE-SUPELEC, Paris, France, in 1991. Since then, he has worked for the THALES Group where he is now Senior Expert in the Advanced Radar Concept Business Unit of Thales Air Systems and Representative at the Board of KTD PCC (Key Technology Domain: Processing, Cognition and Control) for the Global Business Unit Thales Land & Air Systems. He has been an Emeritus Member of SEE since 2011 and he was awarded the Aymé Poirson Prize (for application of sciences to industry) by the French Academy of Sciences in 2014, the SEE Ampere Medal in 2007, the Thévenin Prize in 2014 and the NATO SET Lecture Award in 2012. He is President of SEE Technical Club ISIC "Engineering of Information and Communications Systems" and a member of the SEE administrative board. He is member of the administrative board of SMAI and GRETSI. He was an invited lecturer for UNESCO on "Advanced School and Workshop on Matrix Geometries and Applications" in Trieste at the ITCP in June 2013. He is the General Co-chairman of the new international conference GSI "Geometric Sciences of Information". He was co-editor of MDPI Entropy Book "Information, Entropy and Their Geometric Structures". He has co-organized the CIRM seminar TGSI'17 "Topological and Geometrical Structures of Information".

Frank Nielsen received his PhD (1996) on computational geometry and his habilitation in visual computing (2006) from the university of Nice-Sophia Antipolis (France). After the french national service, he joined Sony CSL (Japan) in 1997. He is currently professor in the computer science department of Ecole polytechnique (France). He co-organizes with Frédéric Barbaresco the biannual Geometric Sciences of Information (GSI, gsi2017.org) conference, and is co-editor of the newly launched Springer journal of Information Geometry and an associate editor of MDPI Entropy. He edited several books including *Computational Information Geometry* (Springer, 2017), and wrote several textbooks including *Introduction to HPC with MPI for Data Science* (Springer, 2016). His research interests focus on computational information geometry with applications to machine learning and visual computing.

Preface to "Differential Geometrical Theory of Statistics"

This Special Issue "Differential Geometrical Theory of Statistics" collates selected invited and contributed talks presented during the conference GSI'15 on "Geometric Science of Information" which was held at the Ecole Polytechnique, Paris-Saclay Campus, France, in October 2015 (Conference web site: http://www.see.asso.fr/gsi2015).

Let us first start with a short historical review on the birth of the interplay of probability with geometry and computing, which is rooted in the 17th century.

1. Preamble: *Aleae Geometria*, the Geometry of Chance by Blaise Pascal

The "calculation of probabilities" began four years after the death of René Descartes, in 1654, in a correspondence between Blaise Pascal and Pierre Fermat. They exchanged letters on elementary problems of gambling, in this case a problem of dice and a problem of "parties". Pascal and Fermat were particularly interested by this problem and succeeded in "Party rule" by two different methods. One understands the legitimate pride of Pascal in his address of the same year at the Académie Parisienne created by Mersenne, to which he presented "the ripe fruit of our Geometry" (*"les fruits mûrs de notre Géométrie" in French*), an entirely new treaty about an absolutely unexplored matter, the distribution of chance in games. In the same way, Pascal, in his introduction to "Les Pensées", wrote that under the influence of what Méré has given to the game, he throws the bases of the calculation of probabilities and composes the Treatise of the Arithmetical Triangle. If Pascal appears at first sight as the initiator of the calculation of probabilities, looking a little closer, its role in the emergence of this theory is more complex. However, there is no trace of word probabilities in Pascal's work. To designate what might resemble what we now call calculation of probabilities, one does not even find the word in such a context. The only occurrences of probability are found in "Les

Provinciales" where he referred to the doctrine of the Jesuits, or in "Les Pensées". In Pascal's writings, we do not find the words of "Doctrine des chances" or "Calcul des chances", but only "Géométrie du hasard" (geometry of chance). In 1654, Blaise Pascal submitted a short paper to "Celeberrimae matheseos Academiae Parisiensi" (ancestor of the French Royal Academy of Sciences founded in 1666), with the title "***Aleae Geometria***" (Geometry of Chance) or "*De compositione aleae in ludis ipsi subjectis*", which was the seminal paper founding Probability as a new discipline in Science. In this paper, Pascal said "… et sic matheseos demonstrationes cum aleae incertitudine jugendo, et quae contraria videntur conciliando, ab utraque nominationem suam accipiens, stupendum hunc titulum jure sibi arrogat: Aleae Geometria" that we can translate as "*By the union thus realized between the demonstrations of mathematics and the uncertainty of chance, and by the conciliation of apparent contradictions, it can derive its name from both sides and arrogate to itself this astonishing title: Geometry of Chance*" (« *… par l'union ainsi réalisée entre les démonstrations des mathématiques et l'incertitude du hasard, et par la conciliation entre les contraires apparents, elle peut tirer son nom de part et d'autre et s'arroger à bon droit ce titre étonnant: Géométrie du Hasard* ». We can observe that Blaise Pascal attached a geometrical sense to probabilities in this seminal paper. Like Jacques Bernoulli, we can also provide references to another Blaise Pascal document entitled "*Art de penser*" (the "Logique" of Port-Royal), published the year of his death (1662), the last chapters of which contain elements on the calculus of probabilities applied to history, medicine, miracles, literary criticism, and life events, etc.

(¹) « **Novissima autem ac penitus intractatae materiae tractatio,** « scilicet de compositione aleae in ludis ipsi subjectis (quod « gallico nostro idiomate dicitur *faire les partis des jeux*) : ubi « anceps fortuna aequitate rationis ita reprimitur ut utrique « lusorum *quod jure competit* exacte semper assignetur. Quod « quidem eo fortius ratiocinando quaerendum, *quo minus tentando* « *investigari possit* : ambigui enim sortis eventus fortuitae contin- « gentiae potius quam naturali necessitati merito tribuuntur. Ideo « res hactenus erravit incerta; nunc autem quae *experimento* « *rebellis fuerat*, rationis dominium effugere non potuit : eam « quippe tanta securitate in artem per geometriam reduximus, « ut, certitudinis ejus particeps facta, jam audacter prodeat; et « sic, matheseos demonstrationes cum aleae incertitudine jungendo, « et quae contraria videntur conciliando, ab utraque nominationem « suam accipiens, stupendum hunc titulum jure sibi arrogat : « *aleae geometria*. » (*OEuvres de Pascal*, t. IV, p. 358 de l'édition de 1819.)

Figure 1. Blaise Pascal and His Seminal Text on « Aleae Geometria »

In "De l'esprit géométrique », the use of reason for knowledge is thought on a geometric model. In geometry, the first principles are given by the natural lights common to all men, and there is no need to define them. Other principles are clearly defined by definitions of names such that it is always possible to mentally substitute the definition for the defined. These definitions of names are completely free, the only condition to be respected is univocity and invariability. Judging his solution as one of his most important contributions to science, Pascal envisioned the drafting of a small treatise entitled "Géométrie du Hasard" (Geometry of Chance). He would never write it. Inspired by this, Christian Huygens wrote the first treatise on the calculation of chances, the "*De ratiociniis in ludo aleae*" ("On calculation in games of chance", 1657). We can conclude this preamble by observing that Blaise Pascal's seminal work on Probability was inspired by *Geometry*. The objective of this edited book is to come back to this initial idea that we can *geometrize statistics* in a rigorous way.

We can also make reference to Blaise Pascal for this book on computing geometrical statistics, because he was the inventor of the computer with his "Pascaline" machine. The introduction of Pascaline marks the beginning of the development of mechanical calculus in Europe. This development, which will traverse from the calculating machines to the electrical and electronic calculators of the following centuries, will culminate with the invention of the microprocessor. However, it was also Charles Babbage who conceived his analytical machine from 1834 to 1837, a programmable calculating machine which was the ancestor of the computers of the 1940s, combining the inventions of Blaise Pascal and Jacquard's machine, with instructions written on perforated cards. One of the descendants of the Pascaline, this was the first machine which performed with the intelligence of man.

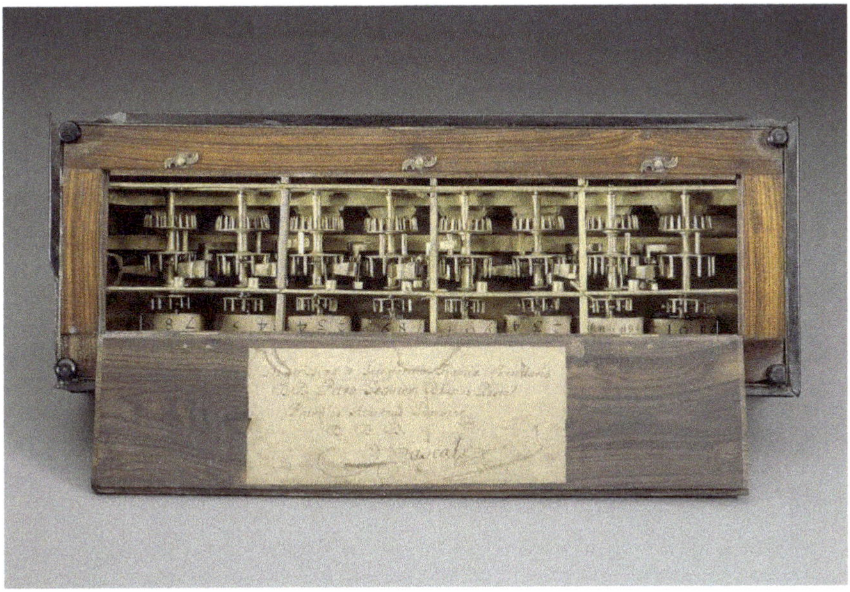

Figure 2. La « pascaline », Computing Machine, Blaise Pascal 1645

Before introducing the chapters of this book, let us recall that the modern birth of information geometry in the 20th century started with the differential-geometric modeling of parametric family of distributions in the pioneer work of Professor Harold Hotelling in 1929 and in Prodessor Maurice Fréchet Lecture at IHP (Institut Henri Poincaré, Paris) during Winter 1939.

Professor Hotelling spent half a year collaborating with Sir Ronald A. Fisher on setting the firm foundation of mathematical statistics in Rothamsted Research (UK) [20–22]. He submitted a groundbreaking note entitled "Spaces of Statistical Parameters" to the American Mathematical Society (AMS) meeting in 1929. Since he did not join the meeting, the note was nevertheless read by Prof. O. Ore. In this work, he introduced the Fisher information metric and the induced Riemannian geometry for modeling parametric family of distributions. C. R. Rao later independently introduced this geometric structure in his celebrated paper entitled "Information and the accuracy attainable in the estimation of statistical parameters" (1945). This paper is truly exceptional since it introduces three key results: (1) Cramér-Rao lower bound, (2) Riemannian geometry of statistical spaces, and (3) Rao-Blackwellization of estimators.

In 1943, Maurice Fréchet wrote a seminal paper (developing elements of his Winter 1939 Lecture at Institut Henri Poincaré in Paris) [23,24] introducing what was then called the Cramer-Rao bound. This paper contains in fact much more than this important discovery. In particular, Maurice Fréchet introduces more general notions relative to "distinguished functions", densities with estimator reaching the bound, defined with a function, solution of Clairaut's equation. The solutions "envelope of the Clairaut's equation" are related to standard Legendre transform and basic structures of Information Geometry. This Fréchet's analysis can also be revisited on the basis of Jean-Louis Koszul works as seminal foundation of "Information Geometry" based on Legendre-Clairaut equation.

We can also make references to De Moivre and Leibnitz contributions in seminal development of Probability [25–27] and give reference to papers written on History of probability [28–32].

We thank all the contributors of this edited book for further pushing the envelope of the geometrization of statistics in novel directions.

This edited book is organized in six chapters as follows:

Chapter I: Geometric Thermodynamics of Jean-Marie Souriau

This first chapter introduces and develops Jean-Marie Souriau's (1922-2012) model of Lie group thermodynamics and relativistic thermodynamics of continua. The contributions are listed below:

- **From Tools in Symplectic and Poisson Geometry to J.-M. Souriau's Theories of Statistical Mechanics and Thermodynamics** by Charles-Michel Marle
- **Geometric Theory of Heat from Souriau Lie Groups Thermodynamics and Koszul Hessian Geometry: Applications in Information Geometry for Exponential Families** by Frédéric Barbaresco
- **Link between Lie Group Statistical Mechanics and Thermodynamics of Continua** by Géry de Saxcé

Chapter II: Koszul-Vinberg Model of Hessian Information Geometry

The second chapter deals with Jean-Louis Koszul's model of Hessian Information Geometry based on Koszul-Vinberg's characteristic function and the homology theory of Koszul-Vinberg algebroids and their modules (KV homology). The two contributions are:

- **Foliations-Webs-Hessian Geometry-Information Geometry-Entropy and Cohomology** (IN MEMORIAM OF ALEXANDER GROTHENDIECK) by Michel Nguiffo Boyom
- **Explicit Formula of Koszul–Vinberg Characteristic Functions for a Wide Class of Regular Convex Cones** by Hideyuki Ishi

Chapter III: Divergence Geometry and Information Geometry

The third chapter develops new algorithms related to the area of divergence geometry (minimum divergence estimator, Rényi divergence) and Information Geometry: Mixture of densities, Expectations on q-Exponential Family, Sparse Goodness-of-Fit Testing.

The five contributions are:

- **A Proximal Point Algorithm for Minimum Divergence Estimators with Application to Mixture Models** by Diaa Al Mohamad and Michel Broniatowski
- **Geometry Induced by a Generalization of Rényi Divergence** by David C. de Souza, Rui F. Vigelis and Charles C. Cavalcante

- **Guaranteed Bounds on Information-Theoretic Measures of Univariate Mixtures Using Piecewise Log-Sum-Exp Inequalities** by Frank Nielsen and Ke Sun
- **A Sequence of Escort Distributions and Generalizations of Expectations on q-Exponential Family** by Hiroshi Matsuzoe
- **The Information Geometry of Sparse Goodness-of-Fit Testing** by Paul Marriott, Radka Sabolová, Germain Van Bever and Frank Critchley

Chapter IV : Density of Probability on manifold and metric space

The fourth Chapter proposes new approaches to estimate parametric and non-parametric probabilities densities for structured covariance matrices (Toeplitz and Block-Toeplitz Hermitian Positive Definite Matrix, Symmetric Positive Definite Matrix).

- **Kernel Density Estimation on the Siegel Space with an Application to Radar Processing** by Emmanuel Chevallier, Thibault Forget, Frédéric Barbaresco and Jesus Angulo
- **Riemannian Laplace Distribution on the Space of Symmetric Positive Definite Matrices** by Hatem Hajri, Ioana Ilea, Salem Said, Lionel Bombrun and Yannick Berthoumieu

Chapter V: Statistics on Paths and on Riemannian Manifolds

The fifth chapter describes new methods to introduce statistical tools for paths and for data on Riemannian manifolds, with the following three contributions:

- **Entropy Minimizing Curves with Application to Flight Path Design and Clustering** by Stéphane Puechmorel and Florence Nicol
- **Anisotropically Weighted and Nonholonomically Constrained Evolutions on Manifolds** by Stefan Sommer
- **Non-Asymptotic Confidence Sets for Circular Means** by Thomas Hotz, Florian Kelma and Johannes Wieditz

Chapter VI: Entropy and Complexity in Linguistic

The sixth chapter concludes this edited book with new perspectives for defining topological structures, entropy and complexity in linguistics with the following contribution:

- **Syntactic Parameters and a Coding Theory Perspective on Entropy and Complexity of Language Families** by Matilde Marcolli

Frédéric Barbaresco and Frank Nielsen
Guest Editors

References

1. BARBARESCO, F. & DJAFARI, A., "Information, Entropy and Their Geometric Structures", MDPI Entropy, September 2015; http://www.mdpi.com/books/pdfview/book/127
2. BAYES, Th., «An essay towards solving a problem in the doctrine of chance», Philosophical Transactions of the Royal Society of London, 53 (1763), trad. J.-P. Cléro, Cahiers d'histoire et de philosophie des sciences, n° 18, 1988.
3. BERNOULLI, J., Ars conjectandi (1713), die Werke von Jakob Bernoulli, 3 vols., Basel, 1969-1975.
4. BYRNE, E., Probability and Opinion: A Study in the Medieval Pre-suppositions of Post-Medieval Theories of probability, La Haye, Martinus Nijhoff, 1968.
5. CARDANO, De ludo aleae (ca. 1520), Opera Omnia, 10 vols., Stuttgart, 1966.

6. CARDANO, The Book on Games of Chance, trad. S. H. Gould, New York, 1961.
7. DASTON, L., Probability in the Enlightenment, Princeton, 1988.
8. DAVID, F. N., Games, Gods and Gambling, A History of Probability and Statistical Ideas, London, Charles Griffin & Co, 1962.
9. ABOUT, P.J., BOY, M., «La correspondance de Blaise Pascal et de Pierre de Fermat», Cahiers de Fontenay, n° 32, p. 59-73.
10. DAVIDSON, H. M., Pascal and the Arts of the Mind, Cambridge, Cambridge University Press, 1993.
11. EDWARDS, A. W. F., «Pascal and the Problem of Points», International Statistical Review, t. 51, 1983, p. 259-266.
12. EDWARDS, A. W. F., «Pascal's Problem: The Gambler's Ruin», International Statistical Review, t. 50, 1982, p. 73-79.
13. GODFROY-GÉNIN A.S., Pascal la Géométrie du Hasard, Math. & Sci. hum., (38e année, n° 150, 2000, p. 7-39
14. ORE, O., Cardano, the gambling scholar, Princeton, 1953.
15. «Pascal et les probabilités», Cahiers Pédagogiques de philosophie et d'histoire des mathématiques, fascicule 4, IREM et CRDP de Rouen, 1993.
16. TODHUNTER, I., A History of Mathematical Theory of Probability from the Time of Pascal to that of Laplace, Cambridge et Londres, Macmillan, 1865.
17. PASCAL, B., Les Provinciales, Paris, Le Guern éd., 1987.
18. PASCAL, B., Oeuvres complètes, J. Mesnard éd., 4 volumes publiés, 1964-1970.
19. PASCAL, B., Pensées de Pascal, Paris, Ph. Sellier éd., 1991.
20. HACKING, I., The Emergence of Probability, Cambridge, 1975.
21. KENDALL, M. G., PEARSON, E. S., (eds.)., Studies in the History of Statistics and Probability, 2 vols., London, 1970-1977.
22. PEARSON, K., The History of Statistics in the 17th and 18th Centuries, London, E.S. Pearson, ed., 1978.
23. BARBARESCO, F., "Les densités de probabilité « distinguées » et l'équation d'Alexis Clairaut: regards croisés de Maurice Fréchet et de Jean Louis Koszul », Colloque GRETSI'17, Juan-Les-Pins-September 2017
24. FRECHET M., Sur l'extension de certaines évaluations statistiques au cas de petits échantillons. Revue de l'Institut International de Statistique 1943, vol. 11, n° 3/4, pp. 182–205.
25. DE MOIVRE, A., The Doctrine of Chances, 3rd edition, London, 1756.
26. LEIBNIZ, G. W., «Nouveaux essais sur l'entendement humain», Sämtliche Schriften und Briefe, Berlin, 1962-1980, réed. Garnier-Flammarion, 1966.
27. LEIBNIZ, G. W., Opuscules et fragments inédits, Couturat, ed., Paris, 1961.
28. SCHNEIDER, I., «Why do we find the origin of a calculus of probabilities in the seventeenth century ?», Pisa Conference Proceedings, vol. 2, Dordrecht and Boston, J.Hintikka, D. Gruender, E. Agazzi eds., 1980.
29. SCHNEIDER, I., Die Entwicklung des Wahrscheinlichkeitsbegriff in des Mathematik von Pascal bis Laplace, Munich, 1972.
30. SHEYNIN, O., «On the early history of the law of large numbers», Studies in the History of Statistics and Probability, vol. 1, Paerson and Kendall eds., 1970.
31. SHEYNIN, O., «On the prehistory of the theory of probability», Archives for History of Exact Sciences 12, 1974.
32. STIGLER, S., The History of Statistics: The measurement of Uncertainety Before 1900, Cambridge (Mass.), The Belknap Press of Harvard University Press, 1986.

Chapter 1: Geometric Thermodynamic of Jean-Marie Souriau

entropy MDPI

Article

From Tools in Symplectic and Poisson Geometry to J.-M. Souriau's Theories of Statistical Mechanics and Thermodynamics †

Charles-Michel Marle

Institut de Mathématiques de Jussieu, Université Pierre et Marie Curie, 4, Place Jussieu, 75252 Paris Cedex 05, France; charles-michel.marle@math.cnrs.fr

† In memory of Jean-Marie Souriau (1922–2012).

Academic Editors: Frédéric Barbaresco and Frank Nielsen
Received: 28 July 2016; Accepted: 5 October 2016; Published: 19 October 2016

Abstract: I present in this paper some tools in symplectic and Poisson geometry in view of their applications in geometric mechanics and mathematical physics. After a short discussion of the Lagrangian an Hamiltonian formalisms, including the use of symmetry groups, and a presentation of the Tulczyjew's isomorphisms (which explain some aspects of the relations between these formalisms), I explain the concept of manifold of motions of a mechanical system and its use, due to J.-M. Souriau, in statistical mechanics and thermodynamics. The generalization of the notion of thermodynamic equilibrium in which the one-dimensional group of time translations is replaced by a multi-dimensional, maybe non-commutative Lie group, is fully discussed and examples of applications in physics are given.

Keywords: Lagrangian formalism; Hamiltonian formalism; symplectic manifolds; Poisson structures; symmetry groups; momentum maps; thermodynamic equilibria; generalized Gibbs states

---

1. Introduction

1.1. Contents of the Paper, Sources and Further Reading

This paper presents tools in symplectic and Poisson geometry in view of their application in geometric mechanics and mathematical physics. The Lagrangian formalism and symmetries of Lagrangian systems are discussed in Sections 2 and 3, the Hamiltonian formalism and symmetries of Hamiltonian systems in Sections 4 and 5. Section 6 introduces the concepts of Gibbs state and of thermodynamic equilibrium of a mechanical system, and presents several examples. For a monoatomic classical ideal gas, eventually in a gravity field, or a monoatomic relativistic gas the Maxwell–Boltzmann and Maxwell–Jüttner probability distributions are derived. The Dulong and Petit law which governs the specific heat of solids is obtained. Finally Section 7 presents the generalization of the concept of Gibbs state, due to Jean-Marie Souriau, in which the group of time translations is replaced by a (multi-dimensional and eventually non-Abelian) Lie group.

Several books [1–11] discuss, much more fully than in the present paper, the contents of Sections 2–5. The interested reader is referred to these books for detailed proofs of results whose proofs are only briefly sketched here. The recent paper [12] contains detailed proofs of most results presented here in Sections 4 and 5.

The main sources used for Sections 6 and 7 are the book and papers by Jean-Marie Souriau [13–17] and the beautiful small book by Mackey [18].

The Euler–Poincaré equation, which is presented with Lagrangian symmetries at the end of Section 3, is not really related to symmetries of a Lagrangian system, since the Lie algebra which acts on the configuration space of the system is not a Lie algebra of symmetries of the Lagrangian. Moreover in its intrinsic form that equation uses the concept of Hamiltonian momentum map presented later, in Section 5. Since the Euler–Poincaré equation is not used in the following sections, the reader can skip the corresponding subsection at his or her first reading.

1.2. Notations

The notations used are more or less those generally used now in differential geometry. The tangent and cotangent bundles to a smooth manifold M are denoted by TM and T^*M, respectively, and their canonical projections by $\tau_M : TM \to M$ and $\pi_M : T^*M \to M$. The vector spaces of k-multivectors and k-forms on M are denoted by $A^k(M)$ and $\Omega^k(M)$, respectively, with $k \in \mathbb{Z}$ and, of course, $A^k(M) = \{0\}$ and $\Omega^k(M) = \{0\}$ if $k < 0$ and if $k > \dim M$, k-multivectors and k-forms being skew-symmetric. The exterior algebras of multivectors and forms of all degrees are denoted by $A(M) = \oplus_k A^k(M)$ and $\Omega(M) = \oplus_k \Omega^k(M)$, respectively. The exterior differentiation operator of differential forms on a smooth manifold M is denoted by $\mathrm{d} : \Omega(M) \to \Omega(M)$. The interior product of a differential form $\eta \in \Omega(M)$ by a vector field $X \in A^1(M)$ is denoted by $\mathrm{i}(X)\eta$.

Let $f : M \to N$ be a smooth map defined on a smooth manifold M, with values in another smooth manifold N. The pull-back of a form $\eta \in \Omega(N)$ by a smooth map $f : M \to N$ is denoted by $f^*\eta \in \Omega(M)$.

A smooth, time-dependent vector field on the smooth manifold M is a smooth map $X : \mathbb{R} \times M \to TM$ such that, for each $t \in \mathbb{R}$ and $x \in M$, $X(t,x) \in T_xM$, the vector space tangent to M at x. When, for any $x \in M$, $X(t,x)$ does not depend on $t \in \mathbb{R}$, X is a smooth vector field in the usual sense, i.e., an element in $A^1(M)$. Of course a time-dependent vector field can be defined on an open subset of $\mathbb{R} \times M$ instead than on the whole $\mathbb{R} \times M$. It defines a differential equation

$$\frac{\mathrm{d}\varphi(t)}{\mathrm{d}t} = X\big(t, \varphi(t)\big), \tag{1}$$

said to be *associated* to X. The (full) *flow* of X is the map Ψ^X, defined on an open subset of $\mathbb{R} \times \mathbb{R} \times M$, taking its values in M, such that for each $t_0 \in \mathbb{R}$ and $x_0 \in M$ the parametrized curve $t \mapsto \Psi^X(t, t_0, x_0)$ is the maximal integral curve of Equation (1) satisfying $\Psi(t_0, t_0, x_0) = x_0$. When t_0 and $t \in \mathbb{R}$ are fixed, the map $x_0 \mapsto \Psi^X(t, t_0, x_0)$ is a diffeomorphism, defined on an open subset of M (which may be empty) and taking its values in another open subset of M, denoted by $\Psi^X_{(t,t_0)}$. When X is in fact a vector field in the usual sense (not dependent on time), $\Psi^X_{(t,t_0)}$ only depends on $t - t_0$. Instead of the full flow of X we can use its *reduced flow* Φ^X, defined on an open subset of $\mathbb{R} \times M$ and taking its values in M, related to the full flow Ψ^X by

$$\Phi^X(t, x_0) = \Psi^X(t, 0, x_0), \quad \Psi^X(t, t_0, x_0) = \Phi^X(t - t_0, x_0).$$

For each $t \in \mathbb{R}$, the map $x_0 \mapsto \Phi^X(t, x_0) = \Psi^X(t, 0, x_0)$ is a diffeomorphism, denoted by Φ^X_t, defined on an open subset of M (which may be empty) onto another open subset of M.

When $f : M \to N$ is a smooth map defined on a smooth manifold M, with values in another smooth manifold N, there exists a smooth map $Tf : TM \to TN$ called the *prolongation of f to vectors*, which for each fixed $x \in M$ linearly maps T_xM into $T_{f(x)}N$. When f is a diffeomorphism of M onto N, Tf is an isomorphism of TM onto TN. That property allows us to define the *canonical lifts* of a vector field X in $A^1(M)$ to the tangent bundle TM and to the cotangent bundle T^*M. Indeed, for each $t \in \mathbb{R}$, Φ^X_t is a diffeomorphism of an open subset of M onto another open subset of M. Therefore $T\Phi^X_t$ is a diffeomorphism of an open subset of TM onto another open subset of TM. It turns out that when t takes all possible values in $\mathbb{R}$ the set of all diffeomorphisms $T\Phi^X_t$ is the reduced flow of a vector field $\overline{X}$ on TM, which is the *canonical lift* of X to the tangent bundle TM.

Similarly, the transpose $(T\Phi^X_{-t})^T$ of $T\Phi^X_{-t}$ is a diffeomorphism of an open subset of the cotangent bundle T^*M onto another open subset of T^*M, and when t takes all possible values in $\mathbb{R}$ the set of all diffeomorphisms $(T\Phi^X_{-t})^T$ is the reduced flow of a vector field $\widehat{X}$ on T^*M, which is the *canonical lift* of X to the cotangent bundle T^*M.

The canonical lifts of a vector field to the tangent and cotangent bundles are used in Sections 3 and 5. They can be defined too for time-dependent vector fields.

2. The Lagrangian Formalism

2.1. The Configuration Space and the Space of Kinematic States

The principles of mechanics were stated by the great English mathematician *Isaac Newton* (1642–1727) in his book *Philosophia Naturalis Principia Mathematica* published in 1687 [19]. On this basis, a little more than a century later, *Joseph Louis Lagrange* (1736–1813) in his book *Mécanique analytique* [20] derived the equations (today known as the *Euler–Lagrange equations*) which govern the motion of a mechanical system made of any number of material points or rigid material bodies interacting between them by very general forces, and eventually submitted to external forces.

In modern mathematical language, these equations are written on the *configuration space* and on the *space of kinematic states* of the considered mechanical system. The *configuration space* is a smooth n-dimensional manifold N whose elements are all the possible configurations of the system (a configuration being the position in space of all parts of the system). The *space of kinematic states* is the tangent bundle TN to the configuration space, which is $2n$-dimensional. Each element of the space of kinematic states is a vector tangent to the configuration space at one of its elements, i.e., at a configuration of the mechanical system, which describes the velocity at which this configuration changes with time. In local coordinates a configuration of the system is determined by the n coordinates $x^1, \dots, x^n$ of a point in N, and a kinematic state by the $2n$ coordinates $x^1, \dots, x^n, v^1, \dots v^n$ of a vector tangent to N at some element in N.

2.2. The Euler–Lagrange Equations

When the mechanical system is *conservative*, the Euler–Lagrange equations involve a single real valued function L called the *Lagrangian* of the system, defined on the product of the real line $\mathbb{R}$ (spanned by the variable t representing the time) with the manifold TN of *kinematic states* of the system. In local coordinates, the Lagrangian L is expressed as a function of the $2n+1$ variables, $t, x^1, \dots, x^n, v^1, \dots, v^n$ and the Euler–Lagrange equations have the remarkably simple form

$$\frac{\mathrm{d}}{\mathrm{d}t}\left(\frac{\partial L}{\partial v^i}\big(t, x(t), v(t)\big)\right) - \frac{\partial L}{\partial x^i}\big(t, x(t), v(t)\big) = 0\,, \quad 1 \leq i \leq n\,,$$

where $x(t)$ stands for $x^1(t), \dots, x^n(t)$ and $v(t)$ for $v^1(t), \dots, v^n(t)$ with, of course,

$$v^i(t) = \frac{\mathrm{d}x^i(t)}{\mathrm{d}t}\,, \quad 1 \leq i \leq n\,.$$

2.3. Hamilton's Principle of Stationary Action

The great Irish mathematician *William Rowan Hamilton* (1805–1865) observed [21,22] that the Euler–Lagrange equations can be obtained by applying the standard techniques of *Calculus of Variations*, due to *Leonhard Euler* (1707–1783) and *Joseph Louis Lagrange*, to the *action integral* (Lagrange observed that fact before Hamilton, but in the last edition of his book he chose to derive the Euler–Lagrange equations by application of the *principle of virtual works*, using a very clever evaluation of the virtual work of inertial forces for a smooth infinitesimal variation of the motion).

$$I_L(\gamma) = \int_{t_0}^{t_1} L\big(t, x(t), v(t)\big)\,\mathrm{d}t\,, \quad \text{with } v(t) = \frac{\mathrm{d}x(t)}{\mathrm{d}t}\,,$$

where $\gamma : [t_0, t_1] \to N$ is a smooth curve in N parametrized by the time t. These equations express the fact that the action integral $I_L(\gamma)$ is *stationary* with respect to any smooth infinitesimal variation of γ with fixed end-points $(t_0, \gamma(t_0))$ and $(t_1, \gamma(t_1))$. This fact is today called *Hamilton's principle of stationary action*. The reader interested in Calculus of Variations and its applications in mechanics and physics is referred to the books [23–25].

2.4. The Euler-Cartan Theorem

The *Lagrangian formalism* is the use of Hamilton's principle of stationary action for the derivation of the equations of motion of a system. It is widely used in mathematical physics, often with more general Lagrangians involving more than one independent variable and higher order partial derivatives of dependent variables. For simplicity I will consider here only the Lagrangians of (maybe time-dependent) conservative mechanical systems.

An intrinsic geometric expression of the Euler–Lagrange equations, wich does not use local coordinates, was obtained by the great French mathematician *Élie Cartan* (1869–1951). Let us introduce the concepts used by the statement of this theorem.

Definition 1. *Let N be the configuration space of a mechanical system and let its tangent bundle TN be the space of kinematic states of that system. We assume that the evolution with time of the state of the system is governed by the Euler–Lagrange equations for a smooth, maybe time-dependent Lagrangian $L : \mathbb{R} \times TN \to \mathbb{R}$.*

1. *The cotangent bundle T^*N is called the phase space of the system.*
2. *The map $\mathcal{L}_L : \mathbb{R} \times TN \to T^*N$*

$$\mathcal{L}_L(t, v) = \mathrm{d}_{\mathrm{vert}} L(t, v) , \quad t \in \mathbb{R} , \, v \in TN ,$$

   where $\mathrm{d}_{\mathrm{vert}} L(t, v)$ is the vertical differential of L at (t, v), i.e., the differential at v of the the map $w \mapsto L(t, w)$, with $w \in \tau_N^{-1}(\tau_N(v))$, is called the Legendre map associated to L.
3. *The map $E_L : \mathbb{R} \times TN \to \mathbb{R}$ given by*

$$E_L(t, v) = \langle \mathcal{L}_L(t, v), v \rangle - L(t, v) , \quad t \in \mathbb{R} , \, v \in TN ,$$

   is called the the energy function associated to L.
4. *The 1-form on $\mathbb{R} \times TN$*

$$\widehat{\varpi}_L = \mathcal{L}_L^* \theta_N - E_L(t, v) \mathrm{d}t ,$$

   where θ_N is the Liouville 1-form on T^*N, is called the Euler–Poincaré 1-form.

Theorem 1 (Euler-Cartan Theorem). *A smooth curve $\gamma : [t_0, t_1] \to N$ parametrized by the time $t \in [t_0, t_1]$ is a solution of the Euler–Lagrange equations if and only if, for each $t \in [t_0, t_1]$ the derivative with respect to t of the map $t \mapsto \left(t, \dfrac{\mathrm{d}\gamma(t)}{\mathrm{d}t} \right)$ belongs to the kernel of the 2-form $\mathrm{d}\widehat{\varpi}_L$, in other words if and only if*

$$\mathrm{i} \left(\frac{\mathrm{d}}{\mathrm{d}t} \left(t, \frac{\mathrm{d}\gamma(t)}{\mathrm{d}t} \right) \right) \mathrm{d}\widehat{\varpi}_L \left(t, \frac{\mathrm{d}\gamma(t)}{\mathrm{d}t} \right) = 0 .$$

The interested reader will find the proof of that theorem in [26], (Theorem 2.2, Chapter IV, p. 262) or, for hyper-regular Lagrangians (an additional assumption which in fact, is not necessary) in [27], Chapter IV, Theorem 2.1, p. 167.

Remark 1. *In his book [14], Jean-Marie Souriau uses a slightly different terminology: for him the odd-dimensional space $\mathbb{R} \times TN$ is the evolution space of the system, and the exact 2-form $\mathrm{d}\widehat{\varpi}_L$ on that space is the Lagrange form. He defines that 2-form in a setting more general than that of the Lagrangian formalism.*

3. Lagrangian Symmetries

3.1. Assumptions and Notations

In this section N is the configuration space of a conservative Lagrangian mechanical system with a smooth, maybe time dependent Lagrangian $L : \mathbb{R} \times TN \to \mathbb{R}$. Let $\widehat{\varpi}_L$ be the Poincaré-Cartan 1-form on the evolution space $\mathbb{R} \times TN$.

Several kinds of symmetries can be defined for such a system. Very often, they are special cases of *infinitesimal symmetries of the Poincaré-Cartan form*, which play an important part in the famous *Noether theorem*.

Definition 2. *An infinitesimal symmetry of the Poincaré-Cartan form $\widehat{\varpi}_L$ is a vector field Z on $\mathbb{R} \times TN$ such that*

$$\mathcal{L}(Z)\widehat{\varpi}_L = 0\,,$$

$\mathcal{L}(Z)$ denoting the Lie derivative of differential forms with respect to Z.

Example 1.

1. *Let us assume that the Lagrangian L does not depend on the time $t \in \mathbb{R}$, i.e., is a smooth function on TN. The vector field on $\mathbb{R} \times TN$ denoted by $\dfrac{\partial}{\partial t}$, whose projection on $\mathbb{R}$ is equal to 1 and whose projection on TN is 0, is an infinitesimal symmetry of $\widehat{\varpi}_L$.*
2. *Let X be a smooth vector field on N and $\overline{X}$ be its canonical lift to the tangent bundle TN. We still assume that L does not depend on the time t. Moreover we assume that $\overline{X}$ is an infinitesimal symmetry of the Lagrangian L, i.e., that $\mathcal{L}(\overline{X})L = 0$. Considered as a vector field on $\mathbb{R} \times TN$ whose projection on the factor $\mathbb{R}$ is 0, $\overline{X}$ is an infinitesimal symmetry of $\widehat{\varpi}_L$.*

3.2. The Noether Theorem in Lagrangian Formalism

Theorem 2 (E. Noether's Theorem in Lagrangian Formalism). *Let Z be an infinitesimal symmetry of the Poincaré-Cartan form $\widehat{\varpi}_L$. For each possible motion $\gamma : [t_0, t_1] \to N$ of the Lagrangian system, the function $\mathrm{i}(Z)\widehat{\varpi}_L$, defined on $\mathbb{R} \times TN$, keeps a constant value along the parametrized curve $t \mapsto \left(t, \dfrac{\mathrm{d}\gamma(t)}{\mathrm{d}t}\right)$.*

Proof. Let $\gamma : [t_0, t_1] \to N$ be a motion of the Lagrangian system, i.e., a solution of the Euler–Lagrange equations. The Euler-Cartan Theorem 1 proves that, for any $t \in [t_0, t_1]$,

$$\mathrm{i}\left(\frac{\mathrm{d}}{\mathrm{d}t}\left(t, \frac{\mathrm{d}\gamma(t)}{\mathrm{d}t}\right)\right)\mathrm{d}\widehat{\varpi}_L\left(t, \frac{\mathrm{d}\gamma(t)}{\mathrm{d}t}\right) = 0\,.$$

Since Z is an infinitesimal symmetry of $\widehat{\varpi}_L$,

$$\mathcal{L}(Z)\widehat{\varpi}_L = 0\,.$$

Using the well known formula relating the Lie derivative, the interior product and the exterior derivative

$$\mathcal{L}(Z) = \mathrm{i}(Z) \circ \mathrm{d} + \mathrm{d} \circ \mathrm{i}(Z)$$

we can write

$$\frac{\mathrm{d}}{\mathrm{d}t}\left(\mathrm{i}(Z)\widetilde{\varpi}_L\left(t,\frac{\mathrm{d}\gamma(t)}{\mathrm{d}t}\right)\right) = \left\langle \mathrm{d}\mathrm{i}(Z)\widehat{\varpi}_L, \frac{\mathrm{d}}{\mathrm{d}t}\left(t,\frac{\mathrm{d}\gamma(t)}{\mathrm{d}t}\right)\right\rangle$$
$$= -\left\langle \mathrm{i}(Z)\mathrm{d}\widehat{\varpi}_L, \frac{\mathrm{d}}{\mathrm{d}t}\left(t,\frac{\mathrm{d}\gamma(t)}{\mathrm{d}t}\right)\right\rangle$$
$$= 0.$$

□

Example 2. *When the Lagrangian L does not depend on time, application of Emmy Noether's theorem to the vector field* $\dfrac{\partial}{\partial t}$ *shows that the energy* E_L *remains constant during any possible motion of the system, since* $\mathrm{i}\left(\dfrac{\partial}{\partial t}\right)\widehat{\varpi}_L = -E_L$.

Remark 2.

1. *Theorem 2 is due to the German mathematician Emmy Noether (1882–1935), who proved it under much more general assumptions than those used here. For a very nice presentation of Emmy Noether's theorems in a much more general setting and their applications in mathematical physics, interested readers are referred to the very nice book by Yvette Kosmann-Schwarzbach [28].*
2. *Several generalizations of the Noether theorem exist. For example, if instead of being an infinitesimal symmetry of* $\widehat{\varpi}_L$*, i.e., instead of satisfying* $\mathcal{L}(Z)\widehat{\varpi}_L = 0$ *the vector field Z satisfies*

$$\mathcal{L}(Z)\widehat{\varpi}_L = \mathrm{d}f\,,$$

where $f : \mathbb{R}\times TM \to \mathbb{R}$ *is a smooth function, which implies of course* $\mathcal{L}(Z)(\mathrm{d}\widehat{\varpi}_L) = 0$*, the function*

$$\mathrm{i}(Z)\widehat{\varpi}_L - f$$

keeps a constant value along $t \mapsto \left(t, \dfrac{\mathrm{d}\gamma(t)}{\mathrm{d}t}\right)$.

3.3. *The Lagrangian Momentum Map*

The Lie bracket of two infinitesimal symmetries of $\widehat{\varpi}_L$ is too an infinitesimal symmetry of $\widehat{\varpi}_L$. Let us therefore assume that there exists a finite-dimensional Lie algebra of vector fields on $\mathbb{R}\times TN$ whose elements are infinitesimal symmetries of $\widehat{\varpi}_L$.

Definition 3. *Let* $\psi : \mathcal{G} \to A^1(\mathbb{R}\times TN)$ *be a Lie algebras homomorphism of a finite-dimensional real Lie algebra* $\mathcal{G}$ *into the Lie algebra of smooth vector fields on* $\mathbb{R}\times TN$ *such that, for each* $X \in \mathcal{G}$, $\psi(X)$ *is an infinitesimal symmetry of* $\widehat{\varpi}_L$*. The Lie algebras homomorphism* ψ *is said to be a Lie algebra action on* $\mathbb{R}\times TN$ *by infinitesimal symmetries of* $\widehat{\varpi}_L$*. The map* $K_L : \mathbb{R}\times TN \to \mathcal{G}^*$*, which takes its values in the dual* $\mathcal{G}^*$ *of the Lie algebra* $\mathcal{G}$*, defined by*

$$\langle K_L(t,v), X\rangle = \mathrm{i}(\psi(X))\widehat{\varpi}_L(t,v)\,,\quad X\in\mathcal{G}\,,\quad (t,v)\in\mathbb{R}\times TN\,,$$

is called the Lagrangian momentum of the Lie algebra action ψ.

Corollary 1 (of E. Noether's Theorem). *Let* $\psi : \mathcal{G} \to A^1(\mathbb{R}\times TM)$ *be an action of a finite-dimensional real Lie algebra* $\mathcal{G}$ *on the evolution space* $\mathbb{R}\times TN$ *of a conservative Lagrangian system, by infinitesimal symmetries of the Poincaré-Cartan form* $\widehat{\varpi}_L$*. For each possible motion* $\gamma : [t_0, t_1] \to N$ *of that system, the Lagrangian momentum map* K_L *keeps a constant value along the parametrized curve* $t \mapsto \left(t, \dfrac{\mathrm{d}\gamma(t)}{\mathrm{d}t}\right)$.

Proof. Since for each $X \in \mathcal{G}$ the function $(t,v) \mapsto \langle K_L(t,v), X\rangle$ keeps a constant value along the parametrized curve $t \mapsto \left(t, \dfrac{d\gamma(t)}{dt}\right)$, the map K_L itself keeps a constant value along that parametrized curve. □

Example 3. *Let us assume that the Lagrangian L does not depend explicitly on the time t and is invariant by the canonical lift to the tangent bundle of the action on N of the six-dimensional group of Euclidean diplacements (rotations and translations) of the physical space. The corresponding infinitesimal action of the Lie algebra of infinitesimal Euclidean displacements (considered as an action on $\mathbb{R} \times TN$, the action on the factor $\mathbb{R}$ being trivial) is an action by infinitesimal symmetries of $\widehat{\omega}_L$. The six components of the Lagrangian momentum map are the three components of the total linear momentum and the three components of the total angular momentum.*

Remark 3. *These results are valid without any assumption of hyper-regularity of the Lagrangian.*

3.4. The Euler–Poincaré Equation

In a short Note [29] published in 1901, the great french mathematician *Henri Poincaré* (1854–1912) proposed a new formulation of the equations of mechanics.

Let N be the configuration manifold of a conservative Lagrangian system, with a smooth Lagrangian $L : TN \to \mathbb{R}$ which does not depend explicitly on time. Poincaré assumes that there exists an homomorphism ψ of a finite-dimensional real Lie algebra $\mathcal{G}$ into the Lie algebra $A^1(N)$ of smooth vector fields on N, such that for each $x \in N$, the values at x of the vector fields $\psi(X)$, when X varies in $\mathcal{G}$, completely fill the tangent space T_xN. The action ψ is then said to be *locally transitive.*

Of course these assumptions imply $\dim \mathcal{G} \geq \dim N$.

Under these assumptions, Henri Poincaré proved that the equations of motion of the Lagrangian system could be written on $N \times \mathcal{G}$ or on $N \times \mathcal{G}^*$, where $\mathcal{G}^*$ is the dual of the Lie algebra $\mathcal{G}$, instead of on the tangent bundle TN. When $\dim \mathcal{G} = \dim N$ (which can occur only when the tangent bundle TN is trivial) the obtained equation, called the *Euler–Poincaré equation,* is *perfectly equivalent to the Euler–Lagrange equations* and may, in certain cases, be easier to use. But when $\dim \mathcal{G} > \dim N$, the system made by the Euler–Poincaré equation is *underdetermined.*

Let $\gamma : [t_0, t_1] \to N$ be a smooth parametrized curve in N. Poincaré proves that there exists a smooth curve $V : [t_0, t_1] \to \mathcal{G}$ in the Lie algebra $\mathcal{G}$ such that, for each $t \in [t_0, t_1]$,

$$\psi\big(V(t)\big)\big(\gamma(t)\big) = \frac{d\gamma(t)}{dt}. \tag{2}$$

When $\dim \mathcal{G} > \dim N$ the smooth curve V in $\mathcal{G}$ is not uniquely determined by the smooth curve γ in N. However, instead of writing the second-order Euler–Lagrange differential equations on TN satisfied by γ when this curve is a possible motion of the Lagrangian system, Poincaré derives a *first order differential equation for the curve V* and proves that it is satisfied, together with Equation (2), *if and only if γ is a possible motion of the Lagrangian system.*

Let $\varphi : N \times \mathcal{G} \to TN$ and $\overline{L} : N \times \mathcal{G} \to \mathbb{R}$ be the maps

$$\varphi(x, X) = \psi(X)(x), \quad \overline{L}(x, X) = L \circ \varphi(x, X).$$

We denote by $d_1\overline{L} : N \times \mathcal{G} \to T^*N$ and by $d_2\overline{L} : N \times \mathcal{G} \to \mathcal{G}^*$ the partial differentials of $\overline{L} : N \times \mathcal{G} \to \mathbb{R}$ with respect to its first variable $x \in N$ and with respect to its second variable $X \in \mathcal{G}$.

The map $\varphi : N \times \mathcal{G} \to TN$ is a *surjective vector bundles morphism* of the trivial vector bundle $N \times \mathcal{G}$ into the tangent bundle TN. Its *transpose* $\varphi^T : T^*N \to N \times \mathcal{G}^*$ is therefore an *injective vector bundles morphism,* which can be written

$$\varphi^T(\xi) = \big(\pi_N(\xi), J(\xi)\big),$$

where $\pi_N : T^*N \to N$ is the canonical projection of the cotangent bundle and $J : T^*N \to \mathcal{G}^*$ is a smooth map whose restriction to each fibre $T^*_x N$ of the cotangent bundle is linear, and is the transpose of the map $X \mapsto \varphi(x, X) = \psi(X)(x)$.

Remark 4. *The homomorphism ψ of the Lie algebra $\mathcal{G}$ into the Lie algebra $A^1(N)$ of smooth vector fields on N is an action of that Lie algebra, in the sense defined below Definition 11. That action can be canonically lifted into a Hamiltonian action of $\mathcal{G}$ on T^*N, endowed with its canonical symplectic form $d\theta_N$ Definition 13. The map J is in fact a Hamiltonian momentum map for that Hamiltonian action Proposition 5.*

Let $\mathcal{L}_L = d_{\mathrm{vert}}L : TN \to T^*N$ be the *Legendre map* defined in Definition 1.

Theorem 3 (Euler–Poincaré Equation). *With the above defined notations, let $\gamma : [t_0, t_1] \to N$ be a smooth parametrized curve in N and $V : [t_0, t_1] \to \mathcal{G}$ be a smooth parametrized curve such that, for each $t \in [t_0, t_1]$,*

$$\psi\big(V(t)\big)\big(\gamma(t)\big) = \frac{d\gamma(t)}{dt} . \tag{3}$$

The curve γ is a possible motion of the Lagrangian system if and only if V satisfies the equation

$$\left(\frac{d}{dt} - \mathrm{ad}^*_{V(t)}\right)\Big(J \circ \mathcal{L}_L \circ \varphi\big(\gamma(t), V(t)\big)\Big) - J \circ d_1\overline{L}\big(\gamma(t), V(t)\big) = 0 . \tag{4}$$

The interested reader will find a proof of that theorem in local coordinates in the original Note by Poincaré [29]. More intrinsic proofs can be found in [12,30]. Another proof is possible, in which that theorem is deduced from the Euler-Cartan Theorem 1.

Remark 5. *Equation (3) is called the compatibility condition and Equation (4) is the Euler–Poincaré equation. It can be written under the equivalent form*

$$\left(\frac{d}{dt} - \mathrm{ad}^*_{V(t)}\right)\Big(d_2\overline{L}\big(\gamma(t), V(t)\big)\Big) - J \circ d_1\overline{L}\big(\gamma(t), V(t)\big) = 0 . \tag{5}$$

Examples of applications of the Euler–Poincaré equation can be found in [5,6,12,30] and, for an application in thermodynamics, [31].

4. The Hamiltonian Formalism

The Lagrangian formalism can be applied to any smooth Lagrangian. Its application yields *second order differential equations* on $\mathbb{R} \times N$ (in local coordinates, the *Euler–Lagrange equations*) which in general *are not solved with respect to the second order derivatives of the unknown functions with respect to time*. The classical existence and unicity theorems for the solutions of differential equations (such as the *Cauchy-Lipschitz theorem*) therefore *cannot be applied to these equations.*

Under the additional assumption that the Lagrangian is *hyper-regular*, a very clever change of variables discovered by *William Rowan Hamilton* (*Lagrange* obtained however Hamilton's equations before *Hamilton*, but only in a special case, for the slow "variations of constants" such as the orbital parameters of planets in the solar system [32,33]). Hamilton [21,22] allows a new formulation of these equations in the framework of *symplectic geometry*. The *Hamiltonian formalism* discussed below is the use of these new equations. It was later generalized independently of the Lagrangian formalism.

4.1. Hyper-Regular Lagrangians

Assumptions Made in this Section

We consider in this section a smooth, maybe time-dependent Lagrangian $L : \mathbb{R} \times TN \to \mathbb{R}$, which is such that the Legendre map Definition 1 $\mathcal{L}_L : \mathbb{R} \times TN \to T^*N$ satisfies the following property: for each fixed value of the time $t \in \mathbb{R}$, the map $v \mapsto \mathcal{L}_L(t,v)$ is a smooth diffeomorphism of the tangent bundle TN onto the cotangent bundle T^*N. An equivalent assumption is the following: the map $(\mathrm{id}_{\mathbb{R}}, \mathcal{L}_L) : (t,v) \mapsto (t, \mathcal{L}_L(t,v))$ is a smooth diffeomorphism of $\mathbb{R} \times TN$ onto $\mathbb{R} \times T^*N$. The Lagrangian L is then said to be *hyper-regular*. The equations of motion can be written on $\mathbb{R} \times T^*N$ instead of $\mathbb{R} \times TN$.

Definition 4. *Under the assumption Section 4.1, the function $H_L : \mathbb{R} \times T^*N \to \mathbb{R}$ given by*

$$H_L(t,p) = E_L \circ (\mathrm{id}_{\mathbb{R}}, \mathcal{L}_L)^{-1}(t,p)\,, \quad t \in \mathbb{R}\,,\ p \in T^*N\,,$$

($E_L : \mathbb{R} \times TN \to \mathbb{R}$ being the energy function defined in Definition 1) is called the Hamiltonian associated to the hyper-regular Lagrangian L.

The 1-form defined on $\mathbb{R} \times T^*N$

$$\widehat{\varpi}_{H_L} = \theta_N - H_L \mathrm{d}t\,,$$

where θ_N is the Liouville 1-form on T^*N, is called the Poincaré-Cartan 1-form in the Hamiltonian formalism.

Remark 6. *The Poincaré-Cartan 1-form $\widehat{\varpi}_L$ on $\mathbb{R} \times TN$, defined in Definition 1, is the pull-back, by the diffeomorphism $(\mathrm{id}_{\mathbb{R}}, \mathcal{L}_L) : \mathbb{R} \times TN \to \mathbb{R} \times T^*N$, of the Poincaré-Cartan 1-form $\widehat{\varpi}_{H_L}$ in the Hamiltonian formalism on $\mathbb{R} \times T^*N$ defined above.*

4.2. Presymplectic Manifolds

Definition 5. *A presymplectic form on a smooth manifold M is a 2-form ω on M which is closed, i.e., such that $\mathrm{d}\omega = 0$. A manifold M equipped with a presymplectic form ω is called a* presymplectic manifold *and denoted by (M, ω). The kernel* $\ker\omega$ *of a presymplectic form ω defined on a smooth manifold M is the set of vectors $v \in TM$ such that $\mathrm{i}(v)\omega = 0$.*

Remark 7. *A symplectic form ω on a manifold M is a presymplectic form which, moreover, is non-degenerate, i.e., such that for each $x \in M$ and each non-zero vector $v \in T_xM$, there exists another vector $w \in T_xM$ such that $\omega(x)(v,w) \neq 0$. Or in other words, a presymplectic form ω whose kernel is the set of null vectors.*

The kernel of a presymplectic form ω on a smooth manifold M is a vector sub-bundle of TM if and only if for each $x \in M$, the vector subspace T_xM of vectors $v \in T_xM$ which satisfy $\mathrm{i}(v)\omega = 0$ is of a fixed dimension, the same for all points $x \in M$. A presymplectic form which satisfies that condition is said to be of constant rank.

Proposition 1. *Let ω be a presymplectic form of constant rank Remark 7 on a smooth manifold M. The kernel* $\ker\omega$ *of ω is a completely integrable vector sub-bundle of TM, which defines a foliation $\mathcal{F}_\omega$ of M into connected immersed submanifolds which, at each point of M, have the fibre of* $\ker\omega$ *at that point as tangent vector space.*

We now assume in addition that this foliation is simple, i.e., such that the set of leaves of $\mathcal{F}_\omega$, denoted by $M/\ker\omega$, has a smooth manifold structure for which the canonical projection $p : M \to M/\ker\omega$ (which associates to each point $x \in M$ the leaf which contains x) is a smooth submersion. There exists on $M/\ker\omega$ a unique symplectic form ω_r such that

$$\omega = p^*\omega_r\,.$$

Proof. Since $d\omega = 0$, the fact that $\ker \omega$ is completely integrable is an immediate consequence of the Frobenius' theorem ([27], Chapter III, Theorem 5.1, p. 132). The existence and unicity of a symplectic form ω_r on $M/\ker\omega$ such that $\omega = p^*\omega_r$ results from the fact that $M/\ker\omega$ is built by quotienting M by the kernel of ω. □

Presymplectic Manifolds in Mechanics

Let us go back to the assumptions and notations of Section 4.1. We have seen in Remark 6 that the Poincaré-Cartan 1-form in Hamiltonian formalism $\widehat{\varpi}_{H_L}$ on $\mathbb{R}\times T^*N$ and the Poincaré-Cartan 1-form in Lagrangian formalism $\widehat{\varpi}_L$ on $\mathbb{R}\times TN$ are related by

$$\widehat{\varpi}_L = (\mathrm{id}_{\mathbb{R}}, \mathcal{L}_L)^*\widehat{\varpi}_{H_L}\,.$$

Their exterior differentials $d\widehat{\varpi}_L$ and $d\widehat{\varpi}_{H_L}$ both are *presymplectic 2-forms* on the odd-dimensional manifolds $\mathbb{R}\times TN$ and $\mathbb{R}\times T^*N$, respectively. At any point of these manifolds, the kernels of these closed 2-forms are one-dimensional. They therefore Proposition 1 determine *foliations into smooth curves* of these manifolds. The Euler-Cartan Theorem 1 shows that each of these curves is a possible *motion* of the system, described either in the Lagrangian formalism, or in the Hamiltonian formalism, respectively.

The set of all possible motions of the system, called by Jean-Marie Souriau the *manifold of motions* of the system, is described by the quotient $(\mathbb{R}\times TN)/\ker d\widehat{\varpi}_L$ in the Lagrangian formalism, and by the quotient $(\mathbb{R}\times T^*N)/\ker d\widehat{\varpi}_{H_L}$ in the Hamiltonian formalism. Both are (maybe non-Hausdorff) *symplectic manifolds*, the projections on these quotient manifolds of the presymplectic forms $d\widehat{\varpi}_L$ and $d\widehat{\varpi}_{H_L}$ both being symplectic forms. Of course the diffeomorphism $(\mathrm{id}_{\mathbb{R}}, \mathcal{L}_L) : \mathbb{R}\times TN \to \mathbb{R}\times T^*N$ projects onto a symplectomorphism between the Lagrangian and Hamiltonian descriptions of the manifold of motions of the system.

4.3. The Hamilton Equation

Proposition 2. *Let N be the configuration manifold of a Lagrangian system whose Lagrangian $L : \mathbb{R}\times TN \to \mathbb{R}$, maybe time-dependent, is smooth and hyper-regular, and $H_L : \mathbb{R}\times T^*N \to \mathbb{R}$ be the associated Hamiltonian Definition 4. Let $\varphi : [t_0, t_1] \to N$ be a smooth curve parametrized by the time $t\in[t_0,t_1]$, and let $\psi : [t_0,t_1] \to T^*N$ be the parametrized curve in T^*N*

$$\psi(t) = \mathcal{L}_L\left(t, \frac{d\gamma(t)}{dt}\right)\,,\quad t\in[t_0,t_1]\,,$$

where $\mathcal{L}_L : \mathbb{R}\times TN \to T^*N$ is the Legendre map Definition 1.

The parametrized curve $t\mapsto\gamma(t)$ is a motion of the system if and only if the parametrized curve $t\mapsto\psi(t)$ satisfies the equatin, called the Hamilton equation,

$$\mathrm{i}\left(\frac{d\psi(t)}{dt}\right) d\theta_N = -dH_{Lt}\,,$$

where $dH_{Lt} = dH_L - \dfrac{\partial H_L}{\partial t}dt$ is the differential of the function $H_{Lt} : T^*N \to \mathbb{R}$ in which the time t is considered as a parameter with respect to which there is no differentiation.

When the parametrized curve ψ satisfies the Hamilton equation stated above, it satisfies too the equation, called the energy equation

$$\frac{d}{dt}\Big(H_L(t,\psi(t))\Big) = \frac{\partial H_L}{\partial t}(t,\psi(t))\,.$$

Proof. These results directly follow from the Euler-Cartan Theorem 1. □

Remark 8. *The 2-form* $\mathrm{d}\theta_N$ *is a symplectic form on the cotangent bundle* T^*N*, called its canonical symplectic form. We have shown that when the Lagrangian* L *is hyper-regular, the equations of motion can be written in three equivalent manners:*

1. *as the Euler–Lagrange equations on* $\mathbb{R} \times TM$*,*
2. *as the equations given by the kernels of the presymplectic forms* $\mathrm{d}\widehat{\omega}_L$ *or* $\mathrm{d}\widehat{\omega}_{H_L}$ *which determine the foliations into curves of the evolution spaces* $\mathbb{R} \times TM$ *in the Lagrangian formalism, or* $\mathbb{R} \times T^*M$ *in the Hamiltonian formalism,*
3. *as the Hamilton equation associated to the Hamiltonian* H_L *on the symplectic manifold* $(T^*N, \mathrm{d}\theta_N)$*, often called the phase space of the system.*

4.3.1. The Tulczyjew Isomorphisms

Around 1974, Tulczyjew [34,35] discovered (β_N was probably known long before 1974, but I believe that α_N, much more hidden, was noticed by Tulczyjew for the first time) two remarkable vector bundles isomorphisms $\alpha_N : TT^*N \to T^*TN$ and $\beta_N : TT^*N \to T^*T^*N$.

The first one α_N is an isomorphism of the bundle $(TT^*N, T\pi_N, TN)$ onto the bundle (T^*TN, π_{TN}, TN), while the second β_N is an isomorphism of the bundle $(TT^*N, \tau_{T^*N}, T^*N)$ onto the bundle $(T^*T^*N, \pi_{T^*N}, T^*N)$. The diagram below is commutative.

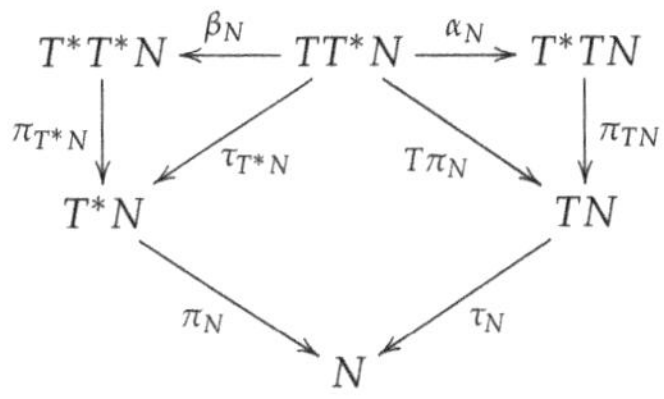

Since they are the total spaces of cotangent bundles, the manifolds T^*TN and T^*T^*N are endowed with the Liouville 1-forms θ_{TN} and θ_{T^*N}, and with the canonical symplectic forms $\mathrm{d}\theta_{TN}$ and $\mathrm{d}\theta_{T^*N}$, respectively. Using the isomorphisms α_N and β_N, we can therefore define on TT^*N two 1-forms $\alpha_N^*\theta_{TN}$ and $\beta_N^*\theta_{T^*N}$, and two symplectic 2-forms $\alpha_N^*(\mathrm{d}\theta_{TN})$ and $\beta_N^*(\mathrm{d}\theta_{T^*N})$. The very remarkable property of the isomorphisms α_N and β_N is that *the two symplectic forms so obtained on* TT^*N *are equal*:

$$\alpha_N^*(\mathrm{d}\theta_{TN}) = \beta_N^*(\mathrm{d}\theta_{T^*N}) .$$

The 1-forms $\alpha_N^*\theta_{TN}$ and $\beta_N^*\theta_{T^*N}$ are not equal, their difference is the differential of a smooth function.

4.3.2. Lagrangian Submanifolds

In view of applications to implicit Hamiltonian systems, let us recall here that a Lagrangian submanifold of a symplectic manifold (M, ω) is a submanifold N whose dimension is half the dimension of M, on which the form induced by the symplectic form ω is 0.

Let $L : TN \to \mathbb{R}$ and $H : T^*N \to \mathbb{R}$ be two smooth real valued functions, defined on TN and on T^*N, respectively. The graphs $\mathrm{d}L(TN)$ and $\mathrm{d}H(T^*N)$ of their differentials are Lagrangian submanifolds of the symplectic manifolds $(T^*TN, \mathrm{d}\theta_{TN})$ and $(T^*T^*N, \mathrm{d}\theta_{T^*N})$. Their pull-backs $\alpha_N^{-1}(\mathrm{d}L(TN))$ and $\beta_N^{-1}(\mathrm{d}H(T^*N))$ by the symplectomorphisms α_N and β_N are therefore two Lagrangian submanifolds of the manifold TT^*N endowed with the symplectic form $\alpha_N^*(\mathrm{d}\theta_{TN})$, which is equal to the symplectic form $\beta_N^*(\mathrm{d}\theta_{T^*N})$.

The following theorem enlightens some aspects of the relationships between the Hamiltonian and the Lagrangian formalisms.

Theorem 4 (W. M. Tulczyjew). *With the notations specified above Section 4.3.2, let $X_H : T^*N \to TT^*N$ be the Hamiltonian vector field on the symplectic manifold $(T^*N, \mathrm{d}\theta_N)$ associated to the Hamiltonian $H : T^*N \to \mathbb{R}$, defined by $\mathrm{i}(X_H)\mathrm{d}\theta_N = -\mathrm{d}H$. Then*

$$X_H(T^*N) = \beta_N^{-1}\big(\mathrm{d}H(T^*N)\big).$$

Moreover, the equality

$$\alpha_N^{-1}\big(\mathrm{d}L(TN)\big) = \beta_N^{-1}\big(\mathrm{d}H(T^*N)\big)$$

holds if and only if the Lagrangian L is hyper-regular and such that

$$\mathrm{d}H = \mathrm{d}\big(E_L \circ \mathcal{L}_L^{-1}\big),$$

where $\mathcal{L}_L : TN \to T^*N$ is the Legendre map and $E_L : TN \to \mathbb{R}$ the energy associated to the Lagrangian L.

The interested reader will find the proof of that theorem in the works of Tulczyjew ([34,35]).

When L is not hyper-regular, $\alpha_N^{-1}\big(\mathrm{d}L(TN)\big)$ still is a Lagrangian submanifold of the symplectic manifold $\big(TT^*N, \alpha_N^*(\mathrm{d}\theta_{TN})\big)$, but it is no more the graph of a smooth vector field X_H defined on T^*N. Tulczyjew proposes to consider this Lagrangian submanifold as an *implicit Hamilton equation* on T^*N.

These results can be extended to Lagrangians and Hamiltonians which may depend on time.

4.4. The Hamiltonian Formalism on Symplectic and Poisson Manifolds

4.4.1. The Hamilton Formalism on Symplectic Manifolds

In pure mathematics as well as in applications of mathematics to mechanics and physics, symplectic manifolds other than cotangent bundles are encountered. A theorem due to the french mathematician *Gaston Darboux* (1842–1917) asserts that any symplectic manifold (M, ω) is of even dimension $2n$ and is locally isomorphic to the cotangent bundle to a n-dimensional manifold: in a neighbourhood of each of its point there exist local coordinates $(x^1, \dots, x^n, p_1, \dots, p_n)$, called *Darboux coordinates* with which the symplectic form ω is expressed exactly as the canonical symplectic form of a cotangent bundle:

$$\omega = \sum_{i=1}^{n} \mathrm{d}p_i \wedge \mathrm{d}x^i.$$

Let (M, ω) be a symplectic manifold and $H : \mathbb{R} \times M \to \mathbb{R}$ a smooth function, said to be a *time-dependent Hamiltonian*. It determines a *time-dependent Hamiltonian vector field* X_H on M, such that

$$\mathrm{i}(X_H)\omega = -\mathrm{d}H_t,$$

$H_t : M \to \mathbb{R}$ being the function H in which the variable t is considered as a parameter with respect to which no differentiation is made.

The *Hamilton equation* determined by H is the differential equation

$$\frac{\mathrm{d}\psi(t)}{\mathrm{d}t} = X_H\big(t, \psi(t)\big).$$

The Hamiltonian formalism can therefore be applied to any smooth, maybe time dependent Hamiltonian on M, even when there is no associated Lagrangian.

The Hamiltonian formalism is not limited to symplectic manifolds: it can be applied, for example, to *Poisson manifolds* [36], *contact manifolds* and *Jacobi manifolds* [37]. For simplicity I will consider only Poisson manifolds. Readers interested in Jacobi manifolds and their generalizations are referred to the papers by Lichnerowicz quoted above and to the very important paper by Kirillov [38].

Definition 6. *A Poisson manifold is a smooth manifold P whose algebra of smooth functions $C^\infty(P,\mathbb{R})$ is endowed with a bilinear composition law, called the Poisson bracket, which associates to any pair (f,g) of smooth functions on P another smooth function denoted by $\{f,g\}$, that composition satisfying the three properties*

1. *it is skew-symmetric,*
$$\{g,f\} = -\{f,g\},$$
2. *it satisfies the Jacobi identity*
$$\{f,\{g,h\}\} + \{g,\{h,f\}\} + \{h,\{f,g\}\} = 0,$$
3. *it satisfies the Leibniz identity*
$$\{f,gh\} = \{f,g\}h + g\{f,h\}.$$

Example 4.

1. *On the vector space of smooth functions defined on a symplectic manifold (M,ω), there exists a composition law, called the Poisson bracket, which satisfies the properties stated in Definition 6. Let us recall briefly its definition. The symplectic form ω allows us to associate, to any smooth function $f \in C^\infty(M,\mathbb{R})$, a smooth vector field $X_f \in A^1(M,\mathbb{R})$, called the Hamiltonian vector field associated to f, defined by*
$$\mathrm{i}(X_f)\omega = -\mathrm{d}f\,.$$
The Poisson bracket $\{f,g\}$ of two smooth functions f and $g \in C^\infty(M,\mathbb{R})$ is defined by the three equivalent equalities
$$\{f,g\} = \mathrm{i}(X_f)\mathrm{d}g = -\mathrm{i}(X_g)\mathrm{d}f = \omega(X_f,X_g)\,.$$
Any symplectic manifold is therefore a Poisson manifold.
The Poisson bracket of smooth functions defined on a symplectic manifold (when that symplectic manifold is a cotangent bundle) was discovered by Siméon Denis Poisson (1781–1840) [39].
2. *Let $\mathcal{G}$ be a finite-dimensional real Lie algebra, and let $\mathcal{G}^*$ be its dual vector space. For each smooth function $f \in C^\infty(\mathcal{G}^*,\mathbb{R})$ and each $\zeta \in \mathcal{G}^*$, the differential $\mathrm{d}f(\zeta)$ is a linear form on $\mathcal{G}^*$, in other words an element of the dual vector space of $\mathcal{G}^*$. Identifying with $\mathcal{G}$ the dual vector space of $\mathcal{G}^*$, we can therefore consider $\mathrm{d}f(\zeta)$ as an element in $\mathcal{G}$. With this identification, we can define the Poisson bracket of two smooth functions f and $g \in C^\infty(\mathcal{G}^*,\mathbb{R})$ by*
$$\{f,g\}(\zeta) = \big[\mathrm{d}f(\zeta),\mathrm{d}g(\zeta)\big]\,,\quad \zeta \in \mathcal{G}^*\,,$$
the bracket in the right hand side being the bracket in the Lie algebra $\mathcal{G}$. The Poisson bracket of functions in $C^\infty(\mathcal{G}^*,\mathbb{R})$ so defined satifies the properties stated in Definition 6. The dual vector space of any finite-dimensional real Lie algebra is therefore endowed with a Poisson structure, called its canonical Lie-Poisson structure or its Kirillov-Kostant-Souriau Poisson structure. Discovered by Sophus Lie, this structure was indeed rediscovered independently by Alexander Kirillov, Bertram Kostant and Jean-Marie Souriau.
3. *A symplectic cocycle of a finite-dimensional, real Lie algebra $\mathcal{G}$ is a skew-symmetric bilinear map Θ : $\mathcal{G}\times\mathcal{G} \to \mathcal{G}^*$ which satisfies, for all X, Y and $Z \in \mathcal{G}$,*
$$\Theta([X,Y],Z) + \Theta([Y,Z],X) + \Theta([Z,X],Y) = 0\,.$$
The canonical Lie-Poisson bracket of two smooth functions f and $g \in C^\infty(\mathcal{G}^*,\mathbb{R})$ can be modified by means of the symplectic cocycle Θ, by setting
$$\{f,g\}_\Theta(\zeta) = \big[\mathrm{d}f(\zeta),\mathrm{d}g(\zeta)\big] - \Theta\big(\mathrm{d}f(\zeta),\mathrm{d}g(\zeta)\big)\,,\quad \zeta \in \mathcal{G}^*\,.$$

This bracket still satifies the properties stated in Definition 6, therefore defines on $\mathcal{G}^*$ a Poisson structure called its canonical Lie-Poisson structure modified by Θ.

4.4.2. Properties of Poisson Manifolds

The interested reader will find the proofs of the properties recalled here in [8–11].

1. On a Poisson manifold P, the Poisson bracket $\{f,g\}$ of two smooth functions $f$and g can be expressed by means of a smooth field of bivectors Λ:

$$\{f,g\} = \Lambda(\mathrm{d}f, \mathrm{d}g), \quad f \text{ and } g \in C^\infty(P, \mathbb{R}),$$

called the *Poisson bivector field* of P. The considered Poisson manifold is often denoted by (P, Λ). The Poisson bivector field Λ identically satisfies

$$[\Lambda, \Lambda] = 0,$$

the bracket $[\ ,\ ]$ in the left hand side being the *Schouten-Nijenhuis bracket*. That bivector field determines a vector bundle morphism $\Lambda^\sharp : T^*P \to TP$, defined by

$$\Lambda(\eta, \zeta) = \langle \zeta, \Lambda^\sharp(\eta) \rangle,$$

where η and $\zeta \in T^*P$ are two covectors attached to the same point in P.

Readers interested in the Schouten-Nijenhuis bracket will find thorough presentations of its properties in [40,41].

2. Let (P, Λ) be a Poisson manifold. A (maybe time-dependent) vector field on P can be associated to each (maybe time-dependent) smooth function $H : \mathbb{R} \times P \to \mathbb{R}$. It is called the *Hamiltonian vector field* associated to the *Hamiltonian* H, and denoted by X_H. Its expression is

$$X_H(t, x) = \Lambda^\sharp(x)\big(\mathrm{d}H_t(x)\big),$$

where $\mathrm{d}H_t(x) = \mathrm{d}H(t, x) - \dfrac{\partial H(t, x)}{\partial t}\mathrm{d}t$ is the differential of the function deduced from H by considering t as a parameter with respect to which no differentiation is made.

The *Hamilton equation* determined by the (maybe time-dependent) Hamiltonian H is

$$\frac{\mathrm{d}\varphi(t)}{\mathrm{d}t} = X_H\big((t, \varphi(t)\big) = \Lambda^\sharp(\mathrm{d}H_t)\big(\varphi(t)\big).$$

3. Any Poisson manifold is foliated, by a generalized foliation whose leaves may not be all of the same dimension, into immersed connected symplectic manifolds called the *symplectic leaves* of the Poisson manifold. The value, at any point of a Poisson manifold, of the Poisson bracket of two smooth functions only depends on the restrictions of these functions to the symplectic leaf through the considered point, and can be calculated as the Poisson bracket of functions defined on that leaf, with the Poisson structure associated to the symplectic structure of that leaf. This property was discovered by Alan Weinstein, in his very thorough study of the local structure of Poisson manifolds [42].

5. Hamiltonian Symmetries

5.1. *Presymplectic, Symplectic and Poisson Maps and Vector Fields*

Let M be a manifold endowed with some structure, which can be either

- a *presymplectic structure*, determined by a presymplectic form, i.e., a 2-form ω which is closed ($\mathrm{d}\omega = 0$),

- a *symplectic structure,* determined by a symplectic form ω, i.e., a 2-form ω which is both closed ($d\omega = 0$) and nondegenerate ($\ker \omega = \{0\}$),
- a *Poisson structure,* determined by a smooth Poisson bivector field Λ satisfying $[\Lambda, \Lambda] = 0$.

Definition 7. *A presymplectic (resp. symplectic, resp. Poisson) diffeomorphism of a presymplectic (resp., symplectic, resp. Poisson) manifold (M, ω) (resp. (M, Λ)) is a smooth diffeomorphism $f : M \to M$ such that $f^*\omega = \omega$ (resp. $f^*\Lambda = \Lambda$).*

Definition 8. *A smooth vector field X on a presymplectic (resp. symplectic, resp. Poisson) manifold (M, ω) (resp. (M, Λ)) is said to be a presysmplectic (resp. symplectic, resp. Poisson) vector field if $\mathcal{L}(X)\omega = 0$ (resp. if $\mathcal{L}(X)\Lambda = 0$), where $\mathcal{L}(X)$ denotes the Lie derivative of forms or mutivector fields with respect to X.*

Definition 9. *Let (M, ω) be a presymplectic or symplectic manifold. A smooth vector field X on M is said to be Hamiltonian if there exists a smooth function $H : M \to \mathbb{R}$, called a* Hamiltonian *for X, such that*

$$\mathrm{i}(X)\omega = -\mathrm{d}H\,.$$

Not any smooth function on a presymplectic manifold can be a Hamiltonian.

Definition 10. *Let (M, Λ) be a Poisson manifold. A smooth vector field X on M is said to be Hamiltonian if there exists a smooth function $H \in C^\infty(M, \mathbb{R})$, called a Hamiltonian for X, such that $X = \Lambda^\sharp(\mathrm{d}H)$. An equivalent definition is that*

$$\mathrm{i}(X)\mathrm{d}g = \{H, g\} \quad \textit{for any } g \in C^\infty(M, \mathbb{R})\,,$$

where $\{H, g\} = \Lambda(\mathrm{d}H, \mathrm{d}g)$ denotes the Poisson bracket of the functions H and g.

On a symplectic or a Poisson manifold, any smooth function can be a Hamiltonian.

Proposition 3. *A Hamiltonian vector field on a presymplectic (resp. symplectic, resp. Poisson) manifold automatically is a presymplectic (resp. symplectic, resp. Poisson) vector field.*

The proof of this result, which is easy, can be found in any book on symplectic and Poisson geoemetry, for example [8–10].

5.2. Lie Algebras and Lie Groups Actions

Definition 11. *An action on the left (resp. an action on the right) of a Lie group G on a smooth manifold M is a smooth map $\Phi : G \times M \to M$ (resp. a smooth map $\Psi : M \times G \to M$) such that*

- *for each fixed $g \in G$, the map $\Phi_g : M \to M$ defined by $\Phi_g(x) = \Phi(g, x)$ (resp. the map $\Psi_g : M \to M$ defined by $\Psi_g(x) = \Psi(x, g)$) is a smooth diffeomorphism of M,*
- *$\Phi_e = \mathrm{id}_M$ (resp. $\Psi_e = \mathrm{id}_M$), e being the neutral element of G,*
- *for each pair $(g_1, g_2) \in G \times G$, $\Phi_{g_1} \circ \Phi_{g_2} = \Phi_{g_1 g_2}$ (resp. $\Psi_{g_1} \circ \Psi_{g_2} = \Psi_{g_2 g_1}$).*

An action *of a Lie algebra $\mathcal{G}$ on a smooth manifold M is a* Lie algebras morphism *of $\mathcal{G}$ into the Lie algebra $A^1(M)$ of smooth vector fields on M, i.e., a linear map $\psi : \mathcal{G} \to A^1(M)$ which associates to each $X \in \mathcal{G}$ a smooth vector field $\psi(X)$ on M such that for each pair $(X, Y) \in \mathcal{G} \times \mathcal{G}$, $\psi([X, Y]) = [\psi(X), \psi(Y)]$.*

Proposition 4. *An action Ψ, either on the left or on the right, of a Lie group G on a smooth manifold M, automatically determines an action ψ of its Lie algebra $\mathcal{G}$ on that manifold, which associates to each $X \in \mathcal{G}$ the vector field $\psi(X)$ on M, often denoted by X_M and called the fundamental vector field on M associated to X. It is defined by*

$$\psi(X)(x) = X_M(x) = \frac{\mathrm{d}}{\mathrm{d}s}\big(\Psi_{\exp(sX)}(x)\big)\big|_{s=0}\,, \quad x \in M\,,$$

with the following convention: ψ is a Lie algebras homomorphism when we take for Lie algebra $\mathcal{G}$ of the Lie group G the Lie algebra or right invariant vector fields on G if Ψ is an action on the left, and the Lie algebra of left invariant vector fields on G if Ψ is an action on the right.

Proof. If Ψ is an action of G on M on the left (respectively, on the right), the vector field on G which is right invariant (respectively, left invariant) and whose value at e is X, and the associated fundamental vector field X_M on M, are compatible by the map $g \mapsto \Psi_g(x)$. Therefore the map $\psi : \mathcal{G} \to A^1(M)$ is a Lie algebras homomorphism, if we take for definition of the bracket on $\mathcal{G}$ the bracket of right invariant (respectively, left invariant) vector fields on G. □

Definition 12. *When M is a presymplectic (or a symplectic, or a Poisson) manifold, an action Ψ of a Lie group G (respectively, an action ψ of a Lie algebra $\mathcal{G}$) on the manifold M is called a presymplectic (or a symplectic, or a Poisson) action if for each $g \in G$, Ψ_g is a presymplectic, or a symplectic, or a Poisson diffeomorphism of M (respectively, if for each $X \in \mathcal{G}$, $\psi(X)$ is a presymplectic, or a symplectic, or a Poisson vector field on M.*

Definition 13. *An action ψ of a Lie algeba $\mathcal{G}$ on a presymplectic or symplectic manifold (M,ω), or on a Poisson manifold (M,Λ), is said to be Hamiltonian if for each $X \in \mathcal{G}$, the vector field $\psi(X)$ on M is Hamiltonian.*

An action Ψ (either on the left or on the right) of a Lie group G on a presymplectic or symplectic manifold (M,ω), or on a Poisson manifold (M,Λ), is said to be Hamiltonian *if that action is presymplectic, or symplectic, or Poisson (according to the structure of M), and if in addition the associated action of the Lie algebra $\mathcal{G}$ of G is Hamiltonian.*

Remark 9. *A Hamiltonian action of a Lie group, or of a Lie algebra, on a presymplectic, symplectic or Poisson manifold, is automatically a presymplectic, symplectic or Poisson action. This result immediately follows from Proposition 3.*

5.3. Momentum Maps of Hamiltonian Actions

Proposition 5. *Let ψ be a Hamiltonian action of a finite-dimensional Lie algebra $\mathcal{G}$ on a presymplectic, symplectic or Poisson manifold (M,ω) or (M,Λ). There exists a smooth map $J : M \to \mathcal{G}^*$, taking its values in the dual space $\mathcal{G}^*$ of the Lie algebra $\mathcal{G}$, such that for each $X \in \mathcal{G}$ the Hamiltonian vector field $\psi(X)$ on M admits as Hamiltonian the function $J_X : M \to \mathbb{R}$, defined by*

$$J_X(x) = \langle J(x), X \rangle \,, \quad x \in M \,.$$

The map J is called a momentum map for the Lie algebra action ψ. When ψ is the action of the Lie algebra $\mathcal{G}$ of a Lie group G associated to a Hamiltonian action Ψ of a Lie group G, J is called a momentum map for the Hamiltonian Lie group action Ψ.

The proof of that result, which is easy, can be found for example in [8–10].

Remark 10. *The momentum map J is not unique:*

- *when (M,ω) is a connected symplectic manifold, J is determined up to addition of an arbitrary constant element in $\mathcal{G}^*$;*
- *when (M,Λ) is a connected Poisson manifold, the momentum map J is determined up to addition of an arbitrary $\mathcal{G}^*$-valued smooth map which, coupled with any $X \in \mathcal{G}$, yields a Casimir of the Poisson algebra of (M,Λ), i.e., a smooth function on M whose Poisson bracket with any other smooth function on that manifold is the function identically equal to 0.*

5.4. Noether's Theorem in Hamiltonian Formalism

Theorem 5 (Noether's Theorem in Hamiltonian Formalism). *Let X_f and X_g be two Hamiltonian vector fields on a presymplectic or symplectic manifold (M,ω), or on a Poisson manifold (M,Λ), which admit as*

Hamiltonians, respectively, the smooth functions f and g on the manifold M. The function f remains constant on each integral curve of X_g if and only if g remains constant on each integral curve of X_f.

Proof. The function f is constant on each integral curve of X_g if and only if $\mathrm{i}(X_g)\mathrm{d}f = 0$, since each integral curve of X_g is connected. We can use the Poisson bracket, even when M is a presymplectic manifold, since the Poisson bracket of two Hamiltonians on a presymplectic manifold still can be defined. So we can write

$$\mathrm{i}(X_g)\mathrm{d}f = \{g,f\} = -\{f,g\} = -\mathrm{i}(X_f)\mathrm{d}g\,.$$ □

Corollary 2 (of Noether's Theorem in Hamiltonian Formalism). *Let $\psi : \mathcal{G} \to A^1(M)$ be a Hamiltonian action of a finite-dimensional Lie algebra $\mathcal{G}$ on a presymplectic or symplectic manifold (M,ω), or on a Poisson manifold (M,Λ), and let $J : M \to \mathcal{G}^*$ be a momentum map of this action. Let X_H be a Hamiltonian vector field on M admitting as Hamiltonian a smooth function H. If for each $X \in \mathcal{G}$ we have $\mathrm{i}\big(\psi(X)\big)(\mathrm{d}H) = 0$, the momentum map J remains constant on each integral curve of X_H.*

Proof. This result is obtained by applying Theorem 5 to the pairs of Hamiltonian vector fields made by X_H and each vector field associated to an element of a basis of $\mathcal{G}$. □

5.5. Symplectic Cocycles

Theorem 6 (J. M. Souriau [14]). *Let Φ be a Hamiltonian action (either on the left or on the right) of a Lie group G on a connected symplectic manifold (M,ω) and let $J : M \to \mathcal{G}^*$ be a momentum map of this action. There exists an affine action A (either on the left or on the right) of the Lie group G on the dual $\mathcal{G}^*$ of its Lie algebra $\mathcal{G}$ such that the momentum map J is equivariant with respect to the actions Φ of G on M and A of G on $\mathcal{G}^*$, i.e., such that*

$$J \circ \Phi_g(x) = A_g \circ J(x) \quad \text{for all } g \in G\,,\ x \in M\,.$$

The action A can be written, with $g \in G$ and $\xi \in \mathcal{G}^*$,

$$\begin{cases} A(g,\xi) = \mathrm{Ad}^*_{g^{-1}}(\xi) + \theta(g) & \text{if } \Phi \text{ is an action on the left,}\\ A(\xi,g) = \mathrm{Ad}^*_{g}(\xi) - \theta(g^{-1}) & \text{if } \Phi \text{ is an action on the right.}\end{cases}$$

Proof. Let us assume that Φ is an action on the left. The fundamental vector field X_M associated to each $X \in \mathcal{G}$ is Hamiltonian, with the function $J_X : M \to \mathbb{R}$, given by

$$J_X(x) = \langle J(x), X\rangle\,, \quad x \in M\,,$$

as Hamiltonian. For each $g \in G$ the direct image $(\Phi_{g^{-1}})_*(X_M)$ of X_M by the symplectic diffeomorphism $\Phi_{g^{-1}}$ is Hamiltonian, with $J_X \circ \Phi_g$ as Hamiltonian. An easy calculation shows that this vector field is the fundamental vector field associated to $\mathrm{Ad}_{g^{-1}}(X) \in \mathcal{G}$. The function

$$x \mapsto \langle J(x), \mathrm{Ad}_{g^{-1}}(X)\rangle = \langle \mathrm{Ad}^*_{g^{-1}} \circ J(x), X\rangle$$

is therefore a Hamiltonian for that vector field. These two functions defined on the connected manifold M, which both are admissible Hamiltonians for the same Hamiltonian vector field, differ only by a constant (which may depend on $g \in G$). We can set, for any $g \in G$,

$$\theta(g) = J \circ \Phi_g(x) - \mathrm{Ad}^*_{g^{-1}} \circ J(x)$$

and check that the map $A : G \times \mathcal{G}^* \to \mathcal{G}^*$ defined in the statement is indeed an action for which J is equivariant.

A similar proof, with some changes of signs, holds when Φ is an action on the right. □

Proposition 6. *Under the assumptions and with the notations of Theorem 6, the map $\theta : G \to \mathcal{G}^*$ is a cocycle of the Lie group G with values in $\mathcal{G}^*$, for the coadjoint representation. It means that it satisfies, for all g and $h \in G$,*

$$\theta(gh) = \theta(g) + \mathrm{Ad}^*_{g^{-1}}\big(\theta(h)\big).$$

More precisely θ is a symplectic cocycle. It means that its differential $T_e\theta : T_eG \equiv \mathcal{G} \to \mathcal{G}^*$ at the neutral element $e \in G$ can be considered as a skew-symmetric bilinear form on $\mathcal{G}$:

$$\Theta(X,Y) = \big\langle T_e\theta(X), Y\big\rangle = -\big\langle T_e\theta(Y), X\big\rangle.$$

The skew-symmetric bilinear form Θ is a symplectic cocycle of the Lie algebra $\mathcal{G}$. It means that it is skew-symmetric and satisfies, for all X, Y and $Z \in \mathcal{G}$,

$$\Theta\big([X,Y],Z\big) + \Theta\big([Y,Z],X\big) + \Theta\big([Z,X],Y\big) = 0.$$

Proof. These properties easily follow from the fact that when Φ is an action on the left, for g and $h \in G$, $\Phi_g \circ \Phi_h = \Phi_{gh}$ (and a similar equality when Φ is an action on the right). The interested reader will find more details in [9,12,14]. □

Proposition 7. *Still under the assumptions and with the notations of Theorem 6, the composition law which associates to each pair (f,g) of smooth real-valued functions on $\mathcal{G}^*$ the function $\{f,g\}_\Theta$ given by*

$$\{f,g\}_\Theta(x) = \big\langle x, [df(x), dg(x)]\big\rangle - \Theta\big(df(x), dg(x)\big), \quad x \in \mathcal{G}^*,$$

($\mathcal{G}$ being identified with its bidual $\mathcal{G}^{**}$), determines a Poisson structure on $\mathcal{G}^*$, and the momentum map $J : M \to \mathcal{G}^*$ is a Poisson map, M being endowed with the Poisson structure associated to its symplectic structure.

Proof. The fact that the bracket $(f,g) \mapsto \{f,g\}_\Theta$ on $C^\infty(\mathcal{G}^*, \mathbb{R})$ is a Poisson bracket was already indicated in Example 4. It can be verified by easy calculations. The fact that J is a Poisson map can be proven by first looking at linear functions on $\mathcal{G}^*$, i.e., elements in $\mathcal{G}$. The reader will find a detailed proof in [12]. □

Remark 11. *When the momentum map J is replaced by another momentum map $J_1 = J + \mu$, where $\mu \in \mathcal{G}^*$ is a constant, the symplectic Lie group cocycle θ and the symplectic Lie algebra cocycle Θ are replaced by θ_1 and Θ_1, respectively, given by*

$$\theta_1(g) = \theta(g) + \mu - \mathrm{Ad}^*_{g^{-1}}(\mu), \quad g \in G,$$
$$\Theta_1(X,Y) = \Theta(X,Y) + \big\langle \mu, [X,Y]\big\rangle, \quad X \text{ and } Y \in \mathcal{G}.$$

These formulae show that $\theta_1 - \theta$ and $\Theta_1 - \Theta$ are symplectic coboundaries of the Lie group G and the Lie algebra $\mathcal{G}$. In other words, the cohomology classes of the cocycles θ and Θ only depend on the Hamiltonian action Φ of G on the symplectic manifold (M, ω).

5.6. The Use of Symmetries in Hamiltonian Mechanics

5.6.1. Symmetries of the Phase Space

Hamiltonian Symmetries are often used for the search of solutions of the equations of motion of mechanical systems. The symmetries considered are those of the *phase space* of the mechanical system. This space is very often a *symplectic manifold*, either the cotangent bundle to the configuration space with its canonical symplectic structure, or a more general symplectic manifold. Sometimes, after some simplifications, the phase space is a *Poisson manifold*.

The *Marsden-Weinstein reduction procedure* [43,44] or one of its generalizations [10] is the method most often used to facilitate the determination of solutions of the equations of motion. In a first step, a possible value of the momentum map is chosen and the subset of the phase space on which the momentum map takes this value is determined. In a second step, that subset (when it is a smooth manifold) is quotiented by its isotropic foliation. The quotient manifold is a symplectic manifold of a dimension smaller than that of the original phase space, and one has an easier to solve Hamiltonian system on that reduced phase space.

When Hamiltonian symmetries are used for the reduction of the dimension of the phase space of a mechanical system, the symplectic cocycle of the Lie group of symmetries action, or of the Lie algebra of symmetries action, is almost always the *zero cocycle*.

For example, if the group of symmetries is the canonical lift to the cotangent bundle of a group of symmetries of the configuration space, not only the *canonical symplectic form*, but the *Liouville* 1*-form* of the cotangent bundle itself remains invariant under the action of the symmetry group, and this fact implies that the symplectic cohomology class of the action is zero.

5.6.2. Symmetries of the Space of Motions

A completely different way of using symmetries was initiated by Jean-Marie Souriau, who proposed to consider the symmetries of the *manifold of motions* of the mechanical system. He observed that the Lagrangian and Hamiltonian formalisms, in their usual formulations, *involve the choice of a particular reference frame*, in which the motion is described. This choice *destroys a part of the natural symmetries of the system*.

For example, in classical (non-relativistic) mechanics, the natural symmetry group of an isolated mechanical system must contain the symmetry group of the *Galilean space-time*, called the *Galilean group*. This group is of *dimension 10*. It contains not only the *group of Euclidean displacements of space* which is of *dimension 6* and the *group of time translations* which is of *dimension 1*, but the *group of linear changes of Galilean reference frames* which is of *dimension 3*.

If we use the *Lagrangian formalism* or the *Hamiltonian formalism*, the Lagrangian or the Hamiltonian of the system *depends on the reference frame*: *it is not invariant with respect to linear changes of Galilean reference frames*.

It may seem strange to consider the set of all possible motions of a system, which is unknown as long as we have not determined all these possible motions. One may ask if it is really useful when we want to determine not all possible motions, but only one motion with prescribed initial data, since that motion is just one point of the (unknown) manifold of motion!

Souriau's answers to this objection are the following.

1. We know that the manifold of motions has a *symplectic structure*, and very often many things are known about its *symmetry properties*.
2. In classical (non-relativistic) mechanics, there exists a natural mathematical object which *does not depend on the choice of a particular reference frame* (even if the decriptions given to that object by different observers depend on the reference frame used by these observers): it is the *evolution space* of the system.

The knowledge of the equations which govern the system's evolution allows the full mathematical description of the *evolution space*, even when these equations are not yet solved.

Moreover, the symmetry properties of the *evolution space* are the same as those of the manifold of motions.

For example, the *evolution space* of a classical mechanical system with configuration manifold N is

1. in the Lagrangian formalism, the space $\mathbb{R} \times TN$ endowed with the presymplectic form $d\widehat{\varpi}_L$, whose kernel is of dimension 1 when the Lagrangian L is hyper-regular,
2. in the Hamiltonian formalism, the space $\mathbb{R} \times T^*N$ with the presymplectic form $d\widehat{\varpi}_H$, whose kernel too is of dimension 1.

The Poincaré-Cartan 1-form $\widehat{\varpi}_L$ in the Lagrangian formalism, or $\widehat{\varpi}_H$ in the Hamiltonian formalism, depends on the choice of a particular reference frame, made for using the Lagrangian or the Hamiltonian formalism. But their exterior differentials, the presymplectic forms $\mathrm{d}\widehat{\varpi}_L$ or $\mathrm{d}\widehat{\varpi}_H$, *do not depend on that choice*, modulo a simple change of variables in the evolution space.

Souriau defined this presymplectic form in a framework more general than those of Lagrangian or Hamiltonian formalisms, and called it the *Lagrange form*. In this more general setting, it may not be an exact 2-form. Souriau proposed as a new *Principle*, the assumption that it always projects on the space of motions of the systems as a *symplectic form*, even in relativistic mechanics in which the definition of an evolution space is not clear. He called this new principle the *Maxwell Principle*.

Bargmann proved that the symplectic cohomology of the Galilean group is of dimension 1, and Souriau proved that the cohomology class of its action on the manifold of motions of an isolated classical (non-relativistic) mechanical system can be identified with the *total mass* of the system [14], Chapter III, p. 153.

Readers interested in the Galilean group and momentum maps of its actions are referred to the recent book by de Saxcé and Vallée [45].

6. Statistical Mechanics and Thermodynamics

6.1. Basic Concepts in Statistical Mechanics

During the XVIII–th and XIX–th centuries, the idea that material bodies (fluids as well as solids) are assemblies of a very large number of small, moving particles, began to be considered by some scientists, notably Daniel Bernoulli (1700–1782), Rudolf Clausius (1822–1888), James Clerk Maxwell (1831–1879) and Ludwig Eduardo Boltzmann (1844–1906), as a reasonable possibility. Attemps were made to explain the nature of some measurable macroscopic quantities (for example the temperature of a material body, the pressure exerted by a gas on the walls of the vessel in which it is contained), and the laws which govern the variations of these macroscopic quantities, by application of the laws of classical mechanics to the motions of these very small particles. Described in the framework of the Hamiltonian formalism, the material body is considered as a Hamiltonian system whose phase space is a very high dimensional symplectic manifold (M, ω), since an element of that space gives a perfect information about the positions and the velocities of all the particles of the system. The experimental determination of the exact state of the system being impossible, one only can use the probability of presence, at each instant, of the state of the system in various parts of the phase space. Scientists introduced the concept of a *statistical state*, defined below.

Definition 14. *Let (M, ω) be a symplectic manifold. A statistical state is a probability measure μ on the manifold M.*

6.1.1. The Liouville Measure on a Symplectic Manifold

On each symplectic manifold (M, ω), with $\dim M = 2n$, there exists a positive measure λ_ω, called the *Liouville measure*. Let us briefly recall its definition. Let (U, φ) be a Darboux chart of (M, ω) Section 4.4.1. The open subset U of M is, by means of the diffeomorphism φ, identified with an open subset $\varphi(U)$ of $\mathbb{R}^{2n}$ on which the coordinates (Darboux coordinates) will be denoted by $(p_1, \ldots, p_n, x^1, \ldots, x^n)$. With this identification, the Liouville measure (restricted to U) is simply the Lebesgue measure on the open subset $\varphi(U)$ of $\mathbb{R}^{2n}$. In other words, for each Borel subset A of M contained in U, we have

$$\lambda_\omega(A) = \int_{\varphi(A)} \mathrm{d}p_1 \ldots \mathrm{d}p_n \, \mathrm{d}x^1 \ldots \mathrm{d}x^n \,.$$

One can easily check that this definition does not depend on the choice of the Darboux coordinates $(p_1, \ldots, p_n, x^1, \ldots, x^n)$ on $\varphi(A)$. By using an atlas of Darboux charts on (M, ω), one can easily define $\lambda_\omega(A)$ for any Borel subset A of M.

Definition 15. *A statistical state μ on the symplectic manifold (M,ω) is said to be* continuous *(respectively, is said to be smooth) if it has a continuous (respectively, a smooth) density with respect to the Liouville measure λ_ω, i.e., if there exists a continuous function (respectively, a smooth function) $\rho : M \to \mathbb{R}$ such that, for each Borel subset A of M*

$$\mu(A) = \int_A \rho \mathrm{d}\lambda_\omega \, .$$

Remark 12. *The density ρ of a continuous statistical state on (M,ω) takes its values in $\mathbb{R}^+$ and of course satisfies*

$$\int_M \rho \mathrm{d}\lambda_\omega = 1 \, .$$

For simplicity we only consider in what follows continuous, very often even smooth statistical states.

6.1.2. Variation in Time of a Statistical State

Let H be a smooth time independent Hamiltonian on a symplectic manifold (M,ω), X_H the associated Hamiltonian vector field and Φ^{X_H} its reduced flow. We consider the mechanical system whose time evolution is described by the flow of X_H.

If the state of the system at time t_0, assumed to be perfectly known, is a point $z_0 \in M$, its state at time t_1 is the point $z_1 = \Phi^{X_H}_{t_1-t_0}(z_0)$.

Let us now assume that the state of the system at time t_0 is not perfectly known, but that a continuous probability measure on the phase space M, whose density with respect to the Liouville measure λ_ω is ρ_0, describes the probability distribution of presence of the state of the system at time t_0. In other words, ρ_0 is the density of the statistical state of the system at time t_0. For any other time t_1, the map $\Phi^{X_H}_{t_1-t_0}$ is a symplectomorphism, therefore leaves invariant the Liouville measure λ_ω. The probability density ρ_1 of the statistical state of the system at time t_1 therefore satisfies, for any $x_0 \in M$ for which $x_1 = \Phi^{X_H}_{t_1-t_0}(x_0)$ is defined,

$$\rho_1(x_1) = \rho_1\big(\Phi^{X_H}_{t_1-t_0}(x_0)\big) = \rho_0(x_0) \, .$$

Since $\left(\Phi^{X_H}_{t_1-t_0}\right)^{-1} = \Phi^{X_H}_{t_0-t_1}$, we can write

$$\rho_1 = \rho_0 \circ \Phi^{X_H}_{t_0-t_1} \, .$$

Definition 16. *Let ρ be the density of a continuous statistical state μ on the symplectic manifold (M,ω). The number*

$$s(\rho) = \int_M \rho \log\left(\frac{1}{\rho}\right) \mathrm{d}\lambda_\omega$$

is called the entropy of the statistical state μ or, with a slight abuse of language, the entropy of the density ρ.

Remark 13.

1. *By convention we state that 0 log0 = 0. With that convention the function $x \mapsto x \log x$ is continuous on $\mathbb{R}^+$. If the integral on the right hand side of the equality which defines $s(\rho)$ does not converge, we state that $s(\rho) = -\infty$. With these conventions, $s(\rho)$ exists for any continuous probability density ρ.*
2. *The above Definition 16 of the entropy of a statistical state, founded on ideas developed by Boltzmann in his Kinetic Theory of Gases [46], specially in the derivation of his famous (and controversed) Theorem Êta, is too related with the ideas of Claude Shannon [47] on information theory. The use of information theory in thermodynamics was more recently proposed by Jaynes [48,49] and Mackey [18]. For a very nice discussion of the use of probability concepts in physics and application of information theory in quantum mechanics, the reader is referred to the paper by Balian [50].*

The entropy $s(\rho)$ of a probability density ρ has very remarkable variational properties discussed in the following definitions and proposition.

Definition 17. *Let ρ be the density of a smooth statistical state on a symplectic manifold (M, ω).*

1. *For each function f defined on M, taking its values in $\mathbb{R}$ or in some finite-dimensional vector space, such that the integral on the right hand side of the equality*
$$\mathcal{E}_\rho(f) = \int_M f\rho \mathrm{d}\lambda_\omega$$
converges, the value $\mathcal{E}_\rho(f)$ of that integral is called the mean value of f with respect to ρ.
2. *Let f be a smooth function on M, taking its values in $\mathbb{R}$ or in some finite-dimensional vector space, satisfying the properties stated above. A smooth infinitesimal variation of ρ with fixed mean value of f is a smooth map, defined on the product $]-\varepsilon, \varepsilon[\times M$, with values in $\mathbb{R}^+$, where $\varepsilon > 0$,*
$$(\tau, z) \mapsto \rho(\tau, z), \quad \tau \in]-\varepsilon, \varepsilon[, \; z \in M,$$
such that
   - *for $\tau = 0$ and any $z \in M$, $\rho(0, z) = \rho(z)$,*
   - *for each $\tau \in]-\varepsilon, \varepsilon[$, $z \mapsto \rho_\tau(z) = \rho(\tau, z)$ is a smooth probability density on M such that*
$$\mathcal{E}_{\rho_\tau}(f) = \int_M \rho_\tau f \mathrm{d}\lambda_\omega = \mathcal{E}_\rho(f).$$
3. *The entropy function s is said to be stationary at the probability density ρ with respect to smooth infinitesimal variations of ρ with fixed mean value of f, if for any smooth infinitesimal variation $(\tau, z) \mapsto \rho(\tau, z)$ of ρ with fixed mean value of f*
$$\frac{\mathrm{d}s(\rho_\tau)}{\mathrm{d}\tau}\Big|_{\tau=0} = 0.$$

Proposition 8. *Let $H : M \to \mathbb{R}$ be a smooth Hamiltonian on a symplectic manifold (M, ω) and ρ be the density of a smooth statistical state on M such that the integral defining the mean value $\mathcal{E}_\rho(H)$ of H with respect to ρ converges. The entropy function s is stationary at ρ with respect to smooth infinitesimal variations of ρ with fixed mean value of H, if and only if there exists a real $b \in \mathbb{R}$ such that, for all $z \in M$,*
$$\rho(z) = \frac{1}{P(b)} \exp(-bH(z)), \quad \textit{with} \quad P(b) = \int_M \exp(-bH) \mathrm{d}\lambda_\omega.$$

Proof. Let $\tau \mapsto \rho_\tau$ be a smooth infinitesimal variation of ρ with fixed mean value of H. Since $\int_M \rho_\tau \mathrm{d}\lambda_\omega$ and $\int_M \rho_\tau H \mathrm{d}\lambda_\omega$ do not depend on τ, it satisfies, for all $\tau \in]-\varepsilon, \varepsilon[$,
$$\int_M \frac{\partial\rho(\tau, z)}{\partial\tau} \mathrm{d}\lambda_\omega(z) = 0, \int_M \frac{\partial\rho(\tau, z)}{\partial\tau} H(z) \mathrm{d}\lambda_\omega(z) = 0.$$
Moreover an easy calculation leads to
$$\frac{\mathrm{d}s(\rho_\tau)}{\mathrm{d}\tau}\Big|_{\tau=0} = -\int_M \frac{\partial\rho(\tau, z)}{\partial\tau}\Big|_{\tau=0} (1 + \log(\rho(z)) \mathrm{d}\lambda_\omega(z).$$

A well known result in calculus of variations shows that the entropy function s is stationary at ρ with respect to smooth infinitesimal variations of ρ with fixed mean value of H, if and only if there exist two real constants a and b, called *Lagrange multipliers*, such that, for all $z \in M$,
$$1 + \log(\rho) + a + bH = 0,$$

which leads to

$$\rho = \exp(-1 - a - bH)\,.$$

By writing that $\int_M \rho \mathrm{d}\lambda_\omega = 1$, we see that a is determined by b:

$$\exp(1 + a) = P(b) = \int_M \exp(-bH)\mathrm{d}\lambda_\omega\,.$$

□

Definition 18. *Let $H : M \to \mathbb{R}$ be a smooth Hamiltonian on a symplectic manifold (M, ω). For each $b \in \mathbb{R}$ such that the integral on the right side of the equality*

$$P(b) = \int_M \exp(-bH)\mathrm{d}\lambda_\omega$$

converges, the smooth probability measure on M with density (with respect to the Liouville measure)

$$\rho(b) = \frac{1}{P(b)} \exp(-bH)$$

is called the Gibbs statistical state associated to b. The function $P : b \mapsto P(b)$ is called the partition function.

The following proposition shows that the entropy function, not only is stationary at any Gibbs statistical state, but in a certain sense attains at that state a strict maximum.

Proposition 9. *Let $H : M \to \mathbb{R}$ be a smooth Hamiltonian on a symplectic manifold (M, ω) and $b \in \mathbb{R}$ be such that the integral defining the value $P(b)$ of the partition function P at b converges. Let*

$$\rho_b = \frac{1}{P(b)} \exp(-bH)$$

be the probability density of the Gibbs statistical state associated to b. We assume that the Hamiltonian H is bounded by below, i.e., that there exists a constant m such that $m \leq H(z)$ for any $z \in M$. Then the integral defining

$$\mathcal{E}_{\rho_b}(H) = \int_M \rho_b H \mathrm{d}\lambda_\omega$$

converges. For any other smooth probability density ρ_1 such that

$$\mathcal{E}_{\rho_1}(H) = \mathcal{E}_{\rho_b}(H)\,,$$

we have

$$s(\rho_1) \leq s(\rho_b)\,,$$

and the equality $s(\rho_1) = s(\rho_b)$ holds if and only if $\rho_1 = \rho_b$.

Proof. Since $m \leq H$, the function $\rho_b \exp(-bH)$ satisfies $0 \leq \rho_b \exp(-bH) \leq \exp(-mb)\rho_b$, therefore is integrable on M. Let ρ_1 be any smooth probability density on M satisfying $\mathcal{E}_{\rho_1}(H) = \mathcal{E}_{\rho_b}(H)$. The function defined on $\mathbb{R}^+$

$$x \mapsto h(x) = \begin{cases} x \log\left(\dfrac{1}{x}\right) & \text{if } x > 0 \\ 0 & \text{if } x = 0 \end{cases}$$

being convex, its graph is below the tangent at any of its points $\big(x_0, h(x_0)\big)$. We therefore have, for all $x > 0$ and $x_0 > 0$,

$$h(x) \leq h(x_0) - (1 + \log x_0)(x - x_0) = x_0 - x(1 + \log x_0)\,.$$

With $x = \rho_1(z)$ and $x_0 = \rho_b(z)$, z being any element in M, that inequality becomes

$$h(\rho_1(z)) = \rho_1(z)\log\left(\frac{1}{\rho_1(z)}\right) \leq \rho_b(z) - (1 + \log\rho_b(z))\rho_1(z)\,.$$

By integration over M, using the fact that ρ_b is the probability density of the Gibbs state associated to b, we obtain

$$s(\rho_1) \leq 1 - 1 - \int_M \rho_1 \log\rho_b \mathrm{d}\lambda_\omega = s(\rho_b)\,.$$

We have proven the inequality $s(\rho_1) \leq s(\rho_b)$. If $\rho_1 = \rho_b$, we have of course the equality $s(\rho_1) = s(\rho_b)$. Conversely if $s(\rho_1) = s(\rho_b)$, the functions defined on M

$$z \mapsto \varphi_1(z) = \rho_1(z)\log\left(\frac{1}{\rho_1(z)}\right) \quad \text{and} \quad z \mapsto \varphi(z) = \rho_b(z) - (1 + \log\rho_b(z))\rho_1(z)$$

are continuous on M except, maybe, for φ, at points z at which $\rho_b(z) = 0$ and $\rho_1(z) \neq 0$, but the set of such points is of measure 0 since φ is integrable. They satisfy the inequality $\varphi_1 \leq \varphi$. Both are integrable on M and have the same integral. The function $\varphi - \varphi_1$ is everywhere ≥ 0, is integrable on M and its integral is 0. That function is therefore everywhere equal to 0 on M. We can write, for any $z \in M$,

$$\rho_1(z)\log\left(\frac{1}{\rho_1(z)}\right) = \rho_b(z) - (1 + \log\rho_b(z))\rho_1(z)\,. \tag{6}$$

For each $z \in M$ such that $\rho_1(z) \neq 0$, we can divide that equality by $\rho_1(z)$. We obtain

$$\frac{\rho_b(z)}{\rho_1(z)} - \log\left(\frac{\rho_b(z)}{\rho_1(z)}\right) = 1\,.$$

Since the function $x \mapsto x - \log x$ reaches its minimum, equal to 1, for a unique value of $x > 0$, that value being 1, we see that for each $z \in M$ at which $\rho_1(z) > 0$, we have $\rho_1(z) = \rho_b(z)$. At points $z \in M$ at which $\rho_1(z) = 0$, Equation (6) shows that $\rho_b(z) = 0$. Therefore $\rho_1 = \rho_b$. □

Remark 14. *The maximality property of the entropy function $\rho \mapsto s(\rho)$ at a Gibbs state density ρ_b proven in Proposition 9 of course implies the stationarity of that function at ρ_b with respect to smooth infinitesimal variations of ρ with fixed mean value of H, proven in Proposition 8. That Proposition therefore could be omitted. We chose to keep it because its proof is much easier than that of Proposition 9, and explains why it is interesting to look at probability densities proportional to $\exp(-bH)$ for some $b \in \mathbb{R}$.*

The following proposition shows that a Gibbs statistical state remains invariant under the flow of the Hamiltonian vector field X_H. One can therefore say that a Gibbs state is a statistical equilibrium state. Of course there exist statistical equilibrium states other than Gibbs states.

Proposition 10. *Let H be a smooth Hamiltonian bounded by below on a symplectic manifold (M, ω), $b \in \mathbb{R}$ be such that the integral defining the value $P(b)$ of the partition function P at b converges. The Gibbs state associated to b remains invariant under the flow of of the Hamiltonian vector field X_H.*

Proof. The density ρ_b of the Gibbs state associated to b, with respect to the Liouville measure λ_ω, is

$$\rho_b = \frac{1}{P(b)}\exp(-bH)\,.$$

Since H is constant along each integral curve of X_H, ρ_b too is constant along each integral curve of X_H. Moreover, the Liouville measure λ_ω remains invariant under the flow of X_H. Therefore the Gibbs probability measure associated to b too remains invariant under that flow. □

6.2. Thermodynamic Equilibria and Thermodynamic Functions

6.2.1. Assumptions Made in this Section.

Any Hamiltonian H defined on a symplectic manifold (M,ω) considered in this section will be assumed to be smooth, bounded by below and such that for any real $b > 0$, each one of the three functions, defined on M, $z \mapsto \exp(-bH(z))$, $z \mapsto |H(z)|\exp(-bH(z))$ and $z \mapsto (H(z))^2\exp(-bH(z))$ is everywhere smaller than some function defined on M integrable with respect to the Liouville measure λ_ω. The integrals which define

$$P(b) = \int_M \exp(-bH)\mathrm{d}\lambda_\omega \quad \text{and} \quad \mathcal{E}_{\rho_b}(H) = \int_M H\exp(-bH)\mathrm{d}\lambda_\omega$$

therefore converge.

Proposition 11. *Let H be a Hamiltonian defined on a symplectic manifold (M,ω) satisfying the assumptions indicated in Section 6.2.1. For any real $b > 0$ let*

$$P(b) = \int_M \exp(-bH)\mathrm{d}\lambda_\omega \quad \textit{and} \quad \rho_b = \frac{1}{P(b)}\exp(-bH)$$

be the value at b of the partition function P and the probability density of the Gibbs statistical state associated to b, and

$$E(b) = \mathcal{E}_{\rho_b}(H) = \frac{1}{P(b)}\int_M H\exp(-bH)\mathrm{d}\lambda_\omega$$

be the mean value of H with respect to the probability density ρ_b. The first and second derivatives with respect to b of the partition function P exist, are continuous functions of b given by

$$\frac{\mathrm{d}P(b)}{\mathrm{d}b} = -P(b)E(b)\,, \quad \frac{\mathrm{d}^2P(b)}{\mathrm{d}b^2} = \int_M H^2\exp(-bH)\mathrm{d}\lambda_\omega = P(b)\mathcal{E}_{\rho_b}(H^2)\,.$$

The derivative with respect to b of the function E exists and is a continuous function of b given by

$$\frac{\mathrm{d}E(b)}{\mathrm{d}b} = -\frac{1}{P(b)}\int_M \big(H - \mathcal{E}_{\rho_b}(H)\big)^2\mathrm{d}\lambda_\omega = -\mathcal{E}_{\rho_b}\Big(\big(H - \mathcal{E}_{\rho_b}(H)\big)^2\Big)\,.$$

Let $S(b)$ be the entropy $s(\rho_b)$ of the Gibbs statistical state associated to b. The function S can be expressed in terms of P and E as

$$S(b) = \log\big(P(b)\big) + bE(b)\,.$$

Its derivative with respect to b exists and is a continuous function of b given by

$$\frac{\mathrm{d}S(b)}{\mathrm{d}b} = b\frac{\mathrm{d}E(b)}{\mathrm{d}b}\,.$$

Proof. Using the assumptions Section 6.2.1, we see that the functions $b \mapsto P(b)$ and $b \mapsto \mathcal{E}_{\rho_b}(H) = E(b)$, defined by integrals on M, have a derivative with respect to b which is continuous and which can be calculated by derivation under the sign $\int_M$. The indicated results easily follow, if we observe that for any function f on M such that $\mathcal{E}_{\rho_b}(f)$ and $\mathcal{E}_{\rho_b}(f^2)$ exist, we have the formula, well known in Probability theory,

$$\mathcal{E}_{\rho_b}(f^2) - \big(\mathcal{E}_{\rho_b}(f)\big)^2 = \mathcal{E}_{\rho_b}\Big(\big(f - \mathcal{E}_{\rho_b}(f)\big)^2\Big)\,.$$

□

6.2.2. Physical Meaning of the Introduced Functions

Let us consider a physical system, for example a gas contained in a vessel bounded by rigid, thermally insulated walls, at rest in a Galilean reference frame. We assume that its evolution can be mathematically described by means of a Hamiltonian system on a symplectic manifold (M, ω) whose Hamiltonian H satisfies the assumptions Section 6.2.1. For physicists, a Gibbs statistical state, i.e., a probability measure of density $\rho_b = \dfrac{1}{P(b)} \exp(-bH)$ on M, is a *thermodynamic equilibrium* of the physical system. The set of possible thermodynamic equilibria of the system is therefore indexed by a real parameter $b > 0$. The following argument will show what physical meaning can have that parameter.

Let us consider two similar physical systems, mathematically described by two Hamiltonian systems, of Hamiltonians H_1 on the symplectic manifold (M_1, ω_1) and H_2 on the symplectic manifold (M_2, ω_2). We first assume that they are independent and both in thermodynamic equilibrium, with different values b_1 and b_2 of the parameter b. We denote by $E_1(b_1)$ and $E_2(b_2)$ the mean values of H_1 on the manifold M_1 with respect to the Gibbs state of density ρ_{1,b_1} and of H_2 on the manifold M_2 with respect to the Gibbs state of density ρ_{2,b_2}. We assume now that the two systems are coupled in a way allowing an exchange of energy. For example, the two vessels containing the two gases can be separated by a wall allowing a heat transfer between them. Coupled together, they make a new physical system, mathematically described by a Hamiltonian system on the symplectic manifold $(M_1 \times M_2, p_1^*\omega_1 + p_2^*\omega_2)$, where $p_1 : M_1 \times M_2 \to M_1$ and $p_2 : M_1 \times M_2 \to M_2$ are the canonical projections. The Hamiltonian of this new system can be made as close to $H_1 \circ p_1 + H_2 \circ p_2$ as one wishes, by making very small the coupling between the two systems. The mean value of the Hamiltonian of the new system is therefore very close to $E_1(b_1) + E_2(b_2)$. When the total system will reach a state of thermodynamic equilibrium, the probability densities of the Gibbs states of its two parts, $\rho_{1,b'}$ on M_1 and $\rho_{2,b'}$ on M_2 will be indexed by the same real number $b' > 0$, which must be such that

$$E_1(b') + E_2(b') = E_1(b_1) + E_2(b_2)\,.$$

By Proposition 11, we have, for all $b > 0$,

$$\frac{dE_1(b)}{db} \leq 0\,, \quad \frac{dE_2(b)}{db} \leq 0\,.$$

Therefore b' must lie between b_1 and b_2. If, for example, $b_1 < b_2$, we see that $E_1(b') \leq E_1(b_1)$ and $E_2(b') \geq E_2(b_2)$. In order to reach a state of thermodynamic equilibrium, energy must be transferred from the part of the system where b has the smallest value, towards the part of the system where b has the highest value, until, at thermodynamic equilibrium, b has the same value everywhere. Everyday experience shows that thermal energy flows from parts of a system where the temperature is higher, towards parts where it is lower. For this reason physicists consider the real variable b as a way to appreciate the temperature of a physical system in a state of thermodynamic equilibrium. More precisely, they state that

$$b = \frac{1}{kT}$$

where T is the absolute temperature and k a constant depending on the choice of units of energy and temperature, called *Boltzmann's constant* in honour of the great Austrian scientist Ludwig Eduard Boltzmann (1844–1906).

For a physical system mathematically described by a Hamiltonian system on a symplectic manifold (M, ω), with H as Hamiltonian, in a state of thermodynamic equilibrium, $E(b)$ and $S(b)$ are the *internal energy* and the *entropy* of the system.

6.2.3. Towards Thermodynamic Equilibrium

Everyday experience shows that a physical system, when submitted to external conditions which remain unchanged for a sufficiently long time, very often reaches a state of thermodynamic equilibrium. At first look, it seems that Lagrangian or Hamiltonian systems with time-independent Lagrangians or Hamiltonians cannot exhibit a similar behaviour. Let us indeed consider a mechanical system whose configuration space is a smooth manifold N, described in the Lagrangian formalism by a smooth time-independent hyper-regular Lagarangian $L : TN \to \mathbb{R}$ or, in the Hamiltonian formalism, by the associated Hamiltonian $H_L : T^*N \to \mathbb{R}$. Let $t \mapsto \overrightarrow{x(t)}$ be a motion of that system, $\overrightarrow{x_0} = \overrightarrow{x(t_0)}$ and $\overrightarrow{x_1} = \overrightarrow{x(t_0)}$ be the configurations of the system for that motion at times t_0 and t_1. There exists another motion $t \mapsto \overrightarrow{x'(t)}$ of the system for which $\overrightarrow{x'(t_0)} = \overrightarrow{x_1}$ and $\overrightarrow{x'(t_1)} = \overrightarrow{x_0}$: since the equations of motion are invariant by time reversal, the motion $t \mapsto \overrightarrow{x'(t)}$ is obtained simply by taking as initial condition at time t_0 $\overrightarrow{x'(t_0)} = \overrightarrow{x(t_1)}$ and $\left.\dfrac{d\overrightarrow{x'(t)}}{dt}\right|_{t=t_0} = -\left.\dfrac{d\overrightarrow{x(t)}}{dt}\right|_{t=t_1}$. Another more serious argument against a kind of thermodynamic behaviour of Lagarangian or Hamiltonian systems rests on the famous recurrence theorem due to Poincaré [51]. This theorem asserts indeed that when the useful part of the phase space of the system is of a finite total measure, almost all points in an arbitrarily small open subset of the phase space are recurrent, i.e., the motion starting of such a point at time t_0 repeatedly crosses that open subset again and again, infinitely many times when $t \to +\infty$.

Let us now consider, instead of perfectly defined states, i.e., points in phase space, statistical states, and ask the question: When at time $t = t_0$ a Hamiltonian system on a symplectic manifold (M, ω) is in a statistical state given by some probability measure of density ρ_0 with respect to the Liouville measure λ_ω, does its statistical state converge, when $t \to +\infty$, towards the probability measure of a Gibbs state? This question should be made more precise by specifying what physical meaning has a statistical state and in what mathematical sense a statistical state can converge towards the probability measure of a Gibbs state. A positive partial answer was given by Ludwig Boltzmann when, developing his kinetic theory of gases, he proved his famous (but controversed) *Êta theorem* stating that the entropy of the statistical state of a gas of small particles is a monotonously increasing function of time. This question, linked with time irreversibility in physics, is still the subject of important researches, both by physicists and by mathematicians. The reader is referred to the paper [50] by Balian for a more thorough discussion of that question.

6.3. *Examples of Thermodynamic Equilibria*

6.3.1. Classical Monoatomic Ideal Gas

In classical mechanics, a dilute gas contained in a vessel at rest in a Galilean reference frame is mathematically described by a Hamiltonian system made by a large number of very small massive particles, which interact by very brief collisions between themselves or with the walls of the vessel, whose motions between two collisions are free. Let us first assume that these particles are material points and that no external field is acting on them, other than that describing the interactions by collisions with the walls of the vessel.

The Hamiltonian of one particle in a part of the phase space in which its motion is free is simply

$$\frac{1}{2m}\|\overrightarrow{p}\|^2 = \frac{1}{2m}(p_1^2 + p_2^2 + p_3^2), \quad \text{with} \quad \overrightarrow{p} = m\overrightarrow{v},$$

where m is the mass of the particle, $\overrightarrow{v}$ its velocity vector and $\overrightarrow{p}$ its linear momentum vector (in the considered Galilean reference frame), p_1, p_2 and p_3 the components of $\overrightarrow{p}$ in a fixed orhtonormal basis of the physical space.

Let N be the total number of particles, which may not have all the same mass. We use a integer $i \in \{1, 2, \ldots, N\}$ to label the particles and denote by m_i, $\overrightarrow{x_i}$, $\overrightarrow{v_i}$, $\overrightarrow{p_i}$ the mass and the vectors position, velocity and linear momentum of the i-th particle.

The Hamiltonian of the gas is therefore

$$H = \sum_{i=1}^{N} \frac{1}{2m_i}\|\overrightarrow{p_i}\|^2 + \text{ terms involving the collisions between particles and with the walls}.$$

Interactions of the particles with the walls of the vessel are essential for allowing the motions of particles to remain confined. Interactions between particles are essential to allow the exchanges between them of energy and momentum, which play an important part in the evolution with time of the statistical state of the system. However it appears that while these terms are very important to determine the system's evolution with time, they can be neglected, when the gas is dilute enough, if we only want to determine the final statistical state of the system, once a thermodynamic equilibrium is established. The Hamiltonian used will therefore be

$$H = \sum_{i=1}^{N} \frac{1}{2m_i} \| \overrightarrow{p_i} \|^2 .$$

The partition function is

$$P(b) = \int_M \exp(-bH)\mathrm{d}\lambda_\omega = \int_D \exp\left(-b\sum_{i=1}^{N}\frac{1}{2m_i}\|\overrightarrow{p}_i\|^2\right)\prod_{i=1}^{N}(\mathrm{d}\overrightarrow{x_i}\,\mathrm{d}\overrightarrow{p_i}),$$

where D is the domain of the $6N$-dimensional space spanned by the position vectors $\overrightarrow{x_i}$ and linear momentum vectors $\overrightarrow{p_i}$ of the particles in which all the $\overrightarrow{x_i}$ lie within the vessel containing the gas. An easy calculation leads to

$$P(b) = V^N \left(\frac{2\pi}{b}\right)^{3N/2} \prod_{i=1}^{N}({m_i}^{3/2}) = \prod_{i=1}^{N}\left[V\left(\frac{2\pi m_i}{b}\right)^{3/2}\right],$$

where V is the volume of the vessel which contains the gas. The probability density of the Gibbs state associated to b, with respect to the Liouville measure, therefore is

$$\rho_b = \prod_{i=1}^{N}\left[\frac{1}{V}\left(\frac{b}{2\pi m_i}\right)^{3/2} \exp\left(\frac{-b\|\overrightarrow{p_i}\|^2}{2m_i}\right)\right].$$

We observe that ρ_b is the product of the probability densities $\rho_{i,b}$ for the i-th particle

$$\rho_{i,b} = \frac{1}{V}\left(\frac{b}{2\pi m_i}\right)^{3/2} \exp\left(\frac{-b\|\overrightarrow{p_i}\|^2}{2m_i}\right).$$

The $2N$ stochastic vectors $\overrightarrow{x_i}$ and $\overrightarrow{p_i}$, $i = 1, \ldots, N$ are therefore independent. The position $\overrightarrow{x_i}$ of the i-th particle is uniformly distributed in the volume of the vessel, while the probability measure of its linear momentum $\overrightarrow{p_i}$ is the classical *Maxwell–Boltzmann probability distribution of linear momentum* for an ideal gas of particles of mass m_i, first obtained by Maxwell in 1860. Moreover we see that the three components $p_{i\,1}$, $p_{i\,2}$ and $p_{i\,3}$ of the linear momentum $\overrightarrow{p_i}$ in an orhonormal basis of the physical space are independent stochastic variables.

By using the formulae given in Proposition 11 the internal energy $E(b)$ and the entropy $S(b)$ of the gas can be easily deduced from the partition function $P(b)$. Their expressions are

$$E(b) = \frac{3N}{2b}, \quad S(b) = \frac{3}{2}\sum_{i=1}^{N}\log m_i + \left(\frac{3}{2}(1+\log(2\pi)) + \log V\right)N - \frac{3N}{2}\log b .$$

We see that each of the N particles present in the gas has the same contribution $\dfrac{3}{2b}$ to the internal energy $E(b)$, which does not depend on the mass of the particle. Even more: each degree of freedom of each particle, i.e., each of the the three components of the the linear momentum of the particle on the three axes of an orthonormal basis, has the same contribution $\dfrac{1}{2b}$ to the internal energy $E(b)$. This result is known in physics under the name *Theorem of equipartition of the energy at a thermodynamic equilibrium*. It can be easily generalized for polyatomic gases, in which a particle may carry, in addition to the kinetic energy due to the velocity of its centre of mass, a kinetic energy due to the particle's rotation around its centre of mass. The reader can consult the books by Souriau [14] and Mackey [18] where the kinetic theory of polyatomic gases is discussed.

The pressure in the gas, denoted by $\Pi(b)$ because the notation $P(b)$ is already used for the partition function, is due to the change of linear momentum of the particles which occurs at a collision of the particle with the walls of the vessel containing the gas (or with a probe used to measure that pressure). A classical argument in the kinetic theory of gases (see for example [52,53]) leads to

$$\Pi(b) = \frac{2}{3}\frac{E(b)}{V} = \frac{N}{Vb}.$$

This formula is the well known *equation of state* of an ideal monoatomic gas relating the number of particles by unit of volume, the pressure and the temperature.

With $b = \dfrac{1}{kT}$, the above expressions are exactly those used in classical thermodynamics for an ideal monoatomic gas.

6.3.2. Classical Ideal Monoatomic Gas in a Gravity Field

Let us now assume that the gas, contained in a cylindrical vessel of section Σ and length h, with a vertical axis, is submitted to the vertical gravity field of intensity g directed downwards. We choose Cartesian coordinates x, y, z, the z axis being vertical directed upwards, the bottom of the vessel being in the horizontal surface $z = 0$. The Hamiltonian of a free particle of mass m, position and linear momentum vectors $\overrightarrow{x}$ (components x, y, z) and $\overrightarrow{p}$ (components p_x, p_y and p_z) is

$$\frac{1}{2m}(p_x^2 + p_y^2 + p_z^2) + mgz.$$

As in the previous section we neglect the parts of the Hamiltonian of the gas corresponding to collisions between the particles, or between a particle and the walls of the vessel. The Hamiltonian of the gas is therefore

$$H = \sum_{i=1}^{N}\left(\frac{1}{2m_i}(p_{ix}^2 + p_{iy}^2 + p_{iz}^2) + m_i g z_i\right).$$

Calculations similar to those of the previous section lead to

$$P(b) = \prod_{i=1}^{N}\left[\Sigma\left(\frac{2\pi m_i}{b}\right)^{3/2}\frac{1-\exp(-m_i g b h)}{m_i g b}\right],$$

$$\rho_b = \frac{1}{P(b)}\exp\left[-b\sum_{i=1}^{N}\left(\frac{\|\overrightarrow{p_i}\|^2}{2m_i} + m_i g z_i\right)\right].$$

The expression of ρ_b shows that the $2N$ stochastic vectors $\overrightarrow{x_i}$ and $\overrightarrow{p_i}$ still are independent, and that for each $i \in \{1, \ldots, N\}$, the probability law of each stochastic vector $\overrightarrow{p_i}$ is the same as in the absence of gravity, for the same value of b. Each stochastic vector $\overrightarrow{x_i}$ is no more uniformly distributed in the vessel containing the gas: its probability density is higher at lower altitudes z, and this nonuniformity is more important for the heavier particles than for the lighter ones.

As in the previous section, the formulae given in Proposition 11 allow the calculation of $E(b)$ and $S(b)$. We observe that $E(b)$ now includes the potential energy of the gas in the gravity field, therefore should no more be called the internal energy of the gas.

6.3.3. Relativistic Monoatomic Ideal Gas

In a Galilean reference frame, we consider a relativistic point particle of rest mass m, moving at a velocity $\overrightarrow{v}$. We denote by v the modulus of $\overrightarrow{v}$ and by c the modulus of the velocity of light.

The motion of the particle can be mathematically described by means of the Euler–Lagrange equations, with the Lagrangian

$$L = -mc^2\sqrt{1-\frac{v^2}{c^2}}\,.$$

The components of the linear momentum $\overrightarrow{p}$ of the particle, in an orthonormal frame at rest in the considered Galilean reference frame, are

$$p_i = \frac{\partial L}{\partial v^i} = \frac{mv^i}{\sqrt{1-\dfrac{v^2}{c^2}}}\,, \quad \text{therefore} \quad \overrightarrow{p} = \frac{m\overrightarrow{v}}{\sqrt{1-\dfrac{v^2}{c^2}}}\,.$$

Denoting by p the modulus of $\overrightarrow{p}$, the Hamiltonian of the particle is

$$H = \overrightarrow{p}\cdot\overrightarrow{v} - L = \frac{mc^2}{\sqrt{1-\dfrac{v^2}{c^2}}} = c\sqrt{p^2+m^2c^2}\,.$$

Let us consider a relativistic gas, made of N point particles indexed by $i \in \{1,\dots,N\}$, m_i being the rest mass of the i-th particle. With the same assumptions as those made in Section 6.3.1, we can take for Hamiltonian of the gas

$$H = c\sum_{i=1}^{N}\sqrt{p_i{}^2+m^2c^2}\,.$$

With the same notations as those of Section 6.3.1, the partition function P of the gas takes the value, for each $b>0$,

$$P(b) = \int_D \exp\left(-bc\sum_{i=1}^{N}\sqrt{(p_i)^2+m^2c^2}\right)\prod_{i=1}^{N}(\mathrm{d}\overrightarrow{x_i}\,\mathrm{d}\overrightarrow{p_i})\,.$$

This integral can be expressed in terms of the Bessel function K_2, whose expression is, for each $x>0$,

$$K_2(x) = x\int_0^{+\infty}\exp(-x\operatorname{ch}\chi)\operatorname{sh}^2\chi\operatorname{ch}\chi\,\mathrm{d}\chi\,.$$

We have

$$P(b) = \left(\frac{4\pi Vc}{b}\right)^N\prod_{i=1}^{N}\left(m_i{}^2K_2(m_ibc^2)\right),$$

$$\rho_b = \frac{1}{P(b)}\exp\left(-bc\sum_{i=1}^{N}\sqrt{p_i{}^2+m_i{}^2c^2}\right)\,.$$

This probability density of the Gibbs state shows that the $2N$ stochastic vectors $\overrightarrow{x_i}$ and $\overrightarrow{p_i}$ are independent, that each $\overrightarrow{x_i}$ is uniformly distributed in the vessel containing the gas and that the probability density of each $\overrightarrow{p_i}$ is exactly the probability distribution of the linear momentum of particles in a relativistic gas called the *Maxwell–Jüttner distribution*, obtained by Ferencz Jüttner (1878–1958) in 1911, discussed in the book by the Irish mathematician and physicist Synge [54].

Of course, the formulae given in Proposition 11 allow the calculation of the internal energy $E(b)$, the entropy $S(b)$ and the pressure $\Pi(b)$ of the relativistic gas.

6.3.4. Relativistic IDeal Gas of Massless Particles

We have seen in the previous Chapter that in an inertial reference frame, the Hamiltonian of a relativistic point particle of rest mass m is $c\sqrt{p^2+m^2c^2}$, where p is the modulus of the linear momentum vector $\overrightarrow{p}$ of the particle in the considered reference frame. This expression still has a meaning when the rest mass m of the particle is 0. In an orthonormal reference frame, the equations of motion of a particle whose motion is mathematically described by a Hamiltonian system with Hamiltonian

$$H = cp = c\sqrt{p_1{}^2 + p_2{}^2 + p_3{}^2}$$

are

$$\begin{cases} \dfrac{\mathrm{d}x^i}{\mathrm{d}t} = \dfrac{\partial H}{\partial p_i} = c\,\dfrac{p_i}{p} \\ \dfrac{\mathrm{d}p_i}{\mathrm{d}t} = -\dfrac{\partial H}{\partial x^i} = 0\,, \end{cases} \quad (1 \le i \le 3)\,,$$

which shows that the particle moves on a straight line at the velocity of light c. It seems therefore reasonable to describe a gas of N photons in a vessel of volume V at rest in an inertial reference frame by a Hamiltonian system, with the Hamiltonian

$$H = c\sum_{i=1}^{N} \|\overrightarrow{p_i}\| = c\sum_{i=1}^{N} \sqrt{p_{i1}{}^2 + p_{i2}{}^2 + p_{i3}{}^2}\,.$$

With the same notations as those used in the previous section, the partition function P of the gas takes the value, for each $b > 0$,

$$P(b) = \int_D \exp\left(-bc\sum_{i=1}^{N} \|\overrightarrow{p_i}\|\right) \prod_{i=1}^{N} (\mathrm{d}\overrightarrow{x_i}\,\mathrm{d}\overrightarrow{p_i}) = \left(\frac{8\pi V}{c^3 b^3}\right)^N\,.$$

The probability density of the corresponding Gibbs state, with respect to the Liouville measure $\lambda_\omega = \prod_{i=1}^{N} (\mathrm{d}\overrightarrow{x_i}\,\mathrm{d}\overrightarrow{p_i})$, is

$$\rho_b = \prod_{i=1}^{N} \left(\frac{c^3 b^3}{8\pi V}\right) \exp(-bc\|\overrightarrow{p_i}\|)\,.$$

This formula appears in the books by Synge [54] and Souriau [14]. Physicists consider it as not adequate for the description of a gas of photons contained in a vessel at thermal equilibrium because the number of photons in the vessel, at any given temperature, cannot be imposed: it results from the processes of absorption and emission of photons by the walls of the vessel, heated at the imposed temperature, which spontaneously occur. In other words, this number is a stochastic function whose probability law is imposed by Nature. Souriau proposes, in his book [14], a way to account for the possible variation of the number of photons. Instead of using the *phase space* of the system of N massless relativistic particles contained in a vessel, he uses the *manifold of motions* M_N of that system (which is symplectomorphic to its phase space). He considers that the manifold of motions M of a system of photons in the vessel is the disjoint union

$$M = \bigcup_{N \in \mathbb{N}} M_N\,,$$

of all the manifolds of motions M_N of a system of N massless relativistic particles in the vessel, for all possible values of $N \in \mathbb{N}$. Fo $N = 0$ the manifold M_0 is reduced to a singleton with, as Liouville measure, the measure which takes the value 1 on the only non empty part of that manifold (the whole manifold M_0). Moreover, since any photon cannot be distinguished from any other photon, two motions of the system with the same number N of massless particles which only differ by the labelling of these particles must be considered as identical. Souriau considers too that since the number N of photons freely adjusts itself, the value of the parameter $b = \dfrac{1}{kT}$ must, at thermodynamic equilibrium, be the same in all parts M_N of the system, $N \in \mathbb{N}$. He uses too the fact that a photon can have two different states of (circular) polarization. With these assumptions the value at any b of the partition function of the system is

$$P(b) = \sum_{N=0}^{+\infty} \frac{1}{N!} \left(\frac{16\pi V}{c^3 b^3}\right)^N = \exp\left(\frac{16\pi V}{c^3 b^3}\right).$$

The number N of photons in the vessel at thermodynamic equilibrium is a stochastic function which takes the value n with the probability

$$\text{Probability}([N = n]) = \frac{1}{n!} \left(\frac{16\pi V}{c^3 b^3}\right)^n \exp\left(-\frac{16\pi V}{c^3 b^3}\right).$$

The expression of the partition function P allows the calculation of the internal energy, the entropy and all other thermodynamic functions of the system. However, the formula so obtained for the distribution of photons of various energies at a given temperature does not agree with the law, in very good agreement with experiments, obtained by Max Planck (1858–1947) in 1900. An assembly of photons in thermodynamic equilibrium evidently cannot be described as a classical Hamiltonian system. This fact played an important part for the development of quantum mechanics.

6.3.5. Specific Heat of Solids

The motion of a one-dimensional harmonic oscillator can be described by a Hamiltonian system with, as Hamiltonian,

$$H(p,q) = \frac{p^2}{2m} + \frac{\mu q^2}{2}.$$

The idea that the heat energy of a solid comes from the small vibrations, at a microscopic scale, of its constitutive atoms, lead physicists to attempt to mathematically describe a solid as an assembly of a large number N of three-dimensional harmonic oscillators. By dealing separately with each proper oscillation mode, the solid can even be described as an assembly of $3N$ one-dimensional harmonic oscillators. Exanges of energy between these oscillators is allowed by the existence of small couplings between them. However, for the determination of the thermodynamic equilibria of the solid we will, as in the previous section for ideal gases, consider as negligible the energy of interactions between the oscillators. We therefore take for Hamiltonian of the solid

$$H = \sum_{i=1}^{3N} \left(\frac{{p_i}^2}{2m_i} + \frac{\mu_i {q_i}^2}{2}\right).$$

The value of the paritition function P, for any $b > 0$, is

$$P(b) = \int_{\mathbb{R}^{6N}} \exp\left[-b \sum_{i=1}^{3N} \left(\frac{{p_i}^2}{2m_i} + \frac{\mu_i {q_i}^2}{2}\right)\right] \prod_{i=1}^{3N} (\mathrm{d}p_i \mathrm{d}q_i) = \prod_{i=1}^{3N} \left(\frac{1}{\nu_i}\right) b^{-3N},$$

where

$$\nu_i = \frac{1}{2\pi} \sqrt{\frac{\mu_i}{m_i}}$$

is the frequency of the i-th harmonic oscillator.

The internal energy of the solid is

$$E(b) = -\frac{\mathrm{d} \log P(b)}{\mathrm{d}b} = \frac{3N}{b}.$$

We observe that it only depends on the the temperature and on the number of atoms in the solid, not on the frequencies ν_i of the harmonic oscillators. With $b = \dfrac{1}{kT}$ this result is in agreement with the empirical law for the specific heat of solids, in good agreement with experiments at high temperature, discovered in 1819 by the French scientists Pierre Louis Dulong (1785–1838) and Alexis Thérèse Petit (1791–1820).

7. Generalization for Hamiltonian Actions

7.1. Generalized Gibbs States

In his book [15] and in several papers [13,16,17], Souriau extends the concept of a Gibbs state for a Hamiltonian action of a Lie group G on a symplectic manifold (M, ω). Usual Gibbs states defined in Section 6 for a smooth Hamiltonian H on a symplectic manifold (M, ω) appear as special cases, in which the Lie group is a one-parameter group. If the symplectic manifold (M, ω) is the *phase space* of the Hamiltonian system, that one-parameter group, whose parameter is the time t, is the group of evolution, as a function of time, of the state of the system, starting from its state at some arbitrarily chosen initial time t_0. If (M, ω) is the symplectic manifold of all the *motions* of the system, that one-parameter group, whose parameter is a real $\tau \in \mathbb{R}$, is the transformation group which maps one motion of the system with some initial state at time t_0 onto the motion of the system with the same initial state at another time $(t_0 + \tau)$. We discuss below this generalization.

Notations and Conventions

In this section, $\Phi : G \times M \to M$ is a Hamiltonian action (for example on the left) of a Lie group G on a symplectic manifold (M, ω). We denote by $\mathcal{G}$ the Lie algebra of G, by $\mathcal{G}^*$ its dual space and by $J : M \to \mathcal{G}^*$ a momentum map of the action Φ.

Definition 19. *Let $b \in \mathcal{G}$ be such that the integrals on the right hand sides of the equalities*

$$P(b) = \int_M \exp\bigl(-\langle J, b\rangle\bigr)\,\mathrm{d}\lambda_\omega \quad \textit{and}$$
$$E_J(b) = \mathcal{E}_{\rho_b}(J) = \frac{1}{P(b)}\int_M J\exp\bigl(-\langle J, b\rangle\bigr)\,\mathrm{d}\lambda_\omega$$

converge. The smooth probability measure on M with density (with respect to the Liouville measure λ_ω on M)

$$\rho_b = \frac{1}{P(b)}\exp\bigl(-\langle J, b\rangle\bigr)$$

is called the generalized Gibbs statistical state associated to b. The functions $b \mapsto P(b)$ and $b \mapsto E_J(b)$ so defined on the subset of $\mathcal{G}$ made by elements b for which the integrals defining $P(b)$ and $E_J(b)$ converge are called the partition function associated to the momentum map J and the mean value of J at generalized Gibbs states.

The following Proposition generalizes 9.

Proposition 12. *Let $b \in \mathcal{G}$ be such that the integrals defining $P(b)$ and $E_J(b)$ in Definition 19 converge, and ρ_b be the density of the generalized Gibbs state associated to b. The entropy $s(\rho_b)$, which will be denoted by $S(b)$, exists and is given by*

$$S(b) = \log\bigl(P(b)\bigr) + \langle E_J(b), b\rangle = \log\bigl(P(b)\bigr) - \Bigl\langle D\bigl(\log P(b)\bigr), b\Bigr\rangle. \tag{7}$$

Moreover, for any other smooth probability density ρ_1 such that

$$\mathcal{E}_{\rho_1}(J) = \mathcal{E}_{\rho_b}(J) = E_J(b)\,,$$

we have

$$s(\rho_1) \leq s(\rho_b)\,,$$

and the equality $s(\rho_1) = s(\rho_b)$ holds if and only if $\rho_1 = \rho_b$.

Proof. Equation (7) follows from $\log\left(\dfrac{1}{\rho_b}\right) = \log(P(b)) + \langle J, b\rangle$, and $D(\log P(b)) = -E_J(b)$. The remaining of the proof is the same as that of Proposition 9. □

Remark 15.

1. *The second part of Equation (7), $S(b) = \log(P(b)) - \Big\langle D(\log P(b)), b\Big\rangle$, expresses the fact that the functions $\log(P(b))$ and $-S(b)$ are Legendre transforms of each other: they are linked by the same relation as the relation which links a smooth Lagrangian L and the associated energy E_L.*
2. *The Liouville measure λ_ω remains invariant under the Hamiltonian action Φ, since the symplectic form ω itself remains invariant under that action. However, we have not a full analogue of Proposition 10 because the momentum map J does not remain invariant under the action Φ. We only have the partial anologue stated below.*
3. *Legendre transforms were used by Massieu in thermodynamics in his very early works [55,56], more systematically presented in [57], in which he introduced his characteristic functions (today called thermodynamic potentials) allowing the determination of all the thermodynamic functions of a physical system by partial derivations of a suitably chosen characteristic function. For a modern presentation of that subject the reader is referred to [58,59], Chapter 5, pp. 131–152.*

Proposition 13. *Let $b \in \mathcal{G}$ be such that the integrals defining $P(b)$ and $E_J(b)$ in Definition 19 converge. The generalized Gibbs state associated to b remains invariant under the restriction of the Hamiltonian action Φ to the one-parameter subgroup of G generated by b, $\{\exp(\tau b) \mid \tau \in \mathbb{R}\}$.*

Proof. The orbits of the action on M of the subgroup $\{\exp(\tau b) \mid \tau \in \mathbb{R}\}$ of G are the integral curves of the Hamiltonian vector field whose Hamiltonian is $\langle J, b\rangle$, which of course is constant on each of these curves. Therefore the proof of Proposition 10 is valid for that subgroup. □

7.2. Generalized Thermodynamic Functions

Assumptions Made in this Section

Notations and conventions being the same as in Section 7.1, let Ω be the largest open subset of the Lie algebra $\mathcal{G}$ of G containing all $b \in \mathcal{G}$ satisfying the following properties:

- the functions defined on M, with values, respectively, in $\mathbb{R}$ and in the dual $\mathcal{G}^*$ of $\mathcal{G}$,

$$z \mapsto \exp\Big(-\langle J(z), b\rangle\Big) \quad \text{and} \quad z \mapsto J(z)\exp\Big(-\langle J(z), b\rangle\Big)$$

  are integrable on M with respect to the Liouville measure λ_ω;
- moreover their integrals are differentiable with respect to b, their differentials are continuous and can be calculated by differentiation under the sign $\int_M$.

It is assumed in this section that the considered Hamiltonian action Φ of the Lie group G on the symplectic manifold (M, ω) and its momentum map J are such that the open subset Ω of $\mathcal{G}$ is not empty. This condition is not always satisfied when (M, ω) is a cotangent bundle, but of course it is satisfied when it is a compact manifold.

Proposition 14. *Let $\Phi : G \times M \to M$ be a Hamiltonian action of a Lie group G on a symplectic manifold (M, ω) satisfying the assumptions indicated in Section 7.2. The partition function P associated to the momentum map J and the mean value E_J of J for generalized Gibbs states Definition 19 are defined and continuously differentiable on the open subset Ω of $\mathcal{G}$. For each $b \in \Omega$, the differentials at b of the functions P and $\log P$ (which are linear maps defined on $\mathcal{G}$, with values in $\mathbb{R}$, in other words elements of $\mathcal{G}^*$) are given by*

$$DP(b) = -P(b)E_J(b), \quad D(\log P)(b) = -E_J(b).$$

For each $b \in \Omega$, the differential at b of the map E_J (which is a linear map defined on $\mathcal{G}$, with values in its dual $\mathcal{G}^*$) is given by

$$\big\langle DE_J(b)(Y), Z\big\rangle = \big\langle E_J(b), Y\big\rangle\big\langle E_J(b), Z\big\rangle - \mathcal{E}_{\rho_b}\big(\langle J, Y\rangle\langle J, Z\rangle\big)\,, \quad \textit{with } Y \textit{ and } Z \in \mathcal{G}\,,$$

where we have written, as in Definition 17,

$$\mathcal{E}_{\rho_b}\big(\langle J, Y\rangle\langle J, Z\rangle\big) = \frac{1}{P(b)}\int_M \langle J, Y\rangle\langle J, Z\rangle \exp\big(-\langle J, b\rangle\big)\mathrm{d}\lambda_\omega\,.$$

At each $b \in \Omega$, the differential of the entropy function S Proposition 12, which is a linear map defined on $\mathcal{G}$, with values in $\mathbb{R}$, in other words an element of $\mathcal{G}^*$, is given by

$$\big\langle DS(b), Y\big\rangle = \big\langle DE_J(b)(Y), b\big\rangle\,, \quad Y \in \mathcal{G}\,.$$

Proof. By assumptions Section 7.2, the differentials of P and E_J can be calculated by differentiation under the sign $\int_M$. Easy (but tedious) calculations lead to the indicated results. □

Corollary 3. *With the same assumptions and notations as those in Proposition 14, for any $b \in \Omega$ and $Y \in \mathcal{G}$,*

$$\big\langle DE_J(b)(Y), Y\big\rangle = -\frac{1}{P(b)}\int_M \big\langle J - E_J(b), Y\big\rangle^2 \mathrm{d}\lambda_\omega \leq 0\,.$$

Proof. This result follows from the well known result in Probability theory already used in the proof of Proposition 11. □

The momentum map J of the Hamiltonian action Φ is not uniquely determined: for any constant $\mu \in \mathcal{G}^*$, $J_1 = J + \mu$ too is a momentum map for Φ. The following proposition indicates how the generalized thermodynamic functions P, E_J and S change when J is replaced by J_1.

Proposition 15. *With the same assumptions and notations as those in Proposition 14, let $\mu \in \mathcal{G}^*$ be a constant. When the momentum map J is replaced by $J_1 = J + \mu$, the open subset Ω of $\mathcal{G}$ remains unchanged, while the generalized thermodynamic functions P, E_J and S, are replaced, respectively, by P_1, E_{J_1} and S_1, given by*

$$P_1(b) = \exp\big(-\langle \mu, b\rangle\big)P(b), \quad E_{J_1}(b) = E_J(b) + \mu\,, \quad S_1(b) = S(b)\,.$$

The Gibbs satistical state and its density ρ_b with respect to the Liouville measure λ_ω remain unchanged.

Proof. We have

$$\exp\big(-\langle J + \mu, b\rangle\big) = \exp\big(-\langle \mu, b\rangle\big)\exp\big(-\langle J, b\rangle\big)\,.$$

The indicated results follow by easy calculations. □

The following proposition indicates how the generalized thermodynamic functions P, E_J and S vary along orbits of the adjoint action of the Lie group G on its Lie algebra $\mathcal{G}$.

Proposition 16. *The assumptions and notations are the same as those in Proposition 14. The open subset Ω of $\mathcal{G}$ is an union of orbits of the adjoint action of G on $\mathcal{G}$. In other words, for each $b \in \Omega$ and each $g \in G$, $\mathrm{Ad}_g b \in \Omega$. Moreover, let $\theta : G \to \mathcal{G}^*$ be the symplectic cocycle of G for the coadjoin action of G on $\mathcal{G}^*$ such that, for any $g \in G$,*

$$J \circ \Phi_g = \mathrm{Ad}^*_{g^{-1}} \circ J + \theta(g)\,.$$

Then for each $b \in \Omega$ and each $g \in G$

$$\begin{aligned} P(\mathrm{Ad}_g b) &= \exp\Big(\langle\theta(g^{-1}), b\rangle\Big)P(b) = \exp\Big(-\langle \mathrm{Ad}_g^* \theta(g), b\rangle\Big)P(b)\,,\\ E_J(\mathrm{Ad}_g b) &= \mathrm{Ad}_{g^{-1}}^* E_J(b) + \theta(g)\,,\\ S(\mathrm{Ad}_g b) &= S(b)\,. \end{aligned}$$

Proof. We have

$$\begin{aligned} P(\mathrm{Ad}_g b) &= \int_M \exp(-\langle J, \mathrm{Ad}_g b\rangle) d\lambda_\omega = \int_M \exp(-\langle \mathrm{Ad}_g^* J, b\rangle) d\lambda_\omega\\ &= \int_M \exp\Big(-\langle J \circ \Phi_{g^{-1}} - \theta(g^{-1}, b\rangle\Big) d\lambda_\omega\\ &= \exp\Big(\langle\theta(g^{-1}), b\rangle\Big)P(b) = \exp\Big(-\langle \mathrm{Ad}_g^* \theta(g), b\rangle\Big)P(b)\,, \end{aligned}$$

since $\theta(g^{-1}) = -\mathrm{Ad}_g^* \theta(g)$. By using Propositions 14 and 12, the other results easily follow. □

Remark 16. *The equality*

$$E_J(\mathrm{Ad}_g b) = \mathrm{Ad}_{g^{-1}}^* E_J(b) + \theta(g)$$

means that the map $E_J : \Omega \to \mathcal{G}^*$ is equivariant with respect to the adjoint action of G on the open subset Ω of its Lie algebra $\mathcal{G}$ and its affine action on the left on $\mathcal{G}^*$

$$(g, \xi) \mapsto \mathrm{Ad}_{g^{-1}}^* \xi + \theta(g)\,, \quad g \in G\,, \quad \xi \in \mathcal{G}^*\,.$$

Proposition 17. *The assumptions and notations are the same as those in Proposition 14. For each $b \in \Omega$ and each $X \in \mathcal{G}$, we have*

$$\begin{aligned} \big\langle E_J(b), [X, b]\big\rangle &= \langle \Theta(X), b\rangle\,,\\ DE_J(b)\big([X, b]\big) &= -\mathrm{ad}_X^* E_J(b) + \Theta(X)\,, \end{aligned}$$

where $\Theta = T_e\theta : \mathcal{G} \to \mathcal{G}^*$ is the 1-cocycle of the Lie algebra $\mathcal{G}$ associated to the 1-cocycle θ of the Lie group G.

Proof. Let us set $g = \exp(\tau X)$ in the first equality in Proposition 16, derive that equality with respect to τ, and evaluate the result at $\tau = 0$. We obtain

$$DP(b)\big([X, b]\big) = -P(b)\langle \Theta(X), b\rangle\,.$$

Since, by the first equality of Proposition 14, $DP(b) = -P(b)E_J(b)$, the first stated equality follows.

Let us now set $g = \exp(\tau X)$ in the second equality in Proposition 16, derive that equality with respect to τ, and evaluate the result at $\tau = 0$. We obtain the second equality stated. □

Corollary 4. *With the assumptions and notations of Proposition 17, let us define, for each $b \in \Omega$, a linear map $\Theta_b : \mathcal{G} \to \mathcal{G}^*$ by setting*

$$\Theta_b(X) = \Theta(X) - \mathrm{ad}_X^* E_J(b)\,.$$

The map Θ_b is a symplectic 1-cocycle of the Lie algebra $\mathcal{G}$ for the coadjoint representation, which satisfies

$$\Theta_b(b) = 0\,.$$

Moreover if we replace the momentum map J by $J_1 = J + \mu$, with $\mu \in \mathcal{G}^*$ constant, the 1-cocycle Θ_b remains unchanged.

Proof. For X, Y and Z in $\mathcal{G}$, we have since Θ is a 1-cocycle, $\sum_{\text{circ}(X,Y,Z)}$ meaning a sum over circular permutations of X, Y and Z, using the Jacobi identity in $\mathcal{G}$, we have

$$\begin{aligned}\sum_{\text{circ}(X,Y,Z)} \langle \Theta_b(X), [Y,Z] \rangle &= \sum_{\text{circ}(X,Y,Z)} \langle -\operatorname{ad}^*_X E_J(b), [Y,Z] \rangle \\ &= \sum_{\text{circ}(X,Y,Z)} \langle -E_J(b), [X,[Y,Z]] \rangle \\ &= 0 .\end{aligned}$$

The linear map Θ_b is therefore a 1 cocycle, even a symplectic 1-cocycle since for all X and $Y \in \mathcal{G}$, $\langle \Theta_b(X), Y \rangle = -\langle \Theta_b(Y), X \rangle$.

Using the first equality stated in Proposition 17, we have for any $X \in \mathcal{G}$

$$\langle \Theta_b(b), X \rangle = \langle \Theta(b) - \operatorname{ad}^*_b E_J(b), X \rangle = -\langle \Theta(X), b \rangle + \langle E_J(b), [X,b] \rangle = 0 .$$

If we replace J by $J_1 = J + \mu$, the map $X \mapsto \Theta(X)$ is replaced by $X \mapsto \Theta_1(X) = \Theta(X) + \operatorname{ad}^*_X \mu$ and $E_J(b)$ by $E_{J_1}(b) = E_J(b) + \mu$, therefore Θ_b remains unchanged. □

The following lemma will allow us to define, for each $b \in \Omega$, a remarkable symmetric bilinear form on the vector subspace $[b,\mathcal{G}] = \{[b,X]\,;X \in \mathcal{G}\}$ of the Lie algebra $\mathcal{G}$.

Lemma 1. *Let Ξ be a 1-cocycle of a finite-dimensional Lie algebra $\mathcal{G}$ for the coadjoint representation. For each $b \in \ker \Xi$, let $F_b = [\mathcal{G}, b]$ be the set of elements $X \in \mathcal{G}$ which can be written $X = [X_1, b]$ for some $X_1 \in \mathcal{G}$. Then F_b is a vector subspace of $\mathcal{G}$, and the value of the right hand side of the equality*

$$\Gamma_b(X,Y) = \langle \Xi(X_1), Y \rangle , \quad \textit{with } X_1 \in \mathcal{G}\,,\ X = [X_1,b] \in F_b\,,\ Y \in F_b\,,$$

depends only on X and Y, not on the choice of $X_1 \in \mathcal{G}$ such that $X = [X_1, b]$. That equality defines a bilinear form Γ_b on F_b which is symmetric, i.e., satisfies

$$\Gamma_b(X,Y) = \Gamma_b(Y,X) \quad \textit{for all } X \textit{ and } Y \in F_b\,.$$

Proof. Let X_1 and $X'_1 \in \mathcal{G}$ be such that $[X_1,b] = [X'_1,b] = X$. Let $Y_1 \in \mathcal{G}$ be such that $[Y_1,b] = Y$. We have

$$\begin{aligned}\langle \Xi(X_1 - X'_1), Y \rangle &= \langle \Xi(X_1 - X'_1), [Y_1, b] \rangle \\ &= -\langle \Xi(Y_1), [b, X_1 - X'_1] \rangle - \langle \Xi(b), [X_1 - X'_1, Y_1] \rangle \\ &= 0\end{aligned}$$

since $\Xi(b) = 0$ and $[b, X_1 - X'_1] = 0$. We have shown that $\langle \Xi(X_1), Y \rangle = \langle \Xi(X'_1), Y \rangle$. Therefore Γ_b is a bilinear form on F_b. Similarly

$$\langle \Xi(X_1), Y \rangle = \langle \Xi(X_1), [Y_1,b] \rangle = -\langle \Xi(Y_1), [b,X_1] \rangle - \langle \Xi(b), [X_1,Y_1] \rangle = \langle \Xi(Y_1), X \rangle ,$$

which proves that Γ_b is symmetric. □

Theorem 7. *The assumptions and notations are the same as those in Proposition 14. For each $b \in \Omega$, there exists on the vector subspace $F_b = [\mathcal{G}, b]$ of elements $X \in \mathcal{G}$ which can be written $X = [X_1, b]$ for some $X_1 \in \mathcal{G}$, a symmetric negative bilinear form Γ_b given by*

$$\Gamma_b(X,Y) = \langle \Theta_b(X_1), Y \rangle , \quad \textit{with } X_1 \in \mathcal{G}\,,\ X = [X_1,b] \in F_b\,,\ Y \in F_b\,,$$

where $\Theta_b : \mathcal{G} \to \mathcal{G}^*$ *is the symplectic* 1*-cocycle defined in Corollary 4.*

Proof. We have seen in Corollary 4 that $b \in \ker \Theta_b$. The fact that the equality given in the statement above defines indeed a symmetric bilinear form on F_b directly follows from Lemma 1. We only have to prove that this symmetric bilinear form is negative. Let $X \in F_b$ and $X_1 \in \mathcal{G}$ such that $X = [X_1, b]$. Using Proposition 17 and Corollary 3, we have

$$\Gamma_b(X,X) = \langle \Theta_b(X_1), [X_1,b] \rangle = \langle \Theta(X_1) - \mathrm{ad}^*_{X_1} E_J(b), [X_1,b] \rangle = \langle DE_J(b)[X_1,b], [X_1,b] \rangle \\ \leq 0\,.$$

The symmetric bilinear form Γ_b on F_b is therefore negative. □

Remark 17. *The symmetric negative bilinear forms encountered in Theorem 7 and Corollary 3 seem to be linked with the Fisher metric in information geometry discussed in [31,60,61].*

7.3. Examples of Generalized Gibbs States

7.3.1. Action of the Group of Rotations on a Sphere

The symplectic manifold (M, ω) considered here is the two-dimensional sphere of radius R centered at the origin O of a three-dimensional oriented Euclidean vector space $\overrightarrow{E}$, equipped with its area element as symplectic form. The group G of rotations around the origin (isomorphic to SO(3)) acts on the sphere M by a Hamiltonian action. The Lie algebra $\mathcal{G}$ of G can be identified with $\overrightarrow{E}$, the fundamental vector field on M associated to an element $\overrightarrow{b}$ in $\mathcal{G} \equiv \overrightarrow{E}$ being the vector field on M whose value at a point $m \in M$ is given by the vector product $\overrightarrow{b} \times \overrightarrow{Om}$. The dual $\mathcal{G}^*$ of $\mathcal{G}$ will be too identified with $\overrightarrow{E}$, the coupling by duality being given by the Euclidean scalar product. The momentum map $J : M \to \mathcal{G}^* \equiv \overrightarrow{E}$ is given by

$$J(m) = -R\,\overrightarrow{Om}\,, \quad m \in M\,.$$

Therefore, for any $\overrightarrow{b} \in \mathcal{G} \equiv \overrightarrow{E}$,

$$\langle J(m), \overrightarrow{b} \rangle = -R\,\overrightarrow{Om} \cdot \overrightarrow{b}\,.$$

Let $\overrightarrow{b}$ be any element in $\mathcal{G} \equiv \overrightarrow{E}$. To calculate the partition function $P(\overrightarrow{b})$ we choose an orthonormal basis $(\overrightarrow{e_x}, \overrightarrow{e_y}, \overrightarrow{e_z})$ of $\overrightarrow{E}$ such that $\overrightarrow{b} = \|\overrightarrow{b}\|\overrightarrow{e_z}$, with $\|\overrightarrow{b}\| \in \mathbb{R}^+$, and we use angular coordinates (φ, θ) on the sphere M. The coordinates of a point $m \in M$ are

$$x = R\cos\theta\cos\varphi\,, \quad y = R\cos\theta\sin\varphi\,, \quad z = R\sin\theta\,.$$

We have

$$P(\overrightarrow{b}) = \int_0^{2\pi} \left(\int_{-\pi/2}^{\pi/2} R^2 \exp(R\|\overrightarrow{b}\|\sin\theta\,\mathrm{d}\theta \right) \mathrm{d}\varphi = \frac{4\pi R}{\|\overrightarrow{b}\|}\,\mathrm{sh}(R\|\overrightarrow{b}\|)\,.$$

The probability density (with respect to the natural area measure on the sphere M) of the generalized Gibbs state associated to $\overrightarrow{b}$ is

$$\rho_b(m) = \frac{1}{P(\overrightarrow{b})}\exp(\overrightarrow{Om} \cdot \overrightarrow{b})\,, \quad m \in M\,.$$

We observe that ρ_b reaches its maximal value at the point $m \in M$ such that $\overrightarrow{Om} = \dfrac{R\overrightarrow{b}}{\|\overrightarrow{b}\|}$ and its minimal value at the diametrally opposed point.

7.3.2. The Galilean Group, Its Lie Algebra and Its Actions

In view of the presentation, made below, of some physically meaningful generalized Gibbs states for Hamiltonian actions of subgroups of the Galilean group, we recall in this section some notions about the space-time of classical (non-relativistic) mechanics, the Galilean group, its Lie algebra and its Hamiltonian actions. The interested reader will find a much more detailed treatment on these subjects in the book by Souriau [14] or in the recent book by de Saxcé and Vallée [45]. The paper [62] presents a nice application of Galilean invariance in thermodynamics.

The space-time of classical mechanics is a four-dimensional real affine space which, once an inertial reference frame, units of length and time, orthonormal bases of space and time are chosen, can be identified with $\mathbb{R}^4 \equiv \mathbb{R}^3 \times \mathbb{R}$ (coordinates x, y, z, t). The first three coordinates x, y and z can be considered as the three components of a vector $\overrightarrow{r} \in \mathbb{R}^3$, therefore an element of space-time can be denoted by $(\overrightarrow{r}, t)$. However, as the action of the Galilean group will show, the splitting of space-time into space and time is not uniquely determined, it depends on the choice of an inertial reference frame. In classical mechanics, there exists an absolute time, but no absolute space. There exists instead a space (which is an Euclidean affine three-dimensional space) for each value of the time. The spaces for two distinct values of the time should be considered as disjoint.

The space-time being identified with $\mathbb{R}^3 \times \mathbb{R}$ as explained above, the Galilean group G can be identified with the set of matrices of the form

$$\begin{pmatrix} A & \overrightarrow{b} & \overrightarrow{d} \\ 0 & 1 & e \\ 0 & 0 & 1 \end{pmatrix}, \quad \text{with } A \in \mathrm{SO}(3), \ \overrightarrow{b} \text{ and } \overrightarrow{d} \in \mathbb{R}^3, \ e \in \mathbb{R}, \tag{8}$$

the vector space $\mathbb{R}^3$ being oriented and endowed with its usual Euclidean structure, the matrix $A \in \mathrm{SO}(3)$ acting on it.

The action of the Galilean group G on space-time, identified as indicated above with $\mathbb{R}^3 \times \mathbb{R}$, is the affine action

$$\begin{pmatrix} \overrightarrow{r} \\ t \\ 1 \end{pmatrix} \mapsto \begin{pmatrix} A & \overrightarrow{b} & \overrightarrow{d} \\ 0 & 1 & e \\ 0 & 0 & 1 \end{pmatrix} \begin{pmatrix} \overrightarrow{r} \\ t \\ 1 \end{pmatrix} = \begin{pmatrix} A\overrightarrow{r} + t\overrightarrow{b} + \overrightarrow{d} \\ t + e \\ 1 \end{pmatrix}.$$

The Lie algebra $\mathcal{G}$ of the Galilean group G can be identified with the space of matrices of the form

$$\begin{pmatrix} j(\overrightarrow{\omega}) & \overrightarrow{\beta} & \overrightarrow{\delta} \\ 0 & 0 & \varepsilon \\ 0 & 0 & 0 \end{pmatrix}, \quad \text{with } \overrightarrow{\omega}, \ \overrightarrow{\beta} \text{ and } \overrightarrow{\delta} \in \mathbb{R}^3, \ \varepsilon \in \mathbb{R}. \tag{9}$$

We have denoted by $j(\overrightarrow{\omega})$ the 3×3 skew-symmetric matrix

$$j(\overrightarrow{\omega}) = \begin{pmatrix} 0 & -\omega_z & \omega_y \\ \omega_z & 0 & -\omega_x \\ -\omega_y & \omega_x & 0 \end{pmatrix}.$$

The matrix $j(\overrightarrow{\omega})$ is an element in the Lie algebra $\mathfrak{so}(3)$, and its action on a vector $\overrightarrow{r} \in \mathbb{R}^3$ is given by the vector product

$$j(\overrightarrow{\omega})\overrightarrow{r} = \overrightarrow{\omega} \times \overrightarrow{r}.$$

Let us consider a mechanical system made by a point particle of mass m whose position and velocity at time t, in the reference frame allowing the identification of space-time with $\mathbb{R}^3 \times \mathbb{R}$, are the vectors $\overrightarrow{r}$ and $\overrightarrow{v} \in \mathbb{R}^3$. The action of an element of the Galilean group on $\overrightarrow{r}, \overrightarrow{v}$ and t can be written as

$$\begin{pmatrix} \overrightarrow{r} & \overrightarrow{v} \\ t & 1 \\ 1 & 0 \end{pmatrix} \mapsto \begin{pmatrix} A & \overrightarrow{b} & \overrightarrow{d} \\ 0 & 1 & e \\ 0 & 0 & 1 \end{pmatrix} \begin{pmatrix} \overrightarrow{r} & \overrightarrow{v} \\ t & 1 \\ 1 & 0 \end{pmatrix} = \begin{pmatrix} A\overrightarrow{r} + t\overrightarrow{b} + \overrightarrow{d} & A\overrightarrow{v} + \overrightarrow{b} \\ t+e & 1 \\ 1 & 0 \end{pmatrix}.$$

Souriau has shown in his book [14] that this action is Hamiltonian, with the map J, defined on the evolution space of the particle, with value in the dual $\mathcal{G}^*$ of the Lie algebra $\mathcal{G}$ of the Galilean group, as momentum map

$$J(\overrightarrow{r}, t, \overrightarrow{v}, m) = m\left(\overrightarrow{r} \times \overrightarrow{v},\ \overrightarrow{r} - t\overrightarrow{v},\ \overrightarrow{v},\ \frac{1}{2}\|\overrightarrow{v}\|^2\right).$$

Let $b = \begin{pmatrix} j(\overrightarrow{\omega}) & \overrightarrow{\beta} & \overrightarrow{\delta} \\ 0 & 0 & \varepsilon \\ 0 & 0 & 0 \end{pmatrix}$ be an element in $\mathcal{G}$. Its coupling with $J(\overrightarrow{r}, t, \overrightarrow{v}, m) \in \mathcal{G}^*$ is given by the formula

$$\langle J(\overrightarrow{r}, t, \overrightarrow{v}, m), b\rangle = m\Big(\overrightarrow{\omega} \cdot (\overrightarrow{r} \times \overrightarrow{v}) - (\overrightarrow{r} - t\overrightarrow{v}) \cdot \overrightarrow{\beta} + \overrightarrow{v} \cdot \overrightarrow{\delta} - \frac{1}{2}\|\overrightarrow{v}\|^2 \varepsilon\Big).$$

7.3.3. One-Parameter Subgroups of the Galilean Group

In his book [14], Souriau has shown that when the considered Lie group action is the action of the full Galilean group on the space of motions of an isolated mechanical system, the open subset Ω of the Lie algebra $\mathcal{G}$ of the Galilean group on which the conditions specified in Section 7.2 are satisfied is empty. In other words, generalized Gibbs states of the full Galilean group do not exist. However, generalized Gibbs states for one-parameter subgroups of the Galilean group do exist which have an interesting physical meaning.

Let us consider an element b of $\mathcal{G}$ such that in its matrix expression (expression (9) above) we have $\varepsilon \neq 0$. The one-parameter subgroup G_1 of the Galilean group generated by b is the set of matrices $\exp(\tau b)$, with $\tau \in \mathbb{R}$. We have

$$\exp(\tau b) = \begin{pmatrix} A(\tau) & \overrightarrow{b}(\tau) & \overrightarrow{d}(\tau) \\ 0 & 1 & \tau\varepsilon \\ 0 & 0 & 1 \end{pmatrix},$$

with

$$\begin{aligned} A(\tau) &= \exp\big(\tau j(\overrightarrow{\omega})\big), \\ \overrightarrow{b}(\tau) &= \left(\sum_{n=1}^{\infty} \frac{\tau^n}{n!} \big(j(\overrightarrow{\omega})\big)^{n-1}\right) \overrightarrow{\beta}, \\ \overrightarrow{d}(\tau) &= \left(\sum_{n=1}^{\infty} \frac{\tau^n}{n!} \big(j(\overrightarrow{\omega})\big)^{n-1}\right) \overrightarrow{\delta} + \varepsilon \left(\sum_{n=2}^{\infty} \frac{\tau^n}{n!} \big(j(\overrightarrow{\omega})\big)^{n-2}\right) \overrightarrow{\beta}, \end{aligned}$$

with the usual convention that $\big(j(\overrightarrow{\omega})\big)^0$ is the unit matrix.

The physical meaning of this one-parameter subgroup of the Galilean group can be understood as follows. Let us call *fixed* the affine Euclidean reference frame of space $(O, \overrightarrow{e_x}, \overrightarrow{e_y}, \overrightarrow{e_z})$ used to represent, at time $t = 0$, a point in space by a vector $\overrightarrow{r}$ or by its three components x, y and z. Let us set $\tau = \dfrac{t}{\varepsilon}$.

For each time $t \in \mathbb{R}$, the action of $A(\tau) = A\left(\frac{t}{\varepsilon}\right)$ maps the fixed reference frame $(O, \overrightarrow{e_x}, \overrightarrow{e_y}, \overrightarrow{e_z})$ onto another affine Euclidean reference frame $\big(O(t), \overrightarrow{e_x}(t), \overrightarrow{e_y}(t), \overrightarrow{e_z}(t)\big)$, which we call the *moving* reference frame. The velocity and the acceleration of the relative motion of the moving reference frame with respect to the fixed reference frame is given, at time $t = 0$, by the fundamental vector field associated to the element b of the Lie algebra $\mathcal{G}$ of the Galilean group: we see that each point in space has a motion composed of a rotation around the axis through O parallel to $\overrightarrow{\omega}$, at an angular velocity $\frac{\|\overrightarrow{\omega}\|}{\varepsilon}$, and simultaneously a uniformly accelerated motion of translation at an initial velocity $\frac{\overrightarrow{\delta}}{\varepsilon}$ and acceleration $\frac{\overrightarrow{\beta}}{\varepsilon}$. At time t, the velocity and acceleration of the moving reference frame with respect to its instantaneous position at that time can be described in a similar manner, but instead of O, $\overrightarrow{\omega}$, $\overrightarrow{\beta}$ and $\overrightarrow{\delta}$ we must use the corresponding transformed elements by the action of $A(\tau) = A\left(\frac{t}{\varepsilon}\right)$.

7.3.4. A Gas Contained in a Moving Vessel

We consider a mechanical system made by a gas of N point particles, indexed by $i \in \{1, 2, \ldots, N\}$, contained in a vessel with rigid, undeformable walls, whose motion in space is given by the action of the one-parameter subgroup G_1 of the Galilean group made by the $A\left(\frac{t}{\varepsilon}\right)$, with $t \in \mathbb{R}$, above described. We denote by m_i, $\overrightarrow{r_i}(t)$ and $\overrightarrow{v_i}(t)$ the mass, position vector and velocity vector, respectively, of the i-th particle at time t. Since the motion of the vessel containing the gas is precisely given by the action of G_1, the boundary conditions imposed to the system are invariant by that action, which leaves invariant the evolution space of the mechanical system, is Hamiltonian and projects onto a Hamiltonian action of G_1 on the symplectic manifold of motions of the system. We can therefore consider the generalized Gibbs states of the system, as discussed in Section 7.1. We must evaluate the momentum map J of that action and its coupling with the element $b \in \mathcal{G}$. As in Section 6.3.1 we will neglect, for that evaluation, the contributions of the collisions of the particles between themselves and with the walls of the vessel. The momentum map can therefore be evaluated as if all particles were free, and its coupling $\langle J, b\rangle$ with b is the sum $\sum_{i=1}^{N} \langle J_i, b\rangle$ of the momentum map J_i of the i-th particle, considered as free, with b. We have

$$\big\langle J_i(\overrightarrow{r_i}, t, \overrightarrow{v_i}, m_i), b\big\rangle = m_i\Big(\overrightarrow{\omega} \cdot (\overrightarrow{r_i} \times \overrightarrow{v_i}) - (\overrightarrow{r_i} - t\overrightarrow{v_i}) \cdot \overrightarrow{\beta} + \overrightarrow{v_i} \cdot \overrightarrow{\delta} - \frac{1}{2}\|\overrightarrow{v_i}\|^2 \varepsilon\Big).$$

Following Souriau [14], Chapter IV, pp. 299–303, we observe that $\langle J_i, b\rangle$ is invariant by the action of G_1. We can therefore define $\overrightarrow{r_{i0}}$, t_0 and $\overrightarrow{v_{i0}}$ by setting

$$\begin{pmatrix} \overrightarrow{r_{i0}} & \overrightarrow{v_{i0}} \\ t_0 & 1 \\ 1 & 0 \end{pmatrix} = \exp\left(-\frac{t}{\varepsilon} b\right) \begin{pmatrix} \overrightarrow{r_i} & \overrightarrow{v_i} \\ t & 1 \\ 1 & 0 \end{pmatrix}$$

and write

$$\big\langle J_i(\overrightarrow{r_i}, t, \overrightarrow{v_i}, m_i), b\big\rangle = \big\langle J_i(\overrightarrow{r_{i0}}, t_0, \overrightarrow{v_{i0}}, m_i), b\big\rangle.$$

The vectors $\overrightarrow{r_{i0}}$ and $\overrightarrow{v_{i0}}$ have a clear physical meaning: they are the vectors $\overrightarrow{r_i}$ and $\overrightarrow{v_i}$ as seen by an observer moving with the moving affine Euclidean reference frame $\big(O(t), \overrightarrow{e_x}(t), \overrightarrow{e_y}(t), \overrightarrow{e_z}(t)\big)$. Moreover, as can be easily verified, $t_0 = 0$ of course. We therefore have

$$\begin{aligned} \big\langle J_i(\overrightarrow{r_i}, t, \overrightarrow{v_i}, m_i), b\big\rangle &= m_i\Big(\overrightarrow{\omega} \cdot (\overrightarrow{r_{i0}} \times \overrightarrow{v_{i0}}) - \overrightarrow{r_{i0}} \cdot \overrightarrow{\beta} + \overrightarrow{v_{i0}} \cdot \overrightarrow{\delta} - \frac{1}{2}\|\overrightarrow{v_{i0}}\|^2 \varepsilon\Big) \\ &= m_i\Big(\overrightarrow{v_{i0}} \cdot (\overrightarrow{\omega} \times \overrightarrow{r_{i0}} + \overrightarrow{\delta}) - \overrightarrow{r_{i0}} \cdot \overrightarrow{\beta} - \frac{1}{2}\|\overrightarrow{v_{i0}}\|^2 \varepsilon\Big) \end{aligned}$$

where we have used the well known property of the mixed product

$$\overrightarrow{\omega} \cdot (\overrightarrow{r_{i0}} \times \overrightarrow{v_{i0}}) = \overrightarrow{v_{i0}} \cdot (\overrightarrow{\omega} \times \overrightarrow{r_{i0}}) .$$

Let us set

$$\overrightarrow{U}^* = \frac{1}{\varepsilon}(\overrightarrow{\omega} \times \overrightarrow{r_{i0}} + \overrightarrow{\delta}) .$$

Using $\overrightarrow{v_{i0}} - \overrightarrow{U}^*$ and $\overrightarrow{U}^*$ instead of $\overrightarrow{v_{i0}}$, we can write

$$\langle J_i(\overrightarrow{r_i}, t, \overrightarrow{v_i}, m_i), b\rangle = m_i\varepsilon\left(-\frac{1}{2}\|\overrightarrow{v_{i0}} - \overrightarrow{U}^*\|^2 - \overrightarrow{r_{i0}} \cdot \frac{\overrightarrow{\beta}}{\varepsilon} + \frac{1}{2}\|\overrightarrow{U}^*\|^2\right) .$$

We observe that the vector $\overrightarrow{U}^*$ only depends on ε, $\overrightarrow{\omega}$, $\overrightarrow{\delta}$, which are constants once the element $b \in \mathcal{G}$ is chosen, and of $\overrightarrow{r_{i0}}$, not on $\overrightarrow{v_{i0}}$. It has a clear physical meaning: it is the value of the velocity of the moving affine reference frame with respect to the fixed affine reference frame, at point $\overrightarrow{r_{i0}}$ seen by an observer linked to the moving reference frame. Therefore the vector $\overrightarrow{w_{i0}} = \overrightarrow{v_{i0}} - \overrightarrow{U}^*$ is the *relative velocity* of the i-th particle with respect to the moving affine reference frame, seen by an observer linked to the moving reference frame.

The three components of $\overrightarrow{r_{i0}}$ and the three components of $\overrightarrow{p_{i0}} = m_i\overrightarrow{w_{i0}}$ make a system of Darboux coordinates on the six-dimensional symplectic manilold (M_i, ω_i) of motions of the i-th particle. With a slight abuse of notations, we can consider the momentum map J_i as defined on the space of motions of the i-th particle, instead of being defined on the evolution space of this particle, and write

$$\langle J_i(\overrightarrow{r_{i0}}, \overrightarrow{p_{i,0}}), b\rangle = -\varepsilon\left(\frac{1}{2m_i}\|\overrightarrow{p_{i0}}\|^2 + m_i f_i(\overrightarrow{r_{i0}})\right) , \ \overrightarrow{p_{i0}} = m_i\overrightarrow{w_{i0}} = m_i(\overrightarrow{v_{i0}} - \overrightarrow{U}^*) , \tag{10}$$

and

$$f_i(\overrightarrow{r_{i0}}) = \overrightarrow{r_{i0}} \cdot \frac{\overrightarrow{\beta}}{\varepsilon} - \frac{1}{2\varepsilon^2}\|\overrightarrow{\omega} \times \overrightarrow{r_{i0}}\|^2 - \frac{\overrightarrow{\delta}}{\varepsilon} \cdot \left(\frac{\overrightarrow{\omega}}{\varepsilon} \times \overrightarrow{r_{i0}}\right) - \frac{1}{2\varepsilon^2}\|\overrightarrow{\delta}\|^2 .$$

Equation (10) is well suited for the determination of generalized Gibbs states of the system. Let us set

$$P_i(b) = \int_{M_i} \exp(-\langle J_i, b\rangle) \mathrm{d}\lambda_{\omega_i} , \quad E_{J_i}(b) = \frac{1}{P_i(b)} \int_{M_i} J_i \exp(-\langle J_i, b\rangle) \mathrm{d}\lambda_{\omega_i} .$$

The integrals in the right hand sides of these equalities converge if and only if $\varepsilon < 0$. It means that the matrix b belongs to the subset Ω of the one-dimensional Lie algebra of the considered one-parameter subgroup G_1 of the Galilean group on which generalized Gibbs states can be defined if and only if $\varepsilon < 0$. Assuming that condition satisfied, we can use Definitions 19. The generalized Gibbs state determined by b has the smooth density, with respect to the Liouville measure $\prod_{i=1}^N \lambda_{\omega_i}$ on the symplectic manifold of motions $\Pi_{i=1}^N (M_i, \omega_i)$,

$$\rho(b) = \prod_{i=1}^N \rho_i(b) , \quad \text{with } \rho_i(b) = \frac{1}{P_i(b)} \exp(-\langle J_i, b\rangle) .$$

The partition function, whose expression is

$$P(b) = \prod_{i=1}^N P_i(b) ,$$

can be used, with the help of the formulae given in Section 7.2, to determine all the generalized thermodynamic functions of the gas in a generalized thermodynamic equilibrium state.

Remark 18.

1. *The physical meaning of the parameter ε which appears in the expression of the matrix b is clearly apparent in expression (10) of $\langle J_i, b\rangle$:*
$$\varepsilon = -\frac{1}{kT},$$
T being the absolute temperature and k the Boltzmann's constant.
2. *The same expression (10) shows that the relative motion of the gas with respect to the moving vessel in which it is contained, seen by an observer linked to that moving vessel, is described by a Hamiltonian system in which the kinetic and potential energies of the i-th particle are, respectively, $\frac{1}{2m_i}\|\overrightarrow{p_{i0}}\|^2$ and $m_i f_i(\overrightarrow{r_{i0}})$. This result can be obtained in another way: by deriving the Hamiltonian which governs the relative motion of a mechanical system with respect to a moving frame, as used by Jacobi [63] to determine the famous Jacobi integral of the restricted circular three-body problem (in which two big planets move on concentric circular orbits around their common center of mass, and a third planet of negligible mass moves in the gravitational field created by the two big planets).*
3. *The generalized Gibbs state of the system imposes to the various parts of the system, i.e., to the various particles, to be at the same temperature $T = -\frac{1}{k\varepsilon}$ and to be statistically at rest in the same moving reference frame.*

7.3.5. Three Examples

1. Let us set $\overrightarrow{\omega} = 0$ and $\overrightarrow{\beta} = 0$. The motion of the moving vessel containing the gas (with respect to the so called *fixed reference frame*) is a translation at a constant velocity $\frac{\overrightarrow{\delta}}{\varepsilon}$. The function $f_i(\overrightarrow{r_{i0}})$ is then a constant. In the moving reference frame, which is an inertial frame, we recover the thermodynamic equilibrium state of a monoatomic gas discussed in Section 6.3.1.
2. Let us set now $\overrightarrow{\omega} = 0$ and $\overrightarrow{\delta} = 0$. The motion of the moving vessel containing the gas (with respect to the so called *fixed reference frame*) is now an uniformly accelerated translation, with acceleration $\frac{\overrightarrow{\beta}}{\varepsilon}$. The function $f_i(\overrightarrow{r_{i0}})$ now is
$$f_i(\overrightarrow{r_{i0}}) = \overrightarrow{r_{i0}} \cdot \frac{\overrightarrow{\beta}}{\varepsilon}.$$
In the moving reference frame, which is no more inertial, we recover the thermodynamic equilibrium state of a monoatomic gas in a gravity field $\overrightarrow{g} = -\frac{\overrightarrow{\beta}}{\varepsilon}$ discussed in Section 6.3.2.
3. Let us now set $\overrightarrow{\omega} = \omega\overrightarrow{e_z}$, $\overrightarrow{\beta} = 0$ and $\overrightarrow{\delta} = 0$. The motion of the moving vessel containing the gas (with respect to the so called *fixed reference frame*) is now a rotation around the coordinate z axis at a constant angular velocity $\frac{\omega}{\varepsilon}$. The function $f_i(\overrightarrow{r_{i0}})$ is now
$$f_i(\overrightarrow{r_{i0}}) = -\frac{\omega^2}{2\varepsilon^2}\|\overrightarrow{e_z} \times \overrightarrow{r_{i0}}\|^2.$$
The length $\Delta = \|\overrightarrow{e_z} \times \overrightarrow{r_{i,0}}\|$ is the distance between the i-th particle and the axis of rotation of the moving frame (the coordinate z axis). Moreover, we have seen that $\varepsilon = \frac{-1}{kT}$. Therefore in the generalized Gibbs state, the probability density $\rho_i(b)$ of presence of the i-th particle in its symplectic manifold of motion M_i, ω_i, with respect to the Liouville measure λ_{ω_i}, is
$$\rho_i(b) = \frac{1}{P_i(b)}\exp\left(-\langle J_i, b\rangle\right) = \text{Constant}\cdot\exp\left(-\frac{1}{2m_i kT}\|\overrightarrow{p_{i0}}\|^2 + \frac{m_i}{2kT}\left(\frac{\omega}{\varepsilon}\right)^2\Delta^2\right).$$

This formula describes the behaviour of a gas made of point particles of various masses in a centrifuge rotating at a constant angular velocity $\frac{\omega}{\varepsilon}$: the heavier particles concentrate farther from the rotation axis than the lighter ones.

7.3.6. Other Applications of Generalized Gibbs States

Applications of generalized Gibbs states in thermodynamics of continua, with the use of affine tensors, are presented in the papers by de Saxcé [64,65].

Several applications of generalized Gibbs states of subgroups of the Poincaré group were considered by Souriau. For example, he presents in his book [14], Chapter IV, p. 308, a generalized Gibbs which describes the behaviour of a gas in a relativistic centrifuge, and in his papers [15,16], very nice applications of such generalized Gibbs states in Cosmology.

Acknowledgments: I address my thanks to Alain Chenciner for his interest and his help to study the works of Claude Shannon, to Roger Balian for his comments and his explanations about thermodynamic potentials, and to Frédéric Barbaresco for his kind invitation to participate in the GSI 2015 conference and his encouragements. My warmest thanks to the anonymous referees whose very careful and benevolent reading of my work allowed me to correct several mistakes and to improve this paper.

Conflicts of Interest: The author declares no conflict of interest.

References

1. Abraham, R.; Marsden, J.E. *Foundations of Mechanics*, 2nd ed.; American Chemical Society: Washington, DC, USA, 1978.
2. Arnold, V.I. *Mathematical Methods of Classical Mechanics*, 2nd ed.; Springer: Berlin/Heidelberg, Germany, 1978.
3. Cannas da Silva, A. *Lectures on Symplectic Geometry*; Springer: Berlin/Heidelberg, Germany, 2001.
4. Guillemin, V.; Sternberg, S. *Symplectic Techniques in Physics*; Cambridge University Press: Cambridge, UK, 1984.
5. Holm, D. *Geometric Mechanics, Part I: Dynamics ans Symmetry*; World Scientific: Singapore, 2008.
6. Holm, D. *Geometric Mechanics, Part II: Rotating, Translating and Rolling*; World Scientific: Singapore, 2008.
7. Iglesias, P. *Symétries et Moment*; Éditions Hermann: Paris, France, 2000. (In French)
8. Laurent-Gengoux, C.; Pichereau, A.; Vanhaecke, P. *Poisson Structures*; Springer: Berlin/Heidelberg, Germany, 2013.
9. Libermann, P.; Marle, C.-M. *Symplectic Geometry and Analytical Mechanics*; Springer: Berlin/Heidelberg, Germany, 1987.
10. Ortega, J.-P.; Ratiu, T.-S. *Momentum Maps and Hamiltonian Reduction*; Birkhäuser: Boston, MA, USA; Basel, Switzerland; Berlin, Germany, 2004.
11. Vaisman, I. *Lectures on the Geometry of Poisson Manifolds*; Springer: Berlin/Heidelberg, Germany, 1994.
12. Marle, C.-M. Symmetries of hamiltonian systems on symplectic and poisson manifolds. In *Similarity and Symmetry Methods, Applications in Elasticity and Mechanics of Materials*; Ganghoffer, J.-F., Mladenov, I., Eds.; Springer: Berlin/Heidelberg, Germany, 2014; pp. 183–269.
13. Souriau, J.-M. Définition covariante des équilibres thermodynamiques. *Supplemento al Nuovo Cimento* **1966**, *4*, 203–216. (In French)
14. Souriau, J.-M. *Structure des Systèmes Dynamiques*; Dunod: Malakoff, France, 1969. (In French)
15. Souriau, J.-M. Mécanique Statistique, Groupes de Lie et Cosmologie. In *Géométrie Symplectique et Physique Mathématique*; CNRS Éditions: Paris, France, 1974; pp. 59–113. (In French)
16. Souriau, J.-M. Géométrie symplectique et Physique mathématique. In Deux Conférences de Jean-Marie Souriau, Colloquium de la Société Mathématique de France, Paris, France, 19 February–12 November 1975. (In French)
17. Souriau, J.-M. *Mécanique Classique et Géométrie Symplectique*; Dunod: Malakoff, France, 1984. (In French)
18. Mackey, G.W. *The Mathematical Foundations of Quantum Mechanics*; W. A. Benjamin, Inc.: New York, NY, USA, 1963.

19. Newton, I. *Philosophia Naturalis Principia Mathematica*; Translated in French by Émilie du Chastelet (1756); London, UK, 1687. (In French)
20. Lagrange, J.L. *Mécanique Analytique*, 1st ed.; La veuve de Saint-Pierre: Paris, France, 1808; reprinted by Jacques Gabay: Paris, France, 1989. (In French)
21. Hamilton, W.R. On a general method in Dynamics. In *Sir William Rowan Hamilton Mathematical Works, Volume II*; Cambridge University Press: Cambridge, UK, 1940; pp. 247–308.
22. Hamilton, W.R. Second essay on a general method in Dynamics. In *Sir William Rowan Hamilton Mathematical Works, Volume II*; Cambridge University Press: Cambridge, UK, 1940; pp. 95–144.
23. Bérest, P. *Calcul des Variations Application à la Mécanique et à la Physique*; Ellipses/Éditions Marketing: Paris, France, 1997. (In French)
24. Bourguignon, J.-P. *Calcul Variationnel*; Éditions de l'École Polytechnique: Paris, France, 1991. (In French)
25. Lanczos, C.S. *The Variational Principles of Mechanics*, 4th ed.; Reprinted by Dover, New York, 1970; University of Toronto Press: Toronto, ON, Canada, 1970.
26. Malliavin, P. *Géométrie Différentielle Intrinsèque*; Éditions Hermann: Paris, France, 1972. (In French)
27. Sternberg, S. *Lectures on Differential Geometry*; Prentice-Hall: Upper Saddle River, NJ, USA, 1964.
28. Kosmann-Schwarzbach, Y. *The Noether Theorems*; Springer: Berlin/Heidelberg, Germany, 2011.
29. Poincaré, H. Sur une forme nouvelle des équations de la Méanique. *C. R. Acad. Sci.* **1901**, *7*, 369–371.
30. Marle, C.-M. On Henri Poincaré's note "Sur une forme nouvelle des équations de la Mécanique". *J. Geom. Symmetry Phys.* **2013**, *29*, 1–38.
31. Barbaresco, F. Symplectic structure of information geometry: Fisher metric and euler-poincaré equation of souriau lie group thermodynamics. In *Geometric Science of Information: Second International Conference, GSI 2015, Proceedings*; Nielsen, F., Barbaresco, F., Eds.; Lecture Notes in Computer Science; Springer: Berlin/Heidelberg, Germany, 2015; Volume 9389, pp. 529–540. (In French)
32. Lagrange, J.-L. *Mémoire sur la Théorie Générale de la Variation des Constantes Arbitraires Dans Tous les Problèmes de Mécanique*; Lu le 13 mars 1809 à l'Institut de France; Dans *Œuvres de Lagrange*; Gauthier-Villars: Paris, France, 1877; Volume VI, pp. 771–805. (In French)
33. Lagrange, J.-L. *Second Mémoire sur la Théorie de la Variation des Constantes Arbitraires Dans les Problèmes de Mécanique*; Gauthier-Villars: Paris, France, 1877; Volume VI, pp. 809–816. (In French)
34. Tulczyjew, W.M. Hamiltonian systems, Lagrangian systems and the Legendre transformation. *Symp. Math.* **1974**, *14*, 247–258.
35. Tulczyjew, W.M. *Geometric Formulations of Physical Theories*; Monographs and Textbooks in Physical Science; Bibliopolis: Napoli, Italy, 1989.
36. Lichnerowicz, A. Les variétés de Poisson et leurs algèbres de Lie associées. *J. Differ. Geom.* **1977**, *12*, 253–300. (In French)
37. Lichnerowicz, A. Les variétés de Jacobi et leurs algèbres de Lie associées. *J. Math. Pures Appl.* **1979**, *57*, 453–488. (In French)
38. Kirillov, A. Local lie algebras. *Russ. Math. Surv.* **1976**, *31*, 55–75.
39. Poisson, S.D. Sur la variation des constantes arbitraires dans les questions de mécanique. Mémoire lu le 16 octobre 1809 à l'Institut de France. *Journal de L'École Polytechnique* quinzième cahier, tome VIII, 266–344. (In French)
40. Koszul, J.-L. Crochet de Schouten-Nijenhuis et cohomologie. In *É. Cartan et les Mathématiques D'aujourd'hui*; Astérisque, numéro hors série, Société Mathématique de France: Paris, France, 1985; pp. 257–271. (In French)
41. Marle, C.-M. Calculus on Lie algebroids, Lie groupoids and Poisson manifolds. *Dissertationes Mathematicae 457, Institute of Mathematics, Polish Academy of Sciences (Warszawa)*. **2008**, arXiv:0806.0919.
42. Weinstein, A. The local structure of Poisson manifolds. *J. Differ. Geom.* **1983**, *18*, 523–557.
43. Marsden, J.E.; Weinstein, A. Reduction of symplectic manifolds with symmetry. *Rep. Math. Phys.* **1974**, *5*, 121–130.
44. Meyer, K. Symmetries and integrals in mechanics. In *Dynamical Systems*; Peixoto, M., Ed.; Academic Press, New York, NY, USA, 1973; pp. 259–273.
45. De Saxcé, G.; Vallée, C. *Galilean Mechanics and Thermodynamics of Continua*; John Wiley & Sons: Hoboken, NJ, USA, 2016.
46. Boltzmann, L.E. Leçons sur la Théorie des gaz. Available online: http://iris.univ-lille1.fr/handle/1908/1523 (accessed on 11 October 2016). (In French)

47. Shannon, C.E. A mathematical theory of communication. *Bell Syst. Tech. J.* **1948**, *27*, 379–423, 623–656.
48. Jaynes, E.T. Information theory and statistical mechanics. *Phys. Rev.* **1957**, *106*, 620–630.
49. Jaynes, E.T. Information theory and statistical mechanics II. *Phys. Rev.* **1957**, *108*, 171–190.
50. Balian, R. Information in statistical physics. *Stud. Hist. Philos. Mod. Phys. Part B* **2005**, *36*, 323–353.
51. Poincaré, H. Sur le problème des trois corps et les équations de la dynamique. *Acta Math.* **1890**, doi:10.1007/BF02392506. (In French)
52. Kinetic Theory. Available online: http://hyperphysics.phy-astr.gsu.edu/hbase/kinetic/kinthe.html (accessed on 11 October 2016).
53. Gastebois, G. Théorie Cinétique des Gaz. Available online: http://gilbert.gastebois.pagesperso-orange.fr/java/gaz/gazparfait/theorie_gaz.pdf (accessed on 11 October 2016). (In French)
54. Synge, J.L. *The Relativistic Gas*; North Holland Publishing Company: Amsterdam, The Netherlands, 1957.
55. Massieu, F. Sur les Fonctions caractéristiques des divers fluides. *C. R. Acad. Sci. Paris* **1869**, *69*, 858–862. (In French)
56. Massieu, F. Addition au précédent Mémoire sur les Fonctions caractéristiques. *C. R. Acad. Sci. Paris* **1869**, *69*, 1057–1061. (In French)
57. Massieu, F. *Thermodynamique. Mémoire sur les Fonctions Caractéristiques des Divers Fluides et sur la Théorie des Vapeurs*; Académie des Sciences: Paris, France, 1876; pp. 1–92. (In French).
58. Balian, R. *François Massieu et les Potentiels Thermodynamiques*; Évolution des Disciplines et Histoire des Découvertes; Académie des Sciences: Paris, France, 2015. (In French)
59. Callen, H.B. *Thermodynamics and an Introduction to Thermostatics*, 2nd ed.; John Wiley and Sons: New York, NY, USA, 1985.
60. Barbaresco, F. Koszul information geometry and Souriau geometric temperature/capacity of lie group thermodynamics. *Entropy* **2014**, *16*, 4521–4565.
61. Barbaresco, F. Geometric theory of heat from Souriau lie groups thermodynamics and koszul hessian geometry: Applications in information geometry for exponential families. *Entropy* **2016**, doi:10.20944/preprints201608.0078.v1.
62. De Saxcé, G.; Vallée, C. Bargmann group, momentum tensor and Galilean invariance of Clausius-Duhem inequality. *Int. J. Eng. Sci.* **2012**, *50*, 216–232.
63. Jacobi, C.G.J. Sur le mouvement d'un point et sur un cas particulier du problème des trois corps. *C. R. Acad. Sci.* **1836**, *3*, 59–61. (In French)
64. De Saxcé, G. Entropy and structure for the thermodynamic systems. In *Geometric Science of Information, Second International Conference GSI 2015 Proceedings*; Nielsen, F., Barbaresco, F., Eds.; Lecture Notes in Computer Science; Springer: Berlin/Heidelberg, Germany, 2015; Volume 9389, pp. 519–528.
65. De Saxcé, G. Link between lie group statistical mechanics and thermodynamics of continua. *Entropy* **2016**, doi:10.3390/e18070254.

entropy

MDPI

Article

Geometric Theory of Heat from Souriau Lie Groups Thermodynamics and Koszul Hessian Geometry: Applications in Information Geometry for Exponential Families

Frédéric Barbaresco

Advanced Radar Concepts Business Unit, Thales Air Systems, Limours 91470, France; frederic.barbaresco@thalesgroup.com

Academic Editor: Adom Giffin

Received: 4 August 2016; Accepted: 27 September 2016; Published: 4 November 2016

Abstract: We introduce the symplectic structure of information geometry based on Souriau's Lie group thermodynamics model, with a covariant definition of Gibbs equilibrium via invariances through co-adjoint action of a group on its moment space, defining physical observables like energy, heat, and moment as pure geometrical objects. Using geometric Planck temperature of Souriau model and symplectic cocycle notion, the Fisher metric is identified as a Souriau geometric heat capacity. The Souriau model is based on affine representation of Lie group and Lie algebra that we compare with Koszul works on G/K homogeneous space and bijective correspondence between the set of G-invariant flat connections on G/K and the set of affine representations of the Lie algebra of G. In the framework of Lie group thermodynamics, an Euler-Poincaré equation is elaborated with respect to thermodynamic variables, and a new variational principal for thermodynamics is built through an invariant Poincaré-Cartan-Souriau integral. The Souriau-Fisher metric is linked to KKS (Kostant–Kirillov–Souriau) 2-form that associates a canonical homogeneous symplectic manifold to the co-adjoint orbits. We apply this model in the framework of information geometry for the action of an affine group for exponential families, and provide some illustrations of use cases for multivariate gaussian densities. Information geometry is presented in the context of the seminal work of Fréchet and his Clairaut-Legendre equation. The Souriau model of statistical physics is validated as compatible with the Balian gauge model of thermodynamics. We recall the precursor work of Casalis on affine group invariance for natural exponential families.

Keywords: Lie group thermodynamics; moment map; Gibbs density; Gibbs equilibrium; maximum entropy; information geometry; symplectic geometry; Cartan-Poincaré integral invariant; geometric mechanics; Euler-Poincaré equation; Fisher metric; gauge theory; affine group

---

> Lorsque le fait qu'on rencontre est en opposition avec une théorie régnante, il faut accepter le fait et abandonner la théorie, alors même que celle-ci, soutenue par de grands noms, est généralement adoptée
>
> —Claude Bernard in "Introduction à l'Étude de la Médecine Expérimentale" [1]

> Au départ, la théorie de la stabilité structurelle m'avait paru d'une telle ampleur et d'une telle généralité, qu'avec elle je pouvais espérer en quelque sorte remplacer la thermodynamique par la géométrie, géométriser en un certain sens la thermodynamique, éliminer des considérations thermodynamiques tous les aspects à caractère mesurable et stochastiques pour ne conserver que la caractérisation géométrique correspondante des attracteurs.
>
> —René Thom in "Logos et théorie des Catastrophes" [2]

1. Introduction

This MDPI Entropy Special Issue on "Differential Geometrical Theory of Statistics" collects a limited number of selected invited and contributed talks presented during the GSI'15 conference on "*Geometric Science of Information*" in October 2015. This paper is an extended version of the paper [3] "*Symplectic Structure of Information Geometry: Fisher Metric and Euler-Poincaré Equation of Souriau Lie Group Thermodynamics*" published in GSI'15 Proceedings. At GSI'15 conference, a special session was organized on "*lie groups and geometric mechanics/thermodynamics*", dedicated to Jean-Marie Souriau's works in statistical physics, organized by Gery de Saxcé and Frédéric Barbaresco, and an invited talk on "*Actions of Lie groups and Lie algebras on symplectic and Poisson manifolds. Application to Lagrangian and Hamiltonian systems*" by Charles-Michel Marle, addressing "*Souriau's thermodynamics of Lie groups*". In honor of Jean-Marie Souriau, who died in 2012 and Claude Vallée [4–6], who passed away in 2015, this Special Issue will publish three papers on Souriau's thermodynamics: Marle's paper on "*From Tools in Symplectic and Poisson Geometry to Souriau's Theories of Statistical Mechanics and Thermodynamics*" [7], de Saxcé's paper on "*Link between Lie Group Statistical Mechanics and Thermodynamics of Continua*" [8] and this publication by Barbaresco. This paper also proposes new developments, compared to paper [9] where relations between Souriau and Koszul models have been initiated.

This paper, similar to the goal of the papers of Marle and de Saxcé in this Special Issue, is intended to honor the memory of the French Physicist Jean-Marie Souriau and to popularize his works, currently little known, on statistical physics and thermodynamics. Souriau is well known for his seminal and major contributions in geometric mechanics, the discipline he created in the 1960s, from previous Lagrange's works that he conceptualized in the framework of symplectic geometry, but very few people know or have exploited Souriau's works contained in Chapter IV of his book "*Structure des systèmes dynamiques*" published in 1970 [10] and only translated into English in 1995 in the book "*Structure of Dynamical Systems: A Symplectic View of Physics*" [11], in which he applied the formalism of geometric mechanics to statistical physics. The personal author's contribution is to place the work of Souriau in the broader context of the emerging "*Geometric Science of Information*" [12] (addressed in GSI'15 conference), for which the author will show that the Souriau model of statistical physics is particularly well adapted to generalize "*information geometry*", that the author illustrates for exponential densities family and multivariate gaussian densities. The author will observe that the Riemannian metric introduced by Souriau is a generalization of Fisher metric, used in "*information geometry*", as being identified to the hessian of the logarithm of the generalized partition function (Massieu characteristic function), for the case of densities on homogeneous manifolds where a non-abelian group acts transively. For a group of time translation, we recover the classical thermodynamics and for the Euclidean space, we recover the classical Fisher metric used in Statistics. The author elaborates a new Euler-Poincaré equation for Souriau's thermodynamics, action on "geometric heat" variable Q (element of dual Lie algebra), and parameterized by "geometric temperature" (element of Lie algebra). The author will integrate Souriau thermodynamics in a variational model by defining an extended Cartan-Poincaré integral invariant defined by Souriau "geometric characteristic function" (the logarithm of the generalized Souriau partition function parameterized by geometric temperature). These results are illustrated for multivariate Gaussian densities, where the associated group is identified to compute a Souriau moment map and reduce the Euler-Poincaré equation of geodesics. In addition, the symplectic cocycle and Souriau-Fisher metric are deduced from a Lie group thermodynamics model.

The main contributions of the author in this paper are the following:

- The Souriau model of Lie group thermodynamics is presented with standard notations of Lie group theory, in place of Souriau equations using less classical conventions (that have limited understanding of his work by his contemporaries).
- We prove that Souriau Riemannian metric introduced with symplectic cocycle is a generalization of Fisher metric (called Souriau-Fisher metric in the following) that preserves the property

to be defined as a hessian of partition function logarithm $g_\beta = -\frac{\partial^2\Phi}{\partial\beta^2} = \frac{\partial^2 \log\psi_\Omega}{\partial\beta^2}$ as in classical information geometry. We then establish the equality of two terms, the first one given by Souriau's definition from Lie group cocycle Θ and parameterized by "geometric heat" Q (element of dual Lie algebra) and "geometric temperature" β (element of Lie algebra) and the second one, the hessian of the characteristic function $\Phi(\beta) = -\log\psi_\Omega(\beta)$ with respect to the variable β:

$$g_\beta([\beta, Z_1], [\beta, Z_2]) = \langle \Theta(Z_1), [\beta, Z_2]\rangle + \langle Q, [Z_1, [\beta, Z_2]]\rangle = \frac{\partial^2 \log\psi_\Omega}{\partial\beta^2} \tag{1}$$

A dual Souriau-Fisher metric, the inverse of this last one, could be also elaborated with the hessian of "geometric entropy" $s(Q)$ with respect to the variable Q: $\frac{\partial^2 s(Q)}{\partial Q^2}$ For the maximum entropy density (Gibbs density), the following three terms coincide: $\frac{\partial^2 \log\psi_\Omega}{\partial\beta^2}$ that describes the convexity of the log-likelihood function, $I(\beta) = -E\left[\frac{\partial^2 \log p_\beta(\xi)}{\partial\beta^2}\right]$ the Fisher metric that describes the covariance of the log-likelihood gradient, whereas $I(\beta) = E\left[(\xi - Q)(\xi - Q)^T\right] = Var(\xi)$ that describes the covariance of the observables.

- This Souriau-Fisher metric is also identified to be proportional to the first derivative of the heat $g_\beta = -\frac{\partial Q}{\partial\beta}$, and then comparable by analogy to geometric "specific heat" or "calorific capacity".
- We observe that the Souriau metric is invariant with respect to the action of the group $I(Ad_g(\beta)) = I(\beta)$, due to the fact that the characteristic function $\Phi(\beta)$ after the action of the group is linearly dependent to β. As the Fisher metric is proportional to the hessian of the characteristic function, we have the following invariance:

$$I(Ad_g(\beta)) = -\frac{\partial^2\left(\Phi - \langle\theta(g^{-1}), \beta\rangle\right)}{\partial\beta^2} = -\frac{\partial^2\Phi}{\partial\beta^2} = I(\beta) \tag{2}$$

- We have proposed, based on Souriau's Lie group model and on analogy with mechanical variables, a variational principle of thermodynamics deduced from Poincaré-Cartan integral invariant. The variational principle holds on $\mathfrak{g}$ the Lie algebra, for variations $\delta\beta = \dot\eta + [\beta, \eta]$, where $\eta(t)$ is an arbitrary path that vanishes at the endpoints, $\eta(a) = \eta(b) = 0$:

$$\delta\int_{t_0}^{t_1} \Phi(\beta(t)) \cdot dt = 0 \tag{3}$$

where the Poincaré-Cartan integral invariant $\int_{C_a} \Phi(\beta) \cdot dt = \int_{C_b} \Phi(\beta) \cdot dt$ is defined with $\Phi(\beta)$, the Massieu characteristic function, with the 1-form $\omega = \Phi(\beta) \cdot dt = (\langle Q, \beta\rangle - s) \cdot dt = \langle Q, (\beta \cdot dt)\rangle - s \cdot dt$

- We have deduced Euler-Poincaré equations for the Souriau model:

$$\frac{dQ}{dt} = ad^*_\beta Q \text{ and } \begin{cases} s(Q) = \langle \beta, Q\rangle - \Phi(\beta) \\ \beta = \frac{\partial s(Q)}{\partial Q} \in \mathfrak{g},\ Q = \frac{\partial\Phi(\beta)}{\partial\beta} \in \mathfrak{g}^* \end{cases} \text{ and } \frac{d}{dt}\left(Ad^*_g Q\right) = 0$$
$$\text{with } \begin{cases} \mathfrak{g}^* : \text{ dual Lie algebra} \\ ad^*_X Y : \text{ Coadjoint operator} \end{cases} \tag{4}$$

where Q is the Souriau geometric heat (element of dual Lie algebra) and β is the Souriau geometric temperature (element of the Lie algebra). The second equation is linked to the result of Souriau based on the moment map that a symplectic manifold is always a coadjoint orbit, affine of its group of Hamiltonian transformations (a symplectic manifold homogeneous under the action of a Lie group, is isomorphic, up to a covering, to a coadjoint orbit; symplectic leaves are the orbits of the affine action that makes the moment map equivariant).

- We have established that the affine representation of Lie group and Lie algebra by Jean-Marie Souriau is equivalent to Jean-Louis Koszul's affine representation developed in the framework of hessian geometry of convex sharp cones. Both Souriau and Koszul have elaborated equations requested for Lie group and Lie algebra to ensure the existence of an affine representation. We have compared both approaches of Souriau and Koszul in a table.
- We have applied the Souriau model for exponential families and especially for multivariate Gaussian densities.
- We have applied the Souriau-Koszul model Gibbs density to compute the maximum entropy density for symmetric positive definite matrices, using the inner product $\langle \eta, \xi \rangle = Tr\left(\eta^T \xi\right), \forall \eta, \xi \in Sym(n)$ given by Cartan-Killing form. The Gibbs density (generalization of Gaussian law for theses matrices and defined as maximum entropy density):

$$p_{\hat{\xi}}(\xi) = e^{-\langle \Theta^{-1}(\hat{\xi}), \xi \rangle + \Phi(\Theta^{-1}(\hat{\xi}))} = \psi_\Omega\left(I_d\right) \cdot \left[\det\left(\alpha \hat{\xi}^{-1}\right)\right] \cdot e^{-Tr(\alpha \hat{\xi}^{-1} \xi)}$$
$$\text{with } \alpha = \frac{n+1}{2} \quad (5)$$

- For the case of multivariate Gaussian densities, we have considered $GA(n)$ a sub-group of affine group, that we defined by a $(n+1) \times (n+1)$ embedding in matrix Lie group G_{aff}, and that acts for multivariate Gaussian laws by:

$$\begin{bmatrix} Y \\ 1 \end{bmatrix} = \begin{bmatrix} R^{1/2} & m \\ 0 & 1 \end{bmatrix} \begin{bmatrix} X \\ 1 \end{bmatrix} = \begin{bmatrix} R^{1/2}X + m \\ 1 \end{bmatrix}, \left\{ \begin{array}{l} (m, R) \in R^n \times Sym^+(n) \\ M = \begin{bmatrix} R^{1/2} & m \\ 0 & 1 \end{bmatrix} \in G_{aff} \end{array} \right. \quad (6)$$
$$X \approx \aleph(0, I) \rightarrow Y \approx \aleph(m, R)$$

- For multivariate Gaussian densities, as we have identified the acting sub-group of affine group M, we have also developed the computation of the associated Lie algebras η_L and η_R, adjoint and coadjoint operators, and especially the Souriau "moment map" Π_R:

$$\langle n_L, M^{-1} n_R M \rangle = \langle \Pi_R, n_R \rangle$$
$$\text{with } M = \begin{bmatrix} R^{1/2} & m \\ 0 & 1 \end{bmatrix}, n_L = \begin{bmatrix} R^{-1/2}\dot{R}^{1/2} & R^{-1/2}\dot{m} \\ 0 & 0 \end{bmatrix} \text{ and } \eta_R = \begin{bmatrix} R^{-1/2}\dot{R}^{1/2} & \dot{m} - R^{-1/2}\dot{R}^{1/2} m \\ 0 & 0 \end{bmatrix} \quad (7)$$
$$\Rightarrow \Pi_R = \begin{bmatrix} R^{-1/2}\dot{R}^{1/2} + R^{-1}\dot{m}m^T & R^{-1}\dot{m} \\ 0 & 0 \end{bmatrix}$$

Using Souriau Theorem (geometrization of Noether theorem), we use the property that this moment map Π_R is constant (its components are equal to Noether invariants):

$$\frac{d\Pi_R}{dt} = 0 \Rightarrow \left\{ \begin{array}{l} R^{-1}\dot{R} + R^{-1}\dot{m}m^T = B = cste \\ R^{-1}\dot{m} = b = cste \end{array} \right. \quad (8)$$

to reduce the Euler-Lagrange equation of geodesics between two multivariate Gaussian densities:

$$\left\{ \begin{array}{l} \ddot{R} + \dot{m}\dot{m}^T - \dot{R}R^{-1}\dot{R} = 0 \\ \ddot{m} - \dot{R}R^{-1}\dot{m} = 0 \end{array} \right. \quad (9)$$

to this reduced equation of geodesics:

$$\begin{cases} \dot{m} = Rb \\ \dot{R} = R\left(B - bm^T\right) \end{cases} \tag{10}$$

that we solve by "geodesic shooting" technic based on Eriksen equation of exponential map.

- For the families of multivariate Gaussian densities, that we have identified as homogeneous manifold with the associated sub-group of the affine group $\begin{bmatrix} R^{1/2} & m \\ 0 & 1 \end{bmatrix}$, we have considered the elements of exponential families, that play the role of geometric heat Q in Souriau Lie group thermodynamics, and β the geometric (Planck) temperature:

$$Q = \hat{\xi} = \begin{bmatrix} E[z] \\ E\left[zz^T\right] \end{bmatrix} = \begin{bmatrix} m \\ R + mm^T \end{bmatrix}, \ \beta = \begin{bmatrix} -R^{-1}m \\ \frac{1}{2}R^{-1} \end{bmatrix} \tag{11}$$

We have considered that these elements are homeomorph to the $(n + 1) \times (n + 1)$ matrix elements:

$$Q = \hat{\xi} = \begin{bmatrix} R + mm^T & m \\ 0 & 0 \end{bmatrix} \in \mathfrak{g}^*, \ \beta = \begin{bmatrix} \frac{1}{2}R^{-1} & -R^{-1}m \\ 0 & 0 \end{bmatrix} \in \mathfrak{g} \tag{12}$$

to compute the Souriau symplectic cocycle of the Lie group:

$$\theta(M) = \hat{\xi}\left(Ad_M(\beta)\right) - Ad_M^*\hat{\xi} \tag{13}$$

where the adjoint operator is equal to:

$$Ad_M\beta = \begin{bmatrix} \frac{1}{2}\Omega^{-1} & -\Omega^{-1}n \\ 0 & 0 \end{bmatrix} \text{ with } \Omega = R'^{1/2}RR'^{-1/2} \text{ and } n = \left(\frac{1}{2}m' + R'^{1/2}m\right) \tag{14}$$

with

$$\hat{\xi}\left(Ad_M(\beta)\right) = \begin{bmatrix} \Omega + nn^T & n \\ 0 & 0 \end{bmatrix} \tag{15}$$

and the co-adjoint operator:

$$Ad_M^*\hat{\xi} = \begin{bmatrix} R + mm^T - mm'^T & R'^{1/2}m \\ 0 & 0 \end{bmatrix} \tag{16}$$

- Finally, we have computed the Souriau-Fisher metric $g_\beta\left([\beta, Z_1], [\beta, Z_2]\right) = \widetilde{\Theta}_\beta\left(Z_1, [\beta, Z_2]\right)$ for multivariate Gaussian densities, given by:

$$\begin{aligned} g_\beta\left([\beta, Z_1], [\beta, Z_2]\right) &= \widetilde{\Theta}_\beta\left(Z_1, [\beta, Z_2]\right) = \widetilde{\Theta}\left(Z_1, [\beta, Z_2]\right) + \left\langle \hat{\xi}, [Z_1, [\beta, Z_2]]\right\rangle \\ &= \left\langle \Theta\left(Z_1\right), [\beta, Z_2]\right\rangle + \left\langle \hat{\xi}, [Z_1, [\beta, Z_2]]\right\rangle \end{aligned} \tag{17}$$

with element of Lie algebra given by $Z = \begin{bmatrix} \frac{1}{2}\Omega^{-1} & -\Omega^{-1}n \\ 0 & 0 \end{bmatrix}$.

The plan of the paper is as follows. After this introduction in Section 1, we develop in Section 2 the position of Souriau symplectic model of statistical physics in the historical developments of thermodynamic concepts. In Section 3, we develop and revisit the Lie group thermodynamics model of Jean-Marie Souriau in modern notations. In Section 4, we make the link between Souriau Riemannian metric and Fisher metric defined as a geometric heat capacity of Lie group thermodynamics. In Section 5, we elaborate Euler-Lagrange equations of Lie group thermodynamics and a variational model based on Poincaré-Cartan integral invariant. In Section 6, we explore Souriau affine representation of Lie group and Lie algebra (including the notions of: affine representations and cocycles, Souriau moment map and cocycles, equivariance of Souriau moment map, action of Lie group on a symplectic manifold and dual spaces of finite-dimensional Lie algebras) and we analyze the link and parallelisms with Koszul affine representation, developed in another context (comparison is synthetized in a table). In Section 7, we illustrate Koszul and Souriau Lie group models of information geometry for multivariate Gaussian densities. In Section 8, after identifying the affine group acting for these densities, we compute the Souriau moment map to obtain the Euler-Poincaré equation, solved by geodesic shooting method. In Section 9, Souriau Riemannian metric defined by cocycle for multivariate Gaussian densities is computed. We give a conclusion in Section 10 with research prospects in the framework of affine Poisson geometry [13], Bismut stochastic mechanics [14] and second order extension of the Gibbs state [15,16]. We have three appendices: Appendix A develops the Clairaut(-Legendre) equation of Maurice Fréchet associated to "distinguished functions" as a seminal equation of information geometry; Appendix B is about a Balian Gauge model of thermodynamics and its compliance with the Souriau model; Appendix C is devoted to the link of Casalis-Letac's works on affine group invariance for natural exponential families with Souriau's works.

2. Position of Souriau Symplectic Model of Statistical Physics in Historical Developments of Thermodynamic Concepts

In this Section, we will explain the emergence of thermodynamic concepts that give rise to the generalization of the Souriau model of statistical physics. To understand Souriau's theoretical model of heat, we have to consider first his work in geometric mechanics where he introduced the concept of "moment map" and "symplectic cohomology". We will then introduce the concept of "characteristic function" developed by François Massieu, and generalized by Souriau on homogeneous symplectic manifolds. In his statistical physics model, Souriau has also generalized the notion of "heat capacity" that was initially extended by Pierre Duhem as a key structure to jointly consider mechanics and thermodynamics under the umbrella of the same theory. Pierre Duhem has also integrated, in the corpus, the Massieu's characteristic function as a thermodynamic potential. Souriau's idea to develop a covariant model of Gibbs density on homogeneous manifold was also influenced by the seminal work of Constantin Carathéodory that axiomatized thermodynamics in 1909 based on Carnot's works. Souriau has adapted his geometric mechanical model for the theory of heat, where Henri Poincaré did not succeed in his paper on attempts of mechanical explanation for the principles of thermodynamics.

Lagrange's works on "mécanique analytique (analytic mechanics)" has been interpreted by Jean-Marie Souriau in the framework of differential geometry and has initiated a new discipline called after Souriau, "mécanique géométrique (geometric mechanics)" [17–19]. Souriau has observed that the collection of motions of a dynamical system is a manifold with an antisymmetric flat tensor that is a symplectic form where the structure contains all the pertinent information of the state of the system (positions, velocities, forces, etc.). Souriau said: *"Ce que Lagrange a vu, que n'a pas vu Laplace, c'était la structure symplectique (What Lagrange saw, that Laplace didn't see, was the symplectic structure"* [20]. Using the symmetries of a symplectic manifold, Souriau introduced a mapping which he called the "moment map" [21–23], which takes its values in a space attached to the group of symmetries (in the dual space of its Lie algebra). He [10] called dynamical groups every dimensional group of symplectomorphisms (an isomorphism between symplectic manifolds, a transformation of phase space that is volume-preserving), and introduced Galileo group for classical mechanics and Poincaré group for relativistic mechanics (both are sub-groups of affine group [24,25]). For instance, a Galileo group could be represented in a matrix form by (with A rotation, b the boost, c space translation and e time translation):

$$\begin{bmatrix} x' \\ t \\ 1 \end{bmatrix} = \underbrace{\begin{bmatrix} A & b & c \\ 0 & 1 & e \\ 0 & 0 & 1 \end{bmatrix}}_{\text{GALILEO GROUP}} \begin{bmatrix} x \\ t \\ 1 \end{bmatrix} \text{ with } \begin{cases} A \in SO(3) \\ b, c \in R^3 \\ e \in R \end{cases}, \text{ Lie Algebra } \begin{bmatrix} \omega & \eta & \gamma \\ 0 & 0 & \varepsilon \\ 0 & 0 & 0 \end{bmatrix} \text{ with } \begin{cases} \omega \in so(3) \\ \eta, \gamma \in R^3 \\ \varepsilon \in R^+ \end{cases} \quad (18)$$

Souriau associated to this moment map, the notion of symplectic cohomology, linked to the fact that such a moment is defined up to an additive constant that brings into play an algebraic mechanism (called cohomology). Souriau proved that the moment map is a constant of the motion, and provided geometric generalization of Emmy Noether invariant theorem (invariants of E. Noether theorem are the components of the moment map). For instance, Souriau gave an ontological definition of mass in classical mechanics as the measure of the symplectic cohomology of the action of the Galileo group (the mass is no longer an arbitrary variable but a characteristic of the space). This is no longer true for Poincaré group in relativistic mechanics, where the symplectic cohomology is null, explaining the lack of conservation of mass in relativity. All the details of classical mechanics thus appear as geometric necessities, as ontological elements. Souriau has also observed that the symplectic structure has the property to be able to be reconstructed from its symmetries alone, through a 2-form (called Kirillov–Kostant–Souriau form) defined on coadjoint orbits. Souriau said that the different versions of mechanical science can be classified by the geometry that each implies for space and time; geometry is determined by the covariance of group theory. Thus, Newtonian mechanics is covariant by the group of Galileo, the relativity by the group of Poincaré; General relativity by the "smooth" group (the group of diffeomorphisms of space-time). However, Souriau added "*However, there are some statements of mechanics whose covariance belongs to a fourth group rarely considered: the affine group, a group shown in the following diagram for inclusion. How is it possible that a unitary point of view (which would be necessarily a true thermodynamics), has not yet come to crown the picture? Mystery...*" [26]. See Figure 1.

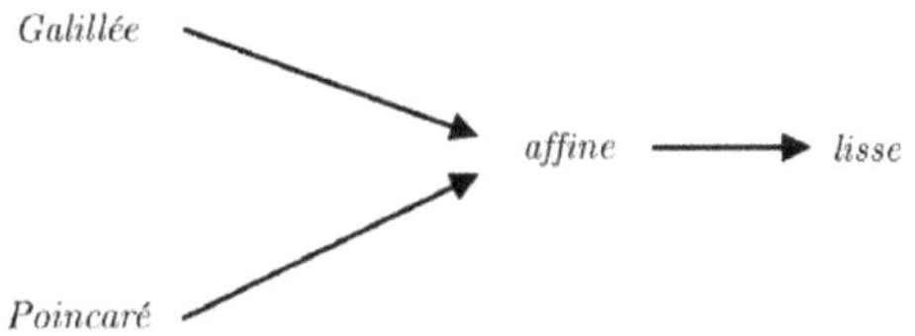

Figure 1. Souriau Scheme about mysterious "affine group" of a true thermodynamics between Galileo group of classical mechanics, Poincaré group of relativistic mechanics and Smooth group of general relativity.

As early as 1966, Souriau applied his theory to statistical mechanics, developed it in the Chapter IV of his book "*Structure of Dynamical Systems*" [11], and elaborated what he called a "Lie group thermodynamics" [10,11,27–37]. Using Lagrange's viewpoint, in Souriau statistical mechanics, a statistical state is a probability measure on the manifold of motions (and no longer in phase space [38]). Souriau observed that Gibbs equilibrium [39] is not covariant with respect to dynamic groups of Physics. To solve this braking of symmetry, Souriau introduced a new "geometric theory of heat" where the equilibrium states are indexed by a parameter β with values in the Lie algebra of the group, generalizing the Gibbs equilibrium states, where β plays the role of a geometric (Planck) temperature. The invariance with respect to the group, and the fact that the entropy s is a convex function of this geometric temperature β, imposes very strict, universal conditions (e.g., there exists necessarily a critical temperature beyond which no equilibrium can exist). Souriau observed that the group of time translations of the classical thermodynamics [40,41] is not a normal subgroup of the Galilei group, proving that if a dynamical system is conservative in an inertial reference frame, it need not be conservative in another. Based on this fact, Souriau generalized the formulation of the Gibbs principle to become compatible with Galileo relativity in classical mechanics and with Poincaré relativity in relativistic mechanics. The maximum entropy principle [42–51] is preserved, and the Gibbs density is given by the density of maximum entropy (among the equilibrium states for which the average value of the energy takes a prescribed value, the Gibbs measures are those which have the largest entropy), but with a new principle "*If a dynamical system is invariant under a Lie subgroup G′ of the Galileo group, then the natural equilibria of the system forms the Gibbs ensemble of the dynamical group G′*" [10]. The classical notion of Gibbs canonical ensemble is extended for a homogneous symplectic manifold on which a Lie group (dynamic group) has a symplectic action. When the group is not abelian (non-commutative group), the symmetry is broken, and new "cohomological" relations should be verified in Lie algebra of the group [52–55]. A natural equilibrium state will thus be characterized by an element of the Lie algebra of the Lie group, determining the equilibrium temperature β. The entropy $s(Q)$, parametrized by Q the geometric heat (mean of energy U, element of the dual Lie algebra) is defined by the Legendre transform [56–59] of the Massieu potential $\Phi(\beta)$ parametrized by β ($\Phi(\beta)$ is the minus logarithm of the partition function $\psi_\Omega(\beta)$):

$$s(Q) = \langle \beta, Q \rangle - \Phi(\beta) \text{ with } \begin{cases} Q = \dfrac{\partial \Phi}{\partial \beta} \in \mathfrak{g}^* \\ \beta = \dfrac{\partial s}{\partial Q} \in \mathfrak{g} \end{cases} \tag{19}$$

$$\begin{gathered} p_{Gibbs}(\xi) = e^{\Phi(\beta) - \langle \beta, U(\xi) \rangle} = \frac{e^{-\langle \beta, U(\xi) \rangle}}{\int\limits_M e^{-\langle \beta, U(\xi) \rangle} d\omega}, \\ Q = \frac{\partial \Phi(\beta)}{\partial \beta} = \frac{\int\limits_M U(\xi) e^{-\langle \beta, U(\xi) \rangle} d\omega}{\int\limits_M e^{-\langle \beta, U(\xi) \rangle} d\omega} = \int\limits_M U(\xi) p(\xi) d\omega \quad \text{with } \Phi(\beta) = -\log \int\limits_M e^{-\langle \beta, U(\xi) \rangle} d\omega \end{gathered} \tag{20}$$

Souriau completed his "geometric heat theory" by introducing a 2-form in the Lie algebra, that is a Riemannian metric tensor in the values of adjoint orbit of β, $[\beta, Z]$ with Z an element of the Lie algebra. This metric is given for (β, Q):

$$g_\beta\left([\beta, Z_1], [\beta, Z_2]\right) = \langle \Theta(Z_1), [\beta, Z_2] \rangle + \langle Q, [Z_1, [\beta, Z_2]] \rangle \tag{21}$$

where Θ is a cocycle of the Lie algebra, defined by $\Theta = T_e\theta$ with θ a cocycle of the Lie group defined by $\theta(M) = Q\left(Ad_M(\beta)\right) - Ad_M^* Q$. We have observed that this metric g_β is also given by the hessian of the Massieu potential $g_\beta = -\dfrac{\partial^2 \Phi}{\partial \beta^2} = \dfrac{\partial \log \psi_\Omega}{\partial \beta^2}$ as Fisher metric in classical information geometry theory [60], and so this is a generalization of the Fisher metric for homogeneous manifold. We call this new metric the Souriau-Fisher metric. As $g_\beta = -\dfrac{\partial Q}{\partial \beta}$, Souriau compared it by analogy with classical thermodynamics to a "geometric specific heat" (geometric calorific capacity).

The potential theory of thermodynamics and the introduction of "characteristic function" (previous function $\Phi(\beta) = -\log \psi_\Omega(\beta)$ in Souriau theory) was initiated by François Jacques Dominique Massieu [61–64]. Massieu was the son of Pierre François Marie Massieu and Thérèse Claire Castel. He married in 1862 with Mlle Morand and had 2 children. He graduated from Ecole Polytechnique in 1851 and Ecole des Mines de Paris in 1956, he has integrated "Corps des Mines". He defended his Ph.D. in 1861 on "*Sur les intégrales algébriques des problèmes de mécanique*" and on "*Sur le mode de propagation des ondes planes et la surface de l'onde élémentaire dans les cristaux biréfringents à deux axes*" [65] with the jury composed of Lamé, Delaunay et Puiseux. In 1870, François Massieu presented his paper to the French Academy of Sciences on "*characteristic functions of the various fluids and the theory of vapors*" [61]. The design of the characteristic function is the finest scientific title of Mr. Massieu. A prominent judge, Joseph Bertrand, do not hesitate to declare, in a statement read to the French Academy of Sciences 25 July 1870, that "*the introduction of this function in formulas that summarize all the possible consequences of the two fundamental theorems seems, for the theory, a similar service almost equivalent to that Clausius has made by linking the Carnot's theorem to entropy"* [66]. The final manuscript was published by Massieu in 1873, *"Exposé des principes fondamentaux de la théorie mécanique de la chaleur (Note destinée à servir d'introduction au Mémoire de l'auteur sur les fonctions caractéristiques des divers fluides et la théorie des vapeurs)"* [63].

Massieu introduced the following potential $\Phi(\beta)$, called "characteristic function", as illustrated in Figure 2, that is the potential used by Souriau to generalize the theory: $s(Q) = \langle \beta, Q \rangle - \Phi(\beta) \underset{\beta = \frac{1}{T}}{\Rightarrow} \Phi = \frac{Q}{T} - S$. However, in his third paper, Massieu was influenced by M. Bertrand, as illustrated in Figure 3, to replace the variable $\beta = \frac{1}{T}$ (that he used in his two first papers) by T. We have then to wait 50 years more for the paper of Planck, who introduced again the good variable $\beta = \frac{1}{T}$, and then generalized by Souriau, giving to Planck temperature β an ontological and geometric status as element of the Lie algebra of the dynamic group.

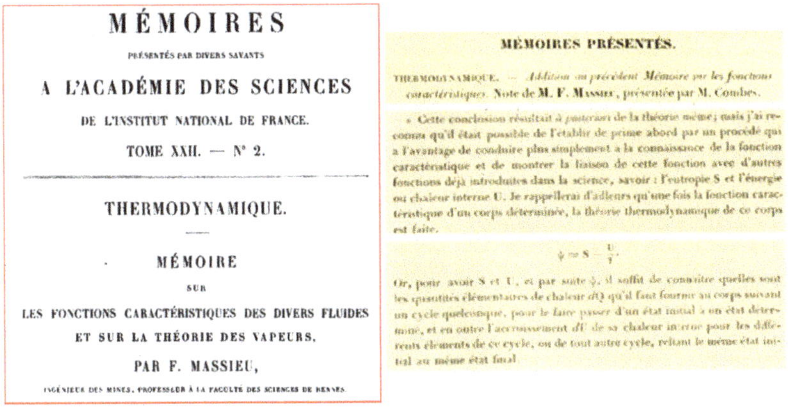
MÉMOIRES
PRÉSENTÉS PAR DIVERS SAVANTS
A L'ACADÉMIE DES SCIENCES
DE L'INSTITUT NATIONAL DE FRANCE.
TOME XXII. — N° 2.

THERMODYNAMIQUE.

MÉMOIRE
SUR
LES FONCTIONS CARACTÉRISTIQUES DES DIVERS FLUIDES
ET SUR LA THÉORIE DES VAPEURS,
PAR F. MASSIEU,
INGÉNIEUR DES MINES, PROFESSEUR À LA FACULTÉ DES SCIENCES DE RENNES.

MÉMOIRES PRÉSENTÉS.

THERMODYNAMIQUE. — *Addition au précédent Mémoire sur les fonctions caractéristiques.* Note de M. F. Massieu, présentée par M. Combes.

« Cette conclusion résultait à *posteriori* de la théorie même; mais j'ai reconnu qu'il était possible de l'établir de prime abord par un procédé qui a l'avantage de conduire plus simplement à la connaissance de la fonction caractéristique et de montrer la liaison de cette fonction avec d'autres fonctions déjà introduites dans la science, savoir : l'entropie S et l'énergie ou chaleur interne U. Je rappellerai d'ailleurs qu'une fois la fonction caractéristique d'un corps déterminée, la théorie thermodynamique de ce corps est faite.

$$\psi = S - \frac{U}{T}.$$

Or, pour avoir S et U, et par suite ψ, il suffit de connaître quelles sont les quantités élémentaires de chaleur dQ qu'il faut fournir au corps suivant un cycle quelconque, pour le faire passer d'un état initial à un état déterminé, et en outre l'accroissement dU de sa chaleur interne pour les différents éléments de ce cycle, ou de tout autre cycle, reliant le même état initial au même état final

Figure 2. Extract from the second paper of François Massieu to the French Academy of Sciences [61,62].

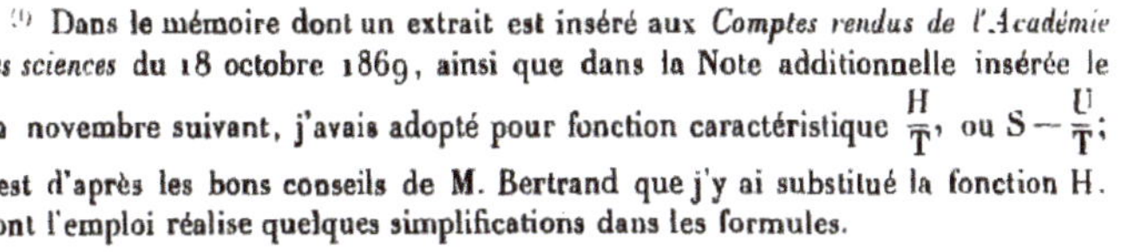
(1) Dans le mémoire dont un extrait est inséré aux *Comptes rendus de l'Académie des sciences* du 18 octobre 1869, ainsi que dans la Note additionnelle insérée le 22 novembre suivant, j'avais adopté pour fonction caractéristique $\frac{H}{T}$, ou $S - \frac{U}{T}$; c'est d'après les bons conseils de M. Bertrand que j'y ai substitué la fonction H, dont l'emploi réalise quelques simplifications dans les formules.

Figure 3. Remark of Massieu in 1876 paper [64], where he explained why he took into account the "good advice" of Bertrand to replace variable $1/T$, used in his initial paper of 1869, by the variable T.

This Lie group thermodynamics of Souriau is able to explain astronomical phenomenon (rotation of celestial bodies: the Earth and the stars rotating about themselves). The geometric temperature β can be also interpreted as a space-time vector (generalization of the temperature vector of Planck), where the temperature vector and entropy flux are in duality unifying heat conduction and viscosity (equations of Fourier and Navier). In case of centrifuge system (e.g., used for enrichment of uranium), the Gibbs Equilibrium state [60,67] are given by Souriau equations as the variation in concentration of the components of an inhomogeneous gas. Classical statistical mechanics corresponds to the dynamical group of time translations, for which we recover from Souriau equations the concepts and principles of classical thermodynamics (temperature, energy, heat, work, entropy, thermodynamic potentials) and of the kinetic theory of gases (pressure, specific heats, Maxwell's velocity distribution, etc.).

Souriau also studied continuous medium thermodynamics, where the "temperature vector" is no longer constrained to be in Lie algebra, but only contrained by phenomenologic equations (e.g., Navier equations, etc.). For thermodynamic equilibrium, the "temperature vector" is then a Killing vector of Space-Time. For each point X, there is a "temperature vector" $\beta(X)$, such it is an infinitesimal conformal transform of the metric of the universe g_{ij}. Conservation equations can then be deduced for components of impulsion-energy tensor T^{ij} and entropy flux S^j with $\hat{\partial}_i T^{ij} = 0$ and $\partial_i S^j = 0$. Temperature and metric are related by the following equations:

$$\begin{cases} \hat{\partial}_i \beta_j + \hat{\partial}_j \beta_i = \lambda g_{ij} \\ \partial_i \beta_j + \partial_j \beta_i - 2\Gamma_{ij}^k \beta_k = \lambda g_{ij} \end{cases} \text{ with } \begin{cases} \hat{\partial}_i . : \text{ covariant derivative} \\ \beta_j : \text{ component of Temperature vector} \end{cases} \quad (22)$$

$$\lambda = 0 \Rightarrow \text{Killing Equation}$$

Leon Brillouin made the link between Boltzmann entropy and Negentropie of information theory [68–71], but before Jean-Marie Souriau, only Constantin Carathéodory and Pierre Duhem [72–75] initiated first theoretical works to generalize thermodynamics.

After three years as lecturer at Lille university, Duhem published a paper in the official revue of the Ecole Normale Supérieure, in 1891, "On general equations of thermodynamics" [72] (Sur les équations générales de la Thermodynamique) in Annales Scientifiques de l'Ecole Normale Supérieure. Duhem generalized the concept of "virtual work" under the action of "external actions" by taking into account both mechanical and thermal actions. In 1894, the design of a generalized mechanics based on thermodynamics was further developed: ordinary mechanics had already become "a particular case of a more general science". Duhem writes "*We made dynamics a special case of thermodynamics, a science that embraces common principles in all changes of state bodies, changes of places as well as changes in physical qualities*" (*Nous avons fait de la dynamique un cas particulier de la thermodynamique, une Science qui embrasse dans des principes communs tous les changements d'état des corps, aussi bien les changements de lieu que les changements de qualités physiques*). In the equations of his generalized mechanics-thermodynamics, some new terms had to be introduced, in order to account for the intrinsic viscosity and friction of the system. As observed by Stefano Bordoni, Duhem aimed at widening the scope of physics: the new physics could not confine itself to "*local motion*" but had to describe what Duhem qualified "*motions of modification*". If Boltzmann had tried to proceed from "local motion" to attain the explanation of more complex transformations, Duhem was trying to proceed from general laws concerning general transformation in order to reach "*local motion*" as a simplified specific case. Four scientists were credited by Duhem with having carried out "the most important researches on that subject": Massieu had managed to derive thermodynamics from a "characteristic function and its partial derivatives"; Gibbs had shown that Massieu's functions "could play the role of potentials in the determination of the states of equilibrium" in a given system; von Helmholtz had put forward "similar ideas"; von Oettingen had given "an exposition of thermodynamics of remarkable generality" based on general duality concept in "Die thermodynamischen Beziehungen antithetisch entwickelt" published at St. Petersburg in 1885. Duhem took into account a system whose elements had the same temperature and where the state of the system could be completely specified by giving its temperature and n other independent quantities. He then introduced some "external forces", and held the system in equilibrium. A virtual work corresponded to such forces, and a set of $n + 1$ equations corresponded to the condition of equilibrium of the physical system. From the thermodynamic point of view, every infinitesimal transformation involving the generalized displacements had to obey to the first law, which could be expressed in terms of the $(n + 1)$ generalized Lagrangian parameters. The amount of heat could be written as a sum of $(n + 1)$ terms. The new alliance between mechanics and thermodynamics led to a sort of symmetry between thermal and mechanical quantities. The $n + 1$ functions played the role of *generalized thermal capacities*, and the last term was nothing other than the ordinary *thermal capacity*. The knowledge of the "*equilibrium equations of a system*" allowed Duhem to compute the partial derivatives of the thermal capacity with regard to all the parameters which described the state of the system, apart from its derivative with regard to temperature. The thermal capacities were therefore known "*except for an unspecified function of temperature*".

The axiomatic approach of thermodynamics was published in 1909 in Mathematische Annalen [76] under the title "*Examination of the Foundations of Thermodynamics*" (*Untersuchungen überdie Grundlagen der Thermodynamik*) by Constantin Carathéodory based on Carnot's works [77]. Carathéodory introduced entropy through a mathematical approach based on the geometric behavior of a certain class of partial differential equations called Pfaffians. Carathéodory's investigations start by revisiting the first law and reformulating the second law of thermodynamics in the form of two axioms. The first axiom applies to a multiphase system change under adiabatic conditions (axiom of classical thermodynamics due to Clausius [78,79]). The second axiom assumes that in the neighborhood of any equilibrium state of a system (of any number of thermodynamic coordinates), there exist states that are inaccessible by reversible adiabatic processes. In the book of Misha Gromov "*Metric Structures for Riemannian and Non-Riemannian Spaces*", written and edited by Pierre Pansu and Jacques Lafontaine, a new metric is introduced called "*Carnot-Carathéodory metric*". In one of his papers, Misha Gromov [80,81] gives historical remarks "*This result (which seems obvious by the modern standards) appears (in a more general form) in the 1909-paper by Carathéorody on formalization of the classical thermodynamics where horizontal curves roughly correspond to adiabatic processes. In fact, the above proof may be performed in the language of Carnot (cycles) and for this reason the metris distH were christened 'Carnot-Carathéodory' in Gromov-Lafontaine-Pansu book*" [82]. When I ask this question to Pierre Pansu, he gave me the answer that "*The section 4 of [76], entitled Hilfsatz aus der Theorie des Pfaffschen Gleichungen (Lemma from the theory of Pfaffian equations) opens with a statement relating to the differential 1-forms. Carathéodory says, If a Pfaffian equation dx0 + X1 dx1 + X2 dx2 + ... + Xn dxn = 0 is given, in which the Xi are finite, continuous, differentiable functions of the xi, and one knows that in any neighborhood of an arbitrary point P of the space of xi there is a point that one cannot reach along a curve that satisfies this equation then the expression must necessarily possess a multiplier that makes it into a complete differential*". This is confirmed in the introduction of his paper [76], where Carathéodory said "*Finally, in order to be able to treat systems with arbitrarily many degrees of freedom from the outset, instead of the Carnot cycle that is almost always used, but is intuitive and easy to control only for systems with two degrees of freedom, one must employ a theorem from the theory of Pfaffian differential equations, for which a simple proof is given in the fourth section*".

We have also to make reference to Henri Poincaré [83] that published the paper "*On attempts of mechanical explanation for the principles of thermodynamics (Sur les tentatives d'explication mécanique des principes de la thermodynamique)*" at the Comptes rendus de l'Académie des sciences in 1889 [84], in which he tried to consolidate links between mechanics and thermomechanics principles. These elements were also developed in Poincaré's lecture of 1892 [85] on "*thermodynamique*" in Chapter XVII "*Reduction of thermodynamics principles to the general principles of mechanics (Réduction des principes de la Thermodynamique aux principes généraux de la mécanique)*". Poincaré writes in his book [85] "*It is otherwise with the second law of thermodynamics. Clausius was the first to attempt to bring it to the principles of mechanics, but not succeed satisfactorily. Helmholtz in his memoir on the principle of least actions developed a theory much more perfect than that of Clausius. However, it cannot account for irreversible phenomena. (Il en est autrement du second principe de la thermodynamique. Clausius, a le premier, tenté de le ramener aux principes de la Mécanique, mais sans y réussir d'une manière satisfaisante. Helmoltz dans son mémoire sur le principe de la moindre action, a développé une théorie beaucoup plus parfaite que celle de Clausius; cependant elle ne peut rendre compte des phénomènes irréversibles.)*". About Helmoltz work, Poincaré observes [85] "*It follows from these examples that the Helmholtz hypothesis is true in the case of body turning around an axis; So it seems applicable to vortex motions of molecules (Il résulte de ces exemples que l'hypothèse d'Helmoltz est exacte dans le cas de corps tournant autour d'un axe; elle parait donc applicable aux mouvements tourbillonnaires des molecules.)*", but he adds in the following that the Helmoltz model is also true in the case of vibrating motions as molecular motions. However, he finally observes that the Helmoltz model cannot explain the increasing of entropy and concludes [85] "*All attempts of this nature must be abandoned; the only ones that have any chance of success are those based on the intervention of statistical laws, for example, the kinetic theory of gases. This view, which I cannot develop here, can be summed up in a somewhat vulgar way as follows: Suppose we want to place a grain of oats in the middle of a heap of wheat; it will be easy; then suppose we wanted to find it and remove it; we cannot achieve it. All irreversible phenomena, according to some physicists, would be built on this model (Toutes les tentatives de cette nature doivent donc être abandonnées; les seules qui aient quelque chance de succès sont celles qui sont fondées sur l'intervention des lois statistiques comme, par exemple, la théorie cinétique des gaz. Ce point de vue, que je ne puis développer ici, peut se résumer d'une façon un peu vulgaire comme il suit: Supposons que nous voulions placer un grain d'avoine au milieu d'un tas de blé; cela sera facile; supposons que nous voulions ensuite l'y retrouver et l'en retirer; nous ne pourrons y parvenir. Tous les phénomènes irréversibles, d'après certains physiciens, seraient construits sur ce modèle)". In Poincaré's lecture, Massieu has greatly influenced Poincaré to introduce Massieu characteristic function in probability [86]. As we have observed, Poincaré has introduced characteristic function in probability lecture after his lecture on thermodynamics where he discovered in its second edition [85], the Massieu's characteristic function. We can read that* "Since from the functions of Mr. Massieu one can deduce other functions of variables, all equations of thermodynamics can be written so as to only contain these functions and their derivatives; it will thus result in some cases, a great simplification (Puisque des fonctions de M. Massieu on peut déduire les autres fonctions des variables, toutes les équations de la Thermodynamique pourront s'écrire de manière à ne plus renfermer que ces fonctions et leurs dérivées; il en résultera donc, dans certains cas, une grande simplification)." *[85]. He [85] added* "MM. Gibbs von Helmholtz, Duhem have used this function H = U − TS assuming that T and V are constant. Mr. von Helmotz has called it 'free energy' and also proposes to give him the name of "kinetic potential"; Duhem called it 'the thermodynamic potential at constant volume'; this is the most justified naming (MM. Gibbs, von Helmoltz, Duhem ont fait usage de cette function H = TS − U en y supposant T et V constants. M. von Helmotz l'a appellée énergie libre et a propose également de lui donner le nom de potential kinetique; M. Duhem la nomme potentiel thermodynamique à volume constant; c'est la dénomination la plus justifiée)". *In 1906, Henri Poincaré also published a note [87]* "Reflection on The kinetic theory of gases" (Réflexions sur la théorie cinétique des gaz), *where he said that:* "The kinetic theory of gases leaves awkward points for those who are accustomed to mathematical rigor ... One of the points which embarrassed me most was the following one: it is a question of demonstrating that the entropy keeps decreasing, but the reasoning of Gibbs seems to suppose that having made vary the outside conditions we wait that the regime is established before making them vary again. Is this supposition essential, or in other words, we could arrive at opposite results to the principle of Carnot by making vary the outside conditions too fast so that the permanent regime has time to become established?".

Jean-Marie Souriau has elaborated a disruptive and innovative "*théorie géométrique de la chaleur (geometric theory of heat)*" [88] after the works of his predecessors as illustrated in Figure 4: "*théorie analytique de la chaleur (analytic theory of heat)*" by Jean Baptiste Joseph Fourier [88], "*théorie mécanique de la chaleur (mechanic theory of heat)*" by François Clausius [89] and François Massieu and "*théorie mathématique de la chaleur (mathematic theory of heat)*" by Siméon-Denis Poisson [90,91], as illustrated in this figure:

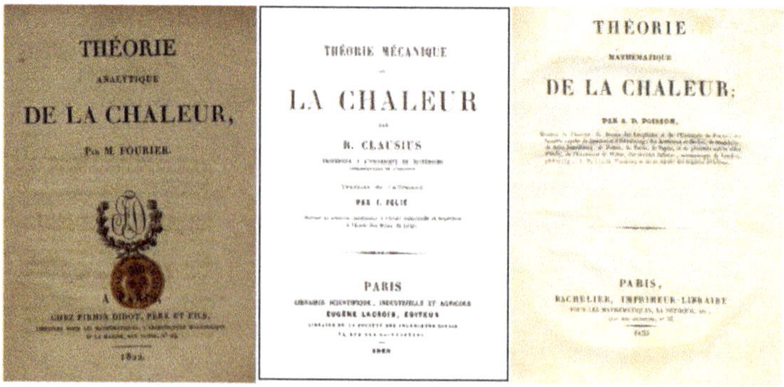

Figure 4. "Théorie analytique de la chaleur (analytic theory of heat)" by Jean Baptiste Joseph Fourier [88], "théorie mécanique de la chaleur (mechanic theory of heat)" by François Clausius [89] and "théorie mathématique de la chaleur (mathematic theory of heat)" by Siméon-Denis Poisson [90].

3. Revisited Souriau Symplectic Model of Statistical Physics

In this Section, we will revisit the Souriau model of thermodynamics but with modern notations, replacing personal Souriau conventions used in his book of 1970 by more classical ones.

In 1970, Souriau introduced the concept of co-adjoint action of a group on its momentum space (or *"moment map"*: mapping induced by symplectic manifold symmetries), based on the orbit method works, that allows to define physical observables like energy, heat and momentum or moment as pure geometrical objects (the moment map takes its values in a space determined by the group of symmetries: the dual space of its Lie algebra). The moment(um) map is a constant of the motion and is associated to symplectic cohomology (assignment of algebraic invariants to a topological space that arises from the algebraic dualization of the homology construction). Souriau introduced the moment map in 1965 in a lecture notes at Marseille University and published it in 1966. Souriau gave the formal definition and its name based on its physical interpretation in 1967. Souriau then studied its properties of equivariance, and formulated the coadjoint orbit theorem in his book in 1970. However, in his book, Souriau also observed in Chapter IV that Gibbs equilibrium states are not covariant by dynamical groups (Galileo or Poincaré groups) and then he developed a covariant model that he called *"Lie group thermodynamics"*, where equilibriums are indexed by a *"geometric (Planck) temperature"*, given by a vector β that lies in the Lie algebra of the dynamical group. For Souriau, all the details of classical mechanics appear as geometric necessities (e.g., mass is the measure of the symplectic cohomology of the action of a Galileo group). Based on this new covariant model of thermodynamic Gibbs equilibrium, Souriau has formulated statistical mechanics and thermodynamics in the framework of symplectic geometry by use of symplectic moments and distribution-tensor concepts, giving a geometric status for temperature, heat and entropy.

There is a controversy about the name "momentum map" or "moment map". Smale [92] referred to this map as the "angular momentum", while Souriau used the French word "moment". Cushman and Duistermaat [93] have suggested that the proper English translation of Souriau's French word was "momentum" which fit better with standard usage in mechanics. On the other hand, Guillemin and Sternberg [94] have validated the name given by Souriau and have used "moment" in English. In this paper, we will see that name "moment" given by Souriau was the most appropriate word. In his Chapter IV of his book [10], studying statistical mechanics, Souriau [10] has ingeniously observed that moments of inertia in mechanics are equivalent to moments in probability in his new geometric model of statistical physics. We will see that in Souriau Lie group thermodynamic model, these statistical moments will be given by the energy and the heat defined geometrically by Souriau, and will be associated with "moment map" in dual Lie algebra.

This work has been extended by Claude Vallée [5,6] and Gery de Saxcé [4,8,95,96]. More recently, Kapranov has also given a thermodynamical interpretation of the moment map for toric varieties [97] and Pavlov, thermodynamics from the differential geometry standpoint [98].

The conservation of the moment of a Hamiltonian action was called by Souriau the *"symplectic or geometric Noether theorem"*. Considering phases space as symplectic manifold, cotangent fiber of configuration space with canonical symplectic form, if Hamiltonian has Lie algebra, then the moment map is constant along the system integral curves. Noether theorem is obtained by considering independently each component of the moment map.

In a first step to establish new foundations of thermodynamics, Souriau [10] has defined a Gibbs canonical ensemble on a symplectic manifold M for a Lie group action on M. In classical statistical mechanics, a state is given by the solution of Liouville equation on the phase space, the partition function. As symplectic manifolds have a completely continuous measure, invariant by diffeomorphisms, the Liouville measure λ, all statistical states will be the product of the Liouville measure by the scalar function given by the generalized partition function $e^{\Phi(\beta)-\langle\beta,U(\xi)\rangle}$ defined by the energy U (defined in the dual of the Lie algebra of this dynamical group) and the geometric temperature β, where Φ is a normalizing constant such the mass of probability is equal to 1,

$\Phi(\beta) = -\log \int\limits_M e^{-\langle\beta, U(\xi)\rangle} d\lambda$ [99]. Jean-Marie Souriau then generalizes the Gibbs equilibrium state to all symplectic manifolds that have a dynamical group. To ensure that all integrals that will be defined could converge, *the canonical Gibbs ensemble is the largest open proper subset (in Lie algebra) where these integrals are convergent. This canonical Gibbs ensemble is convex.* The derivative of Φ, $Q = \frac{\partial \Phi}{\partial \beta}$ (thermodynamic heat) is equal to the mean value of the energy U. The minus derivative of this generalized heat Q, $K = -\frac{\partial Q}{\partial \beta}$ is symmetric and positive (this is a geometric heat capacity). Entropy s is then defined by Legendre transform of Φ, $s = \langle \beta, Q \rangle - \Phi$. If this approach is applied for the group of time translation, this is the classical thermodynamics theory. However, *Souriau* [10] *has observed that if we apply this theory for non-commutative group (Galileo or Poincaré groups), the symmetry has been broken. Classical Gibbs equilibrium states are no longer invariant by this group.* This symmetry breaking provides new equations, discovered by Souriau [10].

We can read in his paper this prophetical sentence *"This Lie group thermodynamics could be also of first interest for mathematics (Peut-être cette Thermodynamique des groups de Lie a-t-elle un intérêt mathématique)"* [30]. He explains that for the dynamic Galileo group with only one axe of rotation, this thermodynamic theory is the theory of centrifuge where the temperature vector dimension is equal to 2 (sub-group of invariance of size 2), used to make "uranium 235" and "ribonucleic acid" [30]. The physical meaning of these two dimensions for vector-valued temperature is "thermic conduction" and "viscosity". Souriau said that the model unifies "heat conduction" and "viscosity" (Fourier and Navier equations) in the same theory of irreversible process. Souriau has applied this theory in detail for relativistic ideal gas with the Poincaré group for the dynamical group.

Before introducing the Souriau Model of Lie group thermodynamics, we will first remind readers of the classical notation of Lie group theory in their application to Lie group thermodynamics:

- The coadjoint representation of G is the contragredient of the adjoint representation. It associates to each $g \in G$ the linear isomorphism $Ad_g^* \in GL(\mathfrak{g}^*)$, which satisfies, for each $\xi \in \mathfrak{g}^*$ and $X \in \mathfrak{g}$:

$$\left\langle Ad^*_{g^{-1}}(\xi), X \right\rangle = \left\langle \xi, Ad_{g^{-1}}(X) \right\rangle \quad (23)$$

- The adjoint representation of the Lie algebra $\mathfrak{g}$ is the linear representation of $\mathfrak{g}$ into itself which associates, to each $X \in \mathfrak{g}$, the linear map $ad_X \in gl(\mathfrak{g})$. ad Tangent application of Ad at neutral element e of G:

$$\begin{aligned} &ad = T_e Ad : T_e G \to End(T_e G) \\ &X, Y \in T_e G \mapsto ad_X(Y) = [X, Y] \end{aligned} \quad (24)$$

- The coadjoint representation of the Lie algebra $\mathfrak{g}$ is the contragredient of the adjoint representation. It associates, to each $X \in \mathfrak{g}$, the linear map $ad_X^* \in gl(\mathfrak{g}^*)$ which satisfies, for each $\xi \in \mathfrak{g}^*$ and $X \in \mathfrak{g}$:

$$\langle ad^*_{-X}(\xi), Y \rangle = \langle \xi, Ad_{-X}(Y) \rangle \quad (25)$$

We can illustrate for group of matrices for $G = GL_n(K)$ with $K = R$ or C.

$$T_e G = M_n(K),\ X \in M_n(K), g \in G\ \ Ad_g(X) = gXg^{-1} \quad (26)$$

$$X, Y \in M_n(K)\ ad_X(Y) = (T_e Ad)_X(Y) = XY - YX = [X, Y] \quad (27)$$

Then, the curve from $e = I_d = c(0)$ tangent to $X = c(1)$ is given by $c(t) = \exp(tX)$ and transform by Ad: $\gamma(t) = Ad\exp(tX)$

$$ad_X(Y) = (T_e Ad)_X(Y) = \frac{d}{dt}\gamma(t)Y\Big|_{t=0} = \frac{d}{dt}\exp(tX)Y\exp(tX)^{-1}\Big|_{t=0} = XY - YX \quad (28)$$

For each temperature β, element of the Lie algebra $\mathfrak{g}$, Souriau has introduced a tensor $\widetilde{\Theta}_\beta$, equal to the sum of the cocycle $\widetilde{\Theta}$ and the heat coboundary (with [.,.] Lie bracket):

$$\widetilde{\Theta}_\beta(Z_1,Z_2) = \widetilde{\Theta}(Z_1,Z_2) + \langle Q, ad_{Z_1}(Z_2)\rangle \text{ with } ad_{Z_1}(Z_2) = [Z_1,Z_2] \tag{29}$$

This tensor $\widetilde{\Theta}_\beta$ has the following properties:

- $\widetilde{\Theta}(X,Y) = \langle \Theta(X), Y\rangle$ where the map Θ is the one-cocycle of the Lie algebra $\mathfrak{g}$ with values in $\mathfrak{g}^*$, with $\Theta(X) = T_e\theta(X(e))$ where θ the one-cocycle of the Lie group G. $\widetilde{\Theta}(X,Y)$ is constant on M and the map $\widetilde{\Theta}(X,Y) : \mathfrak{g}\times\mathfrak{g} \to \Re$ is a skew-symmetric bilinear form, and is called the *symplectic cocycle of Lie algebra* $\mathfrak{g}$ associated to the *moment map* J, with the following properties:

$$\widetilde{\Theta}(X,Y) = J_{[X,Y]} - \{J_X, J_Y\} \text{ with } \{.,.\} \text{ Poisson Bracket and } J \text{ the Moment Map} \tag{30}$$

$$\widetilde{\Theta}([X,Y],Z) + \widetilde{\Theta}([Y,Z],X) + \widetilde{\Theta}([Z,X],Y) = 0 \tag{31}$$

where J_X linear application from $\mathfrak{g}$ to differential function on M: $\begin{array}{l}\mathfrak{g}\to C^\infty(M,R)\\ X\to J_X\end{array}$ and the associated differentiable application J, called moment(um) map:

$$\begin{array}{ll} J : M \to \mathfrak{g}^* & \text{such that } J_X(x) = \langle J(x), X\rangle,\ X\in\mathfrak{g} \\ x \mapsto J(x) & \end{array} \tag{32}$$

If instead of J we take the following moment map: $J'(x) = J(x) + Q\,,\ x\in M$

where $Q\in\mathfrak{g}^*$ is constant, the symplectic cocycle θ is replaced by $\theta'(g) = \theta(g) + Q - Ad_g^*Q$

where $\theta' - \theta = Q - Ad_g^*Q$ is one-coboundary of G with values in $\mathfrak{g}^*$. We also have properties $\theta(g_1g_2) = Ad_{g_1}^*\theta(g_2) + \theta(g_1)$ and $\theta(e) = 0$.

- The geometric temperature, element of the algebra $\mathfrak{g}$, is in the thekernel of the tensor $\widetilde{\Theta}_\beta$:

$$\beta \in Ker\,\widetilde{\Theta}_\beta, \text{ such that } \widetilde{\Theta}_\beta(\beta,\beta) = 0\,,\ \forall\beta\in\mathfrak{g} \tag{33}$$

- The following symmetric tensor g_β, defined on all values of $ad_\beta(.) = [\beta,.]$ is positive definite:

$$g_\beta([\beta,Z_1],[\beta,Z_2]) = \widetilde{\Theta}_\beta(Z_1,[\beta,Z_2]) \tag{34}$$

$$g_\beta([\beta,Z_1],Z_2) = \widetilde{\Theta}_\beta(Z_1,Z_2)\ ,\ \forall Z_1\in\mathfrak{g}, \forall Z_2 \in \mathrm{Im}\left(ad_\beta(.)\right) \tag{35}$$

$$g_\beta(Z_1,Z_2) \geq 0\,,\ \forall Z_1,Z_2 \in \mathrm{Im}\left(ad_\beta(.)\right) \tag{36}$$

where the linear map $ad_X \in gl(\mathfrak{g})$ is the adjoint representation of the Lie algebra $\mathfrak{g}$ defined by $X,Y\in\mathfrak{g}(=T_eG)\mapsto ad_X(Y) = [X,Y]$, and the co-adjoint representation of the Lie algebra $\mathfrak{g}$ the linear map $ad_X^* \in gl(\mathfrak{g}^*)$ which satisfies, for each $\xi\in\mathfrak{g}^*$ and $X,Y\in\mathfrak{g}$:$\langle ad_X^*(\xi),Y\rangle = \langle\xi, -ad_X(Y)\rangle$*These equations are universal*, because they are not dependent on the symplectic manifold but only on the dynamical group G, the symplectic cocycle Θ, the temperature β and the heat Q. Souriau called this model "*Lie groups thermodynamics*".

We will give the main theorem of Souriau for this "Lie group thermodynamics":

Theorem 1 (Souriau Theorem of Lie Group Thermodynamics). *Let Ω be the largest open proper subset of g, Lie algebra of G, such that $\int_M e^{-\langle\beta,U(\xi)\rangle}d\lambda$ and $\int_M \xi\cdot e^{-\langle\beta,U(\xi)\rangle}d\lambda$ are convergent integrals, this set Ω is convex and is invariant under every transformation $Ad_g(.)$, where $g\mapsto Ad_g(.)$ is the adjoint representation of G, such that $Ad_g = T_e i_g$ with $i_g : h\mapsto ghg^{-1}$. Let $a : G\times\mathfrak{g}^*\to\mathfrak{g}^*$ a unique affine action a such that linear* *part is a coadjoint representation of G, that is the contragradient of the adjoint representation. It associates to each $g \in G$ the linear isomorphism $Ad_g^* \in GL(\mathfrak{g}^*)$, satisfying, for each:*

$$\xi \in \mathfrak{g}^* \text{ and } X \in \mathfrak{g} : \left\langle Ad_g^*(\xi), X \right\rangle = \left\langle \xi, Ad_{g^{-1}}(X) \right\rangle.$$

Then, the fundamental equations of Lie group thermodynamics are given by the action of the group:

- *Action of Lie group on Lie algebra:*

$$\beta \to Ad_g(\beta) \tag{37}$$

- *Transformation of characteristic function after action of Lie group:*

$$\Phi \to \Phi - \left\langle \theta\left(g^{-1}\right), \beta \right\rangle \tag{38}$$

- *Invariance of entropy with respect to action of Lie group:*

$$s \to s \tag{39}$$

- *Action of Lie group on geometric heat, element of dual Lie algebra:*

$$Q \to a(g, Q) = Ad_g^*(Q) + \theta(g) \tag{40}$$

Souriau equations of Lie group thermodynamics are summarized in the following Figures 5 and 6:

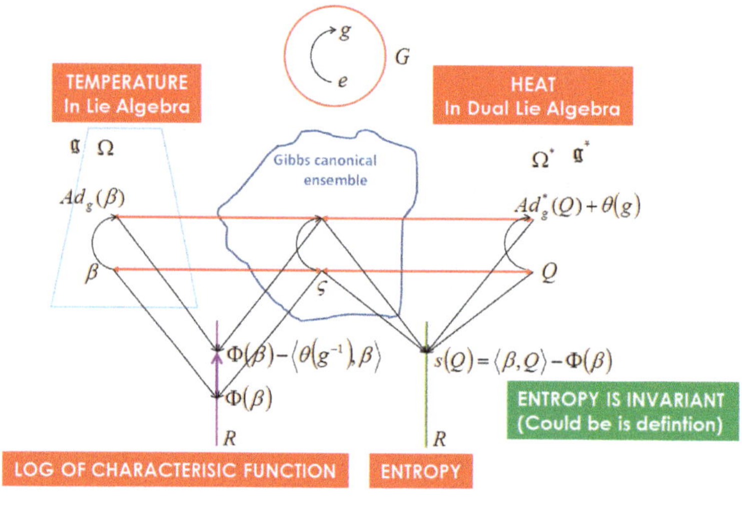

Figure 5. Global Souriau scheme of Lie group thermodynamics.

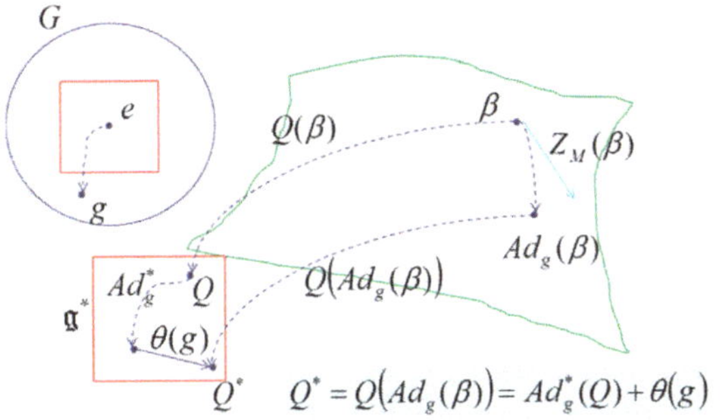

Figure 6. Broken symmetry on geometric heat Q due to adjoint action of the group on temperature β as an element of the Lie algebra.

For Hamiltonian, actions of a Lie group on a connected symplectic manifold, the equivariance of the moment map with respect to an affine action of the group on the dual of its Lie algebra has been studied by Marle and Libermann [100] and Lichnerowics [101,102]:

Theorem 2 (Marle Theorem on Cocycles). *Let G be a connected and simply connected Lie group, $R : G \to GL(E)$ be a linear representation of G in a finite-dimensional vector space E, and $r : \mathfrak{g} \to gl(E)$ be the associated linear representation of its Lie algebra $\mathfrak{g}$. For any one-cocycle $\Theta : \mathfrak{g} \to E$ of the Lie algebra $\mathfrak{g}$ for the linear representation r, there exists a unique one-cocycle $\theta : G \to E$ of the Lie group G for the linear representation R such that $\Theta(X) = T_e\theta\,(X(e))$, which has Θ as associated Lie algebra one-cocycle. The Lie group one-cocycle θ is a Lie group one-coboundary if and only if the Lie algebra one-cocycle Θ is a Lie algebra one-coboundary.*

Let G be a Lie group whose Lie algebra is $\mathfrak{g}$. The skew-symmetric bilinear form $\widetilde{\Theta}$ on $\mathfrak{g} = T_eG$ can be extended into a closed differential two-form on G, since the identity on $\widetilde{\Theta}$ means that its exterior differential $d\widetilde{\Theta}$ vanishes. In other words, $\widetilde{\Theta}$ is a 2-cocycle for the restriction of the de Rham cohomology of G to left invariant differential forms. In the framework of Lie group action on a symplectic manifold, equivariance of moment could be studied to prove that there is a unique action $a(.,.)$ of the Lie group G on the dual $\mathfrak{g}^*$ of its Lie algebra for which the moment map J is equivariant, that means for each $x \in M$:

$$J\left(\Phi_g(x)\right) = a(g, J(x)) = Ad^*_g\left(J(x)\right) + \theta(g) \tag{41}$$

where $\Phi : G \times M \to M$ is an action of Lie group G on differentiable manifold M, the fundamental field associated to an element X of Lie algebra $\mathfrak{g}$ of group G is the vectors field X_M on M:

$$X_M(x) = \frac{d}{dt}\Phi_{\exp(-tX)}\left(x\right)\Big|_{t=0} \tag{42}$$

with $\Phi_{g_1}\left(\Phi_{g_2}(x)\right) = \Phi_{g_1g_2}(x)$ and $\Phi_e(x) = x$. Φ is Hamiltonian on a symplectic manifold M, if Φ is symplectic and if for all $X \in \mathfrak{g}$, the fundamental field X_M is globally Hamiltonian. The cohomology class of the symplectic cocycle θ only depends on the Hamiltonian action Φ, and not on J.

In Appendix B, we observe that Souriau Lie group thermodynamics is compatible with Balian gauge theory of thermodynamics [103], that is obtained by symplectization in dimension $2n + 2$ of contact manifold in dimension $2n + 1$. All elements of the Souriau geometric temperature vector are multiplied by the same gauge parameter.

We conclude this section by this Bourbakiste citation of Jean-Marie Souriau [34]:

It is obvious that one can only define average values on objects belonging to a vector (or affine) space; Therefore—so this assertion may seem Bourbakist—that we will observe and measure average values only as quantity belonging to a set having physically an affine structure. It is clear that this structure is necessarily unique—if not the average values would not be well defined. (Il est évident que l'on ne peut définir de valeurs moyennes que sur des objets appartenant à un espace vectoriel (ou affine); donc—si bourbakiste que puisse sembler cette affirmation—que l'on n'observera et ne mesurera de valeurs moyennes que sur des grandeurs appartenant à un ensemble possédant physiquement une structure affine. Il est clair que cette structure est nécessairement unique—sinon les valeurs moyennes ne seraient pas bien définies.).

4. The Souriau-Fisher Metric as Geometric Heat Capacity of Lie Group Thermodynamics

We observe that Souriau Riemannian metric, introduced with symplectic cocycle, is a generalization of the Fisher metric, that we call the Souriau-Fisher metric, that preserves the property to be defined as a hessian of the partition function logarithm $g_\beta = -\frac{\partial^2 \Phi}{\partial \beta^2} = \frac{\partial^2 \log \psi_\Omega}{\partial \beta^2}$ as in classical information geometry. We will establish the equality of two terms, between Souriau definition based on Lie group cocycle Θ and parameterized by "geometric heat" Q (element of dual Lie algebra) and "geometric temperature" β (element of Lie algebra) and hessian of characteristic function $\Phi(\beta) = -\log \psi_\Omega(\beta)$ with respect to the variable β:

$$g_\beta\left([\beta, Z_1], [\beta, Z_2]\right) = \langle \Theta(Z_1), [\beta, Z_2] \rangle + \langle Q, [Z_1, [\beta, Z_2]] \rangle = \frac{\partial^2 \log \psi_\Omega}{\partial \beta^2} \tag{43}$$

If we differentiate this relation of Souriau theorem $Q\left(Ad_g(\beta)\right) = Ad_g^*(Q) + \theta(g)$, this relation occurs:

$$\frac{\partial Q}{\partial \beta}\left(-[Z_1, \beta], .\right) = \widetilde{\Theta}\left(Z_1, [\beta, .]\right) + \langle Q, Ad_{.Z_1}([\beta, .]) \rangle = \widetilde{\Theta}_\beta\left(Z_1, [\beta, .]\right) \tag{44}$$

$$-\frac{\partial Q}{\partial \beta}\left([Z_1, \beta], Z_2.\right) = \widetilde{\Theta}\left(Z_1, [\beta, Z_2]\right) + \langle Q, Ad_{.Z_1}([\beta, Z_2]) \rangle = \widetilde{\Theta}_\beta\left(Z_1, [\beta, Z_2]\right) \tag{45}$$

$$\Rightarrow -\frac{\partial Q}{\partial \beta} = g_\beta\left([\beta, Z_1], [\beta, Z_2]\right) \tag{46}$$

As the entropy is defined by the Legendre transform of the characteristic function, this Souriau-Fisher metric is also equal to the inverse of the hessian of "geometric entropy" $s(Q)$ with respect to the variable Q: $\frac{\partial^2 s(Q)}{\partial Q^2}$

For the maximum entropy density (Gibbs density), the following three terms coincide: $\frac{\partial^2 \log \psi_\Omega}{\partial \beta^2}$ that describes the convexity of the log-likelihood function, $I(\beta) = -E\left[\frac{\partial^2 \log p_\beta(\xi)}{\partial \beta^2}\right]$ the Fisher metric that describes the covariance of the log-likelihood gradient, whereas $I(\beta) = E\left[(\xi - Q)(\xi - Q)^T\right] = Var(\xi)$ that describes the covariance of the observables.

We can also observe that the Fisher metric $I(\beta) = -\frac{\partial Q}{\partial \beta}$ is exactly the Souriau metric defined through symplectic cocycle:

$$I(\beta) = \widetilde{\Theta}_\beta\left(Z_1, [\beta, Z_2]\right) = g_\beta\left([\beta, Z_1], [\beta, Z_2]\right) \tag{47}$$

The Fisher metric $I(\beta) = -\frac{\partial^2 \Phi(\beta)}{\partial \beta^2} = -\frac{\partial Q}{\partial \beta}$ has been considered by Souriau as a *generalization of "heat capacity"*. Souriau called it K the *"geometric capacity"*.

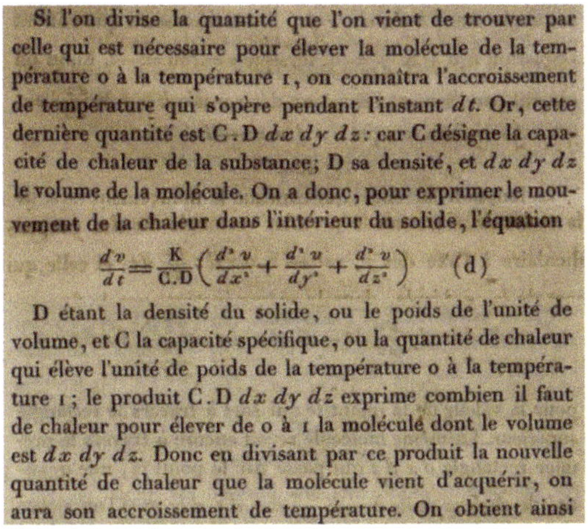
Si l'on divise la quantité que l'on vient de trouver par celle qui est nécessaire pour élever la molécule de la température o à la température 1, on connaîtra l'accroissement de température qui s'opère pendant l'instant dt. Or, cette dernière quantité est C.D $dx\,dy\,dz$: car C désigne la capacité de chaleur de la substance; D sa densité, et $dx\,dy\,dz$ le volume de la molécule. On a donc, pour exprimer le mouvement de la chaleur dans l'intérieur du solide, l'équation

$$\frac{dv}{dt}=\frac{K}{C.D}\left(\frac{d^2v}{dx^2}+\frac{d^2v}{dy^2}+\frac{d^2v}{dz^2}\right) \quad (d)$$

D étant la densité du solide, ou le poids de l'unité de volume, et C la capacité spécifique, ou la quantité de chaleur qui élève l'unité de poids de la température o à la température 1; le produit C.D $dx\,dy\,dz$ exprime combien il faut de chaleur pour élever de o à 1 la molécule dont le volume est $dx\,dy\,dz$. Donc en divisant par ce produit la nouvelle quantité de chaleur que la molécule vient d'acquérir, on aura son accroissement de température. On obtient ainsi

Figure 7. Fourier heat equation in seminal manuscript of Joseph Fourier [88].

For $\beta = \frac{1}{kT}$, $K = -\frac{\partial Q}{\partial \beta} = -\frac{\partial Q}{\partial T}\left(\frac{\partial(1/kT)}{\partial T}\right)^{-1} = kT^2\frac{\partial Q}{\partial T}$ linking the geometric capacity to calorific capacity, then Fisher metric can be introduced in Fourier heat equation (see Figure 7):

$$\frac{\partial T}{\partial t} = \frac{\kappa}{C \cdot D}\Delta T \text{ with } \frac{\partial Q}{\partial T} = C \cdot D \Rightarrow \frac{\partial \beta^{-1}}{\partial t} = \kappa\left[\left(\beta^2/k\right) \cdot I_{Fisher}(\beta)\right]^{-1}\Delta\beta^{-1} \tag{48}$$

We can also observe that Q is related to the mean, and K to the variance of U:

$$K = I(\beta) = -\frac{\partial Q}{\partial \beta} = \text{var}(U) = \int_M U(\xi)^2 \cdot p_\beta(\xi)d\omega - \left(\int_M U(\xi) \cdot p_\beta(\xi)d\omega\right)^2 \tag{49}$$

We observe that the entropy s is unchanged, and Φ is changed but with linear dependence to β, with the consequence that Fisher Souriau metric is invariant:

$$s\left[Q\left(Ad_g(\beta)\right)\right] = s(Q(\beta)) \text{ and } I\left(Ad_g(\beta)\right) = -\frac{\partial^2\left(\Phi - \left\langle\theta\left(g^{-1}\right), \beta\right\rangle\right)}{\partial\beta^2} = -\frac{\partial^2\Phi}{\partial\beta^2} = I(\beta) \tag{50}$$

We have observed that the concept of "heat capacity" is important in the Souriau model because it gives a geometric meaning to its definition. The notion of "heat capacity" has been generalized by Pierre Duhem in his general equations of thermodynamics.

Souriau [34] proposed to define a thermometer (θερμόσ) device principle that could measure this geometric temperature using "relative ideal gas thermometer" based on a theory of dynamical group thermometry and has also recovered the (geometric) Laplace barometric law

5. Euler-Poincaré Equations and Variational Principle of Souriau Lie Group Thermodynamics

When a Lie algebra acts locally transitively on the configuration space of a Lagrangian mechanical system, Henri Poincaré proved that the Euler-Lagrange equations are equivalent to a new system of differential equations defined on the product of the configuration space with the Lie algebra. Marle has written about the Euler-Poincaré equations [104], under an intrinsic form, without any reference to a particular system of local coordinates, proving that they can be conveniently expressed in terms of the Legendre and moment maps of the lift to the cotangent bundle of the Lie algebra action on the configuration space. The Lagrangian is a smooth real valued function L defined on the tangent bundle

TM. To each parameterized continuous, piecewise smooth curve $\gamma : [t_0, t_1] \to M$, defined on a closed interval $[t_0, t_1]$, with values in M, one associates the value at γ of the action integral:

$$I(\gamma) = \int_{t_0}^{t_1} L\left(\frac{d\gamma(t)}{dt}\right) dt \tag{51}$$

The partial differential of the function $L : M \times \mathfrak{g} \to \Re$ with respect to its second variable $d_2\overline{L}$, which plays an important part in the Euler-Poincaré equation, can be expressed in terms of the moment and Legendre maps: $d_2\overline{L} = p_{\mathfrak{g}^*} \circ \varphi^t \circ \mathrm{L} \circ \varphi$ with $J = p_{\mathfrak{g}^*} \circ \varphi^t (\Rightarrow d_2\overline{L} = J \circ \mathrm{L} \circ \varphi)$ the moment map, $p_{\mathfrak{g}^*} : M \times \mathfrak{g}^* \to \mathfrak{g}^*$ the canonical projection on the second factor, $\mathrm{L} : TM \to T^*M$ the Legendre transform, with:

$$\varphi : M \times \mathfrak{g} \to TM / \varphi(x, X) = X_M(x) \text{ and } \varphi^t : T^*M \to M \times \mathfrak{g}^* / \varphi^t(\xi) = (\pi_M(\xi), J(\xi)) \tag{52}$$

The Euler-Poincaré equation can therefore be written under the form:

$$\left(\frac{d}{dt} - ad^*_{V(t)}\right) (J \circ \mathrm{L} \circ \varphi\,(\gamma(t), V(t))) = J \circ d_1\overline{L}\,(\gamma(t), V(t)) \text{ with } \frac{d\gamma(t)}{dt} = \varphi\,(\gamma(t), V(t)) \tag{53}$$

with

$$H(\xi) = \left\langle \xi, \mathrm{L}^{-1}(\xi) \right\rangle - L\left(\mathrm{L}^{-1}(\xi)\right),\ \xi \in T^*M,\ \mathrm{L} : TM \to T^*M,\ H : T^*M \to R\,. \tag{54}$$

Following the remark made by Poincaré at the end of his note [105], the most interesting case is when the map $\overline{L} : M \times \mathfrak{g} \to R$ only depends on its second variable $X \in \mathfrak{g}$. The Euler-Poincaré equation becomes:

$$\left(\frac{d}{dt} - ad^*_{V(t)}\right) \left(d\overline{L}\,(V(t))\right) = 0 \tag{55}$$

We can use analogy of structure when the convex Gibbs ensemble is homogeneous [106]. We can then apply Euler-Poincaré equation for Lie group thermodynamics. Considering Clairaut's equation:

$$s\,(Q) = \langle \beta, Q \rangle - \Phi(\beta) = \left\langle \Theta^{-1}(Q), Q \right\rangle - \Phi\left(\Theta^{-1}(Q)\right) \tag{56}$$

with $Q = \Theta(\beta) = \dfrac{\partial \Phi}{\partial \beta} \in \mathfrak{g}^*$, $\beta = \Theta^{-1}(Q) \in \mathfrak{g}$, a *Souriau-Euler-Poincaré equation* can be elaborated for Souriau Lie group thermodynamics:

$$\frac{dQ}{dt} = ad^*_\beta Q \tag{57}$$

or

$$\frac{d}{dt}\left(Ad^*_g Q\right) = 0. \tag{58}$$

The first equation, the Euler-Poincaré equation is a reduction of Euler-Lagrange equations using symmetries and especially the fact that a group is acting homogeneously on the symplectic manifold:

$$\frac{dQ}{dt} = ad^*_\beta Q \text{ and } \left\{ \begin{array}{l} s(Q) = \langle \beta, Q \rangle - \Phi(\beta) \\ \beta = \frac{\partial s(Q)}{\partial Q} \in \mathfrak{g},\ Q = \frac{\partial \Phi(\beta)}{\partial \beta} \in \mathfrak{g}^* \end{array} \right. \tag{59}$$

Back to Koszul model of information geometry, we can then deduce an equivalent of the Euler-Poincaré equation for statistical models

$$\frac{dx^*}{dt} = ad_x^* x^* \text{ and } \begin{cases} \Phi^*(x^*) = \langle x, x^* \rangle - \Phi(x) \\ x = \frac{\partial \Phi^*(x^*)}{\partial x} \in \Omega \,,\; x^* = \frac{\partial \Phi(x)}{\partial x} \in \Omega^* \end{cases} \tag{60}$$

We can use this Euler-Poincaré equation to deduce an associated equation on entropy: $\frac{ds}{dt} = \left\langle \frac{d\beta}{dt}, Q \right\rangle + \left\langle \beta, ad_\beta^* Q \right\rangle - \frac{d\Phi}{dt}$ that reduces to

$$\frac{ds}{dt} = \left\langle \frac{d\beta}{dt}, Q \right\rangle - \frac{d\Phi}{dt} \tag{61}$$

due to $\langle \xi, ad_V X \rangle = -\langle ad_V^* \xi, X \rangle \Rightarrow \left\langle \beta, ad_\beta^* Q \right\rangle = \langle Q, ad_\beta \beta \rangle = 0$.

With these new equation of thermodynamics $\frac{dQ}{dt} = ad_\beta^* Q$ and $\frac{d}{dt}(Ad_g^* Q) = 0$, we can observe that the new important notion is related to co-adjoint orbits, that are associated to a symplectic manifold by Souriau with KKS 2-form.

We will then define the Poincaré-Cartan integral invariant for Lie group thermodynamics. Classically in mechanics, the Pfaffian form $\omega = p \cdot dq - H \cdot dt$ is related to Poincaré-Cartan integral invariant [107]. Dedecker has observed, based on the relation [108]:

$$\omega = \partial_{\dot q} L \cdot dq - \left(\partial_{\dot q} L \cdot \dot q - L \right) \cdot dt = L \cdot dt + \partial_{\dot q} L \varpi \text{ with } \varpi = dq - \dot q \cdot dt \tag{62}$$

that the property that among all forms $\chi \equiv L \cdot dt \mathrm{mod} \varpi$ the form $\omega = p \cdot dq - H \cdot dt$ is the only one satisfying $d\chi \equiv 0 \mathrm{mod} \varpi$, is a particular case of more general Lepage congruence.

Analogies between geometric mechanics and geometric Lie group thermodynamics, provides the following similarities of structures:

$$\begin{cases} \dot q \leftrightarrow \beta \\ p \leftrightarrow Q \end{cases}, \begin{cases} L(\dot q) \leftrightarrow \Phi(\beta) \\ H(p) \leftrightarrow s(Q) \\ H = p \cdot \dot q - L \leftrightarrow s = \langle Q, \beta \rangle - \Phi \end{cases}$$
$$\text{and} \begin{cases} \dot q = \frac{dq}{dt} = \frac{\partial H}{\partial p} \leftrightarrow \beta = \frac{\partial s}{\partial Q} \\ p = \frac{\partial L}{\partial \dot q} \leftrightarrow Q = \frac{\partial \Phi}{\partial \beta} \end{cases} \tag{63}$$

We can then consider a similar *Poincaré-Cartan-Souriau Pfaffian form*:

$$\omega = p \cdot dq - H \cdot dt \leftrightarrow \omega = \langle Q, (\beta \cdot dt) \rangle - s \cdot dt = (\langle Q, \beta \rangle - s) \cdot dt = \Phi(\beta) \cdot dt \tag{64}$$

This analogy provides an associated *Poincaré-Cartan-Souriau integral invariant*. Poincaré-Cartan integral invariant $\int_{C_a} p \cdot dq - H.dt = \int_{C_b} p \cdot dq - H \cdot dt$ is given for Souriau thermodynamics by:

$$\int_{C_a} \Phi(\beta) \cdot dt = \int_{C_b} \Phi(\beta) \cdot dt \tag{65}$$

We can then deduce an *Euler-Poincaré-Souriau variational principle* for thermodynamics: The variational principle holds on $\mathfrak{g}$, for variations $\delta\beta = \dot\eta + [\beta, \eta]$, where $\eta(t)$ is an arbitrary path that vanishes at the endpoints, $\eta(a) = \eta(b) = 0$:

$$\delta \int_{t_0}^{t_1} \Phi\left(\beta(t)\right) \cdot dt = 0 \quad (66)$$

6. Souriau Affine Representation of Lie Group and Lie Algebra and Comparison with the Koszul Affine Representation

This affine representation of Lie group/algebra used by Souriau has been intensively studied by Marle [7,100,109,110]. Souriau called the mechanics deduced from this model, "affine mechanics". We will explain affine representations and associated notions as cocycles, Souriau moment map and cocycles, equivariance of Souriau moment map, action of Lie group on a symplectic manifold and dual spaces of finite-dimensional Lie algebras. We have observed that these tools have been developed in parallel by Jean-Louis Koszul. We will establish close links and synthetize the comparisons in a table of both approaches.

6.1. *Affine Representations and Cocycles*

Souriau model of Lie group thermodynamics is linked with affine representation of Lie group and Lie algebra. We will give in the following main elements of this affine representation.

Let G be a Lie group and E a finite-dimensional vector space. A map $A : G \to Aff(E)$ can always be written as:

$$A(g)(x) = R(g)(x) + \theta(g) \text{ with } g \in G, x \in E \quad (67)$$

where the maps $R : G \to GL(E)$ and $\theta : G \to E$ are determined by A. The map A is *an affine representation of* G in E.

The map $\theta : G \to E$ is a one-cocycle of G with values in E, for the linear representation R; it means that θ is a smooth map which satisfies, for all $g, h \in G$:

$$\theta(gh) = R(g)(\theta(h)) + \theta(g) \quad (68)$$

The linear representation R is called the linear part of the affine representation A, and θ is called the one-cocycle of G associated to the affine representation A. A one-coboundary of G with values in E, for the linear representation R, is a map $\theta : G \to E$ which can be expressed as:

$$\theta(g) = R(g)(c) - c \,,\ g \in G \quad (69)$$

where c is a fixed element in E and then there exist an element $c \in E$ such that, for all $g \in G$ and $x \in E$:

$$A(g)(x) = R(g)(x + c) - c \quad (70)$$

Let $\mathfrak{g}$ be a Lie algebra and E a finite-dimensional vector space. A linear map $a : \mathfrak{g} \to aff(E)$ always can be written as:

$$a(X)(x) = r(X)(x) + \Theta(X) \text{ with } X \in \mathfrak{g}, x \in E \quad (71)$$

where the linear maps $r : \mathfrak{g} \to gl(E)$ and $\Theta : \mathfrak{g} \to E$ are determined by a. The map a is an affine representation of G in E. The linear map $\Theta : \mathfrak{g} \to E$ is a one-cocycle of G with values in E, for the linear representation r; it means that Θ satisfies, for all $X, Y \in \mathfrak{g}$:

$$\Theta\left([X, Y]\right) = r(X)\left(\Theta(Y)\right) - r(Y)\left(\Theta(X)\right) \quad (72)$$

Θ is called the one-cocycle of $\mathfrak{g}$ associated to the affine representation a. A one-coboundary of $\mathfrak{g}$ with values in E, for the linear representation r, is a linear map $\Theta : \mathfrak{g} \to E$ which can be expressed as: $\Theta(X) = r(X)(c)\,,\ X \in \mathfrak{g}$ where c is a fixed element in E., and then there exist an element $c \in E$ such that, for all $X \in \mathfrak{g}$ and $x \in E$:

$$a(X)(x) = r(X)(x + c)$$

Let $A : G \rightarrow Aff(E)$ be an affine representation of a Lie group $\mathfrak{g}$ in a finite-dimensional vector space E, and $\mathfrak{g}$ be the Lie algebra of G. Let $R : G \rightarrow GL(E)$ and $\theta : G \rightarrow E$ be, respectively, the linear part and the associated cocycle of the affine representation A. Let $a : \mathfrak{g} \rightarrow aff(E)$ be the affine representation of the Lie algebra $\mathfrak{g}$ associated to the affine representation $A : G \rightarrow Aff(E)$ of the Lie group G. The linear part of a is the linear representation $r : \mathfrak{g} \rightarrow gl(E)$ associated to the linear representation $R : G \rightarrow GL(E)$, and the associated cocycle $\Theta : \mathfrak{g} \rightarrow E$ is related to the one-cocycle $\theta : G \rightarrow E$ by:

$$\Theta(X) = T_e\theta\left(X(e)\right), X \in \mathfrak{g} \tag{73}$$

This is deduced from:

$$\left.\frac{dA\left(\exp(tX)\right)(x)}{dt}\right|_{t=0} = \left.\frac{d\left(R(\exp(tX))(x) + \theta(\exp(tX))\right)}{dt}\right|_{t=0} \Rightarrow a(X)(x) = r(X)(x) + T_e\theta(X) \tag{74}$$

Let G be a connected and simply connected Lie group, $R : G \rightarrow GL(E)$ be a linear representation of G in a finite-dimensional vector space E, and $r : \mathfrak{g} \rightarrow gl(E)$ be the associated linear representation of its Lie algebra $\mathfrak{g}$. For any one-cocycle $\Theta : \mathfrak{g} \rightarrow E$ of the Lie algebra $\mathfrak{g}$ for the linear representation r, there exists a unique one-cocycle $\theta : G \rightarrow E$ of the Lie group G for the linear representation R such that:

$$\Theta(X) = T_e\theta\left(X(e)\right) \tag{75}$$

in other words, which has Θ as associated Lie algebra one-cocycle. The Lie group one-cocycle θ is a Lie group one-coboundary if and only if the Lie algebra one-cocycle Θ is a Lie algebra one-coboundary.

$$\left.\frac{d\theta\left(g\exp(tX)\right)}{dt}\right|_{t=0} = \left.\frac{d\left(\theta(g) + R(g)\left(\theta(\exp(tX))\right)\right)}{dt}\right|_{t=0} \Rightarrow T_g\theta\left(TL_g(X)\right) = R(g)\left(\Theta(x)\right) \tag{76}$$

which proves that if it exists, the Lie group one-cocycle θ such that $T_e\theta = \Theta$ is unique.

6.2. Souriau Moment Map and Cocycles

Souriau first introduced the moment map in his book. We will give the link with previous cocycles of affine representation.

There exist J_X linear application from $\mathfrak{g}$ to differential function on M:

$$\begin{aligned} &\mathfrak{g} \rightarrow C^\infty(M, R) \\ &X \rightarrow J_X \end{aligned} \tag{77}$$

We can then associate a differentiable application J, called moment(um) map for the Hamiltonian Lie group action Φ:

$$\begin{aligned} &J : M \rightarrow \mathfrak{g}^* \\ &\quad x \mapsto J(x) \text{ such that } J_X(x) = \langle J(x), X\rangle\,,\ X \in \mathfrak{g} \end{aligned} \tag{78}$$

Let J moment map, for each $(X,Y) \in \mathfrak{g} \times \mathfrak{g}$, we associate a smooth function $\tilde{\Theta}\left(X,Y\right) : M \rightarrow \Re$ defined by:

$$\tilde{\Theta}(X,Y) = J_{[X,Y]} - \{J_X, J_Y\} \text{ with } \{.,.\} : \text{Poisson Bracket} \tag{79}$$

It is a Casimir of the Poisson algebra $C^\infty(M,\Re)$, that satisfies:

$$\tilde{\Theta}([X,Y],Z) + \tilde{\Theta}([Y,Z],X) + \tilde{\Theta}([Z,X],Y) = 0 \tag{80}$$

When the Poisson manifold is a connected symplectic manifold, the function $\tilde{\Theta}\left(X,Y\right)$ is constant on M and the map:

$$\widetilde{\Theta}(X,Y):\mathfrak{g}\times\mathfrak{g}\rightarrow\Re \quad (81)$$

is a skew-symmetric bilinear form, and is called the symplectic Cocycle of Lie algebra $\mathfrak{g}$ associated to the moment map J.

Let $\Theta:\mathfrak{g}\rightarrow\mathfrak{g}^*$ be the map such that for all:

$$X,Y\in\mathfrak{g}:\ \langle\Theta(X),Y\rangle=\widetilde{\Theta}(X,Y) \quad (82)$$

The map Θ is therefore the one-cocycle of the Lie algebra $\mathfrak{g}$ with values in $\mathfrak{g}^*$ for the coadjoint representation $X\mapsto ad_X^*$ of $\mathfrak{g}$ associated to the affine action of $\mathfrak{g}$ on its dual:

$$a_\Theta(X)(\xi)=ad^*_{-X}(\xi)+\Theta(X)\ ,\ X\in\mathfrak{g}\ ,\ \xi\in\mathfrak{g}^* \quad (83)$$

Let G be a Lie group whose Lie algebra is $\mathfrak{g}$. The skew-symmetric bilinear form $\widetilde{\Theta}$ on $\mathfrak{g}=T_eG$ can be extended into a closed differential two-form on G, since the identity on $\widetilde{\Theta}$ means that its exterior differential $d\widetilde{\Theta}$ vanishes. In other words, $\widetilde{\Theta}$ is a 2-cocycle for the restriction of the de Rham cohomology of G to left (or right) invariant differential forms.

6.3. Equivariance of Souriau Moment Map

There exists a unique affine action a such that the linear part is a coadjoint representation:

$$\begin{aligned} &a:G\times\mathfrak{g}^*\rightarrow\mathfrak{g}^* \\ &a(g,\xi)=Ad^*_{g^{-1}}\xi+\theta(g) \end{aligned} \quad (84)$$

with $\left\langle Ad^*_{g^{-1}}\xi,X\right\rangle=\langle\xi,Ad_{g^{-1}}X\rangle$ and that induce equivariance of moment J.

6.4. Action of Lie Group on a Symplectic Manifold

Let $\Phi:G\times M\rightarrow M$ be an action of Lie group G on differentiable manifold M, the fundamental field associated to an element X of Lie algebra $\mathfrak{g}$ of group G is the vectors field X_M on M:

$$X_M(x)=\frac{d}{dt}\Phi_{\exp(-tX)}(x)\Big|_{t=0}\quad \text{With } \Phi_{g_1}(\Phi_{g_2}(x))=\Phi_{g_1g_2}(x) \text{ and } \Phi_e(x)=x \quad (85)$$

Φ is Hamiltonian on a symplectic manifold M, if Φ is symplectic and if for all $X\in\mathfrak{g}$, the fundamental field X_M is globally Hamiltonian.

There is a unique action a of the Lie group G on the dual $\mathfrak{g}^*$ of its Lie algebra for which the moment map J is equivariant, that means satisfies for each $x\in M$

$$J(\Phi_g(x))=a(g,J(x))=Ad^*_{g^{-1}}(J(x))+\theta(g) \quad (86)$$

$\theta:G\rightarrow\mathfrak{g}^*$ is called cocycle associated to the differential $T_e\theta$ of 1-cocyle θ associated to J at neutral element e:

$$\langle T_e\theta(X),Y\rangle=\widetilde{\Theta}(X,Y)=J_{[X,Y]}-\{J_X,J_Y\} \quad (87)$$

If instead of J we take the moment map $J'(x)=J(x)+\mu\ ,\ x\in M$, where $\mu\in\mathfrak{g}^*$ is constant, the symplectic cocycle θ is replaced by:

$$\theta'(g)=\theta(g)+\mu-Ad^*_g\mu \quad (88)$$

where $\theta'-\theta=\mu-Ad^*_g\mu$ is one-coboundary of G with values in $\mathfrak{g}^*$.

Therefore, the cohomology class of the symplectic cocycle θ only depends on the Hamiltonian action Φ, not on the choice of its moment map J. We have also:

$$\tilde{\Theta}'(X,Y) = \tilde{\Theta}(X,Y) + \langle \mu, [X,Y] \rangle \tag{89}$$

This property is used by Jean-Marie Souriau [10] to offer a very nice cohomological interpretation of the total mass of a classical (nonrelativistic) isolated mechanical system. He [10] proves that the space of all possible motions of the system is a symplectic manifold on which the Galilean group acts by a Hamiltonian action. The dimension of the symplectic cohomology space of the Galilean group (the quotient of the space of symplectic one-cocycles by the space of symplectic one-coboundaries) is equal to 1. The cohomology class of the symplectic cocycle associated to a moment map of the action of the Galilean group on the space of motions of the system is interpreted as the total mass of the system.

For Hamiltonian actions of a Lie group on a connected symplectic manifold, the equivariance of the moment map with respect to an affine action of the group on the dual of its Lie algebra has been proved by Marle [110]. Marle [110] has also developed the notion of symplectic cocycle and has proved that given a Lie algebra symplectic cocycle, there exists on the associated connected and simply connected Lie group a unique corresponding Lie group symplectic cocycle. Marle [104] has also proved that there exists a two-parameter family of deformations of these actions (the Hamiltonian actions of a Lie group on its cotangent bundle obtained by lifting the actions of the group on itself by translations) into a pair of mutually symplectically orthogonal Hamiltonian actions whose moment maps are equivariant with respect to an affine action involving any given Lie group symplectic cocycle. Marle [104] has also explained why a reduction occurs for Euler-Poncaré equation mainly when the Hamiltonian can be expressed as the moment map composed with a smooth function defined on the dual of the Lie algebra; the Euler-Poincaré equation is then equivalent to the Hamilton equation written on the dual of the Lie algebra.

6.5. Dual Spaces of Finite-Dimensional Lie Algebras

Let $\mathfrak{g}$ be a finite-dimensional Lie algebra, and $\mathfrak{g}^*$ its dual space. The Lie algebra $\mathfrak{g}$ can be considered as the dual of $\mathfrak{g}^*$, that means as the space of linear functions on $\mathfrak{g}^*$, and the bracket of the Lie algebra $\mathfrak{g}$ is a composition law on this space of linear functions. This composition law can be extended to the space $C^\infty(\mathfrak{g}^*, \Re)$ by setting:

$$\{f,g\}(x) = \langle x, [df(x), dg(x)] \rangle \ , \ f \text{ and } g \in C^\infty(\mathfrak{g}^*, \Re), \ x \in \mathfrak{g}^* \tag{90}$$

If we apply this formula for Souriau Lie group thermodynamics, and for entropy $s(Q)$ depending on geometric heat Q:

$$\{s_1, s_2\}(Q) = \langle Q, [ds_1(Q), ds_2(Q)] \rangle \ , \ s_1 \text{ and } s_2 \in C^\infty(\mathfrak{g}^*, \Re), \ Q \in \mathfrak{g}^* \tag{91}$$

This bracket on $C^\infty(\mathfrak{g}^*, \Re)$ defines a Poisson structure on $\mathfrak{g}^*$, called its canonical Poisson structure. It implicitly appears in the works of Sophus Lie, and was rediscovered by Alexander Kirillov [111], Bertram Kostant and Jean-Marie Souriau.

The above defined canonical Poisson structure on $\mathfrak{g}^*$ can be modified by means of a symplectic cocycle $\tilde{\Theta}$ by defining the new bracket:

$$\{f,g\}_{\tilde{\Theta}}(x) = \langle x, [df(x), dg(x)] \rangle - \tilde{\Theta}(df(x), dg(x)) \tag{92}$$

with $\tilde{\Theta}$ a symplectic cocycle of the Lie algebra $\mathfrak{g}$ being a skew-symmetric bilinear map $\tilde{\Theta} : \mathfrak{g} \times \mathfrak{g} \to \Re$ which satisfies:

$$\tilde{\Theta}([X,Y],Z) + \tilde{\Theta}([Y,Z],X) + \tilde{\Theta}([Z,X],Y) = 0 \tag{93}$$

This Poisson structure is called the modified canonical Poisson structure by means of the symplectic cocycle $\tilde{\Theta}$. The symplectic leaves of $\mathfrak{g}^*$ equipped with this Poisson structure are the orbits of an affine action whose linear part is the coadjoint action, with an additional term determined by $\tilde{\Theta}$.

6.6. Koszul Affine Representation of Lie Group and Lie Algebra

Previously, we have developed Souriau's works on the affine representation of a Lie group used to elaborate the Lie group thermodynamics. We will study here another approach of affine representation of Lie group and Lie algebra introduced by Jean-Louis Koszul. We consolidate the link of Jean-Louis Koszul work with Souriau model. This model uses an affine representation of a Lie group and of a Lie algebra in a finite-dimensional vector space, seen as special examples of actions.

Since the work of Henri Poincare and Elie Cartan, the theory of differential forms has become an essential instrument of modern differential geometry [112–115] used by Jean-Marie Souriau for identifying the space of motions as a symplectic manifold. However, as said by Paulette Libermann [116], except Henri Poincaré who wrote shortly before his death a report on the work of Elie Cartan during his application for the Sorbonne University, the French mathematicians did not see the importance of Cartan's breakthroughs. Souriau followed lectures of Elie Cartan in 1945. The second student of Elie Cartan was Jean-Louis Koszul. Koszul introduced the concepts of affine spaces, affine transformations and affine representations [117–124]. More especially, we are interested by Koszul's definition for affine representations of Lie groups and Lie algebras. Koszul studied symmetric homogeneous spaces and defined relation between invariant flat affine connections to affine representations of Lie algebras, and characterized invariant Hessian metrics by affine representations of Lie algebras [117–124]. Koszul provided correspondence between symmetric homogeneous spaces with invariant Hessian structures by using affine representations of Lie algebras, and proved that a simply connected symmetric homogeneous space with invariant Hessian structure is a direct product of a Euclidean space and a homogeneous self-dual regular convex cone [117–124]. Let *G* be a connected Lie group and let *G/K* be a homogeneous space on which *G* acts effectively, Koszul gave a bijective correspondence between the set of *G*-invariant flat connections on *G/K* and the set of a certain class of affine representations of the Lie algebra of *G* [117–124]. The main theorem of Koszul is: let *G/K* be a homogeneous space of a connected Lie group *G* and let $\mathfrak{g}$ and k be the Lie algebras of *G* and *K*, assuming that *G/K* is endowed with a *G*-invariant flat connection, then $\mathfrak{g}$ admits an affine representation (*f*,*q*) on the vector space *E*. Conversely, suppose that *G* is simply connected and that $\mathfrak{g}$ is endowed with an affine representation, then *G/K* admits a *G*-invariant flat connection.

Koszul has proved the following [117–124]. Let Ω be a convex domain in R^n containing no complete straight lines, and an associated convex cone $V(\Omega) = \{(\lambda x, x) \in R^n \times R / x \in \Omega, \lambda \in R^+\}$. Then there exists an affine embedding:

$$\ell : x \in \Omega \mapsto \begin{bmatrix} x \\ 1 \end{bmatrix} \in V(\Omega) \tag{94}$$

If we consider η the group of homomorphism of $A(n,R)$ into $GL(n+1,R)$ given by:

$$s \in A(n,R) \mapsto \begin{bmatrix} \mathbf{f}(s) & \mathbf{q}(s) \\ 0 & 1 \end{bmatrix} \in GL(n+1,R) \tag{95}$$

and associated affine representation of Lie algebra:

$$\begin{bmatrix} f & q \\ 0 & 0 \end{bmatrix} \tag{96}$$

with $A(n,R)$ the group of all affine transformations of R^n. We have $\eta\,(G(\Omega)) \subset G\,(V(\Omega))$ and the pair (η,ℓ) of the homomorphism $\eta : G(\Omega) \to G\,(V(\Omega))$ and the map $\ell : \Omega \to V(\Omega)$ is equivariant.

A Hessian structure (D, g) on a homogeneous space G/K is said to be an invariant Hessian structure if both D and g are G-invariant. A homogeneous space G/K with an invariant Hessian structure (D, g) is called a homogeneous Hessian manifold and is denoted by $(G/K, D, g)$. Another result of Koszul is that a homogeneous self-dual regular convex cone is characterized as a simply connected symmetric homogeneous space admitting an invariant Hessian structure that is defined by the positive definite second Koszul form (we have identified in a previous paper that this second Koszul form is related to the Fisher metric). In parallel, Vinberg [125,126] gave a realization of a homogeneous regular convex domain as a real Siegel domain. Koszul has observed that regular convex cones admit canonical Hessian structures, improving some results of Pyateckii-Shapiro that studied realizations of homogeneous bounded domains by considering Siegel domains in connection with automorphic forms. Koszul defined a characteristic function ψ_Ω of a regular convex cone Ω, and showed that $\psi_\Omega = Dd\log\psi_\Omega$ is a Hessian metric on Ω invariant under affine automorphisms of Ω. If Ω is a homogeneous self dual cone, then the gradient mapping is a symmetry with respect to the canonical Hessian metric, and is a symmetric homogeneous Riemannian manifold. More information on Koszul Hessian geometry can be found in [127–136].

We will now focus our attention to Koszul affine representation of Lie group/algebra. Let G a connex Lie group and E a real or complex vector space of finite dimension, Koszul has introduced an affine representation of G in E such that [117–124]:

$$\begin{aligned} &E \to E \\ &a \mapsto sa \ \forall s \in G \end{aligned} \tag{97}$$

is an affine transformation. We set $A(E)$ the set of all affine transformations of a vector space E, a Lie group called affine transformation group of E. The set $GL(E)$ of all regular linear transformations of E, a subgroup of $A(E)$.

We define a linear representation from G to $GL(E)$:

$$\begin{aligned} &\mathbf{f} : G \to GL(E) \\ &s \mapsto \mathbf{f}(s)a = sa - so \ \forall a \in E \end{aligned} \tag{98}$$

and an application from G to E:

$$\begin{aligned} &\mathbf{q} : G \to E \\ &s \mapsto \mathbf{q}(s) = so \ \forall s \in G \end{aligned} \tag{99}$$

Then we have $\forall s, t \in G$:

$$\mathbf{f}(s)\mathbf{q}(t) + \mathbf{q}(s) = \mathbf{q}(st) \tag{100}$$

deduced from $\mathbf{f}(s)\mathbf{q}(t) + \mathbf{q}(s) = s\mathbf{q}(t) - so + so = s\mathbf{q}(t) = sto = \mathbf{q}(st)$.

On the contrary, if an application q from G to E and a linear representation $\mathbf{f}$ from G to $GL(E)$ verify previous equation, then we can define an affine representation of G in E, written $(\mathbf{f}, \mathbf{q})$:

$$Aff(s) : a \mapsto sa = \mathbf{f}(s)a + \mathbf{q}(s) \ \forall s \in G, \forall a \in E \tag{101}$$

The condition $\mathbf{f}(s)\mathbf{q}(t) + \mathbf{q}(s) = \mathbf{q}(st)$ is equivalent to requiring the following mapping to be an homomorphism:

$$Aff : s \in G \mapsto Aff(s) \in A(E) \tag{102}$$

We write f the linear representation of Lie algebra $\mathfrak{g}$ of G, defined by $\mathbf{f}$ and q the restriction to $\mathfrak{g}$ of the differential to $\mathbf{q}$ (f and q the differential of $\mathbf{f}$ and $\mathbf{q}$ respectively), Koszul has proved that:

$$\begin{gathered} f(X)q(Y) - f(Y)q(X) = q([X,Y]) \ \forall X,Y \in \mathfrak{g} \\ \text{with } f : \mathfrak{g} \to gl(E) \text{ and } q : \mathfrak{g} \mapsto E \end{gathered} \tag{103}$$

where $gl(E)$ the set of all linear endomorphisms of E, the Lie algebra of $GL(E)$.

Using the computation,

$$q(Ad_s Y) = \left. \frac{d\mathbf{q}(s \cdot e^{tY} \cdot s^{-1})}{dt} \right|_{t=0} = \mathbf{f}(s)f(Y)\mathbf{q}(s^{-1}) + \mathbf{f}(s)q(Y) \tag{104}$$

We can obtain:

$$q([X,Y]) = \left. \frac{d\mathbf{q}(Ad_{e^{tX}} Y)}{dt} \right|_{t=0} = f(X)q(Y)\mathbf{q}(e) + \mathbf{f}(e)f(Y)(-q(X)) + f(X)q(Y) \tag{105}$$

where e is the unit element in G. Since $\mathbf{f}(e)$ is the identity mapping and $\mathbf{q}(e) = 0$, we have the equality: $f(X)q(Y) - f(Y)q(X) = q([X,Y])$.

A pair (f,q) of a linear representation f of a Lie algebra $\mathfrak{g}$ on E and a linear mapping q from $\mathfrak{g}$ to E is an affine representation of $\mathfrak{g}$ on E, if it satisfies $f(X)q(Y) - f(Y)q(X) = q([X,Y])$.

Conversely, if we assume that $\mathfrak{g}$ admits an affine representation (f,q) on E, using an affine coordinate system $\{x^1,...,x^n\}$ on E, we can express an affine mapping $v \mapsto f(X)v + q(Y)$ by an $(n+1) \times (n+1)$ matrix representation:

$$aff(X) = \begin{bmatrix} f(X) & q(X) \\ 0 & 0 \end{bmatrix} \tag{106}$$

where $f(X)$ is a $n \times n$ matrix and $q(X)$ is a n row vector.

$X \mapsto aff(X)$ is an injective Lie algebra homomorphism from $\mathfrak{g}$ in the Lie algebra of all $(n+1) \times (n+1)$ matrices, $gl(n+1,R)$:

$$\left| \begin{array}{c} \mathfrak{g} \to gl(n+1,R) \\ X \mapsto aff(X) \end{array} \right. \tag{107}$$

If we denote $\mathfrak{g}_{aff} = aff(\mathfrak{g})$, we write G_{aff} the linear Lie subgroup of $GL(n+1,R)$ generated by $\mathfrak{g}_{aff}$. An element of $s \in G_{aff}$ is expressed by:

$$Aff(s) = \begin{bmatrix} \mathbf{f}(s) & \mathbf{q}(s) \\ 0 & 1 \end{bmatrix} \tag{108}$$

Let M_{aff} be the orbit of G_{aff} through the origin o, then $M_{aff} = \mathbf{q}(G_{aff}) = G_{aff}/K_{aff}$ where $K_{aff} - \left\{ s \in G_{aff}/\mathbf{q}(s) = 0 \right\} = Ker(\mathbf{q})$.

Example. Let Ω be a convex domain in R^n containing no complete straight lines, we define a convex cone $V(\Omega)$ in $R^{n+1} = R^n \times R$ by $V(\Omega) = \{(\lambda x, x) \in R^n \times R / x \in \Omega, \lambda \in R^+\}$. Then there exists an affine embedding:

$$\ell : x \in \Omega \mapsto \begin{bmatrix} x \\ 1 \end{bmatrix} \in V(\Omega) \tag{109}$$

If we consider η the group of homomorphism of $A(n,R)$ into $GL(n+1,R)$ given by:

$$s \in A(n,R) \mapsto \begin{bmatrix} \mathbf{f}(s) & \mathbf{q}(s) \\ 0 & 1 \end{bmatrix} \in GL(n+1,R) \tag{110}$$

with $A(n,R)$ the group of all affine transformations of R^n. We have $\eta\left(G(\Omega)\right) \subset G\left(V(\Omega)\right)$ and the pair (η,ℓ) of the homomorphism $\eta : G(\Omega) \to G\left(V(\Omega)\right)$ and the map $\ell : \Omega \to V(\Omega)$ is equivariant:

$$\ell \circ s = \eta(s) \circ \ell \text{ and } d\ell \circ s = \eta(s) \circ d\ell \tag{111}$$

6.7. Comparison of Koszul and Souriau Affine Representation of Lie Group and Lie Algebra

We will compare, in the following Table 1, affine representation of Lie group and Lie algebra from Souriau and Koszul approaches:

Table 1. Table comparing Souriau and Koszul affine representation of Lie group and Lie algebra.

Souriau Model of Affine Representation of Lie Groups and Algebra	Koszul Model of Affine Representation of Lie Groups and Algebra
$A(g)(x) = R(g)(x) + \theta(g)$ with $g \in G, x \in E$ $R : G \to GL(E)$ and $\theta : G \to E$	$Aff(s) : a \mapsto sa = \mathbf{f}(s)a + \mathbf{q}(s) \quad \forall s \in G, \forall a \in E$ $\mathbf{f} : G \to GL(E)$ $s \mapsto \mathbf{f}(s)a = sa - so \quad \forall a \in E$ $\mathbf{q} : G \to E$ $s \mapsto \mathbf{q}(s) = so \quad \forall s \in G$
$\theta(gh) = R(g)(\theta(h)) + \theta(g)$ with $g,h \in G$ $\theta : G \to E$ is a one-cocycle of G with values in E,	$\mathbf{q}(st) = \mathbf{f}(s)\mathbf{q}(t) + \mathbf{q}(s)$
$a(X)(x) = r(X)(x) + \Theta(X)$ with $X \in \mathfrak{g}, x \in E$ The linear map $\Theta : \mathfrak{g} \to E$ is a one-cocycle of G with values in E: $\Theta(X) = T_e\theta\left(X(e)\right), X \in \mathfrak{g}$	$v \mapsto f(X)v + q(Y)$ f and q the differential of $\mathbf{f}$ and $\mathbf{q}$ respectively
$\Theta\left([X,Y]\right) = r(X)\left(\Theta(Y)\right) - r(Y)\left(\Theta(X)\right)$	$q\left([X,Y]\right) = f(X)q(Y) - f(Y)q(X)\ \forall X,Y \in \mathfrak{g}$ with $f : \mathfrak{g} \to gl(E)$ and $q : \mathfrak{g} \mapsto E$
none	$aff(X) = \begin{bmatrix} f(X) & q(X) \\ 0 & 0 \end{bmatrix}$
none	$Aff(s) = \begin{bmatrix} \mathbf{f}(s) & \mathbf{q}(s) \\ 0 & 1 \end{bmatrix}$

6.8. Additional Elements on Koszul Affine Representation of Lie Group and Lie Algebra

Let $\left\{x^1, x^2, ..., x^n\right\}$ be a local coordinate system on M, the Christoffel's symbols Γ_{ij}^k of the connection D are defined by:

$$D_{\frac{\partial}{\partial x^i}} \frac{\partial}{\partial x^j} = \sum_{k=1}^{n} \Gamma_{ij}^k \frac{\partial}{\partial x^k} \tag{112}$$

The torsion tensor T of D is given by:

$$T(X,Y) = D_X Y - D_Y X - [X,Y] \tag{113}$$

$$T\left(\frac{\partial}{\partial x^i}, \frac{\partial}{\partial x^j}\right) = \sum_{k=1}^{n} T_{ij}^k \frac{\partial}{\partial x^k} \text{ with } T_{ij}^k = \Gamma_{ij}^k - \Gamma_{ji}^k \tag{114}$$

The curvature tensor R of D is given by:

$$R(X,Y)Z = D_X D_Y Z - D_Y D_X Z - D_{[X,Y]} Z \tag{115}$$

$$R\left(\frac{\partial}{\partial x^k},\frac{\partial}{\partial x^l}\right)\frac{\partial}{\partial x^j}=\sum_i R^i_{jkl}\frac{\partial}{\partial x^i} \text{ with } R^i_{jkl}=\frac{\partial\Gamma^i_{lj}}{\partial x^k}-\frac{\partial\Gamma^i_{kj}}{\partial x^l}+\sum_m\left(\Gamma^m_{lj}\Gamma^i_{km}-\Gamma^m_{kj}\Gamma^i_{lm}\right) \tag{116}$$

The Ricci tensor Ric of D is given by:

$$Ric\,(Y,Z)=Tr\,\{X\to R\,(X,Y)\,Z\} \tag{117}$$

$$R_{jk}=Ric\left(\frac{\partial}{\partial x^j},\frac{\partial}{\partial x^k}\right)=\sum_i R^i_{kij} \tag{118}$$

In the following, we will consider a homogeneous space G/K endowed with a G-invariant flat connection D (homogeneous flat manifold) written $(G/K, D)$. Koszul has proved a bijective correspondence between the set of G-invariant flat connections on G/K and the set of affine representations of the Lie algebra of G. Let (G, K) be the pair of connected Lie group G and its closed subgroup K. Let $\mathfrak{g}$ the Lie algebra of G and k be the Lie subalgebra of $\mathfrak{g}$ corresponding to K. X^* is defined as the vector field on $M=G/K$ induced by the 1-parameter group of transformation e^{-tX}. We denote $A_{X^*}=L_{X^*}-D_{X^*}$, with L_{X^*} the Lie derivative.

Let V be the tangent space of G/K at $o=\{K\}$ and let consider, the following values at o:

$$f(X)=A_{X^*,o} \tag{119}$$

$$q(X)=X_o^* \tag{120}$$

where $A_{X^*}Y^*=-D_{Y^*}X^*$ (where D is a locally flat linear connection: its torsion and curvature tensors vanish identically), then:

$$f\left([X,Y]\right)=[f(X),f(Y)] \tag{121}$$

$$f(X)q(Y)-f(Y)q(X)=q\left([X,Y]\right) \tag{122}$$

where $\ker(\mathrm{k})=q$, and (f,q) an affine representation of the Lie algebra $\mathfrak{g}$:

$$\forall X\in g,\ X_a=\sum_i\left(\sum_j f(X)^j_i x^i+q(X)^i\right)\frac{\partial}{\partial x^i} \tag{123}$$

The 1-parameter transformation group generated by X_a is an affine transformation group of V, with linear parts given by $e^{-t.f(X)}$ and translation vector parts:

$$\sum_{n=1}^{\infty}\frac{(-t)^n}{n!}f(X)^{n-1}q(X) \tag{124}$$

These relations are proved by using:

$$\begin{cases} A_{X^*}Y^*-A_{Y^*}X^*=[X^*,Y^*] \\ [A_{X^*},A_{Y^*}]=A_{[X^*,Y]^*} \end{cases} \text{ with } A_{X^*}Y^*=-D_{Y^*}X^* \tag{125}$$

based on the property that the connection D is locally flat and there is local coordinate systems on M such that $D_{\frac{\partial}{\partial x^i}}\frac{\partial}{\partial x^j}=0$ with a vanishing torsion and curvature:

$$T\,(X,Y)=0\Rightarrow D_XY-D_YX=[X,Y] \tag{126}$$

$$R\,(X,Y)\,Z=0\Rightarrow D_XD_YZ-D_YD_XZ=D_{[X,Y]}Z \tag{127}$$

deduced from the fact the a locally flat linear connection (vanishing of torsion and curvature).

Let ω be an invariant volume element on G/K in an affine local coordinate system $\{x^1, x^2, ..., x^n\}$ in a neighborhood of o:

$$\omega = \Phi \cdot dx^1 \wedge ... \wedge dx^n \tag{128}$$

We can write $X^* = \sum_i \chi^i \frac{\partial}{\partial x^i}$ and develop the Lie derivative of the volume element ω:

$$L_{X^*}\omega = (L_{X^*}\Phi).dx^1 \wedge ... \wedge dx^n + \sum_j \Phi.dx^1 \wedge \cdots \wedge L_{X^*}dx^j \wedge \cdots \wedge dx^n = \left(X^*\Phi + \left(\sum_j \frac{\partial \chi^j}{\partial x^j}\right)\Phi\right) dx^1 \wedge ... \wedge dx^n \tag{129}$$

Since the volume element ω is invariant by G:

$$L_{X^*}\omega = 0 \Rightarrow X^*\Phi + \left(\sum_j \frac{\partial \chi^j}{\partial x^j}\right)\Phi = 0 \Rightarrow X^*\log\Phi = -\sum_j \frac{\partial \chi^j}{\partial x^j} \tag{130}$$

By using $A_{X^*}Y^* = -D_{Y^*}X^*$, we have:

$$\left(D_{\frac{\partial}{\partial x^i}}(A_{X^*})\right)\left(\frac{\partial}{\partial x^j}\right) = D_{\frac{\partial}{\partial x^i}}\left(A_{X^*}\left(\frac{\partial}{\partial x^j}\right)\right) - A_{X^*}\left(D_{\frac{\partial}{\partial x^i}}\frac{\partial}{\partial x^j}\right) = -D_{\frac{\partial}{\partial x^i}}D_{\frac{\partial}{\partial x^j}}\left(\sum_k \chi^k \frac{\partial}{\partial x^k}\right) = -\sum_k \frac{\partial^2 \chi^k}{\partial x^i \partial x^j}\frac{\partial}{\partial x^k} \tag{131}$$

But as D is locally flat and X^* is an infinitesimal affine transformation with respect to D:

$$D_{\frac{\partial}{\partial x^i}}(A_{X^*}) = 0 \Rightarrow \frac{\partial^2 \chi^k}{\partial x^i \partial x^j} = 0 \tag{132}$$

The Koszul form and canonical bilinear form are given by:

$$\alpha = \sum_i \frac{\partial \log\Phi}{\partial x^i} dx^i = D\log\Phi \tag{133}$$

$$D\alpha = \sum_{i,j} \frac{\partial^2 \log\Phi}{\partial x^i \partial x^j} dx^i dx^j = Dd\log\Phi \tag{134}$$

$$L_{X^*}\alpha = L_{X^*}D\log\Phi = DL_{X^*}\log\Phi = DX^*\log\Phi = -D\left(\sum_j \frac{\partial \chi^j}{\partial x^j}\right) = -\sum_{,j} \frac{\partial^2 \chi^j}{\partial x^i \partial x^j} dx^i = 0 \tag{135}$$

Then, $L_{X^*}\alpha = 0 \; \forall X \in g$.

By using $X^*\log\Phi = -\sum_j \frac{\partial \chi^j}{\partial x^j}$, we can obtain:

$$\alpha(X^*) = (D\log\Phi)(X^*) \underset{L_{X^*}\alpha = 0}{\Rightarrow} D_{X^*}\log\Phi = -\sum_j \frac{\partial \chi^j}{\partial x^j} \tag{136}$$

By using $A_{X^*}Y^* = -D_{Y^*}X^*$, we can develop:

$$A_{X^*}\left(\frac{\partial}{\partial x^j}\right) = -D_{\frac{\partial}{\partial x^j}}X^* = -\sum_i \frac{\partial \chi^i}{\partial x^j}\frac{\partial}{\partial x^i} \tag{137}$$

As $f(X) = A_{X^*,o}$ and $q(X) = X_o^*$:

$$Tr(f(X)) = Tr(A_{X^*,o}) = -\sum_i \frac{\partial \chi^i}{\partial x^i}(o) = \alpha(X_0^*) = \alpha_0(q(X)) \tag{138}$$

If we use that $L_{X^*}\alpha = 0 \; \forall X \in \mathfrak{g}$, then we obtain:

$$(D\alpha)(X^*, Y^*) = (D_{Y^*}\alpha)(X^*) = -(A_{Y^*}\alpha)(X^*) = -A_{Y^*}(\alpha(X^*)) + \alpha(A_{Y^*}X^*) = \alpha(A_{Y^*}X^*) \tag{139}$$

$$D\alpha_0\left(q(X),q(Y)\right) = \alpha_0\left(f(Y)q(X)\right) \tag{140}$$

To synthetize the result proved by Jean-Louis Koszul, if α_o and $D\alpha_o$ are the values of α and $D\alpha$ at o, then:

$$\alpha_o\left(q(X)\right) = Tr\left(f(X)\right)\ \forall X \in g \tag{141}$$

$$D\alpha_o\left(q(X),q(Y)\right) = \langle q(X),q(Y)\rangle_o = \alpha_0\left(f(X)q(Y)\right)\ \forall X,Y \in \mathfrak{g} \tag{142}$$

Jean-Louis Koszul has also proved that the inner product $\langle .,. \rangle$ on V, given by the Riemannian metric g_{ij}, satisfies the following conditions:

$$\langle f(X)q(Y),q(Z)\rangle + \langle q(Y),f(X)q(Z)\rangle = \langle f(Y)q(X),q(Z)\rangle + \langle q(X),f(Y)q(Z)\rangle \tag{143}$$

To make the link with Souriau model of thermodynamics, the first Koszul form $\alpha = D\log\Phi = Tr\left(f(X)\right)$ will play the role of the geometric heat Q and the second koszul form $D\alpha = Dd\log\Phi = \langle q(X),q(Y)\rangle_o$ will be the equivalent of Souriau-Fisher metric that is G-invariant.

Koszul theory is wider and integrates "information geometry" in its corpus. Koszul [117–124] has proved general results, for example: on a complex homogeneous space, an invariant volume defines with the complex structure, an invariant Hermitian form. If this space is a bounded domain, then this hermitian form is positive definite and coincides with the classical Bergman metric of this domain. During his stay at Institute for Advanced Study in Princeton, Koszul [117–124] has also demonstrated the reciprocal for a class of complex homogeneous spaces, defined by open orbits of complex affine transformation groups. Koszul and Vey [137,138] have also developed extended results with the following theorem for connected hessian manifolds:

Theorem 3 (Koszul-Vey Theorem). *Let M be a connected hessian manifold with hessian metric g. Suppose that M admits a closed 1-form α such that $D\alpha = g$ and there exists a group G of affine automorphisms of M preserving α:*

- *If* M/G is quasi-compact, then the universal covering manifold of M is affinely isomorphic to a convex domain Ω of an affine space not containing any full straight line.
- *If M/G is compact, then Ω is a sharp convex cone.*

On this basis, Koszul has given a Lie group construction of a homogeneous cone that has been developed and applied in information geometry by Shima and Boyom in the framework of Hessian geometry. The results of Koszul are also fundamental in the framework of Souriau thermodynamics.

7. Souriau Lie Group Model and Koszul Hessian Geometry Applied in the Context of Information Geometry for Multivariate Gaussian Densities

We will enlighten Souriau model with Koszul hessian geometry applied in information geometry [117–124], recently studied in [3,9,139]. We have previously shown that information geometry could be founded on the notion of Koszul-Vinberg characteristic function $\psi_\Omega(x) = \int_{\Omega^*} e^{-\langle x,\xi\rangle}d\xi,\ \forall x \in \Omega$ where Ω is a convex cone and Ω^* the dual cone with respect to Cartan-Killing inner product $\langle x,y\rangle = -B\left(x,\theta(y)\right)$ invariant by automorphisms of Ω, with $B\left(.,.\right)$ the Killing form and $\theta(.)$ the Cartan involution. We can develop the Koszul characteristic function:

$$\psi_\Omega(x+\lambda u) = \psi_\Omega(x) - \lambda\langle x^*,u\rangle + \frac{\lambda^2}{2}\langle K(x)u,u\rangle + \ldots \tag{144}$$

$$\text{with } x^* = \frac{d\Phi(x)}{dx},\ \Phi(x) = -\log\psi_\Omega(x) \text{ and } K(x) = \frac{d^2\Phi(x)}{dx^2} \tag{145}$$

This characteristic function is at the cornerstone of modern concept of information geometry, defining Koszul density by solution of maximum Koszul-Shannon entropy [140]:

$$\underset{p}{Max}\left[-\int_{\Omega^*} p_{\hat{\xi}}(\xi)\log p_{\hat{\xi}}(\xi)\cdot d\xi\right] \text{ such that} \int_{\Omega^*} p_{\hat{\xi}}(\xi)d\xi = 1 \text{ and} \int_{\Omega^*} \xi\cdot p_{\hat{\xi}}(\xi)d\xi = \hat{\xi} \tag{146}$$

$$\begin{aligned} & p_{\hat{\xi}}(\xi) = \frac{e^{-\langle\Theta^{-1}(\hat{\xi}),\xi\rangle}}{\int_{\Omega^*} e^{-\langle\Theta^{-1}(\hat{\xi}),\xi\rangle}.d\xi}\hat{\xi} = \Theta(\beta) = \frac{\partial\Phi(\beta)}{\partial\beta} \text{ where } \Phi(\beta) = -\log\psi_\Omega(\beta) \\ & \psi_\Omega(\beta) = \int_{\Omega^*} e^{-\langle\beta,\xi\rangle}d\xi\ ,\ S(\hat{\xi}) = -\int_{\Omega^*} p_{\hat{\xi}}(\xi)\log p_{\hat{\xi}}(\xi)\cdot d\xi \text{ and } \beta = \Theta^{-1}(\hat{\xi}) \\ & S(\hat{\xi}) = \langle\hat{\xi},\beta\rangle - \Phi(\beta) \end{aligned} \tag{147}$$

This last relation is a Legendre transform between the logarithm of characteristic function and the entropy:

$$\begin{aligned} & \log p_{\hat{\xi}}(\xi) = -\langle\xi,\beta\rangle + \Phi(\beta) \\ & S(\bar{\xi}) = -\int_{\Omega^*} p_{\hat{\xi}}(\xi)\cdot\log p_{\hat{\xi}}(\xi)\cdot d\xi = -E\left[\log p_{\hat{\xi}}(\xi)\right] \\ & S(\bar{\xi}) = \langle E[\xi],\beta\rangle - \Phi(\beta) = \langle\hat{\xi},\beta\rangle - \Phi(\beta) \end{aligned} \tag{148}$$

The inversion $\Theta^{-1}(\hat{\xi})$ is given by the Legendre transform based on the property that the Koszul-Shannon entropy is given by the Legendre transform of minus the logarithm of the characteristic function:

$$S(\hat{\xi}) = \langle\beta,\hat{\xi}\rangle - \Phi(\beta) \text{ with } \Phi(\beta) = -\log\int_{\Omega^*} e^{-\langle\xi,\beta\rangle}d\xi \quad \forall\beta\in\Omega \text{ and } \forall\xi,\hat{\xi}\in\Omega^* \tag{149}$$

We can observe the fundamental property that $E[S(\xi)] = S(E[\xi])\ ,\ \xi\in\Omega^*$, and also as observed by Maurice Fréchet that "distinguished functions" (densities with estimator reaching the Fréchet-Darmois bound) are solutions of the *Alexis Clairaut equation* introduced by Clairaut in 1734 [141], as illustrated in Figure 8:

$$S(\hat{\xi}) = \left\langle\Theta^{-1}(\hat{\xi}),\hat{\xi}\right\rangle - \Phi\left[\Theta^{-1}(\hat{\xi})\right] \forall\hat{\xi}\in\{\Theta(\beta)/\beta\in\Omega\} \tag{150}$$

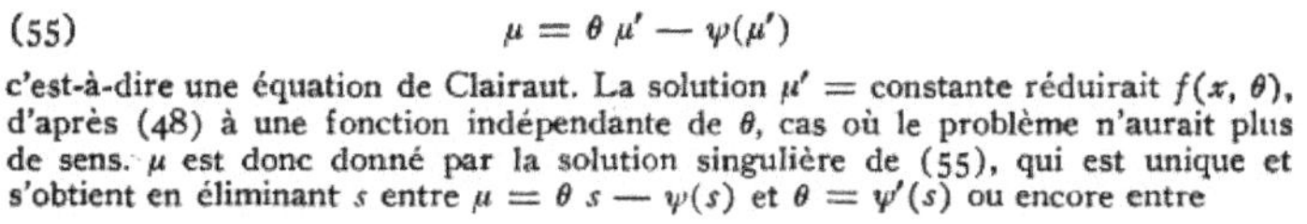
(55) $\mu = \theta\,\mu' - \psi(\mu')$

c'est-à-dire une équation de Clairaut. La solution μ' = constante réduirait $f(x, \theta)$, d'après (48) à une fonction indépendante de θ, cas où le problème n'aurait plus de sens. μ est donc donné par la solution singulière de (55), qui est unique et s'obtient en éliminant s entre $\mu = \theta\,s - \psi(s)$ et $\theta = \psi'(s)$ ou encore entre

Figure 8. Clairaut-Legendre equation introduced by Maurice Fréchet in his 1943 paper [141].

Details of Fréchet elaboration for this Clairaut(-Legendre) equation for "distinguished function" is given in Appendix A, and other elements are available on Fréchet's papers [141–144].

In this structure, the Fisher metric $I(x)$ makes appear naturally a *Koszul hessian geometry* [145,146], if we observe that

$$\log p_{\hat{\xi}}(\xi) = -\langle \xi, \beta \rangle + \Phi(\beta)$$
$$S(\bar{\xi}) = -\int_{\Omega^*} p_{\hat{\xi}}(\xi) \cdot \log p_{\hat{\xi}}(\xi) \cdot d\xi = -E\left[\log p_{\hat{\xi}}(\xi)\right] \tag{151}$$
$$S(\bar{\xi}) = \langle E[\xi], \beta \rangle - \Phi(\beta) = \langle \hat{\xi}, \beta \rangle - \Phi(\beta)$$

Then we can recover the relation with Fisher metric:

$$I(\beta) = -E\left[\frac{\partial^2 \log p_\beta(\xi)}{\partial \beta^2}\right] = -E\left[\frac{\partial^2 \left(-\langle \xi, \beta \rangle + \Phi(\beta)\right)}{\partial \beta^2}\right] = -\frac{\partial^2 \Phi(\beta)}{\partial \beta^2}$$
$$\hat{\xi} = \frac{\partial \Phi(\beta)}{\partial \beta} \tag{152}$$
$$I(\beta) = E\left[\frac{\partial \log p_\beta(\xi)}{\partial \beta} \frac{\partial \log p_\beta(\xi)}{\partial \beta}^T\right] = E\left[\left(\xi - \hat{\xi}\right)\left(\xi - \hat{\xi}\right)^T\right] = E\left[\xi^2\right] - E\left[\xi\right]^2 = Var(\xi)$$

with Crouzeix relation established in 1977 [147,148], $\frac{\partial^2 \Phi}{\partial \beta^2} = \left[\frac{\partial^2 S}{\partial \hat{\xi}^2}\right]^{-1}$ giving the dual metric, in dual space, where entropy S and (minus) logarithm of characteristic function, Φ, are dual potential functions.

The first metric of information geometry [149,150], the Fisher metric is given by the hessian of the characteristic function logarithm:

$$I(\beta) = -E\left[\frac{\partial^2 \log p_\beta(\xi)}{\partial \beta^2}\right] = -\frac{\partial^2 \Phi(\beta)}{\partial \beta^2} = \frac{\partial^2 \log \psi_\Omega(\beta)}{\partial \beta^2} \tag{153}$$

$$ds_g^2 = d\beta^T I(\beta) d\beta = \sum_{ij} g_{ij} d\beta_i d\beta_j \text{ with } g_{ij} = [I(\beta)]_{ij} \tag{154}$$

The second metric of information geometry is given by hessian of the Shannon entropy:

$$\frac{\partial^2 S(\hat{\xi})}{\partial \hat{\xi}^2} = \left[\frac{\partial^2 \Phi(\beta)}{\partial \beta^2}\right]^{-1} \text{ with } S(\hat{\xi}) = \langle \hat{\xi}, \beta \rangle - \Phi(\beta) \tag{155}$$

$$ds_h^2 = d\hat{\xi}^T \left[\frac{\partial^2 S(\hat{\xi})}{\partial \hat{\xi}^2}\right] d\hat{\xi} = \sum_{ij} h_{ij} d\hat{\xi}_i d\hat{\xi}_j \text{ with } h_{ij} = \left[\frac{\partial^2 S(\hat{\xi})}{\partial \hat{\xi}^2}\right]_{ij} \tag{156}$$

Both metrics will provide the same distance:

$$ds_g^2 = ds_h^2 \tag{157}$$

From the Cartan inner product, we can generate logarithm of the Koszul characteristic function, and its Legendre transform to define Koszul entropy, Koszul density and Koszul metric, as explained in the following Figure 9:

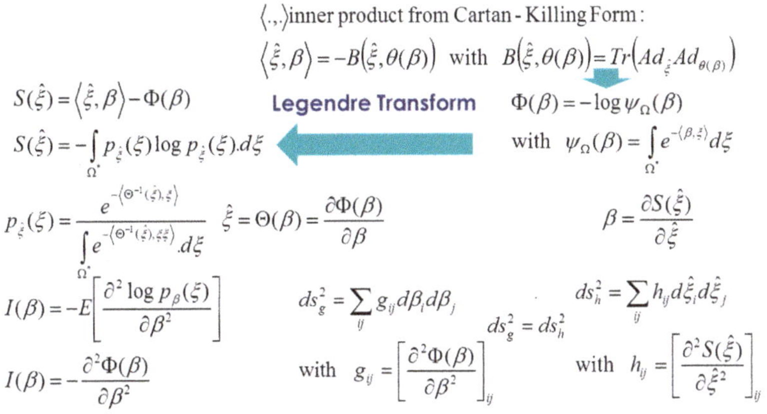

$\langle .,. \rangle$ inner product from Cartan - Killing Form:

$$\left\langle \hat{\xi},\beta \right\rangle = -B\left(\hat{\xi},\theta(\beta)\right) \text{ with } B\left(\hat{\xi},\theta(\beta)\right) = Tr\left(Ad_{\hat{\xi}}Ad_{\theta(\beta)}\right)$$

$$S(\hat{\xi}) = \left\langle \hat{\xi},\beta \right\rangle - \Phi(\beta) \qquad \textbf{Legendre Transform} \qquad \Phi(\beta) = -\log \psi_\Omega(\beta)$$

$$S(\hat{\xi}) = -\int_{\Omega^*} p_{\hat{\xi}}(\xi)\log p_{\hat{\xi}}(\xi).d\xi \qquad \text{with } \psi_\Omega(\beta) = \int_{\Omega^*} e^{-\langle \beta,\xi \rangle} d\xi$$

$$p_{\hat{\xi}}(\xi) = \frac{e^{-\left\langle \Theta^{-1}(\hat{\xi}),\xi \right\rangle}}{\int_{\Omega^*} e^{-\left\langle \Theta^{-1}(\hat{\xi}),\xi \right\rangle}.d\xi} \qquad \hat{\xi} = \Theta(\beta) = \frac{\partial \Phi(\beta)}{\partial \beta} \qquad \beta = \frac{\partial S(\hat{\xi})}{\partial \hat{\xi}}$$

$$I(\beta) = -E\left[\frac{\partial^2 \log p_\beta(\xi)}{\partial \beta^2}\right] \qquad ds_g^2 = \sum_{ij} g_{ij} d\beta_i d\beta_j \qquad ds_g^2 = ds_h^2 \qquad ds_h^2 = \sum_{ij} h_{ij} d\hat{\xi}_i d\hat{\xi}_j$$

$$I(\beta) = -\frac{\partial^2 \Phi(\beta)}{\partial \beta^2} \qquad \text{with } g_{ij} = \left[\frac{\partial^2 \Phi(\beta)}{\partial \beta^2}\right]_{ij} \qquad \text{with } h_{ij} = \left[\frac{\partial^2 S(\hat{\xi})}{\partial \hat{\xi}^2}\right]_{ij}$$

Figure 9. Generation of Koszul elements from Cartan inner product.

This information geometry has been intensively studied for structured matrices [151–166] and in statistics [167] and is linked to the seminal work of Siegel [168] on symmetric bounded domains.

We can apply this Koszul geometry framework for cones of symmetric positive definite matrices. Let the inner product $\langle \eta, \xi \rangle = Tr\left(\eta^T \xi\right), \forall \eta, \xi \in Sym(n)$ given by Cartan-Killing form, Ω be the set of symmetric positive definite matrices is an open convex cone and is self-dual $\Omega^* = \Omega$.

$$\begin{aligned} &\langle \eta, \xi \rangle = Tr\left(\eta^T \xi\right), \forall \eta, \xi \in Sym(n) \\ &\psi_\Omega(\beta) = \int_{\Omega^*} e^{-\langle \beta,\xi \rangle} d\xi = \det(\beta)^{-\frac{n+1}{2}} \psi_\Omega(I_d) \\ &\hat{\xi} = \frac{\partial \Phi(\beta)}{\partial \beta} = \frac{\partial(-\log\psi_\Omega(\beta))}{\partial \beta} = \frac{n+1}{2}\beta^{-1} \end{aligned} \tag{158}$$

$$\begin{aligned} &p_{\hat{\xi}}(\xi) = e^{-\langle \Theta^{-1}(\hat{\xi}),\xi \rangle + \Phi(\Theta^{-1}(\hat{\xi}))} = \psi_\Omega(I_d) \cdot \left[\det\left(\alpha \hat{\xi}^{-1}\right)\right] \cdot e^{-Tr(\alpha \hat{\xi}^{-1} \xi)} \\ &\text{with } \alpha = \frac{n+1}{2} \end{aligned} \tag{159}$$

We will in the following illustrate information geometry for multivariate Gaussian density [169]:

$$p_{\hat{\xi}}(\xi) = \frac{1}{(2\pi)^{n/2} \det(R)^{1/2}} e^{-\frac{1}{2}(z-m)^T R^{-1}(z-m)} \tag{160}$$

If we develop:

$$\begin{aligned} \frac{1}{2}(z-m)^T R^{-1}(z-m) &= \frac{1}{2}\left[z^T R^{-1} z - m^T R^{-1} z - z^T R^{-1} m + m^T R^{-1} m\right] \\ &= \frac{1}{2} z^T R^{-1} z - m^T R^{-1} z + \tfrac{1}{2} m^T R^{-1} m \end{aligned} \tag{161}$$

We can write the density as a Gibbs density:

$$\begin{aligned} &p_{\hat{\xi}}(\xi) = \frac{1}{(2\pi)^{n/2} \det(R)^{1/2} e^{\frac{1}{2} m^T R^{-1} m}} e^{-\left[-m^T R^{-1} z + \frac{1}{2} z^T R^{-1} z\right]} = \frac{1}{Z} e^{-\langle \xi, \beta \rangle} \\ &\xi = \begin{bmatrix} z \\ zz^T \end{bmatrix} \text{ and } \beta = \begin{bmatrix} -R^{-1} m \\ \frac{1}{2} R^{-1} \end{bmatrix} = \begin{bmatrix} a \\ H \end{bmatrix} \\ &\text{with } \langle \xi, \beta \rangle = a^T z + z^T H z = Tr\left[z a^T + H^T z z^T\right] \end{aligned} \tag{162}$$

We can then rewrite density with canonical variables:

$$
\begin{aligned}
& p_{\hat{\xi}}(\xi) = \frac{1}{\int\limits_{\Omega^*} e^{-\langle \xi,\beta\rangle}.d\xi} e^{-\langle \xi,\beta\rangle} = \frac{1}{Z} e^{-\langle \xi,\beta\rangle} \text{ with } \log(Z) = n\log(2\pi) + \frac{1}{2}\log\det(R) + \frac{1}{2} m^T R^{-1} m \\
& \xi = \begin{bmatrix} z \\ zz^T \end{bmatrix}, \hat{\xi} = \begin{bmatrix} E[z] \\ E[zz^T] \end{bmatrix} = \begin{bmatrix} m \\ R + mm^T \end{bmatrix}, \beta = \begin{bmatrix} a \\ H \end{bmatrix} = \begin{bmatrix} -R^{-1}m \\ \frac{1}{2}R^{-1} \end{bmatrix} \\
& \text{with } \langle \xi,\beta\rangle = Tr\left[za^T + H^T zz^T \right] \\
& R = E\left[(z-m)(z-m)^T \right] = E\left[zz^T - mz^T - zm^T + mm^T \right] = E\left[zz^T \right] - mm^T
\end{aligned}
\tag{163}
$$

The first potential function (free energy/logarithm of characteristic function) is given by:

$$
\begin{aligned}
& \psi_\Omega(\beta) = \int\limits_{\Omega^*} e^{-\langle \xi,\beta\rangle} \cdot d\xi \\
& \text{and } \Phi(\beta) = -\log\psi_\Omega(\beta) = \frac{1}{2}\left[-Tr\left[H^{-1}aa^T \right] + \log\left[(2)^n \det H \right] - n\log(2\pi) \right]
\end{aligned}
\tag{164}
$$

We verify the relation between the first potential function and moment:

$$
\begin{aligned}
& \frac{\partial\Phi(\beta)}{\partial\beta} = \frac{\partial\left[-\log\psi_\Omega(\beta) \right]}{\partial\beta} = \int\limits_{\Omega^*} \xi \frac{e^{-\langle \xi,\beta\rangle}}{\int\limits_{\Omega^*} e^{-\langle \xi,\beta\rangle} \cdot d\xi} \cdot d\xi = \int\limits_{\Omega^*} \xi \cdot p_{\hat{\xi}}(\xi) \cdot d\xi = \hat{\xi} \\
& \frac{\partial\Phi(\beta)}{\partial\beta} = \begin{bmatrix} \dfrac{\partial\Phi(\beta)}{\partial a} \\ \dfrac{\partial\Phi(\beta)}{\partial H} \end{bmatrix} = \begin{bmatrix} m \\ R + mm^T \end{bmatrix} = \hat{\xi}
\end{aligned}
\tag{165}
$$

The second potential function (Shannon entropy) is given as a Legendre transform of the first one:

$$
\begin{aligned}
& S(\hat{\xi}) = \langle \hat{\xi},\beta\rangle - \Phi(\beta) \text{ with } \frac{\partial\Phi(\beta)}{\partial\beta} = \hat{\xi} \text{ and } \frac{\partial S(\hat{\xi})}{\partial\hat{\xi}} = \beta \\
& S(\hat{\xi}) = -\int\limits_{\Omega^*} \frac{e^{-\langle \xi,\beta\rangle}}{\int\limits_{\Omega^*} e^{-\langle \xi,\beta\rangle}.d\xi} \log \frac{e^{-\langle \xi,\beta\rangle}}{\int\limits_{\Omega^*} e^{-\langle \xi,\beta\rangle}.d\xi} \cdot d\xi = -\int\limits_{\Omega^*} p_{\hat{\xi}}(\xi) \log p_{\hat{\xi}}(\xi) \cdot d\xi
\end{aligned}
\tag{166}
$$

$$
S(\hat{\xi}) = -\int\limits_{\Omega^*} p_{\hat{\xi}}(\xi) \log p_{\hat{\xi}}(\xi) \cdot d\xi = \frac{1}{2}\left[\log(2)^n \det\left[H^{-1} \right] + n\log(2\pi \cdot e) \right] = \frac{1}{2}\left[\log\det[R] + n\log(2\pi \cdot e) \right] \tag{167}
$$

This remark was made by Jean-Souriau in his book [10] as soon as 1969. He has observed, as illustrated in Figure 10 that if we take vector with tensor components $\xi = \begin{pmatrix} z \\ z \otimes z \end{pmatrix}$, components of $\hat{\xi}$ will provide moments of the first and second order of the density of probability $p_{\hat{\xi}}(\xi)$. He used this change of variable $z' = H^{1/2}z + H^{-1/2}a$, to compute the logarithm of the characteristic function $\Phi(\beta)$:

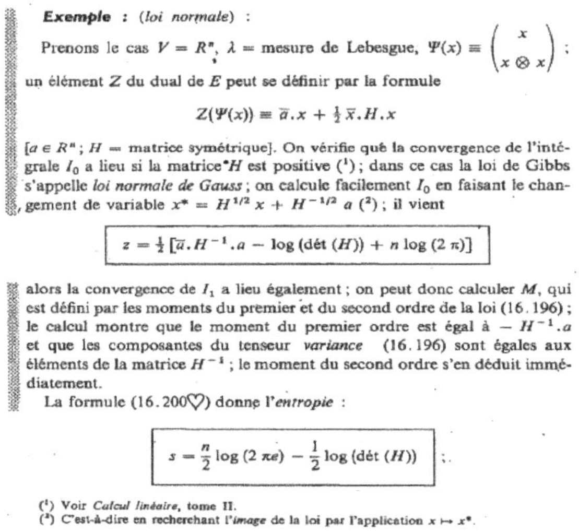

Exemple : (*loi normale*) :

Prenons le cas $V = R^n$, λ = mesure de Lebesgue, $\Psi(x) \equiv \begin{pmatrix} x \\ x \otimes x \end{pmatrix}$;
un élément Z du dual de E peut se définir par la formule

$$Z(\Psi(x)) \equiv \bar{a}.x + \tfrac{1}{2}\bar{x}.H.x$$

[$a \in R^n$; H = matrice symétrique]. On vérifie que la convergence de l'intégrale I_0 a lieu si la matrice H est positive (¹) ; dans ce cas la loi de Gibbs s'appelle *loi normale de Gauss* ; on calcule facilement I_0 en faisant le changement de variable $x^* = H^{1/2}\,x + H^{-1/2}\,a$ (²) ; il vient

$$z = \tfrac{1}{2}\,[\bar{a}.H^{-1}.a - \log(\text{dét}(H)) + n\log(2\pi)]$$

alors la convergence de I_1 a lieu également ; on peut donc calculer M, qui est défini par les moments du premier et du second ordre de la loi (16.196) ; le calcul montre que le moment du premier ordre est égal à $-H^{-1}.a$ et que les composantes du tenseur *variance* (16.196) sont égales aux éléments de la matrice H^{-1} ; le moment du second ordre s'en déduit immédiatement.

La formule (16.200♡) donne l'*entropie* :

$$s = \frac{n}{2}\log(2\pi e) - \frac{1}{2}\log(\text{dét}(H))\ ;.$$

(¹) Voir *Calcul linéaire*, tome II.
(²) C'est-à-dire en recherchant l'*image* de la loi par l'application $x \mapsto x^*$.

Figure 10. Introduction of potential function for multivariate Gaussian law in Souriau book [10].

We can finally compute the metric from the matrix g_{ij}:

$$ds^2 = \sum_{ij} g_{ij} d\theta_i d\theta_j = dm^T R^{-1} dm + \frac{1}{2} Tr\left[\left(R^{-1}dR\right)^2\right] \tag{168}$$

and from classical expression of the Euler-Lagrange equation:

$$\sum_{i=1}^{n} g_{ik}\ddot{\theta}_i + \sum_{i,j=1}^{n} \Gamma_{ijk}\dot{\theta}_i\dot{\theta}_j = 0\,,\ k = 1, \ldots, n \text{ with } \Gamma_{ijk} = \frac{1}{2}\left[\frac{\partial g_{jk}}{\partial \theta_i} + \frac{\partial g_{jk}}{\partial \theta_j} + \frac{\partial g_{ij}}{\partial \theta_k}\right] \tag{169}$$

That is explicitely given by [170]:

$$\begin{cases} \ddot{R} + \dot{m}\dot{m}^T - \dot{R}R^{-1}\dot{R} = 0 \\ \ddot{m} - \dot{R}R^{-1}\dot{m} = 0 \end{cases} \tag{170}$$

We cannot integrate this Euler-Lagrange equation. We will see that Lie group theory will provide new reduced equation, Euler-Poincaré equation, using Souriau theorem.

We make reference to the book of Deza that gives a survey about distance and metric space [171].

The case of Natural Exponential families that are invariant by an affine group has been studied by Casalis (in 1999 paper and in her Ph.D. thesis) [172–178] and by Letac [179–181]. We give the details of Casalis' development in Appendix C. Barndorff-Nielsen has also studied transformation models for exponential families [182–186]. In this section, we will only consider the case of multivariate Gaussian densities.

8. Affine Group Action for Multivariate Gaussian Densities and Souriau's Moment Map: Computation of Geodesics by Geodesic Shooting

To more deeply understand Koszul and Souriau Lie group models of information geometry, we will illustrate their tools for multivariate Gaussian densities.

Consider the general linear group $GL(n)$ consisting of the invertible $n \times n$ matrices, that is a topological group acting linearly on R^n by:

$$\begin{aligned} GL(n) \times R^n &\to R^n \\ (A, x) &\mapsto Ax \end{aligned} \tag{171}$$

The group $GL(n)$ is a Lie group, is a subgroup of the general affine group $GA(n)$, composed of all pairs (A, v) where $A \in GL(n)$ and $v \in R^n$, the group operation given by:

$$(A_1, v_1)(A_2, v_2) = (A_1 A_2, A_1 v_2 + v_1) \tag{172}$$

$GL(n)$ is an open subset of R^{n^2}, and may be considered as n^2-dimensional differential manifold with the same differentiable structure than R^{n^2}. Multiplication and inversion are infinitely often differentiable mappings. Consider the vector space $gl(n)$ of real $n \times n$ matrices and the commutator product:

$$\begin{aligned} gl(n) \times gl(n) &\to gl(n) \\ (A, B) &\mapsto AB - BA = [A, B] \end{aligned} \tag{173}$$

This is a Lie product making $gl(n)$ into a Lie algebra. The exponential map is then the mapping defined by:

$$\begin{aligned} \exp: gl(n) &\to GL(n) \\ A &\mapsto \exp(A) = \sum_{n=0}^{\infty} \frac{A^n}{n!} \end{aligned} \tag{174}$$

Restricting A to have positive determinant, one obtains the positive general affine group $GA_+(n)$ that acts transitively on R^n by:

$$((A, v), x) \mapsto Ax + v \tag{175}$$

In case of symmetric positive definite matrices $Sym^+(n)$, we can use the Cholesky decomposition:

$$R = LL^T \tag{176}$$

where L is a lower triangular matrix with real and positive diagonal entries, and L^T denotes the transpose of L, to define the square root of R.

Given a positive semidefinite matrix R, according to the spectral theorem, the continuous functional calculus can be applied to obtain a matrix $R^{1/2}$ such that $R^{1/2}$ is itself positive and $R^{1/2}R^{1/2} = R$. The operator $R^{1/2}$ is the unique non-negative square root of R.

$N_n = \{\aleph(\mu, \Sigma) / \mu \in R^n, \Sigma \in Sym^+{}_n\}$ the class of regular multivariate normal distributions, where μ is the mean vector and Σ is the (symmetric positive definite) covariance matrix, is invariant under the transitive action of $GA(n)$. The induced action of $GA(n)$ on $R^n \times Sym^+{}_n$ is then given by:

$$\begin{aligned} GA(n) \times (R^n \times Sym^+ n) &\to R^n \times Sym^+ n \\ ((A, v), (\mu, \Sigma)) &\mapsto (A\mu + v, A\Sigma A^T) \end{aligned} \tag{177}$$

and

$$\begin{aligned} GA(n) \times R^n &\to R^n \\ ((A, v), x) &\mapsto Ax + v \end{aligned} \tag{178}$$

As the isotropy group of $(0, I_n)$ is equal to $O(n)$, we can observe that:

$$N_n = GA(n)/O(n) \tag{179}$$

N_n is an open subset of the vector space $T_n = \{(\eta, \Omega) / \eta \in R^n, \Omega \in Sym_n\}$ and is a differentiable manifold, where the tangent space at any point may be identified with T_n.

The Fisher information defines a metric given to N_n a Riemannian manifold structure. The inner product of two tangent vectors $(\eta_1, \Omega_1) \in T_n$, $(\eta_2, \Omega_2) \in T_n$ at the point $(\mu, \Sigma) \in N_n$ is given by:

$$g_{(\mu,\Sigma))}\left((\eta_1, \Omega_1), (\eta_1, \Omega_1)\right) = \eta_1^T \Sigma^{-1} \eta_2 + \frac{1}{2} Tr\left(\Sigma^{-1}\Omega_1\Sigma^{-1}\Omega_2\right) \tag{180}$$

Niels Christian Bang Jesperson has proved that the transformation model on R^n with parameter set $R^n \times Sym^+{}_n$ are exactly those of the form $p_{\mu,\Sigma} = f_{\mu,\Sigma}\lambda$ where λ is the Lebesque measure, where $f_{\mu,\Sigma}(x) = h\left((x-\mu)^T \Sigma^{-1}(x-\mu)\right)/\det(\Sigma)^{1/2}$ and $h : [0, +\infty[\to R^+$ is a continuous function with $\int_0^{+\infty} h(s) s^{\frac{n}{2}-1} ds < +\infty$. Distributions with densities of this form are called elliptic distributions.

To improve understanding of tools, we will consider $GA(n)$ as a sub-group of affine group, that could be defined by a matrix Lie group G_{aff}, that acts for multivariate Gaussian laws, as illustrated in Figure 11:

$$\begin{bmatrix} Y \\ 1 \end{bmatrix} = \begin{bmatrix} R^{1/2} & m \\ 0 & 1 \end{bmatrix} \begin{bmatrix} X \\ 1 \end{bmatrix} = \begin{bmatrix} R^{1/2}X + m \\ 1 \end{bmatrix}, \begin{cases} (m,R) \in R^n \times Sym^+(n) \\ M = \begin{bmatrix} R^{1/2} & m \\ 0 & 1 \end{bmatrix} \in G_{aff} \end{cases} \tag{181}$$

$$X \approx \aleph(0,I) \to Y \approx \aleph(m,R)$$

We can verify that M is a Lie group with classical properties, that product of M preserves the structure, the associativity, the non-commutativity, and the existence of neutral element:

$$\left.\begin{aligned} M_1 \cdot M_2 &= \begin{bmatrix} R_1^{1/2} & m_1 \\ 0 & 1 \end{bmatrix} \begin{bmatrix} R_2^{1/2} & m_2 \\ 0 & 1 \end{bmatrix} = \begin{bmatrix} R_1^{1/2}R_2^{1/2} & R_1^{1/2}m_2 + m_1 \\ 0 & 1 \end{bmatrix} \\ M_2 \cdot M_1 &= \begin{bmatrix} R_2^{1/2} & m_2 \\ 0 & 1 \end{bmatrix} \begin{bmatrix} R_1^{1/2} & m_1 \\ 0 & 1 \end{bmatrix} = \begin{bmatrix} R_2^{1/2}R_1^{1/2} & R_2^{1/2}m_1 + m_2 \\ 0 & 1 \end{bmatrix} \end{aligned}\right\} \\ \Rightarrow \begin{cases} M_1 \cdot M_2 \in G_{aff} \\ M_2 \cdot M_1 \in G_{aff} \\ M_1 \cdot M_2 \neq M_2 \cdot M_1 \\ M_1 \cdot (M_2 \cdot M_3) = (M_1 \cdot M_2) \cdot M_3 \\ M_1 \cdot I = M_1 \end{cases} \tag{182}$$

We can also observe that the inverse preserves the structure:

$$M = \begin{bmatrix} R^{1/2} & m \\ 0 & 1 \end{bmatrix} \Rightarrow M_R^{-1} = M_L^{-1} = M^{-1} = \begin{bmatrix} R^{-1/2} & -R^{-1/2}m \\ 0 & 1 \end{bmatrix} \in G_{aff} \tag{183}$$

To this Lie group we can associate a Lie algebra whose underlying vector space is the tangent space of the Lie group at the identity element and which completely captures the local structure of the group. This Lie group acts smoothly on the manifold, and acts on the vector fields. Any tangent vector at the identity of a Lie group can be extended to a left (respectively right) invariant vector field by left (respectively right) translating the tangent vector to other points of the manifold. This identifies the tangent space at the identity $\mathfrak{g} = T_I(G)$ with the space of left invariant vector fields, and therefore makes the tangent space at the identity into a Lie algebra, called the Lie algebra of G.

$$L_G : \begin{cases} G_{aff} \to G_{aff} \\ M \mapsto L_M N = M \cdot N \end{cases} \text{ and } R_G : \begin{cases} G_{aff} \to G_{aff} \\ M \mapsto R_M N = N \cdot M \end{cases} \tag{184}$$

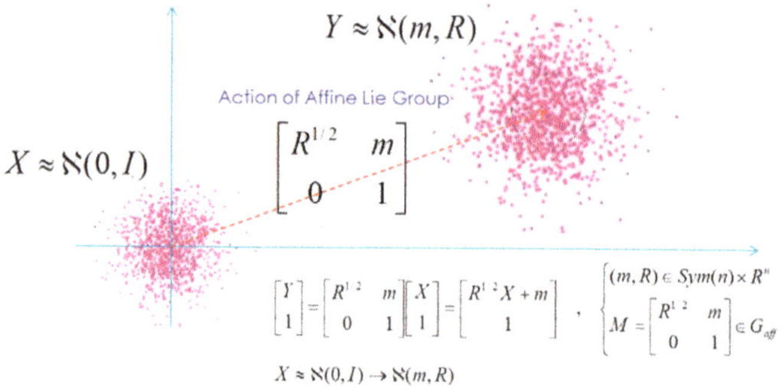

Figure 11. Affine Lie group action for multivariate Gaussian law.

Considering the curve $\gamma(t)$ and its derivative $\dot{\gamma}(t)$:

$$\gamma(t) = \begin{bmatrix} R^{1/2}(t) & m(t) \\ 0 & 1 \end{bmatrix} \text{ and } \dot{\gamma}(t) = \begin{bmatrix} \dot{R}^{1/2}(t) & \dot{m}(t) \\ 0 & 0 \end{bmatrix} \tag{185}$$

We can consider the curve with the point $\gamma(0)$ moved at the identity element on the left or on the right. Then, the tangent plan at identity element provides the Lie algebra:

$$\Gamma_L(t) = L_{M^{-1}}(\gamma(t)) = \begin{bmatrix} R^{-1/2}R^{1/2}(t) & R^{-1/2}(m(t) - m) \\ 0 & 1 \end{bmatrix} \tag{186}$$

$$\dot{\Gamma}_L(t)\Big|_{t=0} = \begin{bmatrix} R^{-1/2}\dot{R}^{1/2}(0) & R^{-1/2}\dot{m}(0) \\ 0 & 1 \end{bmatrix} = \frac{d}{dt}\left(L_{M^{-1}}(\gamma(t))\right)\Big|_{t=0} = dL_{M^{-1}}\dot{\gamma}(0) = dL_{M^{-1}}\dot{M} \tag{187}$$

Lie algebra on the right and on the left is the defined by:

$$dL_{M^{-1}} : T_M(G) \to \mathfrak{g}_L$$

$$\dot{M} \mapsto \Omega_L = dL_{M^{-1}}\dot{M} = M^{-1}\dot{M} = \begin{bmatrix} R^{-1/2}\dot{R}^{1/2} & R^{-1/2}\dot{m} \\ 0 & 0 \end{bmatrix} \tag{188}$$

$$dR_{M^{-1}} : T_M(G) \to \mathfrak{g}_R$$

$$\dot{M} \mapsto \Omega_R = dR_{M^{-1}}\dot{M} = \dot{M}M^{-1} = \begin{bmatrix} R^{-1/2}\dot{R}^{1/2} & \dot{m} - R^{-1/2}\dot{R}^{1/2}m \\ 0 & 0 \end{bmatrix} \tag{189}$$

We can then observe the velocities in two different ways, either by placing in a fixed outside frame, either by putting in place of the element in the process of moving by placing in the reference frame of the element.

$$\begin{bmatrix} X(t) \\ 1 \end{bmatrix} = M\begin{bmatrix} x \\ 1 \end{bmatrix} \Rightarrow \begin{bmatrix} \dot{X}(t) \\ 0 \end{bmatrix} = \Omega_R\begin{bmatrix} X(t) \\ 1 \end{bmatrix} \text{ with } x \text{ fixed} \tag{190}$$

$$\begin{bmatrix} x(t) \\ 1 \end{bmatrix} = M^{-1}\begin{bmatrix} X \\ 1 \end{bmatrix} \Rightarrow \begin{bmatrix} \dot{x}(t) \\ 0 \end{bmatrix} = -\Omega_L\begin{bmatrix} X \\ 1 \end{bmatrix} \text{ with } X \text{ fixed} \tag{191}$$

In the following, we will complete the global view by the operators which will allow to link algebra (from the left or the right) between them and also connect to their dual. We will first consider the automorphisms, the action by conjugation of the Lie group on itself that allows this operator to carry a member of the group.

$$\begin{aligned} &AD : G \times G \to G \\ &M, N \mapsto AD_M N = M.N.M^{-1} \end{aligned} \tag{192}$$

$$\begin{cases} M_1 = \begin{bmatrix} R_1^{1/2} & m_1 \\ 0 & 1 \end{bmatrix}, \; M_2 = \begin{bmatrix} R_2^{1/2} & m_2 \\ 0 & 1 \end{bmatrix} \\ AD_{M_1} M_2 = \begin{bmatrix} R_2^{1/2} & -R_2^{1/2} m_1 + R_1^{1/2} m_2 + m_1 \\ 0 & 1 \end{bmatrix} \end{cases} \tag{193}$$

If now we consider a curve *N(t)* curve on the manifold via the identity at *t* = 0. Its image by the previous operator will be then curve $\gamma = M \cdot N(t) \cdot M^{-1}$ passing through identity element at *t* = *0*. As $\dot{N}(0)$ is an element of the Lie algebra and its image by previous conjugation operator is called the Adjoint operator:

$$\begin{aligned} &Ad : G \times \mathfrak{g} \to \mathfrak{g} \\ &M, n \mapsto Ad_M n = M.n.M^{-1} = \left.\frac{d}{dt}\right|_{t=0} (AD_M N(t)) \text{ with } \begin{cases} N(0) = I \\ \dot{N}(0) = n \in g \end{cases} \end{aligned} \tag{194}$$

We can then compute the Adjoint operator for the previous Lie group:

$$\begin{cases} n_{2L} = \begin{bmatrix} R_2^{-1/2} \dot{R}_2^{1/2} & R_2^{-1/2} \dot{m}_2 \\ 0 & 0 \end{bmatrix}, \; n_{2R} = \begin{bmatrix} R_2^{-1/2} \dot{R}_2^{1/2} & -R_2^{-1/2} \dot{R}_2^{1/2} m_2 + \dot{m}_2 \\ 0 & 0 \end{bmatrix} \\ Ad_{M_1} n_{2L} = n_{2R} \text{ and } Ad_{M2} n_{2R} = \begin{bmatrix} R_2^{-1/2} \dot{R}_2^{1/2} & -R_2^{-1/2} \dot{R}_2^{1/2} m_2 + \dot{R}_2^{1/2} m_2 + R_2^{1/2} \dot{m}_2 \\ 0 & 0 \end{bmatrix}, \; Ad_{M_1^{-1}} n_{2R} = n_{2L} \end{cases} \tag{195}$$

We recall that the Lie algebra has been defined as the tangent space at the identity of a Lie group. We will then introduce a Lie bracket $[.,.]$, the expression of the operator associated with the combined action of the Lie algebra on itself, called an adjoint operator. The adjoint operator represents the action by conjugation of the Lie algebra on itself and is defined by:

$$\begin{aligned} &ad : \mathfrak{g} \times \mathfrak{g} \to \mathfrak{g} \\ &n, m \mapsto ad_m n = m \cdot n - n \cdot m = \left.\frac{d}{dt}\right|_{t=0} (Ad_M n(t)) = [m, n] \text{ with } \begin{cases} \dot{N}(0) = n \in g \\ \dot{M}(0) = m \in g \end{cases} \end{aligned} \tag{196}$$

We can then compute this operator for our use case:

$$n_{1L} = \begin{bmatrix} R_1^{-1/2} \dot{R}_1^{1/2} & R_1^{-1/2} \dot{m}_1 \\ 0 & 0 \end{bmatrix}, \; n_{2L} = \begin{bmatrix} R_2^{-1/2} \dot{R}_2^{1/2} & R_2^{-1/2} \dot{m}_2 \\ 0 & 0 \end{bmatrix} \tag{197}$$

$$ad_{n_{1L}} n_{2L} = [n_{1L}, n_{2L}] = \begin{bmatrix} 0 & R_1^{-1/2} \left(\dot{R}_1^{1/2} \dot{m}_2 - \dot{R}_2^{1/2} \dot{m}_1 \right) R_2^{-1/2} \\ 0 & 0 \end{bmatrix} \tag{198}$$

$$ad_{n_{1R}} n_{2R} = [n_{1R}, n_{2R}] = \begin{bmatrix} 0 & R_1^{-1/2} \dot{R}_1^{1/2} \left(-R_2^{-1/2} \dot{R}_2^{1/2} m_2 + \dot{m}_2 \right) - R_2^{-1/2} \dot{R}_2^{1/2} \left(-R_1^{-1/2} \dot{R}_1^{1/2} m_1 + \dot{m}_1 \right) \\ 0 & 0 \end{bmatrix} \tag{199}$$

To study the geodesic trajectories of the group, we consider the Lagrangian from the total kinetic energy (a quadratic form on speeds). It may therefore in particular be written in the left algebra "left", with the scalar product associated with the metric.

$$E_L = \frac{1}{2}\langle n_L, n_L\rangle = \frac{1}{2}Tr\left[n_L^T n_L\right] \tag{200}$$

If we consider as scalar product:

$$\begin{aligned}\langle .,.\rangle : \mathfrak{g}^* \times \mathfrak{g} \to R \\ k, n \mapsto \langle k, n\rangle = Tr\left(k^T n\right)\end{aligned} \tag{201}$$

and left algebra:

$$n_L = \begin{bmatrix} R^{-1/2}\dot{R}^{1/2} & R^{-1/2}\dot{m} \\ 0 & 0 \end{bmatrix} \tag{202}$$

we obtain for the total kinetic energy

$$E_L = \frac{1}{2}\left(Tr\left(R^{-1}\dot{R}\right) + \dot{m}^T R^{-1}\dot{m}\right) \tag{203}$$

We will then introduce the coadjoint operator that will enable us to work on the elements of the dual algebra of the Lie algebra defined above. Like algebra, which is physically the space of instantaneous speeds, the dual algebra is the space of moments. For the dual of left algebra, the moment is given by:

$$\Pi_L = \frac{\partial E_L}{\partial n_L} = n_L \tag{204}$$

Where E_L is the kinetic energy of the system and is currently associated with Π_L is an element of the left algebra. The moment space is the dual algebra, denoted $\mathfrak{g}^*$, associated with the Lie algebra $\mathfrak{g}$. This value is deduced from the computation:

$$\begin{aligned}&\left\langle \frac{\partial E_L}{\partial n_L}, \delta U\right\rangle = \underset{\varepsilon\to 0}{Lim}\frac{E_L(n_L + \varepsilon\cdot\delta U) - E_L(n_L)}{\varepsilon} \\ &\text{with } E_L(n_L + \varepsilon\cdot\delta U) = \frac{1}{2}\langle n_L + \varepsilon.\delta U, n_L + \varepsilon\cdot\delta U\rangle = \frac{1}{2}(n_L + \varepsilon\cdot\delta U)^T(n_L + \varepsilon\cdot\delta U) \\ &\left\langle \frac{\partial E_L}{\partial n_L}, \delta U\right\rangle = 2\cdot\frac{1}{2}tr\left(\eta_L^T\delta U\right) = \langle n_L, \delta U\rangle \Rightarrow \frac{\partial E_L}{\partial n_L} = n_L\end{aligned} \tag{205}$$

Then the moment map is given by:

$$\begin{aligned}\alpha_M : \mathfrak{g} \to \mathfrak{g}^* \\ n_L \mapsto \Pi_L = \eta_L\end{aligned} \tag{206}$$

We can observe that the application that turns left algebra into dual algebra is the identity application but, physically, the first are moments and the seconds are instantaneous speeds.

We can also define the moment Π_R associated to the right algebra η_R by:

$$\langle \Pi_L, n_L\rangle = \left\langle \Pi_L, M^{-1}n_R M\right\rangle = \langle \Pi_R, n_R\rangle \tag{207}$$

But as $\Pi_L = n_L$, we can deduce that:

$$\langle n_L, M^{-1}n_RM\rangle = \langle \Pi_R, n_R\rangle$$

$$\text{with } M = \begin{bmatrix} R^{1/2} & m \\ 0 & 1 \end{bmatrix}, n_L = \begin{bmatrix} R^{-1/2}\dot{R}^{1/2} & R^{-1/2}\dot{m} \\ 0 & 0 \end{bmatrix} \text{ and } \eta_R = \begin{bmatrix} R^{-1/2}\dot{R}^{1/2} & \dot{m} - R^{-1/2}\dot{R}^{1/2}\dot{m} \\ 0 & 0 \end{bmatrix} \tag{208}$$

$$\Rightarrow \Pi_R = \begin{bmatrix} R^{-1/2}\dot{R}^{1/2} + R^{-1}\dot{m}m^T & R^{-1}\dot{m} \\ 0 & 0 \end{bmatrix}$$

Then, the operator that transform the right algebra to its dual algebra is given by:

$$\beta_M : \mathfrak{g} \to \mathfrak{g}^*$$
$$n_R = \begin{bmatrix} \eta_{R1} & \eta_{R2} \\ 0 & 0 \end{bmatrix} \mapsto \Pi_R = \begin{bmatrix} \eta_{R1}\left(1 + m^TR^{-1}m\right) + \eta_{R2}m^TR^{-1} & \eta_{R1}R^{-1}m + R^{-1}\eta_{R2} \\ 0 & 0 \end{bmatrix} \tag{209}$$

There is an operator to change the view of algebra. Therefore, there is an operator that did the same to the dual algebra. This is called the co-adjoint operator and it is the conjugate action of the Lie group on its dual algebra:

$$\begin{cases} Ad^* : G \times \mathfrak{g}^* \to \mathfrak{g} \\ M, \eta \mapsto Ad^*_M\eta \end{cases} \text{ with } \langle Ad^*_M\eta, n\rangle = \langle \eta, Ad_Mn\rangle \text{ where } n \in \mathfrak{g} \tag{210}$$

We can then develop this expression for our use in the case of an affine sup-group. We find:

$$\begin{cases} M = \begin{bmatrix} A & b \\ 0 & 1 \end{bmatrix} \in G \\ \eta = \begin{bmatrix} \eta_1 & \eta_2 \\ 0 & 0 \end{bmatrix} \in \mathfrak{g}^* \\ n = \begin{bmatrix} n_1 & n_2 \\ 0 & 0 \end{bmatrix} \in \mathfrak{g} \end{cases} \Rightarrow \begin{cases} \langle Ad^*_M\eta, n\rangle = \langle \eta, Ad_Mn\rangle = \langle \eta, MnM^{-1}\rangle \\ \langle Ad^*_M\eta, n\rangle = \left\langle \begin{bmatrix} \eta_1 - \eta_2b^T & A\eta_2 \\ 0 & 0 \end{bmatrix}, n\right\rangle \end{cases} \Rightarrow Ad^*_M\eta = \begin{bmatrix} \eta_1 - \eta_2b^T & A\eta_2 \\ 0 & 0 \end{bmatrix} \tag{211}$$

and we can also observe that:

$$Ad^*_{M^{-1}}\eta = \begin{bmatrix} \eta_1 + A\eta_2b^T & A\eta_2 \\ 0 & 0 \end{bmatrix} \tag{212}$$

Similarly there exists the following relation between the left and the right algebras:

$$Ad^*_M\Pi_R = \Pi_L \text{ and } Ad^*_{M^{-1}}\Pi_L = \Pi_R \tag{213}$$

As we have defined a commutator on the Lie algebra, it is possible to define one on its dual algebra. This commutator on the dual algebra can also be defined using the operator expressing the combined action of the algebra of its dual algebra. This operator is called the co-adjoint operator:

$$\begin{cases} ad^* : \mathfrak{g} \times \mathfrak{g}^* \to \mathfrak{g}^* \\ n, \eta \mapsto ad^*_n\eta \end{cases} \text{ with } \langle ad^*_n\eta, \kappa\rangle = \langle \eta, ad_n\kappa\rangle \text{ where } \kappa \in \mathfrak{g} \tag{214}$$

We can develop this co-adjoint operator on its dual algebra for our use-case:

$$\begin{cases} \kappa = \begin{bmatrix} \kappa_1 & \kappa_2 \\ 0 & 0 \end{bmatrix} \in G \\ \eta = \begin{bmatrix} \eta_1 & \eta_2 \\ 0 & 0 \end{bmatrix} \in \mathfrak{g}^* \\ n = \begin{bmatrix} n_1 & n_2 \\ 0 & 0 \end{bmatrix} \in \mathfrak{g} \end{cases} \Rightarrow \begin{cases} \langle ad_n^* \eta, \kappa \rangle = \langle \eta, ad_n \kappa \rangle = \langle \eta, n\kappa - \kappa n \rangle \\ \langle ad_n^* \eta, \kappa \rangle = \left\langle \begin{bmatrix} -\eta_2 n_2^T & n_1 \eta_2 \\ 0 & 0 \end{bmatrix}, \kappa \right\rangle \end{cases} \Rightarrow \begin{cases} ad_n^* \eta = \begin{bmatrix} -\eta_2 n_2^T & n_1 \eta_2 \\ 0 & 0 \end{bmatrix} \\ ad_n^* \eta = \{n, \eta\} \end{cases} \tag{215}$$

This co-adjoint operator will give the Euler-Poincaré equation. While the Euler-Lagrange equations is defined on the tangent bundle (union of the tangent spaces at each point) of the manifold and give the geodesics, the Euler-Poincaré equation gives a differential system on the dual Lie algebra of the group associated with the manifold.

We can also complete these maps by using additional ones. First, $p \in T_M^* G$ the moment associated with $\dot{M} \in T_M G$ in tangent space of G at M and also two other moments map the element of the dual algebra in dual tangent space, respectively on the left and on the right:

$$\begin{cases} \langle \Pi_L, n_L \rangle = \left\langle dL_{M^{-1}}^* \Pi_L, \dot{M} \right\rangle \\ \left\langle \Pi_L, dL_{M^{-1}} \dot{M} \right\rangle = \left\langle \Pi_L, M^{-1} \dot{M} \right\rangle \end{cases} \Rightarrow p = \left(M^{-1} \right)^T \Pi_L \tag{216}$$

where

$$\begin{array}{ccc} dL_{M^{-1}}^* : \mathfrak{g}_L^* \to T_M^* G & & dR_{M^{-1}}^* : \mathfrak{g}_R^* \to T_M^* G \\ \Pi_L \mapsto p = \left(M^{-1} \right)^T \Pi_L & \text{and} & \Pi_R \mapsto p = \Pi_R \left(M^{-1} \right)^T \end{array} \tag{217}$$

From these relations, we can also observe that:

$$\begin{aligned} &\Pi_L = n_L = M^{-1} \dot{M} \\ &\Rightarrow \begin{cases} p = \left(M^{-1} \right)^T M^{-1} \dot{M} \\ p = \Xi_M \cdot \dot{M} \text{ with } \Xi_M = \left(M^{-1} \right)^T M^{-1} \end{cases} \end{aligned} \tag{218}$$

All these maps could be summarized in the following Figure 12:

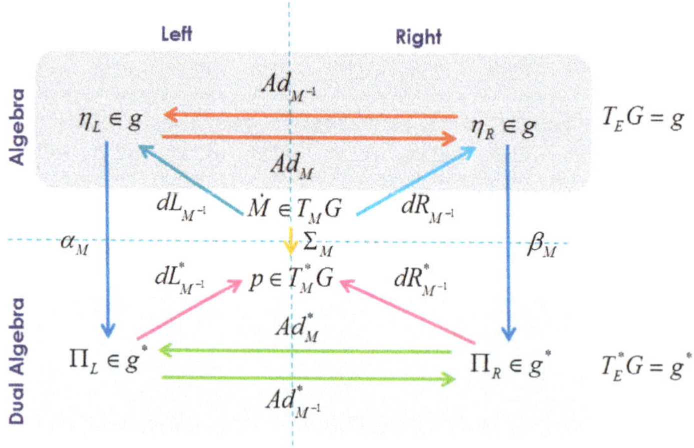

Figure 12. Maps between algebras.

Heni Poincaré proved that when a Lie algebra acts locally and transitively on the configuration space of a Lagrangian mechanical system, the Euler-Lagrange equations are equivalent to a new system of differential equations defined on the product of the configuration space with the Lie algebra.

If we consider that the following function is stationary for a Lagragian $l(.)$ invariant with respect to the action of a group on the left:

$$S(\eta_L) = \int_a^b l(\eta_L)dt \text{ with } \delta S(\eta_L) = 0 \text{ and } l : \mathfrak{g} \to R \tag{219}$$

The solution is given by the Euler-Poincaré equation:

$$\begin{gathered} \frac{d}{dt}\frac{\delta l}{\delta \eta_L} = ad^*_{\eta_L}\frac{\delta l}{\delta \eta_L} \\ \delta\eta_L = \dot{\Gamma} + ad_{\eta_L}\Gamma \text{ where } \Gamma(t) \in \mathfrak{g} \end{gathered} \tag{220}$$

If we take for the function $l(.)$, the total kinetic energy E_L, using $\Pi_L = M^{-1}\dot{M} = \frac{\partial E_L}{\partial n_L} \in \mathfrak{g}_L$, then the Euler-Poincaré equation is given by:

$$\frac{d\Pi_L}{dt} = ad^*_{n_L}\Pi_L \text{ with } \frac{\delta l}{\delta \eta_L} = \frac{\partial E_L}{\partial n_L} = \Pi_L \in \mathfrak{g}_L \tag{221}$$

The following quantities are conserved:

$$\frac{d\Pi_R}{dt} = 0 \tag{222}$$

With this second theorem, it is possible to write the geodesic not from its coordinate system but from the quantity of motion, and in addition to determine explicitly what the conserved quantities along the geodesic are (conservations are related to the symmetries of the variety and hence the invariance of the Lagrangian under the action of the group).

For our use-case, the Euler-Poincaré equation is given by:

$$\begin{cases} \dot{\eta}_{L1} = -\eta_{L2}\eta_{L2}^T \\ \dot{\eta}_{L2} = \eta_{L2}\eta_{L1} \end{cases} \text{with} \begin{cases} \eta_{L1} = R^{-1/2}\dot{R}^{1/2} \\ \eta_{L2} = R^{-1/2}\dot{m} \end{cases} \Rightarrow \begin{cases} \left(R^{-1/2}\dot{R}^{1/2}\right)^{\bullet} = -R^{-1/2}\dot{m}\dot{m}^T R^{-1/2} \\ \left(R^{-1/2}\dot{m}\right)^{\bullet} = \dot{R}^{-1/2}\dot{R}^{1/2}R^{-1/2}\dot{m} \end{cases} \tag{223}$$

If we remark that we have $R^{-1/2}\dot{R}^{1/2} = R^{-1/2}\left(R^{-1/2}\dot{R}\right) = R^{-1}\dot{R}$, then the conserved Souriau moment could be given by:

$$\Pi_R = \begin{bmatrix} R^{-1/2}\dot{R}^{1/2} + R^{-1}\dot{m}m^T & R^{-1}\dot{m} \\ 0 & 0 \end{bmatrix} = \begin{bmatrix} R^{-1}\dot{R} + R^{-1}\dot{m}m^T & R^{-1}\dot{m} \\ 0 & 0 \end{bmatrix} \tag{224}$$

Components of the Souriau moment give the conserved quantities that are the classical elements given by Emmy Noether Theorem (Souriau moment is a geometrization of Emmy Noether Theorem):

$$\frac{d\Pi_R}{dt} = \begin{bmatrix} \frac{d\left(R^{-1}\dot{R}+R^{-1}\dot{m}m^T\right)}{dt} & \frac{d\left(R^{-1}\dot{m}\right)}{dt} \\ 0 & 0 \end{bmatrix} = 0 \Rightarrow \begin{cases} R^{-1}\dot{R} + R^{-1}\dot{m}m^T = B = cste \\ R^{-1}\dot{m} = b = cste \end{cases} \tag{225}$$

From this constant, we can obtain a reduced equation of geodesic:

$$\begin{cases} \dot{m} = Rb \\ \dot{R} = R\left(B - bm^T\right) \end{cases} \tag{226}$$

This is the Euler-Poincaré equation of geodesic. We can observe that we have obtained a reduction of the following Euler-Lagrange equation [27,156,187]: $\begin{cases} \ddot{R} + \dot{m}\dot{m}^T - \dot{R}R^{-1}\dot{R} = 0 \\ \ddot{m} - \dot{R}R^{-1}\dot{m} = 0 \end{cases}$ associated to the information geometry metric $ds^2 = dm^T R^{-1} dm + \frac{1}{2} Tr\left(\left(R^{-1}dR\right)^2\right)$.

The Fisher information defines a metric turning $N_n = \{(m, R) \in R^n \times Sym^+(n)\}$ into a Riemannian manifold. The inner product of two tangent vectors $(m_1, R_1) \in T_n$ and $(m_2, R_2) \in T_n$ at the point $(\mu, \Sigma) \in N_n$ is given by:

$$g_{(\mu,\Sigma)}\left((m_1, R_1), (m_2, R_2)\right) = m_1^T \Sigma^{-1} m_2 + \frac{1}{2} tr\left(\Sigma^{-1} R_1 \Sigma^{-1} R_2\right) \tag{227}$$

and the geodesic is given by:

$$l(\chi) = \int_{t_0}^{t_1} \sqrt{g_{\chi(t)}\left(\dot{\chi}(t), \dot{\chi}(t)\right)} dt \tag{228}$$

We can also observe that the manifold of multivariate Gaussian is homogeneous with respect to positive affine group $GA^+(n)$:

$$ds_Y^2 = ds_X^2 \text{ for } Y = \Sigma^{1/2} X + \mu \text{ with } \mathrm{GA}^+(n) = \{(\mu, \Sigma) \in R \times GL(R) / \det(\Sigma) > 0\} \tag{229}$$

characterized by the action of the group $(m, R) \mapsto \rho.(m, R) = \left(\Sigma^{1/2} m + \mu, \Sigma^{1/2} R \Sigma^{1/2T}\right), \rho \in GA^+(n)$

$$\text{with } \begin{bmatrix} Y \\ 1 \end{bmatrix} = \begin{bmatrix} \Sigma^{1/2} & \mu \\ 0 & 1 \end{bmatrix} \begin{bmatrix} X \\ 1 \end{bmatrix} \tag{230}$$

$$\begin{aligned} ds_Y^2 &= d\left(\Sigma^{1/2} m + \mu\right)^T \left(\Sigma^{1/2} R \Sigma^{1/2T}\right)^{-1} d\left(\Sigma^{1/2} m + \mu\right) + \frac{1}{2} Tr\left(\left(\left(\Sigma^{1/2} R \Sigma^{1/2T}\right)^{-1} d\left(\Sigma^{1/2} R \Sigma^{1/2T}\right)\right)^2\right) \\ ds_Y^2 &= dm^T R^{-1} dm + \frac{1}{2} Tr\left(\left(R^{-1} dR\right)^2\right) = ds_X^2 \end{aligned} \tag{231}$$

Since the special orthogonal group $SO(n) = \{\delta \in GL(R) / \det(\delta) = 1\}$ is the stabilizer subgroup of $(0, I_n)$, we have the following isomorphism:

$$\begin{aligned} & GA^+(n)/SO(n) \to N_n = \{(m, R) \in R^n \times Sym^+(n)\} \\ & \rho = (\mu, \Sigma) \mapsto \rho.(0, I_n) = \left(\mu, \Sigma^{1/2} \Sigma^{1/2T}\right) = (\mu, \Sigma) \end{aligned} \tag{232}$$

We can then restrict the computation of the geodesic from $(0, I_n)$ and then we can partially integrate the system of equations:

$$\begin{cases} \dot{m} = Rb \\ \dot{R} = R\left(B - bm^T\right) \end{cases} \tag{233}$$

where $\left(R^{-1}(0)\dot{m}(0), R^{-1}(0)\left(\dot{R}(0) + \dot{m}(0)m(0)^T\right)\right) = (b, B) \in R^n \times Sym_n(R)$ are the integration constants.

From this Euler-Poincaré equation, we can compute geodesics by geodesic shooting [188–191] using classical Eriksen equations [192–195], by the following change of parameters:

$$\begin{cases} \Delta(t) = R^{-1}(t) \\ \delta(t) = R^{-1}(t)m(t) \end{cases} \Rightarrow \begin{cases} \dot{\Delta} = -B\Delta + bm^T \\ \dot{\delta} = -B\delta + \left(1 + \delta^T \Delta^{-1} \delta\right) b \\ \Delta(0) = I_p, \delta(0) = 0 \end{cases} \text{ with } \begin{cases} \dot{\Delta}(0) = -B \\ \dot{\delta}(0) = b \end{cases} \tag{234}$$

The initial speed of the geodesic is given by $\left(\dot{\delta}(0), \dot{\Delta}(0)\right)$. The geodesic shooting is given by the exponential map:

$$\Lambda(t) = \exp(tA) = \sum_{n=0}^{\infty} \frac{(tA)^n}{n!} = \begin{pmatrix} \Delta & \delta & \Phi \\ \delta^T & \varepsilon & \gamma^T \\ \Phi^T & \gamma & \Gamma \end{pmatrix} \text{ with } A = \begin{pmatrix} -B & b & 0 \\ b^T & 0 & -b^T \\ 0 & -b & B \end{pmatrix} \tag{235}$$

This equation can be interpreted by group theory. A could be considered as an element of Lie algebra $so(n+1,n)$ of the special Lorentz group $SO_O(n+1,n)$ and more specifically as the element p of Cartan Decomposition l + p where l is the Lie algebra of a maximal compact sub-group $K = S(O(n+1) \times O(n))$ of the group $G = SO_O(n+1,n)$. We know that its exponential map defines a geodesic on Riemannian Symetric space G/K.

This equation can be established by the following developments:

$$\dot{\Lambda}(t) = A.\Lambda(t) \Rightarrow \begin{pmatrix} \dot{\Delta} & \dot{\delta} & \dot{\Phi} \\ \dot{\delta}^T & \dot{\varepsilon} & \dot{\gamma}^T \\ \dot{\Phi}^T & \dot{\gamma} & \dot{\Gamma} \end{pmatrix} = \begin{pmatrix} -B & b & 0 \\ b^T & 0 & -b^T \\ 0 & -b & B \end{pmatrix} . \begin{pmatrix} \Delta & \delta & \Phi \\ \delta^T & \varepsilon & \gamma^T \\ \Phi^T & \gamma & \Gamma \end{pmatrix} \tag{236}$$

We can then deduce that:

$$\begin{cases} \dot{\Delta} = -B\Delta + b\delta^T \\ \dot{\delta} = -B\delta + \varepsilon b \end{cases} \tag{237}$$

If $\varepsilon = 1 + \delta^T\Delta^{-1}\delta$, then (Δ, δ) is solution to the geodesic equation previously defined. Since $\varepsilon(0) = 1$, it suffices to demonstrate that $\dot{\varepsilon} = \dot{\tau}$ where $\tau = \delta^T\Delta^{-1}\delta$.

From $\dot{\Lambda}(t) = \Lambda(t).A$, using that $\dot{\delta}^T = b^T\Delta - b^T\Phi^T$, we can deduce:

$$\begin{cases} \dot{\varepsilon} = b^T\delta - b^T\gamma \\ \dot{\tau} = b^T\delta - b^T\left((\tau - \varepsilon)\Delta^{-1}\delta + \Phi^T\Delta^{-1}\delta\right) \end{cases} \tag{238}$$

Then $\dot{\varepsilon} = \dot{\tau}$, if $\gamma = (\tau - \varepsilon)\Delta^{-1}\delta + \Phi\Delta^{-1}\delta$, that could be verified using relation $\Lambda.\Lambda^{-1} = I$, by observing that:

$$\Lambda^{-1} = \exp(-tA) = \Lambda(-t) = \begin{bmatrix} \Gamma & \gamma & \Phi^T \\ \gamma^T & \varepsilon & \delta^T \\ \Phi & \delta & \Delta \end{bmatrix} \tag{239}$$

$$\Lambda.\Lambda^{-1} = I \Rightarrow \begin{cases} \Delta\gamma + \varepsilon\delta + \Phi\delta = 0 \\ \Delta\Phi^T + \delta\delta^T + \Phi\Delta = 0 \end{cases} \Rightarrow \begin{cases} \gamma = -\varepsilon\Delta^{-1}\delta - \Delta^{-1}\Phi\delta \\ \Phi^T\Delta^{-1} + \Delta^{-1}\delta\delta^T\Delta^{-1} + \Delta^{-1}\Phi = 0 \end{cases} \Rightarrow \begin{cases} \gamma = -\varepsilon\Delta^{-1}\delta - \Delta^{-1}\Phi\delta \\ \Phi^T\Delta^{-1}\delta + \tau\Delta^{-1}\delta + \Delta^{-1}\Phi\delta = 0 \end{cases} \tag{240}$$

We can then compute γ from two last equations:

$$\gamma = (\tau - \varepsilon)\Delta^{-1}\delta + \Phi^T\Delta^{-1}\delta \tag{241}$$

As $\dot{\tau} = b^T\delta - b^T\left((\tau - \varepsilon)\Delta^{-1}\delta + \Phi^T\Delta^{-1}\delta\right)$ then we can deduce that $\dot{\tau} = b^T\delta - b^T\gamma$ and then $\dot{\tau} = \dot{\varepsilon}$.

To interpret elements of Λ, $(\Gamma(t), \gamma(t)) = (\Delta(-t), \delta(-t))$, opposite points to $(\Delta(t), \delta(t))$, and $\varepsilon = 1 + \delta^T\Delta^{-1}\delta = 1 + \gamma^T\Gamma^{-1}\gamma$.

Then the geodesic that goes through the origin $(0, I_n)$ with initial tangent vector $(b, -B)$ is the curve given by $(\delta(t), \Delta(t))$. Then the distance computation is reduced to estimate the initial tangent vector space related by $\left(R^{-1}(0)\dot{m}(0), R^{-1}(0)\left(\dot{R}(0)+\dot{m}(0)m(0)^T\right)\right)=(b,B)\in R^n\times Sym_n(R)$The distance will be then given by the initial tangent vector:

$$d=\sqrt{\dot{m}(0)^T R^{-1}(0)\dot{m}(0)+\frac{1}{2}Tr\left[\left(R^{-1}(0)\dot{R}(0)\right)^2\right]} \tag{242}$$

This initial tangent vector will be identified by "Geodesic Shooting". Let $V=\log_A B$:

$$\begin{cases} \dfrac{dV_m}{dt}=\dfrac{1}{2}\left(\dfrac{dR}{dt}\right)R^{-1}V_m+\dfrac{1}{2}V_R R^{-1}\left(\dfrac{dm}{dt}\right) \\ \dfrac{dV_R}{dt}=\dfrac{1}{2}\left(\left(\dfrac{dR}{dt}\right)R^{-1}V_m+V_R R^{-1}\left(\dfrac{dR}{dt}\right)\right)-\dfrac{1}{2}\left(\left(\dfrac{dm}{dt}\right)V_m^T+V_m^T\left(\dfrac{dm}{dt}\right)\right) \end{cases} \tag{243}$$

Geodesic Shooting is corrected by using Jacobi Field J and parallel transport: $J(t)=\left.\frac{\partial\chi_\alpha(t)}{\partial\alpha}\right|_{t=0}$ solution to $\frac{d^2J(t)}{dt^2}+R\left(J(t),\dot{\chi}(t)\right)\dot{\chi}(t)=0$ with R the Riemann Curvarture tensor.

We consider a geodesic χ between θ_0 and θ_1 with an initial tangent vector V, and we suppose that V is perturbated by W, to $V+W$. The variation of the final point θ_1 can be determined thanks to the Jacobi field with $J(0)=0$ and $\dot{J}(0)=W$. In term of the exponential map, this could be written:

$$J(t)=\frac{d}{d\alpha}\exp_{\theta_0}\left(t\left(V+\alpha W\right)\right)\Big|_{\alpha=0} \tag{244}$$

This could be illustrated in the Figure 13:

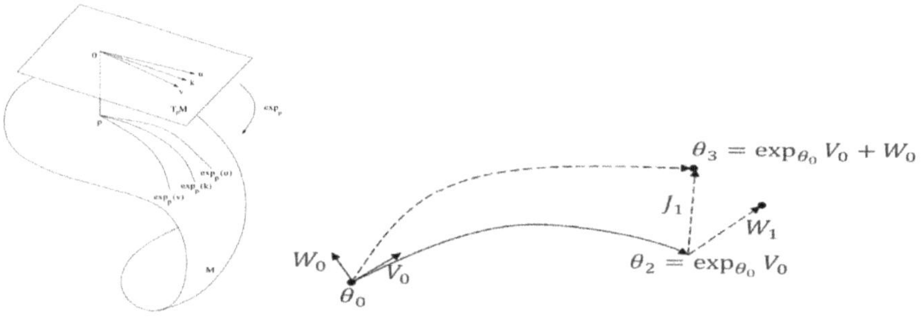

Figure 13. Geodesic shooting principle.

We give some illustration, in Figure 14, of geodesic shooting to compute the distance between multivariate Gaussian density for the case n = 2:

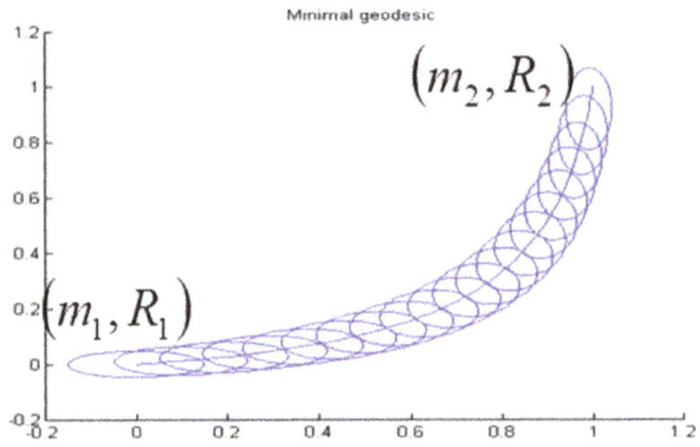

Figure 14. GeodesicsShooting between two multivariate Gaussian in case $n = 2$.

9. Souriau Riemannian Metric for Multivariate Gaussian Densities

To illustrate the Souriau-Fisher metric, we will consider the family of multivariate Gaussian densities and will develop some elements that we have previously developed purely theoretically.

For the families of multivariate Gaussian densities, that we have identified as homogeneous manifold with the associated sub-group of the affine group $\begin{bmatrix} R^{1/2} & m \\ 0 & 1 \end{bmatrix}$, we have seen that if we consider them as elements of exponential families, we can write $\hat{\xi}$ (element of the dual Lie algebra) that play the role of geometric heat Q in Souriau Lie group thermodynamics, and β the geometric (Planck) temperature.

$$\hat{\xi} = \begin{bmatrix} E[z] \\ E\left[zz^T\right] \end{bmatrix} = \begin{bmatrix} m \\ R + mm^T \end{bmatrix}, \ \beta = \begin{bmatrix} -R^{-1}m \\ \frac{1}{2}R^{-1} \end{bmatrix} \tag{245}$$

These elements are homeomorphic to the matrix elements in matrix Lie algebra and dual Lie algebra:

$$\hat{\xi} = \begin{bmatrix} R + mm^T & m \\ 0 & 0 \end{bmatrix} \in \mathfrak{g}^*, \ \beta = \begin{bmatrix} \frac{1}{2}R^{-1} & -R^{-1}m \\ 0 & 0 \end{bmatrix} \in \mathfrak{g} \tag{246}$$

If we consider $M = \begin{bmatrix} R'^{1/2} & m' \\ 0 & 1 \end{bmatrix}$, then we can compute the co-adjoint operator:

$$Ad^*_M \hat{\xi} = \begin{bmatrix} R + mm^T - mm'^T & R'^{1/2}m \\ 0 & 0 \end{bmatrix} \tag{247}$$

We can also compute the adjoint operator:

$$\begin{gathered} Ad_M\beta = M \cdot \beta \cdot M^{-1} = \begin{bmatrix} R'^{1/2} & m' \\ 0 & 1 \end{bmatrix} \begin{bmatrix} \frac{1}{2}R^{-1} & -R^{-1}m \\ 0 & 0 \end{bmatrix} \begin{bmatrix} R'^{-1/2} & -R'^{-1/2}m' \\ 0 & 1 \end{bmatrix} \\ Ad_M\beta = \begin{bmatrix} \frac{1}{2}R'^{1/2}R^{-1}R'^{-1/2} & -\frac{1}{2}R'^{1/2}R^{-1}R'^{-1/2}m' - R'^{1/2}R^{-1}m \\ 0 & 0 \end{bmatrix} \end{gathered} \tag{248}$$

We can rewrite $Ad_M\beta$ with the following identification:

$$Ad_M\beta = \begin{bmatrix} \frac{1}{2}\Omega^{-1} & -\Omega^{-1}n \\ 0 & 0 \end{bmatrix} \quad \text{with } \Omega = R'^{1/2}RR'^{-1/2} \text{ and } n = \left(\frac{1}{2}m' + R'^{1/2}m\right) \tag{249}$$

We have then to develop $\hat{\xi}\left(Ad_M(\beta)\right)$, that is to say $\hat{\xi}(\beta)$ after action of the group on the Lie algebra for β, given by $Ad_M(\beta)$. By analogy of structure between $\hat{\xi}(\beta)$ and β, we can write:

$$\left.\begin{array}{l} \beta = \begin{bmatrix} \frac{1}{2}R^{-1} & -R^{-1}m \\ 0 & 0 \end{bmatrix} \\ \hat{\xi}(\beta) = \begin{bmatrix} R + mm^T & m \\ 0 & 0 \end{bmatrix} \end{array}\right\} \Rightarrow \left\{\begin{array}{l} Ad_M\beta = \begin{bmatrix} \frac{1}{2}\Omega^{-1} & -\Omega^{-1}n \\ 0 & 0 \end{bmatrix} \\ \hat{\xi}\left(Ad_M(\beta)\right) = \begin{bmatrix} \Omega + nn^T & n \\ 0 & 0 \end{bmatrix} \end{array}\right. \tag{250}$$

We have then to identify the cocycle $\theta(M)$ from $\hat{\xi}\left(Ad_M(\beta)\right) = Ad_M^*(\hat{\xi}) + \theta(M)$ $\Rightarrow \theta(M) = \hat{\xi}\left(Ad_M(\beta)\right) - Ad_M^*\hat{\xi}$ where:

$$Ad_M^*\hat{\xi} = \begin{bmatrix} R + mm^T - mm'^T & R'^{1/2}m \\ 0 & 0 \end{bmatrix} \tag{251}$$

$$\hat{\xi}\left(Ad_M(\beta)\right) = \begin{bmatrix} R'^{1/2}RR'^{-1/2} + \left(\frac{1}{2}m' + R'^{1/2}m\right)\left(\frac{1}{2}m' + R'^{1/2}m\right)^T & \left(\frac{1}{2}m' + R'^{1/2}m\right) \\ 0 & 0 \end{bmatrix} \tag{252}$$

The cocycle is then given by:

$$\begin{array}{l} \theta(M) = \begin{bmatrix} R'^{1/2}RR'^{-1/2} + \left(\frac{1}{2}m' + R'^{1/2}m\right)\left(\frac{1}{2}m' + R'^{1/2}m\right)^T & \left(\frac{1}{2}m' + R'^{1/2}m\right) \\ 0 & 0 \end{bmatrix} - \begin{bmatrix} R + mm^T - mm'^T & R'^{1/2}m \\ 0 & 0 \end{bmatrix} \\ \theta(M) = \begin{bmatrix} \left(R'^{1/2}RR'^{-1/2} - R\right) + \left(R'^{1/2}mm^TR'^{1/2T} - mm^T\right) + \left(\frac{1}{2}m'm^TR'^{1/2T} + \frac{1}{2}R'^{1/2}mm'^T - mm'^T\right) & \frac{1}{2}m' \\ 0 & 0 \end{bmatrix} \end{array} \tag{253}$$

From $\theta(M) = \hat{\xi}\left(Ad_M(\beta)\right) - Ad_M^*\hat{\xi}$, we can compute cocycle in Lie algebra

$$\Theta = T_e\theta \tag{254}$$

used to define the tensor:

$$\begin{array}{l} \widetilde{\Theta}(X,Y) : \mathfrak{g} \times \mathfrak{g} \to \Re \\ X, Y \mapsto \langle \Theta(X), Y \rangle \end{array} \tag{255}$$

In this second part, we will compute the Souriau-Fisher metric given by:

$$g_\beta\left([\beta, Z_1], [\beta, Z_2]\right) = \widetilde{\Theta}_\beta\left(Z_1, [\beta, Z_2]\right) \tag{256}$$

with

$$\widetilde{\Theta}_\beta(Z_1, Z_2) = \widetilde{\Theta}(Z_1, Z_2) + \langle \hat{\xi}, ad_{Z_1} Z_2 \rangle = \langle \Theta(Z_1), Z_2 \rangle + \langle \hat{\xi}, [Z_1, Z_2] \rangle \tag{257}$$

$$\begin{aligned} g_\beta\left([\beta, Z_1], [\beta, Z_2]\right) &= \widetilde{\Theta}_\beta\left(Z_1, [\beta, Z_2]\right) = \widetilde{\Theta}\left(Z_1, [\beta, Z_2]\right) + \langle \hat{\xi}, [Z_1, [\beta, Z_2]] \rangle \\ &= \langle \Theta(Z_1), [\beta, Z_2] \rangle + \langle \hat{\xi}, [Z_1, [\beta, Z_2]] \rangle \end{aligned} \tag{258}$$

where

$$\beta = \begin{bmatrix} \frac{1}{2}R^{-1} & -R^{-1}m \\ 0 & 0 \end{bmatrix} \text{ and } \hat{\xi} = \begin{bmatrix} R + mm^T & m \\ 0 & 0 \end{bmatrix} \quad (259)$$

$$\text{If we set } Z_1 = \begin{bmatrix} \frac{1}{2}\Omega_1^{-1} & -\Omega_1^{-1}n_1 \\ 0 & 0 \end{bmatrix} \text{ and } Z_2 = \begin{bmatrix} \frac{1}{2}\Omega_2^{-1} & -\Omega_2^{-1}n_2 \\ 0 & 0 \end{bmatrix} \quad (260)$$

With $\langle ..., ... \rangle$ the inner product given by

$$\langle \xi, \beta \rangle = Tr\left[ba^T + H^T L\right] \text{ with } \xi = \begin{bmatrix} L & b \\ 0 & 0 \end{bmatrix}, \beta = \begin{bmatrix} H & a \\ 0 & 0 \end{bmatrix} \quad (261)$$

$$\begin{aligned} [\beta, Z_2] = \beta Z_2 - Z_2\beta &= \begin{bmatrix} \frac{1}{2}R^{-1} & -R^{-1}m \\ 0 & 0 \end{bmatrix} \begin{bmatrix} \frac{1}{2}\Omega_2^{-1} & -\Omega_2^{-1}n_2 \\ 0 & 0 \end{bmatrix} - \begin{bmatrix} \frac{1}{2}\Omega_2^{-1} & -\Omega_2^{-1}n_2 \\ 0 & 0 \end{bmatrix} \begin{bmatrix} \frac{1}{2}R^{-1} & -R^{-1}m \\ 0 & 0 \end{bmatrix} \\ [\beta, Z_2] &= \begin{bmatrix} \frac{1}{4}\left(R^{-1}\Omega_2^{-1} - \Omega_2^{-1}R^{-1}\right) & -\frac{1}{2}\left(R^{-1}\Omega_2^{-1}n_2 - \Omega_2^{-1}R^{-1}m\right) \\ 0 & 0 \end{bmatrix} \end{aligned} \quad (262)$$

$$\begin{aligned} [Z_1, [\beta, Z_2]] &= \begin{bmatrix} \frac{1}{2}\Omega_1^{-1} & -\Omega_1^{-1}n_1 \\ 0 & 0 \end{bmatrix} \begin{bmatrix} \frac{1}{4}\left(R^{-1}\Omega_2^{-1} - \Omega_2^{-1}R^{-1}\right) & -\frac{1}{2}\left(R^{-1}\Omega_2^{-1}n_2 - \Omega_2^{-1}R^{-1}m\right) \\ 0 & 0 \end{bmatrix} \\ &\quad - \begin{bmatrix} \frac{1}{4}\left(R^{-1}\Omega_2^{-1} - \Omega_2^{-1}R^{-1}\right) & -\frac{1}{2}\left(R^{-1}\Omega_2^{-1}n_2 - \Omega_2^{-1}R^{-1}m\right) \\ 0 & 0 \end{bmatrix} \begin{bmatrix} \frac{1}{2}\Omega_1^{-1} & -\Omega_1^{-1}n_1 \\ 0 & 0 \end{bmatrix} \\ &= \begin{bmatrix} \frac{1}{8}\left(\Omega_1^{-1}\left(R^{-1}\Omega_2^{-1} - \Omega_2^{-1}R^{-1}\right) - \left(R^{-1}\Omega_2^{-1} - \Omega_2^{-1}R^{-1}\right)\Omega_1^{-1}\right) & -\frac{1}{4}\left(\Omega_1^{-1}\left(R^{-1}\Omega_2^{-1}n_2 - \Omega_2^{-1}R^{-1}m\right) - \left(R^{-1}\Omega_2^{-1} - \Omega_2^{-1}R^{-1}\right)\Omega_1^{-1}n_1\right) \\ 0 & 0 \end{bmatrix} \end{aligned} \quad (263)$$

We can then compute:

$$\begin{aligned} \langle \hat{\xi}, [Z_1, [\beta, Z_2]] \rangle &= Tr\left[\frac{1}{4}m\left(\left(R^{-1}\Omega_2^{-1} - \Omega_2^{-1}R^{-1}\right)\Omega_1^{-1}n_1 - \Omega_1^{-1}\left(R^{-1}\Omega_2^{-1}n_2 - \Omega_2^{-1}R^{-1}m\right)\right)^T\right] \\ &+ Tr\left[\left(\frac{1}{8}\left(\Omega_1^{-1}\left(R^{-1}\Omega_2^{-1} - \Omega_2^{-1}R^{-1}\right) - \left(R^{-1}\Omega_2^{-1} - \Omega_2^{-1}R^{-1}\right)\Omega_1^{-1}\right)\right)\left(R + mm^T\right)\right] \end{aligned} \quad (264)$$

The Souriau-Fisher metric is defined in Lie algebra $g_\beta\left([\beta, Z_1], [\beta, Z_2]\right)$ where:

$$\begin{aligned} &[\beta, Z_1] = \begin{bmatrix} \frac{1}{4}\left(R^{-1}\Omega_1^{-1} - \Omega_1^{-1}R^{-1}\right) & -\frac{1}{2}\left(R^{-1}\Omega_1^{-1}n_1 - \Omega_1^{-1}R^{-1}m\right) \\ 0 & 0 \end{bmatrix} = \begin{bmatrix} \frac{1}{2}G_1^{-1} & -G_1^{-1}g_1 \\ 0 & 0 \end{bmatrix} \\ &\text{with } G_1 = 2\left(\Omega_1 R - R\Omega_1\right) \text{ and } g_1 = (I - R\Omega_1 R^{-1}\Omega_1^{-1})n_1 + (\Omega_1 R\Omega_1^{-1}R^{-1} - I)m \\ &[\beta, Z_2] = \begin{bmatrix} \frac{1}{4}\left(R^{-1}\Omega_2^{-1} - \Omega_2^{-1}R^{-1}\right) & -\frac{1}{2}\left(R^{-1}\Omega_2^{-1}n_2 - \Omega_2^{-1}R^{-1}m\right) \\ 0 & 0 \end{bmatrix} = \begin{bmatrix} \frac{1}{2}G_2^{-1} & -G_2^{-1}g_2 \\ 0 & 0 \end{bmatrix} \\ &\text{with } G_2 = 2\left(\Omega_2 R - R\Omega_2\right) \text{ and } g_2 = (I - R\Omega_2 R^{-1}\Omega_2^{-1})n_2 + (\Omega_2 R\Omega_2^{-1}R^{-1} - I)m \end{aligned} \quad (265)$$

and

$$\beta = \begin{bmatrix} \frac{1}{2}R^{-1} & -R^{-1}m \\ 0 & 0 \end{bmatrix} \quad (266)$$

Another approach to develop the Souriau-Fisher metric $g_\beta\left([\beta, Z_1], [\beta, Z_2]\right)$ is to compute the tensor $\widetilde{\Theta}(X, Y)$ from the moment map J:

$$\widetilde{\Theta}(X, Y) = J_{[X,Y]} - \{J_X, J_Y\} \text{ with } \{.,.\} \text{ Poisson Bracket and } J \text{ the Moment Map} \quad (267)$$

$$\widetilde{\Theta}(X, Y) : \mathfrak{g} \times \mathfrak{g} \to \Re \quad (268)$$

We can then write the Souriau-Fisher metric as:

$$\widetilde{\Theta}_{\beta}(Z_1, Z_2) = J_{[Z_1,Z_2]} - \{J_{Z_1}, J_{Z_2}\} + \langle \hat{\xi}, [Z_1, Z_2] \rangle \tag{269}$$

Where the associated differentiable application J, called moment map is:

$$\begin{array}{ll} J : M \to g^* & \text{such that } J_X(x) = \langle J(x), X \rangle, \; X \in g \\ \quad x \mapsto J(x) & \end{array} \tag{270}$$

This moment map could be identified with the operator that transforms the right algebra to an element of its dual algebra given by:

$$\begin{array}{l} \beta_M : \mathfrak{g} \to \mathfrak{g}^* \\ Z = \begin{bmatrix} \mathrm{N} & \eta \\ 0 & 0 \end{bmatrix} \mapsto J = \begin{bmatrix} \mathrm{N}\left(1 + m^T R^{-1} m\right) + \eta m^T R^{-1} & \mathrm{N}R^{-1}m + R^{-1}\eta \\ 0 & 0 \end{bmatrix} \end{array} \tag{271}$$

10. Conclusions

In this paper, we have developed a Souriau model of Lie group thermodynamics that recovers the symmetry broken by lack of covariance of Gibbs density in classical statistical mechanics with respect to dynamic groups action in physics (Galileo and Poincaré groups, sub-group of affine group). The ontological model of Souriau gives geometric status to (Planck) temperature (element of Lie alebra), heat (element of dual Lie algebra) and entropy. Souriau said in one of his papers [30] on this new "Lie group thermodynamics" that "*these formulas are universal, in that they do not involve the symplectic manifold, but only group G, the symplectic cocycle. Perhaps this Lie group thermodynamics could be of interest for mathematics*".

For this new covariant thermodynamics, the fundamental notion is the coadjoint orbit that is linked to positive definite KKS (Kostant–Kirillov–Souriau) 2-form [196]:

$$\omega_w(X, Y) = \langle w, [U, V] \rangle \text{ with } X = ad_w U \in T_w\mathrm{M} \text{ and } Y = ad_w V \in T_w\mathrm{M} \tag{272}$$

that is the Kähler-form of a G-invariant kähler structure compatible with the canonical complex structure of M, and determines a canonical symplectic structure on M. When the cocycle is equal to zero, the KKS and Souriau-Fisher metric are equal. This 2-form introduced by Jean-Marie Souriau is linked to the coadjoint action and the coadjoint orbits of the group on its moment space. Souriau provided a classification of the homogeneous symplectic manifolds with this moment map. The coadjoint representation of a Lie group G is the dual of the adjoint representation. If $\mathfrak{g}$ denotes the Lie algebra of G, the corresponding action of G on $\mathfrak{g}^*$, the dual space to $\mathfrak{g}$, is called the coadjoint action. Souriau proved based on the moment map that a symplectic manifold is always a coadjoint orbit, affine of its group of Hamiltonian transformations, deducing that coadjoint orbits are the universal models of symplectic manifolds: a symplectic manifold homogeneous under the action of a Lie group, is isomorphic, up to a covering, to a coadjoint orbit. So the link between Souriau-Fisher metric and KKS 2-form will provide a symplectic structure and foundation to information manifolds. For Souriau thermodynamics, the Souriau-Fisher metric is the canonical structure linked to KKS 2-form, modified by the cocycle (its symplectic leaves are the orbits of the affine action that makes equivariant the moment map). This last property allows us to determine all homogeneous spaces of a Lie group admitting an invariant symplectic structure by the action of this group: for example, there are the orbits of the coadjoint representation of this group or of a central extension of this group (the central extension allowing suppressing the cocycle). For affine coadjoint orbits, we make reference to Alice Tumpach Ph.D. [197–199] who has developed previous works of Neeb [200], Biquard and Gauduchon [201–204].

Other promising domains of research are theory of generating maps [205–208] and the link with Poisson geometry through affine Poisson group. As observed by Pierre Dazord [209] in his paper *"Groupe de Poisson Affines"*, the extension of a Poisson group to an affine Poisson group due to Drinfel'd [210] includes the affine structures of Souriau on dual Lie algebra. For an affine Poisson group, its universal covering could be identified to a vector space with an associated affine structure. If this vector space is an abelian affine Poisson group, we can find the affine structure of Souriau. For the abelian group $(R^3,+)$, affine Poisson groups are the affine structures of Souriau.

Souriau model of Lie group thermodynamics could be a promising way to achieve René Thom's dream to replace thermodynamics by geometry [211,212], and could be extended to the second order extension of the Gibbs state [213,214].

We could explore the links between *"stochastic mechanics"* (*mécanique alétoire*) developed by Jean-Michel Bismut based on Malliavin Calculus (stochastic calculus of variations) and Souriau "Lie group thermodynamics", especially to extend covariant Souriau Gibbs density on the stochastic symplectic manifold (e.g., to model centrifuge with random vibrating axe and the Gibbs density).

We have seen that Souriau has replaced classical Maximum Entropy approach by replacing Lagrange parameters by only one geometric "temperature vector" as element of Lie algebra. In parallel, as refered in [15], Ingarden has introduced [213,214] second and higher order temperature of the Gibbs state that could be extended to Souriau theory of thermodynamics. Ingarden higher order temperatures could be defined in the case when no variational is considered, but when a probability distribution depending on more than one parameter. It has been observed that Ingarden can fail if the following assumptions are not fulfilled: the number of components of the sum goes to infinity and the components of the sum are stochastically independent. Gibbs hypothesis can also fail if stochastic interactions with the environment are not sufficiently weak. In all these cases, we never observe absolute thermal equilibrium of Gibbs type but only flows or turbulence. Nonequilibrium thermodynamics could be indirectly addressed by means of the concept of high order temperatures. Momentum $Q = \frac{\partial \Phi(\beta)}{\partial \beta}$ should be replaced by higher order moments given by the relation $Q_k = \frac{\partial \Phi(\beta_1,\ldots,\beta_n)}{\partial \beta_k} = \frac{\int\limits_M U^k(\xi)\cdot e^{-\sum\limits_{k=1}^{n}\langle \beta_k, U^k(\xi)\rangle} d\omega}{\int\limits_M e^{-\sum\limits_{k=1}^{n}\langle \beta_k, U^k(\xi)\rangle} d\omega}$ defined by extended Massieu characteristic function $\Phi(\beta_1,\ldots,\beta_n) = -\log \int\limits_M e^{-\sum\limits_{k=1}^{n}\langle \beta_k, U^k(\xi)\rangle} d\omega$. Entropy is defined by Legendre transform of this Massieu characteristic function $S(Q_1,\ldots,Q_n) = \sum\limits_{k=1}^{n}\langle \beta_k, Q_k\rangle - \Phi(\beta_1,\ldots,\beta_n)$ where $\beta_k = \frac{\partial S(Q_1,\ldots,Q_n)}{\partial Q_k}$. We are able also to define high order thermal capacities given by $K_k = -\frac{\partial Q_k}{\partial \beta_k}$. The Gibbs density could be then extended with respect to high order temperatures by $p_{Gibbs}(\xi) = e^{\sum\limits_{k=1}^{n}\langle \beta_k, U^k(\xi)\rangle - \Phi(\beta_1,\ldots,\beta_n)} = \frac{e^{-\sum\limits_{k=1}^{n}\langle \beta_k, U^k(\xi)\rangle}}{\int\limits_M e^{-\sum\limits_{k=1}^{n}\langle \beta_k, U^k(\xi)\rangle} d\omega}$.

We also have to make reference to the works of Streater [16], Nencka [215] and Burdet [216]. Nencka and Streater [215], for certain unitary representations of a Lie algebra $\mathfrak{g}$, define the statistical manifold $\mathcal{M}$ of states as the convex cone of $X \in \mathfrak{g}$ for which the partition function $Z = Tr\left[\exp(-X)\right]$ is finite. The Hessian of $\log Z$ defines a Riemannian metric g on dual Lie algebra $\mathfrak{g}^*$. They observe that $\mathfrak{g}^*$ foliates into the union of coadjoint orbits, each of which can be given a complex Kostant structure (that of Kostant).

To conclude, we will make reference to Alain Berthoz [217] at College de France who has studied brain coding of movement. The most recent studies on this topic, by Alexandre Afgoustidis Ph.D. [218] "*Invariant Harmonic Analysis and Geometry in the Workings of the Brain*" supervised by Daniel Bennequin, Afgoustidis [218] consolidate the idea that brain vestibular channels and otolithes code Lie algebra of the homogeneous Galileo group as illustrated in the following Figure 15.

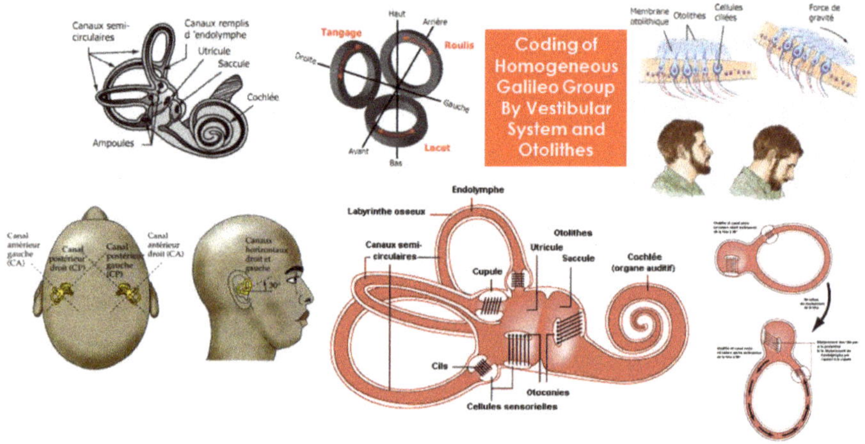

Figure 15. Coding of homogeneous Galileo algebra by vestibular system and otolithes.

Souriau gave the same ideas in this direction regarding how the brain could code invariants [219]:

Lorsque il y un tremblement de terre, nous assistons à la mort de l'Espace. ... Nous vivons avec nos habitudes que nous pensons universelles. ... La neuroscience s'occupe rarement de la géométrie ... Pour les singes qui vivent dans les arbres, certaines propriétés du groupe d'Euclide sont mieux câblées dans leurs cerveaux (When there is an earthquake, we are witnessing the death of Space ... We live with our habits that we think are universal.... Neuroscience rarely is interested in geometry ... For the monkeys that live in trees, some properties of the Euclid group are better coded in their brains).

Souriau added anecdotes from a discussion with a student of Bohr that [220]:

L'élève demanda à Bohr qu'il ne comprenait pas le principe de correspondance. Bohr lui demanda de s'assoir et il tourna autour de lui. Bohr lui dit tu dois commencer à avoir mal au cœur, c'est que tu commences à comprendre ce qu'est le principe de correspondance (The student said to Bohr that he *did not understand the principle of correspondence. Bohr asked him to sit and he turned around. Bohr said, you should start to be seasick, it is then that you begin to understand what the correspondence principle is.).*

Acknowledgments: I would like to thank Charles-Michel Marle and Gery de Saxcé for the fruitful discussions on Souriau model of statistical physics that help me to understand the fundamental notion of affine representation of Lie group and algebra, moment map and coadjoint orbits. I would also like to thank Michel Boyom that introduce me to Jean-Louis Koszul works on affine representation of Lie group and Lie algebra.

Si on ajoute que la critique qui accoutume l'esprit, surtout en matière de faits, à recevoir de simples probabilités pour des preuves, est, par cet endroit, moins propre à le former, que ne le doit être la géométrie qui lui fait contracter l'habitude de n'acquiescer qu'à l'évidence; nous répliquerons qu'à la rigueur on pourrait conclure de cette différence même, que la critique donne, au contraire, plus d'exercice à l'esprit que la géométrie: parce que l'évidence, qui est une et absolue, le fixe au premier aspect sans lui laisser ni la liberté de douter, ni le mérite de choisir; au lieu que les probabilités étant susceptibles du plus et du moins, il faut, pour se mettre en état de prendre un parti, les comparer ensemble, les discuter et les peser. Un genre d'étude qui rompt, pour ainsi dire, l'esprit à cette opération, est certainement d'un usage plus étendu que celui où tout est soumis à l'évidence; parce que les occasions de se déterminer sur des vraisemblances ou probabilités, sont plus fréquentes que celles qui exigent qu'on procède par démonstrations: pourquoi ne dirions –nous pas que souvent elles tiennent aussi à des objets beaucoup plus importants?

—Joseph de Maistre in L'Espit de Finesse [221]

Le cadavre qui s'acoutre se méconnait et imaginant l'éternité s'en approrie l'illusion ... C'est pourquoi j'abandonnerai ces frusques et jetant le masque de mes jours, je fuirai le temps où, de concert avec les autres, je m'éreinte à me trahir.

—Emile Cioran in Précis de decomposition [222]

Conflicts of Interest: The author declares no conflict of interest.

Appendix A. Clairaut(-Legendre) Equation of Maurice Fréchet Associated to "Distinguished Functions" as Fundamental Equation of Information Geometry

Before Rao [223,224], in 1943, Maurice Fréchet [141] wrote a seminal paper introducing what was then called the Cramer-Rao bound. This paper contains in fact much more that this important discovery. In particular, Maurice Fréchet introduces more general notions relative to "distinguished functions", densities with estimator reaching the bound, defined with a function, solution of Clairaut's equation. The solutions "envelope of the Clairaut's equation" are equivalent to standard Legendre transform without convexity constraints but only smoothness assumption. This Fréchet's analysis can be revisited on the basis of Jean-Louis Koszul's works as a seminal foundation of "information geometry".

We will use Maurice Fréchet notations, to consider the estimator:

$$T = H\left(X_1, ..., X_n\right) \tag{A1}$$

and the random variable

$$A(X) = \frac{\partial \log p_\theta(X)}{\partial \theta} \tag{A2}$$

that are associated to:

$$U = \sum_i A\left(X_i\right) \tag{A3}$$

The normalizing constraint $\int\limits_{-\infty}^{+\infty} p_\theta(x)dx = 1$ implies that: $\int\limits_{-\infty}^{+\infty} ... \int\limits_{-\infty}^{+\infty} \prod\limits_i p_\theta(x_i)dx_i = 1$

If we consider the derivative if this last expression with respect to θ, then

$$\int\limits_{-\infty}^{+\infty} ... \int\limits_{-\infty}^{+\infty} \left[\sum_i A(x_i)\right] \prod_i p_\theta(x_i)dx_i = 0 \text{ gives} : E_\theta\left[U\right] = 0 \tag{A4}$$

Similarly, if we assume that $E_\theta\left[T\right] = \theta$, then $\int\limits_{-\infty}^{+\infty} ... \int\limits_{-\infty}^{+\infty} H\left(x_1, ..., x_n\right) \prod\limits_i p_\theta(x_i)dx_i = \theta$, and we obtain by derivation with respect to θ:

$$E\left[\left(T - \theta\right) U\right] = 1 \tag{A5}$$

But as $E\left[T\right] = \theta$ and $E\left[U\right] = 0$, we immediately deduce that:

$$E\left[\left(T - E\left[T\right]\right)\left(U - E\left[U\right]\right)\right] = 1 \tag{A6}$$

From Schwarz inequality, we can develop the following relations:

$$\begin{gathered} \left[E\left(ZT\right)\right]^2 \leq E\left[Z^2\right] E\left[T^2\right] \\ 1 \leq E\left[\left(T - E\left[T\right]\right)^2\right] E\left[\left(U - E\left[U\right]\right)^2\right] = \left(\sigma_T \sigma_U\right)^2 \end{gathered} \tag{A7}$$

U being the summation of independent variables, Bienaymé equality could be applied:

$$(\sigma_U)^2 = \sum_i \left[\sigma_{A(X_i)}\right]^2 = n\,(\sigma_A)^2 \tag{A8}$$

From which, Fréchet deduced the bound, rediscovered by Cramer and Rao 2 years later:

$$(\sigma_T)^2 \geq \frac{1}{n\,(\sigma_A)^2} \tag{A9}$$

Fréchet [141] observed that it is a remarkable inequality where the second member is independent of the choice of the function H defining the "empirical value" T, where the first member can be taken to any empirical value $T = H\,(X_1, ..., X_n)$ subject to the unique condition $E_\theta\,[T] = \theta$ regardless is θ.

The classic condition that the Schwarz inequality becomes an equality helps us to determine when σ_T reaches its lower bound $\frac{1}{\sqrt{n}\sigma_n}$.

The previous inequality becomes an equality if there are two numbers α and β (not random and not both zero) such that $\alpha\,(H' - \theta) + \beta U = 0$, with H' being a particular function among eligible H such that we have an equality. This equality is rewritten $H' = \theta + \lambda' U$ with λ' being a non-random number.

If we use the previous equation, then:

$$E\left[(T - E\,[T])\,(U - E\,[U])\right] = 1 \Rightarrow E\left[(H' - \theta)\,U\right] = \lambda' E_\theta\left[U^2\right] = 1 \tag{A10}$$

We obtain:

$$U = \sum_i A\,(X_i) \Rightarrow \lambda' n E_\theta\left[A^2\right] = 1 \tag{A11}$$

From which we obtain λ' and the form of the associated estimator H':

$$\lambda' = \frac{1}{nE\,[A^2]} \Rightarrow H' = \theta + \frac{1}{nE\,[A^2]}\sum_i \frac{\partial \mathrm{log} p_\theta(X_i)}{\partial\theta} \tag{A12}$$

It is therefore deduced that the estimator that reaches the terminal is of the form:

$$H' = \theta + \frac{\sum_i \frac{\partial \mathrm{log} p_\theta(X_i)}{\partial\theta}}{n\int_{-\infty}^{+\infty}\left[\frac{\partial p_\theta(x)}{\partial\theta}\right]^2 \frac{dx}{p_\theta(x)}} \tag{A13}$$

with $E\,[H'] = \theta + \lambda' E\,[U] = \theta$.

H' would be one of the eligible functions, if H' would be independent of θ. Indeed, if we consider $E_{\theta_0}\,[H'] = \theta_0$, $E\left[(H' - \theta_0)^2\right] \leq E_{\theta_0}\left[(H - \theta_0)^2\right]$ $\forall H$ such that $E_{\theta_0}\,[H] = \theta_0$.

$H = \theta_0$ satisfies the equation and inequality shows that it is almost certainly equal to θ_0.

So to look for θ_0, we should know beforehand θ_0.

At this stage, Fréchet [141] looked for "distinguished functions" ("densités distinguées" in French), as any probability density $p_\theta(x)$ such that the function:

$$h(x) = \theta + \frac{\frac{\partial \mathrm{log} p_\theta(x)}{\partial\theta}}{\int_{-\infty}^{+\infty}\left[\frac{\partial p_\theta(x)}{\partial\theta}\right]^2 \frac{dx}{p_\theta(x)}} \tag{A14}$$

is independent of θ. The objective of Fréchet is then to determine the minimizing function $T = H'\,(X_1, ..., X_n)$ that reaches the bound. We can deduce from previous relations that:

$$\lambda(\theta)\frac{\partial \mathrm{log} p_\theta(x)}{\partial\theta} = h(x) - \theta \tag{A15}$$

But as $\lambda(\theta) > 0$, *we can consider* $\frac{1}{\lambda(\theta)}$ *as the second derivative of a function* $\Phi(\theta)$ such that:

$$\frac{\partial \log p_\theta(x)}{\partial \theta} = \frac{\partial^2 \Phi(\theta)}{\partial \theta^2}\left[h(x) - \theta\right] \tag{A16}$$

From which we deduce that:

$$\ell(x) = \log p_\theta(x) - \frac{\partial \Phi(\theta)}{\partial \theta}\left[h(x) - \theta\right] - \Phi(\theta) \tag{A17}$$

Is an independent quantity of θ. *A distinguished function* will be then given by:

$$p_\theta(x) = e^{\frac{\partial \Phi(\theta)}{\partial \theta}[h(x)-\theta]+\Phi(\theta)+\ell(x)} \tag{A18}$$

With the normalizing constraint $\int_{-\infty}^{+\infty} p_\theta(x)dx = 1$.

These two conditions are sufficient. Indeed, reciprocally, let three functions $\Phi(\theta)$, $h(x)$ and $\ell(x)$ that we have, for any

$$\theta : \int_{-\infty}^{+\infty} e^{\frac{\partial \Phi(\theta)}{\partial \theta}[h(x)-\theta]+\Phi(\theta)+\ell(x)} dx = 1 \tag{A19}$$

Then the function is distinguished:

$$\theta + \frac{\frac{\partial \log p_\theta(x)}{\partial \theta}}{\int_{-\infty}^{+\infty} \left[\frac{\partial p_\theta(x)}{\partial \theta}\right]^2 \frac{dx}{p_\theta(x)}} = \theta + \lambda(x)\frac{\partial^2 \Phi(\theta)}{\partial \theta^2}\left[h(x) - \theta\right] \tag{A20}$$

$$\text{If } \lambda(x)\frac{\partial^2 \Phi(\theta)}{\partial \theta^2} = 1, \text{ when } \frac{1}{\lambda(x)} = \int_{-\infty}^{+\infty} \left[\frac{\partial \log p_\theta(x)}{\partial \theta}\right]^2 p_\theta(x)dx = (\sigma_A)^2 \tag{A21}$$

The function is reduced to $h(x)$ and then is not dependent of θ.

We have then the following relation:

$$\frac{1}{\lambda(x)} = \int_{-\infty}^{+\infty} \left(\frac{\partial^2 \Phi(\theta)}{\partial \theta^2}\right)^2 \left[h(x) - \theta\right]^2 e^{\frac{\partial \Phi(\theta)}{\partial \theta}(h(x)-\theta)+\Phi(\theta)+\ell(x)} dx \tag{A22}$$

The relation is valid for any θ, we can derive prefious equation with respect with θ:

$$\int_{-\infty}^{+\infty} e^{\frac{\partial \Phi(\theta)}{\partial \theta}(h(x)-\theta)+\Phi(\theta)+\ell(x)} \left(\frac{\partial^2 \Phi(\theta)}{\partial \theta^2}\right)\left[h(x) - \theta\right] dx = 0 \tag{A23}$$

We can divide by $\frac{\partial^2 \Phi(\theta)}{\partial \theta^2}$ because it does not depend on x.

If we derive again with respect to θ, we will have:

$$\int_{-\infty}^{+\infty} e^{\frac{\partial \Phi(\theta)}{\partial \theta}(h(x)-\theta)+\Phi(\theta)+\ell(x)} \left(\frac{\partial^2 \Phi(\theta)}{\partial \theta^2}\right)\left[h(x) - \theta\right]^2 dx = \int_{-\infty}^{+\infty} e^{\frac{\partial \Phi(\theta)}{\partial \theta}(h(x)-\theta)+\Phi(\theta)+\ell(x)} dx = 1 \tag{A24}$$

Combining this relation with that of $\frac{1}{\lambda(x)}$, we can deduce that $\lambda(x)\frac{\partial^2\Phi(\theta)}{\partial\theta^2} = 1$ and as $\lambda(x) > 0$ then $\frac{\partial^2\Phi(\theta)}{\partial\theta^2} > 0$.

Fréchet emphasizes at this step [141], another way to approach the problem. We can select arbitrarily $h(x)$ and $l(x)$ and then $\Phi(\theta)$ is determined by:

$$\int_{-\infty}^{+\infty} e^{\frac{\partial\Phi(\theta)}{\partial\theta}[h(x)-\theta]+\Phi(\theta)+\ell(x)}dx = 1 \tag{A25}$$

That could be rewritten:

$$e^{\theta.\frac{\partial\Phi(\theta)}{\partial\theta}-\Phi(\theta)} = \int_{-\infty}^{+\infty} e^{\frac{\partial\Phi(\theta)}{\partial\theta}h(x)+\ell(x)}dx \tag{A26}$$

If we then fixed arbitrarily $h(x)$ and $l(x)$ and let s an arbitrary variable, the following function will be an explicit positive function given by $e^{\Psi(s)}$:

$$\int_{-\infty}^{+\infty} e^{s.h(x)+\ell(x)}dx = e^{\Psi(s)} \tag{A27}$$

Fréchet obtained finally the function $\Phi(\theta)$ *as solution of the equation* [141]:

$$\Phi(\theta) = \theta\cdot\frac{\partial\Phi(\theta)}{\partial\theta} - \Psi\left(\frac{\partial\Phi(\theta)}{\partial\theta}\right) \tag{A28}$$

Fréchet noted that this is the Alexis Clairaut equation [141].

The case $\frac{\partial\Phi(\theta)}{\partial\theta} = cste$ would reduce the density to a function that would be independent of θ, and so $\Phi(\theta)$ is given by a singular solution of this Clairaut equation, which is unique and could be computed by eliminating the variable s between:

$$\Phi = \theta\cdot s - \Psi(s) \text{ and } \theta = \frac{\partial\Psi(s)}{\partial s} \tag{A29}$$

Or between:

$$e^{\theta\cdot s-\Phi(\theta)} = \int_{-\infty}^{+\infty} e^{s\cdot h(x)+\ell(x)}dx \text{ and } \int_{-\infty}^{+\infty} e^{s\cdot h(x)+\ell(x)}\,[h(x)-\theta]\,dx = 0 \tag{A30}$$

$\Phi(\theta) = -\log\int_{-\infty}^{+\infty} e^{s\cdot h(x)+\ell(x)}dx + \theta\cdot s$ where s is given implicitly by $\int_{-\infty}^{+\infty} e^{s\cdot h(x)+\ell(x)}\,[h(x)-\theta]\,dx = 0$.

Then we know the distinguished function, H' among functions $H(X_1, ..., X_n)$ verifying $E_\theta[H] = \theta$ and such that σ_H reaches for each value of θ, an absolute minimum, equal to $\frac{1}{\sqrt{n}\sigma_A}$.

For the previous equation:

$$h(x) = \theta + \frac{\frac{\partial \log p_\theta(x)}{\partial\theta}}{\int_{-\infty}^{+\infty}\left[\frac{\partial p_\theta(x)}{\partial\theta}\right]^2\frac{dx}{p_\theta(x)}} \tag{A31}$$

We can rewrite the estimator as:

$$H'(X_1, ..., X_n) = \frac{1}{n}\left[h\left(X_1\right) + ... + h\left(X_n\right)\right] \tag{A32}$$

and compute the associated empirical value:

$$t = H'(x_1, ..., x_n) = \frac{1}{n}\sum_i h(x_i) = \theta + \lambda(\theta)\sum_i \frac{\partial \log p_\theta(x_i)}{\partial \theta}$$

If we take $\theta = t$, we have as $\lambda(\theta) > 0$:

$$\sum_i \frac{\partial \log p_t(x_i)}{\partial t} = 0 \tag{A33}$$

When $p_\theta(x)$ is a distinguished function, the empirical value t of θ corresponding to a sample $x_1, ..., x_n$ is a root of previous equation in t. This equation has a root and only one when X is a distinguished variable. Indeed, as we have:

$$p_\theta(x) = e^{\frac{\partial \Phi(\theta)}{\partial \theta}[h(x)-\theta]+\Phi(\theta)+\ell(x)} \tag{A34}$$

$$\sum_i \frac{\partial \log p_t(x_i)}{\partial t} = \frac{\partial^2 \Phi(t)}{\partial t^2}\left[\frac{\sum_i h(x_i)}{n} - t\right] \text{ with } \frac{\partial^2 \Phi(t)}{\partial t^2} > 0 \tag{A35}$$

We can then recover the unique root: $t = \frac{\sum_i h(x_i)}{n}$.

This function $T \equiv H'\left(X_1, ..., X_n\right) = \frac{1}{n}\sum_i h\left(X_i\right)$ can have an arbitrary form, that is a sum of functions of each only one of the quantities and it is even the arithmetic average of N values of a same auxiliary random variable $Y = h(X)$. The dispersion is given by:

$$\left(\sigma_{T_n}\right)^2 = \frac{1}{n\left(\sigma_A\right)^2} = \frac{1}{n\int\limits_{-\infty}^{+\infty}\left[\frac{\partial p_\theta(x)}{\partial \theta}\right]^2 \frac{dx}{p_\theta(x)}} = \frac{1}{n\frac{\partial^2 \Phi(\theta)}{\partial \theta^2}} \tag{A36}$$

and T_n follows the probability density:

$$p_\theta(t) = \sqrt{n}\frac{1}{\sigma_A\sqrt{2\pi}}e^{-\frac{n(t-\theta)^2}{2\cdot\sigma_A^2}} \text{ with } \left(\sigma_A\right)^2 = \frac{\partial^2 \Phi(\theta)}{\partial \theta^2} \tag{A37}$$

Clairaut Equation and Legendre Transform

We have just observed that Fréchet shows that distinguished functions depend on a function $\Phi(\theta)$, solution of the Clairaut equation:

$$\Phi(\theta) = \theta \cdot \frac{\partial \Phi(\theta)}{\partial \theta} - \Psi\left(\frac{\partial \Phi(\theta)}{\partial \theta}\right) \tag{A38}$$

Or given by the Legendre transform:

$$\Phi = \theta \cdot s - \Psi\left(s\right) \text{ and } \theta = \frac{\partial \Psi\left(s\right)}{\partial s} \tag{A39}$$

Fréchet also observed that this function $\Phi(\theta)$ could be rewritten:

$\Phi(\theta) = -\log \int\limits_{-\infty}^{+\infty} e^{s\cdot h(x)+\ell(x)}dx + \theta \cdot s$ where s is given implicitly by $\int\limits_{-\infty}^{+\infty} e^{s\cdot h(x)+\ell(x)}\left[h(x) - \theta\right]dx = 0$.

This equation is the fundamental equation of information geometry.

The "Legendre" transform was introduced by Adrien-Marie Legendre in 1787 [225] to solve a minimal surface problem Gaspard Monge in 1784. Using a result of Jean Baptiste Meusnier, a student of Monge, it solves the problem by a change of variable corresponding to the transform which now entitled with his name. Legendre wrote: *"I have just arrived by a change of variables that can be useful in other occasions."* About this transformation, Darboux [226] in his book gives an interpretation of Chasles: *"This comes after a comment by Mr. Chasles, to substitute its polar reciprocal on the surface compared to a paraboloïd."* The equation of Clairaut was introduced 40 years earlier in 1734 by Alexis Clairaut [225]. Solutions "envelope of the Clairaut equation" are equivalent to the Legendre transform with unconditional convexity, but only under differentiability constraint. Indeed, for a non-convex function, Legendre transformation is not defined where the Hessian of the function is canceled, so that the equation of Clairaut only makes the hypothesis of differentiability. The portion of the strictly convex function g in Clairaut equation $y = px - g(p)$ to the function f giving the envelope solutions by the formula $y = f(x)$ is precisely the Legendre transformation. The approach of Fréchet may be reconsidered in a more general context on the basis of the work of Jean-Louis Koszul.

Appendix B. Balian Gauge Model of Thermodynamics and its Compliance with Souriau Model

Supported by Industial group TOTAL (previously Elf-Aquitaine), Roger Balian has introduced a Gauge theory of thermodynamics [103] and has also developed information geometry in statistical physics and quantum physics [103,227–235]. Balian has observed that the entropy S (we use Balian notation, contrary with previous section where we use $-S$ as neg-entropy) can be regarded as an extensive variable $q^0 = S\left(q^1, ..., q^n\right)$, with $q^i(i = 1, ..., n)$, n independent quantities, usually extensive and conservative, characterizing the system. The n intensive variables γ_i are defined as the partial derivatives:

$$\gamma_i = \frac{\partial S(q^1, ..., q^n)}{\partial q^i} \tag{B1}$$

Balian has introduced a non-vanishing gauge variable p_0, without physical relevance, which multiplies all the intensive variables, defining a new set of variables:

$$p_i = -p_0 . \gamma_i \, , \, i = 1, ..., n \tag{B2}$$

The $2n$ + 1-dimensional space is thereby extended into a $2n$ + 2-dimensional thermodynamic space T spanned by the variables $p_i \, , q^i$ with $i = 0, 1, ..., n$, where the physical system is associated with a n + 1-dimensional manifold M in T, parameterized for instance by the coordinates $q^1, ..., q^n$ and p_0. A gauge transformation which changes the extra variable p_0 while keeping the ratios $p_i / p_0 = -\gamma_i$ invariant *is not observable*, so that a state of the system is represented by any point of a one-dimensional ray lying in M, along which the physical variables $q^0, ..., q^n, \gamma_1, ..., \gamma_n$ are fixed. Then, the relation between contact and canonical transformations is a direct outcome of this gauge invariance: the contact structure $\tilde{\omega} = dq^0 \quad \sum_{i=1}^{n} \gamma_i \cdot dq^i$ in n + 1 dimension can be embedded into a symplectic structure in $2n$ + 2 dimension, with 1-form:

$$\omega = \sum_{i=0}^{n} p_i \cdot dq^i \tag{B3}$$

as *symplectization*, with geometric interpretation in the theory of fiber bundles.

The n + 1-dimensional thermodynamic manifolds M are characterized by the vanishing of this form $\omega = 0$. The 1-form induces then *a symplectic structure on* T:

$$d\omega = \sum_{i=0}^{n} dp_i \wedge dq^i \tag{B4}$$

Any thermodynamic manifold M belongs to the set of the so-called Lagrangian manifolds in T, which are the integral submanifolds of $d\omega$ with maximum dimension $(n + 1)$. Moreover, M *is gauge invariant*, which is implied by $\omega = 0$. The extensivity of the entropy function $S\left(q^1, ..., q^n\right)$ is expressed by the Gibbs-Duhem relation $S = \sum_{i=1}^{n} q^i \frac{\partial S}{\partial q^i}$, rewritten with previous relation $\sum_{i=0}^{n} p_i q^i = 0$, defining a $2n +$ 1-dimensional extensivity sheet in T, where the thermodynamic manifolds M should lie. Considering an infinitesimal canonical transformation, generated by the Hamiltonian $h(q^0, q^1, ..., q^n, p_0, p_1, ..., p_n)$, $\dot{q}_i = \frac{\partial h}{\partial p_i}$ and $\dot{p}_i = \frac{\partial h}{\partial q^i}$, the Hamilton's equations are given by Poisson bracket:

$$\dot{g} = \{g, h\} = \sum_{i=0}^{n} \frac{\partial g}{\partial q^i} \frac{\partial h}{\partial p_i} - \frac{\partial h}{\partial q_i} \frac{\partial g}{\partial p_i} \tag{B5}$$

The concavity of the entropy $S\left(q^1, ..., q^n\right)$, as function of the extensive variables, expresses the stability of equilibrium states. This property produces constraints on the physical manifolds M in the $2n + 2$-dimensional space. It entails the existence of a metric structure in the n-dimensional space q_i relying on the quadratic form:

$$ds^2 = -d^2 S = -\sum_{i,j=1}^{n} \frac{\partial^2 S}{\partial q^i \partial q^j} dq^i dq^j \tag{B6}$$

which defines a distance between two neighboring thermodynamic states.

$$\text{As } d\gamma_i = \sum_{j=1}^{n} \frac{\partial^2 S}{\partial q^i \partial q^j} dq^j, \text{ then: } ds^2 = -\sum_{i=1}^{n} d\gamma_i dq_i = \frac{1}{p_0} \sum_{i=0}^{n} dp_i dq^i \tag{B7}$$

The factor $1/p_0$ *ensures gauge invariance.* In a continuous transformation generated by h, the metric evolves according to:

$$\frac{d}{d\tau}(ds^2) = \frac{1}{p_0} \frac{\partial h}{\partial q^0} ds^2 + \frac{1}{p_0} \sum_{i,j=0}^{n} \left(\frac{\partial^2 h}{\partial q^i \partial p_j} dp_i dp_j - \frac{\partial^2 h}{\partial q^i \partial q^j} dq^i dq^j \right) \tag{B8}$$

We can observe that *this gauge theory of thermodynamics is compatible with Souriau Lie groupTthermodynamics*, where we have to consider the Souriau vector $\beta = \begin{bmatrix} \gamma_1 \\ \vdots \\ \gamma_n \end{bmatrix}$, transformed in a new vector:

$$p_i = -p_0 . \gamma_i,\ p = \begin{bmatrix} -p_0 \gamma_1 \\ \vdots \\ -p_0 \gamma_n \end{bmatrix} = -p_0 \cdot \beta \tag{B9}$$

Appendix C. Casalis-Letac Affine Group Invariance for Natural Exponential Families

The characterization of the natural exponential families of R^{d} which are preserved by a group of affine transformations has been examined by Muriel Casalis in her Ph.D. [173] and her different papers [172,174–178]. Her method has consisted of translating the invariance property of the family into a property concerning the measures which generate it, and to characterize such measures.

Let E a vector space of finite size, E^* its dual. $\langle \theta, x \rangle$ duality bracket with $(\theta, x) \in E^* \times E$. μ positive Radon measure on E, Laplace transform is:

$$L_\mu : E^* \to [0, \infty] \text{ with } \theta \mapsto L_\mu(\theta) = \int_E e^{\langle \theta, x \rangle} \mu(dx) \tag{C1}$$

Let transformation $k_\mu(\theta)$ defined on $\Theta(u)$ interior of $D_\mu = \{\theta \in E^*, L_\mu < \infty\}$:

$$k_\mu(\theta) = \log L_\mu(\theta) \tag{C2}$$

natural exponential families are given by:

$$F(\mu) = \left\{P(\theta,\mu)(dx) = e^{\langle\theta,x\rangle - k_\mu(\theta)}\mu(dx), \theta \in \Theta(\mu)\right\} \tag{C3}$$

with injective function (domain of means):

$$k'_\mu(\theta) = \int_E xP(\theta,\mu)\,\mu(dx) \tag{C4}$$

the inverse function:

$$\psi_\mu : M_F \to \Theta(\mu) \text{ with } M_F = \operatorname{Im}\left(k'_\mu\left(\Theta(\mu)\right)\right) \tag{C5}$$

and the Covariance operator:

$$V_F(m) = k''_\mu\left(\psi_\mu(m)\right) = \left(\psi'_\mu(m)\right)^{-1}, \; m \in M_F \tag{C6}$$

Measure generetad by a family F is then given by:

$$F(\mu) = F(\mu') \Leftrightarrow \exists(a,b) \in E^* \times R, \text{such that } \mu'(dx) = e^{\langle a,x\rangle + b}\mu(dx) \tag{C7}$$

Let F an exponential family of E generated by μ and $\varphi : x \mapsto g_\varphi x + v_\varphi$ with $g_\varphi \in GL(E)$ automorphisms of E and $v_\varphi \in E$, then the family $\varphi(F) = \{\varphi(P(\theta,\mu)), \theta \in \Theta(\mu)\}$ is an exponential familly of E generated by $\varphi(\mu)$

Definition C1. *An exponential family F is invariant by a group G (affine group of E), if*

$$\forall\varphi \in G, \varphi(F) = F : \forall\mu, F(\varphi(\mu)) = F(\mu) \tag{C8}$$

(the contrary could be false)

Then Muriel Casalis has established the following theorem:

Theorem C1 (Casalis). *Let $F = F(\mu)$ an exponential family of E and G affine group of E, then F is invariant by G if and only:*

$$\begin{gathered}
\exists a : G \to E^*, \exists b : G \to R, \text{ such that:} \\
\forall(\varphi,\varphi') \in G^2, \begin{cases} a(\varphi\varphi') = {}^t g_\varphi^{-1} a(\varphi') + a(\varphi) \\ b(\varphi\varphi') = b(\varphi) + b(\varphi') - \left\langle a(\varphi'), g_\varphi^{-1} v_\varphi \right\rangle \end{cases} \\
\forall\varphi \in G, \varphi(\mu)(dx) = e^{\langle a(\varphi),x\rangle + b(\varphi)}\mu(dx)
\end{gathered} \tag{C9}$$

When G is a linear subgroup, b is a character of G and a could be obtained by the help of cohomology of Lie groups.

If we define action of G on E^* by:

$$g \cdot x = {}^t g^{-1} x, g \in G, x \in E^* \tag{C10}$$

It can be verified that:

$$a(g_1 g_2) = g_1 \cdot a(g_2) + a(g_1) \tag{C11}$$

the action a is an inhomogeneous 1-cocycle:

$\forall n>0$, let the set of all functions from G^n to E^*, $\Im\left(G^n,E^*\right)$ called inhomogenesous n-cochains, then we can define the operators $d^n:\Im\left(G^n,E^*\right)\to\Im\left(G^{n+1},E^*\right)$ by:

$$\begin{aligned} d^nF\left(g_1,\cdots,g_{n+1}\right) &= g_1.F\left(g_2,\cdots,g_{n+1}\right)+\sum_{i=1}^{n}(-1)^iF\left(g_1,g_2,\cdots,g_ig_{i+1},\cdots,g_n\right) \\ &+(-1)^{n+1}F\left(g_1,g_2,\cdots,g_n\right) \end{aligned} \tag{C12}$$

Let $Z^n\left(G,E^*\right)=Ker\left(d^n\right),B\left(G,E^*\right)=Im\left(d^{n-1}\right)$, with Z^n inhomogneous n-cocycles, the quotient:

$$H^n\left(G,E^*\right)=Z^n\left(G,E^*\right)/B^n\left(G,E^*\right) \tag{C13}$$

is the Cohomology group of G with value in E^*. We have:

$$\begin{aligned} d^0:E^*&\to\Im\left(G,E^*\right) \\ x&\mapsto\left(g\mapsto g\cdot x-x\right) \end{aligned} \tag{C14}$$

$$Z^0=\left\{x\in E^*;g\cdot x=x,\forall g\in G\right\} \tag{C15}$$

$$\begin{aligned} &d^1:\Im\left(G,E^*\right)\to\Im\left(G^2,E^*\right) \\ &F\mapsto d^1F\ ,\ d^1F\left(g_1,g_2\right)=g_1\cdot F(g_2)-F\left(g_1g_2\right)+F(g_1) \end{aligned} \tag{C16}$$

$$Z^1=\left\{F\in\Im\left(G,E^*\right);F\left(g_1g_2\right)=g_1\cdot F(g_2)+F(g_1),\forall\left(g_1,g_2\right)\in G^2\right\} \tag{C17}$$

$$B^1=\left\{F\in\Im\left(G,E^*\right);\exists x\in E^*,F(g)=g\cdot x-x\right\} \tag{C18}$$

When the Cohomology group $H^1\left(G,E^*\right)=0$ then:

$$Z^1\left(G,E^*\right)=B^1\left(G,E^*\right) \tag{C19}$$

Then if $F=F(\mu)$ is an exponential family invariant by G, μ verifies:

$$\forall g\in G,g\left(\mu\right)\left(dx\right)=e^{\langle c,x\rangle-\langle c,g^{-1}x\rangle+b(g)}\mu(dx) \tag{C20}$$

$$\forall g\in G,g\left(e^{\langle c,x\rangle}\mu(dx)\right)=e^{b(g)}e^{\langle c,x\rangle}\mu(dx)\text{ with }\mu_0(dx)=e^{\langle c,x\rangle}\mu(dx) \tag{C21}$$

For all compact group, $H^1\left(G,E^*\right)=0$ and we can express a:

$$\begin{aligned} &A:G\to GA(E) \\ &g\mapsto A_g\ ,\ A_g(\theta)={}^tg^{-1}\theta+a(g) \end{aligned} \tag{C22}$$

$$\begin{gathered} \forall\left(g,g'\right)\in G^2,A_{gg'}=A_gA_{g'} \\ A(G)\text{ compact sub}-\text{group of }GA(E) \end{gathered} \tag{C23}$$

$$\exists\text{fixed point}\Rightarrow\forall g\in G,A_g(c)={}^tg^{-1}c+a(g)=c\Rightarrow a(g)=\left(I_d-{}^tg^{-1}\right)c \tag{C24}$$

References

1. Bernard, C. Introduction à l'Étude de la Médecine Expérimentale. Available online: http://classiques.uqac.ca/classiques/bernard_claude/intro_etude_medecine_exp/intro_medecine_exper.pdf (accessed on 17 October 2016).
2. Thom, R. *Logos et Théorie des Catastrophes*; Editions Patiño: Genève, Switzerland, 1988.

3. Barbaresco, F. Symplectic structure of information geometry: Fisher metric and Euler-Poincaré equation of souriau lie group thermodynamics. In *Geometric Science of Information, Second International Conference GSI 2015*; Nielsen, F., Barbaresco, F., Eds.; Springer: Berlin/Heidelberg, Germany, 2015; Volume 9389, pp. 529–540.
4. De Saxcé, G.; Vallée, C. *Galilean Mechanics and Thermodynamics of Continua*; Wiley-ISTE: London, UK, 2016.
5. Vallée, C. Relativistic thermodynamics of continua. *Int. J. Eng. Sci.* **1981**, *19*, 589–601. [CrossRef]
6. Vallée, C.; Lerintiu, C. Convex analysis and entropy calculation in statistical mechanics. *Proc. A Razmadze Math. Inst.* **2005**, *137*, 111–129.
7. Marle, C.M. From Tools in Symplectic and Poisson Geometry to J.-M. Souriau's Theories of Statistical Mechanics and Thermodynamics. *Entropy* **2016**, *18*, 370. [CrossRef]
8. De Saxcé, G. Link between lie group statistical mechanics and thermodynamics of continua. In *Special Issue MDPI Entropy "Differential Geometrical Theory of Statistics"*; MDPI: Basel, Switzerland, 2016; Volume 18, p. 254.
9. Barbaresco, F. Koszul information geometry and souriau geometric temperature/capacity of lie group thermodynamics. *Entropy* **2014**, *16*, 4521–4565. [CrossRef]
10. Souriau, J.M. *Structure des Systèmes Dynamiques*; Editions Jacques Gabay: Paris, France, 1970. (In French)
11. Souriau, J.M. Structure of Dynamical Systems, volume 149 of Progress in Mathematics. In *A Symplectic View of Physics*; Birkhäuser: Basel, Switzerland, 1997.
12. Nielsen, F.; Barbaresco, F. *Geometric Science of Information*; Springer: Berlin/Heidelberg, Germany, 2015.
13. Kosmann-Schwarzbach, Y. La géométrie de Poisson, création du XXième siècle. In *Siméon-Denis Poisson*; Ecole Polytechnique: Paris, France, 2013; pp. 129–172.
14. Bismut, J.M. *Mécanique Aléatoire*; Springer: Berlin/Heidelberg, Germany, 1981; Volume 866.
15. Casas-Vázquez, J.; Jou, D. Temperature in non-equilibrium states: A review of open problems and current proposals. *Rep. Prog. Phys.* **2003**, *66*, 1937–2023. [CrossRef]
16. Streater, R.F. The information manifold for relatively bounded potentials. *Tr. Mat. Inst. Steklova* **2000**, *228*, 217–235.
17. Arnold, V.I. Sur la géométrie différentielle des groupes de Lie de dimension infinie et ses applications à l'hydrodynamique des fluides parfaits. *Ann. Inst. Fourier* **1966**, *16*, 319–361. [CrossRef]
18. Arnold, V.I.; Givental, A.B. Symplectic geometry. In *Dynamical Systems IV: Symplectic Geometry and Its Applications, Encyclopaedia of Mathematical Sciences*; Arnol'd, V.I., Novikov, S.P., Eds.; Springer: Berlin, Germany, 1990; Volume 4, pp. 1–136.
19. Marle, C.M.; de Saxcé, G.; Vallée, C. L'oeuvre de Jean-Marie Souriau, Gazette de la SMF, Hommage à Jean-Marie Souriau. 2012. Published by SMF, Paris.
20. Patrick Iglesias, Itinéraire d'un Mathématicien: Un entretien Avec Jean-Marie Souriau, Le Journal de Maths des Elèves. Available online: http://www.lutecium.fr/jp-petit/science/gal_port/interview_Souriau.pdf (accessed on 27 October 2016). (In French)
21. Iglesias, P. *Symétries et Moment*; Hermann: Paris, France, 2000.
22. Kosmann-Schwarzbach, Y. *Groupes et Symmetries*; Ecole Polytechnique: Paris, France, 2006.
23. Kosmann-Schwarzbach, Y. En homage à Jean-Marie Souriau, quelques souvenirs. *Gazette des Mathématiciens* **2012**, *133*, 105–106.
24. Ghys, E. Actions localement libres du groupe affine. *Invent. Math.* **1985**, *82*, 479–526. [CrossRef]
25. Rais, M. La representation coadjointe du groupe affine. *Ann. Inst. Fourier* **1978**, *28*, 207–237. (In French) [CrossRef]
26. Souriau, J.M. Mécanique des états condensés de la matière. In Proceedings of the 1st International Seminar of Mechanics Federation of Grenoble, Grenoble, France, 19–21 May 1992. (In French)
27. Souriau, J.M. Géométrie de l'espace de phases. *Commun. Math. Phys.* **1966**, *374*, 1–30. (In French)
28. Souriau, J.M. Définition covariante des équilibres thermodynamiques. *Nuovo Cimento* **1966**, *1*, 203–216. (In French)
29. Souriau, J.M. *Mécanique Statistique, Groupes de Lie et Cosmologie*; Colloques Internationaux du CNRS Numéro 237: Paris, France, 1974; pp. 59–113. (In French)
30. Souriau, J.M. *Géométrie Symplectique et Physique Mathématique*; Éditions du C.N.R.S.: Paris, France, 1975. (In French)
31. Souriau, J.M. *Thermodynamique Relativiste des Fluides*; Centre de Physique Théorique: Marseille, France, 1977. (In French)

32. Souriau, J.M. *Interpretation Géometrique des Etatsquantiques*; Springer: Berlin/Heidelberg, Germany, 1977; Volume 570. (In French)
33. Souriau, J.M. Thermodynamique et géométrie. In *Differential Geometrical Methods in Mathematical Physics II*; Bleuler, K., Reetz, A., Petry, H.R., Eds.; Springer: Berlin/Heidelberg, Germany, 1978; pp. 369–397. (In French)
34. Souriau, J.M. *Dynamic Systems Structure*; Chapters 16–19; Unpublished work, 1980.
35. Souriau, J.M.; Iglesias, P. *Heat Cold and Geometry. Differential Geometry and Mathematical Physics, Mathematical Physics Studies Volume*; Springer: Amsterdam, The Netherlands, 1983; pp. 37–68.
36. Souriau, J.M. Mécanique classique et géométrie symplectique. CNRS Marseille. *Cent. Phys. Théor.* Report ref. CPT-84/PE-1695 1984. (In French)
37. Souriau, J.M. On Geometric Mechanics. *Discret. Cont. Dyn. Syst. J.* **2007**, *19*, 595–607. [CrossRef]
38. Laplace, P.S. Mémoire sur la probabilité des causes sur les évènements. In *Mémoires de Mathématique et de Physique*; De l'Imprimerie Royale: Paris, France, 1774. (In French)
39. Gibbs, J.W. Elementary principles in statistical mechanics. In *The Rational Foundation of Thermodynamics*; Scribner: New York, NY, USA, 1902.
40. Ruelle, D.P. *Thermodynamic Formalism*; Addison-Wesley: New York, NY, USA, 1978.
41. Ruelle, D.P. Extending the definition of entropy to nonequilibrium steady states. *Proc. Natl. Acad. Sci. USA* **2003**, *100*, 3054–3058. [CrossRef] [PubMed]
42. Jaynes, E.T. Information theory and statistical mechanics. *Phys. Rev.* **1957**, *106*, 620–630. [CrossRef]
43. Jaynes, E.T. Information theory and statistical mechanics II. *Phys. Rev.* **1957**, *108*, 171–190. [CrossRef]
44. Jaynes, E.T. Prior probabilities. *IEEE Trans. Syst. Sci. Cybern.* **1968**, *4*, 227–241. [CrossRef]
45. Jaynes, E.T. The well-posed problem. *Found. Phys.* **1973**, *3*, 477–493. [CrossRef]
46. Jaynes, E.T. Where do we stand on maximum entropy? In *The Maximum Entropy Formalism*; Levine, R.D., Tribus, M., Eds.; MIT Press: Cambridge, MA, USA, 1979; pp. 15–118.
47. Jaynes, E.T. The minimum entropy production principle. *Annu. Rev. Phys. Chem.* **1980**, *31*, 579–601. [CrossRef]
48. Jaynes, E.T. On the rationale of maximum entropy methods. *IEEE Proc.* **1982**, *70*, 939–952. [CrossRef]
49. Jaynes, E.T. *Papers on Probability, Statistics and Statistical Physics*; Reidel: Dordrecht, The Netherlands, 1982.
50. Ollivier, Y. Aspects de l'entropie en Mathématiques et en Physique (Théorie de l'information, Systèmes Dynamiques, Grandes Déviations, Irréversibilité). Available online: http://www.yann-ollivier.org/entropie/entropie.pdf (accessed on 7 August 2015). (In French)
51. Villani, C. (Ir)rréversibilité et Entropie. Available online: http://www.bourbaphy.fr/villani.pdf (accessed on 5 August 2015). (In French)
52. Godement, R. *Introduction à la Théorie des Groupes de Lie*; Springer: Berlin/Heidelberg, Germany, 2004.
53. Guichardet, A. *Cohomologie des Groups Topologiques et des Algèbres de Lie*; Cedic/Fernand Nathan: Paris, France, 1980.
54. Guichardet, A. La method des orbites: Historiques, principes, résultats. In *Leçons de Mathématiques D'aujourd'hui*; Cassini: Paris, France, 2010; Volume 4, pp. 33–59.
55. Guichardet, A. *Le Problème de Kepler, Histoire et Théorie*; Ecole Polytechnique: Paris, France, 2012.
56. Dubois, J.G.; Dufour, J.P. La théorie des catastrophes. V. Transformées de Legendre et thermodynamique. In *Annales de l'IHP Physique Théorique*; Institut Henri Poincaré: Paris, France, 1978; Volume 29, pp. 1–50.
57. Monge, G. *Sur le Calcul Intégral des Equations aux Différences Partielles*; Mémoires de l'Académie des Sciences: Paris, France, 1784; pp. 118–192. (In French)
58. Moreau, J.J. Fonctions convexes duales et points proximaux dans un espace hilbertien. *C. R. Acad. Sci.* **1962**, *255*, 2897–2899. (In French)
59. Plastino, A.; Plastino, A.R. On the Universality of thermodynamics' Legendre transform structure. *Phys. Lett. A* **1997**, *226*, 257–263. [CrossRef]
60. Friedrich, T. Die fisher-information und symplectische strukturen. *Math. Nachr.* **1991**, *153*, 273–296. (In German) [CrossRef]
61. Massieu, F. Sur les Fonctions caractéristiques des divers fluides. *C. R. Acad. Sci.* **1869**, *69*, 858–862. (In French)
62. Massieu, F. Addition au précédent Mémoire sur les Fonctions caractéristiques. *C. R. Acad. Sci.* **1869**, *69*, 1057–1061. (In French)

63. Massieu, F. *Exposé des Principes Fondamentaux de la Théorie Mécanique de la Chaleur (note Destinée à Servir D'introduction au Mémoire de L'auteur sur les Fonctions Caractéristiques des Divers Fluides et la Théorie des Vapeurs)*; Académie des Sciences: Paris, France, 1873; p. 31. (In French)
64. Massieu, F. *Thermodynamique: Mémoire sur les Fonctions Caractéristiques des Divers Fluides et sur la Théorie des Vapeurs*; Académie des Sciences: Paris, France, 1876; p. 92. (In French)
65. Massieu, F. Sur les Intégrales Algébriques des Problèmes de Mécanique. Suivie de Sur le Mode de Propagation des Ondes Planes et la Surface de L'onde Elémentaire dans les Cristaux Biréfringents à Deux Axes. Ph.D. Thesis, Faculté des Sciences de Paris, Paris, France, 1861.
66. Nivoit, E. Notice sur la vie et les Travaux de M. Massieu, Inspecteur Général des Mines. Available online: http://facultes19.ish-lyon.cnrs.fr/fiche.php?indice=1153 (accessed 27 October).
67. Gibbs, J.W. Graphical Methods in the Thermodynamics of Fluids. In *The Scientific Papers of J. Willard Gibbs*; Dover: New York, NY, USA, 1961.
68. Brillouin, L. *Science and Information Theory*; Academic Press: New York, NY, USA, 1956.
69. Brillouin, L. Maxwell's demon cannot operate: Information and entropy. *J. Appl. Phys.* **1951**, *22*, 334–337. [CrossRef]
70. Brillouin, L. Physical entropy and information. *J. Appl. Phys.* **1951**, *22*, 338–343. [CrossRef]
71. Brillouin, L. Negentropy principle of information. *J. Appl. Phys.* **1953**, *24*, 1152–1163. [CrossRef]
72. Duhem, P. Sur les équations générales de la thermodynamique. In *Annales scientifiques de l'École Normale Supérieure*; Ecole Normale Supérieure: Paris, France, 1891; Volume 8, pp. 231–266. (In French)
73. Duhem, P. Commentaire aux principes de la Thermodynamique—Première partie. *J. Math. Appl.* **1892**, *8*, 269–330. (In French)
74. Duhem, P. Commentaire aux principes de la Thermodynamique—Troisième partie. *J. Math. Appl.* **1894**, *10*, 207–286. (In French)
75. Duhem, P. Les théories de la chaleur. *Revue des deux Mondes* **1895**, *130*, 851–868.
76. Carathéodory, C. Untersuchungen über die Grundlagen der Thermodynamik (Examination of the foundations of thermodynamics). *Math. Ann.* **1909**, *67*, 355–386. [CrossRef]
77. Carnot, S. *Réflexions sur la Puissance Motrice du feu*; Dover: New York, NY, USA, 1960.
78. Clausius, R. *On the Mechanical Theory of Heat*; Browne, W.R., Translator; Macmillan: London, UK, 1879.
79. Darrigol, O. The Origins of the Entropy Concept. Available online: http://www.bourbaphy.fr/darrigol.pdf (accessed on 5 August 2015). (In French)
80. Gromov, M. In a Search for a Structure, Part 1: On Entropy. Available online: http://www.ihes.fr/~gromov/PDF/structre-serch-entropy-july5-2012.pdf (accessed on 6 August 2015).
81. Gromov, M. Six Lectures on Probability, Symmetry, Linearity. Available online: http://www.ihes.fr/~gromov/PDF/probability-huge-Lecture-Nov-2014.pdf (accessed on 6 August 2015).
82. Gromov, M. Metric Structures for Riemannian and Non-Riemannian Spaces (Modern Birkhäuser Classics), 3rd ed.Lafontaine, J., Pansu, P., Eds.; Birkhäuser: Basel, Switzerland, 2006.
83. Kozlov, V.V. Heat equilibrium by Gibbs and poincaré. *Dokl. RAN* **2002**, *382*, 602–606. (In French)
84. Poincaré, H. Sur les tentatives d'explication mécanique des principes de la thermodynamique. *C. R. Acad. Sci.* **1889**, *108*, 550–553.
85. Poincaré, H. *Thermodynamique, Cours de Physique Mathématique*. Available online: http://gallica.bnf.fr/ark:/12148/bpt6k2048983 (accessed on 24 October 2016). (In French)
86. Poincaré, H. *Calcul des Probabilités*; Gauthier-Villars: Paris, France, 1896. (In French)
87. Poincaré, H. Réflexions sur la théorie cinétique des gaz. *J. Phys. Theor. Appl.* **1906**, *5*, 369–403. [CrossRef]
88. Fourier, J. *Théorie Analytique de la Chaleur*; Chez Firmin Didot: Paris, France, 1822. (In French)
89. Clausius, R. *Théorie Mécanique de la Chaleur*; Lacroix: Paris, France, 1868. (In French)
90. Poisson, S.D. *Théorie Mathématique de la Chaleur*; Bachelier: Paris, France, 1835. (In French)
91. Kosmann-Schwarzbach, Y. *Siméon-Denis Poisson: Les Mathématiques au Service de la Science*; Ecole Polytechnique: Paris, France, 2013. (In French)
92. Smale, S. Topology and Mechanics. *Invent. Math.* **1970**, *10*, 305–331. [CrossRef]
93. Cushman, R.; Duistermaat, J.J. The quantum mechanical spherical pendulum. *Bull. Am. Math. Soc.* **1988**, *19*, 475–479. [CrossRef]
94. Guillemin, V.; Sternberg, S. The moment map and collective motion. *Ann. Phys.* **1980**, *1278*, 220–253. [CrossRef]

95. De Saxcé, G.; Vallée, C. Bargmann group, momentum tensor and Galilean invariance of Clausius-Duhem Inequality. *Int. J. Eng. Sci.* **2012**, *50*, 216–232. [CrossRef]
96. De Saxcé, G. Entropy and structure for the thermodynamic systems. In *Geometric Science of Information, Second International Conference GSI 2015 Proceedings*; Nielsen, F., Barbaresco, F., Eds.; Lecture Notes in Computer Science; Springer: Berlin/Heidelberg, Germany, 2015; Volume 9389, pp. 519–528.
97. Kapranov, M. Thermodynamics and the moment map. 2011; arXiv:1108.3472v1.
98. Pavlov, V.P.; Sergeev, V.M. Thermodynamics from the differential geometry standpoint. *Theor. Math. Phys.* **2008**, *157*, 1484–1490. [CrossRef]
99. Cartier, P.; DeWitt-Morette, C. *Functional Integration. Action and Symmetries*; Cambridge University Press: Cambridge, UK, 2004.
100. Libermann, P.; Marle, C.M. *Symplectic Geometry and Analytical Mechanics*; Springer Science & Business Media: Berlin/Heidelberg, Germany, 1987.
101. Lichnerowicz, A. Espaces homogènes Kähleriens. In *Colloque de Géométrie Différentielle*; CNRSP: Paris, France, 1953; pp. 171–184. (In French)
102. Lichnerowicz, A. *Représentation Coadjointe Quotient et Espaces Homogènes de Contact, cours du Collège de France*; Springer: Berlin/Heidelberg, Germany, 1986. (In French)
103. Balian, R.; Valentin, P. Hamiltonian structure of thermodynamics with gauge. *Eur. Phys. J. B* **2001**, *21*, 269–282. [CrossRef]
104. Marle, C.M. On Henri Poincaré's note "Sur une forme nouvelle des équations de la mécanique". *J. Geom. Symmetry Phys.* **2013**, *29*, 1–38.
105. Poincaré, H. Sur une forme nouvelle des équations de la Mécanique. *C. R. Acad. Sci.* **1901**, *7*, 369–371. (In French)
106. Sternberg, S. Symplectic homogeneous spaces. *Trans. Am. Math. Soc.* **1975**, *212*, 113–130. [CrossRef]
107. Bourguignon, J.P. *Calcul Variationnel*; Ecole Polytechnique: Paris, France, 2007. (In French)
108. Dedecker, P. A property of differential forms in the calculus of variations. *Pac. J. Math.* **1957**, *7*, 1545–1549. [CrossRef]
109. Marle, C.M. On mechanical systems with a Lie group as configuration space. In *Jean Leray '99 Conference Proceedings*; De Gosson, M., Ed.; Springer: Berlin/Heidelberg, Germany, 2003; pp. 183–203.
110. Marle, C.M. Symmetries of Hamiltonian systems on symplectic and poisson manifolds. In *Similarity and Symmetry Methods*; Springer: Berlin/Heidelberg, Germany, 2014; pp. 185–269.
111. Kirillov, A.A. Merits and demerits of the orbit method. *Bull. Am. Math. Soc.* **1999**, *36*, 433–488. [CrossRef]
112. Cartan, E. La structure des groupes de transformations continus et la théorie du trièdre mobile. *Bull. Sci. Math.* **1910**, *34*, 250–284. (In French)
113. Cartan, E. *Leçons sur les Invariants Intégraux*; Hermann: Paris, France, 1922. (In French)
114. Cartan, E. Les récentes généralisations de la notion d'espace. *Bull. Sci. Math.* **1924**, *48*, 294–320. (In French)
115. Cartan, E. *Le rôle de la Théorie des Groupes de Lie dans L'évolution de la Géométrie Modern*; C.R. Congrès International: Oslo, Norway, 1936; Volume 1, pp. 92–103. (In French)
116. Libermann, P. La géométrie différentielle d'Elie Cartan à Charles Ehresmann et André Lichnerowicz. In *Géométrie au XXe siècle, 1930-2000: Histoire et Horizons*; Hermann: Paris, France, 2005. (In French)
117. Koszul, J.L. Sur la forme hermitienne canonique des espaces homogènes complexes. *Can. J. Math.* **1955**, *7*, 562–576. (In French) [CrossRef]
118. Koszul, J.L. *Exposés sur les Espaces Homogènes Symétriques*; Publicação da Sociedade de Matematica de São Paulo: São Paulo, Brazil, 1959. (In French)
119. Koszul, J.L. Domaines bornées homogènes et orbites de groupes de transformations affines. *Bull. Soc. Math. Fr.* **1961**, *89*, 515–533. (In French)
120. Koszul, J.L. Ouverts convexes homogènes des espaces affines. *Math. Z.* **1962**, *79*, 254–259. (In French) [CrossRef]
121. Koszul, J.L. Variétés localement plates et convexité. *Osaka J. Math.* **1965**, *2*, 285–290. (In French)
122. Koszul, J.L. *Lectures on Groups of Transformations*; Tata Institute of Fundamental Research: Bombay, India, 1965.
123. Koszul, J.L. Déformations des variétés localement plates. *Ann. Inst. Fourier* **1968**, *18*, 103–114. (In French) [CrossRef]
124. Koszul, J.L. Trajectoires convexes de groupes affines unimodulaires. In *Essays on Topology and Related Topics*; Springer: Berlin, Germany, 1970; pp. 105–110.

125. Vinberg, E.B. The theory of homogeneous convex cones. *Trudy Moskovskogo Matematicheskogo Obshchestva* **1963**, *12*, 303–358.
126. Vinberg, E.B. Structure of the group of automorphisms of a homogeneous convex cone. *Trudy Moskovskogo Matematicheskogo Obshchestva* **1965**, *13*, 56–83. (In Russian)
127. Byande, P.M.; Ngakeu, F.; Boyom, M.N.; Wolak, R. KV-cohomology and differential geometry of affinely flat manifolds. Information geometry. *Afr. Diaspora J. Math.* **2012**, *14*, 197–226.
128. Byande, P.M. *Des Structures Affines à la Géométrie de L'information*; Omniscriptum: Saarbrücken, France, 2012.
129. Nguiffo Boyom, M. Sur les structures affines homotopes à zéro des groupes de Lie. *J. Differ. Geom.* **1990**, *31*, 859–911. (In French)
130. Nguiffo Boyom, M. Structures localement plates dans certaines variétés symplectiques. *Math. Scand.* **1995**, *76*, 61–84. (In French) [CrossRef]
131. Nguiffo Boyom, M. The cohomology of Koszul-Vinberg algebras. *Pac. J. Math.* **2006**, *225*, 119–153. [CrossRef]
132. Nguiffo Boyom, M. *Some Lagrangian Invariants of Symplectic Manifolds, Geometry and Topology of Manifolds*; Banach Center Institute of Mathematics, Polish Academy of Sciences: Warsaw, Poland, 2007; Volume 76, pp. 515–525.
133. Nguiffo Boyom, M. Métriques kählériennes affinement plates de certaines variétés symplectiques. I. *Proc. Lond. Math. Soc.* **1993**, *2*, 358–380. (In French) [CrossRef]
134. Nguiffo Boyom, M.; Byande, P.M. *KV Cohomology in Information Geometry Matrix Information Geometry*; Springer: Heidelberg, Germany, 2013; pp. 69–92.
135. Nguiffo Boyom, M. Transversally Hessian foliations and information geometry I. *Am. Inst. Phys. Proc.* **2014**, *1641*, 82–89.
136. Nguiffo Boyom, M.; Wolak, R. Transverse Hessian metrics information geometry MaxEnt 2014. *AIP. Conf. Proc. Am. Inst. Phys.* **2015**. [CrossRef]
137. Vey, J. *Sur une Notion D'hyperbolicité des Variables Localement Plates. Thèse de Troisième Cycle de Mathématiques Pures*; Faculté des Sciences de l'université de Grenoble: Grenoble, France, 1969. (In French)
138. Vey, J. Sur les Automorphismes affines des ouverts convexes saillants. Annali della scuola normale superiore di pisa. *Classe Sci.* **1970**, *24*, 641–665. (In French)
139. Barbaresco, F. Koszul information geometry and Souriau Lie group thermodynamics. In AIP Conference Proceedings, Proceedings of MaxEnt'14 Conference, Amboise, France, 21–26 September 2014.
140. Lesne, A. Shannon entropy: A rigorous notion at the crossroads between probability, information theory, dynamical systems and statistical physics. *Math. Struct. Comput. Sci.* **2014**, *24*, e240311. [CrossRef]
141. Fréchet, M.R. Sur l'extension de certaines évaluations statistiques au cas de petits échantillons. *Rev. Inst. Int. Stat.* **1943**, *11*, 182–205. (In French) [CrossRef]
142. Fréchet, M.R. Les espaces abstraits topologiquement affines. *Acta Math.* **1925**, *47*, 25–52. [CrossRef]
143. Fréchet, M.R. Les éléments aléatoires de nature quelconque dans un espace distancié. *Ann. Inst. Henri Poincaré* **1948**, *10*, 215–310.
144. Fréchet, M.R. Généralisations de la loi de probabilité de Laplace. *Ann. Inst. Henri Poincaré* **1951**, *12*, 1–29. (In French)
145. Shima, H. *The Geometry of Hessian Structures*; World Scientific: Singapore, 2007.
146. Shima, H. Geometry of Hessian Structures. In *Springer Lecture Notes in Computer Science*; Nielsen, F., Frederic, B., Eds.; Springer: Berlin/Heidelberg, Germany, 2013; Volume 8085, pp. 37–55.
147. Crouzeix, J.P. A relationship between the second derivatives of a convex function and of its conjugate. *Math. Program.* **1977**, *3*, 364–365. [CrossRef]
148. Hiriart-Urruty, J.B. A new set-valued second-order derivative for convex functions. In *Mathematics for Optimization*; Elsevier: Amsterdam, The Netherlands, 1986.
149. Bakhvalov, N.S. Memorial: Nikolai Nikolaevitch Chentsov. *Theory Probab. Appl.* **1994**, *38*, 506–515. [CrossRef]
150. Chentsov, N.N. *Statistical Decision Rules and Optimal Inference*; American Mathematical Society: Providence, RI, USA, 1982.
151. Berezin, F. Quantization in complex symmetric spaces. *Izv. Akad. Nauk SSSR Ser. Math.* **1975**, *9*, 363–402. [CrossRef]
152. Bhatia, R. *Positive Definite Matrices*; Princeton University Press: Princeton, NJ, USA, 2007.
153. Bhatia, R. The bipolar decomposition. *Linear Algebra Appl.* **2013**, *439*, 3031–3037. [CrossRef]

154. Bini, D.A.; Garoni, C.; Iannazzo, B.; Capizzano, S.S.; Sesana, D. Asymptotic Behaviour and Computation of Geometric-Like Means of Toeplitz Matrices, SLA14 Conference, Kalamata, Greece, September 2014; Available online: http://noether.math.uoa.gr/conferences/sla2014/sites/default/files/Iannazzo.pdf (accessed on 8–12 September 2014).
155. Bini, D.A.; Garoni, C.; Iannazzo, B.; Capizzano, S.S. Geometric means of toeplitz matrices by positive parametrizations. **2016**, in press.
156. Calvo, M.; Oller, J.M. An explicit solution of information geodesic equations for the multivariate normal model. *Stat. Decis.* **1991**, *9*, 119–138. [CrossRef]
157. Calvo, M.; Oller, J.M. A distance between multivariate normal distributions based in an embedding into the Siegel group. *J. Multivar. Anal. Arch.* **1990**, *35*, 223–242. [CrossRef]
158. Calvo, M.; Oller, J.M. A distance between elliptical distributions based in an embedding into the Siegel group. *J. Comput. Appl. Math.* **2002**, *145*, 319–334. [CrossRef]
159. Chevallier, E.; Barbaresco, F.; Angulo, J. Probability density estimation on the hyperbolic space applied to radar processing. In *Geometric Science of Information Proceedings*; Lecture Notes in Computer Science; Springer: Berlin/Heidelberg, Germany, 2015; Volume 9389, pp. 753–761.
160. Chevallier, E.; Forget, T.; Barbaresco, F.; Angulo, J. Kernel Density Estimation on the Siegel Space Applied to Radar Processing. Available online: https://hal-ensmp.archives-ouvertes.fr/hal-01344910/document (accessed on 24 October 2016).
161. Costa, S.I.R.; Santosa, S.A.; Strapasson, J.E. Fisher information distance: A geometrical reading. *Discret. Appl. Math.* **2015**, *197*, 59–69. [CrossRef]
162. Jeuris, B.; Vandebril, R.; Vandereycken, B. A survey and comparison of contemporary algorithms for computing the matrix geometric mean. *Electron. Trans. Numer. Anal.* **2012**, *39*, 379–402.
163. Jeuris, B. Riemannian Optimization for Averaging Positive Definite Matrices. Ph.D. Thesis, Katholieke Universiteit Leuven, Leuven, Belgium, 2015.
164. Jeuris, B.; Vandebril, R. *The Kähler Mean of Block-Toeplitz Matrices with Toeplitz Structured Blocks*; Department of Computer Science, KU Leuven: Leuven, Belgium, 2015.
165. Maliavin, P. Invariant or quasi-invariant probability measures for infinite dimensional groups, Part II: Unitarizing measures or Berezinian measures. *Jpn. J. Math.* **2008**, *3*, 19–47. [CrossRef]
166. Strapasson, J.E.; Porto, J.P.S.; Costa, S.I.R. On bounds for the Fisher-Rao distance between multivariate normal distributions. *AIP Conf. Proc.* **2015**, *1641*, 313–320.
167. Hua, L.K. *Harmonic Analysis of Functions of Several Complex Variables in the Classical Domains*; American Mathematical Society: Providence, RI, USA, 1963.
168. Siegel, C.L. Symplectic geometry. *Am. J. Math.* **1943**, *65*, 1–86. [CrossRef]
169. Yoshizawa, S.; Tanabe, K. Dual differential geometry associated with the Kullback-Leibler information on the Gaussian distributions and its 2-parameters deformations. *SUT J. Math.* **1999**, *35*, 113–137.
170. Skovgaard, L.T. *A Riemannian Geometry of the Multivariate Normal Model*; Technical Report for Stanford University: Stanford, CA, USA, April 1981.
171. Deza, M.M.; Deza, E. *Encyclopedia of Distances*, 3rd ed.; Springer: Berlin/Heidelberg, Germany, 2013; p. 242.
172. Casalis, M. Familles exponentielles naturelles invariantes par un groupe de translations. *C. R. Acad. Sci. Ser. I Math.* **1988**, *307*, 621–623. (In French)
173. Casalis, M. Familles Exponentielles Naturelles Invariantes par un Groupe. Ph.D. Thesis, Thèse de l'Université Paul Sabatier, Toulouse, France, 1990. (In French)
174. Casalis, M. Familles exponentielles naturelles sur rd invariantes par un groupe. *Int. Stat. Rev.* **1991**, *59*, 241–262. (In French) [CrossRef]
175. Casalis, M. Les familles exponentielles à variance quadratique homogène sont des lois de Wishart sur un cône symétrique. *C. R. Acad. Sci. Ser. I Math.* **1991**, *312*, 537–540. (In French)
176. Casalis, M.; Letac, G. Characterization of the Jørgensen set in generalized linear models. *Test* **1994**, *3*, 145–162. [CrossRef]
177. Casalis, M.; Letac, G. The Lukacs-Olkin-Rubin characterization of the Wishart distributions on symmetric cone. *Ann. Stat.* **1996**, *24*, 763–786. [CrossRef]
178. Casalis, M. The 2d + 4 simple quadratic natural exponential families on Rd. *Ann. Stat.* **1996**, *24*, 1828–1854.
179. Letac, G. A characterization of the Wishart exponential families by an invariance property. *J. Theor. Probab.* **1989**, *2*, 71–86. [CrossRef]

180. Letac, G. *Lectures on Natural Exponential Families and Their Variance Functions, Volume 50 of Monografias de Matematica (Mathematical Monographs)*; Instituto de Matematica Pura e Aplicada (IMPA): Rio de Janeiro, Brazil, 1992.
181. Letac, G. Les familles exponentielles statistiques invariantes par les groupes du Cône et du paraboloïde de revolution. In *Journal of Applied Probability, Volume 31, Studies in Applied Probability*; Takacs, L., Galambos, J., Gani, J., Eds.; Applied Probability Trust: Sheffield, UK, 1994; pp. 71–95.
182. Barndorff-Nielsen, O.E. Differential geometry and statistics: Some mathematical aspects. *Indian J. Math.* **1987**, *29*, 335–350.
183. Barndorff-Nielsen, O.E.; Jupp, P.E. Yokes and symplectic structures. *J. Stat. Plan Inference* **1997**, *63*, 133–146. [CrossRef]
184. Barndorff-Nielsen, O.E.; Jupp, P.E. Statistics, yokes and symplectic geometry. *Annales de la Faculté des sciences de Toulouse: Mathématiques* **1997**, *6*, 389–427. [CrossRef]
185. Barndorff-Nielsen, O.E. *Information and Exponential Families in Stattistical Theory*; Wiley: New York, NY, USA, 2014.
186. Jespersen, N.C.B. On the structure of transformation models. *Ann. Stat.* **1999**, *17*, 195–208.
187. Skovgaard, L.T. A Riemannian geometry of the multivariate normal model. *Scand. J. Stat.* **1984**, *11*, 211–223.
188. Han, M.; Park, F.C. DTI segmentation and fiber tracking using metrics on multivariate normal distributions. *J. Math. Imaging Vis.* **2014**, *49*, 317–334. [CrossRef]
189. Imai, T.; Takaesu, A.; Wakayama, M. Remarks on geodesics for multivariate normal models. *J. Math. Ind.* **2011**, *3*, 125–130.
190. Inoue, H. Group theoretical study on geodesics for the elliptical models. In *Geometric Science of Information Proceedings*; Lecture Notes in Computer Science; Springer: Berlin/Heidelberg, Germany, 2015; Volume 9389, pp. 605–614.
191. Pilté, M.; Barbaresco, F. Tracking quality monitoring based on information geometry and geodesic shooting. In Proceedings of the 17th International Radar Symposium (IRS), Krakow, Poland, 10–12 May 2016; pp. 1–6.
192. Eriksen, P.S. *(k, 1)* Exponential transformation models. *Scand. J. Stat.* **1984**, *11*, 129–145.
193. Eriksen, P. *Geodesics Connected with the Fisher Metric on the Multivariate Normal Manifold*; Technical Report 86-13; Institute of Electronic Systems, Aalborg University: Aalborg, Denmark, 1986.
194. Eriksen, P.S. Geodesics connected with the Fisher metric on the multivariate normal manifold. In Proceedings of the GST Workshop, Lancaster, UK, 28–31 October 1987.
195. Feragen, A.; Lauze, F.; Hauberg, S. Geodesic exponential kernels: When curvature and linearity conflict. In Proceedings of the IEEE Conference on Computer Vision and Pattern Recognition (CVPR), 8–10 June 2015; pp. 3032–3042.
196. Besse, A.L. *Einstein Manifolds, Ergebnisse der Mathematik und ihre Grenzgebiete*; Springer: Berlin/Heidelberg, Germany, 1986.
197. Tumpach, A.B. Infinite-dimensional hyperkähler manifolds associated with Hermitian-symmetric affine coadjoint orbits. *Ann. Inst. Fourier* **2009**, *59*, 167–197. [CrossRef]
198. Tumpach, A.B. Classification of infinite-dimensional Hermitian-symmetric affine coadjoint orbits. *Forum Math.* **2009**, *21*, 375–393. [CrossRef]
199. Tumpach, A.B. Variétés Kählériennes et Hyperkählériennes de Dimension Infinie. Ph.D. Thesis, Ecole Polytechnique, Paris, France, 26 July 2005.
200. Neeb, K.-H. Infinite-dimensional groups and their representations. In *Lie Theory*; Birkhäuser: Basel, Switzerland, 2004.
201. Gauduchon, P. Calabi's Extremal Kähler Metrics: An Elementary Introduction. Available online: germanio.math.unifi.it/wp-content/uploads/2015/03/dercalabi.pdf (accessed on 27 October 2016).
202. Biquard, O.; Gauduchon, P. Hyperkähler Metrics on Cotangent Bundles of Hermitian Symmetric Spaces. Available online: https://www.math.ens.fr/~biquard/aarhus96.pdf (accessed on 27 October 2016).
203. Biquard, O.; Gauduchon, P. La métrique hyperkählérienne des orbites coadjointes de type symétrique d'un groupe de Lie complexe semi-simple. *Comptes Rendus de l'Académie des Sciences* **1996**, *323*, 1259–1264. (In French)
204. Biquard, O.; Gauduchon, P. Géométrie hyperkählérienne des espaces hermitiens symétriques complexifiés. *Séminaire de Théorie Spectrale et Géométrie* **1998**, *16*, 127–173. [CrossRef]

205. Chaperon, M. *Jets, Transversalité, Singularités: Petite Introduction aux Grandes Idées de René Thom;* Kouneiher, J., Flament, D., Nabonnand, P., Szczeciniarz, J.-J., Eds.; Géométrie au Vingtième Siècle, Histoire et Horizons: Hermann, Paris, 2005; pp. 246–256.
206. Chaperon, M. Generating maps, invariant manifolds, conjugacy. *J. Geom. Phys.* **2015**, *87*, 76–85. [CrossRef]
207. Viterbo, C. Symplectic topology as the geometry of generating functions. *Math. Ann.* **1992**, *292*, 685–710. [CrossRef]
208. Viterbo, C. Generating functions, symplectic geometry and applications. In Proceedings of the International Congress of Mathematics, Zürich, Switzerland, 3–11 August 1994.
209. Dazord, P.; Weinstein, A. *Symplectic, Groupoids, and Integrable Systems;* Springer: Berlin/Heidelberg, Germany, 1991; pp. 99–128.
210. Drinfeld, V.G. Hamiltonian structures on Lie groups. *Sov. Math. Dokl.* **1983**, *27*, 68–71.
211. Thom, R. Une théorie dynamique de la Morphogenèse. In *Towards a Theoretical Biology I;* Waddington, C.H., Ed.; University of Edinburgh Press: Edinburgh, UK, 1966; pp. 52–166.
212. Thom, R. *Stabilité Structurelle et Morphogénèse*, 2nd ed.; Inter Editions: Paris, France, 1977.
213. Ingarden, R.S.; Nakagomi, T. The second order extension of the Gibbs state. *Open Syst. Inf. Dyn.* **1992**, *1*, 243–258. [CrossRef]
214. Ingarden, R.S.; Meller, J. Temperatures in Linguistics as a Model of Thermodynamics. *Open Syst. Inf. Dyn.* **1994**, *2*, 211–230. [CrossRef]
215. Nencka, H.; Streater, R.F. Information Geometry for some Lie algebras. *Infin. Dimens. Anal. Quantum Probab. Relat. Top.* **1999**, *2*, 441–460. [CrossRef]
216. Burdet, G.; Perrin, M.; Perroud, M. Generating functions for the affine symplectic group. *Comm. Math. Phys.* **1978**, *3*, 241–254. [CrossRef]
217. Berthoz, A. *Le Sens du Movement;* Odile Jacob Edirot: Paris, France, 1997. (In French)
218. Afgoustidis, A. Invariant Harmonic Analysis and Geometry in the Workings of the Brain. Available online: https://hal-univ-diderot.archives-ouvertes.fr/tel-01343703 (accessed on 17 October 2016).
219. Souriau, J.M. Innovaxiom—Interview of Jean-Marie Souriau. Available online: https://www.youtube.com/watch?v=Lb_TWYqBUS4 (accessed on 27 October 2016).
220. Souriau, J.M. Quantique ? Alors c'est Géométrique. Available online: http://www.ahm.msh-paris.fr/Video.aspx?domain=84fa1a68-95c0-4c74-aed7-06055edaca16&language=fr&metaDescriptionId=dd3bd275-8372-4130-976b-847c36156a83&mediatype=VideoWithShots (accessed on 27 October 2016).
221. Masseau, D. *Les marges des Lumières Françaises (1750–1789);* Dix-huitième Siècle Année: Paris, France, 2005; Volume 37, pp. 638–639. (In French)
222. Cioran, E. *Précis de Décomposition Poche;* Gallimard: Paris, France, 1977.
223. Rao, C.R. Information and the accuracy attainable in the estimation of statistical parameters. *Bull. Calcutta Math. Soc.* **1945**, *37*, 81–91.
224. Burbea, J.; Rao, C.R. Entropy differential metric, distance and divergence measures in probability spaces: A unified approach. *J. Multivar. Anal.* **1982**, *12*, 575–596. [CrossRef]
225. Legendre, A.M. *Mémoire Sur L'intégration de Quelques Equations aux Différences Partielles;* Mémoires de l'Académie des Sciences: Paris, France, 1787; pp. 309–351. (In French)
226. Darboux, G. *Leçons sur la Théorie Générale des Surfaces et les Applications Géométriques du Calcul Infinitésimal: Premiere Partie (Généralités, Coordonnées Curvilignes, Surface Minima);* Gauthier-Villars: Paris, France, 1887. (In French)
227. Balian, R.; Alhassid, Y.; Reinhardt, H. Dissipation in many-body systems: A geometric approach based on information theory. *Phys. Rep.* **1986**, *131*, 1–146. [CrossRef]
228. Balian, R.; Balazs, N. Equiprobability, inference and entropy in quantum theory. *Ann. Phys.* **1987**, *179*, 97–144. [CrossRef]
229. Balian, R. On the principles of quantum mechanics. *Am. J. Phys.* **1989**, *57*, 1019–1027. [CrossRef]
230. Balian, R. *From Microphysics to Macrophysics: Methods and Applications of Statistical Physics;* Springer: Heidelberg, Germany, 1991 & 1992; Volumes I and II.
231. Balian, R. Incomplete descriptions and relevant entropies. *Am. J. Phys.* **1999**, *67*, 1078–1090. [CrossRef]
232. Balian, R. Entropy, a protean concept. In *Poincaré Seminar 2003;* Dalibard, J., Duplantier, B., Rivasseau, V., Eds.; Birkhauser: Basel, Switzerland, 2004; pp. 119–144.

233. Balian, R. Information in statistical physics. In *Studies in History and Philosophy of Modern Physics, Part B*; Elsevier: Amsterdam, The Netherlands, 2005.
234. Balian, R. The entropy-based quantum metric. *Entropy* **2014**, *16*, 3878–3888. [CrossRef]
235. Balian, R. *François Massieu et les Potentiels Thermodynamiques, Évolution des Disciplines et Histoire des Découvertes*; Académie des Sciences: Avril, France, 2015.

entropy
MDPI

Article

Link between Lie Group Statistical Mechanics and Thermodynamics of Continua

Géry de Saxcé

Laboratoire de Mécanique de Lille, CNRS FRE 3723, Université de Lille 1, Villeneuve d'Ascq F59655, France; gery.desaxce@univ-lille1.fr; Tel.: +33-3-2033-7172

Academic Editors: Frédéric Barbaresco and Frank Nielsen
Received: 29 April 2016; Accepted: 7 July 2016; Published: 12 July 2016

Abstract: In this work, we consider the value of the momentum map of the symplectic mechanics as an affine tensor called momentum tensor. From this point of view, we analyze the underlying geometric structure of the theories of Lie group statistical mechanics and relativistic thermodynamics of continua, formulated by Souriau independently of each other. We bridge the gap between them in the classical Galilean context. These geometric structures of the thermodynamics are rich and we think they might be a source of inspiration for the geometric theory of information based on the concept of entropy.

Keywords: Lie groups; symplectic geometry; affine tensors; continuum thermodynamics; statistical mechanics

1. Introduction

In [1], Souriau proposes to revisit mechanics, emphasizing its affine nature. It is this viewpoint that we will adopt here, starting from a generalization of the concept of momentum under the form of an affine object [2]. Our starting point is closely related to Souriau's approach on the basis of two key ideas: a new definition of momenta and the crucial part played by the affine group of $\mathbb{R}^n$. This group proposes an intentionally poor geometrical structure. Indeed, this choice is guided by the fact that it contains both Galileo and Poincaré groups [3,4], which allows the simultaneous involvement of the Galilean and relativistic mechanics. In the follow-up, we shall detail only the applications to classical mechanics and thermodynamics.

A class of tensors corresponds to each group. The components of these tensors are transformed according to the action of the considered group. The standard tensors discussed in the literature are those of the linear group of $\mathbb{R}^n$. We will call them linear tensors. A fruitful standpoint consists of considering the class of the affine tensors, corresponding to the affine group [2,5]. This viewpoint is closely related to symplectic mechanics [3,4,6] in the sense that the values of the momentum map *are just* the components of the momentum tensors.

The present paper is structured as follows. In Section 2, we present briefly the affine tensors, starting with the most simple ones: the *points* of an affine space which are 1-contravariant and the real affine functions on this space or *affine forms* which are 1-covariant. As a subgroup G of the affine group of $\mathbb{R}^n$ naturally acts onto the affine tensors by restriction to G of their transformation law, we define the corresponding G-tensor. In Section 3, we use this framework, defining the momentum as a mixed 1-covariant and 1-contravariant affine tensor. If G is a Lie group, we demonstrate the important fact that its transformation law *is nothing other* than the coadjoint representation of G in the dual $\mathfrak{g}^*$ of its Lie algebra. In Section 4, we recall classical tools of symplectic mechanics around the concept of symplectic action and a momentum map. An important result called the Kirillov–Kostant–Souriau theorem reveals the orbit symplectic structure. In Section 5, we recall shortly the main concepts of the Lie group statistical mechanics proposed by Souriau in [3,4], using geometric tools. In Section 6, we present briefly the cornerstone results of the Galilean version of a thermodynamics of continua compatible with general relativity proposed by Souriau in [7,8] independently of his statistical mechanics. In Section 7, we reveal the link between the previous relativistic thermodynamics of continua and Lie group statistical mechanics in the classical Galilean context, working in seven steps.

2. Affine Tensors

Points of an affine space. Let $A\mathcal{T}$ be an affine space associated to a linear space $\mathcal{T}$ of finite dimension n. By the choice of an affine frame f composed of a basis of $\mathcal{T}$ and an origin $\boldsymbol{a}_0$, we can associate to each point $\boldsymbol{a}$ a set of n (affine) components V^i gathered in the n-column $V \in \mathbb{R}^n$. For a change of affine frames, the transformation law for the components of a point reads:

$$V = C + P\,V' \,, \tag{1}$$

which is an affine representation of the affine group of $\mathbb{R}^n$ denoted $\mathbb{A}ff(n)$. It is clearly different from the usual transformation law of vectors $V = P\,V'$.

Affine forms. The affine maps $\boldsymbol{\Psi}$ from $A\mathcal{T}$ into $\mathbb{R}$ are called affine forms and their set is denoted $A^*\mathcal{T}$. In an affine frame, $\boldsymbol{\Psi}$ is represented by an affine function Ψ from $\mathbb{R}^n$ into $\mathbb{R}$. Hence, it holds:

$$\boldsymbol{\Psi}(\boldsymbol{a}) = \Psi(V) = \chi + \Phi\,V \,,$$

where $\chi = \Psi(0) = \boldsymbol{\Psi}(\boldsymbol{a}_0)$ and $\Phi = lin(\Psi)$ is a n-row. We call $\Phi_1, \Phi_2, \cdots, \Phi_n, \chi$ the components of $\boldsymbol{\Psi}$ or, equivalently, the couple of χ and the row Φ collecting the Φ_α. The set $A^*\mathcal{T}$ is a linear space of dimension $(n+1)$ called the vector dual of $A\mathcal{T}$. If we change the affine frame, the components of an affine form are modified according to the induced action of $\mathbb{A}ff(n)$, that leads to, taking into account (1):

$$\chi' = \chi - \Phi\,P^{-1}C, \qquad \Phi' = \Phi\,P^{-1} \,, \tag{2}$$

which is a linear representation of $\mathbb{A}ff(n)$.

Affine tensors. We can generalize this construction and define an affine tensor as an object:

- that assigns a set of components to each affine frame f of an affine space $A\mathcal{T}$ of finite dimension n,
- with a transformation law, when changing of frames, which is an affine or a linear representation of $\mathbb{A}ff(n)$.

With this definition, the affine tensors are a natural generalization of the classical tensors that we shall call linear tensors, these last ones being trivial affine tensors for which the affine transformation $a = (C, P)$ acts through its linear part $P = lin(a)$. An affine tensor can be constructed as a map which is affine or linear with respect to each of its arguments. Similar to the linear tensors, the affine ones can be classified in three families: covariants, contravariant and mixed. The most simple affine tensors are the points which are 1-contravariant and the affine forms which are 1-covariant but we can construct more complex ones having a strong physical meaning: the *torsors* (proposed in [5]), the *co-torsors* and the *momenta* extensively detailled in [2]. For more details on the affine dual space, affine tensor product, affine wedge product and affine tangent bundles, the reader interested in this topic is referred to the so-called AV-differential geometry [9].

G-tensors. A subgroup G of $\mathbb{A}ff(n)$ naturally acts onto the affine tensors by restriction to G of their transformation law. Let F_G be a set of affine frames of which G is a transformation group. The elements of F_G are called G-frames. A G-tensor is an object:

- that assigns a set of components to each G-frame f,
- with a transformation law, when changing of frames, which is an affine or a linear representation of G.

For instance, if G is the group of Euclidean transformations, we recover the classical Euclidean tensors. Hence, each G-tensor can be identified with an orbit of G within the space of the tensor components.

3. Momentum as Affine Tensor

Let $\mathcal{M}$ be a differential manifold of dimension n and G a Lie subgroup of $\mathbb{A}ff(n)$. In the applications to physics, $\mathcal{M}$ will be for us typically the space-time and G a subgroup of $\mathbb{A}ff(n)$ with a physical meaning in the framework of classical mechanics (Galileo's group) or relativity (Poincaré's group). The points of the space-time $\mathcal{M}$ are events of which the coordinate X^0 is the time t and $X^i = x^i$ for i running from 1 to 3 gives the position.

The tangent space to $\mathcal{M}$ at the point $\boldsymbol{X}$ equipped with a structure of affine space is called the affine tangent space and is denoted $AT_{\boldsymbol{X}}\mathcal{M}$. Its elements are called tangent points at $\boldsymbol{X}$. The set of affine forms on the affine tangent space is denoted $A^*T_{\boldsymbol{X}}\mathcal{M}$. We call momentum a bilinear map $\boldsymbol{\mu}$:

$$\boldsymbol{\mu} : T_{\boldsymbol{X}}\mathcal{M} \times A^*T_{\boldsymbol{X}}\mathcal{M} \to \mathbb{R} : (\overrightarrow{\boldsymbol{V}}, \boldsymbol{\Psi}) \mapsto \boldsymbol{\mu}(\overrightarrow{\boldsymbol{V}}, \boldsymbol{\Psi})$$

It is a mixed 1-covariant and 1-contravariant affine tensor. Taking into account the bilinearity, it is represented in an affine frame f by:

$$\boldsymbol{\mu}(\overrightarrow{\boldsymbol{V}}, \boldsymbol{\Psi}) = (\chi\, K_\beta + \Phi_\alpha L^\alpha_\beta)\, V^\beta$$

where K_β and L^α_β are the components of $\boldsymbol{\mu}$ in the affine frame f or, equivalently, the couple $\mu = (K, L)$ of the row K collecting the K_β and the $n \times n$ matrix L of elements L^α_β. Owing to (2), the transformation law is given by the induced action of $\mathbb{A}ff(n)$:

$$K' = K\, P^{-1}, \qquad L' = (P\, L + C\, K)\, P^{-1} \tag{3}$$

If the action is restricted to the subgroup G, the momentum $\boldsymbol{\mu}$ is a G-tensor.

On the other hand, have a look to the Lie algebra $\mathfrak{g}$ of G, that is the set of infinitesimal generators $Z = da = (dC, dP)$ with $a \in G$. Let us identify the space of the momentum components $\mu = (K, L)$ to the dual $\mathfrak{g}^*$ of the Lie algebra thanks to the dual pairing:

$$\mu\, Z = \mu\, da = (K, L)\,(dC, dP) = K\, dC + Tr(L\, dP) \tag{4}$$

We know that the group acts on its Lie algebra by the adjoint representation:

$$Ad(a) : \mathfrak{g} \to \mathfrak{g} : Z' \mapsto Z = Ad(a)\, Z' = a\, Z'\, a^{-1}\,.$$

As G is a group of affine transformations, any infinitesimal generator Z is represented by:

$$\tilde{Z} = d\tilde{P} = d \begin{pmatrix} 1 & 0 \\ C & P \end{pmatrix} = \begin{pmatrix} 0 & 0 \\ dC & dP \end{pmatrix}.$$

Then $\tilde{Z} = \tilde{P}\, \tilde{Z}'\, \tilde{P}^{-1}$ leads to:

$$dC = P\,(dC' - dP'\, P^{-1}C), \qquad dP = P\, dP'\, P^{-1}\,. \tag{5}$$

This adjoint representation induces the coadjoint representation of G in $\mathfrak{g}^*$ defined by:

$$(Ad^*(a)\, \mu')\, Z = \mu'\,(Ad(a^{-1})\, Z)\,.$$

Taking into account (4), one finds that the coadjoint representation:

$$Ad^*(a) : \mathfrak{g}^* \to \mathfrak{g}^* : \mu' \mapsto \mu = Ad^*(a)\,\mu'$$

is given by:

$$K = K'\,P^{-1}, \qquad L = (P\,L' + C\,K')\,P^{-1}\,.$$

It is noteworthy to observe that the transformation law (3) of momenta *is nothing other* than the coadjoint representation!

However, this mathematical construction is not relevant for all considered physical applications and we need to extend it by considering a map θ from G into $\mathfrak{g}^*$ and a generalized transformation law:

$$\mu = a \cdot \mu' = Ad^*(a)\,\mu' + \theta(a)\,, \tag{6}$$

where θ eventually depends on an invariant of the orbit. It is an affine representation of G in $\mathfrak{g}^*$ (because we wish the momentum to be an affine tensor) provided:

$$\forall a, b \in G, \qquad \theta(ab) = \theta(a) + Ad^*(a)\,\theta(b) \tag{7}$$

Remark 1. *This action induces a structure of affine space on the set of momentum tensors. Let $\pi : \mathcal{F} \to \mathcal{M}$ be a G-principal bundle of affine frames with the free action $(a, f) \mapsto f' = a \cdot f$ on each fiber. Then we can build the associated G-principal bundle:*

$$\hat{\pi} : \mathfrak{g}^* \times \mathcal{F} \to (\mathfrak{g}^* \times \mathcal{F})/G : (\mu, f) \mapsto \boldsymbol{\mu} = orb(\mu, f)$$

for the free action:

$$(a, (\mu, f)) \mapsto (\mu', f') = a \cdot (\mu, f) = (a \cdot \mu, a \cdot f)$$

where the action on $\mathfrak{g}^*$ is (6). Clearly, the orbit $\boldsymbol{\mu} = orb(\mu, f)$ can be identified to the momentum G-tensor $\boldsymbol{\mu}$ of components μ in the G-frame f.

4. Symplectic Action and Momentum Map

Let $(\mathcal{N}, \omega)$ be a symplectic manifold [3,4,6,10]. A Lie group G smoothly left acting on $\mathcal{N}$ and preserving the symplectic form ω is said to be symplectic.The interior product of a vector $\overrightarrow{V}$ and a p-form ω is denoted $\iota(\overrightarrow{V})\,\omega$. A map $\psi : \mathcal{N} \to \mathfrak{g}^*$ such that:

$$\forall \eta \in \mathcal{N}, \quad \forall Z \in \mathfrak{g}, \qquad \iota(Z \cdot \eta)\,\omega = -d(\psi(\eta) Z)\,,$$

is called a momentum map of G. It is the quantity involved in Noether's theorem that claims ψ is constant on each leaf of $\mathcal{N}$. In [3] (Theorem 11.17, p. 109, or its English translation [4]), Souriau proved there exists a smooth map θ from G into $\mathfrak{g}^*$:

$$\theta(a) = \psi(a \cdot \eta) - Ad^*(a)\,\psi(\eta)\,, \tag{8}$$

which is a symplectic cocycle, that is a map $\theta : G \to \mathfrak{g}$ verifying the identity (7) and such that $(D\theta)(e)$ is a 2-form. An important result, called the Kirillov–Kostant–Souriau theorem, reveals the orbit symplectic structure [3] (Theorem 11.34, Pages 116–118). Let G be a Lie group and an orbit of the coadjoint representation $orb\,(\mu) \subset \mathfrak{g}^*$. Then the orbit $orb\,(\mu)$ is a symplectic manifold, G is a symplectic group and any $\mu \in \mathfrak{g}^*$ is its own momentum.

Remark 2. *Replacing η by $a^{-1} \cdot \eta$ in (8), this formula reads:*

$$\psi(\eta) = Ad^*(a)\,\psi'(\eta) + \theta(a)\,,$$

where $\psi \mapsto \psi' = a \cdot \psi$ *is the induced action of the one of G on* $\mathcal{N}$*. It is worth observing it is just (6) with* $\mu = \psi(\eta)$ *and* $\mu' = \psi'(\eta)$*. In this sense, the values of the momentum map are just the components of the momentum G-tensors defined in the previous Section.*

Remark 3. *We saw at Remark of Section 3 that the momentum G-tensor* $\boldsymbol{\mu}$ *is identified to the orbit* $\boldsymbol{\mu} = orb(\mu, f)$ *and, disregarding the frames for simplification, we can identify* $\boldsymbol{\mu}$ *to the component orbit* $orb(\mu)$*.*

5. Lie Group Statistical Mechanics

In order to discover the underlined geometric structure of the statistical mechanics, we are interested in the affine maps $\boldsymbol{\Theta}$ on the affine space of momentum tensors. In an affine frame, $\boldsymbol{\Theta}$ is represented by an affine function Θ from $\mathfrak{g}^*$ into $\mathbb{R}$:

$$\boldsymbol{\Theta}(\boldsymbol{\mu}) = \Theta(\mu) = z + \mu Z \,,$$

where $z = \Theta(0) = \boldsymbol{\Theta}(\boldsymbol{\mu}_0)$ and $Z = lin(\Theta) \in \mathfrak{g}$ are the affine components of $\boldsymbol{\Theta}$. If the components of the momentum tensors are modified according to (6), the change of affine components of $\boldsymbol{\Theta}$ is given by the induced action:

$$z = z' - \theta(a)\, Ad(a)\, Z', \qquad Z = Ad(a)\, Z' \,. \tag{9}$$

Then $\boldsymbol{\Theta}$ is a G-tensors. In [3,4], Souriau proposed a statistical mechanics model using geometric tools. Let $d\lambda$ be a measure on $\boldsymbol{\mu} = orb\,(\mu)$ and a Gibbs probability measure $p\, d\lambda$ with:

$$p = e^{-\Theta(\mu)} = e^{-(z+\mu\, Z)} \,.$$

The normalization condition $\int_{orb(\mu)} p\, d\lambda = 1$ links the components of $\boldsymbol{\Theta}$, allowing to express z in terms of Z:

$$z(Z) = \ln \int_{orb(\mu)} e^{-\mu\, Z}\, d\lambda \,. \tag{10}$$

The corresponding entropy and mean momenta are:

$$\begin{aligned} s(Z) &= -\int_{orb(\mu)} p\, \ln p\, d\lambda = z + M\, Z, \\ M(Z) &= \int_{orb(\mu)} \mu\, p\, d\lambda = -\frac{\partial z}{\partial Z}\,, \end{aligned} \tag{11}$$

satisfying the same transformation law as the one (6) of μ. Hence M are the components of a momentum tensor $\boldsymbol{M}$ which can be identified to the orbit $orb(M)$, that defines a map $\boldsymbol{\mu} \mapsto \boldsymbol{M}$, i.e., a correspondance between two orbits. This construction is formal and, for reasons of integrability, the integrals will be performed only on a subset of the orbit according to an heuristic way explained latter on.

People generally consider that the definition of the entropy is relevant for applications insofar as the number of particles in the system is very huge. For instance, the number of atoms contained in one mole is Avogadro's number equal to 6×10^{23}. It is worth noting that Vallée and Lerintiu proposed a generalization of the ideal gas law based on convex analysis and a definition of entropy which does not require the classical approximations (Stirling's Formula) [11].

6. Relativistic Thermodynamics of Continua

Independently of his statistical mechanics, Souriau proposed in [7,8] a thermodynamics of continua compatible with general relativity. Following in his footsteps, one can quote the works by Iglesias [12] and Vallée [13]. In his Ph.D thesis, Vallée studied the invariant form of constitutive laws in the context of special relativity where the gravitation effects are neglected. In [14], the author and Vallée proposed a Galilean version of this theory of which we recall the cornerstone results. For more details, the reader is referred to [2].

Galileo's group $\mathbb{GAL}$ is a subgroup of the affine group $\mathbb{A}ff(4)$, collecting the Galilean transformations, that is the affine transformations $dX' \mapsto dX = P\,dX' + C$ of $\mathbb{R}^4$ such that:

$$C = \begin{pmatrix} \tau_0 \\ k \end{pmatrix}, \quad P = \begin{pmatrix} 1 & 0 \\ u & R \end{pmatrix}, \tag{12}$$

where $u \in \mathbb{R}^3$ is a Galilean boost, $R \in \mathbb{SO}(3)$ is a rotation, $k \in \mathbb{R}^3$ is a spatial translation and $\tau_0 \in \mathbb{R}$ is a clock change. Hence, Galileo's group is a Lie group of dimension 10. The $\mathbb{GAL}$-tensors will also be called *Galilean tensors*.

$\mathcal{M}$ is the space-time equipped with a symmetric $\mathbb{GAL}$-connection ∇ representing the gravitation, the matter and its evolution is characterized by a line bundle $\pi_0 : \mathcal{M} \mapsto \mathcal{M}_0$. The trajectory of the particle $\mathbf{X}_0 \in \mathcal{M}_0$ is the corresponding fiber $\pi_0^{-1}(\mathbf{X}_0)$. In local charts, $\mathbf{X}_0$ is represented by $s' \in \mathbb{R}^3$ and its position x at time t is given by a map:

$$x = \varphi(t, s')\,. \tag{13}$$

The 4-velocity:

$$\vec{\boldsymbol{U}} = \frac{\overrightarrow{d\boldsymbol{X}}}{dt}\,,$$

is the tangent vector to the fiber parameterized by the time. In a local chart, it is represented by:

$$U = \begin{pmatrix} 1 \\ v \end{pmatrix}, \tag{14}$$

where v is the usual velocity. Conversely, φ can be obtained as the flow of the 4-velocity.

β being the reciprocal temperature, that is $1\,/\,k_B T$ where k_B is Boltzmann's constant and T the absolute temperature, there are five basic tensor fields defined on the space-time $\mathcal{M}$:

- the 4-flux of mass $\vec{\boldsymbol{N}} = \rho\,\vec{\boldsymbol{U}}$ where ρ is the density,
- the 4-flux of entropy $\vec{\boldsymbol{S}} = \rho\, s\,\vec{\boldsymbol{U}} = s\,\vec{\boldsymbol{N}}$ where s is the specific entropy,
- Planck's temperature vector $\vec{\boldsymbol{W}} = \beta\,\vec{\boldsymbol{U}}$,
- its gradient $\mathbf{f} = \nabla\,\vec{\boldsymbol{W}}$ called friction tensor,
- the momentum tensor of a continuum $\boldsymbol{T}$, a linear map from $T_{\mathbf{X}}\mathcal{M}$ into itself.

In local charts, they are respectively represented by two 4-columns N, W and two 4×4 matrices f and T. Then we proved in [14] the following result characterizing the reversible processes:

Theorem 1. *If Planck's potential ζ smoothly depends on s', W and $F = \partial x / \partial s'$ through right Cauchy strains:*

$$C = F^T F\,, \tag{15}$$

then:

$$T = U\,\Pi + \begin{pmatrix} 0 & 0 \\ -\sigma v & \sigma \end{pmatrix} \tag{16}$$

with

$$\Pi = -\rho\,\frac{\partial \zeta}{\partial W}, \qquad \sigma = -\frac{2\rho}{\beta}\,F\,\frac{\partial \zeta}{\partial C}\,F^T\,, \tag{17}$$

represents the momentum tensor of the continuum and is such that:

$$(\nabla \zeta)\,N = -Tr\,(T\,f)\,,$$

Combining this result with the geometric version of the first principle of thermodynamics:

$$\boldsymbol{Div\, T = 0, \qquad Div\, \overrightarrow{N} = 0,}\ , \tag{18}$$

In [7,8], Souriau claimed that the 4-flux of entropy is given by:

$$\overrightarrow{S} = \boldsymbol{T}\,\overrightarrow{W} + \zeta\,\overrightarrow{N}\,, \tag{19}$$

and proved it is divergence free. Moreover the specific entropy s is an integral of the motion [2].

Let us introduce now the 5-temperature $\hat{\overrightarrow{W}}$ represented by the 5-column:

$$\hat{W} = \begin{pmatrix} W \\ \zeta \end{pmatrix}, \tag{20}$$

and the tensor $\hat{\boldsymbol{T}}$ represented by the 4×5 matrix

$$\hat{T} = \begin{pmatrix} T & N \end{pmatrix} \tag{21}$$

which allows gathering Equation (18) in the more compact form

$$Div\, \hat{T} = 0$$

and representing (19) in the more compact form:

$$S = \hat{T}\,\hat{W}\,,$$

local expression of the contracted product of $\hat{\mathbf{T}}$ and $\hat{\overrightarrow{\mathbf{W}}}$:

$$\vec{S} = \hat{\mathbf{T}} \cdot \hat{\vec{W}}\,, \tag{22}$$

It is the cornerstone equation of Souriau's theory. In this form, it can be seen as a geometrization of Clausius' definition of the entropy as state function of a system:

$$\mathcal{S} = \frac{Q}{\theta}\,, \tag{23}$$

where Q is a the amount of heat absorbed in an isothermal process. Scalar quantities are replaced by analogous tensorial ones: $\mathcal{S}$ by its 4-flux $\vec{S}$, Q by $\hat{\mathbf{T}}$ and $\beta = 1\ /\ \theta$ by its 5-flux $\hat{\vec{W}}$. Replacing (19) by (22) is not a purely formal manipulation but it takes a strong meaning when considering Bargmann's group $\mathbb{B}$ [15], a central extension of Galileo's one [16], set of the affine transformations $d\hat{X}' \mapsto d\hat{X} = \hat{P}\,d\hat{X}' + \hat{C}$ of $\mathbb{R}^5$ such that.

$$\hat{P} = \begin{pmatrix} 1 & 0 & 0 \\ u & R & 0 \\ \frac{1}{2} \parallel u \parallel^2 & u^T R & 1 \end{pmatrix}. \tag{24}$$

The $\mathbb{B}$-tensors are called Bargmannian tensors. From this viewpoint, the 5-column (20) represents a Bargmannian vector $\hat{\vec{W}}$ of transformation law:

$$\hat{W} = \hat{P}\,\hat{W}'\,, \tag{25}$$

and the 4×5 matrix (21) represents a Bargmannian 1-covariant and 1-contravariant tensor $\hat{T}$ of transformation law:

$$\hat{T} = P\,\hat{T}'\,\hat{P}^{-1}\,.$$

7. Planck's Potential of a Continuum

Now, let us reveal the link between the previous relativistic thermodynamics of continua and Lie group statistical mechanics in the classical Galilean context and, to simplify, in absence of gravitation. In other words, how to deduce $\boldsymbol{T}$ from $\boldsymbol{M}$ and ζ from z? We work in seven steps:

- *Step 1: defining the orbit.* To begin with, we consider the momentum as an Galilean tensor, i.e., its components ar modified only by the action of Galilean transformations. In order to calculate the integral (10), the orbit is parameterized thanks to a momentum map. Calculating the infinitesimal generators $Z = (dC, dP)$ by differentiation of (12):

$$dC = \begin{pmatrix} d\tau_0 \\ dk \end{pmatrix}, \qquad dP = \begin{pmatrix} 0 & 0 \\ du & j(d\varpi) \end{pmatrix},$$

where $j(d\varpi)\, v = d\varpi \times v$, the dual pairing (4) reads:

$$\mu Z = l \cdot d\varpi - q \cdot du + p \cdot dk - e\, d\tau_0 \,. \tag{26}$$

The most general form of the action (6) itemizes in:

$$p = R\,p' + m\,u, \qquad q = R\,(q' - \tau_0\, p') + m\,(k - \tau_0\, u)\,, \tag{27}$$

$$l = R\,l' - u \times (R\,q') + k \times (R\,p') + m\,k \times u\,, \tag{28}$$

$$e = e' + u \cdot (R\,p') + \frac{1}{2} m \parallel u \parallel^2 \,. \tag{29}$$

where the orbit invariant m occuring in the symplectic cocycle θ is physically interpreted as the particle mass. In [3] (Theorem 11.34, p. 151), the cocycle of Galileo's group is derived from an explicit form of the symplectic form. An alternative method to obtain it using only the Lie group structure is proposed in [2] (Theorem 16.3, p. 329 and Theorem 17.4, p. 374).

Taking into account (3), the transformation law (6) of the Galilean momentum tensor $\boldsymbol{\mu}$ reads:

$$K = K'\,P^{-1} + K_m(C,P), \qquad L = (P\,L' + C\,K')\,P^{-1} + L_m(C,P)\,, \tag{30}$$

where K_m and L_m are the components of θ. In particular, one has:

$$K_m(C,P) = m\left(-\frac{1}{2} \parallel u \parallel^2, u^T\right) . \tag{31}$$

- *Step 2: representing the orbit by equations.* To obtain them, we have to determine a functional basis. The first step is to calculate their number. We start determining the isotropy group of μ. The analysis will be restricted to massive particles: $m \neq 0$. The components p, q, l, e being given, we have to solve the following system:

$$p = R\,p + m\,u\,, \tag{32}$$

$$q = R\,q - \tau_0 (R\,p + m\,u) + m\,k\,, \tag{33}$$

$$l = R\,l - u \times (R\,q) + k \times (R\,p) + m\,k \times u\,, \tag{34}$$

$$u \cdot (R\,p) + \frac{1}{2} m \parallel u \parallel^2 = 0\,, \tag{35}$$

with respect to τ_0, k, R, u. Owing to (32), the boost u can be expressed with respect to the rotation R by:

$$u = \frac{1}{m}\,(p - R\,p)\,, \tag{36}$$

that allows us to satisfy automatically (35). Next, owing to (32), Equation (33) can be simplified as follows:

$$q = R\,q - \tau_0 p + m\,k\,,$$

that allows to determine the spatial translation k with respect to R and the clock change τ_0:

$$k = \frac{1}{m}\,(q - R\,q + \tau_0 p). \tag{37}$$

Finally, because of (32), Equation (34) is simplified as follows:

$$l = R\,l - u \times (R\,q) + k \times p\,.$$

Substituting (37) into the last relation gives:

$$l = R\,l - u \times (R\,q) + \frac{1}{m}\,q \times p - \frac{1}{m}\,(R\,q) \times p\,.$$

Owing to (32) and the definition of the spin angular momentum l_0

$$l_0 = l - q \times p \;/\; m\,,$$

leads to:

$$l_0 = R\,l_0\,. \tag{38}$$

These quantity being given, we have to determine the rotations satisfying the previous relation. It turns out that two cases must be considered.

- *Generic orbits : massive particle with spin or rigid body.* If l_0 does not vanish, the solutions of (38) are the rotations of an arbitrary angle ϑ about the axis l_0. We know by (36) and (37) that u and k are determined in a unique manner with respect to R and τ_0. The isotropy group of μ can be parameterised by ϑ and τ_0. It is a Lie group of dimension 2. The dimension of the orbit of μ is $10 - 2 = 8$. The maximum number of independent invariant functions is $10 - 8 = 2$. A possible functional basis is composed of:

$$s_0 = \parallel l_0 \parallel\,, \tag{39}$$

$$e_0 = e - \frac{1}{2\,m}\,\parallel p \parallel^2\,, \tag{40}$$

  of which the values are constant on the orbit which represents a massive particle with spin or a rigid body (seen from a long way off).
- *Singular orbits : spinless massive particle.* In the particular case $l_0 = 0$, all the rotations of $\mathbb{SO}(3)$ satisfy (38), then the isotropy group is of dimension 4. By similar reasoning to the case of non vanishing l_0, we conclude that dimension of the orbit is 6 and the number of invariant functions is 4. A possible functional basis is composed of e_0 and the three null components of l_0.

For the orbits with $m = 0$, the reader is referred to [6] (pp. 440, 441).

To physically interpret the components of the momentum, let consider a coordinate system X' in which a particle is at rest and characterized by the components $p' = 0$, $q' = 0$, $l' = l_0$ and $e' = e_0$ of the momentum tensor. Let us consider another coordinate system $X = PX' + C$ with a Galilean boost v and a translation of the origin at $k = x_0$ (hence $\tau_0 = 0$ and $R = 1_{\mathbb{R}^3}$), providing the trajectory equation:

$$x = x_0 + v\,t\,, \tag{41}$$

of the particle moving in uniform straight motion at velocity v. Owing (27) and (28), we can determine the new components of the torsor in X:

$$p = m\,v, \quad q = m\,x_0, \quad l = l_0 + q \times v, \quad e = e_0 + \frac{m}{2} \parallel v \parallel^2, \tag{42}$$

The third relation of (42) is the classical *transport law of the angular momentum*. In fact, it is a particular case of the general transformation laws (28) when considering only a Galilean boost. The transformation law reveals the physical meaning of the momentum tensor components:

- The quantity p, proportional to the mass and to the velocity, is the *linear momentum*.
- The quantity q, proportional to the mass and to the initial position, provides the trajectory equation. It is called *passage* because indicating the particle is passing through x_0 at time $t = 0$.
- The quantity l splits into two terms. The second one, $q \times v \ = x \times m\,v \ = x \times p$, is the *orbital angular momentum*. The first one, $l_0 = l - \ q \times p \ / \ m$, is the *spin angular momentum*. Their sum, l, is the *angular momentum*.

- *Step 3: parameterizing the orbit.* If the particle has an internal structure, introducing the moment of inertia matrix $\mathcal{J}$ and the spin ϖ, we have, according to König's theorem:

$$l_0 = \mathcal{J}\,\varpi, \qquad e_0 = \frac{1}{2}\,\varpi \cdot (\mathcal{J}\,\varpi)\,.$$

Hence each orbit defines a particle of mass m, spin s_0, inertia $\mathcal{J}$ and can be parameterized by 8 coordinates, the 3 components of q, the 3 components of p and the 2 components of the unit vector n defining the spin direction, thanks to the momentum map $\mathbb{R}^3 \times \mathbb{R}^3 \times \mathbb{S}^2 \to \mathfrak{g}^* : (q, p, n) \mapsto \mu = \psi(q, p, n)$ such that:

$$l = \frac{1}{m}\,q \times p + s_0 n, \qquad e = \frac{1}{2\,m} \parallel p \parallel^2 + \frac{s_0^2}{2}\,n \cdot (\mathcal{J}^{-1}\,n)\,.$$

The corresponding measure is $d\lambda \ = \ d^3q\,d^3p\,d^2n$. For simplicity, we consider further only a singular orbit of dimension 6 representing a spinless particle of mass m, which corresponds to the particular case $l_0 = 0$ then $n = 0$. It can be parameterized by 6 coordinates, the 3 components of q and the 3 components of p thanks to the map:

$$\psi : \mathbb{R}^3 \times \mathbb{R}^3 \to \mathfrak{g}^* : (q, p) \mapsto \mu = \psi(q, p)\,,$$

such that:

$$l = \frac{1}{m}\,q \times p, \qquad e = \frac{1}{2\,m} \parallel p \parallel^2\,. \tag{43}$$

- *Step 4: modelling the deformation.* Statistical mechanics is essentially based on a set of discrete particles and, in essence, incompatible with continuum mechanics. Thus, according to usual arguments, the passage from the statistical mechanics to continuum mechanics is obtained by equivalence between the set of N particles (in huge number) and a box of finite volume V occupied by them, large with respect to the particle size but so small with respect to the continuous medium that it can be considered as infinitesimal. Let us consider N identical particles contained in V, large with respect to the particles but representing the volume element of the continuum thermodynamics. The motion of the matter being characterized by (13), let us consider the change of coordinate

$$t = t', \qquad x = \varphi(t', s')\,.$$

The jacobean matrix reads:

$$\frac{\partial X}{\partial X'} = P = \begin{pmatrix} 1 & 0 \\ v & F \end{pmatrix} . \tag{44}$$

From then on, the momentum is considered as an affine tensor, i.e., its components are modified by the action of any affine transformation.

Besides, we suppose that the box of initial volume V_0 is at rest in the considered coordinate system ($v = 0$) and the deformation gradient F is uniform in the box, then:

$$dx = F\, ds' .$$

According to (3), the linear momentum is transformed according to:

$$p = F^{-T} p' . \tag{45}$$

For a particle initially at position x, the passage is given by (42):

$$q = m\, x .$$

The measure becomes

$$d\lambda = m^3 d^3 x\, d^3 p\, d^2 n = m^3 d^3 s'\, d^3 p'\, d^2 n .$$

For reasons that will be justified at Step 5, we consider the infinitesimal generator:

$$Z = (-W, 0) .$$

As the box is at rest in the considered coordinate system, the velocity is null and, owing to (14):

$$W = \beta\, U = \begin{pmatrix} \beta \\ 0 \end{pmatrix} . \tag{46}$$

Hence the dual pairing (26) is reduced to:

$$\mu\, Z = \beta\, e ,$$

and, owing to (43), (45) and (15), for a spinless massive particle:

$$\mu\, Z = \frac{\beta}{2\, m} \parallel p \parallel^2 = \frac{\beta}{2\, m} \parallel F^{-T} p' \parallel^2 = \frac{\beta}{2\, m}\, p'^T C^{-1} p' .$$

For reasons of integrability as explained in Section 6, *it is usual to replace the orbit by the subset* $V_0 \times \mathbb{R}^3 \times \mathbb{S}^2 \subsetneq orb(\mu)$. It is worth remarking that, unlike the orbit, this set is not preserved by the action but the integrals in (10) and (11) are invariant. Equation (10) gives for a particle:

$$z = \ln(m^3 I_0 I_1 I_2) ,$$

where:

$$I_0 = \int_{V_0} d^3 s' = V_0 ,$$

$$I_1 = \int_{\mathbb{R}^3} e^{-\frac{\beta}{2m} p'^T C^{-1} p'} d^3 p' ,$$

$$I_2 = \int_{\mathbb{S}^2} d^2 n = 4\, \pi .$$

Finally:

$$z = \frac{1}{2}\ln(\det(C)) - \frac{3}{2}\ln\beta + C^{te}, \tag{47}$$

where the value of the constant is not relevant in the sequel since it does not depend on W and F (through C). It is worth remarking that, unlike $orb(\mu)$, the subset $V_0 \times \mathbb{R}^3 \times \mathbb{S}^2$ is not preserved by the action and depends on the arbitrary choice of V_0. Nevertheless, z—then s and M—depends on V_0 only through $\ln(V_0)$ which is absorbed in the constant and has no influence on the derivatives (17).

As pointed out by Barbaresco [17], there is a puzzling analogy between the integral occuring in (10) and Koszul–Vinberg characteristic function [18,19]:

$$\psi_\Omega(Z) = \int_{\Omega^*} e^{-\mu Z}\, d\lambda\,,$$

where Ω is a sharp open convex cone and Ω^* is the set of linear strictly positive forms on $\bar{\Omega} - \{0\}$. Considering Galileo's group, it is worth remarking that the cone of future directed timelike vectors (i.e., such that $\beta > 0$) [20] is preserved by linear Galilean transformations. The momentum orbits are contained in Ω^* but the integral does not converge on the orbits or on Ω^*.

- *Step 5: identification.* It is based on the following result.

Theorem 2. *The transformation law of the temperature vector $\hat{\vec{W}}$ is the same as the one of affine maps $\boldsymbol{\Theta}$ on the affine space of momentum tensors through the identification:*

$$Z = (-W, 0), \qquad z = m\,\zeta\,,$$

Proof. First of all, let us verify that the form $Z = (-W, 0)$ does not depend on the choice of the affine frame. Indeed, starting from $Z' = (-W', 0)$ and applying the adjoint representation (5) with $dC' = -W'$ and $dP' = 0$, we find that $dC = -W$ and $dP = 0$ with:

$$W = P\,W'\,.$$

Besides, using the notations of (30), Equation (9) gives:

$$z = z' - \theta(a)\,Ad(a)\,Z' = z' + K_m P\,W'\,.$$

On the other hand, let $\hat{W}$ be the 5-column (20) representing the temperature vector:

$$\hat{W} = \begin{pmatrix} W \\ \zeta \end{pmatrix} = \begin{pmatrix} \beta \\ w \\ \zeta \end{pmatrix}.$$

Taking into account (12) and (31), it is easy to verify that its transformation law (25) with the linear Bargmannian transformation (24) can be recast as:

$$\begin{pmatrix} W \\ \zeta \end{pmatrix} = \begin{pmatrix} P & 0 \\ F_1 P & 1 \end{pmatrix} \begin{pmatrix} W' \\ \zeta' \end{pmatrix},$$

which is the transformation law of the affine map $\boldsymbol{\Theta}$ provided $z = m\,\zeta$, that achieves the proof. □

- *Step 6: boost method.* For the box at rest in the coordinate system X, the temperature 4-vector is given by (46):

$$W = \begin{pmatrix} \beta \\ 0 \end{pmatrix} .$$

A new coordinate system $\bar{X}$ in which the box has the velocity v can be deduced from $X = P\bar{X} + C$ by applying a boost $u = -v$ (hence $k = 0$, $\tau_0 = 0$ and $R = 1_{\mathbb{R}^3}$). The transformation law of vectors gives the new components

$$\bar{W} = \begin{pmatrix} \beta \\ \beta v \end{pmatrix} ,$$

and (9) leads to:

$$\bar{z} = z + \frac{m\,\beta}{2} \parallel v \parallel^2 = z + \frac{m}{2\,\beta} \parallel w \parallel^2 .$$

Taking into account (47) and leaving out the bars:

$$z = \frac{1}{2}\ \ln(\det(\mathcal{C})) - \frac{3}{2}\ \ln\beta + \frac{m}{2\,\beta} \parallel w \parallel^2 + C^{te} . \tag{48}$$

It is clear from (11) that s is Legendre conjugate of $-z$, then, introducing the internal energy (which is nothing other than the Galilean invariant (40)):

$$e_{int} = e - \frac{1}{2\,m} \parallel p \parallel^2 ,$$

the entropy is:

$$s = \frac{3}{2}\ \ln e_{int} + \frac{1}{2}\ \ln(\det(\mathcal{C})) + C^{te} ,$$

and, by $Z = \partial s \,/\, \partial M$, we derive the corresponding momenta:

$$\beta = \frac{\partial s}{\partial e} = \frac{3}{2\,e_{int}}, \qquad w = -grad_{\,p}\, s = \frac{3}{2\,e_{int}}\,\frac{p}{m} .$$

As Equation (47), Equation (48) and the expressions of s, β and w are not affected by the arbitrary choice of V_0.

- *Step 7: link between z and ζ.* As z is an extensive quantity, its value for N identical particles is $z_N = N\,z$. Planck's potential ζ being a specific quantity, we claim that:

$$\zeta = \frac{z_N}{N\,m} = \frac{z}{m} = \frac{1}{2\,m}\ \ln(\det(\mathcal{C})) - \frac{3}{2\,m}\ \ln\beta + \frac{1}{2\,\beta} \parallel w \parallel^2 + C^{te} .$$

By (16) and (17), we obtain the linear 4-momentum $\Pi = (\mathcal{H}, -p^T)$ and Cauchy's stresses:

$$\mathcal{H} = \rho\left(\frac{3}{2}\,\frac{k_B T}{m} + \frac{1}{2} \parallel v \parallel^2\right), \qquad p = \rho v, \qquad \sigma = -q\,1_{\mathbb{R}^3} ,$$

where, by the expression of the pressure, we recover the *ideal gas law*:

$$q = \frac{\rho}{m}\,k_B T = \frac{N}{V}\,k_B T .$$

The first principle of thermodynamics (18) reads:

$$\frac{\partial \mathcal{H}}{\partial t} + div\ (\mathcal{H} v - \sigma v) = 0, \qquad \rho\,\frac{dv}{dt} = -grad\, q, \qquad \frac{\partial \rho}{\partial t} + div\,(\rho\, v) = 0 .$$

We recognize the balance of energy, linear momentum and mass.

Remark 4. *The Hessian matrix I of $-z$, considered as function of W through Z, is positive definite [3]. It is Fisher metric of the Information Geometry. For the expression (48), it is easy to verify it:*

$$-\delta M\,\delta Z = \frac{1}{\beta}\left(e_{int}(\delta\beta)^2 + m \parallel \delta w - \frac{\delta\beta}{m}\, p \parallel^2\right) > 0\,,$$

for any non vanishing δZ taking into account $\beta > 0, e_{int} > 0$ and $m > 0$. On this basis, we can construct a thermodynamic length of a path $t \mapsto X(t)$ [21]:

$$\mathcal{L} = \int_{t_0}^{t_1} \sqrt{(\delta W(t))^T I(t)\,\delta W(t)}dt\,,$$

where $\delta W(t)$ is the perturbation of the temperature vector, tangent to the space-time at $X(t)$. We can also define a related quantity, Jensen–Shannon divergence of the path:

$$\mathcal{J} = (t_1 - t_0)\int_{t_0}^{t_1} (\delta W(t))^T I(t)\,\delta W(t)dt\,.$$

8. Conclusions

The above approach is not limited to classical mechanics but can be used as guiding ideas to tackle the relativistic mechanics. Beyond the strict application to physics, it can be taken as source of inspiration to broach other topics such as the science of information from the viewpoint of differential geometry and Lie groups. We hope to have modestly contributed to this aim.

Conflicts of Interest: The authors declare no conflict of interest.

References

1. Souriau, J.-M. Milieux continus de dimension 1, 2 ou 3 : statique et dynamique. Available online: http://jmsouriau.com/Publications/JMSouriau-MilContDim1991.pdf (accessed on 9 July 2016). (In French)
2. De Saxcé, G.; Vallée, C. *Galilean Mechanics and Thermodynamics of Continua*; Wiley-ISTE: London, UK, 2016.
3. Souriau, J.-M. *Structure des Systèmes Dynamiques*; Dunod: Paris, France, 1970. (In French)
4. Souriau, J.-M. *Structure of Dynamical Systems: A Symplectic View of Physics*; Birkhäuser: New York, NY, USA, 1997.
5. De Saxcé, G.; Vallée, C. Affine Tensors in Mechanics of Freely Falling Particles and Rigid Bodies. *Math. Mech. Solid J.* **2011**, *17*, 413–430.
6. Guillemin, V.; Sternberg, S. *Symplectic Techniques in Physics*; Cambridge University Press: Cambridge, MA, USA, 1984.
7. Souriau, J.-M. Thermodynamique et Géométrie. In *Differential Geometrical Methods in Mathematical Physics II*; Springer: Berlin/Heidelberg, Germany, 1976; pp. 369–397. (In French)
8. Souriau, J. M. *Thermodynamique Relativiste des Fluides*; Centre de Phsyique Théorique: Marseille, France, 1977; Volume 35, pp. 21–34. (In French)
9. Tulczyjew, W.; Urbański, P.; Grabowski, J. A pseudocategory of principal bundles. *Atti della Reale Accademia delle Scienze di Torino* **1988**, *122*, 66–72. (In Italian)
10. Libermann, P.; Marle, C.-M. *Symplectic Geometry and Analytical Mechanics*; Springer: Dordrecht, The Netherlands, 1987.
11. Vallée, C.; Lerintiu, C. Convex analysis and entropy calculation in statistical mechanics. *Proc. A. Razmadze Math. Inst.* **2005**, *137*, 111–129.
12. Iglesias, P. Essai de "thermodynamique rationnelle" des milieux continus. *Annales de l'IHP Physique Théorique* **1981**, *34*, 1–24. (In French)
13. Vallée, C. Relativistic thermodynamics of continua. *Int. J. Eng. Sci.* **1981**, *19*, 589–601.

14. De Saxcé, G.; Vallée, C. Bargmann Group, Momentum Tensor and Galilean invariance of Clausius–Duhem Inequality. *Int. J. Eng. Sci.* **2012**, *50*, 216-232.
15. Bargmann, V. On unitary representation of continuous groups. *Ann. Math.* **1954**, *59*, 1–46, doi:10.2307/1969831.
16. Hall, M. *The Theory of Groups*; Macmillan Co.: New York, NY, USA, 1953.
17. Barbaresco, F. Koszul Information Geometry and Souriau Geometric Temperature/Capacity of Lie Group Thermodynamics. *Entropy* **2014**, *16*, 4521–4565.
18. Koszul, J.L. Ouverts convexes homogènes des espaces affines. *Mathematische Zeitschrift* **1962**, *79*, 254–259. (In French)
19. Vinberg, È.B. Structure of the group of automorphisms of a homogeneous convex cone. *Trudy Moskovskogo Matematicheskogo Obshchestva* **1965**, *13*, 56–83. (In Russian)
20. Künzle, H.P. Galilei and Lorentz structures on space-time: Comparison of the corresponding geometry and physics. *Annales de l'IHP Physique Théorique* **1972**, *17*, 337–362.
21. Crooks, G.E. Measuring thermodynamic length. *Phys. Rev. Lett.* **2009**, *99*, 100602.

Chapter 2:
Koszul-Vinberg Model of Hessian Information Geometry

entropy

MDPI

Article

Foliations-Webs-Hessian Geometry-Information Geometry-Entropy and Cohomology †

Michel Nguiffo Boyom

ALEXANDER GROTHENDIECK INSTITUTE, IMAG-UMR CNRS 5149-c.c.051, University of Montpellier, PL. E. Bataillon, F-34095 Montpellier, France; boyom@math.univ-montp2.fr; Tel.: +33-467-143-571

† IN MEMORIAM OF ALEXANDER GROTHENDIECK. THE MAN.

Academic Editors: Frédéric Barbaresco and Frank Nielsen
Received: 1 June 2016; Accepted: 16 November 2016; Published: 2 December 2016

Abstract: Let us begin by considering two book titles: A provocative title, *What Is a Statistical Model? McCullagh (2002)* and an alternative title, *In a Search for Structure. The Fisher Information. Gromov (2012).* It is the richness in open problems and the links with other research domains that make a research topic exciting. Information geometry has both properties. Differential information geometry is the differential geometry of statistical models. The topology of information is the topology of statistical models. This highlights the importance of both questions raised by Peter McCullagh and Misha Gromov. The title of this paper looks like a list of key words. However, the aim is to emphasize the links between those topics. The theory of homology of Koszul-Vinberg algebroids and their modules (KV homology in short) is a useful key for exploring those links. In Part A we overview three constructions of the KV homology. The first construction is based on the pioneering brute formula of the coboundary operator. The second construction is based on the theory of semi-simplicial objects. The third construction is based on the anomaly functions of abstract algebras and their abstract modules. We use the KV homology for investigating links between differential information geometry and differential topology. For instance, "dualistic relation of Amari" and "Riemannian or symplectic Foliations"; "Koszul geometry" and "linearization of webs"; "KV homology" and "complexity of models". Regarding the complexity of a model, the challenge is to measure how far from being an exponential family is a given model. In Part A we deal with the classical theory of models. Part B is devoted to answering both questions raised by McCullagh and B Gromov. A few criticisms and examples are used to support our criticisms and to motivate a new approach. In a given category an outstanding challenge is to find an invariant which encodes the points of a moduli space. In Part B we face four challenges. (1) The introduction of a new theory of statistical models. This re-establishment must answer both questions of McCullagh and Gromov; (2) The search for an characteristic invariant which encodes the points of the moduli space of isomorphism class of models; (3) The introduction of the theory of homological statistical models. This is a pioneering notion. We address its links with Hessian geometry; (4) We emphasize the links between the classical theory of models, the new theory and Vanishing Theorems in the theory of homological statistical models. Subsequently, the differential information geometry has a homological nature. That is another notable feature of our approach. This paper is dedicated to our friend and colleague Alexander Grothendieck.

Keywords: KV cohomoloy; functor of Amari; Riemannian foliation; symplectic foliation; entropy flow; moduli space of statistical models; homological statistical models; geometry of Koszul; localization; vanishing theorem

MSC: 55R10; 55N20; 62B05; 55C12; 55C07

Contents

1. Introduction

1.1. The Notation

Throughout the paper we use tha following notation. N is the set of non negative integers, Z is the ring of integers, R is the field of real numbers, $C^\infty(M)$ is the associative commutative algebra of real valued smooth functions in a smooth manifold M. Let ∇ be a Koszul connection in a manifold M, R^∇ is the curvature tensor of ∇. It is defined by

$$R^\nabla(X,Y) = \nabla_X\nabla_Y - \nabla_Y\nabla_X - \nabla_{[X,Y]}.$$

T^∇ is the torsion tensor of ∇. It is defined by

$$T^\nabla(X,Y) = \nabla_X Y - \nabla_Y X - [X,Y].$$

Let X be a smooth vector field in M. $L_X\nabla$ is the Lie derivative of ∇ in the direction $X \cdot \iota(X)R^\nabla$ is the inner product by X. To a pair of Koszul connections (∇, ∇^*) we assign three differential operators. They are denoted by $D^{\nabla\nabla^*}$, D^∇ and D_∇.

(A.1) $D^{\nabla\nabla^*}$ is a first order differential operator. It is defined in the vector bundle $Hom(TM, TM)$. Its values belong to the vector bundle $Hom(TM^{\otimes 2}, TM)$.

(A.2) D^∇ and D_∇ are 2nd order differential operators. They are defined in the vector bundle TM. Their values belong to the vector bundle $Hom(TM^{\otimes 2}, TM)$. Let X be a section of TM and let ψ be a section of $T^*M \otimes TM$. The differential operators just mentioned are defined by

$$D^{\nabla\nabla^*}(\psi) = \nabla^* \circ \psi - \psi \circ \nabla, \quad (1a)$$

$$D^\nabla(X) = L_X\nabla - \iota(X)R^\nabla, \quad (1b)$$

$$D_\nabla(X) = \nabla^2(X). \quad (1c)$$

Part A of this paper is partially devoted to the global analysis of the differential equation

$$FE(\nabla\nabla^*): \quad D^{\nabla\nabla^*}(\psi) = O.$$

The solutions to $FE(\nabla\nabla^*)$ are useful for addressing the links between the KV homology, the differential topology and the information geometry.

The purpose of a forthcoming paper is the study of the differential equations

$$FE^*(\nabla): \quad D^\nabla(X) = 0,$$

$$FE^{**}(\nabla): \quad D_\nabla(X) = 0.$$

In the Appendix A to this paper we overview the role played by the solutions to $FE^{**}(\nabla)$ in some still open problems.

1.2. Some Explicit Formulas

Let $x = (x_1, ..., x_m)$ be a system of local coordinate functions of M. In those coordinates the Christoffel symbols of both ∇ and ∇^* are denoted by $\Gamma_{ij:k}$ and $\Gamma^*_{ij:k}$ respectively. We use those coordinate functions for presenting an element $\psi \in \mathcal{M}(\nabla\nabla^*)$ as a matrix $[\psi_{ij}]$. Thus by setting $\partial i = \frac{\partial}{\partial x_i}$ one has

$$\nabla_{\partial i}\partial j = \Gamma_{ij:k}\partial k.$$

We focus on $FE(\nabla\nabla^*)$ and of $FE^{**}(\nabla)$. They are equivalent to the following system of partial differential equations

$$[S_{ij:k}]: \quad \frac{\partial \psi_{kj}}{\partial x_i} - \sum_{1\leq \ell \leq m} (\Gamma_{ij:\ell}\psi_{k\ell} - \Gamma^*_{i\ell:k}\psi_{\ell}j) = 0,$$

$$[\Theta^k_{ij}(X)]: \quad \frac{\partial^2 X^k}{\partial x_i \partial x_j} + \sum_{\alpha}[\Gamma^k_{i\alpha}\frac{\partial X^\alpha}{\partial x_j} + \Gamma^k_{j\alpha}\frac{\partial X^\alpha}{\partial x_i} - \Gamma^\alpha_{ij\frac{\partial X^k}{\partial x_\alpha}}] + \sum_{\alpha}[\frac{\partial \Gamma^k_{j\alpha}}{\partial x_i} + \sum_{\beta}[\Gamma^\beta_{j\alpha}\Gamma^k_{i\beta} - \Gamma^\beta_{ij}\Gamma^k_{\beta}\alpha]]X^\alpha = 0.$$

In Part A we address the links between the following topics DTO, HGE, IGE and ENT. Those topics are presented as vertices of a square whose centre is denoted by KVH.

(1) DTO stands for Differential TOpology. In DTO, FWE stands for Foliations and WEbs.
(2) HGE stands for Hessian GEometry. Its sources are the geometry of bounded domains, the topology of bounded domains, the analysis in bounded domains. Among the notable references are [1–3]. Hessian geometry has significant impacts on thermodynamics, see [4,5], About the impacts on other related topics the readers are referred to [6–12].
(3) IGE stands for Information GEometry. That is the geometry of statistical models. More generally its concern is the differential geometry of statistical manifolds. The range of the information geometry is large [13]. Currently, the interest in information geometry is increasing. This comes from the links with many major research domains [14–16]. We address some significant aspects of those links. Non-specialist readers are referred to some fundamental references such as [17,18]. See also [4,19–23]. The information geometry also provides a unifying approach to many problems in differential geometry, see [21,24,25]. The information geometry has a large scope of applications, e.g., physics, chemistry, biology and finance.
(4) ENT stands for ENTropy. The notion of entropy appears in many mathematical topics, in Physics, in thermodynamics and in mechanics. Recent interest in the entropy function arises from its topological nature [14]. In Part B we introduce the entropy flow of a pair of vector fields. The Fisher information is then defined as the Hessian of the entropy flow.
(5) KVH stands for KV Homology. The theory of KV homology was developed in [9]. The motivation was the conjecture of M. Gerstenhaber in the category of locally flat manifolds. In this paper we emphasize other notable roles played by the theory of KV homology. It is also useful for discussing a problem raised by John Milnor in [26].

The conjecture of Gerstenhaber is the following claim.

Every restricted theory of deformation generates its proper cohomology theory [27].

Loosely speaking, in a restricted theory of deformation one has the notion of both infinitesimal deformation and trivial deformation. The challenge is the search for a cochain complex admitting infinitesimal deformations as cocycles. In the present paper, KVH is useful for emphasizing the links between the vertices DTO, HGE, IGE and ENT. That is our reason for devoting a section to KVH.

Warning.

We propose to overview the structure of this paper. The readers are advised to read this paper as through it were a wander around the vertices of the square "DTO-HGE-IGE-ENT". Thus, depending on his interests and his concerns a reader could walk several times across the same vertex. For instance the information geometry appears in many sections, depending on the purpose and on the aims.

1.3. The content of the Paper

This paper is divided into Part A and Part B.

Part A: Sections 1–7.

Section 1 is the Introduction. Section 2 is devoted to algebroids, modules of algebroids and the theory of KV homology of the Koszul-Vinberg algebroids. To introduce the KV cohomology we have adopted three approaches. Each approach is based on its specific machinery. However, the readers will face three cochain complexes which are pairwise quasi isomorphic. The KV cohomology is present throughout this paper. At the end of Part B the reader will see that the theory of statistical models is but a vanishing theorem in the theory of KV hohomology. The first approach is based on the pioneering fundamental brute formula of the coboundary operator. Historically, the brute formula is the first to have been constructed [9].

This first approach is used in many sections of this paper. Regarding the theory of deformation of the Koszul Geometry, the KV cohomology is the solution to the conjecture of Gerstenhaber. In the theory of modules of KV algebroids the role played by the KV cohomology is practically SIMILAR to the role played by the Hocshild cohomology in the category of associative alggebroids and their modules. This last remark holds for the role played by the Chevalley-Eilenberg cohomology in the category of Lie algebroids and their modules. Nevertheless, our comparison fails in the theory of Extension of modules over algebroids. In both categories of extensions of modules over associative algebroids and Lie algebroids the moduli space of equivalence class is encoded by cohomology classes of degree one. In the category of extensions of KV modules the moduli space is encoded by a spectral sequence. That was a unexpected feature in [9]. The pioneering coboundary operator of Nijenhuis [28] may be derived from the total brute coboundary operator introduced in [29].

The second approach is based on the notion of simplecial objects.

The third approach is based on the theory of anomaly functions for abstract algebras and their abstract modules. The idea has emerged from recent correspondences with one of my former teachers. The KV anomaly function of a Koszul connection ∇ may be expressed in terms of the ∇-Hessian operators ∇^2, namely

$$KV_\nabla(X,Y,Z) =< \nabla^2(Z),(X,Y) > - < \nabla^2(Z),(Y,X) > .$$

This approach is a powerful for addressing the relationships between the global analysis, the differential topology and the information geometry. The approach by the anomaly functions suggests many conjectures. Among those conjectures is the following.

Conjecture. Every anomaly function of algebras and of modules yields a theory of cohomology of algebras and modules.

Section 3. This section is devoted to the theory of KV (co)homology of Koszul-Vinberg algebroids. We focus on cohomological data which are used in the paper.

Section 4. This section is devoted the KV algebroids which are defined by structures of locally flat manifold. The KV cohomology theory is used for highlighting the impacts on the differential topology of the information geometry and its methods. We make the most of some relationships between the KV cohomology and the global analysis of the differential equation $FE^*(\nabla\nabla^*)$. We also sketch the global analysis of the differential equation

$$FE^{**}(\nabla).$$

This leads to the function

$$\mathcal{LC} \ni \nabla \to r^b(\nabla) \in \mathbb{Z}.$$

We explain how to interpret r^b as a distance. (See the Appendix A to this paper). For instance, the function r^b gives rise to an numerical invariant $r^b(\mathbb{M})$ which measures how far from being an exponential family is a statistical model $\mathbb{M}$. This result is a significant contribution to the information geometry, see [18,22,24].

Section 5. We are interested in how interact the information geometry, the KV cohomology and the geometry and Koszul. In particular we relate the notion of hyperbolicity and vanishing theorems in the KV cohomology.

Section 6. This section is devoted to the homological version of the geometry of Koszul. Our approach involves the dualistic relation of Amari. The KV cohomology links the dualistic relation with the geometry of Koszul.

Section 7. In this section summarize the highlighting features of Part A.

Part B: Sections 8–14.

Section 8. This is the starting section of the second part B. This Part B is devoted to new insights in the theory of statistical models. On 2002 Peter McCullar raised the provocative question.

What Is a Statistical Model

Across the world (Australia, Canada, Europe, US) the $McCullagh^s$ paper became the object of many criticisms and questions by eminent theoretical and applied statisticians [30].

Part B is aimed at supplying some deficiencies in the current theory of statistical models. We address some criticisms which support the need of re-establishing the theory of statistical model for measurable sets. Those criticisms are used for highlight the lack of both Structure and Relations. Those criticisms also highlight the search of M. Gromov [15]. The need for structures and relations was the intuition of Peter McCullagh. Loosely speaking there is a lack of Intrinsic Geometry in the sense of Erlangen. Subsequently the lack of intrinsic geometry yields other things that are lacking. The problem of the moduli space of models is not studied, although this would be crucial for applied information geometry, and for applied statistics. That might be a key in reading some the controversy about [30].

Section 9. In this section we address the problem of moduli space of statistical models. The problem of moduli space in a category is a major question in Mathematic. It is generally a difficult problem that involves finding a characteristic invariant which encodes the point of the moduli space. Such an invariant is a crucial step toward the geometry and the topology of a moduli space. Among other needs, the problem of encoding the moduli space of models has motivated our need of a new approach, that is to say the need of a theory having nice mathematical structure and relations. In this Part B the problem of the moduli space is solved. To summarize the theorem describing the moduli spaces of statistical models we need the following notation.

A gauge structure in a manifold M is a pair (M, ∇) where ∇ is a Koszul connection in M. The category of gauge structures in M is denoted by $\mathcal{LC}(M)$. We are concerned with the vector bundle $T^{*\otimes 2}M$ of bi-linear forms in the tangent bundle TM. The sheaf of sections of $T^{*\otimes 2}M$ is denoted by $\mathcal{BL}(M)$.

The category of m-dimensional statistical models (to be defined) of a measurable (Ξ, Ω) is denoted by $\mathcal{GM}_m(\Xi, \Omega)$. The category of random functors

$$\mathcal{LC}(M) \times \Xi \rightarrow \mathcal{BL}(M)$$

is denoted by $\mathcal{F}(\mathcal{LC}, \mathcal{BL})(M)$. One of the interesting breakthrough in this Part.B is the following solution to the problem of moduli.

Theorem 1. *There exists a functor*

$$\mathcal{GM}_m(\Xi, \Omega) \ni \mathbb{M} \rightarrow q_{\mathbb{M}} \in \mathcal{BL}(M) \tag{2}$$

which determines a model $\mathbb{M}$ up to isomorphism.
Let p be the probability density of a model $\mathbb{M}$. The mathematical expectation of $q_{\mathbb{M}}(\nabla)$ is defined by

$$E(q_{\mathbb{M}}(\nabla)) = \int_{\Xi} p q_{\mathbb{M}}(\nabla). \tag{3}$$

The quantity $E(q_{\mathbb{M}})(\nabla)$ does not depend on the Koszul connection ∇. It is called the Fisher information of $\mathbb{M}$.

This theorem emphasizes the Search for structure [16].

Section 10. This section is devoted to introduce the category of homological statistical models. This may be interpreted as a variant of the topology of the information. Another approach is to be found in Baudot-Bennequin [31].

The current theory (as in [17]) is called the classical (or local) theory. This means that a statistical model as in [17,18] is derived from the localization of a homological model. Loosely speaking such a model expresses a local vanishing theorem in the theory of homological statistical models.

Section 11. This section is devoted to discussing the links between the geometry of Koszul and the theory of homological statistical models. Those investigations lead to this notable feature.

The Geometry of Koszul, the homological statistical models and the classical information geometry locally look alike.

Section 12. Through Section 9 the framework is the category of equivariant locally trivial fibration. This assumption is weakened in Section 12. We recall the relationships between the Cech cohomology and the theory of locally trivial fiber bundle. We extend the scope of applications of the methods of the information geometry. Those extensions produce some interesting results. Here is an instance.

Theorem 2. *Let M be an oriented compact real analytic manifold and let $C^{\omega}(M^2)$ be the space of real valued analytic functions in M^2. There exists a non trivial map of $C^{\omega}(M^2)$ in the family of (positive) stratified Riemannian foliation in M.*

Sections 13. This Section 13 is a variant of Section 7.

Section 14 is an appendix we have mentioned. It is devoted to overview a few new significant results. Those results are derived from the global analysis of the differential operators

$$\left\{D^{\nabla}, D_{\nabla}, \nabla \in \mathcal{LC}(M)\right\}.$$

The solutions to a few open problems are announced.

2. Algebroids, Moduls of Algebroids, Anomaly Functions

The purpose of this section is to introduce basic notions in the algebraic topology of locally flat manifolds.

2.1. The Algebroids and Modules

Given a smooth fiber bundle

$$B \to M$$

the set of smooth sections of B is denoted by $\Gamma(B)$.

Definition 1. *An (abstract) real algebra is a real vector space $\mathcal{A}$ endowed with a bilinear map*

$$\mathcal{A} \times \mathcal{A} \to \mathcal{A}$$

Definition 2. *An (abstract) real two-sided module of an (abstract) algebra $\mathcal{A}$ is a real vector space W with two bilinear mappings*

$$\mathcal{A} \times W \to W,$$

$$W \times \mathcal{A} \to W$$

Warning.

Here algebra means a multiplication $a \cdot b$ without any rule of calculations. So the product $a \cdot b \cdot c$ is meaningless.

Throughout this paper, the smooth manifolds we deal with are connected and paracompact. In a smooth manifold M all geometrical objects we are interested in are smooth as well.

The vector space of smooth vector fields in a manifold M is denoted by $\mathcal{X}(M)$. It is a left module of the associative commutative algebra $C^\infty(M)$.

Consider a real vector bundle

$$\mathcal{E} \to M.$$

The real vector space of sections of $\mathcal{E}$ is denoted by $\Gamma(\mathcal{E})$.

Definition 3. *A real algebroid over a smooth manifold M is a real vector bundle whose vector space of sections is a real algebra.*

So the vector space of sections of a real algebroid $\mathcal{E}$ is endowed with a $\mathbb{R}$-bilinear map

$$\Gamma(\mathcal{E}) \times \Gamma(\mathcal{E}) \ni (s, s^*) \to s \cdot s^* \in \Gamma(\mathcal{E})$$

To simplify the multiplication of two sections is denoted $s \cdot s*$.

Definition 4. *A two-sided module of an algebroid $\mathcal{E}$ is a vector bundle*

$$\mathcal{V} \to M$$

whose vector space of sections is a two-sided module of the algebra $\Gamma(\mathcal{E})$.

Let s be section $\mathcal{E}$ and let v be a section of $\mathcal{V}$. Both left action s on v and the right action of s on v are denoted by $s \cdot v$ and $v \cdot s$.

Definition 5. *An anchored vector bundle over M is a pair*

$$(\mathcal{E}, b)$$

formed by a real vector bundle $\mathcal{E}$ and a vector bundle homomorphism

$$\mathcal{E} \ni e \to b(e) \in TM.$$

The homomorphism b is called the anchor map.

2.2. Anomaly Functions of Algebroids and of Modules

Let $\mathcal{V}$ be a two-sided module of an algebroid $(\mathcal{E}, b)$.

Definition 6. *An anomaly function of an algebroid $\mathcal{E}$ is a 3-linear map $A_{\mathcal{E}}$ of $\Gamma(\mathcal{E})^3$ in $\Gamma(\mathcal{E})$ whose values $A_{\mathcal{E}}(s_1, s_2, s_3)$ belong to $span_{\mathbb{R}}[(s_i{\cdot}s_j){\cdot}s_k, s_i{\cdot}(s_j{\cdot}s_k); i, j, k \in [1, 2, 3]]$. An anomaly function of an $\mathcal{E}$-module $\mathcal{V}$ is a 3-linear map $A_{\mathcal{E}\mathcal{V}}$ of $\Gamma(\mathcal{E})^2 \times \Gamma(\mathcal{V})$ in $\Gamma(\mathcal{V})$ whose values $A_{\mathcal{E}\mathcal{V}}(s, s^*, v)$ belong to $span_{\mathbb{R}}[(s{\cdot}s^*){\cdot}v, s{\cdot}(s^*{\cdot}v) \forall s$, $s^* \in \Gamma(\mathcal{E}), \forall v \in \Gamma(\mathcal{V})]$.*

In this paper we are interested in some anomaly functions which have strong geometrical impacts. They are defined below.

Definition 7. *Let $\mathcal{E}$ be an algebroid and let $s, s^*, s^{*}* \in \Gamma(\mathcal{E})$.*
(1) *The associator anomaly function of $\mathcal{E}$ is defined by*

$$Ass(s, s^*, s^{**}) = (s \cdot s^*) \cdot s^{**} - s \cdot (s^* \cdot s^{**}).$$

(2) *The Koszul-Vinberg anomaly function of $\mathcal{E}$ is defined by*

$$KV(s, s^*, s^{**}) = Ass(s, s^*, s^{**}) - Ass(s^*, s, s^{**}).$$

(3) *The Jacobi anomaly functions of $\mathcal{E}$ are defined by*

$$J(s, s^*, s^{**}) = (s \cdot s^*) \cdot s^{**} + (s^* \cdot s^{**}) \cdot s + (s^{**} \cdot s) \cdot s^*.$$

Definition 8. *Let v be a section of a two-sided $\mathcal{E}$-module $\mathcal{V}$.*
(1) *The associator anomaly function of a left module $\mathcal{V}$ is defined as*

$$Ass(s, s^*, v) = (s \cdot s^*) \cdot v - s \cdot (s^* \cdot v).$$

(2) *The KV anomaly functions of a two sided module $\mathcal{V}$ are defined as*

$$KV(s, s^*, v) = Ass(s, s^*, v) - Ass(s^*, s, v),$$

$$KV(s, v, s^*) = (s \cdot v) \cdot s^* - s \cdot (v \cdot s^*) - (v \cdot s) \cdot s^* + v \cdot (s \cdot s^*).$$

Definition 9. *We keep the notation used above. Let s, s^* be sections of $\mathcal{E}$, let v be a section of $\mathcal{V}$ and $f \in C^\infty(M)$.*
(1) *The Leibniz anomaly function of an anchored algebroid $\mathcal{E}$ is defined by*

$$L(s, f, s^*) = s \cdot (fs^*) - df(b(s))s^* - fs \cdot s^*.$$

(2) *The Leibniz anomaly function of the $\mathcal{E}$-module $\mathcal{V}$ is defined by*

$$L(s, f, v) = s \cdot (fv) - df(b(s))v - fs \cdot v.$$

A category of algebroids and modules of algebroids is defined by its anomaly functions. The anomaly functions are also used for introducing theories of homology of algebroids.

Some categories of anchored algebroids play important roles in the differential geometry.

Definition 10. *(A1): A Lie algebroid is an anchored algebroid $(\mathcal{E}, b)$ satisfying the identities*

$$s \cdot s* = 0,$$

$$L(s, f, s*) = 0.$$

(B1): A KV algebroid is an anchored algebroid $(\mathcal{E}, b)$ satisfying the identities

$$KV(s, s*, s**) = 0,$$

$$L(s, f, s*) = 0.$$

(B2): A vector bundle $\mathcal{V}$ is a module of Lie algebroid $(\mathcal{E}, b)$ if it satisfies the identities

$$L(s, f, v) = 0,$$

$$(s \cdot s*) \cdot v - s \cdot (s * \cdot v) + s * \cdot (s \cdot v) = 0.$$

A vector bundle $\mathcal{V}$ is a two-sided KV module of a Koszul-Vinberg algebroid $(\mathcal{E}, b)$ if it satisfies the identities

$$L(s, f, v) = 0,$$

$$KV(s, s^*, v) = 0,$$

$$KV(s, v, s^*) = 0.$$

Warning.

Consider a vector V space as the trivial vector bundle

$$V \times O \to 0.$$

Then we get

$$\Gamma(V \times 0) = V.$$

Therefore an algebra is an anchored algebroid over a point; its anchor map of is the zero map. Therefore, the Leibniz anomaly of an algebra is nothing but the bilinearity of the multiplication. So the notion of KV algebra and KV module is clear.

3. The Theory of Cohomology of KV Algebroids and Their Modules

This section is devoted to the cohomology of KV algebroids and KV modules of KV algebroids. KV stands for Koszul-Vinberg. We shall introduce three approaches to the theory of KV cohomology. Each approach has its particular advantage. So, depending on the needs or on the concerns one or other approach may be convenient. The three approaches are called "Version brute formula", "Version semi simplicial objects", "Version anomaly functions". The same graded vector space is common to the three constructions. They differ in their coboundary operators. However, three constructions lead to cohomology complexes which are pairwise quasi isomorphic.

Each construction leads to two cochain complexes. Those complexes are called the KV complex and total KV complex. They are denoted by C^*_{KV} and C^*_τ. In final we obtain six cohomological complexes.

3.1. The Theory of KV Cohomology—Version the Brute Formula of the Coboundary Operator

The geometric framework is the category of real KV algebraoids and their two sided modules. However our machineries only make use of $\mathbb{R}$-multi-linear calculations in the vector spaces of sections of vector bundles. Without any damage we replace the categories of KV algebroids and modules of KV algebroids by the categories of KV algebras and abstract modules of KV algebras.

3.1.1. The Cochain Complex C_{KV}.

Let W be a two-sided module of a KV algebra $\mathcal{A}$.

Definition 11. *The vector subspace $J(W) \subset W$ is defined by*

$$(a \cdot b) \cdot w - a \cdot (b \cdot w) = 0 \quad \forall a, b \in \mathcal{A}$$

We consider the $\mathbb{Z}$-graded vector space

$$C_{KV}(\mathcal{A}, W) = \sum_q C^q_{KV}(\mathcal{A}, W).$$

The homogeneous vector sub-spaces are defined by

$$C^q_{KV}(\mathcal{A}, W) = 0 \quad \forall q < 0,$$
$$C^0_{KV}(\mathcal{A}, W) = J(W),$$
$$C^q_{KV}(\mathcal{A}, W) = Hom_{\mathbb{R}}(\mathcal{A}^{\otimes q}, W) \quad \forall q > 0.$$

Before pursuing we fix the following notation.

Let

$$\xi = a_1 \otimes ... \otimes a_{q+1} \in \mathcal{A}^{\otimes q+1}$$

and let $a \in \mathcal{A}$,

$$\partial_i \xi = a_1 \otimes ...\hat{a}_i... \otimes a_{q+1},$$
$$\partial^2_{i,k+1}\xi = \partial_i(\partial_{k+1}\xi),$$
$$a.\xi = \sum_1^{q+1} a_1 \otimes ...a_{j-1} \otimes a.a_j \otimes a_{j+1}...a_{q+1}.$$

We are going to define the coboundary operator

$$\delta_{KV} : C^q(\mathcal{A}, W) \to C^{q+1}(\mathcal{A}, W).$$

The coboundary operator is a linear map. It is defined by

$$[\delta_{KV}(w)](a) = -a \cdot w + w \cdot a \quad \forall w \in J(W), \tag{4a}$$

$$[\delta_{KV} f](\xi) = \sum_1^q (-1)^i [a_i \cdot f(\partial_i \xi) - f(a_i \cdot \partial_i \xi) + (f(\partial^2_{i,q+1}\xi \otimes a_i)) \cdot a_{q+1}] \forall f \in C^q_{KV}(\mathcal{A}, W),$$
$$\forall \xi \in \mathcal{A}^{\otimes q+1}. \tag{4b}$$

The operator δ_{KV} satisfies the identity

$$\delta^2_{KV} f = 0 \quad \forall f \in C_{KV}(\mathcal{A}, W).$$

Therefore the pair $(C^*_{KV}(\mathcal{A}, W), \delta_{KV})$ is a cochain complex. Its cohomology space is denoted by

$$H_{KV}(\mathcal{A}, W) = \sum_q H^q_{KV}(\mathcal{A}, W).$$

The conjecture of Gerstenhaber: Comments.

A KV algebra $\mathcal{A}$ is a two-sided module of itself. An infinitesimal deformations of $\mathcal{A}$ is a 1-cocycle of $C_{KV}(\mathcal{A}, \mathcal{A})$ [9]. By the conjecture of Gerstenhaber the cohomology complex $C_{KV}(\mathcal{A}, \mathcal{A})$ is generated by the theory of deformations in the category of KV algebras.

The theory of deformation of KV algebras is the algebraic version of the theory of deformation of locally flat manifolds [2]. Therefore, the complex $C_{KV}(\mathcal{A}, \dashv)$ is the solution to the conjecture of Muray Gerstenhaber in the category of locally flat manifolds [27].

Features.

(1) The 2nd cohomology space $H^2_{KV}(\mathcal{A}, \mathcal{A})$ is the space of non trivial deformations of $\mathcal{A}$.

The definition of KV algebra of a locally flat manifold will be given in the next section.

Following [2] every hyperbolic locally flat manifold has non trivial deformations. Thus, if $\mathcal{A}$ is the KV algebra of a hyperbolic locally flat manifold then

$$H^2_{KV}(\mathcal{A}, \mathcal{A}) \neq 0.$$

(2) Let W be a two-sided module of a KV algebra $\mathcal{A}$. We consider W as a trivial KV algebra, viz

$$w \cdot w* = 0 \quad \forall w, w* \in W.$$

Let $EXT_{KV}(\mathcal{A}, W)$ be the set of equivalence classes of short exact sequences of KV algebras

$$0 \to W \to \mathcal{B} \to \mathcal{A} \to 0.$$

An interpretation of the 2nd cohomology space of $C_{KV}(\mathcal{A}, W)$ is the identification

$$H^2_{KV}(\mathcal{A}, W) = EXT_{KV}(\mathcal{A}, W).$$

Let $W, W*$ be two-sided modules of $\mathcal{A}$. Let $EXT_{\mathcal{A}}(W*, W)$ be the set of equivalence classes of exact short sequences of two-sided $\mathcal{A}$-modules

$$0 \to W \to T \to W* \to 0.$$

In both the category of associative algebras and the category of Lie algebras we have

$$HH^1(A, Hom_{\mathbb{R}}(W*, W)) = EXT_A(W*, W),$$

$$H^1_{CE}(A, Hom_{\mathbb{R}}(W*, W)) = EXT_A(W*, W).$$

Here $HH(A, -)$ stands for Hochschild cohomology of an associative algebra A and $H_{CE}(A, -)$ stands for cohomology of Chevalley-Eilenberg of a Lie algebra A.

Unfortunately in the category of KV modules of KV algebras this interpretation of the first cohomology space fails. Loosely speaking in the category of KV algebras the set $H^1(\mathcal{A}, Hom(W*, W))$ is not canonically isomorphic to set $EXT_{\mathcal{A}}(W*, W)$ [9].

3.1.2. The Total Cochain Complex C_τ.

The purpose is the total complex

$$C_\tau(\mathcal{A}, W) = \sum_q C^q_\tau(\mathcal{A}, W).$$

Its homogeneous vector subspaces are defined by

$$C^q_\tau(\mathcal{A}, W) = 0 \quad \forall q < 0,$$

$$C^0_\tau(\mathcal{A}, W) = W,$$

$$C^q_\tau(\mathcal{A}, W) = Hom_{\mathbb{R}}(\mathcal{A}^{\otimes q}, W) \quad \forall q > 0.$$

The total coboundary operator is a linear map

$$C^q_\tau(\mathcal{A}, W) \to C^{q+1}_\tau(\mathcal{A}, W).$$

That operator is defined by

$$(1): \quad [\delta_\tau w](a) = -a \cdot w + wa \quad \forall (a, w) \in \mathcal{A} \times W,$$

$$(2): \quad [\delta_\tau f](\xi) = \sum_1^{q+1}(-1)^i[a_i \cdot f(\partial_i\xi) - f(a_i \cdot \partial_i\xi) + (f(\partial^2_{i,q+1}\xi \otimes a_i)) \cdot a_{q+1}] \quad \forall f \in C^q_\tau(\mathcal{A}, W).$$

The pair

$$(C^*_\tau(\mathcal{A}, W), \delta_\tau)$$

is a cochain complex, viz

$$\delta^2_\tau = 0.$$

The derived cohomology space is denoted by

$$H_\tau(\mathcal{A}, W) = \sum_q H^q_\tau(\mathcal{A}, W).$$

It is called the W-valued total KV cohomology of $\mathcal{A}$.

3.2. The Theory of KV Cohomology—Version: the Semi-Simplicial Objects

Let V be a two-sided module of a KV algebra $\mathcal{A}$. Our aim is the construction of semi simplicial $\mathcal{A}$-modules whose derived cochain complex is quasi isomorphic to the KV cochin complex $C_{KV}(\mathcal{A}, V)$.

3.2.1. Extension

We start by considering the vector space

$$\mathcal{B} = \mathcal{A} \oplus \mathbb{R}.$$

Its elements are denoted by $(s + \lambda)$. We endow $\mathcal{B}$ with the multiplication which is defined by

$$(s + \lambda) \cdot (s^* + \lambda^*) = s \cdot s^* + \lambda s^* + \lambda^* s + \lambda\lambda^*.$$

With the multiplication we just defined, $\mathcal{B}$ is a real KV algebra. In other words we have

$$KV(X_1, X_2, X_3) = 0.$$

Here

$$X_j = s_j + \lambda_j.$$

In the $\mathcal{A}$-module V we have a structure of left $\mathcal{B}$-module which is defined by

$$(s + \lambda)\cdot v = s\cdot v + \lambda v \quad \forall (s + \lambda) \in \mathcal{B}, \quad \forall v \in V.$$

3.2.2. Construction

Let $\tilde{\mathcal{B}}$ be the vector space spanned by $\mathcal{A} \times \mathbb{R}$. Its elements are finite linear combinations of $(s, \lambda), s \in \mathcal{A} \times \mathbb{R}$.

The tensor algebra of $\tilde{\mathcal{B}}$ is denoted by $T(\tilde{\mathcal{B}})$. It has a $\mathbb{Z}$-grading. its homogeneous vector sub-spaces are defined by

$$T_q(\tilde{\mathcal{B}}) = \tilde{\mathcal{B}}^{\otimes q}.$$

A monomial element is denoted by

$$\xi = x_1 \otimes x_2 \otimes ... \otimes x_q.$$

Here

$$x_j = (s_j, \lambda_j) \in \mathcal{A} \times \mathbb{R}.$$

The KV algebra $\mathcal{A}$ is a two-sided ideal of the KV algebra $\mathcal{B}$. Thereby, the vector space $\tilde{\mathcal{B}}$ is canonically a left module of $\mathcal{A}$.

We define the natural two-sided action of $\mathbb{R}$ in $\tilde{\mathcal{B}}$ by setting

$$\lambda \cdot (s^*, \lambda^*) = (\lambda s, \lambda\lambda^*),$$

$$(s^*, \lambda^*) \cdot \lambda = (\lambda s^*, \lambda^* \lambda).$$

Thereby every vector subspace $T_q(\tilde{\mathcal{B}})$ is a left KV module of $\mathcal{B}$. Here the left action of $\mathcal{B}$ in $T_q(\tilde{\mathcal{B}})$ is defined

$$(s + \lambda) \cdot \xi = s \cdot \xi + \lambda \xi.$$

Before continuing we recall the (extended) action of $\mathcal{A}$ in tensor space $T_q(\tilde{\mathcal{B}})$,

$$s \cdot (x_1 \otimes x_2 \otimes ... \otimes x_q) = \sum_{j=1}^{q} x_1 \otimes x_2 ... \otimes s \cdot x_j \otimes ... \otimes x_q.$$

We recall a notation which has been used in the last subsections,

$$\partial_j \xi = x_1 \otimes x_2 \otimes ... \hat{x}_j ... \otimes x_q.$$

The symbol $\hat{x}_j$ means that x_j is missing. Let $1 \in \mathbb{R}$ be the unit element, then $\tilde{1}$ stands for $(0, 1) \in \tilde{\mathcal{B}}$. We are going to construct semi simplicial modules of $\mathcal{B}$.

3.2.3. Notation-Definitions

Implicitly we use set isomorphism

$$\tilde{\mathcal{B}} \ni x = (s, \lambda) \rightarrow X^* = s + \lambda \in \mathcal{B}.$$

Then $\forall \xi \in T_q(\tilde{\mathcal{B}})$ one has

$$\tilde{1}^* \cdot \xi = \xi \cdot$$

We go back to the $\mathbb{Z}$-graded $\mathcal{B}$-module

$$T_*(\tilde{\mathcal{B}}) = \sum_q T_q(\tilde{\mathcal{B}}).$$

Definition 12. *Let j, q be two positive integers with $j < q$, let*

$$\xi = x_1 \otimes x_2 ... \otimes x_q.$$

The linear maps

$$d_j : T_q(\tilde{\mathcal{B}}) \rightarrow T_{q-1}(\tilde{\mathcal{B}})$$

and

$$S_j : T_q(\tilde{\mathcal{B}}) \rightarrow T_{q+1}(\tilde{\mathcal{B}})$$

are defined by

$$d_j \xi = X_j^* \cdot \partial_j \xi,$$

$$S_j \xi = e_j(\tilde{1}) \xi$$

The right member of the last equality has the following meaning

$$e_j(\tilde{1}) \xi = x_1 \otimes x_2 ... \otimes x_{j-1} \otimes \tilde{1} \otimes x_j ... \otimes x_q$$

Structure. *The maps d_j and S_j satisfy the following identities*

$$d_i d_j = d_{j-1} d_i \quad if \quad i \leq j, \tag{5a}$$
$$S_i S_j = S_{j+1} S_i \quad if \quad i < j, \tag{5b}$$
$$(S_{j-1} d_i - d_i S_j)(\xi) = e_{j-1}(x_i)\partial_i \xi \quad if \quad 1 < i < j, \tag{5c}$$
$$(d_{i+1} S_j - S_j d_i)(\xi) = e_j(x_i)\partial_i \xi \quad if \quad j+1 < i \leq q, \tag{5d}$$
$$d_i(S_i(\xi)) = \xi \quad if \quad i = j. \tag{5e}$$

Definition 13. *The system $\{T_q(\mathcal{B}), d_i, S_i\}$ is called the canonical semi simplicial module of $\mathcal{B}$.*

3.2.4. The KV Chain Complex

From the canonical simplicial $\mathcal{B}$-module we derive the chain complex $C_*(\mathcal{B})$. it has a $\mathbb{Z}$-grading which is defined by

$$C_q(\mathcal{B}) = 0 \quad if \quad q < 0, \tag{6a}$$
$$C_0(\mathcal{B}) = \mathcal{R}, \tag{6b}$$
$$C_q(\mathcal{B}) = T_q(\tilde{\mathcal{B}}) \quad if \quad q > 0. \tag{6c}$$

Now one defines the (linear) boundary operator

$$d : C_q(\mathcal{B}) \rightarrow C_{q-1}(\mathcal{B})$$

by setting

$$d(C_0(\mathcal{B})) = 0,$$
$$d(C_1(\mathcal{B})) = 0,$$
$$d\xi = \sum_1^q (-1)^j d_j \xi \quad if \quad q > 1.$$

By the virtue of (5a) we have

$$d^2 = 0.$$

3.2.5. The V-Valued KV Homology

We keep the notation used in the preceding sub-subsection. So the vector spaces $\mathcal{A}$, $\mathcal{B}$ and V are the same as in the preceding subsubsection.

We consider the $\mathbb{Z}$-graded vector space

$$C_*(\mathcal{B}, V) = \oplus_q C_q(\mathcal{B}, V).$$

Its homogeneous sub-spaces are defined by

$$C_q(\mathcal{B}, V) = 0 \quad if \quad q < 0,$$
$$C_0(\mathcal{B}, V) = V,$$
$$C_q(\mathcal{B}, V) = T_q(\tilde{\mathcal{B}}) \otimes V \quad if \quad q > 0.$$

Every homogeneous vector subspace $C_q(\mathcal{B}, V)$ is a left module of the KV algebra $\mathcal{B}$. The left action is defined by

$$s \cdot (\xi \otimes v) = s \cdot \xi \otimes v + \xi \otimes s \cdot v.$$

Let j and q be two positive integers such that $j < q$.
Let $\xi = x_1 \otimes x_2 ... \otimes x_q$. To define the linear map

$$d_j : C_q(\mathcal{B}, V) \to C_{q-1}(\mathcal{B}, V)$$

we put

$$d_j(\xi \otimes v) = X_j^* \cdot (\partial_j \xi \otimes v).$$

Henceforth one defines the boundary operator

$$d : C_q(\mathcal{B}, V) \to C_{q-1}(\mathcal{B}, V)$$

by setting

$$d = \sum_1^q (-1)^j d_j.$$

So we obtain a chain complex whose homology space of degree q is denoted by $H_q(\mathcal{B}, V)$.

Definition 14. *The graded vector space*

$$H_*(\mathcal{B}, V) = \sum_q H_q(\mathcal{B}, V)$$

is called the total homology of $\mathcal{B}$ with coefficients in V.

3.2.6. Two Cochain Complexes

We are going to define two cochain complexes. They are denoted by $C_{KV}(\mathcal{B}, V)$ and by $C_\tau(\mathcal{B}, V)$ respectively.
We recall that the vector subspace $J(V) \subset V$ is defined by

$$(s \cdot s*) \cdot v - s \cdot (s * \cdot v) = 0 \quad \forall s \quad s* \in \mathcal{B}.$$

Let us set

$$C^0_{KV}(\mathcal{B}, V) = J(V),$$
$$C^0_\tau(\mathcal{B}, V) = V,$$
$$C^q(\mathcal{B}, V) = Hom_{\mathbb{R}}(T_q(\tilde{\mathcal{B}}) \quad \forall q \geq 1.$$

Let (j, q) be a pair of non negative integers such that $j < q$. We are going to define the linear map

$$d_j : C^q(\mathcal{B}, V) \to C^{q+1}(\mathcal{B}, V).$$

Given $f \in C_q(\mathcal{B}, V)$ and

$$\xi = x_1 \otimes ... \otimes x_{q+1}$$

we put

$$d_j f(\xi) = X_j^* \cdot f(\partial_j \xi) - f(d_j \xi).$$

The family of linear mappings d_j has property $S \cdot 1$, viz

$$d_j d_i = d_i d_{j-1} \quad \forall i, j \quad with \quad i < j.$$

We use these data for constructing two cochain complexes. They are denoted by (C^*_{KV}, d_{KV}) and by (C^*_τ, d_τ) respectively. The underlying graded vector spaces are defined by

$$C_{KV} = J(V) \oplus \sum_{q>0} C^q(\mathcal{B}, V),$$

$$C_\tau = V \oplus \sum_{q>0} C^q(\mathcal{B}, V).$$

Their coboundary operators are defined by

$$(d_{KV}v)(s) = -s \cdot v,$$

$$(d_\tau w)(s) = -sw,$$

$$d_{KV}(f) = \sum_1^q (-1)^j d_j(f) \quad if \quad q > 0,$$

$$d_\tau(f) = \sum_1^{q+1} (-1)^j d_j(f) \quad if \quad q > 0.$$

The simplicial formula (5a) yields the identities

$$d^2_{KV} = 0,$$

$$d^2_\tau = 0.$$

The cohomology space

$$H_{KV}(\mathcal{B}, V) = \sum_q H^q_{KV}(\mathcal{B}, V)$$

is called the V-valued KV cohomology of $\mathcal{B}$.

The cohomology space

$$H_\tau(\mathcal{B}, V) = \sum_q H^q_\tau(\mathcal{B}, V)$$

is called the V-valued total KV cohomology of $\mathcal{B}$.

The algebra $\mathcal{A}$ is a two-sided ideal of the KV algebra $\mathcal{B}$. Mutatis mutandis our construction gives the cohomology spaces $H_{KV}(\mathcal{A}, V)$ and $H_\tau(\mathcal{A}, V)$. They are called the V-valued KV cohomology and the V-valued total KV cohomology of $\mathcal{A}$.

Comments.

Though the spectral sequences are not the purpose of this paper we recall that the pair $(\mathcal{A} \subset \mathcal{B})$ *gives rise to a spectral sequences* E^{ij}_r *[32–34]. The term* E^{ij}_0 *is nothing other than* $H_{KV}(\mathcal{A}, V)$ *[29]. In other words one has*

$$H^q_{KV}(\mathcal{A}, V) = \sum_{0 \le j \le q} E^{j,q-j}_0.$$

3.2.7. Residual Cohomology

Before pursuing we introduce the notion of residual cohomology. It will be used in the section be devoted the homological statistical models.

The machinery we are going to introduce is similar to the machinery of Eilenberg [35]. In particular we introduce the residual cohomology. Our construction leads to an exact cohomology sequence which links the residual cohomology with the equivariant cohomology. We restrict the attention to the category of left modules of KV algebroids. We keep our previous notation.

We recall that for every positive integer $q > 0$ the vector space $C^q(\mathcal{B}, V)$ is a left module of $\mathcal{B}$. The left action of $s \in \mathcal{B}$ is defined by

$$(s \cdot f)(\xi) = s \cdot f(\xi) - f(s \cdot \xi).$$

Definition 15. *A cochain $f \in C^q(\mathcal{B}, V)$ is called a left invariant cochain if*

$$s \cdot f = 0 \quad \forall s \in \mathcal{B} \quad \forall s.$$

A straightforward consequence of this definition is that a left invariant cochain is a cocycle of both C^*_{KV} and C^*_{τ}. The vector subspace of left invariant q-cochains of $\mathcal{B}$ is denoted by $H^q_e(\mathcal{B}, V)$. It is easy to see that

$$Z^q_\tau(\mathcal{B}, V) \cap Z^q_{KV}(\mathcal{B}, V) = H^q_e(\mathcal{B}, V),$$
$$Z^q_\tau(\mathcal{A}, V) \cap Z^q_{KV}(\mathcal{A}, V) = H^q_e(\mathcal{A}, V).$$

Definition 16. *A KV cochain of degree q whose coboundary is left invariant invariant is called a residual KV cocycles.*

(1) The vector subspace of residual KV cocycles of degree q is denoted by Z^q_{KVres}.

(2) The vector subspace of residual coboundaries of degree q is defined by $B^q_{KVres} = H^q_e(\mathcal{B}, V) + d_{KV}(C^{q-1}_{KV}(\mathcal{B}, V))$. The residual KV cohomology space of degree q is the quotient vector space.

(3) $H^q_{KVres}(\mathcal{B}, V) = \frac{Z^q_{KVres}}{B^q_{KVres}}$.

(4) By replacing the KV complex by the total KV complex one defines the vector space of residual total cocycles $Z^q_{\tau res}$ and the space of residual total coboundaries $B^q_{\tau res}$. Therefore we get the residual total KV cohomology space

$$H^q_{\tau res}(\mathcal{A}, V) = \frac{Z^q_{\tau res}}{B^q_{\tau,res}}$$

The definitions above lead to the cohomological exact sequences which is similar to those constructed by Eilenberg machinery [35]. We are going to pay a special attention to two cohomology exact sequences.

(1) At one side the operator d_{KV} yields a canonical linear map

$$H^q_{KVres}(\mathcal{B}, V) \to H^{q+1}_e(\mathcal{B}, V).$$

(2) At another side every KV cocycle is a residual cocycle and every KV coboundary is a residual coboundary as well. Then one has a canonical linear map

$$H^q_{KV}(\mathcal{B}, V) \to H^q_{KVres}(\mathcal{A}, V).$$

Those canonical linear mappings yield the following exact sequences

$$\to H^{q-1}_{KVres}(\mathcal{B}, V) \to H^q_e(\mathcal{B}, V) \to H^q_{KV}(\mathcal{B}, V) \to H^q_{KVres}(\mathcal{B}, V) \to$$

$$\to H^{q-1}_{\tau res}(\mathcal{B}, V) \to H^q_e(\mathcal{B}, V) \to H^q_\tau(\mathcal{B}, V) \to H^q_{\tau res}(\mathcal{B}, V) \to$$

Some Comments.

$(c.1)$: *We replace the KV $\mathcal{B}$ by $\mathcal{A}$. Then we obtain the exact sequences*

$$\to H^{q-1}_{KVres}(\mathcal{A},V) \to H^q_e(\mathcal{A},V) \to H^q_{KV}(\mathcal{A},V) \to H^q_{KVres}(\mathcal{A},V) \to$$

$$\to H^{q-1}_{\tau res}(\mathcal{A},V) \to H^q_e(\mathcal{A},V) \to H^q_{\tau}(\mathcal{A},V) \to H^q_{\tau res}(\mathcal{A},V) \to$$

$(c.2)$: *The KV cohomology difers from the total cohomology. Loosely speaking their intersecttion is the equivariant cohomology $H^*_e(\mathcal{B},V)$ their difference is the residual cohomology. The domain of their efficiency are different as well. Here are two illustrations.*

Example 1.

In the introduction we have stated a conjecture of M. Gerstenhaber, namely *Every Restricted Theory of Deformation Generates Its Proper Theory of Cohomology*.

From the viewpoint of this conjecture, the KV cohomology is the completion a long history [2,9,28]. Besides Koszul and Nijenhuis, other pioneering authors are Vinberg, Richardson, Gerstenhaber, Matsushima, Vey.

The challenge was the search for a theory of cohomology which might be generated by the theory of deformation of locally flat manifolds [8]. The expected theory is the now known KV theory of KV cohomolgy [9].

Example 2.

The total cohomology is close to both the pioneering Nijenhuis work [28,36]. In [29] we have constructed a spectral sequence which relates to [28,36].

From another viewpoint, the total KV cohomology is useful for exploring the relationships between the information geometry and the theory of Riemannian foliations. This purpose will be addressed in the next sections.

3.3. *The Theory of KV Cohomology—Version the Anomaly Functions*

This subsection is devoted to use the KV anomaly functions for introducing the theory of cohomology of KV algebroids and their modules.

This viewpoint leads to an unifying framework for introducing the theory of cohomology of abstract algebras and their abstract two-sided modules. Here are a few examples of cohomology theory which are based on the anomaly functions.

Example 1. The theory of Hochschild cohomology of associative algebras is based on the associator anomaly function.

Example 2. The theory of Chevalley-Eilenberg-Koszul cohomology of Lie algebras is based on the Jacobi anomaly function.

Example 3. The theory of cohomology of Leibniz algebras is based on the Jacobi anomaly function as well.

3.3.1. The General Challenge $CH(\mathbb{D})$

We consider data

$$\mathbb{D} = [(\mathcal{A}, A_{\mathcal{A}}), (V, A_{AV}), Hom(T(\mathcal{A}), V)].$$

Here

(1) V is an (abstract) two sided module of an (abstract) algebra $\mathcal{A}$.
(2) $A_{\mathcal{A}}$ and $A_{\mathcal{A}V}$ are fixed anomaly functions of $\mathcal{A}$ and of V respectively.

(3) $Hom(T(\mathcal{A}), V)$ stands for the $\mathbb{Z}$-graded vector space

$$Hom(T(\mathcal{A}), V) = \oplus_q Hom_{\mathbb{R}}(\mathcal{A}^{\otimes q}, V).$$

Let $A_{\mathbb{D}}$ be the category of (abstract) algebras and (abstract) modules whose structures are defined by the pair $(A_{\mathcal{A}}, A_{\mathcal{AV}})$. So the rules of calculations in the category $\mathbb{A}$ are defined by the identities

$$A_{\mathcal{A}}(a, b, c) = 0,$$

$$A_{\mathcal{AV}}(a, b, v) = 0.$$

The challenge is the search of a particular family of linear maps

$$Hom(\mathcal{A}^{\otimes q}, V) \ni f \to d_q(f) \in Hom(\mathcal{A}^{\otimes q+1}, V).$$

Such a particular family d_q must satisfy a condition that we call the property Δ.

Property Δ

$\forall \xi = a_1 \otimes a_2 ... \otimes a_{q+2} \in \mathcal{A}^{q+2}$, $\forall f \in Hom(\mathcal{A}^{\otimes q}, V)$ *the quantity* $[d_{q+1}(d_q(f))](\xi)$ *depends linearly on the values of the anomaly functions*

$$\{A_{\mathcal{A}}(a_i, a_j, a_k), \quad A_{AV}(a_i, a_j, v)\}$$

Let us assume that a family d_q is a solution to $CH(\mathbb{D})$. Then the category $A_{\mathbb{D}}$ admits a theory of cohomology with coefficients in modules.

The next is devoted to this challenge in the category of KV algebras and KV modules. The geometry version is the category of KV algebroids and KV modules of KV algebroids.

3.3.2. Challenge $CH(\mathcal{D})$ for KV Algebras

Let W be a two-sided module of an abstract algebra $\mathcal{A}$. We assume that the following bilinear mappings are non trivial applications

$$\mathcal{A} \times W \ni (X, w) \to X \cdot w \in W,$$

$$W \times \mathcal{A} \ni (w, X) \to w \cdot X \in W.$$

Let $f \in Hom(\mathcal{A}^{\otimes q}, W)$. We consider a monomial $\xi \in \mathcal{A}^{\otimes q+1}$, so

$$\xi = X_1 \otimes ... \otimes X_{q+1} \in \mathcal{A}^{\otimes q+1}.$$

Our construction is divided into many STEPS.

Step 1.

Let $(i < j)$ be a pair of positive integers with $1 \leq i < j \leq q$. The linear the map

$$S_{[i,j]}(f) \in Hom(\mathcal{A}^{q+1}, V).$$

$S_{[ij]}$ is defined by

$$S_{[i,j]}(f)(X_1 \otimes ... \otimes X_{q+1}) = (-1)^j [X_j \cdot f(X_1 \otimes ... \otimes X_i \otimes ... \hat{X}_j \otimes X_{j+1} ... \otimes X_{q+1})$$
$$+ (f(X_1 \otimes ... \otimes X_i \otimes ... \hat{X}_j \otimes ... \hat{X_{q+1}} \otimes X_j) \cdot X_{q+1}$$
$$- \omega(f) f(X_1 \otimes ... \otimes X_j \cdot X_i \otimes ... \hat{X}_j \otimes ... \otimes X_{q+1})]$$

$$+(-1)^i[X_i \cdot f(X_1 \otimes ...\hat{X}_i \otimes ... \otimes X_j \otimes . \otimes X_{q+1})$$
$$+(f(X_1 \otimes ...\hat{X}_i \otimes ... \otimes X_j \otimes ...\hat{X_{q+1}} \otimes X_i)) \cdot X_{q+1}$$
$$-\omega(f)f(X_1 \otimes ...\hat{X}_i \otimes ... \otimes X_i \cdot X_j \otimes ... \otimes X_{q+1})].$$

In the right side member of $S_{[i,j]}(f)(\xi)$ the coefficient $\omega(f)$ is the degree of f, viz $\omega(f) = q$ for all $f \in Hom(\mathcal{A}^q, W)$.

Step 2.

For every pair $(i, q+1)$ with $1 \leq i \leq q$ we define the map $S_{[i,q+1]}(f)$ by

$$S_{[i,q+1]}(f)(X_1 \otimes ... \otimes X_{q+1}) = (-1)^i[X_i \cdot f(X_1 \otimes ...\hat{X}_i \otimes ... \otimes X_{q+1})$$
$$+(f(X_1 \otimes ...\hat{X}_i \otimes ...\hat{X_{q+1}} \otimes X_i)) \cdot X_{q+1}$$
$$-\omega(f)f(X_1 \otimes ...\hat{X}_i \otimes ...X_i \cdot X_{q+1})].$$

Step 3.

Let $g \in Hom(\mathcal{A}^{\otimes q+1}, W)$ and let

$$\xi = X_1 \otimes ... \otimes X_{q+2} \in \mathcal{A}^{\otimes q+2}.$$

Let i, j, k be three positive integers such that $i < j < q+2; \quad k \leq q+2$. We have already introduced the notation

$$\partial_k \xi = X_1 \otimes ...\hat{X}_k \otimes ... \otimes X_{q+2},$$
$$\partial^2_{k,q+2}\xi = X_1 \otimes ...\hat{X}_k \otimes ... \otimes ...\hat{X}_{q+2}.$$

We define $S^k_{[i,j]}(g) \in Hom(\mathcal{A}^{\otimes q+2}, W)$ by setting

$$S^k_{[i,j]}(g)(\xi) = (-1)^{i+k}[X_k \cdot g(\partial_k \xi) + (g(\partial^2_{k,q+2}\xi \otimes X_k)) \cdot X_{q+2}$$
$$+\omega(g)g(X_1 \otimes ... \otimes X_k \cdot X_i \otimes ...\hat{X}_k \otimes ... \otimes X_{q+2})]$$
$$+(-1)^{j+k}[X_k \cdot g(\partial_k \xi) + (g(\partial^2_{k,q+2}\xi) \otimes X_k) \cdot X_{q+2}$$
$$+\omega(g)g(X_1 \otimes ... \otimes X_k \cdot X_j \otimes ...\hat{X}_k \otimes ... \otimes X_{q+2})].$$

Given a triple (i, j, k) with $i < j < k < q+2$ we put

$$S_{[i,j,k]}(g)(\xi) = S^k_{[i,j]}(g)(\xi) + S^j_{[i,k]}(g)(\xi) + S^i_{[j,k]}(g)(\xi).$$

The proof of the following statement is based on direct calculations.

Lemma 1.

$$(* * **): \quad \sum_{[i<j]} S_{[i,j]}(g)(\xi) = \sum_{[i<j<k]} S_{[i,j,k]}(g)(\xi)$$

Let $f \in Hom(\mathcal{A}^q, W)$. In both the left side and the right side of the equality $(* * **)$ we replace g by $\sum_{i<j} S_{[i,j]}(f)$. Then we obtain a linear mapping

$$Hom(\mathcal{A}^{\text{II}}, W) \ni f \to E^{****}(f) \in Hom(\mathcal{A}^{q+2}, W).$$

Our aim is to evaluate ξ of $E^{****}(f)$ at $\xi \in \mathcal{A}^{\otimes q+2}$. Here

$$\xi = X_1 \otimes ... \otimes X_{q+2}.$$

To calculate $[E^{****}(f)](\xi)$ we take into account both STEP1 and STEP2. Then we obtain

$$[E^{****}(f)](\xi) = \sum_{[i<j<q+2;1\leq k\leq q+2]} [E^{****}_{[ijk]}(f)](\xi).$$

At the right side member

$$[E^{****}_{[ijk]}(f)](\xi) = (-1)^{i+j}[KV(X_i, X_j, f(X_1 \otimes ...\hat{X}_i \otimes ..\hat{X}_j \otimes ...X_k \otimes ... \otimes X_{q+2}))$$
$$+KV(X_i, f(X_1 \otimes ...\hat{X}_i...\hat{X}_j \otimes ... \otimes X_{q+1} \otimes X_j), X_{q+2})$$
$$+KV(X_j, f(X_1 \otimes ...\hat{X}_i...\hat{X}_j \otimes ... \otimes X_{q+1} \otimes X_i), X_{q+2})$$
$$+\omega(f)(\omega(f)+1)f(X_1 \otimes ...\hat{X}_i...\hat{X}_j \otimes ... \otimes KV(X_i, X_j, X_k) \otimes ... \otimes X_{q+2})].$$

Step 4.

We are in position to face $CH(\mathbb{D})$.

Definition 17. *Let $f \in Hom(\mathcal{A}^{\otimes q}, W)$ and $\xi = X_1 \otimes ... \otimes X_{q+1} \in \mathcal{A}^{\otimes q+1}$. We take into account Step 1, Step 2 and Step 3. Therefore, we define the linear map*

$$Hom(\mathcal{A}^{\otimes q}, W) \ni f \rightarrow \partial f \in Hom(\mathcal{A}^{\otimes q+1}, W)$$

by putting

$$[\partial f](\xi) = \sum_{1\leq i<j\leq q+1} S_{[i,j]}(f)(\xi)$$

The following lemma is a straightforward consequence of the machinery in STEP3.

Lemma 2.

$$\partial^2 f(\xi) = \sum_{[i<j<q+2];1\leq k\leq q+2} [E^{****}_{[ijk]}(f)](\xi)$$

Lemma 2 tells us that $\partial^2 f(\xi)$ depends linearly on the values of the KV anomaly functions.

The challenge $CH(\mathbb{D})$ is won in the category of KV algebras and their two-sided KV modules.

We replace the category of KV algebras and their two-sided modules by the category of KV algebroids and their bi-modules. Then we win the geometry version of $CH(\mathbb{D}$.

We use Lemma 2 for introducing a theory of KV homology of KV algebras and their two-sided modules.

3.3.3. The KV Cohomology

Let W be a two sided KV module of a KV algebra $\mathcal{A}$. We consider the graded vector space

$$\mathcal{C}_{KV} = \oplus_q C^q_{KV}.$$

The homogeneous subspaces are defined by $C^q_{KV} = 0$ if q is a negative integer, $C^0_{KV} = J(W)$, $C^q_{KV} = Hom(\mathcal{A}^{\otimes q}, W)$ if q is a positive integer.

We define the linear map

$$C^q_{KV} \ni f \rightarrow \partial_{KV} f \in C^q + 1_{KV}$$

by setting

$$\partial_{KV}(w)(X) = -X \cdot w + w \cdot X \quad if \quad w \in J(W), \tag{7a}$$

$$\partial_{KV} f = \sum_{[i<j]} S_{[i,j]}(f) \quad if \quad q > 0. \tag{7b}$$

By Lemma 2 we obtain the following statement

Theorem 3. *For every two sided KV module W of a KV algebra $\mathcal{A}$ the pair $(C^*_{KV}, \partial_{KV})$ is a cochain complex.*

3.3.4. The Total Cohomology

Let W be a two-sided module of a KV algebra $\mathcal{A}$. Our concern is the $\mathbb{Z}$-graded vector space

$$C_\tau = W + \oplus_{q>0} C^q(\mathcal{A}, W).$$

For our present purpose the maps S_{ij} are not subject the requirement as in Step 2.
We define the coboundary operator ∂_τ by setting

$$\partial_\tau w(X) = -X \cdot w + w \cdot X \quad \forall w in W,$$

$$\partial_\tau f(\xi) = \sum_{1 \leq i < j \leq q+1} S_{[i,j]}(f)(\xi) \quad \forall q > 0.$$

The quantity $(\partial^2_\tau f(\xi)$ depends linearly on the KV anomaly functions of the pair $(\mathcal{A}, W)$. Thus the pair $(C^*_\tau, \partial_\tau)$ is a cochain complex. Its cohomology is called the W-valued total KV cohomology of $\mathcal{A}$. We denote it by $H^*_\tau(\mathcal{A}, W)$.

3.3.5. The Residual Cohomology, Some Exact Sequences, Related Topics, DTO-HEG-IGE-ENT

In the next sections we will see that the links between the information geometry and the differential topology involve the real valued total KV cohomology of KV algebroids. Many relevant relationships are based on the exact sequences

$$\rightarrow H^{q-1}_{KVres}(\mathcal{A}, \mathbb{R}) \rightarrow H^q_e(\mathcal{A}, \mathbb{R}) \rightarrow H^q_{KV}(\mathcal{A}, \mathbb{R}) \rightarrow H^q_{KVres}(\mathcal{A}, \mathbb{R}) \rightarrow$$

$$\rightarrow H^{q-1}_{\tau res}(\mathcal{A}, \mathbb{R}) \rightarrow H^q_e(\mathcal{A}, \mathbb{R}) \rightarrow H^q_\tau(\mathcal{A}, \mathbb{R}) \rightarrow H^q_{\tau res}(\mathcal{A}, \mathbb{R}) \rightarrow$$

Now we are provided with cohomological tools which will be used in the next sections.
We plan to perform KV cohomological methods for studying some links between the vertices of the square "DTO, IGE, ENT HGE" as in Figure 1. We recall basic notions.

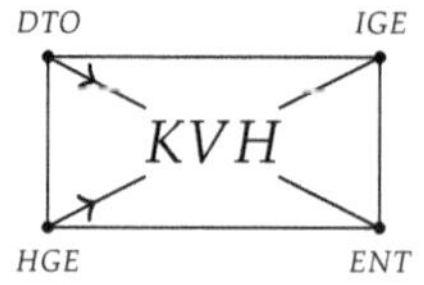

Figure 1. Federation.

DTO stands for Differential TOpology.

The purposes: Riemannian foliations and Riemannian webs. Symplectic foliations and symplectic webs. Linearization of webs.

Our aims: We use cohomological methods for constructing Riemannian foliations, Riemannian webs, linearizable webs.

Nowadays, there does not exist any criterion for deciding whether a manifold supports those differential topological objects. Our aim is to discuss sufficient conditions for a manifold admitting those structures. Our approach leads to notable results. The key tools are the KV cohomology and the dualistic relation of Amari. Both the KV cohomology and the dualistic relation product remarkable split exact sequences. Notable results are based on those exact sequences. *HGE stands for Hessian GEometry.*

The purposes: Hessian structures, geometry of Koszul, hyperbolicity, cohomological vanishing theorems. Our aims: The geometry of Koszul is a cohomological vanishing theorem. Statistical geometry and vanishing theorem, the solution to a hold question of Alexander K Guts (announced).

Theorem 3 as in [2] may be rephrased in the framework of the theory of KV homology. For a compact locally flat manifold (M, ∇) being hyperbolic it is necessary and sufficient that $C^2_{KV}(\mathcal{A}, C^\infty(M))$ contains a positive definite EXACT cocycle. To be hyperbolic is a geometrical-topological property of the developing map of locally flat manifolds. To be hyperbolicitic means that the image of the developing is a convex domain not containing any straight line. This formulation is far from being a homological statement. So the Hessian GEOmetry is a link between the theory of KV homology and the Riemannian Riemannian geometry.

The geometry of Koszul, the geometry of homogeneous bounded domains and related topics have been studied by Vinberg, Piatecci-Shapiro and many other mathematicians [3]. The geometry of Siegel domains belongs to that galaxy [7,12]. Almost all of those studies are closely related to the Hessian geometry.

Among the open problems in the Hessian geometry are two questions we are concerned with. The first is to know whether the metric tensor g of a Riemannian manifold is a Hessian metric. Alexander K. Guts raised this question in a mail (to me) forty years ago. The second question is to know whether a locally flat manifold admits a Hessian tensor metric. The solutions to those two problems are announced in the Appendix A to this paper.

IGE stands for Information GEometry.

The purposes: The differential geometry of statistical models, the complexity of statistical models, ramifications of the information geometry.

Our aims: We revisit the classical theory of statistical models, requests of McCullagh and Gromov. A search of a characteristic invariant. The moduli space of models. The homological nature of the information geometry.

The information geometry is the differential geometry in statistical models for measurable sets. In both the theoretical statistics and the applied statistics the exponential families and their generalizations are optimal statistical models. There are many references, e.g., [17,18,22,37]. Here Murray-Rice 1.15 means Murray-Rice Chapter 1, Section 15. A major problem is to know whether a given statistical model is isomorphic to an exponential model. That is what we call the complexity problem of statistical models. This challenge is a still open problem. It explicitly arises from the purposes which are discussed in [22] here, see also [30]. In the appendix to this paper we present a recently discovered invariant which measures how far from being an exponential family is a given model. That invariant is useful for exploring the differential topology of statistical models. That is particularly important when models are singular, viz models whose Fisher information is not inversible.

ENT stands for ENTropy.

Pierre Baudot and Daniel Bennequin recently discovered that the entropy function has a homological nature [31]. We recall that in 2002 Peter McCullagh raised a fundamental geometric-topological question in the theory of information: What Is a Statistical Model? [30] A few years after Misha Gromov raised a similar request: The Search of Structure. Fisher Information [15,16].

Those two titles are two formulations of the same need.

The paper of McCullagh became the subject of controversy. It gave rise to questions, discussions, criticisms, see [30].

In Part B of this paper we will be addressing this fundamental problem. A reading of the McCullagh paper would be useful for drawing a comparison between our approach and [15,16,30].

4. The KV Topology of Locally Flat Manifolds

4.1. The Total Cohomology and Riemannian Foliations

In this section we focus on the KV algebroids which are defined by structures of locally flat manifolds. To facilitate a continuous reading of this paper we recall fundamental notions which are needed.

Definition 18. *A locally flat manifold is a pair (M, D). Here D is a torsion free Koszul connection whose curvature tensor R^D vanishes identically.*

The pair (M, D) defines a Koszul-Vinberg algebroid

$$A = (TM, D, 1)$$

The anchor map is the identity map of TM. The multiplication of sections is defined by D, viz

$$X \cdot Y = D_X Y$$

forall $X, Y \in \mathcal{X}(M)$.

The KV algebra of (M, D) is the algebra

$$\mathcal{A} := (\mathcal{X}(M), D).$$

The cotangent bundle T^*M is a left module of the KV algebroid $(TM, D, 1)$. For every $(X, Y, \theta) \in \mathcal{X}(M) \times \mathcal{X}(M) \times \Gamma(T^*M)$ the differential 1-form $X \cdot \theta$ is defined by

$$[X \cdot \theta](Y) = [d(\theta(Y))](X) - \theta(X \cdot Y).$$

In the right hand member of the equality above $d(\theta(Y))$ is the exterior derivative of the real valued function $\theta(Y)$.

Let $S^2(T^*M)$ be the vector bundle of symmetric bi-linear forms in M.

The vector space of sections of $S^2(T^*M)$ is denoted by $\mathcal{S}_2(M)$, viz

$$\mathcal{S}_2(M) = \Gamma(S^2(T^*M)).$$

The vector space $\mathcal{S}_2(M)$ is a left module of the KV algebra $\mathcal{A}$. The left action of $\mathcal{A}$ in $\mathcal{S}_2(M)$ is defined by

$$(X \cdot g)(Y, Z) = [dg(Y, Z)](X) - g(X \cdot Y, Z) - g(Y, X \cdot Z).$$

We put

$$\Omega^1(M) = \Gamma(T^*M).$$

The T^*M-valued cohomology of the KV algebroid $(TM, D, 1)$ is but the cohomology of $\mathcal{A}$ with coefficients in $\Omega^1(M)$. The KV cohomology and the total cohomology are denoted by

$$H^*_{KV}(\mathcal{A}, \Omega^1(M)),$$

$$H^*_{\tau}(\mathcal{A}, \Omega^1(M)).$$

Warning.

We observe that elements of $\mathcal{S}_2(M)$ may be regarded as 1-cochains of $\mathcal{A}$ with coefficients in its left module $\Omega^1(M)$. By [29] we have

$$Z^2_\tau(\mathcal{A}, C^\infty(M)) = \mathcal{S}^{\mathcal{A}}_2(M). \tag{8}$$

At another side we have the cohomolgy exact sequence

$$\to H^1_{KVres}(\mathcal{A}, V) \to H^2_{KVe}(\mathcal{A}, V) \to H^2_{KV}(\mathcal{A}, V) \to H^2_{KVres}(\mathcal{A}, V) \to \tag{9}$$

By Equations (8) and (9) we obtain the inclusion maps

$$\mathcal{S}^{\mathcal{A}}_2(M) \subset Z^1_{KV}(\mathcal{A}, \Omega^1(M)) \subset Z^2_{KV}(\mathcal{A}, \mathbb{R}).$$

Mutatis mutandis one also has

$$\mathcal{S}^{\mathcal{A}}_2(M) \subset Z^1_\tau(\mathcal{A}, \Omega^1(M) \cap Z^2_\tau(\mathcal{A}, \mathbb{R}).$$

Remark 1 (Important Remarks). *We give some subtle consequences of (1).*

(R.1) Every exact total 2-cocycle $\omega \in C^2_\tau(\mathcal{A}, \mathbb{R})$ is a skew symmetric bilinear form. Viz one has the identity

$$\omega(X, X) = 0 \quad \forall X \in \mathcal{A}.$$

(R.2) Every symmetric KV 2-cocycle $g \in Z^2_{KV}(\mathcal{A}, \mathbb{R})$ is locally an exact KV cocycle, viz in a neighbourhood of every point there exists a local section $\theta \in \Omega^1(M)$ such that

$$g = \delta_{KV}\theta.$$

(R.3) Every symmetric total 2-cocycle is a left invariant cochain, viz

$$Z^2_\tau(\mathcal{A}, \mathbb{R}) \cap \mathcal{S}_2(M) = \mathcal{S}^{\mathcal{A}}_2(M).$$

By (R.1) and (R.3) we obtain the inclusion map

$$\mathcal{S}^{\mathcal{A}}_2(M) \subset H^2_\tau(\mathcal{A}, \mathbb{R}).$$

Let $H^2_{dR}(M)$ be the second cohomology space of the de Rham complex of M. The following theorem is useful for relating the total KV cohomology and the differential topology.

Theorem 4. *[29] There exists a canonical linear injection of $H^2_{dR}(M)$ in $H^2_\tau(\mathcal{A}, \mathbb{R})$ such that*

$$H^2_\tau(\mathcal{A}, \mathbb{R}) = H^2_{dR}(M) \oplus \mathcal{S}^{\mathcal{A}}_2(M)$$

The theorem above highlights a fruitful link between the total KV cohomology and the differential topology. We are particularly interested in D-geodesic Riemannian foliations in a locally flat manifold (M, D).

Warning.

Throughout this paper a Riemannian metric tensor in a manifold M is a non-degenerate symmetric bilinear form in M.

A positive metric tensor is a positive definite metric tensor.

In the next we use the following definition of Riemannian foliation and symplectic foliation.

Definition 19. *A Riemannian foliation is an element $g \in \mathcal{S}_2(M)$ which has the following properties*

(1.1) *$rank(g) = constant$,*

(1.2) *$L_X g = 0 \quad \forall X \in \Gamma(Ker(g))$. A symplectic foliation is a (de Rham) closed differential 2-form ω which satisfies*

(2.1) *$rank(\omega) = constant$,*

(2.2) *$L_X \omega = 0 \quad \forall X \in \Gamma(Ker(\omega))$.*

Warning.

When g is positive semi-definite our definition is equivalent to the classical definition of Riemannian foliation [38–40].

The complete integrability of Ker(g) *and the conditions to be satisfied by the holonomy of leaves are equivalent to the Property (2.2).*

The set of Riemannian foliations in a manifold M is denoted by $\mathcal{RF}(M)$. The last theorem above yields the inclusion map

$$\frac{H^2_\tau(\mathcal{A},\mathbb{R})}{H^2_{dR}(M)} \subset \mathcal{RF}(M).$$

We often use the notion of affine coordinates functions in a locally flat manifold. For non specialists we recall two definitions and the link between them.

Definition 20. *An m-dimensional affinely flat manifold is an m-dimensional smooth manifold M admitting a complete atlas $\{(U_j, \phi_j)\}$ whose local coordinate changes coincide with affine transformations of the affine space $\mathbb{R}^m$.*

We denoted an affine atlas by

$$\mathbb{A} = \{(U_j, \phi_j)\}.$$

Definition 21. *An affinely flat structure $(M, \mathbb{A})$ and a locally flat structure (M, ∇) are compatible if local coordinate functions of $(M, \mathbb{A})$ are solutions to the Hessian equation*

$$\nabla^2 x_j = O$$

Theorem 5. *For every positive integer m the relation to be compatible with a locally flat manifold is an equivalence between the category of m-dimensional affinely flat manifolds and the category of m-dimensional locally flat manifolds.*

4.2. The General Linearization Problem of Webs

In the framework $\mathcal{RF}(M)$ the inclusion

$$\frac{H^2_\tau(\mathcal{A},\mathbb{R})}{H^2_{dR}(M)} \subset \mathcal{RF}(M)$$

may be rewritten as the exact sequence

$$O \to H^2_{dR}(M) \to H^2_\tau(\mathcal{A}, \mathbb{R}) \to \mathcal{RF}(M).$$

Let (M, ∇) be a locally flat manifold whose KV algebra is denoted by $\mathcal{A}$. Every finite family in $H^2_\tau(\mathcal{A}, \mathbb{R})$ is a family of ∇-geodesic Riemannian foliations.

There does not exist any criterion to know whether a manifold supports Riemannian foliations. The exact cohomology sequences we have been performing provide us with a cohomological method for constructing Riemannian foliations in the category of locally flat manifolds. This is an impact of the theory of KV homology on DTO.

In the next section we will introduce other new ingredients which highlight the impacts on DTO of the information geometry.

Further we will see that those new machineries from the information geometry have a homological nature.

Another major problems in the differential topology is the linearization of webs. Among references are [41–43].

Definition 22. *Consider a finite family of distributions $\mathcal{D}_j \subset TM, j := 1, 2, ..., k$. Those distributions are in general position at a point $x \in M$ if for every subset $J \subset \{1, 2, ..., k\}$ one has*

$$dim(\sum_{j \in J} \mathcal{D}_j(x)) = \min\left\{dim(M), \sum_{j \in J} dim(\mathcal{D}_j(x))\right\}.$$

Definition 23. *A k-web in M is a family of completely integrable distributions which are in general position everywhere in M.*

A Comment.

The distributions belonging to a web may have different dimensions. An example of problem is the symplectic linearization of lagrangian 2-webs.

Let $(\mathcal{D}_j, j := 1, 2)$ be a lagrangian 2-web in a 2n-dimensional symplectic manifold (M, ω). The challenge is the search of special local Darboux coordinate functions

$$(x, y) = (x_1, ..., x_n, y_1, ..., y_n).$$

Those functions must have three properties
(1): $\omega(x, y) = \Sigma_j dx_j \wedge dy_j$; (2) : *The leaves of $\mathcal{D}_1$ are defined by $x = constant$;* (3): *The leaves of $\mathcal{D}_2$ are defined by $y = constant$.*

Definition 24. *An affine web in an affine space is a web whose leaves are affine subspaces.*

Definition 25. *A web in a m-dimensional manifold is linearizable if it is locally diffeomorphic to an affine web in a m-dimensional affine space.*

Example 1. In the symplectic manifold $(\mathbb{R}^2, e^{xy} dx \wedge dy)$ one considers the lagrangian 2-web which is defined by

$$\mathcal{L}_1 = \{(x, y) | x = constant\},$$

$$\mathcal{L}_2 = \{(x, y) | y = constant\}.$$

This lagrangian 2-web is not symplectic linearizable.

Example 2. We keep $(\mathcal{L}_1, \mathcal{L}_2)$ as in example.1. It is symplectic linearizable in $(\mathbb{R}^2, (e^x + e^y) dx \wedge dy)$. The linearization problem for lagrangian 2-webs is closely related to the locally flat geometry [10,44,45].

Example 3. What about the linearization of the 3-web defined by $L_1 := \{(x = constant, y)\}$, $L_2 := \{(x, y = constant)\}$, $L_3 := \{e^{-x}(x+y) = constant\}$, $(x,y) \in \mathbb{R}^2$.

Up to today the question as to whether it is linearizable is subject to controversies, see [42] and references therein.

4.3. The Total KV Cohomology and the Differential Topology Continued

We implement the KV cohomology to address some open problems in the differential topology. For our purpose we recall a few classical notions which are needed.

Definition 26. *A metric vector bundle over a manifold M is a vector bundle $\mathcal{V}$ endowed with a non-degenerate inner product $< v, v^* >$.*

A Koszul connection in a vector bundle $\mathcal{V}$ is a bilinear map

$$\Gamma(TM) \times \Gamma(\mathcal{V}) \ni (X, v) \to \nabla_X v \in \Gamma(\mathcal{V})$$

which has the properties

$$\nabla_{fX} v = f \nabla_X v \forall v, \forall f \in C\infty(M), \tag{10a}$$
$$\nabla_X f v = df(X) v + f \nabla_X v \forall v, \forall f \in C\infty(M). \tag{10b}$$

Definition 27. *A metric connection in $(\mathcal{V}, < -, >)$ is a Koszul connection ∇ which satisfies*

$$d(< v, v^* >)(X) - < \nabla_X v, v^* > - < v, \nabla_X v^* > = 0.$$

Definition 28. *Let $(M, \mathcal{D})$ be a foliation in the usual sense, viz $\mathcal{D}$ has constant rank and is in involution.*
(1): $(M, \mathcal{D})$ is transversally Riemannian if there exists a $g \in \mathcal{S}_2(M)$ such that

$$\mathcal{D} = \mathcal{K}er(g).$$

(2): $(M, \mathcal{D})$ is transversally symplectic if there exists a (de Rham) closed differential 2-form ω such that

$$\mathcal{D} = Ker(\omega)$$

A transversally Riemannian foliation and a transversally symplectic foliation are denoted by

$$(\mathcal{D}, g),$$
$$(\mathcal{D}, \omega).$$

Definition 29. *Given a Koszul connection ∇, a transversally Riemannian foliation $(\mathcal{D}, g)$ (respectively a transversally symplectic foliation $(\mathcal{D}, \omega)$) is called ∇-geodesic if*

$$\nabla g = 0,$$

$$\nabla \omega = 0$$

The notions of transversally Riemannian foliation and transversally symplectic foliation are weaker than the notion of Riemannian foliation and symplectic foliation. However if ∇ a torsion free Koszul connection every ∇geodesic transversally Riemannian foliation is a Riemannian foliation. Every ∇-geodesic transversally symplectic foliation is a symplectic foliation.

For the general theory of Riemannian foliations the readers are referred to [39,40,46], see also the monograph [38] and the references therein.

We have pointed out that criterions for deciding whether a smooth manifold admits Riemannian foliations (respectively symplectic foliations) are missing. Our purpose is to address this existence problem in the category $\mathcal{SLC}$ whose objects are symmetric gauge structures. Such an object is a pair (M, ∇) where ∇ is a torsion free Koszul connection in M. The category of locally flat structure $\mathcal{LF}$ is a subcategory of $\mathcal{SLC}$. The theory of KV homology is useful for discussing geodesic Riemannian foliations in the category $\mathcal{LF}$. In a locally flat manifold (M, D) we have been dealing with the decomposition

$$H^2_\tau(\mathcal{A}, \mathbb{R}) = H^2_{dR}(M) \oplus \mathcal{S}^D_2(M).$$

Here $\mathcal{A}$ is the KV algebra of (M, D).

Let $b_2(M)$ be the second Betti number of M. We define the numerical geometric invariant $r(D)$ by

$$r(D) = dim(H^2_\tau(\mathcal{A}, \mathbb{R})) - b_2(M).$$

Formally $r(D)$ is the codimension of $H^2_{dR}(M) \subset H^2_\tau(\mathcal{A}, \mathbb{R})$, viz

$$r(D) = dim(\frac{H^2_\tau(\mathcal{A}, \mathbb{R})}{H^2_{dR}(M)}).$$

We consider the exact sequences

$$O \to H^2_{dR}(M) \to H^2_\tau(\mathcal{A}, \mathbb{R}) \to \mathcal{S}^A_2(M) \to 0$$

and

$$\to H^2_{\tau,e}(\mathcal{A}, \mathbb{R}) \to H^2_\tau(\mathcal{A}, \mathbb{R}) \to H^2_{\tau,res}(\mathcal{A}, \mathbb{R}) \to H^3_{\tau,e}(\mathcal{R}, \mathbb{R}) \to$$

From those exact sequences, one deduces the equality

$$\frac{H^2_\tau(\mathcal{A}, \mathbb{R})}{H^2_{dR}(M)} = \frac{H^2_{\tau,e}(\mathcal{A}, \mathbb{R})}{H^2_{dR}(M)}.$$

Thus $r(D)$ is formally the dimension of $\mathcal{S}^A_2(M)$.

The present approach leads to the following statement

Proposition 1. *If $r(D) > 0$ then M admits non trivial D-geodesic Riemannian foliations.*

Proof. Let B be a non zero element of $\mathcal{S}^A_2(M)$ and let $\mathcal{D}$ be the kernel of B.

(1) Suppose that

$$0 < rank(\mathcal{D}) < dim(M)$$

Therefore, (M, B) is a D-geodesic Riemannian foliation.

(2) Suppose that

$$rank(\mathcal{D}) = O.$$

Then (M, B) is a Riemannian manifold the Levi-Civita connection of which is D. Therefore, the proposition holds. □

Before proceeding we define three numerical invariants

$$r(M) = \max \{r(D) | D \in \mathcal{LC}(M)\},$$

$$s(M, \mathcal{A}) = \max \left\{rank(B) | B \in \mathcal{S}^A_2(M)\right\},$$

$$s(M) = \max \{s(M, \mathcal{A}) | D \in \mathcal{LF}(M)\}.$$

The non negative integers $r(M)$ and $s(M)$ are global geometric invariants. They connect the total KV cohomology to geodesic Riemannian foliations. By this viewpoint the proposition has an interesting corollary.

Corollary 1. *In an m-dimensional manifold M suppose that the following inequalities are satisfied*

$$0 < s(M) < m.$$

Then the manifold M admits a locally flat structure (M, D^*) which supports a non trivial D^*-geodesic Riemannian foliation.

The integer $s(M)$ is a local characteristic invariant of some class of 2-webs in Hessian manifolds. Let (M, D) be a locally flat manifold whose KV algebra is denoted by $\mathcal{A}$. we recall that a Hessian metric tensor in (M, D) is a inversible cocycle $g \in Z^2_{KV}(\mathcal{A}, \mathbb{R})$.

Theorem 6. *Let (M, D, g) and (M^*, D^*, g^*) be m-dimensional Hessian manifolds. We assume that the following inequalities hold*

$$0 < s(M, D) = s(M^*, D^*) = s < m.$$

Then M and M^* admit linearizable 2-webs which are locally isomorphic.

Proof. The proof is based on methods of the information geometry.

Let $\mathcal{A}$ and $\mathcal{A}*$ be the KV algebras of (M, D) and of (M^*, D^*) respectively. By the hypothesis there exists a pair of geosic Riemannian foliations

$$(B, B^*) \in \mathcal{S}_2^{\mathcal{A}} \times \mathcal{S}_2^{\mathcal{A}^*}$$

such that

$$rank(B) = rank(B^*) = s.$$

By the dualistic relation both M and M^* admit locally flat structures $(M, \tilde{D})$ and $(M^*, \tilde{D}^*)$ defined by

$$g(Y, \tilde{D}_X Z) = Xg(Y, Z) - g(D_X Y, Z),$$
$$g^*(Y, \tilde{D}^*_X Z) = Xg^*(y, Z) - g^*(D^*_X Y, Z).$$

Their KV algebras are denoted by $\tilde{\mathcal{A}}$ and $\tilde{A}^*$.

Step a

There exists a 1-cocycle

$$\psi \in Z^1_\tau(\tilde{\mathcal{A}}, \tilde{A})$$

such that

$$B(X, Y) = g(\psi(X), Y),$$
$$Ker(B) = Ker(\psi).$$

By the definition of $\tilde{D}$ we have

$$TM = Ker(\psi) \oplus im(\psi).$$

Further $im(\psi)$ is $\tilde{D}$-geodesic and $Ker(B)$ is D-geodesic. Therefore, the pair

$$(Ker(\psi), im(\psi))$$

is a 2-web in M.

In (M^*, D^*, g^*) we obtain similar 2-web

$$(Ker(\psi^*), im(\psi^*)).$$

By the choice of B and B^* we have

$$rank(Ker(\psi)) = rank(Ker(\psi^*)) = m - s.$$

Now we perform the following arguments.

(a): The foliation B is D-geodesic. In a neighbourhood of every point $p_0 \in (M, D)$ we linearize B by choosing appropriate local affine coordinate functions

$$(x, y) = (x_1, ..., x_{m-s}, y_1, ..., y_s).$$

The leaves of $Ker(\psi)$ are defined by

$$y = constant.$$

Thereby those leaves are locally isomorphic to affine sub-spaces.

Step b

The distribution $im(\psi)$ is $\tilde{D}$-geodesic. Therefore, near the same point $p_0 \in (M, \tilde{D})$ we linearize $im(\psi$ by choosing appropriate local affine coordiante functions

$$(x^*, y^*) = (x_1^*, ..., y_1^*, ...).$$

The leaves of $im(\psi)$ are defined by

$$x^* = constant.$$

Thus near p_0 the foliation defined by $m(\psi)$ is isomorphic to an linear foliation.

Step c

By both step a and step b we choose a neighbourhood of p_0 which is the domain of systems of appropriate local coordinate functions (x, y) and (x^*, y^*). From those data we pick the local coordinate functions

$$(x, y^*) = (x_1, ..., x_{m-s}, y_1^*, ..., y_s^*).$$

So we linearize the 2-web $(Ker(\psi), im(\psi))$ with the local coordinate functions (x, y^*).

$$(Ker(\psi), im(\psi)).$$

Thus near the p_0 the 2-web $(Ker(\psi), im(\psi))$ is isomorphic to the linear 2-web (L_1, L_2) which is defined in $\mathbb{R}^m$ by

$$\mathbb{R}^m = \mathbb{R}^{m-s} \times \mathbb{R}^s.$$

Step d

At a point p_0^* in M^* we perform the construction as in step a and in steps b and c, then we linearize $(Ker(\psi^*), im(\psi*))$ by choosing appropriate local coordinate functions

$$(x^0, y^{0*}) = (x_1^0, ..., x_{m-s}^0, y_1^{0*}, ..., y_s^{0*}).$$

In final, near the $p_O^* \in M^*$ the web $(Ker(\psi^*), im(\psi^*))$ is diffeomorphic to the affine web whose leaves are parallel to a decomposition

$$\mathbb{R}^m = V^{m-s} \times V^s.$$

Here V^{m-s} and V^s are vector subspaces of $\mathbb{R}^m$. Their dimensions are $m-s$ and s.

Conclusion.

There exists a unique linear transformation ϕ of $\mathbb{R}^m$ such that

$$\phi(\mathbb{R}^{m-s} \times 0) = V^{m-s},$$

$$\phi(0 \times \mathbb{R}^s) = V^s.$$

Thereby there is a local diffeomorphism Φ of M in M^* subject to the requirements

$$\Phi(p_0) = p_0^*,$$

$$(x^0, y^{0*}) \circ \Phi = (x, y^*).$$

The differential of Φ is denoted by Φ_*. We express the properties above by

$$\Phi(p_0) = p_0^*,$$

$$\Phi_*[Ker(\psi), im(\psi)] = [Ker(\psi^*), im(\psi^*)].$$

This ends the sketch of proof of Theorem. □

In the next we use the following definitions.

Definition 30. *A finite family*

$$\{B_J, \quad J \subset \mathbb{Z}\} \subset \mathcal{S}_2^{\mathcal{A}}(M)$$

is in general position if the distributions $\{Ker(B_j), j \in J\}$ *are in general position.*

The following statement is a straight corollary of the theorem we just demonstrated.

Proposition 2. *In a locally flat manifold* (M, D) *with* $r(D) > 0$ *every finite family in general position define a linearizable Riemannian web.*

4.4. The KV Cohomology and Differential Topology Continued

We have seen how the total cohomology and linearizable Riemannian webs are related. More precisely the theory of KV cohomology provides sufficient conditions for a locally flat manifold admitting linearizable Riemannian webs. That approach is based on the split exact sequence

$$0 \to H^2_{dR}(M) \to H^2_\tau(\mathcal{A}, C^\infty(M)) \to \mathcal{S}_2^{\mathcal{A}}(M) \to 0.$$

4.4.1. Kernels of 2-Cocycles and Foliations

Not all locally flat manifolds admit locally flat foliations. The existence of locally flat foliations is related to the linear holomnomy representation, viz the linear component of the affine holonomy representation of the fundamental group. Via the developing map the affine holonomy representation is conjugate to the natural action of the fundamental group in the universal covering. The KV homology is useful for investigating the existence of locally flat foliations. To simplify we work in the analytic category. So our purposes include singular foliations.

For those purposes we focus on an elementary item which has a notable impacts on our request.

Let (M, D) be a locally flat manifold whose KV algebra is denoted by $\mathcal{A}$. Let $g \in C^2(\mathcal{A}, C^\infty(M))$. The left kernel and the right kernel of g are denoted by $Ker(g)$ and $K^0er(g)$ respectively.

$Ker(g)$ is defined by

$$g(X, Y) = 0 \quad \forall Y \in \mathcal{A}.$$

$K^0er(g)$ is defined by

$$g(Y, X) = 0 \quad \forall Y \in \mathcal{A}.$$

The scalar KV 2-cocycles have elementary relevant properties

(1) The left kernel of every KV 2-cocycle is closed under the Poisson bracket of vector fields.
(2) The right kernel of every KV 2-cocycle is a KV subalgebra of the KV algebra $\mathcal{A}$.

We translate those elementary properties in term of the differential topology

Theorem 7. *In an analytic locally flat manifold* (M, D)

(1) The arrow

$$Z^2_{KV}(\mathcal{A}, C^\infty(M)) \ni g \to Ker(g)$$

maps the set of analytic 2-cocycles in the category of analytic stratified foliations M,

(2) The arrow

$$Z^2_{KV}(\mathcal{A}, C^\infty(M)) \ni g \to K^0er(g)$$

maps the set of analytic 2-cocycles in the category of stratified locally flat foliations,

(3) If a 2-cocycle g is a symmetric form then $Ker(g)$ *is a stratified locally flat transversally Riemannian foliation.*

The vector subspace of symmetric 2-cocycles the kernels of which are D-geodesic is denoted by $\tilde{Z}^2_{KV}(\mathcal{A})$. The corresponding cohomology vector subspace is denoted by

$$\tilde{H}^2_{KV}(\mathcal{A}) \subset H^2_{KV}(\mathcal{A}, C^\infty(M)).$$

By the exact sequence

$$O \to H^2_{dR}(M) \to H^2_\tau(\mathcal{A}, C^\infty(M)) \to \mathcal{S}^{\mathcal{A}}_2(M) \to 0$$

we have the inclusion map

$$\frac{H^2_\tau(\mathcal{A}, C^\infty(M))}{H^2_{dR}(M)} \subset \tilde{H}^2_{KV}(\mathcal{A}) \subset \mathcal{RF}(M).$$

5. The Information Geometry, Gauge Homomorphisms and the Differential Topology

We combine the dualistic relation with gauge homomorphisms to relate the total cohomology and two problems.

(i) The first is the existence problem for Riemannian foliations.
(ii) The second is the linearization of webs.

Those relationships highlight other roles played by the total KV cohomology. Through this section we use the brute coboundary operator.

5.1. The Dualistic Relation

We are interest in the foliation counterpart of the reduction in statistical models. The statistical reduction theorem is Theorem 3.5 as in [18]. We recall the notions which are needed.

Definition 31. *A dual pair is a quadruple (M,g,D,D^*) where (M,g) is a Riemannian manifold, D and D^* are Koszul connections in M which are related to the metric tensor g by*

$$Xg(Y,Z) = g(D_XY,Z) + g(Y,D^*_XZ) \quad \forall X,Y,Z.$$

We recall that a Riemannian tensor is a non degenerate symmetric bilinear 2-form.

The dualistic relation between linear connections plays a central role in the information geometry [17,18,47,48].

Definition 32. *Let (M,g) be a Riemannian manifold.*

(1) A dual pair (M,g,D,D^*) is called a flat pair if the connection D is flat, viz $R^\nabla = 0$.
(2) A flat pair (M,g,D,D^*) is called a dually flat pair if both (M,D) and (M,D^*) are locally flat manifolds.

Given a dual pair (M,D,D^*) let us set $A = D - D^*$. Here are the relationships between the torsion tensors T^D and T^{D^*} (respectively the relationship between the curvature tensors R^D and R^{D^*})

$$g(R^D(X,Y)\cdot Z,T) + g(Z,R^D(X,Y)\cdot T) = 0,$$

$$g(T^D(X,Y),Z) - g(T^{D^*}(X,Y),Z) = g(Y,A(X,Z)) - g(X,A(Y,Z)).$$

Proposition 3. *Given a flat pair (M,g,D,D^*), the following assertions are equivalent.*

(1) Both D and D^* are torsion free.
(2) D is torsion free and A is symmetric, viz

$$A(X,Y) = A(Y,X).$$

(3) D^* is torsion free and the metric tensor g a is KV cocycle of the KV algebra $\mathcal{A}^*$ of the locally flat manifold (M,D^*).
(4) The flat pair (M,g,D,D^*) is a dually flat pair.

Proof. Let us prove that 1 implies (2)
If both T^D and T^{D^*} vanish identically then A is symmetric, viz $A(X,Y) = A(Y,X)$.

Let us prove that (2) implies (3).

Since D is a flat connection, (2) implies that both the torsion tensor and the curvature tensor of D vanish identically. Then (M,D) is a locally flat manifold whose KV complex is denoted by $(C^*(\mathcal{A},\mathbb{R}),\delta_{KV})$. Using the dualistic relation of the pair (M,g,D,D^*) one obtains the identity

$$\delta_{KV}g(X,Y,Z) = g(A(X,Y) - A(Y,X),Z) = g(T_{D^*}(X,Y),Z),$$

therefore (2) implies (3).

Let us prove that (3) implies (4).

The assertion (3) implies that (M,D^*) is a locally flat manifold. Since g is δ_{KV}-closed D is torsion free. Thereby (M,g,D,D^*) is a dually flat pair.

Let us prove that (4) implies (1).

This implication derives directly from the definition of dually flat pair. □

A Comment.

From the proposition just proved arises a relationship between the dually flatness and the KV cohomology.

Indeed let (M, D^0) *be a fixed locally flat manifold whose KV algebra is denoted by* $\mathcal{A}^0$. *Let* $C^*_{KV}(\mathcal{A}^0, \mathbb{R})$ *be the KV complex of* $\tilde{\mathbb{R}}$*-valued cochains of the KV algebroid* $(TM, D^0, 1)$. *We know that every* $g \in \mathcal{R}ie(M)$ *yields a flat pair* (M, g, D^0, D^g).

Here D^g *is the flat Koszul connection defined by*

$$g(D^g_X Y, Z) = Xg(Y, Z) - g(Y, D^0_X Z).$$

Proposition 4. *The following assertions are equivalent.*

(1) (M, g, D^0, D^g) *is a dually flat pair.*

(2) $\delta^0_{KV}(g) = 0$

The scalar KV cohomology of a fixed locally flat manifold (M, D^0) provides a way of constructing new locally flat structures in M. Indeed let us set

$$\mathcal{H}es(M, D^0) = Z^2_{KV}(\mathcal{A}^0, \mathbb{R}) \cap \mathcal{R}ie(M).$$

For every $g \in \mathcal{H}es(M, D^0)$ there is a unique $D^g \in \mathcal{LF}(M)$ such that (M, g, D^0, D^g) is a dually flat pair.

So the dualistic relation leads to the map

$$\mathcal{H}es(M, D^0) \ni g \to D^g \in \mathcal{LF}(M).$$

We recall that a gauge map in TM is a vector bundle morphism of TM in TM which projects on the identity map of M. The readers interested in others topological studies involving connections and gauge transformations are referred to [49].

Given two symmetric cocycles $g, g^* \in \mathcal{H}es(M, D^0)$ there is a unique gauge transformation

$$\phi^* : TM \to TM$$

such that

$$g^*(X, Y) = g(\phi^*(X), Y).$$

The following properties are equivalent

$$\phi(D^0_X Y) = D^0_X \phi(Y), \tag{11a}$$

$$D^g = D^{g^*}. \tag{11b}$$

We fix a metric tensor $g* \in \mathcal{H}es(M, D^0)$. A gauge transformation ϕ is called g-symmetric if we have

$$g(\phi(X), Y) = g(X, \phi(Y)) \quad \forall (X, Y).$$

Every g-symmetric gauge transformation ϕ defines the metric tensor

$$g_\phi(X, Y) = g(\phi(X), Y).$$

This gives rise to the flat pair

$$(M, g_\phi, D^0, D^{g_\phi}).$$

To simplify we set

$$D^\phi = D^{g_\phi}.$$

We note $Sym(g)$ the subset of g-symmetric gauge transformations ϕ such that the following assertions are equivalent

(1) $\phi \in Sym(g)$.
(2) (M, g_ϕ, D^0, D^ϕ) is a dually flat pair.

The Lie group of D^0-preserving gauge transformations of TM is denoted by G^0. It is easy to see that for every $\phi \in Sym(g)$ the following assertions are equivalent

(1) $\phi \in G^0$,
(2) $g_\phi \in \mathcal{H}es(M, D^0)$.

Henceforth we deal with a fixed $g^* \in \mathcal{H}es(M, D^0$. The triple (M, g^*, D^0) leads to the dually flat pair (M, g, D^0, D^{g^*}). We set

$$D^* = D^{g^*}.$$

The tangent bundle TM is regarded as a left KV module of the KV algebroid $(TM, D^*, 1)$.

The KV algebras of (M, D^0) and of (M, D^*) are denoted by $\mathcal{A}^0$ and by $\mathcal{A}^*$ respectively. Their coboundary operators are noted δ^0 and δ^* respectively.

We focus on the role played by the total KV cohomology of the algebroid $(M, D^*, 1)$.

Let ϕ be a g^*-symmetric gauge transformation. Then ϕ gives rise to the metric tensor g_ϕ which is defined by

$$g_\phi(X, Y) = g^*(\phi(X), Y).$$

Lemma 3. *The following assertions are equivalent,*

(1) $g_\phi \in \mathcal{H}es(M, D^0)$,
(2) $\phi \in Z^1_\tau(\mathcal{A}^*, \mathcal{A}^*)$.

Hint.

Use the following formula

$$\delta^0_{KV} g_\phi(X, Y, Z) = g^*(\delta^*_\tau \phi(X, Y), Z).$$

Following the pioneering definition as in [2] a hyperbolic locally flat manifold is a positive exact Hessian manifold $(M, D, \delta_{KV}\theta)$. We extend the notion of hyperbolicity by deleting the condition that $\delta_{KV}\theta$ is positive. Now denote by $\mathcal{H}yp(M, D^0)$ the set of exact Hessian structures in (M, D^0).

A hyperbolic structure is defined by a triple (M, D, θ) where (M, D) is a locally flat manifold and θ is a de Rham closed differential 1-form such that the symmetric bilinear $\delta_{KV}\theta$ is definite.

The following statement is a straightforward consequence of Lemma 3.

Corollary 2. *The following statements are equivalent.*

(1) $g_\phi \in \mathcal{H}yp(M, D^0)$,
(2) $\phi \in B^1\tau(\mathcal{A}^*, \mathcal{A}^*)$

Proof of Corollary. By (1) there exists a (de Rham) closed differential 1-form θ such that

$$g_\phi(X, Y) = X\theta(Y) - \theta(D^0_X Y).$$

Let ξ be the unique vector field such that

$$\theta = \iota_\xi g^*.$$

Therefore one has

$$g^*(\phi(X), Y) = Xg^*(\xi, Y) - g^*(\xi, D^0_X Y).$$

Since the quadruple

$$(M, g^*, D^0, D^*)$$

is a dually flat pair one has the identity

$$g^*(\phi(X), Y) = g^*(D^*_X \xi, Y).$$

Thus we get the expected conclusion, viz

$$\phi(X) = D^*_X \xi.$$

Conversely let us assume that there exists a vector ξ satisfying the identity

$$\phi(X) = D^*_X \xi.$$

That leads to the identity

$$g^*(D^*_X \xi, Y) = Xg^*(\xi, Y) - g(\xi, D^0_X Y).$$

In other words one has

$$g_\phi \in \mathcal{H}yp(M, D^0).$$

This ends the proof of Corollary 2. □

The set of $g*$-symmetric gauge transformation is denoted by $\Sigma(g^*)$.
We have the canonical isomorphism

$$\Sigma(g^*) \ni \phi \to g_\phi \in \mathcal{R}ie(M). \tag{12}$$

Now we define the sets

$$\tilde{Z}^1_\tau(\mathcal{A}^*, \mathcal{A}^*) = \Sigma(g^*) \cap Z^1_\tau(\mathcal{A}^* \mathcal{A}^*),$$
$$\tilde{B}^1_\tau(\mathcal{A}^*, \mathcal{A}^*) = \Sigma(g^*) \cap B^1_\tau(\mathcal{A}^*, \mathcal{A}^*).$$

Combining Lemma 3 and its corollary with the isomorphism Equation (12). Then we obtain the identifications

$$\tilde{Z}^1_\tau(\mathcal{A}^*, \mathcal{A}^*) = \mathcal{H}es(M, D^*),$$
$$\tilde{B}^1_\tau(\mathcal{A}^*, \mathcal{A}^*) = \mathcal{H}yp(M, D^*).$$

Reminder.

We recall that a hyperbolic manifold (or a Koszul manifold) is δ_{KV}-exact Hessian manifold (M, g, D).

It is easily seen that the set of positive hyperbolic structures in a locally flat manifold (M, D) is a convex subset of $\mathcal{H}es(M, D)$.

So show the Koszul geometry is a vanishing theorem in the theory of KV homology of KV algebroids. The theory of homological statistical model (to be introduced in Part B) is another impact on the information geometry of the KV cohomology.

At the present step we have the relations

$$\frac{\mathcal{H}es(M, D^*)}{\mathcal{H}yp(M, D^*)} \subset H^2_{KV}(\mathcal{A}^*, \mathbb{R}),$$

$$\frac{\tilde{Z}^1_\tau(\mathcal{A}^*, \mathcal{A}^*)}{\tilde{B}^1_\tau(\mathcal{A}^*, \mathcal{A}^*)} = \frac{\mathcal{H}es(M, D^*)}{\mathcal{H}yp(M, D^*)}.$$

Another outstanding result of Koszul is the non rigidity of compact positive hyperbolic manifolds [2]. The non rigidity means that every open neighborhood of a positive Hyperbolic locally flat manifold $(M, D, \delta_{KV}\theta)$ contain another positive hyperbolic locally flat structure which is not isomorphic to (M, D). This non rigidity property may be expressed with the Maurer–Cartan polynomial function P^A_{MC} of (M, D) (see the local convexity theorem in [29]. In the next sub-subsection we revisit the notion of dual pair of foliations as in [18].

5.1.1. Statistcal Reductions

The statistical reduction theorem is the following statement.

Theorem 8 ([18])**.** *Let (M, g, D, D^*) be a dually flat pair and let N be a submanifold of M. Assume that N is either D-geodesic or D^*-geodesic. Then N inherits a structure of dually flat pair which is either (N, g_N, D, D^*_N) or (N, g_N, D_N, D^*)).*

The foliation counterpart of the reduction theorem is of great interest in the differential topology of statistical models see [18]. In the preceding sections we have addressed a cohomological aspect of this purpose. The matter will be more extensively studied in a forthcoming paper (See the Appendix A).

In mathematical physics a principal connection 1-form is called a gauge field.

In the differential geometry a principal connection 1-form in a bundle of linear frames is called a linear connection.

In the category of vector bundle Koszul connections are algebroid counterpart of principal connection 1-forms.

In a tangent bundle TM, depending on concerns and needs Koszul connections may called linear connections or linear gauges.

Definition 33. *Let $D, D^* \in \mathcal{LC}(M)$. A vector bundle homomorphism*

$$\psi : TM \to TM$$

is called a gauge homomorphism of (M, D) in (M, D^*) if for all pairs of vector fields (X, Y) one has

$$D^*_X\psi(Y) = \psi(D_XY).$$

The vector space of gauge homomorphisms of (M, D) in (M, D^*) is denoted by $\mathcal{M}(D, D^*)$. The vector space $\mathcal{M}(D, D^*)$ is not a $C^\infty(M)$-module.

5.1.2. A Uselful Complex

In this subsubsection we fix a dually flat pair (M, g, D, D^*) whose KV algebras are denoted by $\mathcal{A}$ and by $\mathcal{A}^*$. The tangent bundle TM is endowed the structure left module of the anchored KV algebroids $(TM, D, 1)$ and $(TM, D^*, 1)$. This means that each of the KV algebras $\mathcal{A}$ or $\mathcal{A}^*$ is regarded as a left module of itself.

We consider the tensor product

$$C = C^*_\tau(\mathcal{A}^*, \mathcal{A}^*) \otimes C^*_\tau(\mathcal{A}, \mathbb{R}).$$

We endow C with the $\mathbb{Z}$ bi-grading.

$$C^{i,0} = C^i_\tau(\mathcal{A}^*, \mathcal{A}^*) \otimes C^\infty(M),$$
$$C^{0,j} = \mathcal{A}^* \otimes C^j_\tau(\mathcal{A}, \mathbb{R}),$$

$$C^{i,j} = C^i_\tau(\mathcal{A}^*, \mathcal{A}^*) \otimes C^j_{KV}(\mathcal{A}, \mathbb{R}).$$

We recall that $C^*(\mathcal{A}, \mathbb{R})$ stands for $C^*(\mathcal{A}, C^\infty(M))$.
For every non negative integer q we set

$$C^q = \Sigma_{i+j=q} C^{i,j}.$$

We defines the linear map

$$\delta_{i,j} : C^{i,j} \to C^{i+1,j} \oplus C^{i,j+1}$$

by

$$\delta_{i,j} = \delta_\tau \otimes 1 + (-1)^i \otimes \delta_\tau.$$

So we obtain a linear map

$$C^q \to C^{q+1}$$

Therefore, we consider the bi-graded differential vector space

$$C := (C^{**}, \delta_{**}).$$

That is a bi-graded cochain complex whose q^{th}*cohomology* is denoted by $H^q(C)$. The cohomology inherits the bi-grading

$$H^q(C) = \sum_{[i+j=q]} H^{i,j}(C).$$

Here

$$H^{i,j}(C) = \frac{C^{i,j} \cap [Z^i_\tau(\mathcal{A}^*, \mathcal{A}^*) \otimes Z^j_\tau(\mathcal{A}, \mathbb{R})]}{im(\delta_{i-1,j}) + im(\delta_i, j-1)}$$

In the next subsubsection we shall discuss the impacts of this cohomology.

Remark 2. *The pair (C^{**}, δ_{**}) generates a spectral sequence [34]. That spectral sequence is a useful tool for simultaneously computing both the KV cohomology and the total KV cohomology of KV algebroids. Those matters are not the purpose of this paper.*

5.1.3. The Homological Nature of Gauge Homomorphisms

Giving a dually flat pair (M, g, D, D^*) one considers the linear map

$$C^{1,0}_\tau(\mathcal{A}^*, \mathcal{A}^*) \ni \psi \to \psi \otimes q_\psi \in C^{1,2}.$$

Here the symmetric 2-form q_ψ is defined by

$$q_\psi(X, Y) = \frac{1}{2}[g(\psi(X), Y) + g(X, \psi(Y))].$$

To relate the bi-complex (C^{**}, δ^{**}) and the space of gauge homomorphisms we use the following statement.

Theorem 9. *Given a gauge morphism*

$$\psi : TM \to TM$$

the following statements are equivalent

(1) $\psi \in \mathcal{M}(D, D^*)$,
(2) $\delta_{1,2}(\psi \otimes q_\psi) = 0$

Proof. (1) implies (2).

Suppose that $\psi \in \mathcal{M}(D, D^*)$. Then we have

$$D^*_X\psi(Y) = \psi(D_X Y) \quad \forall(X, Y).$$

Since both D and D^* are torsion free one has the identity

$$D^*_X.\psi(Y) - \psi(D^*_X Y) - D^*_Y\psi(X) + \psi(D^*_Y X) = 0.$$

Thus ψ is a (1,0)-cocycle of the total KV complex (C^{**}, δ_{**}).
At another side the relation $D^*_X \circ \psi = \psi \circ D_X$ leads to the identity

$$D_X q_\psi = 0.$$

So q_ψ is a (0,2)-cocycle of complex (C^{**}, δ_{**}). We conclude that

$$\delta_{1,2}(\psi \otimes q_\psi) = 0, \quad QED.$$

(2) implies (1).
We recall the formula

$$\delta_{1,2}(\psi \otimes q_\psi) = (\delta_\tau \psi) \otimes q_\psi - \psi \otimes \delta_\tau q_\psi.$$

By this formula

$$\delta_{1,2}(\psi \otimes q_\psi) \in C^{2,2} \oplus C^{1,3}$$

Thus the statement (2) is equivalent to the system

$$\delta_\tau \psi = 0,$$

$$\delta_\tau q_\psi = 0. \quad \square$$

To continue the proof we perform the following lemma.

Lemma 4 ([29]). *For every symmetric cochain $B \in C^{0,2}$, viz*

$$B(X, Y) = B(Y, X)$$

the following identities are equivalent

$$\delta_\tau B = 0, \tag{13a}$$
$$\nabla B = 0, \tag{13b}$$

By Lemma 4 the bilinear form q_ψ is D-parallel. Thereby we get the identity

$$Xq_\psi(Y, Z) - q_\psi(D_X Y, Z) - q_\psi(Y, D_X Z) = 0.$$

To usefully interpret this identity we involve the dualistic relation

$$Xg(Y, Z) = g(D_X Y, Z) + g(Y, D^*_X Z).$$

This expression leads to the identity

$$g(D^*_X\psi(Y) - \psi(D_X Y), Z) + g(Y, D^*_X\psi(Z) - \psi(D_X Z)) = 0. \tag{14}$$

A highlighting consequence is the identity

$$D_X^*\psi(Y) - \psi(D_XY) = D_Y^*\psi(X) - \psi(D_YX). \tag{15}$$

To every vector field X we assign the linear map

$$Y \to S_X(Y) = D_X^*Y - \psi(D_XY).$$

Then we rewrite Equations (14) and (15) as

$$g(S_X(Y), Z) + g(Y, S_X(Z)) = 0,$$

$$S_X(Y) = S_Y(X).$$

We consider the last identities in the framework of the Sternberg geometry [50,51].
Since the application

$$(X, Y) \to S_X(Y)$$

is $C^\infty(M)$-bi-linear it belongs to the first Kuranishi-Spencer prolongation of the orthogonal Lie algebra $so(g)$. Thereby $S_X(Y)$ vanishes identically. In other words we have

$$\psi \in \mathcal{M}(D, D^*).$$

This ends the proof of Theorem

A Comment.

The Sternberg geometry is the algebraic counterpart of the global analysis on manifolds. It has been introduced by Shlomo Sternberg and Victor Guillemin. It is an algebraic model for transitive differential geometry [50]. In that approach the Riemannian geometry is a geometry of type one. All of its Kuranishi-Spencer prolongations are trivial. The unique relevant geometrical invariant of the Riemnnian geometry is the curvature tensor of the Levi-Civita connection. Except the connection of Levi-Civita the other metric connections have been of few interest. Really other metric connections may have outstanding impacts on the differential topology. I shall address this purpose in a forthcoming paper.

5.1.4. The Homological Nature of the Equation $FE^{\nabla\nabla^*}$

Before proceeding we plan to discuss some homological ingredients which are connected to the differential equation

$$FE^{\nabla\nabla^*}: \quad D^{\nabla\nabla^*}(\psi) = 0.$$

Let us consider a dually flat pair (M, g^*, D, D^*) and the KV complex

$$\psi \in C^{1,0} = C^1_\tau(\mathcal{A}^*, \mathcal{A}^*).$$

Lemma 4 yields the following corollary.

Corollary 3. *We keep the notation used the preceding sub-subsection. Given a gauge morphism ψ the following statements are equivalent.*

(1) $\psi \otimes q_\psi$ is an exact (1,2)-cocyle,
(2) $\psi \in B^1_\tau(\mathcal{A}^*, \mathcal{A}^*)$.

Proof. Assume that the assertion (2) holds. Then there is $\xi \in \mathcal{A}^*$ satisfying the condition

$$\psi(X) = D_X^*\xi.$$

Thereby one has

$$q_{D^*\xi}\otimes \in Z^2_\tau(\mathcal{A},\mathbb{R}).$$

So one gets

$$D^*\xi \otimes q_{D^*\xi} = \delta_{0,2}[\xi \otimes q_{D^*\xi}].$$

Therefore assumption (2) implies (1).

Conversely assume that (1) holds, viz the (1,2)-cochain $\psi \otimes q_\psi$ is exact.

There exists

$$\xi \otimes \alpha \oplus \psi^* \otimes \beta \in C^{0,2} + C^{1,1}$$

such that

$$\psi \otimes q_\psi = \delta_{0,2}(\xi \otimes \alpha) + \delta 1, 1(\psi^* \otimes \beta).$$

Thus for vector fields Z, X, Y we have

$$\psi(Z) \otimes q_\psi(X,Y) = \delta_\tau \xi \otimes \alpha(X,Y) + \xi \otimes \delta_\tau \alpha(Z,X,Y) + \delta\tau\psi^*(Z,X) \otimes \beta(Y) + \psi^*(Z) \otimes \delta\tau\beta(X,Y).$$

Since

$$\psi \otimes q_\psi \in C^{1,2} = C^1_\tau(\mathcal{A}^*,\mathcal{A}^*) \otimes C^2_\tau(\mathcal{A}, C^\infty(M))$$

the exactness of $\psi \otimes q_\psi$ implies

$$\delta_\tau \alpha = 0,$$

$$\alpha(X,Y) = \alpha(Y,X).$$

Therefore

$$\psi(Z) \otimes q_\psi(X,Y) = \delta\xi(Z) \otimes \alpha(X,Y) + \psi^*(Z) \otimes \delta_\tau\beta(X,Y).$$

Now we observe that

$$\delta\tau\beta(X,Y) + \delta_\tau\beta(Y,X) = 0.$$

In final we get

$$\psi(Z) \otimes q_\psi(X,Y) = \delta\tau\xi(Z) \otimes \alpha(X,Y).$$

So we obtain

$$\psi(Z) = D^*_Z\xi,$$

$$q_\psi(X,Y) = \alpha(X,Y).$$

This end the proof of the corollary. □

From the mapping

$$C^{1,0} \ni \psi \to \psi \otimes q_\psi \in C^{1,2}$$

we deduce the canonical linear map

$$H^1_\tau(\mathcal{A}^*,\mathcal{A}^*) \ni [\psi] \to [\psi \otimes q_\psi] \in H^{1,2}(C).$$

We define another map

$$C^{1,0} \to C^{1,2}$$

by

$$\psi \to \psi \otimes \omega_\psi.$$

Here the differential 2-form ω is defined by

$$\omega_\psi(X,Y) = \frac{1}{2}[g(\psi(X),Y) - g(X,\psi(Y))].$$

This yields a linear map

$$H^1_\tau(\mathcal{A}^*, \mathcal{A}^*) \ni [\psi] \to [\psi \otimes \omega_\psi] \in H^{1,2}(C).$$

Now let (M, g, ∇, ∇^*) be a dually flat pair whose KV algebras are denoted by $\mathcal{A}$ and $\mathcal{A}^*$.
We identify the vector space $\Gamma(Hom(TM, TM))$ with the space $C^1_\tau(\mathcal{A}^*, \mathcal{A}^*)$.
We keep the notation $D^{\nabla\nabla^*}$, C^{ij}, δ_{ij} and q_ψ. Therefore, we can rephrase Lemma 4 as it follows.

Proposition 5. *For every section ψ of $Hom(TM, TM)$ the following assertions are equivalent.*

$$(1): \quad D^{\nabla\nabla^*}(\psi) = 0,$$

$$(2): \quad \delta_{12}(q_\psi) = 0$$

Here is an interesting feature. In a dually flat pair (M, g, ∇, ∇^*) we combine the double complex

$$\left\{C^{ij}, \delta_{ij}\right\}$$

with the correspondence

$$\psi \to q_\psi.$$

That allow us to replace the differential equation

$$FE^{\nabla\nabla^*}: \quad D^{\nabla\nabla^*}(\psi) = 0$$

by the homological equation

$$\delta_{12}(\psi) = 0.$$

That is a relevant impact on the global analysis of combinations of the KV cohomological methods with methods in the information geometry.

5.1.5. Computational Relations. Riemannian Foliations. Symplectic Foliations: Continued

We continue to relate the vector space of gauge homomorphisms and the differential topology. The tools we use are the KV cohomology and the Amari dualistic relation.

Let (M, g, D, D^*) be a dual pair. The vector subspace of g-preserving elements of $\mathcal{M}(D, D^*)$ is denoted by $\mathcal{M}(g, D, D^*)$. Thus every $\psi \in \mathcal{M}(g, D, D^*)$ satisfies the identity

$$g(\psi(X), Y) + g(X, \psi(Y)) = 0.$$

Now we fix a Koszul connection D^0 and we define the map

$$\mathcal{R}ie(M) \ni g \to D^g \in \mathcal{LC}(M).$$

by setting

$$g(D^g_X Y, Z) = Xg(Y, Z) - g(Y, D^0_X Z).$$

We define the non negative integers

$$n_x(D^0) = dim[\frac{\mathcal{M}_x(D^0, D^g)}{\mathcal{M}_x(g, D^0, D^g)}],$$

$$n(D^O) = \min_{x \in M} dim[\frac{\mathcal{M}_x(D^0, D^g)}{\mathcal{M}_x(g, D^0, D^g)}].$$

Lemma 5. *The integer $n(D^0)$ does not depend on the choice of $g \in \mathcal{R}ie(M)$.*

An Idea.

We fix a metric tensor g. For every $g^* \in \mathcal{R}ie(M)$ there is a unique g-symmetric vector bundle morphism $\phi \in \Sigma(g)$ such that

$$g^*(X,Y) = g(\phi(X),Y).$$

Therefore, we have

$$\phi^{-1} \circ \mathcal{M}(D^0, D^g) = \mathcal{M}(g^*, D^0, D^{g^*}),$$
$$\phi^{-1}\mathcal{M}(g, D^0, D^g) = \mathcal{M}(g^*, D^0, D^{g^*}).$$

Now one defines the numerical invariant $n(M)$.

Definition 34.

$$n(M) = \max\{n(D)|D \in \mathcal{SLC}(M)\}.$$

Given a Koszul connection ∇ the vector space of ∇-parallel differential 2-forms is denoted by $\Omega_2^\nabla(M)$.

Every dual pair (M, g, ∇, ∇^*) gives rise to the linear isomorphisms

$$(1): \quad \frac{\mathcal{M}(\nabla,\nabla^*)}{\mathcal{M}(g,\nabla,\nabla^*)} \ni [\psi] \to q_\psi \in \mathcal{S}_2^\nabla(M),$$
$$(2): \quad \mathcal{M}(g,\nabla,\nabla^*) \ni \psi \to \omega_\psi \in \Omega_2^\nabla(M).$$

The isomorphism (1) derives from the linear map

$$(1*): \quad \psi \to q_\psi(X,Y) = \frac{1}{2}[g(\psi(X),Y) + g(X,\psi(Y))].$$

The isomorphism (2) is defined by

$$(2*): \quad \psi \to \omega_\psi(X,Y) = \frac{1}{2}[g(\psi(X),Y) - g(X,\psi(Y))].$$

Proposition 6. *Let (M, g, ∇, ∇^*) be a dual pair. The inclusion map*

$$\mathcal{M}(g,\nabla,\nabla^*) \to \mathcal{M}(\nabla,\nabla^*)$$

induced the split short exact sequence

$$(*****): \quad 0 \to \Omega_2^\nabla(M) \to \mathcal{M}(\nabla,\nabla^*) \to \mathcal{S}_?^\nabla(M) \to 0.$$

Reminder.

According our previous notation elements of $\Omega_2^\nabla(M)$ are ∇-geodesic symplectic foliations. Those of $\mathcal{S}_2^\nabla(M)$ are ∇-geodesic Riemannian foliations. Thus we apply methods of the information geometry to relate the gauge geometry and the differential topolgy.

Digressions.

Our construction may open to new developments. Here are some unexplored perspectives.

(a) *A ∇-geodesic symplectic foliation $\omega \in \Omega^\nabla$ might carry richer structures such as Kahlerian structures.*

(b) *Suppose that the manifold M is compact and suppose that $g \in \mathcal{S}_2^\nabla(M)$ is a positive Riemannian foliation, viz*

$$g(X,X) \geq 0 \quad \forall X.$$

Then the theory of Molino may be applied to study g [38]. Therefore, the structure theorem of Molino tells that g gives rise to a Lie foliation whoses leaves are the adherences leaves of $\bar{g}$ [39].

(c) *In the principal bundle of first order linear frames of M the analog of a Koszul connection ∇ is a principal connection 1-form ω whose curvature form is denoted by Ω. The curvature form is involved in constructing characteristic classes of M, (the formalism of Chern-Weill.)*

At another side ∇-geodesic Riemannian foliations and ∇-geodesic symplectic foliations are Lie algebroids. They have their extrinsic algebraic topology. In particular the theory of integrable systems may be performed in every leaf of $\omega \in \Omega_2^\nabla(M)$. If one considers the α-connections in a statistical model those new insights may be of interest.

Here is an interpretation of the numerical invaraint $n(\nabla)$.

Theorem 10. *We assume there exists $\nabla \in \mathcal{SLC}(M)$ whose linear holonomy group $H(\nabla)$ is neither an orthogonal subgroup nor a symplectic subgroup. If $n(\nabla) > 0$ then the manifold M admits a couple $(\mathcal{F}_r, \mathcal{F}_s)$ formed by a ∇-geodesic Riemannian foliation $\mathcal{F}_r$ and a ∇-geodesic symplectic foliation $\mathcal{F}_s$.*

Proof. Let g be a Riemannian metric tensor in M. Since

$$n(\nabla) \leq dim(\mathcal{S}_2^\nabla(M)(x))$$

for all $x \in M$ there exists $\psi \in \mathcal{M}(\nabla, \nabla(g)$ such that

$$q_\psi \in \mathcal{S}_2^\nabla(M) \setminus \{0\},$$

$$\omega_\psi \in \Omega_2^\nabla(M) \setminus \{0\}.$$

The assumption that the holonomy group $H(\nabla)$ is neither orthogonal nor symlectic implies

$$Ker(q_\psi) \neq 0,$$

$$Ker(\omega_\psi) \neq 0.$$

Both distributions $Ker(q_\psi)$ and $Ker(\omega_\psi)$ are ∇-geodesic. Since ∇ is torsion free those distributions are completely integrable.

For all $X \in \Gamma(Ker(q_\psi))$ we have

$$L_X q_\psi = 0.$$

Mutatis mutandis for all $X \in \Gamma(Ker(\omega_\psi))$ we have

$$L_X \omega_\psi = 0.$$

From those properties we conclude

$(M, Ker(q_\psi), q_\psi)$ is a ∇-geodesic Riemannian foliation, $(M, Ker(\omega_\psi), \omega_\psi)$ is a ∇-geodesic symplectic foliation.

The theorem is proved. □

A Useful Comment.

Let (M, D) be a locally flat manifold whose KV algebra is denoted by $\mathcal{A}$. To every dual pair (M, g, D, D^g) we assign the short split exact sequence

$$0 \to \Omega_2^A(M) \to \mathcal{M}(D, D^g) \to \mathcal{S}_2^A(M) \to 0$$

which is canonically isomorphic to the short exact sequence

$$0 \to \mathcal{M}(g, D, D^g) \to \mathcal{M}(D, D^g) \to \mathcal{S}_2^A(M) \to 0.$$

We have already defined the geometric invariant

$$r(D) = dim(H_\tau^2(\mathcal{A}, \mathbb{R})) - b_2(M).$$

We observe that the integer $n(D)$ is a byproduct of methods of the information geometry while $r(D)$ is a byproduct of homological methods. However the split short exact sequence (****) leads to the equality

$$n(D) = r(D).$$

Here is a straight consequence of the theorem we just proved.

Proposition 7. *Every odd-dimensional manifold M with $n(M) > 0$ admits a geodesic symplectic foliation.*

The dualistic relation of Amari has another significant impact on the differential topology.

Definition 35. *We consider a dual pair (M, g, ∇, ∇^*). Let $\psi \in \mathcal{M}(\nabla, \nabla^*)$.*

(1) *The g-symmetric part of ψ, ψ^+ is defined by*

$$g(\psi^+(X), Y) = \frac{1}{2}[g(\psi(X), Y) + g(X, \psi(Y))].$$

(2) *The g-skew symmetric part of ψ, ψ^- is defined by*

$$g(\psi^-(X), Y) = \frac{1}{2}[g(\psi(X), Y) - g(X, \psi(Y))].$$

Theorem 11. *Let (M, g, ∇, ∇^*) be a dual pair where (M, g) is a positive Riemannian manifold. Let $\psi \in \mathcal{M}(\nabla, \nabla^*)$.*

(1) *The g-symmetric part ψ^+ is an element $\mathcal{M}(\nabla, \nabla^*)$ whose rank is constant.*
(2) *We have the g-orthogonal decomposition*

$$TM = Ker(\psi^+) \oplus im(\psi^+).$$

(3) *If both ∇ and ∇^* are torsion free then $Ker(\psi^+)$ and $Im(\psi^+)$ are completely integrable.*

A Digression.

We recall that a statistical manifold is a torsion free dual pair (M, g, ∇, ∇^*). If the vector space $\mathcal{M}(\nabla, \nabla^*)$ is non-trivial then it plays an outstanding role in the differential topology of M. We define a canonical map of $\mathcal{M}(\nabla, \nabla^*)$ in the category of 2-webs by

$$\mathcal{M}(\nabla, \nabla*) \ni \psi \to Ker(\psi^+) \oplus im(\psi^+).$$

Thus one may regard elements of $\mathcal{M}(\nabla, \nabla^*)$ as orthogonal 2-webs in M.
We keep our previous notation. The we have

$$q_\psi(X,Y) = g(\psi^+(X),Y),$$

$$\omega_\psi(X,Y) = g(\psi^-(X),Y).$$

Now suppose that (M, g, ∇, ∇^*) is a dually flat pair whose KV algebras are noted $\mathcal{A}$ and $\mathcal{A}^*$. We take into account the inclusion

$$\mathcal{M}(\nabla, \nabla^*) \subset Z^1_\tau(\mathcal{A}^*, \mathcal{A}^*).$$

We have a map of $\mathcal{M}(\nabla, \nabla^*)$ in the space of de Rham 2-cocyles which is defined by

$$\mathcal{M}(\nabla, \nabla*) \ni \psi \to \omega_\psi.$$

Assume that the cocycle $\psi \in \mathcal{M}(\nabla, \nabla^*)$ is exact. Then there exists $\xi \in \mathcal{A}^*$ such that

$$\psi(X) = \nabla^*_X \xi \quad \forall X \in \mathcal{A}.$$

By the dualistic relation one easily sees that

$$\omega_\psi = d_{dR}(\iota_\xi g).$$

Therefore one gets a canonical linear map

$$H^1_\tau(\mathcal{A}^*, \mathcal{A}^*) \ni [\psi] \to [\omega_\psi] \in H^2_{dR}(M, \mathbb{R}).$$

The next subsubsection is devoted to a few consequences of items we just discussed.

5.1.6. Riemannian Webs—Symplectic Webs in Statistical Manifolds

We introduce Riemannian webs and symplectic webs and we discuss their impacts on the topology of statistical manifolds. We recall that a Riemannian foliation is a symmetric bilinear form $g \in \mathcal{S}_2(M)$ with the following properties

(a) $rank(g) = constant$,
(b) $L_X g = 0 \forall X \in \Gamma(Ker(g))$.

We put

$$\mathcal{D} = Ker(g).$$

To avoid confusions the pair $(\mathcal{D}, g)$ stands for the Riemannian foliation g.

Definition 36. *A Riemannian k-web is a family of k Riemannian foliations in general position $(\mathcal{D}_j, g_j)$, $j := 1, ..., k$. A symplectic k-web is a family of k symplectic foliations in general position $(\mathcal{D}_j, \omega_j); j := 1, ..., k)$.*

Let (M, g, D, D^*) be a dually flat pair whose KV algebras are denoted by $\mathcal{A}$ and $\mathcal{A}^*$. We recall the inclusion

$$\mathcal{M}(D, D^*) \subset Z^1_\tau(\mathcal{A}^*, \mathcal{A}^*).$$

Consider a statistical manifold (M, g, ∇, ∇^*). By the classical theorem of Frobenius every ∇-parallel differential system in M is completely integrable.

In a statistical manifold (M, g, ∇, ∇^*) we consider a ∇-geodesic Riemannian k-web

$$[\tilde{g}_j \in \mathcal{S}_2^{\nabla}(M); j : 1, ..., k].$$

The distributions

$$\mathcal{D}_j = Ker(\tilde{g}_j)$$

are in general position. We consider the family $\Psi_j^+ \in \Sigma(g)$ defined by

$$g(\Psi_j^+(X), Y) = \tilde{g}_j(X, Y).$$

We get the family of g-orthogonal 2-webs defined by

$$TM = Ker(\Psi_j^+) \oplus im(\Psi_j^+).$$

Mutatis mutandis we can consider a ∇-geodesic symplectic web

$$[\omega_j \in \Omega_2^{\nabla}(M); j := 1, ..., k].$$

There is family of g-skew symmetric gauge morphisms Ψ_j^- defined by

$$g(\Psi_j^-(X), Y) = \omega_j(X, Y).$$

Since ∇ and ∇^* are torsion free $Ker(\Psi_j^-)$ and $im(\Psi_j^-)$ are completely integrable. Since g is positive definite we get the 2-web

$$TM = Ker(\Psi_j^-) \oplus im(\Psi_j^-).$$

Further every leaf of $im(\Psi_j^-)$ is a symplectic manifold.

Definition 37. *Let $\omega \in \Omega_2^{\nabla}(M)$ be a non trivial symplectic foliation in a statistical manifold (M, g, ∇, ∇^*). Consider $\Psi^- \in \mathcal{M}(\nabla, \nabla^*)$ defined by*

$$g(\Psi^-(X), Y) = \omega(X, Y).$$

The differential 2-form ω is called simple if the foliation $Ker(\Psi^-)$ is simple.

In a statistical manifold (M, g, ∇, ∇^*) every non trivial symplectic web

$$[\omega_j; j := 1, ...] \subset \Omega_2^{\nabla}(M)$$

gives rise to a family of g-orthogonal 2-webs. So in this approach the role played by $\mathcal{S}_2^{\nabla}(M)$ is similar to the role played by $\Omega_2^{\nabla}(M)$. Our construction of Riemannian webs and symplectic webs in the category of dually flat pairs holds in the category of statistical manifolds.

At one side, in a dually flat pair (M, g, D, D^*) our approach yields linearizable webs. This property does not hold in all statistical manifolds.

At another side, in a statistical manifold (M, g, ∇, ∇^*) a Riemannian web

$$[\tilde{g}_j, j \in J] \subset \mathcal{S}_2^{\nabla}(M)$$

or symplectic web

$$[\omega_j, j \in J] \subset \Omega_2^{\nabla}(M)$$

gives rise to families of orthogonal 2-webs. This property does not hold in all dually flat pairs.

The considerations we just discussed may have remarkable impacts on the topological-geometrical structure of statistical manifolds.

From our brief discussion we conclude

Theorem 12. *Consider a statistical manifold (M, g, ∇, ∇^*). Every non trivial simple symplectic foliation $\omega \in \Omega_2^{\nabla}(M)$ is defined by a Riemannian submersion on a symplectic manifold.*

Corollary 4. *Every non trivial simple symplectic web $[\omega_j, j \in J] \subset \Omega_2^{\nabla}(M)$ is defined by family of Riemannian submersions on symplectic manifols.*

5.2. The Hessian Information Geometry, Continued

In [52] Shima pointed out that the Fisher informations of many classical statistical models are Hessian metric tensors.

At another side the exponential models (or exponential family) may be considered as optimal Statistical models.

As already mentioned there does not exists any criterion for knowing whether a given statistical model is isomorphic to an exponential model [22], [13]

In the category of regular models, viz models whose Fisher information is a Riemannian metric, it is known that the Fisher information of an exponential model is a Hessian Riemannian metric [18,52].

In this subsection we address the general situation. We give a cohomological characterization of exponential models. We also introduce a new numerical invariant r^b which measures how far from being an exponential family is a given statistical model. See the Appendix A to this paper.

We recall that the metric tensor g of a Hessian structure (M, D, g) is a 2-cocycle of the KV complex $[C^*_{KV}(\mathcal{A}, \mathbb{R}), \delta_{KV}]$.

To non specialists we go to recall some constructions in the geometry of Koszul [2,52], see also [53].

Let $((M, x^*), D)$ be a pointed locally flat manifold whose universal covering is denoted by $(\tilde{M}, \tilde{D})$. Here the topological space $\tilde{M}$ is the set of homotopy class of continuous paths

$$\{([0,1], 0) \rightarrow (M, x^*)\}.$$

Its topology is the compact-open topology. Let c be a continuous path with

$$c(0) = x^*.$$

For $s \in [0,1]$ the parallel transport of $T_{x^*}M$ in $T_{c(s)}M$ is denoted by τ_s. One defines $Q(c) \in T_{x^*}M$ by

$$Q(c) = \int_0^1 \tau_s^{-1}(\frac{dc(s)}{ds})ds.$$

The tangent vector $Q(c)$ depends only on the homotopy class of the path c. Therefore, Q defines a local homeomorphism

$$\tilde{Q} : \tilde{M} \rightarrow T_{x^*}M.$$

Let $\pi_1(x^*)$ be the fundamental group at x^*. Let $[\gamma] \in \pi_1(x^*)$. The natural left action

$$\pi_1(x^*) \times \tilde{M} \rightarrow \tilde{M}$$

is given by the composition of paths, viz

$$[\gamma].[c] = [\gamma \circ c].$$

The parallel transport along a loop $\gamma(t)$ yields a linear action of $\pi_1(x^*)$ in $T_{x^*}M$ which is denoted by $f([\gamma])$.

Let $[\gamma], [\gamma^*] \in \pi_1(x^*)$. The composition of paths leads to the formula

$$Q([\gamma].[\gamma^*]) = f([\gamma])Q([\gamma^*]) + Q([\gamma]).$$

The last relation shows that the pair (f, Q) is an affine representation of $\pi_1(x^*)$ in $T_{x^*}M$. This representation is called the holonomy representation of the locally flat manifold (M, D). The group $(f, Q)(\pi_1(x^*))$ is called the affine holonomy group of (M, D). Its linear component $f(\pi_1(x^*))$ is called the linear holonomy group of (M, D).

Definition 38 ([2]). *An m-dimensional locally flat structure (M, D) is called hyperbolic if $\tilde{Q}(\tilde{M})$ is a convex domain not containing any straight line in $T_{x^*}M$.*

Definition 39. *A locally flat manifold (M, D) is complete if the map $\tilde{Q}$ is a diffeomorphism onto $T_{x^*}M$.*

Among the major open problems in the theory of space groups is the conjecture of Markus.

Conjecure of Markus: a compact locally flat manifold (M, D) whose linear holonomy group is unimodular is complete.

Before pursuing we recall KV cohomological version of Theorem 3 as in [2].

Theorem 13 ([2]). *A necessary condition for a locally flat manifold (M, D) being hyperbolic is that (M, D) carries a positive Hessian structure whose metric tensor is exact in the KV complex of (M, D). This condition is sufficient if M is compact.*

We have already mentioned a notable consequence of this theorem of Koszul. In the category of compact locally flat manifolds the subcategory of hyperbolic locally flat structures is the same thing as the category of positive exact Hessian structures. So The geometry of compact hyperbolic local flat manifolds is an appropriate vanishing theorem.

In the preceding sections the family of Hessian metrics in a locally flat manifold (M, D) is denoted by $\mathcal{H}es(M, D)$. Therefore, $\mathcal{H}^+es(M)$ stands for the sub-family of positive Hessian metric tensors. It is a convex cone in $\mathcal{R}ie(M)$.

We have already used the KV complex for expressing the dually flatness. More precisely let (M, D) be a fixed locally flat manifold whose KV algebra is noted $\mathcal{A}$. A dual pair $(M, g, D, D(g))$ is dually flat if and only if $g \in Z^2_{KV}(\mathcal{A}, \mathbb{R})$. Therefore, every dually flat pair (M, g, D, D^*) yields two cohomology classes

$$[g]_D \in H^2_{KV}(\mathcal{A}, \mathbb{R}),$$

$$[g]_{D^*} \in H^*_{KV}(\mathcal{A}^*, \mathbb{R}).$$

Thereby, we can use methods of the information geometry for rephrasing Theorem 3 as in [2].

Theorem 14. *A necessary condition for (M, D) being hyperbolic is the existence a positive dually flat pair (M, g, D, D^*) such that*

$$[g] = 0 \in H^2_{KV}(\mathcal{A}, \mathbb{R}).$$

If M is compact this (vanishing) condition is sufficient.

About the geometry of Koszul the non specialists are referred to [2,7,8,52] and bibliography therein [12].

About applications of the geometry of Koszul the readers are refereed to [12,13,54,55].

About relationships between the theory of deformation and the theory of cohomology, the readers are referred to [9,27,56].

5.3. The α-Connetions of Chentsov

Still, nowadays, the information geometry deals with models (Θ, P) whose underlying m-dimensional manifold Θ is an open subset of the euclidean space $\mathbb{R}^m$. Further the Fisher information g is assumed to be regular, viz (Θ, g) is a Riemannian manifold. In this paper this classical information geometry is called the local information geometry. This "local nature" will be explained in Part B of this paper.

At the moment we plan to investigate other topological properties of the local statistical models.

Let (Θ, P) be an m-dimensional local statistical model for a measurable set (Ξ, Ω). The manifold Θ is a domain in the Euclidean space $\mathbb{R}^m$. The function P is non negative. It is defined in $\Theta \times \Xi$. We recall the requirements P is subject to.

(1) $\forall \xi \in \Xi$ the function
$$\Theta \ni \theta \to P(\theta, \xi)$$
is smooth.

(2) $\forall \theta \in \Theta$ the triple
$$(\Xi, \Omega, P(\theta, -))$$
is a probability space.

(3) $\forall \theta, \theta^* \in \Theta$ with $\theta \neq \theta^*$ there exists $\xi \in \Xi$ such that
$$P(\theta, \xi) \neq P(\theta', \xi).$$
(with the requirement (3) (Θ, P) is called identifiable.)

(4) The differentiation d_θ commutes with the integration $\int_\Xi$. The Fisher information of a model (Θ, P) is the symmetric bi-linear form g which is defined by
$$g(X, Y)(\theta) = \int_\Xi P(\theta, \xi)[[d_\theta log(P)]^{\otimes 2}(X, Y)](\theta, \xi) d\xi.$$
Here d_θ stands for the differentiation with respect to the argument $\theta \in \Theta$.

(5) The Fisher information is positive definite.

Remark 3. *The Fisher information g can be defined using any Koszul connection ∇ according to the following formula*
$$g(X, Y)(\theta) = -\int_\Xi P(\theta, \xi)[(\nabla^2 log(P))(X, Y)](\theta, \xi)] d\xi.$$

The right member of the last equality does not depend on the choice of the Koszul connection ∇.

From now on, we deal with a generic statistical model. This means that we do not assume the Fisher information g is definite.

Let $\theta = (\theta_1, ..., \theta_m)$ be a system of Euclidean coordinate functions in $\mathbb{R}^m$. To every real number α is assigned the torsion free Koszul connection ∇^α whose Christoffel symbols in the coordinate (θ_j) are

$$\Gamma^\alpha_{ij,k} = \int_\Xi P(\theta, \xi)[(\frac{\partial^2 log(P(\theta, \xi))}{\partial\theta_i \partial\theta_j} + \frac{1-\alpha}{2} \frac{\partial log(P(\theta, \xi))}{\partial\theta_i} \frac{\partial log(P(\theta, \xi))}{\partial\theta_j}) \frac{\partial log(p(\theta, \xi))}{\partial\theta_k}] d\xi.$$

This definition agrees with any affine coordinate change. We put $\partial_i = \frac{\partial}{\partial\theta_i}$. We have

$$\nabla^\alpha_{\partial_i} \partial_j = \sum_k \Gamma^\alpha_{ij,k} \partial_k.$$

Now we assume a model (Θ, P) is regular. Then the Christoffel symbols and the Fisher information are related by the formula

$$\Gamma^{\alpha}_{ij,k} = g(\nabla^{\alpha}_{\partial_i}\partial_j, \partial_k).$$

Further every quadruple $(\Theta, g, \nabla^{\alpha}, \nabla^{-\alpha})$ is a statistical manifold [18,48].

Thus we have a family of splitting short exact sequences

$$0 \to \Omega^{\nabla^{\alpha}}(\Theta) \to \mathcal{M}(\nabla^{-\alpha}, \nabla^{\alpha}) \to \mathcal{S}_2^{\nabla^{\alpha}}(\Theta) \to 0.$$

So the machinery we have developed in the preceding sections can be performed to explore the differential topology of regular local statistical models. For that purpose the crucial tool is the family of vector space

$$\mathcal{S}_2^{\alpha}(\Theta) = \mathcal{S}_2^{\nabla^{\alpha}}(\Theta).$$

We consider the abstract trivial bundle

$$\cup_{\alpha}[\mathcal{S}^{\alpha} \times \{\alpha\}] \to \mathbb{R}$$

whose fiber over $\alpha \in \mathbb{R}$ is $\mathcal{S}^{\alpha}(\Theta)$. To every $B \in \mathcal{S}^{\alpha}(\Theta)$ we assign the unique $\psi^{+} \in \Sigma(g)$ defined by

$$g(\psi^{+\alpha}(X), Y) = B(X, Y).$$

The machinery in the preceding subsection leads to the following proposition.

Proposition 8. *We assume (Θ, P) is regular.*

(1) Every non zero singular section

$$\mathbb{R} \ni \alpha \to B_{\alpha} \in \mathcal{S}^{\alpha}(\Theta)$$

gives rise the family of (g-orthogonal) 2-web

$$T\Theta = Ker(\psi^{+\alpha}) \oplus im(\psi^{+\alpha}).$$

Further according to the notation used previously (B_{α}) is a family of Riemannian foliations as in [39,40].

(2) By replacing $\mathcal{S}^{\alpha}(\Theta)$ by $\Omega_2^{\nabla}(\Theta)$ every non zero singular section

$$\mathbb{R} \ni \alpha \to \omega_{\alpha} \in \Omega_2^{\nabla}(\Theta)$$

yields a family of symplectic foliations ω_{α}.

Reminder.

(i) $\alpha \to B_{\alpha}$ is called a singular section if each B_{α} is non inversible.
(ii) $\alpha \to \omega_{\alpha}$ is called a simple section if each ω_{α} is simple.

We have used some gauge morphisms to construct Riemannian submersions of statistical manifolds over symplectic manifolds. The notions we just introduced lead to similar situations.

Theorem 15. *Let (Θ, P) be a regular statistical model whose Fisher information is denoted by g. Every simple non zero singular section*

$$\mathbb{R} \ni \alpha \to \omega_{\alpha} \in \Omega^{\alpha}(\Theta)$$

defines an α-family of Riemannian submersions of (Θ, g) onto symplectic manifolds.

Proposition 9. *We assume a model (Θ, P) is regular. For every nonzero real number α one has*

$$\mathcal{M}(\nabla^{\alpha}, \nabla^{-\alpha}) \cap \mathcal{M}(\nabla^{-\alpha}, \nabla^{\alpha}) = \mathcal{M}(g, \nabla^{-\alpha}, \nabla^{\alpha}) + \mathcal{M}(g, \nabla^{*}, \nabla).$$

The Sketch of Proof. The proof is based on the short exact sequences

$$0 \to \mathcal{M}(g, \nabla^{\alpha}, \nabla^{-\alpha}) \to \mathcal{M}(\nabla^{\alpha}, \nabla^{-}\alpha) \to \mathcal{S}_2^{\nabla}(\Theta) \to 0,$$

$$0 \to \mathcal{M}(g, \nabla^{-\alpha}, \nabla^{\alpha}) \to \mathcal{M}(\nabla^{-\alpha}, \nabla^{\alpha}) \to \mathcal{S}_2^{-\nabla}(\Theta) \to 0.$$

Let us suppose that the conclusion of the the proposition fails. Then there is a nonzero 2-form $B \in \mathcal{S}_2^{\nabla^{\alpha}}(\Theta) \cap \mathcal{S}_2^{\nabla^{-\alpha}}(\Theta)$.

(1) If $Ker(B) = 0$, then both ∇^{α} and $\nabla^{-\alpha}$ coincide with the Levi-Civita connection of B. This implies $\alpha = 0$, this contradicts our choice of α.
(2) If $Ker(B) \neq 0$ then $Ker(B)$ and $Ker(B)^{\perp}$ are geodesic for both ∇^{α} and $\nabla^{-\alpha}$. Thus the pair $(Ker(B), Ker(B)^{\perp})$ defines a g-orthogonal 2-web.

At one side, by the virtue of the reduction theorem as in [18] every leaf F of $Ker(B)^{\perp}$ inherits a dual pair $(F, g_F, \nabla_F^{\alpha}, \nabla_F^{-\alpha})$.

At another side, B gives rise to the Riemannian structure (F, B). Furthermore both ∇_F^{α} and $\nabla_F^{-\alpha}$ are torsion free metric connections in (F, B). Thereby one gets

$$\nabla_F^{\alpha} = \nabla_F^{-\alpha}$$

The last equality holds if and only if $\alpha = 0$. This contradicts our assumption. The proposition is proved. □

The proposition above is a separation criterion for α-connections in the following sense. For every nonzero real number α the vector subspace $\mathcal{S}_2^{\nabla^{\alpha}}(\Theta)$ is transverse to $\mathcal{S}_2^{\nabla^{-\alpha}}(\Theta)$ in the vector space $\mathcal{S}_2(\Theta)$ of symmetric forms in Θ.

5.4. The Exponential Models and the Hyperbolicity

A challenge is the search of a criterion for deciding whether a model (Θ, P) is an exponential family. That is the challenge in [22]. Still, nowadays, this problem is open.

The Fisher information of a regular exponential model is a Hessian Riemannian metric. We are going to demonstrate that the converse is globally true. Our proof is based on cohomological arguments.

In the Appendix A to this paper we introduce a new numerical invariant $r^b(\Theta, P)$ which measures how far from being an exponential family is a model (Θ, P).

The invariant r^b derives from the global analysis of differential operators

$$D_{\alpha} = D_{\nabla^{\alpha}},$$

$$D^{\alpha} = D^{\nabla^{\alpha}}.$$

Now we are going to provide a cohomological characterization of exponential models.

Before pursuing we recall a definition.

Let $\theta_j, j := 1, ..., m$ be a system of Euclidean coordinate functions of $\mathbb{R}^m$.

Definition 40. *[18] An m-dimensional statistical model (Θ, P) is called an exponential model for (Ξ, Ω) if there exist a map*

$$\Xi \ni \xi \to [C(\xi), F_1(\xi), ..., F_m(\xi)] \in \mathbb{R}^{m+1}$$

and a smooth function

$$\Theta \ni \theta \to \psi(\theta) \in \mathbb{R}$$

such that

$$P(\theta,\xi) = exp(C(\xi) + \sum_1^m F^j(\xi)\theta_j - \psi(\theta_1,...,\theta_m)).$$

Theorem 16. *Let (Ξ,Ω) be a measurable set and let (Θ,P) be an m-dimensional statistical model for (Ξ,Ω). The Fisher information of (Θ,P) is denoted by g. The following statements are equivalent.*

(1) *There exists $\nabla \in \mathcal{LF}(\Theta)$ such that*

$$\delta_{KV} g = 0,$$

(2) *The model (Θ,P) is an exponential model.*

Demonstration.

$(2) \Rightarrow (1)$.

We assume that (2) holds. Then we fix a system of affine coordinate functions

$$\theta = (\theta_1,...,\theta_m).$$

By the virtue of (2) we have

$$P(\theta,\xi) = exp(C(\xi) + \sum_1^m F^j(\xi)\theta_j - \psi(\theta)).$$

Here $\psi \in C^\infty(\Theta)$ and $(C,F)(\xi) = (C(\xi),F^1(\xi),...,F^m(\xi)) \in \mathbb{R}^{m+1}$. Therefore, one has

$$\frac{\partial^2 log(P(\theta,\xi))}{\partial\theta_i\partial\theta_j} = -\frac{\partial^2\psi}{\partial\theta_i\partial\theta_j}.$$

Thereby one can write

$$-\int_\Xi P(\theta,\xi)\frac{\partial^2 log(P(\theta,\xi))}{\partial\theta_i\partial\theta_j} = \frac{\partial^2\psi(\theta,\xi)}{\partial\theta_i\partial\theta_j}.$$

This shows that we have

$$g = \delta_{KV}(d\psi) \in B^2_{KV}(\mathcal{A},\mathbb{R}).$$

The implication $(2) \to (1)$ is proved.

$(1) \Rightarrow (2)$.

We use a strategy similar to that used in [52]. However our arguments do not depend on rank(g). Let $\nabla \in \mathcal{LF}(\Theta)$ whose KV algebra is denoted by $\mathcal{A}$. We assume

$$g \in Z^2_{KV}(\mathcal{A},C^\infty(\Theta)).$$

Thus we have

$$\delta_{KV} g = 0.$$

In (Θ,∇) we fix a system of local affine coordinate functions

$$\{\theta_1,...,\theta_m\}$$

whose domain is a convex open subset U. We write the matrix of Fisher information g in the basis

$$\left\{\frac{\partial}{\partial\theta_j}\right\},$$

namely

$$g = \sum g_{ij} d\theta_i d\theta_j.$$

Here

$$g_{ij} = g(\frac{\partial}{\partial\theta_i}, \frac{\partial}{\partial\theta_j}).$$

The assumption

$$\delta_{KV} g = 0$$

is equivalent to the system

$$\frac{\partial g_{ij}}{\partial\theta_k} - \frac{\partial g_{kj}}{\partial\theta_i} = 0$$

for all i, j, k.

We use a notation which is used in [52]. We consider the differential 1-forms

$$h_j = \sum_i g_{ij} d\theta_i.$$

Every differential 1-form h_j is a de Rham cocycle. By the Lemma of Poincaré the convex open set U supports smooth functions $\phi_j, j =: 1, ..., m$ which have the following property

$$d\phi_j = h_j.$$

We put

$$\omega = \sum_j \phi_j d\theta_j.$$

Then we have

$$(\delta_{KV}\omega)(\frac{\partial}{\partial\theta_i}, \frac{\partial}{\partial\theta_j}) = g_{ij}.$$

Thus the differential 1-form $\sum_j \phi_j d\theta_j$ is de Rham closed.

Since U is convex it supports a local smooth function Ψ such that

$$d\Psi = \sum \phi_j d\theta_j.$$

So we get

$$g(\frac{\partial}{\partial\theta_i}, \frac{\partial}{\partial\theta_j}) = \frac{\partial^2\psi}{\partial\theta_i\partial\theta_j}.$$

To continue we fix $\theta_0 \in U$ and we consider the function

$$\theta \rightarrow a(\theta)$$

which is defined in U by

$$a(\theta) = \int_{\Xi} P(\theta_0, \xi)[\psi(\theta) + log(P(\theta, \xi))]d\xi.$$

Now recall that the integration $\int_{\Xi}$ commutes with the differentiation $\frac{d}{d\theta}$. Therefore, $\forall i, j \leq dim(\Theta)$ one has

$$\frac{\partial^2 a}{\partial\theta_i\partial\theta_j}(\theta) = \int_{\Xi} P(\theta_0, \xi)\frac{\partial^2(\psi + log(P))}{\partial\theta_i\partial\theta_j}(\theta, \xi)d\xi.$$

The identities above show that

$$\frac{\partial^2 a}{\partial\theta_i \partial\theta_j}(\theta_0) = 0$$

for all $\theta_0 \in U$. *Thereby the function*

$$\theta \to \psi(\theta) + log(P(\theta, \xi))$$

depends affinely on $\theta_1, ..., \theta_m$. *So there exists a* $\mathcal{R}^{m+1}$*-valued function*

$$\Xi \ni \xi \to (C(\xi), F_1(\xi), ..., F_m(\xi)) \in \mathcal{R}^{m+1}$$

such that

$$\psi(\theta) + log(P(\theta, \xi)) = C(\xi) + \sum_1^m F_i(\xi)\theta_i.$$

In final we get

$$P(\theta, \xi) = exp(C(\xi) + \sum F_i(\xi)\theta_i - \psi(\theta))$$

for all $(\theta, \xi) \in U \times \Xi$. *So* (Θ, P) *is locally an exponential family.*

Since the exponential function is injective this local property of (Θ, P) *is a global property, in other words the model is globally an exponential model. In final assertion (1) implies assertion(2). This ends the demonstration of the theorem*

Some Comments.

- *(i) It must be noticed that the demonstration above is independent of the rank of the Fisher information* g. *Therefore, the theorem holds in singular statistical models.*
- *(ii) In regular statistical models the theorem above leads to the notion of e-m-flatness as in [18].*
- *(iii) When the Fisher information* g *is semi-definite the dualistic relation is meaningless. However data* $(\Theta, g, \nabla, \nabla^*)$ *may be regarded as data depending on the transversal structure of the distribution* $Ker(g)$.
- *(iv) In the analytic category the Fisher information is a Riemannian foliation. Therefore, both the information geometry and the topology of information are transversal concepts. This may be called the transversal geometry and the transversal topology of Fisher-Riemannian foliations.*
- *(v) The theorem above does not solve the question as how far from being an exponential family is a given model. It only tells us that exponential families are objects of the Hessian geometry.*

The framework for addressing the challenge just mentioned is the theory of invariants. That is the purpose of a forth going work. Some new results are anounced in the Appendix A to this paper.

6. The Similarity Structure and the Hyperbolicity

We consider a dually flat pair (M, g, ∇, ∇^*). Both (M, ∇, g) and (M, g, ∇^*) are locally hyperbolic in the sense of [2]. So they locally support the geometry of Koszul. That is a consequence of the classical Lemma of Poincare.

Every point of M has an open neighborhood U supporting a local de Rham closed differential 1-forms

$$\omega \in C^1_{KV}(\mathcal{A}, \mathbb{R})$$

and

$$\omega^* \in C^1_{KV}(\mathcal{A}^*, \mathbb{R})$$

subject to the following requirements

$$g|U = \delta\omega,$$

$$g|U = \delta^*\omega^*.$$

By the virtue of Theorem 3 as in [2], for both (M, g, ∇) and (M, g, ∇^*) being globally hyperbolic it is necessary that

$$[g] = 0 \in H^2_{KV}(\mathcal{A}, \mathbb{R})$$

and

$$[g] = 0 \in H^2_{KV}(\mathcal{A}^*, \mathbb{R}).$$

Every choice of local differential 1-forms ω and ω^* gives rise to a unique pair of local similarity vector fields (H, H^*), viz

$$\nabla_X H = X,$$

$$\nabla^*_X H^* = X$$

for all vector fields X. The vector fields H and H^* are Riemannian gradients of ω^* and of ω respectively. This means that those differential 1-forms are defined by

$$\omega = \iota_H g,$$

$$\omega^* = \iota_{H^*} g.$$

Here ι_H stands for the inner product by H, viz

$$\iota_H g(X) = g(H, X).$$

This short discussions lead to the following statement

Theorem 17. *Let (M, g, ∇, ∇^*) be a compact dually flat pair whose KV algebras are denoted by $\mathcal{A}$ and by $\mathcal{A}^*$. The following assertions are equivalent*

(1) *The locally flat manifold (M, ∇) is hyperbolic,*
(2) *the locally flat manifold (M, ∇^*) admits a global similarity vector field H^*.*

Definition 41. *Let $\nabla \in \mathcal{LC}(M)$.*

(1) *The gauge structure (M, ∇) is called a similarity structure if ∇ admits a global similarity vector field $H \in \mathcal{X}(M)$.*
(2) *A dual pair (M, g, ∇, ∇^*) is a similarity dual pair if either (M, ∇) or (M, ∇^*) is a similarity structure.*

The following proposition is a straightforward consequence of our definition.

Proposition 10. *If a gauge structure (M, ∇) is flat and is locally a similarity structure, then (M, ∇) is a locally flat manifold*

7. Some Highlighting Conclusions

In this Part A our aim has been to address various purposes involving the theory of KV homology. Doing that we have pointed significant relationships between some major topics in mathematics and the local information geometry. Those relationships might be sources of new investigations.

We summarize some relevant relationships we have been concerned with.

7.1. The Total KV Cohomology and the Differential Topology

We have addressed the existence problem for a few major objects of the differential topology. Riemannian foliations and symplectic foliations. Riemannian webs and their linearization problem.

To those questions we have obtained substantial solutions in the category of locally flat manifolds. The cohomological methods we have used are based on the split exact short cohomology sequence

$$0 \to H^2_{dR}(M,\mathbb{R}) \to H^2_{\tau}(\mathcal{A},\mathbb{R}) \to \mathcal{S}^{\mathcal{A}}_2(M) \to 0.$$

7.2. The KV Cohomology and the Geometry of Koszul

The Hessian Geometry is a byproduct local vanishing Theorems in the theory of KV cohomology. The geometry of Koszul is a byproduct of global vanishing Theorem in the same sitting.

7.3. The KV Cohomology and the Information Geometry

The category of finite dimensional statistical models for a measurable set (Ξ,Ω) contains the subcategory of finite dimensional Hessian manifolds. From this viewpoint the Hessian information geometry is nothing but the exponential information geometry (i.e., the geometry of exponential families and their genelarizations). The framework for those purposes is closely related to vanishing Theorems in the theory of KV cohomology.

At another side cotangent bundles of Hessian manifolds are Kaehlerian manifolds. This aspect has been discussed by many authors, see [52] and the bibliography ibidem.

7.4. The Differential Topology and the Information Geometry

A lot of outstanding links between the differential topology and the information geometry are based on the dualistic relation of Amari. This approach leads to significant results in the category of statistical manifolds. In a statistical manifold (M,g,∇,∇^*) we have introduced the splitting short exact sequence

$$0 \to \Omega^{\nabla}_2(M) \to \mathcal{M}(\nabla,\nabla^*) \to \mathcal{S}^{\nabla}_2(M) \to 0.$$

Here (i) $\Omega^{\nabla}_2(M)$ is the space of ∇-geodesic symplectic foliations in M; (ii) $\mathcal{S}^{\nabla}_2(M)$ is the space of ∇-geodesic Riemannian foliations in M.

The numerical invariant $n(\nabla)$ has outstanding impacts on the differential topology of M. See our results on orthogonal 2-webs and on Riemannian submersions on symplectic manifolds.

7.5. The KV Cohomology and the Linearization Problem for Webs

In a locally flat pair (M,g,∇,∇^*) we consider the short exact sequence

$$O \to \Omega^{\nabla}_2(M) \to \mathcal{M}(\nabla,\nabla*) \to \mathcal{S}^{\nabla}_2(M) \to O.$$

The linearization of webs of is a difficult outstanding problem in the differential topology. $\mathbb{G}^k[\Omega^{\nabla}_2(M)]$ stands for the family formed by

$$[\omega_1,...,\omega_k] \subset \Omega^{\nabla}_2(M)$$

such that

$$dim[\Sigma_j Ker(\omega_j)] = \min[dim(M),\Sigma_j dim(Ker(\omega_j))].$$

$\mathbb{G}^k[\mathcal{S}^{\nabla}_2(M)]$ stands for the family formed by

$$[B_1,...,B_k] \subset \mathcal{S}^{\nabla}_2(M)$$

such that

$$dim[\Sigma_j Ker(B_j)] = \min[dim(M),\Sigma_j dim(Ker(B_j))].$$

(i) Elements of $\mathbb{G}^p[\Omega^{\nabla}_2(M)]$ are LINEARIZABLE symplectic k-webs.
(ii) Elements of $\mathbb{G}^p[\mathcal{S}^{\nabla}_2(M)]$ are LINEARIZABLE Riemannian k-webs.

We have introduced the double complex

$$\left\{C: \quad C^{ij} = C^i_\tau(\mathcal{A}^*, \mathcal{A}^*) \otimes C^j_\tau(\mathcal{A}, C^\infty(M)), \quad \delta_{ij}\right\}.$$

It gives rise to spectral sequences which may be useful for computing the KV cohomologies $H *_\tau (\mathcal{A}^*, \mathcal{A}^*)$ and $H *_{KV} (\mathcal{A}, C^\infty(M))$. That is not the purpose of this paper. However this double complex is useful for replacing the first order differential equation

$$D^{\nabla\nabla^*}(\psi) = 0$$

by the homological equation

$$\delta_{1,2} q_\psi = 0.$$

We have proved the homological nature of the space of gauge homomorphisms $\mathcal{M}(\nabla, \nabla^*)$. This is useful for relating the image of $\mathcal{M}(\nabla, \nabla^*)$ in $H^1_\tau(\mathcal{A}^*, \mathcal{A}^*)$ and the pair $H^2_{dR}(M), H^{1,2}(C)$.

8. B. The Theory of StatisticaL Models

In the introduction of this paper we have recalled the problem raised by Peter McCullagh.

What is a statistical model [30]?

By the way we have recalled a variant request of Misha Gromov.

In a Search for a Structure. The Fisher Information [15,16].

McCullagh and Gromov choose the same framework for addressing their purpose, The theory of category. This Part B is devoted to the same purpose.

Further the moduli space of isomorphism class of objects of a category $\mathcal{C}$ is denoted by $[\mathcal{C}a]$. In general it is difficult to find an invariant inv_a which encodes $[\mathcal{C}a]$. Subsequently to the questions raised by McCullagh and by Gromov the moduli space of isomorphism class of statistical models is discussed in this Part B. Nowadays, there exists a well established theory of statistical models. The classical references are Amari [17], Amari and Nagaoka [18]. Other remarkable references are Barndorff-Nielsen (Indian Journal of Mathematics 29, Ramanujan Centenary Volume) [21,24], Kass and Vos [37], Murray-Rice (Chapter 1, Section 15 in [22]). In Part A of this paper we have been dealing with this current theory. It has been called the local theory. We suggest reading the attempt by McCullagh to establish a conceptually consistent theory of statistical models [30]. In its time, the paper of McCullagh had been the object of controversy and questions.

We are aimed at re-establishing the theory of statistical models. Our motivations have emerged from some criticisms.

The current theory presents some deficiencies that we plan outlining. (i) A weakness of the current theory is its lacking in geometry; (ii) In the literature on the information geometry many references define an m-dimensional statistical model as an open subset of an Euclidean space $\mathbb{R}^m$. Though this definition may be useful for dealing with coordinate functions, it is topologically and geometrically useless. Let Γ be the group of measurable isomorphisms of a measurable set (Ξ, Ω). The information geometry of a statistical model $\mathbb{M}$ includes the geometry in the sense of Erlangen program of the pair $[\mathbb{M}, \Gamma]$.

Let $\mathbb{M}$ and $\mathbb{M}^*$ be m-dimensional statistical models for the same measurable set (Ξ, Ω). An isomorphism of $\mathbb{M}$ on $\mathbb{M}*$ looks like an sufficient statistic. The geometries $[\mathbb{M}, \Gamma]$ and $[\mathbb{M}*, \Gamma]$ provide the same information. So the impact on the applied information geometry of the theory of moduli space is notable. Subsequently the search for characteristic invariants presents a challenge. An invariant is called characteristic if it determine a model up to isomorphism. So a characteristic invariant encodes the moduli space. That increases the interest in the search of both McCullagh and Gromov.

The Fisher information of widely used models are Hessian metrics [52]. This observation is relevant. However the Fisher information is not a characteristic invariant.

We intend to face the following challenges.
Challenge 1. Revisit the theory of geometric statistical models for measurable sets.
Challenge 2. The Search for a geometric characteristic invariant for statistical models. We recall that such an invariant will encode the points the moduli space of models. Before continuing we recall some definitions.

Definition 42. *A geometric invariant of a model for (Ξ, Ω) is a datum which is invariant under the action of the symmetry group $Aut(\Xi, \Omega)$.*

The framework which is useful for re-establishing the theory of statistical models is the category of locally trivial fiber bundles.

As we have mentioned the need for introducing a new theory of statistical model emerges from some criticisms. We recall the definition a statistical model [18,22,24].

Definition 43. *An m-dimensional statistical model for a measurable set (Ξ, Ω) is a pair (Θ, P) having the properties which follow.*

(1) The manifold Θ is an open subset of the m-dimensional Euclidean space $\mathbb{R}^m$.
(2) P is a positive real valued function

$$\Theta \times \Xi \ni (\theta, \xi) \to P(\theta, \xi) \in \mathbb{R}$$

subject to the requirements which follow.
(3) The function $P(\theta, \xi)$ is differentiable with respect to $\theta \in \Theta$.
(4) For every fixed $\theta \in \Theta$ one set

$$P_\theta = P(\theta, -)$$

then the triple

$$(\Xi, \Omega, P_\theta)$$

is a probability space, viz

$$\int_\Xi P_\theta(\xi) d\xi = 1$$

Furthermore the operation of differentiation

$$d_\theta = \frac{d}{d\theta}$$

commutes with the operation of integration $\int_\Xi$.
(5) (Θ, P) is identifiable, viz for $\theta, \theta^* \in \Theta$

$$P_\theta = P_{\theta^*}$$

if and only if

$$\theta = \theta^*$$

(6) The Fisher information

$$g_\theta(X, Y) = \int_\Xi P(\theta, \xi)[d_\theta log(P(\theta, \xi))]^{\otimes 2}(X, Y) d\xi$$

is positive definite.

Some Criticisms.

The First Critique

The first critique arises from requirement (5).

From the viewpoint of fiber bundles the requirement (5) is useless. Consider the Cartesian product

$$\mathcal{E} = \Theta \times \Xi.$$

That is the same thing as the trivial fiber bundle

$$\mathcal{E} \ni (\theta, \xi) \to \pi(\theta, \xi) = \theta \in \Theta.$$

Therefore P_θ is the restriction to the fiber $\mathcal{E}_\theta$ of the function P.

The Second Critique

The second critique emerges from the requirement (1).

This requirement (1) is too restrictive. It excludes many interesting compact manifolds such as flat tori, euclidean sphere, compact Lie groups.

The Third Critique

From the viewpoint of the differential topology the requirement (6) may be damage to the topology of Θ. When the Fisher information g is singular its kernel is in involution. Thus the topological-geometrical information that are contained in g are transverse to the distribution $Ker(g)$. If $Ker(g)$ is completely integrable then topological and geometrical informations which are contained in g are transversal to the foliation $Ker(g)$. See Part A of this paper. This ends the criticisms.

To motivate for deleting the requirement (1) we construct a compact statistical model which satisfies all of the requirements except the requirement (1).

Let $\mathcal{E}$ be the tangent bundle of the circle S^1. E is the trivial line bundle

$$S^1 \times \mathbb{R} \ni (\theta, t) \to \theta \in S^1.$$

We consider the fonctions f, F and P defined by

$$f(\theta, t) = [\sin^2(\frac{t^2\theta}{1+t^2}) \cos^2(\frac{\theta}{4}) e^{-t^2} + \frac{\pi}{e^2} t^2],$$

$$F(\theta) = \int_{-\infty}^{+\infty} e^{-f(\theta,t)} dt,$$

$$P(\theta, t) = \frac{e^{-f(\theta,t)}}{F(\theta)}.$$

The function $P(\theta, t)$ has the following properties

(i) $(i) : P(\theta, t)$ is smooth,
(ii) $P(0, t) = P(2\pi, t) \quad \forall t \in \mathbb{R}$,
(iii) the $\frac{d}{d\theta}$ commutes with $\int_{\mathbb{R}}$,
(iv) $P(\theta, t) \leq 1 \quad \forall (\theta, t) \in S^1 \times \mathbb{R}$,
(v) if $0 < \theta, \theta^* < 2\pi$ then $P_\theta = P^*_\theta$ if and only if $\theta = \theta^*$,
(vi) $\int_{-\infty}^{+\infty} P(\theta, t) dt = 1$.

These properties show that there is a one to one correspondence between the circle S^1 and a subset of probability densities in $\mathbb{R}$. Thus S^1 is a compact 1-dimensional manifold of probabilities in the measurable set $(\mathbb{R}, \beta(\mathbb{R}))$. Here $\beta(\mathbb{R})$ is the family of Borel subsets of $\mathbb{R}$.

So (S^1, P) is a compact parametric model for $(\mathbb{R}, \beta(\mathbb{R}))$.

A Digression.

Let $\{(\Theta_j, P_j), j := 1, 2\}$ *be statistical models for measurable sets* $\{(\Xi_j, \Omega_j), j := 1, 2\}$. *We put*

$$\Theta = \Theta_1 \times \Theta_2,$$

$$(\Xi, \Omega) = (\Xi_1 \times \Xi_2, \Omega_1 \times \Omega_2),$$

$$P = P_1 \otimes P_2.$$

The function P *is defined in* $\Theta \times \Xi$ *by*

$$P((\theta_1, \theta_2), (\xi_1, \xi_2)) = P_1(\theta_1, \xi_1) P_2(\theta_2, \xi_2)$$

The integration on Ξ *is defined by*

$$\int_{\Xi} f((\theta_1, \theta_2), (\xi_1, \xi_2)) d(\xi_1, \xi_2) = \int_{\Xi_1 \times \Xi_2} f((\theta_1, \theta_2), (\xi_1, \xi_2)) d\xi_1 d\xi_2.$$

Thus we get

$$\int_{\Xi} P[(\theta_1, \theta_2), (\xi_1, \xi_2)] d\xi_1 d\xi_2 = \int_{\Xi_1 \times \Xi_2} P_1(\theta_1, \xi_1) P_2(\theta_2, \xi_2) d\xi_1 d\xi_2 = 1.$$

So (Θ, P) *is a statistical model for* $[\Xi_1 \times \Xi_2, \Omega_1 \times \Omega_2]$.
One is in position to prove that every Euclidean torus $\mathbb{T}^m$ *is a statistical model for* $(\mathbb{R}^m, \beta(\mathbb{R}^m))$.

Another Construction.

For every positive integer m *we consider positive real numbers*

$$\alpha_1 < \alpha_2 < ... < \alpha_m$$

and the real functions which are defined by

$$f_j(\theta, t) = \sin^2(\frac{t^2\theta}{1+t^2}) \cos^2(\frac{\theta}{4}) e^{-t^2} + \alpha_j t^2 \quad (\theta, t) \in E,$$

$$F_j(\theta) = \int_{-\infty}^{+\infty} e^{-f_j(\theta,t)} dt,$$

$$P_j(\theta, t) = \frac{e^{-f_j(\theta,t)}}{F_j(\theta)}.$$

Now we consider the tangent bundle of the m-dimensional flat torus $T\mathbb{T}^m$,

$$\mathbb{T}^m = S^1 \times S^1 \times ... \times S^1.$$

Let

$$(\theta, t) = [(\theta_1, t_1), (\theta_2, t_2), ..., (\theta_m, t_m)] \in T\mathbb{T}^m.$$

We put

$$F(\theta) = \int_{\mathbb{R}^m} e^{-\sum_1^m f_j(\theta_j, t_j)} dt_1 dt_2 ... dt_m,$$

$$P(\theta, t) = \frac{e^{-\sum_1^m f_j(\theta_j, t_j)}}{F(\theta)}.$$

The function $P(\theta, t)$ *satisfies the following requirements*

(1) *If $\theta \neq \theta^*$ there exists $t^{**} \in \mathbb{R}^m$ such that*

$$P(\theta, t^{**}) \neq P(\theta^*, t^{**}),$$

(2) $P(\theta, t) \leq 1 \forall (\theta, t) \in T\mathbb{T}^m$,
(3) $\int_{\mathbb{R}^m} P(\theta, t) dt = 1$.

We deduce that the pair $(\mathbb{T}^m, P)$ is an m-dimensional manifold of probability densities in the measurable set $(\mathbb{R}^m, \beta(\mathbb{R}^m))$.

The image of every local chart of $\mathbb{T}^m$ is a local statistical model in the classical sense [17,18,22]. This ends the Digression.

We are motivated for introducing a new theory of statistical models whose localization yields the current theory. The theory we introduce is an answer to McCullagh and to Gromov.

8.1. The Preliminaries

In this Part B we face three major challenges.
Challenge 1. Taking into account the criticisms we have raised our aim is to introduce a new theory of statistical model whose localization leads to the classical theory of statistical models.
Challenge 2. The second challenge is the Search for an invariant which encodes the point of the moduli space of isomorphism class of statistical models.
Challenge 3. We introduce the theory of homological statistical model and we explore the links between this theory and the challenge 2.
Challenge 4. The fourth challenge is to explore the relationships between "challenge 1, challenge 2, challenge 3" and "Vanishing Theorems in the theory of KV homology".

The theory of KV cohomology and the geometry of Koszul play important roles. We introduce the needed definitions.

Let (Ξ, Ω) be a measurable set. Let $Aut(\Xi, \Omega)$ be the group of measurable isomorphisms of Ξ. Let (M, D) be a locally flat manifold whose KV algebra is denoted by $\mathcal{A}$.

We keep the notation used in Part A of this paper. For instance $\mathcal{S}_2(M)$ is the vector space of differentiable symmetric bi-linear forms in M.

Definition 44. *A random Hessian metric in (M, D) is a map*

$$M \times \Xi \ni (x, \xi) \to Q(x, \xi) \in S^2[T_x^* M],$$

which has the following properties

(1) *for every vector field X the real number $Q(x, \xi)[X, X]$ is non negative, furthermore $\forall v \in T_x M \setminus \{0\} \, \exists \xi \in \Xi$ such that*

$$Q(x, \xi)(v, v) > 0,$$

(2) *for every $\xi \in \Xi$, the random KV cochain*

$$(X, Y) \to Q_\xi(X, Y)(x)$$

with

$$Q_\xi(X, Y)(x) = Q(x, \xi)(X_x, Y_x)$$

is a random cocycle of the KV complex $[C^*_{KV}(\mathcal{A}, C^\infty(M)), \delta_{KV}]$.

Let (Ξ, Ω, p) be a probability space. A random Hessian metric Q generates a Hessian structure (M, g_Q, D) whose tensor metric g_Q is defined by

$$g_Q(X,Y)(x) = \int_{\Xi} Q(x,\xi)(X,Y)p(\xi)d\xi.$$

The group $Aut(\Xi, \Omega)$ of measurable isomorphisms of (Ξ, Ω) is denoted by Γ.

Warning.

When Ξ is a topological space elements of Γ are continuous maps. When Ξ is a differentiable manifold elements of Γ are differentiable maps. Let $\mathbb{P}(\Xi)$ be the Boolean algebra of all subsets of Ξ. The abstract group $Aut(\Xi, \mathbb{P}(\Xi))$ is a subgroup of the group $Isom(\Xi)$ of isomorphisms of the set Ξ.

Definition 45. *A measurable set (Ξ, Ω) is called homogeneous if the natural action of Γ in Ξ is transitive.*

Throughout this paper we will be dealing with homogeneous measurable sets. Below we introduce the framework of the theory of statistical models.

8.2. The Category $\mathcal{FB}(\Gamma, \Xi)$

8.2.1. The Objects of $\mathcal{FB}(\Gamma, \Xi)$

Definition 46. *An object of the category $\mathcal{FB}(\Xi,)$ is a datum $[\mathcal{E}, \pi, M, D]$ which is composed as it follows.*

(1) *M is a connected m-dimensional smooth manifold. The map*

$$\pi : \mathcal{E} \to M$$

is a locally trivial fiber bundle whose fibers $\mathcal{E}_x$ are isomorphic to the set Ξ.

(2) *The pair (M, D) is an m-dimensional locally flat manifold.*

(3) *There is a group action*

$$\Gamma \times [\mathcal{E} \times M] \times \mathbb{R}^m \ni (\gamma, [e, x, \theta]) \to [[(\gamma \cdot e), \gamma \cdot x], \tilde{\gamma} \cdot \theta] \in [\mathcal{E} \times M] \times \mathbb{R}^m.$$

That action is subject to the compatibility requirement

$$\pi(\gamma \cdot e) = \gamma \cdot \pi(e) \quad \forall e \in \mathcal{E}.$$

(4) *Every point $x \in M$ has an open neighborhood U which is the domain of a local fiber chart*

$$\Phi_U \times \phi_U : [\mathcal{E}_U \times U] \ni (e_x, x) \to [\Phi_U(e_x), \phi_U(x)] \in [\mathbb{R}^m \times \Xi] \times \mathbb{R}^m.$$

The local charts are subject to the following compatibility relation

- *(U, ϕ_U) is an affine local chart of the locally flat manifold (M, D),*
- *$\phi_U(\pi(e)) = p_1(\Phi_U(e))$.*

(5) *We set*

$$\Phi_U(e) = (\theta_U(e), \xi_U(e)) \in \mathbb{R}^m \times \Xi.$$

Let $(U, \Phi \times \phi)$ and $(U^*, \Phi^* \times \phi^*)$ be two local charts with

$$U \cap U^* \neq \emptyset,$$

then there exists a unique $\gamma_{UU^*} \in \Gamma$ *such that*

$$[\gamma_{UU^*} \cdot \Phi](e) = \Phi^*(e) \quad \forall e \in \mathcal{E}_{U \cap U^*}.$$

Comments. *Requirements (3) and (4) mean that*

$$[\Phi_U(e), \phi_U(\pi(e)] = [[\theta_U(e), \xi_U(e)], \theta_U(e)]$$

Both requirements (4) and (5) yield the following remarks: the following action is differentiable

$$\Gamma \times M \ni (\gamma, x) \to \gamma \cdot x \in M,$$

the following action is an affine action

$$\Gamma \times \mathbb{R}^m \ni (\gamma, \theta) \to \tilde{\gamma} \cdot \theta,$$

both the left side member and the right side member of (5) have the following meaning.

$$\gamma_{UU^*} \cdot [\theta_U(e), \xi_U(e)] = [\theta_{U^*}(e), \xi_{U^*}(e)].$$

Consequently (5) implies that for all $x \in U \cap U^*$ *one has*

$$\tilde{\gamma}_{UU^*} \cdot \phi(x) = \phi^*(x).$$

Therefore we get

$$\gamma_{UU^*} = \phi^* \circ \phi^{-1}.$$

Suppose that U, U^* and U^{**} are domains of local chart with

$$U \cap U^* \cap U^{**} \neq \emptyset$$

then

$$\gamma_{U^*U^{**}} \circ \gamma_{UU^*} = \gamma_{UU^{**}}.$$

The requirement (3) means that the fibration π is Γ equivariant.
The Figure 2 expresses the requirement property (3).

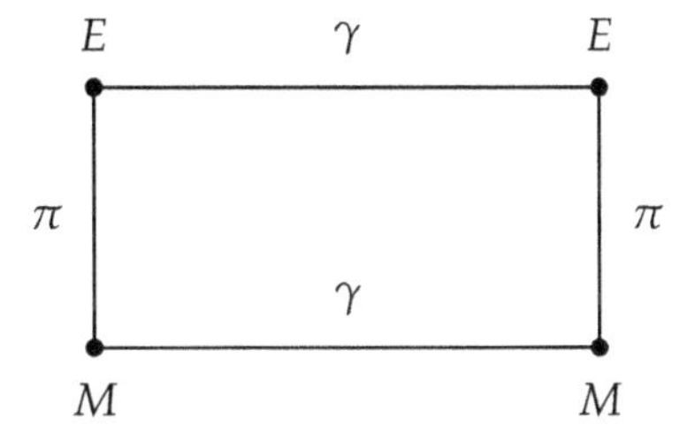

Figure 2. Fibration.

We recall that the group Γ acts in both $\mathcal{E}$ and M. Figure 2 expresses that the projection π of $\mathcal{E}$ on M is Γ-equivariant.

8.2.2. The Morphisms of $\mathcal{FB}(\Gamma, \Xi)$

Let $[\mathcal{E}, \pi, M, D]$ and $[\mathcal{E}^*, \pi^*, M^*, D^*]$ be two objects of $\mathcal{FB}(\Gamma, \Xi)$.
Let $\Psi \times \psi$ be a map

$$[\mathcal{E} \times M] \ni (e, x) \to (\Psi(e), \psi(x)) \in [\mathcal{E}^* \times M^*].$$

Definition 47. *A pair $(\Psi \times \psi)$ is a morphism of the category $\mathcal{FB}(\Gamma, \Xi)$ if the following conditions are satisfied*
(m.1): $\pi^* \circ \Psi = \psi \circ \pi$,
(m.2): both Ψ and ψ are Γ-equivariant isomorphism, that is to say

$$\Psi(\gamma \cdot e) = \gamma \cdot \Psi(e),$$

$$\psi(\gamma \cdot x) = \gamma \cdot \psi(x),$$

(m.3): ψ is an affine map of (M, D) in (M^*, D^*).

The Figure 3 represents the properties (m.1) and (m.2). We are going to define the category of statistical model for (Ξ, Ω). The framework is the category $\mathcal{FB}(\Gamma, \Xi)$.

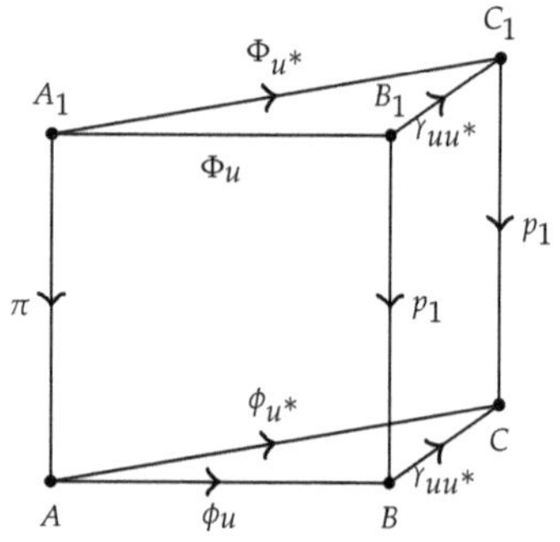

Figure 3. Equivariance.

At one side we recall that the group Γ also acts in $\mathbb{R}^m \times \Xi$. At another side the localizations are made coherent thanks to Cech cocycles γ_{UU^*}. Figure 3 tells two informations. Firstly localizations are Γ-equivariant, secondly thanks to Cech cocycles localizations are coherent.

8.3. *The Category $\mathcal{GM}(\Xi, \Omega)$*

We keep the notation used in the previous subsections. Our purpose is the category of statistical models $\mathcal{GM}(\Xi, \Omega)$.

8.3.1. The Objects of $\mathcal{GM}(\Xi, \Omega)$

Definition 48. *An m-dimensional statistacal model for (Ξ, Ω) is an object of $\mathcal{FB}(\Gamma, \Xi)$, namely*

$$\mathbb{M} = [\mathcal{E}, \pi, M, D]$$

which has the following properties (ρ_*).

$[\rho_1]$: For every local chart $(U, \Phi_U \times \phi_U)$ the subset

$$[\Theta_U \times \Xi] = \Phi_U(\mathcal{E}_U)$$

supports a non negative real valued function P_U *subject to the following requirements.*
$[\rho_{1.1}]$: *For every fixed* $\xi \in \Xi$ *the function*

$$\Theta_U \ni \theta \to P_U(\theta, \xi)$$

is differentiable.
$[\rho_{1.2}]$: *For every fixed* $\theta \in \Theta_U$ *the triple*

$$(\Xi, \Omega, P_U(\theta, -))$$

is a probability space. Further the operation of integration $\int_\Xi$ *commutes with the operation of differentiation* $d_\theta = \frac{d}{d\theta}$.
$[\rho_{1.3}]$: *Let* $(U, \Phi_U \times \phi_U, P_U)$ *and* $(U^*, \Phi_{U^*} \times \phi_{U^*}, P_{U^*})$ *be as in* $[\rho_{1.1}]$ *and in* $[\rho_{1.2}]$.
If $U \cap U^* \neq \emptyset$ *then* P_U, P_{U^*} *and* γ_{UU*} *are related by the formula*

$$P_{U^*} \circ \gamma_{UU^*} = P_U.$$

$[\rho_{1.4}]$: *Let* $U \subset M$ *be an open subset and let* $\gamma \in \Gamma$. *Let us assume that both* U *and* $\gamma \cdot U$ *are domains of local charts*

$$(U, \Phi_U \times \phi_U, P_U)$$

and

$$(\gamma \cdot U, \Phi_{\gamma \cdot U} \times \phi_{\gamma \cdot U}, P_{\gamma \cdot U}).$$

We assume that those local charts satisfy $\rho_{1.1}$, $\rho_{1.2}$ *and* $\rho_{1.3}$. *Then the relations*

$$\Phi_{\gamma \cdot U} \circ \gamma = \gamma \circ \Phi_U,$$

$$\phi_{\gamma \cdot U} \circ \gamma = \gamma \circ \phi_U,$$

implies the equality

$$P_{\gamma \cdot U} \circ \gamma = P_U.$$

A Comment.

Actually, $([\rho_{1.3}])$ *has the following meaning:*

$$P_{U^*}[\tilde{\gamma}_{UU^*} \cdot \theta_U(e), \gamma_{UU^*} \cdot \xi_U(e)] = P_U(\theta_U(e), \xi_U(e))$$

$\forall e \in \mathcal{E}_{U \cap U^*}$.

The Figure 4 represents $(\rho_{1.3})$

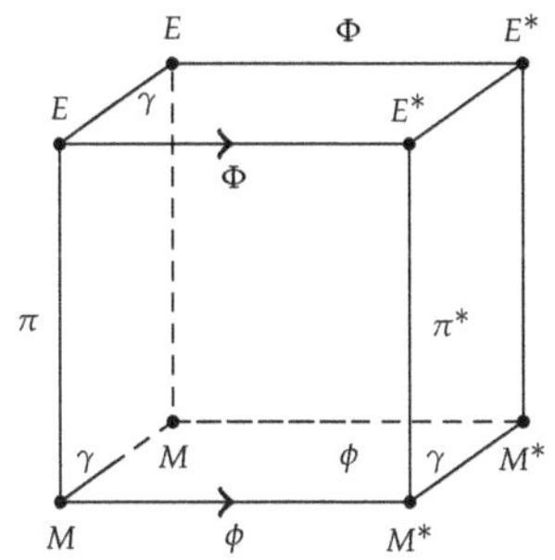

Figure 4. Moduli.

This ends the comment.

Definition 49. *A datum* $[\mathcal{U}, \Phi_{\mathcal{U}} \times \phi_{\mathcal{U}}, P_{\mathcal{U}}, \gamma_{\mathcal{U}\mathcal{U}^*}]$ *as in the last definition is called a local statistical chart of* $[\mathcal{E}, \pi, M, D]$.

Figure 4 is represents what are crucial steps toward the serach of characteristic invariants, viz invariants encoding the points of the moduli space of statistical models. At the present Figure 4 describes the moduli space of the category $\mathcal{FB}(\Gamma, \Xi)$

Before dealing with morphisms of the category $\mathcal{GM}(\Xi, \Omega)$ we introduce a relevant global geometrical invariant.

8.3.2. The Global Probability Density of a Statistical Model

We consider a COMPLETE (or maximal statistical) atlas of an object $[\mathcal{E}, \pi, M, D]$ (of the category $\mathcal{GM}(\Xi, \Omega)$), namely

$$A_{\Phi} = [U_j, \Phi_j, \phi_j, P_j, \gamma_{ij}].$$

The family U_j is an open covering of M. The pair $\mathcal{E}_j \times U_j$ is the domain of the local chart $(\Phi_j \times \phi_j)$. We have

$$\mathcal{E}_j = \mathcal{E}_{U_j}.$$

If $U_i \cap U_j \neq \emptyset$ then one has

$$\phi_j(x) = \tilde{\gamma}_{ji} \cdot \phi_i j(x) \quad \forall x \in U_i \cap U_j.$$

In particular $A = (U_j, \phi_j)$ is an affine atlas of the locally flat manifold (M, D). We have

$$\Phi_j(\mathcal{E}_{y^*}) = \phi_j(y^*) \times \Xi \quad \forall y^* \in U_j.$$

Therefore we set

$$[\mathcal{E}_{y^*}, \Omega_{y^*}] = \Phi_j^{-1}[[\phi_j(y^*) \times \Xi], \Omega].$$

The atlas A_{Φ} satisfies requirements $(\rho_{1.1})$, $(\rho_{1.2})$ and $(\rho_{1.3})$. In $\mathcal{E}_{U_j}$ the local function p_j is defined by

$$p_j = P_j \circ \Phi_j.$$

We suppose that

$$U_i \cap U_j \neq \emptyset.$$

By the virtue of of $[\rho_{1.3}]$ one has

$$p_i(e) = p_j(e)$$

for all $e \in \mathcal{E}_{U_i \cap U_j}$.

Thereby there exists a unique function

$$\mathcal{E} \ni e \to p(e) \in \mathbb{R}$$

whose restriction to $\mathcal{E}_j$ coincides with p_j. The restriction to $\mathcal{E}_x$ is denoted by p_x. The triple

$$(\mathcal{E}_x, \Omega_x, p_x)$$

is a probability space.

Definition 50. *The function*

$$\mathcal{E} \ni e \to p(e) \in \mathbb{R}$$

is called the probability density of the model $[\mathcal{E}, \pi, M, D]$.

Comments.

(i) *We take into account the global probability density p. Then an object of the category* $\mathcal{GM}(\Xi, \Omega)$ *is denoted by*

$$[\mathcal{E}, \pi, M, D, p].$$

(ii) *The function p is* Γ*-equivariant. THIS IS THE GEOMETRY in the sense of Erlangen program.*
(iii) *We have not used any argument depending the dimension of manifolds.*

The Figure 5 expresses coherence to local probability densities We are in position to define the morphisms of the category $\mathcal{GM}(\Xi, \Omega)$.

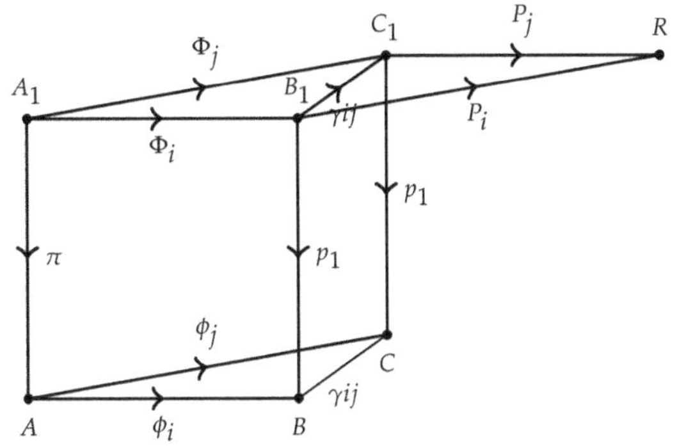

Figure 5. Localisation.

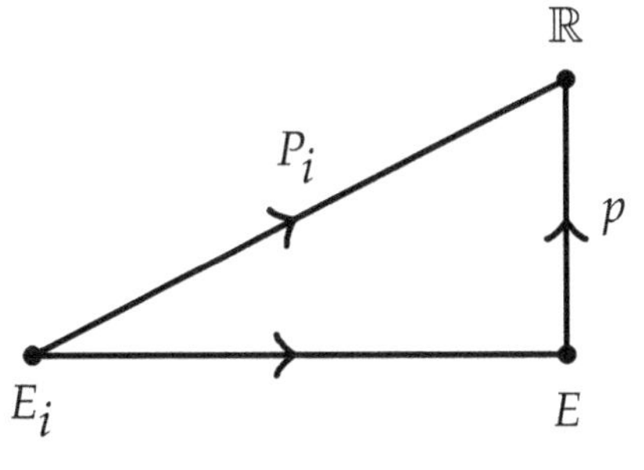

Figure 6. Probability Density.

In Figure 5 one sees that modulo the dynamics of the group Γ in $\mathbb{R}^m \times \Xi$ all localizations look alike. Figure 6 show that local probability densities $\{p_i\}$ are but localizations of a global probability density p

8.3.3. The Morphisms of $\mathcal{GM}(\Xi, \Omega)$

Definition 51. *Let* $\mathbb{M} = [\mathcal{E}, \pi, M, D, p]$ *and* $\mathbb{M}* = [\mathcal{E}^*, \pi^*, M^*, D^*, p^*]$ *be two objects of the category* $\mathcal{GM}(\Xi, \Omega)$*. A* $\mathcal{FB}(\Gamma, \Xi)$*-morphism*

$$(\Psi \times \psi) : [\mathcal{E}, \pi, M, D] \to [\mathcal{E}^*, \pi^*, M^*, D^*]$$

is a morphism of $[\mathcal{E}, \pi, M, D, p]$ *in* $[\mathcal{E}^*, \pi^*, M^*, D^*, p^*]$ *if*

$$p^* \circ \Psi = p.$$

A Comment.

Let $[\mathcal{E}, \pi, M, D]$ *be an object of the category* $\mathcal{FB}(\Gamma, \Xi)$. *Let G be the group of isomorphisms of* $[\mathcal{E}, \pi, M, D]$. *If M is finite dimensional then G is a finite dimensional Lie group. The group G acts in the category M whose objects are probability densities in* $[\mathcal{E}, \pi, M, D,]$.

Definition 52. *The orbit space space*

$$m = \frac{M}{G}$$

is called the moduli space of M.

A Comment.

Every trivialization of

$$\mathbb{M} = [\mathcal{E}, \pi, M, D, p]$$

is a statistical model in the classical sense [18,22]. So we have taken up the challenge 1.

8.3.4. Two Alternative Definitions

We introduce two other presentations of the category $\mathcal{GM}(\Xi, \Omega)$. Those presentations highlight the connection with the searches of McCullagh and Gromov. Those presentation is useful in both the theoretical statistics and the applied statistics [17,18,21,24,55,57,58].

We consider the category $\mathcal{MSE}$ whose objects are a probability spaces (Ξ, Ω, p).

Definition 53. *A morphism of a probability space* (Ξ, Ω, p) *in another probability space* (Ξ^*, Ω^*, p^*) *is a measurable map* Ψ *of* (Ξ, Ω) *in* (Ξ^*, Ω^*) *such that*

$$p = p^* \circ \Psi.$$

Remark 4. *A morphism as in the last definition has a statistical nature. An isomorphisms of* (Ξ, Ω, p) *on* $(\Xi*, \Omega*, p*)$ *is an sufficient statistic. The category* $\mathcal{MSE}$ *is useful for introducing two variant descriptions of the category* $\mathcal{GM}(\Xi, \Omega)$.

Definition 54. *We use the previous notation.*

(1) *A statistical model is a locally trivial fiber bundle over a locally flat manifold*

$$\pi : \mathcal{E} \to M.$$

The fibers of π *are probability spaces.*

(2) *The functor*

$$[\mathcal{E}, p] \to [M, D]$$

is called a $\mathcal{MSE}$*-fibration.*

The category of $\mathcal{MSE}$-fibrations is denoted by $\mathcal{FB}(\mathcal{MSE})$. The morphisms the category $\mathcal{GM}(\Xi, \Omega)$ are called $\mathcal{MSE}$-morphisms.

Definition 55. *A statistical model for a measurable set* (Ξ,Ω) *is a functor of the category* $\mathcal{FB}(\Gamma,\Xi)$ *in the category* $\mathcal{FB}(\mathcal{MSE}$, *namely*

$$[\mathcal{E},\pi,M,D] \to [\mathcal{E},\pi,M,D,p]$$

At the present step it is clear that the information geometry is structured.

8.3.5. Fisher Information in $\mathcal{GM}(\Xi,\Omega)$

We consider a $\mathcal{MSE}$-fibration

$$\mathbb{M} := [\mathcal{E},p] \to (M,D).$$

The Fisher information to be defined is an elementg of $\Gamma(S^2(T^*M))$.
We recall that every $\mathcal{MSE}$-fiber $\mathbb{M}_x, x \in M$ has a structure of probability space

$$\mathbb{M}_x := [\mathcal{E}_x,\Omega_x,p_x].$$

Let X,Y be local vector fields which are defined in a open neighbourhood of $x \in M$.

Definition 56. *The Fisher information at x is defined by*

$$g_x(X,Y) = -\int_{\mathcal{E}_x} p(e)[D^2 log(p(e))](X,Y)d(e)$$

We recall that the horizontal differentiation commutes with the integration along the $\mathcal{MSE}$-fibers, viz

$$d_\theta \circ \int_F = \int_F \circ \frac{\partial}{\partial\theta}.$$

So the Fisher information g is well defined. It has the following properties

(1) g is positive semi-definite,
(2) g is an invariant of the Γ-geometry in $[\mathcal{E},\pi,M,D,p]$.

8.4. Exponential Models

Let $[\mathcal{E},\pi,M,D,p]$ be an object of $\mathcal{GM}(\Xi,\Omega)$. We recall that data which are defined in $\mathcal{E}$ are called random data in the base manifold M. The operation of integration along the $\mathcal{MSE}$-fibers is denoted by $\int_F$. Thus a random datum μ is called smooth if its image $\int_F(\mu)$ is smooth.
Conversely every datum θ^* which is point-wise defined in M is the image of the random datum

$$\theta = \theta^* \circ \pi.$$

So we get

$$\theta^* = \int_F [\theta^* \circ \pi].$$

Thus at every $x \in M$ one has

$$\theta^*(x) = \int_{\mathcal{E}_x} \theta^*(\pi(e))p_x(e)de.$$

A random affine function is a function

$$\mathcal{E} \ni e \to a(e) \in \mathbb{R}$$

subject to the requirement

$$D^2 \int_F a = 0.$$

Definition 57. *An $\mathcal{MSE}$-fibration*

$$[\mathcal{E}, p] \to [M, D]$$

is called an exponential model if the following conditions are satisfied.

(1) *The base manifold M supports a locally flat structure (M, ∇) and a real valued function $\psi \in C^\infty(M)$.*
(2) *The total space $\mathcal{E}$ supports a real valued random function a.*
(3) *The triple $[a, \nabla, \psi]$ is subject to the following requirement*
(4) $\nabla^2 \int_F(a) = O$,
(5) $p(e) = exp[a(e) - \psi(\pi(e))]$.

Remark 5. *At one side Localizations of exact Hessian homological statistical models yield the classical Koszul Information Geometry [55]. That is but the classical Hessian Information Geometry. At another side the KV homology learns that the Hessian information Geometry is the same think as the geometry of exponential famillies see Part A Section 5, Theorem 16.*

Reminder.

In the Appendix A to this paper the reader will find a new invariant $r^b(p)$ measuring how far from being an exponential model is an $\mathcal{MSE}$-fibration

$$[\mathcal{E}, p] \to [M, D].$$

By the virtue of results in Part A, to be an exponential model depends on homological conditions.

8.4.1. The Entropy Flow

We are going to introduce the notion of local entropy flow. Subsequently we will show that the Fisher information of a model $[\mathcal{E}, \pi, M, D, p]$ is the Hessian of the local entropy flow.

To start we consider a $\mathcal{MSE}$-fibration

$$[\mathcal{E}, p] \to [M, D].$$

That is another presentation of the statistical model $[\mathcal{E}, \pi, M, D, p]$.

Let $[U_j, \Phi_j \times \phi_j, \gamma_{ij}, P_j]$ be an atlas of $[\mathcal{E}, \pi, M, D, p]$. We put

$$[\Theta_j, P_j] = [\Phi_j(\mathcal{E}_j), p \circ \Phi_j^{-1}].$$

Then every $[\Theta_j, P_j]$ is a local statistical model for (Ξ, Ω).

Lc X, Y be two vector fields defined in U_j and let $\psi_X(t)$ and $\psi_Y(s)$ be their local flows defined in U_j. Then we set

$$\Phi_j(e) = [\theta_j(e), \xi_j(e)] = [\phi_j(\pi(e)), \xi_j(e)], \quad e \in \mathcal{E}_j,$$
$$\tilde{\psi}_j(t)[\theta_j(e), \xi_j(e)] = ([\phi_j \psi_X(t) \phi_j^{-1}][\theta_j(e)], \xi_j(e)),$$
$$\tilde{\psi}_j(s)[\theta_j(e), \xi_j(e)] = ([\phi_j \psi_Y(s) \phi^{-1}{}_j][\theta_j(e)], \xi_j(e)).$$

Definition 58. *The local entropy flow of the pair (X, Y) is the function $Ent^j_{X,Y}$ defined by*

$$Ent^j_{X,Y}(s,t)(\pi(e)) = \int_\Xi \{P_j[\tilde{\psi}_X(s)(\Phi(e))] log[P_j[\tilde{\psi}_Y(t)(\Phi_j(e))]]\}\, d\xi(e).$$

To pursue we use the compatibility of local charts of the atlas$[U_j, \Phi_j \times \phi_, , \gamma_{ij}, P_j]$. If

$$U_i \cap U_j \neq \emptyset$$

then forall $e \in \mathcal{E}_{U_i \cap U_j}$ we have

$$\Phi_j(e) = \gamma_{ij} \cdot \Phi_i(e),$$

$$\phi_j(\pi(e)) = \gamma_{ij} \cdot \phi_i(\pi(e)).$$

We recall that $\tilde{\psi}(t)$ and $\tilde{\psi}(s)$ are defined by

$$\tilde{\psi}_j(t) = \phi_j \psi_X(t) \phi_j^{-1},$$

$$\tilde{\psi}_j(s) = \phi_j \psi_Y(s) \phi_j^{-1}.$$

Those reminders are useful for concluding that whenever

$$U_i \cap U_j \neq \emptyset$$

we have

$$Ent^i_{X,Y}(s,t)(\pi(e)) = Ent^j_{X,Y}(s,t)(\pi(e)).$$

So the local entropy flow does not depend on local charts.

If both X and Y are complete vector fields then their entropy flow is globally defined. A notable consequence is the following statement.

Theorem 18. *Every $\mathcal{MSE}$-fibration over a compact manifold M admits a globally defined entropy map*

$$\mathcal{X}(M) \times \mathcal{X}(M) \ni (X,Y) \rightarrow Ent_{X,Y} \in C^\infty(\mathbb{R}^2).$$

8.4.2. The Fisher Information as the Hessian of the Local Entropy Flow

we consider the function

$$H_j(s,t,\xi) = P_j[\tilde{\psi}_X(s)(\phi(e))] log[P_j[\tilde{\psi}_j(t)(\Phi_j(e))]].$$

Direct calculations yield

$$[\frac{\partial^2(H_j(s,t))}{\partial s \partial t}][(s,t) = (0,0)] = P_j[\phi_j(e)](X \cdot log[P_j(\Phi_j(e))])(Y \cdot log[P_j(\phi_j(e))]).$$

We know that $\frac{\partial^2}{\partial s \partial t}$ commutes with $\int_\Xi$. Therby we conclude that

$$g\pi(e)(X,Y) = \frac{\partial^2 Ent_j(s,t)(\pi(e))}{\partial s \partial t}[(s,t) = (0,0)].$$

Theorem 19. *We consider an $\mathcal{MSE}$-firation over a compact manifold*

$$\mathbb{M} := [\mathcal{E}, p] \rightarrow [M, D].$$

The Fisher information of $\mathbb{M}$ is the Hessian of the entropy map.

8.4.3. The Amari-Chentsov Connections in $\mathcal{GM}(\Xi, \Omega)$

Let

$$\mathbb{M} = [\mathcal{E}, \pi, M, D, p]$$

be an m-dimensional statistical model for a measurable set (Ξ, Ω). To define the family of α-connections we work in a local chart (Φ_U, Φ_U).

We set

$$\Theta_U = \phi_U(U),$$
$$\Theta \times \Xi = \Phi_U(\mathcal{E}_U).$$

In the base manifold (M, D) the local chart $[U, \phi_u)$ yields a system of local affine coordinate functions

$$\theta = (\theta_1, ..., \theta_m).$$

We use the notation as in [18]. Given a real number α we define the α-connection ∇^α by its Christoffel symbols in the local coordinate functions θ. Those Christoffel are denoted by $\Gamma^\alpha_i j : k$. We proceed as it follows.

Step 1: In the open subset $\Theta_U \subset \mathbb{R}^m$ we put

$$\tilde{\Gamma}^{\alpha,U}_{ij:k}(\theta) = \int_\Xi P_U(\theta, \xi)\left\{[\frac{\partial^2 l_n(\theta,\xi)}{\partial\theta_i\partial\theta_j} + \frac{1+\alpha}{2}\frac{\partial l_n(\theta,\xi)}{\partial\theta_i}\frac{\partial l_n(\theta,\xi)}{\partial\theta_j}]\frac{\partial l_n(\theta,\xi)}{\partial\theta_k}\right\} d\xi.$$

This local definition of $\tilde{\Gamma}^\alpha_{ij:k}$ agrees with affine coordinate change in Θ_U.

Step 2: In the open subset U $\Gamma^{\alpha,U}_{ij:k}$ is defined by

$$\Gamma^{\alpha,U}_{ij:k} = \tilde{\Gamma}^{\alpha,U}_{ij:k} \circ \phi_U.$$

Since the definition of $\tilde{\Gamma}^\alpha_{ij:k}$ agrees with an affine coordinate change we can use an atlas

$$\mathbb{A} = [U_j, \Phi_j \times \phi_j, \gamma_{ij}]$$

for constructing a Koszul connection $\nabla^\alpha(\mathbb{A})$. Since the construction of $\nabla^\alpha(\mathbb{A})$ agrees with affine coordinate change the connection $\nabla^\alpha(\mathbb{A})$ is independent from the choice of $\mathbb{A}$. Every α-connection is torsion free. So an $\mathcal{MSE}$-fibration

$$[\mathcal{E}, p] \to [M, D]$$

gives rise to a ma

$$\mathbb{R} \ni \alpha \to \nabla^\alpha \in \mathcal{SLC}(M).$$

If the Fisher information g is definite then $(M, g, \nabla^\alpha, \nabla^{-\alpha})$ is a dual pair [17,48].

By the virtue of the definition of the Fisher information g a local section of section of $Ker(g)$ is a local vector field $X \in \mathcal{X}(M)$ such that

$$X \cdot p = 0.$$

Therefore, it is easily seen that

$$L_X g = 0.$$

So if data are analytic then g is a stratified Riemannian foliation.

8.4.4. The Homological Nature of the Probability Density

We consider a $\mathcal{MSE}$-fibration

$$\mathbb{M} := [\mathcal{E}, p] \to [M, D].$$

We recall that a random differential q-form in $\mathcal{E}$ is a mapping

$$\mathcal{E} \ni e \to \omega(e) \in \wedge^q(T^*_{\pi(e)})M$$

such
$$\int_F \omega \in \Omega^q(M).$$

The vector space of random differential q-forms is denoted by $\Omega^q(\mathcal{E})$.

Let $\Phi_U \times \phi_U$ be a local chart of $\mathbb{M}$. We set

$$(\Theta_U, \Xi) = (\phi_U(U), \Xi) = \Phi_U(\mathcal{E}_U).$$

We recal that in $\Phi_U(\mathcal{E}_U)$ the partial differentiation $\frac{\partial}{\partial\theta}$ is called the horizontal differentiation in $\mathcal{E}_U$. Therefore we use the relation
$$\int_F \circ \frac{\partial}{\partial\theta} = \frac{d}{d\theta} \circ \int_F$$
for setting the de Rham complex of random differential forms. Namely

$$\Omega(\mathcal{E}): \quad 0 \to \mathbb{R} \to \Omega^0(\mathcal{E}) \to ...\Omega^q(\mathcal{E}) \to \Omega^{q+1}(\mathcal{E})... \to \Omega^m(\mathcal{E}) \to 0.$$

The complex $\Omega(\mathcal{E})$ is a complex of Γ-modules. Here

$$\Gamma = Aut(\Xi, \Omega).$$

Then the cohomology space $H * (\Gamma, \Omega(\mathcal{E}))$ is bigraded,

$$H^{p,q}(\Gamma, \Omega(\mathcal{E})) = H^p(\Gamma, \Omega^q(\mathcal{E})).$$

The probability density p is Γ-invariant. It is an element of $H^{0,0}(\Gamma, \Omega(\mathcal{E}))$.

8.4.5. Another Homological Nature of Entropy

One of main purpose of [14] is the homological nature of the entropy. The classical entropy function of a statistical model $[\mathcal{E}, \pi, M, D, p]$ is defined by

$$E(\pi(e)) = \int_{\mathcal{E}_{\pi(e)}} p(e^*) log(p(e^*)).$$

In the complex $\Omega(\mathcal{E})$ we perform the machinery of Eilenberg [59]. That yields the exact sequence (of random cohomology spaces)

$$\to H^{q-1}_{res}(\mathcal{E}, \mathbb{R}) \to H^q_e(\mathcal{E}, \mathbb{R}) \to H^q_{dR}(\mathcal{E}, \mathbb{R}) \to H^q_{res}(\mathcal{E}, \mathbb{R}) \to$$

We take into account the identities

$$p(\gamma \cdot e) = p(e),$$
$$\gamma \cdot (\pi(e)) = \pi(\gamma \cdot e).$$

Then we have
$$\begin{aligned} E(\gamma \cdot \pi(e)) &= \int_{\mathcal{E}_{\gamma \cdot \pi(e)}} p(\gamma \cdot e^*) log(p(\gamma \cdot e^*)) \\ &= \int_{\mathcal{E}_{\pi(\gamma \cdot e)}} p(e^*) log(p(e^*)) \\ &= E(\pi(e)). \end{aligned}$$

Thus the entropy $E(\pi(e))$ is Γ-equivariant. Therefore, it defines an equivariant cohomology class

$$[E] \in H^0_e(M, \mathbb{R}).$$

This is another topological nature of the entropy. For another viewpoint see [16,31].

Our purpose is to show the theory of statistical models has a homological nature in the category $\mathcal{FB}(\Gamma, \Xi)$.

Definition 59. *A statistical model for a measurable set* (Ξ, Ω) *is couple* $(\mathbb{M}, [p])$ *formed by an object of the category* $(\mathcal{FB} - \Gamma, \Xi)$*, namely*

$$\mathbb{M} = [\mathcal{E}, \pi, M, D]$$

and a smooth Γ*-equivariant random cohomology class*

$$[p] \in H_e^0(\mathcal{E}, \mathbb{R}).$$

Further the to every fiber $p|\mathcal{E}_x$ *is a probability density.*

A Comment.

Let $(U, \Phi \times \phi)$ *be a local chart of* $[\mathcal{E}, \pi, M, D]$ *and let* $x* \in U$*. We set*

$$\Theta_U = \phi(U),$$

$$(\mathcal{E}_{x^*}, \Omega_{x^*}) = \Phi^-1[\{\phi(x^*)\} \times (\Xi, \Omega)].$$

The last definition above says that we obtain the probability space

$$(\mathcal{E}_{x^*}, \Omega_{x^*}, [p]).$$

This property does not depend on the choice of the local chart $(U, \Phi \times \phi)$*. Thus we can regard* $[\mathbb{M}, p]$ *as a special type of homological map*

$$\mathcal{FB}(\Gamma, \Xi) \ni \mathbb{M} \to [p] \in H_e^0(\mathcal{E}, \mathbb{R}).$$

9. The Moduli Space of the Statistical Models

We are going to face another major open problem. The challenge is the search for an invariant which encodes the points of the orbit space

$$m = \frac{M}{G}.$$

That is what is called the problem of moduli space. This problem of moduli space is a major challenge in both the differential geometry and the algebraic geometry (see the theory of Teichmuller). The problem is rather confusedly addressed in [30]. Subsequently it provoked controversies and criticisms.

The Hessian Functor

We consider the category $\mathcal{BF}$ whose objects are pairs $\{M, B\}$ formed by a manifold M equipped bilinear forms $B \in \Gamma(T^{*\otimes 2M})$.

In Part A we have defined the Hessian differential operator of a Koszul connection ∇, namely

$$D_\nabla = \nabla^2.$$

Those operators are useful for addressing the problem of moduli spaces. For our purpose four categories are involved,

(1) The category $\mathcal{LC}$ whose objects are gauge structures (M, ∇),
(2) The category $\mathcal{GM}$ whose objects are statistical models for measurable sets,
(3) the category $\mathcal{BF}$ whose objects are manifolds equipped bilinear forms,

(4) the category $\mathcal{F}(\mathcal{LC}, \mathcal{BF})$ whose objects are functors

$$\mathcal{GM} \to \mathcal{BF}.$$

Definition 60. *The Hessian functor is the functor*

$$\mathcal{GM} \ni \mathbb{M} = [\mathcal{E}, \pi, M, D, p] \to q_{\mathbb{M}} \in \mathcal{F}(\mathcal{LC}, \mathcal{BL})$$

Here $q_{\mathbb{M}}$ is defined by

$$q_{\mathbb{M}}[\nabla] = \nabla^2 log(p).$$

Reminder.

We recall that for vector fields X, Y the bilinear form $q_{\mathbb{M}}[\nabla](X, Y)$ is defined by

$$q_{\mathbb{M}}[\nabla](X, Y) = X \cdot (Y \cdot log(p)) - \nabla_X Y \cdot log(p)$$

The functor $q_{\mathbb{M}}$ is called the Hessian functor of the model

$$\mathbb{M} = [\mathcal{E}, \pi, M, D, p].$$

Our aim is to demonstrate the following claim. Up to isomorphism a statistical model $\mathbb{M}$ is defined by its Hessian functor $q_{\mathbb{M}}$. The functor $q_{\mathbb{M}}$ is an significant contribution to the information geometry.

We fix an object of $\mathcal{FB}(\Gamma, \Xi)$, namely $[\mathcal{E}, \pi, M, D]$. Let $\mathcal{P}(\mathcal{E})$ be the convex set of probability densities in $[\mathcal{E}, \pi, M, D]$. The multiplicative group of positive real valued functions defined in Ξ is denoted by $\mathbb{R}_+^{\Xi}$. The quotient of $\mathcal{P}(\mathcal{E})$ modulo $\mathbb{R}_+^{\Xi}$ is denoted by

$$\mathcal{PRO}(\mathcal{E}) = \frac{\mathcal{P}(\mathcal{E})}{\mathbb{R}_+^{\Xi}}.$$

Lemma 6. *For every $p \in \mathcal{P}(\mathcal{E})$ the image of*

$$\mathbb{M} = [\mathcal{E}, p],$$

namely $q_{\mathbb{M}}$ depends only on the class $[p] \in \mathcal{PRO}(\mathcal{E})$

Proof. We consider

$$\mathbb{M} = [\mathcal{E}, \pi, M, D, p],$$
$$\mathbb{M}^* = [\mathcal{E}, \pi, M, D, p^*].$$

We assume that

$$q_{\mathbb{M}} = q_{\mathbb{M}^*}.$$

Thus in every local trivialization $\Theta_U \times \Xi$ one has the identity

$$X(Ylog(\frac{p^*(x,\xi)}{p(x,\xi)}) - \nabla_X Ylog(\frac{p^*(x,\xi)}{p(x,\xi)})) = 0$$

forall $X, Y \in \Gamma(T\Theta)$, forall $\nabla \in \mathcal{LC}(\Theta)$. That identity holds if and only if the function

$$(x, \xi) \to \frac{p^*(x,\xi)}{p(x,\xi)}$$

belongs to $\mathbb{R}_+^{\Xi}$. This ends the idea. □

A Comment.

The mapping

$$\mathbb{M} \to q_{\mathbb{M}}$$

is a global geometrical invariant in the sense of Erlangen. In other words it is an invariant of the Γ-geometry in $[\mathcal{E}, \pi, M, D, p]$.

Our aim is to demonstrate that

$$\mathbb{M} \to q_{\mathbb{M}}$$

is a characteristic invariant in the category $\mathcal{GM}(\Xi, \Omega)$. In other words the isomorphism class of the model

$$\mathbb{M} = [\mathcal{E}, \pi, (M, D), p]$$

is encoded by the functor

$$\nabla \to q_{\mathbb{M}}[\nabla].$$

The first step is the following lemma.

Lemma 7. *In the same object* $[\mathcal{E}, \pi, M, D]$ *we consider two statistical models*

$$\mathbb{M}_1 = [\mathcal{E}, \pi, M, D, p_1],$$

$$\mathbb{M}_2 = [\mathcal{E}, \pi, M, D, p_2].$$

The following assertions are equivalent

(1) $q_{\mathbb{M}_1} = q_{\mathbb{M}_2}$,
(2) $p_1 = p_2$.

Proof. We work in the domain of a local trivialization of $[\mathcal{E}, \pi, M, D]$. By the virtue of Lemma 6 above we know that

$$q_{p_1} = q_{p_2}$$

if and only if

$$p_1(x, \xi) = \lambda(\xi) p_2(x, \xi)$$

with $\lambda \in \mathbb{R}_+^{\Xi}$. Since both p_1 and p_2 are Γ-equivariant the function

$$\Xi \ni \xi \to \lambda(\xi)$$

is Γ-invariant too. Now we take into account that the natural action of Γ in Ξ is transitive. Therefore the Γ-equivariant function $\lambda(\xi)$ is a constant function. Therefore

$$p_1(x, \xi) = \lambda p_2(x, \xi)$$

The operation of integration along a fiber of π yields

$$\lambda = 1$$

This ends the proof. □

We consider two m-dimensional statistical models for (Ξ, Ω), namely

$$\mathbb{M}_j = [\mathcal{E}_j, \pi_j, M_j, D_j, p_j], \quad j := 1, 2.$$

To simlplify we use the following notation.

$$q_{p_j} = q_{\mathbb{M}_j}.$$

In the category $\mathcal{FB}(\Gamma, \Xi)$ we consider an isomorphism

$$[\mathcal{E}_1 \times M_1] \in (e, x) \to [\Psi(e), \psi(x)] \in \mathcal{E}_2 \times M_2.$$

(1) Let ψ_* be the differential of ψ. For $\nabla \in \mathcal{LC}(M_1)$ the image $\psi_*(\nabla) \in \mathcal{LC}(M_2)$ is defined by

$$[\psi_*(\nabla)]_{X^*} Y^* = \psi_*[\nabla_{\psi_*^{-1}(X^*)} \psi_*^{-1}(Y^*)]$$

for all vector fields $X^*, Y^* \in \mathcal{X}(M_2)$.

(2) It is clear that the datum $[\mathcal{E}_1, \pi, M_1, D_1, p_2 \circ \Psi]$ is an object of the category $\mathcal{GM}(X, \Omega)$. Then for vector fields X, Y in M_1 we calculate (at X, Y) the right hand member of the following equality

$$[q_{p_2 \circ \Psi}(\nabla)] = \nabla^2[log(p_2 \circ \Psi)].$$

Direct calculations yield

$$\begin{aligned}
\nabla^2[log(p_2 \circ \Psi)](X, Y) &= X \cdot [Y \cdot log(p_2 \circ \Psi)] - \nabla_X Y \cdot log(p_2 \circ \Psi) \\
&= X \cdot [Y \cdot log(p_2) \circ \Psi] - \nabla_X Y \cdot [log(p_2) \circ \Psi] \\
&= \psi_*(X) \cdot [\psi_*(Y) \cdot log(p_2)] - \psi_*(\nabla_X Y) \cdot log(p_2) \\
&= [\psi_*(\nabla)^2 log(p_2)](\psi_*(X), \psi_*(Y)).
\end{aligned}$$

Thus for all $\nabla \in \mathcal{LC}(M_1)$ we have

$$q_{[p_2 \circ \Psi]}(\nabla) = q_{p_2}(\psi_*(\nabla)).$$

We summarize the calculations just carried out as it follows

Lemma 8. *Keeping the notation we just used namely p_2 and $\Psi \times \psi$ we have the following equality*

$$q_{[p_2 \circ \Psi]} = q_{p_2} \circ \psi_*$$

We are in position to face the problem of moduli space in the category $\mathcal{GM}(\Xi, \Omega)$.

Theorem 20. *We consider two m-dimensional statistical models*

$$\mathbb{M}_j = [\mathcal{E}_j, \pi_j, M_j, D_j, p_j], \quad j := 1, 2.$$

In the category $\mathcal{FB}(\Gamma, \Xi)$ let $\Psi \times \psi$ be an isomorphism of $[\mathcal{E}_1, \pi_1, M_1, D_1]$ onto $[\mathcal{E}_2, \pi_2, M_2, D_2]$. The following assertions are equivalent.

(1) $q_{p_2} \circ \psi_* = q_{p_1}$,

(2) $p_2 \circ \Psi = p_1$.

Demonstration.

The demonstration is based on Lemmas 7 and 8.

According to our construction one has the following functor

$$q_{\Psi} = q_{p_2} \circ \psi_*.$$

This functor q_{Ψ} is the Hessian functor of the model

$$\mathbb{M}_{\Psi} = [\mathcal{E}_1, \pi, M_1, D_1, p_2 \circ \Psi].$$

Further $\Psi \times \psi$ is an isomorphism of $\mathbb{M}_{\Psi}$ onto $\mathbb{M}_2$.

Let us prove that assertion (2) implies assertion (1).

By of the definition of morphism of models, the pair $\Psi \times \psi$ is an isomorphism of $\mathbb{M}_1$ onto $\mathbb{M}_2$ if and only if

$$p_2 \circ \Psi = p_1.$$

Here we set the explicit formulas. Let $\nabla \in \mathcal{LC}(M_1)$. For all vector fields X, Y in M_2 we have

$$X \cdot (Y \cdot log(p_1)) - \nabla_X Y \cdot log(p_1) = X \cdot (Y \cdot log(p_2 \circ \Psi)) - \nabla_X Y \cdot log(p_2 \circ \Psi)$$

$$= \psi_*(X) \cdot (\psi_*(Y) \cdot log(p_2)) - \psi_*(\nabla_X Y) \cdot log(p_2).$$

now we observe that

$$\psi_*(\nabla_X Y) = [\psi_*(\nabla)]_{[\psi_*(X)]} \psi_*(Y).$$

Therefore (2) impies the equality

$$\psi^*[q_{[p_2]}(\psi_*(\nabla))] = q_{[p_2 \circ \Psi]}(\nabla) = q_{p_1}.$$

This shows the implication (2) $\rightarrow$ (1).

Let us prove that assertion (1) implies assertion (2).

Now we assume that that (1) holds, viz

$$q_{[p_2 \circ \Psi]} = q_{p_1}.$$

Then both $\mathbb{M}_1$ and $\mathbb{M}_{\Psi}$ have the same Hessian functor. By the virtue of Lemma 8 above we deduce that

$$p_2 \circ \Psi = p_1.$$

This ends the demonstration.

Reminder.

(i) *Objects of $\mathcal{GM}(\Gamma, \Xi)$ are quintuplets*

$$\mathbb{M} = [\mathcal{E}, \pi, M, D, p].$$

They are called statistical models for the measurable set (Ξ, Ω).

(ii) *Objects of $\mathcal{FB}(\mathcal{MSE})$ are functors*

$$[\mathcal{E}, p] \rightarrow [M, D].$$

They are called $\mathcal{MSE}$-fibrations.

Those categories are canonically equivalent. Further the actions the group G in those categories lead to the same moduli space

$$m = \frac{\mathcal{GM}(\Xi,\Omega)}{G} = \frac{\mathcal{FB}(\mathcal{MSE})}{G}.$$

We rephrase the theorem on moduli space.

Theorem 21. *The functor*

$$[\mathcal{E}, p] \to q_p \in \mathcal{BF}(M)$$

parametrizes the moduli space m.

This ends the challenge 2.

10. The Homological Statistical Models

In this section we introduce the theory of homological statistical models (HSM in short). We address the links between this theory and the local theory as in [17].

The theory of homological statistical models is useful for strengthening the central role played by the theory of KV homology in the information geometry and in the topology of the information [14,16–18,22,30,37,60,61].

We introduce the theory of localization of homological models. We use it to highlight the role played by local cohomological vanishing theorems as well as the role played by global cohomological vanishing theorems.

The framework is the category $\mathcal{FB}(\Gamma,\Xi)$.

Let $[\mathcal{E},\pi,M,D]$ be an m-dimensional object of the category $\mathcal{FB}(\Gamma,\Xi)$, viz $m = dim(M)$. The KV algebra of (M,D) is denoted by $\mathcal{A}$. The smooth manifold $\mathbb{R}^m$ supports a sheaf of KV algebras $\tilde{\mathcal{A}}$. This sheaf is locally isomorphic to $\mathcal{A}$. The vector space $C^\infty(\mathbb{R}^m)$ is a left module of $\tilde{\mathcal{A}}$. The affine action of Γ in $\mathbb{R}^m$ is $\tilde{\mathcal{A}}$-preserving.

Let $(U,\Phi_U\times\phi_U)$ be a local chart of $[\mathcal{E},\pi,M,D]$. We recall that $d\phi_U$ is the differential of ϕ_U. We have

$$d\phi_U(\mathcal{A}) = \tilde{\mathcal{A}}(\phi_U(U)).$$

Definition 61. *A homological model consists of the following data. The datum $[\mathcal{E},\pi,M,D]$ is an object of the category $\mathcal{FB}(\Gamma,\Xi)$. Every $x\in M$ has an open neighborhood U which is the domain of a local chart of $[\mathcal{E},\pi,M,D]$, namely $(\Phi_U\times\phi_U)$. We set*

$$\Theta_U \times \Xi = \Phi_U(\mathcal{E}_U).$$

Those data are subject to the following requirements.
$HSM.1$: $\Theta\times\Xi$ supports a non negative random symmetric 2-cocycle

$$\Theta_U\times\Xi \ni (\theta,\xi) \to Q_U(\theta,\xi) \in Z^2_{KV}(\tilde{\mathcal{A}},\mathbb{R}).$$

$HSM.2$: Let $[U,\Phi_U\times\phi_U,Q_U]$ and $[U^*,\Phi_{U^*}\times\phi_{U^*},Q_{U^*}]$ as in HSM.1.
If we assume that

$$U\cap U^* \neq \emptyset$$

then there exists $\gamma_{UU^*}\in\Gamma$ such that

$$HSM.2.1 \quad \Phi_{U^*}(e) = \gamma_{UU^*}\cdot\Phi_U(e) \quad \forall e\in\mathcal{E}_{U\cap U^*},$$

$$HSM.2.2 \quad Q_U(\Phi_U(e)) = \gamma^*_{UU^*}\cdot[Q_{U^*}(\Phi_{U^*}(e))].$$

Comments.

The equality

$$Q_U(\Phi_U(e)) = \gamma^*_{UU^*} \cdot [Q_{U^*}(\Phi_{U^*}(e))]$$

has the following meaning. For $v, w \in T_{\phi_U(e)}\Theta_U$ one has

$$Q_U[\theta(e), \xi(e)](v, w) = Q_{U^*}[\gamma_{UU^*} \cdot (\theta(e), \xi(e))](d[\gamma_{UU^*}] \cdot v, d[\gamma_{UU^*}] \cdot w).$$

Morphisms of homological models are defined by replacing the probability P_U by the random cocycle Q_U.

Definition 62. *The category of homological statistical models for a measurable set* (Ξ, Ω) *is denoted by* $\mathcal{HSM}(\Xi, \Omega)$.

10.1. The Cohomology Mapping of $HSM(\Xi, \Omega)$

We consider an m-dimensional object of $\mathcal{HSM}(\Xi, \Omega)$ which is defined by a complete atlas

$$\mathbb{A} = [U_j, \Phi_j \times \phi_j, \gamma_{ij}, Q_j]$$

The underlying object of the atlas $\mathbb{A}$ is denoted by $[\mathcal{E}, \pi, M, D]$. We set

$$\Theta_j = \phi_j(U_j) \subset \mathbb{R}^m.$$

We are not making any difference between $(U_j, \mathcal{A})$ and $(\Theta_j, \tilde{\mathcal{A}})$. We put set

$$\mathcal{E}_{ij} = \mathcal{E}_{U_i \cap U_j}.$$

If we assume that

$$U_i \cap U_j \neq \emptyset$$

then we have

$$\Phi_j(e) = \gamma_{ij} \cdot \Phi_i(e), \quad \forall e \in \mathcal{E}_{ij}$$

and

$$Q_i(\Phi_i(e)) = \gamma^*_{ij} \cdot Q_j(\Phi_j(e)) \quad \forall e \in \mathcal{E}_{ij}.$$

We put

$$q_j(e) = Q_j(\Phi_j(e)) \quad \forall e \in \mathcal{E}_j.$$

Here

$$\mathcal{E}_j = \mathcal{E}_{U_j}.$$

If

$$U_i \cap U_j \neq \emptyset$$

then we know that

$$[\Phi_j \circ \Phi_i^{-1}](\theta_i(e), \xi_i(e)) = \gamma_{ij}(\theta_i(e), \xi_i(e)) \quad \forall e \in \mathcal{E}_{ij}.$$

Therefore we get

$$q_i(e) = q_j(e) \quad \forall e \in \mathcal{E}_{ij}.$$

Therefore q_j is the restriction to $\mathcal{E}_j$ of a (globally defined) map

$$\mathcal{E} \ni e \to Q(e) \in Z^2_{KV}(\mathcal{A}, \mathbb{R}).$$

It is clear that

$$Q_j = Q \circ \Phi_j^{-1}.$$

The action of Γ in $\mathcal{E}$ is Q-preserving. Thus a homological statistical model is a quintuplet

$$[\mathcal{E}, \pi, M, D, Q].$$

Here Q is a map

$$\mathcal{E} \ni e \to Q(e) \in Z^2_{KV}(\mathcal{A}, \mathbb{R}).$$

Thus we get random cohomological map

$$\mathcal{E} \ni e \to [Q](e) = [Q(e)] \in H^2_{KV}(\mathcal{A}, \mathbb{R}).$$

Definition 63. *The mapping $[Q]$ is called cohomology mapping of the homological model $[\mathcal{E}, \pi, M, D, Q]$.*

10.2. An Interpretation of the Equivariant Class [Q]

We intend to interpret the cohomology class $[Q]$ as an obstruction class.

Definition 64. *(1) A homological statistical model whose cohomological map vanishes is called an EXact Homological Statistical Model, (EXHSM); (2) A homological statistical model whose cocycle is a random Hessian metric is called a HEssian Homological Statistical Model (HEHSM); (3) An exact Hessian homological statistical model is called a HYperbolic Homological Statistical Model (HYHSM).*

Given a Hessian Homological model

$$\mathbb{M} = [\mathcal{E}, \pi, M, D, Q]$$

the cohomology map $[Q]$ is the obstruction for $\mathbb{M}$ being an Hyperbolicity model.

The following proposition leads to impacts on the differential topology.

Proposition 11. *The kernel of an exact homological statistical model is in involution. Further if M and all data depending on M are analytic then Q is a stratified transversally Riemannian foliation in M.*

If $[\mathcal{E}, \pi, M, D, Q]$ is exact then there exists a random differential 1-form θ such that

$$Q = \delta_{KV}\theta,$$

viz

$$Q(X, Y) = X \cdot \theta(Y) - \theta(D_X Y) \quad \forall X, Y \in \mathcal{X}(M).$$

That useful for seing that $Ker(Q)$ is in involution.

10.3. Local Vanishing Theorems in the Category $\mathcal{HSM}(\Xi, \Omega)$

Reminder.

The category whose objects are homological statistical models (for (Ξ, Ω)) is denoted by $\mathcal{HSM}(\Xi, \Omega)$. Henceforth we fix an auxiliary structure of probability space (Ξ, Ω, p^*).

Definition 65. *We are interested in random functions defined in $\mathbb{R}^m \times \Xi$.*

(1) *A random function f has the property $p^* - EXP$ if*

$$exp(f(x,\xi)) \leq \int_{\Xi} exp(f(x,\xi))dp^*(\xi) \quad \forall x \in \mathbb{R}^m.$$

(2) *A random closed differential 1-form θ has the property $p^* - EXP$ if every $x \in \mathbb{R}^m$ has an open neighbourhood U satisfying the following conditions, $U \times \Xi$ support a random function f subject to two requirements:*

- $\theta = df$,
- *f has the property $p^* - Exp$.*

(3) *An exact homological statistical model $[\mathcal{E}, \pi, M, D, Q]$ has property $p^* - EXP$ if there exists a random differential 1-form θ satisfying the following conditions*

- *θ has the property $p^* - EXP$,*
- $Q = \delta_{KV}\theta$.

Localization

Our purpose is to explore the relationships between the theory of homological statistical models and the theory of local statistical model as in [18,22], Barndorff-Nielsen 1987

Our aim is to show that the current (local) theory is a byproduct of the localization of homological models. The notion of localization of homological models is but the notion of local vanishing theorem.

Theorem 22. *Let $[\mathcal{E}, \pi, M, D, Q]$ be a homological statistical model.*

(1) *$[\mathcal{E}, \pi, M, D, Q]$ is locally exact.*
(2) *If the $[\mathcal{E}, \pi, M, D, Q]$ has the property $p^* - EXP$ then $[\mathcal{E}, \pi, M, D, Q]$ is locally isomorphic to a classical statistical model (Θ, P) as in [18].*

The Sketch of Proof of (1). Let $(U, \Phi \times \phi)$ be a local chart of $[\mathcal{E}, \pi, M, Q]$. We set

$$\Theta_U = \phi(U).$$

We assume that Θ_U is an open convex subset of $\mathbb{R}^m$. Θ supports a system of affine coordinate functions

$$\theta = (\theta_1, ..., \theta_m).$$

We have

$$Q(\theta, \xi) = \sum Q_{ij}(\theta, \xi) d\theta_i d\theta_j.$$

Since $Q(\theta, \xi)$ is a random KV cocycle of $\tilde{\mathcal{A}}$ we have

$$\delta_{KV} Q = 0.$$

The last equality is equivalent to the following system

$$\frac{\partial Q_{jk}}{\partial \theta_i} - \frac{\partial Q_{ik}}{\partial \theta_j} = 0.$$

We fix $\xi \in \Xi$. For every j the random differential 1-form β_j is defined by

$$\beta_j(\theta, \xi) = \sum_i Q_{ij} d\theta_i.$$

Every $\beta_j(\theta, \xi)$ is a cocycle of the de Rham complex of Θ_U.

By the virtue of the Lemma of Poincaré there exists a local function $h_j(\theta, \xi)$ such that

$$\beta_j = dh_j.$$

Now the differential 1-form $\tilde{\theta}$ is defined by

$$\tilde{\theta} = \sum_j h_j(\theta, \xi) d\theta_j.$$

Direct calculations lead to the following equality

$$Q = \delta_{KV} \tilde{\theta}.$$

This ends the proof of (1). □

The proof of (2). We assume that $\mathbb{M} = [\mathcal{E}, \pi, M, D, Q]$ has the property $p^* - Exp$. We keep the notation we just used.

The random differential 1-form $\tilde{\theta}$ is a (de Rham) cocycle. Therefore $\Theta \times \Xi$ supports a random function $h(\theta, \xi)$ such that

$$\tilde{\theta} = dh.$$

So we have the following conclusion

$$Q(\theta(e), \xi(e)) = D^2 h(\theta(e), \xi(e)) \quad \forall e \in \mathcal{E}_U.$$

Equivalently one gets

$$\frac{\partial^2 h}{\partial \theta_i \partial \theta_j} = Q_{ij}.$$

Since $\mathbb{M}$ has the property $p * -Exp$ we choose a function h has the property $p^* - EXP$. The functions $F(\theta)$ and $P(\theta, \xi)$ are defined by

$$F(\theta) = \int_\Xi exp(h(\theta, \xi)) dp^*(\xi),$$

$$P_Q(\theta, \xi) = \frac{exp(h(\theta, \xi))}{F(\theta)}.$$

By the virtue of the property $p^* - Exp$ the function $P(\theta, \xi)$ satisfies the following requirements

(i) $P_Q \theta, \xi)$ is differentiable with respect to θ,
(ii) P_Q satisfies the following inequalities

$$0 \leq P_Q(\theta, \xi) \leq 1,$$

(iii) P_Q satifies the following identity

$$\int_\Xi P_Q(\theta, \xi) d\xi = 1.$$

Thus the pair (Θ_U, P_Q) is a local statistical model for (Ξ, Ω). This ends the proof of (2). The theorem is demonstrated. □

The pair (Θ_U, P_Q) is called a localization of $\mathbb{M}$.

Definition 66. *A localization (Θ_U, P_Q) is called a Local Vanishing Theorem of $[\mathcal{E}, \pi, M, Q]$.*

Let $[\mathcal{E}, \pi, M, D, p]$ be an object of the category $\mathcal{GM}(\Xi, \Omega)$. We set

$$Q_p = D^2 log(p).$$

Thefore we get the exact homological statistical model

$$\mathbb{M}_p = [\mathcal{E}, \pi, M, D, Q_p]$$

So the notion of vanishing theorem has significant impacts on the information geometry. To simplify an exact models which having the property $p^* - Exp$ (for some probabily space) are is called $Exp - models.$

Theorem 23. *The notation is that used previously.*

(1) *The category $\mathcal{GM}(\Xi, \Omega)$ is a subcategory of the category $\mathcal{EXHSM}(\Xi, \Omega)$.*
(2) *Ojects of $\mathcal{GM}(\Xi, \Omega)$ but homological $Exp - models.$*

Reminder: New Insigts.

(1.1) *The Information GEometry is the geometry of statistical models.*
(1.2) *The Information topology is the topology of statistical models.*
(2.1) *The homological nature of the Information Geometry.*
(2.2) *What is a statistical model? The answer to the question raised by McCullagh should be: A statistical model is a Global Vanishing Theorem in the theory of homological models.*
(2.3) *A local statistical model is a Local Vanishing Theorem in the theory of homological models.*

11. The Homological Statistical Models and the Geometry of Koszul

Our purpose is to to relate the category of homological statistical models and the geometry of Koszul. This relationship is based on the localization of homological statistical models.

Proposition 12. *$\mathcal{EXPHEHSM}(\Xi, \Omega)$ stands for the subcategory whose objects are Hessian Exp-models.*

(1) *The holomogical map leads to the functor of $\mathcal{EXPHEHSM}(ß, \Omega)$ in the category of Hessian structures in (M, D)*

$$[\mathcal{E}, \pi, M, D, Q] \rightarrow (M, D, \tilde{Q}).$$

(2) *If M is compact then the subcategory of exact Hessian homological Exp-models $\mathcal{EXPHYHSM}(\Xi, \Omega)$ is sent in the category of hyperbolic structure in in (M, D).*

12. Examples

This section is devoted to a few examples. The construction involves some basic notions of the differential topology.

Example 1: Dynamics

We consider a triple $[M \times H, p_1, M, \nabla]$. Here (M, ∇) is a compact locally flat manifold, $(H, d\mu)$ is an amenable group. There is an effective affine action

$$H \times (M, \nabla) \rightarrow (M, \nabla).$$

Let $f \in C^\infty(M)$ and $x \in M$. The function

$$f_x : H \rightarrow \mathbb{R}$$

is defined by

$$f_x(h) = f(h \cdot x).$$

Let $L^{\infty}_H(M)$ be the set of $f^* \in C^{\infty}(M)$ such that

$$f^*_x \in L^{\infty}(H),$$

viz

$$\sup_{[h \in H]} |f_x(h)| < \infty \quad \forall x.$$

Now $EXP(L^{\infty}_H(M)$ stands for the set of $f^* \in L^{\infty}_H(M)$ such that

$$exp(f^*_x(h)) \leq \mu(exp(f^*_x)) \quad \forall x.$$

The function $P_{f^*}(x,h)$ is defined by

$$P_{f^*}(x,h) = \frac{exp(f^*(x \cdot h))}{\mu(exp(f^*_x))}.$$

The pair (M, P_{f^*}) is a probability density in H. Now set

$$\tilde{f}^*(x,h) = f^*(x \cdot h)$$

Therefore the datum $[M \times H, p_1, M, \nabla, P_{f^*}]$ is a statistical model for $(H, \mathcal{P}(H))$. Here $\mathcal{P}(M)$ is the boolean algebra of subsets of H and p_1 is the trivial fibration of $M \times H$ over M.

Example 2: Geometry

We focus on an example which plays a significant role in global analysis (and geometry) in some type of bounded domains [2,3]. This example relates the geometry of Koszul and Souriau Lie groups thermodynamics [4] and bibliography therein.

Let $\mathcal{C} \subset \mathcal{R}^m$ be a convex cone and let $\mathcal{C}^*$ be its dual. The characteristic function of $\mathcal{C}$ is defined by

$$\mathcal{C} \ni v \rightarrow \int_{\mathcal{C}^*} exp(- < v, w^* >) dw^*.$$

This gives rise to the following function

$$\mathcal{C} \times \mathcal{C}^* \ni (v, v^*) \rightarrow P(v, v^*) = \frac{exp(- < v, v^* >)}{\int_{\mathcal{C}^*} exp(- < v, w^* >) dw^*}$$

So $(\mathcal{C}, P)$ is a statistical model for $(\mathcal{C}^*, dw^*)$. Here dw^* is the standard Borel measure.

Stratified Analytic Riemannian Foliations

Reminder.

We recall that a (regular) Riemannian foliation M is a symmetric bilinear form $g \in \mathcal{S}_2(M)$ having the following properties

(1) $rank(g) = constant$,
(2) $L_X g = 0 \forall X \in \mathcal{G}(Ker(g))$.

From (2) one easily deduces that Ker(g) is in involution. By the virtue of Theorem of Frobenius (1) and (2) imply that Ker(g) is completely integrable.

In the category of differentiable manifolds, not all involutive singular distributions are completely integrable. Nevertheless, that is true in the category of analytic manifolds [62].

This subsection is mainly devoted to examples of stratified Riemannian foliations in analytic manifolds. For more details about those object the readers are referred to [46,63].

Theorem 24. *Let M be an orientable compact analytic manifold. Let $C^{\omega}(M^2)$ be the space real valued analytic functions defined in M^2. There exists a canonical map of $C^{\omega}(M \times M)$ in the family of analytic stratified Riemannian foliations in M.*

The Idea of Construction.

Let dv be an analytic volume element in M. In M we fix an analytic torsion free Koszul connection ∇. To a function $f \in C^{\omega}(M^2)$ we assign the function $P \in C^{\omega}(M^2)$

$$P_f(x, x^*) = \frac{exp[f(x, x^*)]}{\int_M exp[f(x, x^{**})]dv(x^{**})}.$$

We make the following identification

$$\mathcal{X}(M) = \mathcal{X}(M) \times 0 \subset \mathcal{X}(M^2).$$

The analytic bilinear form $g_f \in \mathcal{S}_2(M)$ is defined by

$$[g_f(x)](X, Y) = -\int_M P_f(x, x^*)[\nabla^2(log(P_f))(X, Y)](x, x^*)dv(x^*).$$

The form g_f has the following properties.

(a) *g_f does not depend on the choice of ∇,*
(b) *g_f is symmetric and positive semi-definite,*
(c) *If X is a section of $Ker(g_f)$ then $L_X g_f = 0$.*

Conclusion.

If

$$rank(g_f) = constant$$

then g_f is a Riemannian foliation as in [38–40,46].

If $rank(g_f)$ is not constant we apply [62]. Thereby g_f is an analytic stratified Riemannian foliation.

Reminder.

The idea of the strafication of g_f.

Step 0

The open subset $U_0 \subset M$ is defined by

$$x \in U_0 \quad iff \quad rank(g_f(x)) = \max_{[x^* \in M]} rank(g_f(x^*)).$$

The closed analytic submanifold $F_1 \subset M$ is defined by

$$F_1 = M \setminus U_0.$$

Step 1

In the pair (F_1, g_f) *the open subset* $U_1 \subset F_1$ *is defined by*

$$x \in U_1 \quad iff \quad rank(g_f(x)) = \max_{[x* \in F_1]} rank(g_f(x*)).$$

Step 2

We iterate this construction. Then we have get a filtration of M

$$... \subset F_n \subset F_{n-1} \subset ... \subset F_1 \subset F_O = M.$$

This filtration has the following properties

(1) $F_{j-1} \setminus F_j$ *is a analytic submanifold of M.*
(2) g_f *defines a regular Riemannian foliation in* $(F_{j-1} \setminus F_j, g_f)$,

Remark 6. *The extrinsic geometry of submanifolds is a particular case of the geometry of singular foliation [25].*

13. Highlighting Conclusions

13.1. Criticisms

In Part B we have raised some criticisms. We have constructed structures of statistical models in flat tori. An m-dimensional flat torus is not homeomorphic to an open subset of $\mathbb{R}^m$. The second criticism is the lack of dynamics. Subsequently, the problem of moduli space is absent from the classical theory. That deficiency is filled in by the characteristic functor

$$\mathbb{M} = [\mathcal{E}, \pi, M, D, p] \to q_{\mathbb{M}}.$$

The current theory requires a model to be identifiable. From the viewpoint of locally trivial fiber bundles, that requirement is useless.

13.2. Complexity

In both the theoretical information geometry and the applied information the exponential models and their generalizations play notable roles. What we call the complexity of a model $[\mathcal{E}, \pi, M, D, p]$ is its distance from the category of exponential models. Up to today there does not exist any INVARIANT which measures how far from being an exponential is a given model. This problem has a homological nature. We have produced a function r^b which fills in that lack. (See the Appendix A below).

13.3. KV Homology and Localization

We have introduced the theory of homological model. Among the notable notions that we have studied is the localization of homological statistical models. It links the theory of homological models and the current theory as in [22]. It may be interpreted as a functor from the theory of homological models to the classical theory of statistical models.

13.4. The Homological Nature of the Information Geometry

$\mathcal{GM}(\Xi, \Omega)$ and $\mathcal{HSM}(\Xi, \Omega)$ are introduced in this Part B. The category of local statistical models for (Ξ, Ω) is denoted by $\mathcal{LM}(\Xi, \Omega)$. On one side, the right arrows below mean subcategory. Then we have

$$\mathcal{LM}(\Xi, \Omega) \to \mathcal{GM}(\Xi, \Omega) \to \mathcal{HSM}(\Xi, \Omega).$$

On another side, the notion of Vanishing Theorem is useful in linking $\mathcal{HSM}(\Xi,\Omega)$ with both $\mathcal{GM}(\Xi,\Omega)$ and $\mathcal{LM}(\Xi,\Omega)$.

(1) The Global Vanishing Theorem is the functor

$$\mathcal{HSM}(\Xi,\Omega) \to \mathcal{GM}(\Xi,\Omega).$$

(2) The Local Vanishing Theorem is the functor

$$\mathcal{HSM}(\Xi,\Omega) \to \mathcal{LM}(\Xi,\Omega).$$

13.5. Homological Models and Hessian Geometry

In the category $\mathcal{HSM}(\Xi,\Omega)$ the Hessian functor is the functor from $\mathcal{HEHSM}(\Xi,\Omega)$ to the category of randon Hessian manifolds.

Furthermore, every structure of probability space (Ξ,Ω,p^*) gives rise to a canonical functor from $\mathcal{HEHSM}(\Xi,\Omega)$ to the category of Hessian manifolds. The canonical functor is defined by

$$[\mathcal{E},\pi,M,D,Q] \to \int_F p^*Q$$

Acknowledgments: The author gratefully thanks the referees for number of comments and suggestions. Their criticisms have been helpful to improve parts of the original manuscript.

Conflicts of Interest: The author declares no conflict of interest.

Appendix A

Usually the appendix is devoted to overview the notions which are used in a paper. In this appendix we announce a few outstanding impacts of Hessian differential operators of Koszul connections.

In the introduction a pair of Koszul connections (∇,∇^*) is used for defining three differential operators

$$X \to D^\nabla(X) = \iota_X R^\nabla - L_X\nabla \quad \forall X \in \Gamma(TM).$$

The differential operator D^∇ is elliptic and involutive in the sense of the global analysis [50,51,64]. Let $\mathcal{J}^\nabla$ be the sheaf of germ of solutions to the equation

$$FE^{**}(\nabla) : D^\nabla(X) = 0.$$

If ∇ torsion free then $FE^{**}(\nabla)$ is a Lie equation.

The non negative integers $r^b(\nabla)$ and $r^b(M)$ are defined by

$$r^b(\nabla) = \min_{[x\in M]} \left\{dim(\mathcal{J}^\nabla(x)\right\},$$

$$r^b(M) = \min_{[\nabla\in\mathcal{SLC}(M)]} \left\{dim(M) - r^b(\nabla)\right\}.$$

Here $\mathcal{SLC}(M)$ is the convex set of torsion free Koszul connections in M. We set the following notation: $\mathcal{R}ie(M)$ is the set of Riemannian metric tensors in M. $\mathcal{LF}(M)$ is the set of locally flat Koszul connection in M. At one side every $g \in \mathcal{R}ie(M)$ gives rise to the map

$$\mathcal{LF}(M) \ni \nabla \to \nabla^* \in \mathcal{LC}(M)$$

which is defined by

$$g(Y,\nabla^*_X Z) = Xg(Y,Z) - g(\nabla_X Y, Z).$$

At another side every $\nabla \in \mathcal{LF}(M)$ gives rise to the map

$$\mathcal{R}i(M) \ni g \rightarrow \nabla^g \in \mathcal{LC}(M)$$

which is defined by

$$g(Y, \nabla^g_X Z) = Xg(Y, Z) - g(\nabla_X Y, Z).$$

In every Riemannian manifold (M, g) we define the following numerical invariants

$$r^b(M, g) = \min_{[\nabla \in \mathcal{LF}(M)]} \left\{ dim(M) - r^b(\nabla^*) \right\},$$

$$r^B(M) = \min_{[g \in \mathcal{R}ie(M)]} \left\{ r^b(M, g) \right\}.$$

In every locally flat manifold (M, ∇) we define the following numerical invariant

$$r^b(M, \nabla) = \min_{[g \in \mathcal{R}ie(M)]} \left\{ r^b(\nabla^g) \right\}$$

The numerical invariants we just defined have notable impacts.

Appendix A.1. The Affinely Flat Geometry

Theorem A1. *In a smooth manifold M the following assertions are equivalent*

(1) $r^b(M) = 0$,
(2) *the manifold M admits locally flat structures.*

Appendix A.2. The Hessian Geometry

Theorem A2 (Answer a hold questions of [65]). *In a Riemannian manifold (M, g) the following assertions are equivalent*

(1) $r^b(M, g) = 0$,
(2) *the Riemannian manifold (M, g) admits Hessian structures (M, g, ∇)*

A Comment.

Assertion (2) has the following meaning.

(i) *(M, ∇) is a locally flat manifold.*
(ii) *every point has an open neighborhood U supporting a system of affine coordinate functions $(x_1, ..., x_m)$ and a local smooth function $h(x_1, ..., x_m)$ such that*

$$g(\frac{\partial}{\partial x_i}, \frac{\partial}{\partial x_j}) = \frac{\partial^2 h}{\partial x_i \partial x_j}.$$

Appendix A.3. The Geometry of Koszul

Theorem A3. *In a locally flat manifold (M, ∇) whose KV algebra is denoted by $\mathcal{A}$ the following assertions are equivalent*

(1) $r^b(M, \nabla) = 0$,
(2) *the KV cohomology space $H^2_{KV}(\mathcal{A}, \mathbb{R})$ contains a metric class $[g]$,*
(3) *the locally flat manifold (M, ∇) admits Hessian structures (M, ∇, g).*

Appendix A.4. The Information Geometry

Let (Ξ, Ω) be a transitive measurable and let

$$\mathbb{M} = [\mathcal{E}, \pi, M, D, p]$$

be an object of $\mathcal{GM}(\Xi, \Omega)$. Let g be the Fisher information of $\mathbb{M}$. Let

$$\{\nabla^{\alpha}, \quad \alpha \in \mathbb{R}\}$$

be the family of α-connections of $\mathbb{M}$. We define the following numerical invariant

$$r^b(\mathbb{M}) = \min_{[\alpha \in \mathbb{R}]} \left\{ dim(M) - r^b(\nabla^{\alpha}) \right\}.$$

Theorem A4. *In $\mathbb{M}$ the following assertions are equivalent.*

(1) $r^b(\mathbb{M}) = 0,$
(2) *$\mathbb{M}$ is an exponential family.*

Corollary A1. *Assume that $\mathbb{M}$ is regular, viz g is positive definite, then the following assertions are equivalent*

(1) $r^b(\mathbb{M}) = 0,$
(2) $r^b(M, g) = 0,$

Appendix A.5. The Differential Topology of a Riemannian Manifold

A Riemannian manifold (M, g) , (whose Levi-Civita connection is denoted by ∇^*), is called special if

$$\mathcal{J}^{\nabla^*} \neq 0$$

Theorem A5. *A special positive Riemannian manifold(M, g) has the following properties*

(1) *(M, g) admits a geodesic flat Hessian foliation*

$$[\mathcal{F}, g|\mathcal{F}, \nabla^*].$$

(2) *The leaves of $\mathcal{F}$ are the orbits of a bi-invariant affine Cartan-Lie group $(\tilde{G}, \tilde{\nabla})$.*
(3) *The bi-invariant affine Cartan-Lie group $(\tilde{G}, \tilde{\nabla})$ is generated by an effective infinitesimal action of a simply connected bi-invariant affine Lie group (G, ∇).*

References

1. Faraut, J.; Koranyi, A. *Analysis on Symmetric Cones*; Oxford University Press: Oxford, UK, 1994.
2. Koszul, J.-L. Déformation des connexions localement plates. *Ann. Inst. Fourier* **1968**, *18*, 103–114. (In French)
3. Vinberg, E.B. The theory of homogeneous convex cones. *Trans. Moscow. Math. Soc.* **1963**, *12*, 303–358.
4. Barbaresco, F. Geometric Theory of Heat from Souriau Lie Groups Thermodynamics and Koszul Geometry: Applications in Information Geometry. *Entropy* **2016**, doi:10.3390/e18110386.
5. Nguiffo Boyom, M.; Wolak, R. Foliations in affinely flat manifolds: Information Geometric. Science of Information. In *Geometric Science of Information*; Springer: Berlin/Heidelberg, Germany, 2013; pp. 283–292.
6. Gindikkin, S.G.; Pjateckii, I.I.; Vinerg, E.B. Homogeneous Kahler manifolds. In *Geometry of Homogeneous Bounded Domains*; Cremonese: Roma, Italy, 1968; pp. 3–87.
7. Kaup, W. Hyperbolische komplexe Rume. *Ann. Inst. Fourier* **1968**, *18*, 303–330.
8. Koszul, J.-L. Variété localement plate et convexité. *Osaka J. Math.* **1965**, *2*, 285–290. (In French)
9. Nguiffo Boyom, M. The Cohomology of Koszul-Vinberg Algebras. *Pac. J. Math.* **2006**, *225*, 119–153.
10. Nguiffo Boyom, M. Réductions Kahlériennes dans les groupes de Lie Résolubles et Applications. *Osaka J. Math.* **2010**, *47*, 237–283. (In French)

11. Shima, H. Homogeneous hessian manifolds. *Ann. Inst. Fourier* **1980**, *30*, 91–128.
12. Vey, J. Une notion d'hyperbolicité sur les variétés localement plates. *C. R. Acad. Sci. Paris* **1968**, *266*, 622–624. (In French)
13. Barbaresco, F. Information geometry of covariance matrix: Cartan-Siegel homogeneous bounded domains. Mostov/Berger fibration and Frechet median. In *Matrix Information Geometry*; Springer: Berlin/Heidelberg, Germany, 2013; pp. 199–255.
14. Baudot, P.; Bennequin, D. Topology forms of informations. *AIP Conf. Proc.* **2014**, *1641*, 213–221.
15. Gromov, M. In a Search for a Structure, Part I: On Entropy. Available online: www.ihes.fr/gromov/PDF/structure-serch-entropy-july5-2012.pdf (accessed on 28 November 2016).
16. Gromov, M. On the structure of entropy. In Proceedings of the 34th International Workshop on Bayesian Inference and Maximum Entropy Methods in Science and Engineering MaxEnt 2014, Amboise, France, 21–26 September 2014.
17. Amari, S.-I. *Differential Geometry Methods in Statistics*; Lecture Notes in Statistics; Springer: Berlin/Heidelberg, Germany, 1990.
18. Amari, S.-I.; Nagaoka, H. *Methods of Information Geometry, Translations of Mathemaical Monographs*; American Mathematical Society: Providence, RI, USA, 2007; Volume 191.
19. Arnaudon, M.; Barbaresco, F.; Yan, L. Medians and Means in Riemannian Geometry: Existence, Uniqueness and Computation. In *Matrix Information Geometry*; Springer: Berlin/Heidelberg, Germany, 2013; pp. 169–197.
20. Armaudon, M.; Nielsen, F. Meadians and means in Fisher geometry. *LMS J. Comput. Math.* **2012**, *15*, 23–37.
21. Barndorff-Nielsen, O.E. Differential geometry and statistics: Some mathematical aspects. *Indian J. Math.* **1987**, *29*, 335–350.
22. Murray, M.K.; Rice, J.W. Monographs on statistics and applied probability. In *Differential Geometry and Statistics*; Chapman and Hall/CRC: Boca Raton, FL, USA, 1993; Volume 48.
23. Nguiffo Boyom, M.; Byande, P.M. KV cohomology in information geometry. In *Matrix Information Geometry*; Springer: Berlin/Heidelberg, Germany, 2013; pp. 69–92.
24. Barndorff-Nielsen, O.E. *Information and Exponential Families in Stattistical Theory*; Wiley: New York, NY, USA, 1978.
25. Nguiffo Boyom, M.; Jamali, M.; Shahid, M.H. Multiply CR warped product Statistical submanifolds of holomorphic statistical space form. In *Geometric Science of Information*; Springer: Berlin/Heidelberg, Germany, 2015; pp. 257–268.
26. Milnor, J. The fundamental groups of complete affinely flat manifolds. *Adv. Math.* **1977**, *25*, 178–187.
27. Gerstenhaber, M. On deformations of Rings and Algebras. *Ann. Math.* **1964**, *79*, 59–103.
28. Nijenhuis, A. Sur une classe de propriétés communes à quelques types différents d'algèbres. *Enseign. Math.* **1968**, *14*, 225–277. (In French)
29. Nguiffo Boyom, M.; Byande, P.M.; Ngakeu, F.; Wolak, R. KV Cohomology and diffrential geometry of locally flat manifolds. Information geometry. *Afr. Diaspora J. Math.* **2012**, *14*, 197–226.
30. McCullagh, P. What is statistical model? *Ann. Stat.* **2002**, *30*, 1225–1310.
31. Baudot, P.; Bennequim, D. The homological nature of Entropy. *Entropy* **2015**, *17*, 3253–3318.
32. Hochschild, G. On the cohomology groups of an associative algebra. *Ann. Math.* **1945**, *46*, 58–67.
33. Hochschild, G.; Serre, J.-P. Cohomology of Lie algebras. *Ann. Math.* **1953**, *57*, 591–603.
34. McCleary, J. *A User's Guide to Spectral Sequences*; Cambridge University Press: Cambridge, UK, 2001.
35. Eilenberg, S. Cohomology of space with operators group. *Trans. Am. Math. Soc* **1949**, *65*, 49–99.
36. Koszul, J.-L. Homologie des complexes de formes différentielles d'ordre supérieur. *Ann. Sci. Ecole Norm. Super.* **1974**, *7*, 139–153. (In French)
37. Kass, R.E.; Vos, P.W. *Geometrical Foundations of Asymptotic Inference*; Wiley: New York, NY, USA, 1997.
38. Moerdijk, I.; Mrcun, J. *Introduction to Foliations and Lie Groupoids*; Cambridge Studies in Advanced Mathematics; Cambridge University Press: Cambridge, UK, 2003.
39. Molino P. *Riemannian, Foliations*; Birkhauser: Boston, MA, USA, 1988.
40. Reinhardt, B.L. Foliations with bundle-like metrics. *Ann. Math.* **1959**, *69*, 119–132.
41. Akivis, M.; Goldberg, V.; Lychagin, V. Linearizability of d-webs, on two-dimensional manifolds. *Sel. Math.* **2004**, *10*, 431–451.
42. Grifone, J.; Saab, J.; Zoltan, M. On the linearization of 3-webs. *Nonlinear Anal.* **2001**, *47*, 2643–2654.
43. Henaut, A. Sur la linéarisation des tissus de C^2. *Topology* **1993**, *32*, 531–542.

44. Hess, H. Connectionson symplectic manifolds and geometric quantization. In *Differential Geometry Methods in Maththematical Physics*; Lectures Notes 836; Springer: Berlin/Heidelberg, Germany, 1980; pp. 153–166
45. Nguiffo Boyom, M. Structures localement plates dans certaines variétés symplectiques. *Math. Scand.* **1995**, *76*, 61–84. (In French)
46. Kamber, F.; Tondeur, P.H. De Rham-Hodge Theory for Riemannian foliations. *Math. Ann.* **1987**, *277*, 415–431.
47. Byande, P.M. *De Structures Affines à la Géométrie de L'information*; Éditions Universitaires Européennes: Saarbrücken, Germany, 2012. (In French)
48. Chentsov, N.N. *Statistical Decision Rules and Optimal Inference*; American Mathematical Society: Providence, RI, USA, 1982.
49. Petrie, T.; Handal, J. *Connections, Definite Forms and Four-Manifolds*; Oxford Mathematcal Monograph, Oxford Science Publications; Clarendon Press: Oxford, UK, 1990.
50. Guillemin, V.; Sternberg, S. An algebraic model for transitive differential geometry. *Bull. Am. Math. Soc.* **1964**, *70*, 16–47.
51. Singer, I.M.; Sternberg, S. The infinte groups and lie and Cartan. *J. Anal. Math. Jerus.* **1965**, *15*, 1–144.
52. Shima, H. *The Geometry of Hessian Structures*; World Scientific: Singapore, 2007.
53. Barbaresco, F. *Symplectic Structure of Information Geometry: Fisher Information and Euler-Poincare Equation of Souriau Lie Group Thermodynamics*; Nielsen, F., Barbaresco, F., Eds.; Springer: Berlin/Heidelberg, Germany, 2015; pp. 529–540.
54. Matsuzoe, H.; Henmi, M. *Hessian Structures on Deformed Exponential Families*; Nielsen, F., Barbaresco, F., Eds.; Springer: Berlin/Heidelberg, Germany, 2013; Volume 8085, pp. 275–282.
55. Barbaresco, F. Koszul Information Geometry and Souriau Geometric Temperature/Capacity of Lie Group Thermodynamics. *Entropy* **2014**, *16*, 4521–4565.
56. Nijenhuis, A.; Richardson, W. Commutative algebra cohomology and deformation of Lie algebras and associative algebras. *J. Algebra* **1968**, *9*, 42–53.
57. Chevallier, E.; Barbaresco, F.; Angulo, J. Kernel Density Estimation on Siegel Space Applied to Radar Processing. *Entropy* **2016**, *18*, 396.
58. Chevallier, E.; Barbaresco, F.; Angulo, J. Probability density estimation on hyperbolic space applied to radar processing. In *Geometric Science of Information*; Springer: Berlin/Heidelberg, Germany, 2015; pp. 753–761.
59. Eilenberg, S. Homology of spaces with operators group I. *Trans. Am. Math. Soc.* **1947**, *62*, 378–417.
60. Nguiffo Boyom, M.; Wolak, R. Transverse Hessian metrics information geometry. In Proceedings of the 34th International Workshop on Bayesian Inference and Maximum Entropy Methods in Science and Engineering, Amboise, France, 21–26 September 2014.
61. Nguiffo Boyom, M.; Wolak, R.A. Transversely Hessian foliations and information geometry. *Int. J. Math.* **2016**, *27*, doi:10.1142/S0129167X16500920.
62. Zeghib, A. On Gromov theory of rigid transformation groups, a dual approach. *Ergod. Theorey Dyn. Syst.* **2000**, *20*, 935–946.
63. Kamber, F.; Tondeur, P.H. *Invariant Operators and the Cohomology of Lie Algebra Sheaves*; Memoirs of the American Matheamtical Society: Providence, RI, USA, 1971.
64. Kumpera, A.; Spencer, D. *Lie Equation*; Princeton University Press: Princeton, NJ, USA, 1972.
65. Guts, A.K. (Omsk State University). Private communication.

entropy

MDPI

Article

Explicit Formula of Koszul–Vinberg Characteristic Functions for a Wide Class of Regular Convex Cones

Hideyuki Ishi

Graduate School of Mathematics, Nagoya University, Nagoya 464-8602, Japan; hideyuki@math.nagoya-u.ac.jp; Tel.: +81-52-789-4877

Academic Editors: Frédéric Barbaresco and Frank Nielsen
Received: 16 September 2016; Accepted: 20 October 2016; Published: 26 October 2016

Abstract: The Koszul–Vinberg characteristic function plays a fundamental role in the theory of convex cones. We give an explicit description of the function and related integral formulas for a new class of convex cones, including homogeneous cones and cones associated with chordal (decomposable) graphs appearing in statistics. Furthermore, we discuss an application to maximum likelihood estimation for a certain exponential family over a cone of this class.

Keywords: convex cone; homogeneous cone; graphical model; Koszul–Vinberg characteristic function

1. Introduction

Let Ω be an open convex cone in a vector space $\mathcal{Z}$. The cone Ω is said to be regular if Ω contains no straight line, which is equivalent to the condition $\overline{\Omega} \cap (-\overline{\Omega}) = \{0\}$. In this paper, we always assume that a convex cone is open and regular. The dual cone Ω^* with respect to an inner product $(\cdot|\cdot)$ on $\mathcal{Z}$ is defined by:

$$\Omega^* := \{\, \xi \in \mathcal{Z}\,;\, (x|\xi) > 0 \ (\forall x \in \overline{\Omega} \setminus \{0\}) \,\}.$$

Then, Ω^* is again a regular open convex cone, and we have $(\Omega^*)^* = \Omega$. The Koszul–Vinberg characteristic function $\varphi_\Omega : \Omega \to \mathbb{R}_{>0}$ defined by:

$$\varphi_\Omega(x) := \int_{\Omega^*} e^{-(x|\xi)}\, d\xi \qquad (x \in \Omega)$$

plays a fundamental role in the theory of regular convex cones [1–4].

In particular, φ_Ω is an important function in the theory of convex programming [5], and it has also been studied recently in connection with thermodynamics [6,7]. There are several (not many) classes of cones for which an explicit formula of the Koszul–Vinberg characteristic function is known. Among them, the class of homogeneous cones [8–10] and the class of cones associated with chordal graphs [11] are particularly fruitful research objects. In this paper, we present a wide class of cones, including both of them, and give an explicit expression of the Koszul–Vinberg characteristic function (Section 3). Moreover, we get integral formulas involving the characteristic functions and the so-called generalized power functions, which are expressed as some product of powers of principal minors of real symmetric matrices (Section 4). After investigating the multiplicative Legendre transform of generalized power functions in Section 5, we study a maximum likelihood estimator for a Wishart-type natural exponential family constructed from the integral formula (Section 6).

A regular open convex cone $\Omega \subset \mathcal{Z}$ is said to be homogeneous if the linear automorphism group $GL(\Omega) := \{\, \alpha \in GL(\mathcal{Z})\,;\, \alpha\Omega = \Omega \,\}$ acts on Ω transitively. The cone $\mathcal{P}_n$ of positive definite $n \times n$ real symmetric matrices is a typical example of homogeneous cones. It is known [12–16] that every homogeneous cone is linearly isomorphic to a cone $\mathcal{P}_n \cap \mathcal{Z}$ with an appropriate subspace $\mathcal{Z}$ of the vector space $\mathrm{Sym}(n, \mathbb{R})$ of all $n \times n$ real symmetric matrices, where $\mathcal{Z}$ admits a specific block decomposition. Based on such results, our matrix realization method [15,17,18] has been developed for the purpose of the efficient study of homogeneous cones. In this paper, we present a generalization of matrix realization dealing with a wide class of convex cones, which turns out to include cones associated with chordal graphs. Actually, it was an enigma for the author that some formulas in [11,19] for the chordal graph resemble the formulas in [8,17] for homogeneous cones so much, and the mystery is now solved by the unified method in this paper to get the formulas. Furthermore, the techniques and ideas in the theory of homogeneous cones, such as Riesz distributions [8,20,21] and homogeneous Hessian metrics [4,18,22], will be applied to various cones to obtain new results in our future research.

Here, we fix some notation used in this paper. We denote by $\mathrm{Mat}(p, q, \mathbb{R})$ the vector space of $p \times q$ real matrices. For a matrix A, we write ${}^{\mathrm{t}}A$ for the transpose of A. The identity matrix of size p is denoted by I_p.

2. New Cones $\mathcal{P}_{\mathcal{V}}$ and $\mathcal{P}_{\mathcal{V}}^*$

2.1. Setting

We fix a partition $n = n_1 + n_2 + \cdots + n_r$ of a positive integer n. Let $\mathcal{V} = \{\mathcal{V}_{lk}\}_{1 \le k < l \le r}$ be a system of vector spaces $\mathcal{V}_{lk} \subset \mathrm{Mat}(n_l, n_k, \mathbb{R})$ satisfying

(V1) $A \in \mathcal{V}_{lk} \Rightarrow A\,{}^{\mathrm{t}}A \in \mathbb{R}I_{n_l} \quad (1 \le k < l \le r)$,
(V2) $A \in \mathcal{V}_{lj},\ B \in \mathcal{V}_{kj} \Rightarrow A\,{}^{\mathrm{t}}B \in \mathcal{V}_{lk} \quad (1 \le j < k < l \le r)$.

The integer r is called the rank of the system $\mathcal{V}$. We denote by n_{lk} the dimension of $\mathcal{V}_{lk}$. Note that some n_{lk} can be zero. Let $\mathcal{Z}_{\mathcal{V}}$ be the space of real symmetric matrices $x \in \mathrm{Sym}(n, \mathbb{R})$ of the form:

$$x = \begin{pmatrix} X_{11} & {}^{\mathrm{t}}X_{21} & \cdots & {}^{\mathrm{t}}X_{r1} \\ X_{21} & X_{22} & & {}^{\mathrm{t}}X_{r2} \\ \vdots & & \ddots & \\ X_{r1} & X_{r2} & \cdots & X_{rr} \end{pmatrix} \qquad \begin{pmatrix} X_{kk} = x_{kk} I_{n_k},\ x_{kk} \in \mathbb{R},\ k = 1, \ldots, r \\ X_{lk} \in \mathcal{V}_{lk},\ 1 \le k < l \le r \end{pmatrix}, \tag{1}$$

and $\mathcal{P}_{\mathcal{V}}$ the subset of $\mathcal{Z}_{\mathcal{V}}$ consisting of positive definite matrices. Then, $\mathcal{P}_{\mathcal{V}}$ is a regular open convex cone in $\mathcal{Z}_{\mathcal{V}}$.

Example 1. *Let $r = 3$, and set $\mathcal{V}_{21} := \left\{ \begin{pmatrix} a & 0 \end{pmatrix} ; a \in \mathbb{R} \right\}$, $\mathcal{V}_{31} := \left\{ \begin{pmatrix} 0 & a \end{pmatrix} ; a \in \mathbb{R} \right\}$, and $\mathcal{V}_{32} := \mathbb{R}$. Then, $\mathcal{Z}_{\mathcal{V}}$ is the space of symmetric matrices x of the form:*

$$x = \begin{pmatrix} x_1 & 0 & x_4 & 0 \\ 0 & x_1 & 0 & x_5 \\ x_4 & 0 & x_2 & x_6 \\ 0 & x_5 & x_6 & x_3 \end{pmatrix}. \tag{2}$$

We shall see later that the cone $\mathcal{P}_{\mathcal{V}} = \mathcal{Z}_{\mathcal{V}} \cap \mathcal{P}_4$ is not homogeneous in this case, but admits various integral formulas, as well as explicit expression of the Koszul–Vinberg characteristic function.

2.2. Inductive Description of $\mathcal{P}_{\mathcal{V}}$

If the system $\mathcal{V} = \{\mathcal{V}_{lk}\}_{1 \le k < l \le r}$ satisfies (V1) and (V2), any subsystem $\mathcal{V}_{\mathcal{I}} := \{\mathcal{V}_{lk}\}_{k,l \in \mathcal{I}}$ with $\mathcal{I} \subset \{1, \ldots, r\}$ also satisfies the same conditions. In particular, the cone corresponding to the subsystem

$\{\mathcal{V}_{lk}\}_{2\le k<l\le r}$ will play an important role in this paper. Let us define $\mathcal{V}' := \{\mathcal{V}'_{lk}\}_{1\le k<l\le r-1}$ by $\mathcal{V}'_{lk} := \mathcal{V}_{l+1,k+1}$. Then, $\mathcal{V}'$ is a system of rank $r-1$. Any $x \in \mathcal{Z}_{\mathcal{V}}$ is written as:

$$x = \begin{pmatrix} x_{11}I_{n_1} & {}^{\mathrm{t}}U \\ U & x' \end{pmatrix} \qquad (x_{11} \in \mathbb{R},\ U \in \mathcal{W},\ x' \in \mathcal{Z}_{\mathcal{V}'}), \tag{3}$$

where:

$$\mathcal{W} := \left\{ U = \begin{pmatrix} X_{21} \\ \vdots \\ X_{r1} \end{pmatrix} ;\ X_{l1} \in \mathcal{V}_{l1}\ (1 < l \le r) \right\}. \tag{4}$$

If $x_{11} \neq 0$, then we have:

$$\begin{pmatrix} x_{11}I_{n_1} & {}^{\mathrm{t}}U \\ U & x' \end{pmatrix} = \begin{pmatrix} I_{n_1} & \\ x_{11}^{-1}U & I_{n-n_1} \end{pmatrix} \begin{pmatrix} x_{11}I_{n_1} & \\ & x' - x_{11}^{-1}U{}^{\mathrm{t}}U \end{pmatrix} \begin{pmatrix} I_{n_1} & x_{11}^{-1}{}^{\mathrm{t}}U \\ & I_{n-n_1} \end{pmatrix}. \tag{5}$$

Note that $U{}^{\mathrm{t}}U$ belongs to $\mathcal{Z}_{\mathcal{V}'}$ thanks to (V1) and (V2). Thus, we deduce the following lemma immediately from (5).

Lemma 1. (i) *Let $x \in \mathcal{Z}_{\mathcal{V}}$ as in (3). Then, $x \in \mathcal{P}_{\mathcal{V}}$ if and only if $x_{11} > 0$ and $x' - x_{11}^{-1}U{}^{\mathrm{t}}U \in \mathcal{P}_{\mathcal{V}'}$.*
(ii) *For $x \in \mathcal{P}_{\mathcal{V}}$, there exist unique $\tilde{U} \in \mathcal{W}$ and $\tilde{x} \in \mathcal{P}_{\mathcal{V}'}$ for which:*

$$\begin{aligned} x &= \begin{pmatrix} I_{n_1} & \\ \tilde{U} & I_{n-n_1} \end{pmatrix} \begin{pmatrix} x_{11}I_{n_1} & \\ & \tilde{x}' \end{pmatrix} \begin{pmatrix} I_{n_1} & {}^{\mathrm{t}}\tilde{U} \\ & I_{n-n_1} \end{pmatrix} \\ &= \begin{pmatrix} x_{11}I_{n_1} & x_{11}{}^{\mathrm{t}}\tilde{U} \\ x_{11}\tilde{U} & \tilde{x}' + x_{11}\tilde{U}{}^{\mathrm{t}}\tilde{U} \end{pmatrix}. \end{aligned} \tag{6}$$

(iii) *The closure $\overline{\mathcal{P}_{\mathcal{V}}}$ of the cone $\mathcal{P}_{\mathcal{V}}$ is described as:*

$$\overline{\mathcal{P}_{\mathcal{V}}} := \left\{ \begin{pmatrix} x_{11}I_{n_1} & x_{11}{}^{\mathrm{t}}\tilde{U} \\ x_{11}\tilde{U} & \tilde{x}' + x_{11}\tilde{U}{}^{\mathrm{t}}\tilde{U} \end{pmatrix} ;\ x_{11} \ge 0,\ \tilde{U} \in \mathcal{W},\ \tilde{x}' \in \overline{\mathcal{P}_{\mathcal{V}'}} \right\}.$$

2.3. *The Dual Cone $\mathcal{P}_{\mathcal{V}}^*$*

We define an inner product on the space $\mathcal{V}_{lk}$ by $(A|B)_{\mathcal{V}_{lk}} := n_l^{-1}\mathrm{tr}\, A^{\mathrm{t}}B$ for $A, B \in \mathcal{V}_{lk}$. Then, we see from (V1) that:

$$A^{\mathrm{t}}B + B^{\mathrm{t}}A = 2(A|B)_{\mathcal{V}_{lk}}I_{n_l}.$$

Gathering these inner products $(\cdot|\cdot)_{\mathcal{V}_{lk}}$, we introduce the standard inner product on the space $\mathcal{Z}_{\mathcal{V}}$ defined by:

$$(x|x') := \sum_{k=1}^{r} x_{kk}x'_{kk} + 2\sum_{1\le k<l\le r} (X_{lk}|X'_{lk})_{\mathcal{V}_{lk}} \tag{7}$$

for $x, x' \in \mathcal{Z}_{\mathcal{V}}$ of the form (1). When $n_1 = n_2 = \cdots = n_r = 1$ (and only in this case), the standard inner product above equals the trace inner product $\mathrm{tr}\,(xx')$.

Let $\widetilde{\mathcal{W}}_k$ $(k = 1, \ldots, r)$ be the vector space of $W \in \mathrm{Mat}(n, n_k, \mathbb{R})$ of the form:

$$W = \begin{pmatrix} 0_{n_1+\cdots+n_{k-1}, n_k} \\ X_{kk} \\ X_{k+1,k} \\ \vdots \\ X_{rk} \end{pmatrix} \qquad (X_{kk} = x_{kk} I_{n_k},\ x_{kk} \in \mathbb{R},\ X_{lk} \in \mathcal{V}_{lk},\ l > k).$$

Clearly, the space $\widetilde{\mathcal{W}}_k$ is isomorphic to $\mathbb{R} \oplus \sum_{l>k} \mathcal{V}_{lk}$, which implies $\dim \widetilde{\mathcal{W}}_k = 1 + q_k$ with $q_k := \sum_{l>k} n_{lk}$. Gathering orthogonal bases of $\mathcal{V}_{lk}$'s, we take a basis of $\widetilde{\mathcal{W}}_k$, so that we have an isomorphism $\widetilde{\mathcal{W}}_k \ni W \mapsto w = \mathrm{vect}(W) \in \mathbb{R}^{1+q_k}$, where the first component w_1 of w is assumed to be x_{kk}. Let us introduce a linear map $\phi_k : \mathcal{Z}_{\mathcal{V}} \to \mathrm{Sym}(1 + q_k, \mathbb{R})$ defined in such a way that:

$$(W^{\mathrm{t}}W|\xi) = {}^{\mathrm{t}}w\phi_k(\xi)w \qquad (\xi \in \mathcal{Z}_{\mathcal{V}},\ W \in \widetilde{\mathcal{W}}_k,\ w = \mathrm{vect}(W) \in \mathbb{R}^{1+q_k}). \tag{8}$$

It is easy to see that $\phi_r(\xi) = \xi_{rr}$ for $\xi \in \mathcal{Z}_{\mathcal{V}}$.

Theorem 1. *The dual cone $\mathcal{P}_{\mathcal{V}}^* \subset \mathcal{Z}_{\mathcal{V}}$ of $\mathcal{P}_{\mathcal{V}}$ with respect to the standard inner product is described as:*

$$\begin{aligned} \mathcal{P}_{\mathcal{V}}^* &= \{\, \xi \in \mathcal{Z}_{\mathcal{V}}\,;\ \phi_k(\xi) \text{ is positive definite for all } k = 1, \ldots, r \,\} \\ &= \{\, \xi \in \mathcal{Z}_{\mathcal{V}}\,;\ \det \phi_k(\xi) > 0 \text{ for all } k = 1, \ldots, r \,\}. \end{aligned} \tag{9}$$

Proof. We shall prove the statement by induction on the rank r. When $r = 1$, we have $\phi_1(\xi) = \xi_{11}$ and $\xi = \xi_{11} I_{n_1}$. Thus, (9) holds in this case.

Let us assume that (9) holds when the rank is smaller than r. In particular, the statement holds for $\mathcal{P}_{\mathcal{V}'}^* \subset \mathcal{Z}_{\mathcal{V}'}$, that is,

$$\begin{aligned} \mathcal{P}_{\mathcal{V}'}^* &= \{\, \xi' \in \mathcal{Z}_{\mathcal{V}'}\,;\ \phi'_k(\xi') \text{ is positive definite for all } k = 1, \ldots, r-1 \,\} \\ &= \{\, \xi' \in \mathcal{Z}_{\mathcal{V}'}\,;\ \det \phi'_k(\xi') > 0 \text{ for all } k = 1, \ldots, r-1 \,\}, \end{aligned}$$

where ϕ'_k is defined similarly to (8) for $\mathcal{Z}_{\mathcal{V}'}$. On the other hand, if:

$$\xi = \begin{pmatrix} \xi_{11} I_{n_1} & {}^{\mathrm{t}}V \\ V & \xi' \end{pmatrix} \qquad (\xi_{11} \in \mathbb{R},\ V \in \mathcal{W},\ \xi' \in \mathcal{Z}_{\mathcal{V}'}), \tag{10}$$

we observe that:

$$\phi_k(\xi) = \phi'_{k-1}(\xi') \qquad (k = 2, \ldots, r).$$

Therefore, in order to prove (9) for $\mathcal{P}_{\mathcal{V}}^*$ of rank r, it suffices to show that:

$$\begin{aligned} \mathcal{P}_{\mathcal{V}}^* &= \{\, \xi \in \mathcal{Z}_{\mathcal{V}}\,;\ \xi' \in \mathcal{P}_{\mathcal{V}'}^* \text{ and } \phi_1(\xi) \text{ is positive definite} \,\} \\ &= \{\, \xi \in \mathcal{Z}_{\mathcal{V}}\,;\ \xi' \in \mathcal{P}_{\mathcal{V}'}^* \text{ and } \det \phi_1(\xi) > 0 \,\}. \end{aligned} \tag{11}$$

If $q_1 = 0$, then any element $\xi \in \mathcal{Z}_{\mathcal{V}}$ is of the form:

$$\xi = \begin{pmatrix} \xi_{11} I_{n_1} & \\ & \xi' \end{pmatrix},$$

which belongs to $\mathcal{P}_{\mathcal{V}}$ if and only if $\xi' \in \mathcal{P}_{\mathcal{V}'}$ and $\phi_1(\xi) = \xi_{11} > 0$, so that (11) holds.

Assume $q_1 > 0$. Keeping in mind that $\widetilde{\mathcal{W}}_1 \simeq \mathbb{R} \oplus \mathcal{W}$ and $\mathcal{W} \simeq \mathbb{R}^{q_1}$ by (4), we have for $\xi \in \mathcal{Z}_{\mathcal{V}}$ as in (10),

$$\phi_1(\xi) = \begin{pmatrix} \xi_{11} & {}^{\mathrm{t}}v \\ v & \psi(\xi') \end{pmatrix} \in \mathrm{Sym}(1+q_1, \mathbb{R}), \tag{12}$$

where $v = \mathrm{vect}(V) \in \mathbb{R}^{q_i}$ and $\psi : \mathcal{Z}_{\mathcal{V}'} \to \mathrm{Sym}(q_i, \mathbb{R})$ is defined in such a way that:

$$(U^{\mathrm{t}}U|\xi') = {}^{\mathrm{t}}u\psi(\xi')u \quad (\xi' \in \mathcal{Z}_{\mathcal{V}'},\, U \in \mathcal{W},\, u = \mathrm{vect}(U) \in \mathbb{R}^{q_1}). \tag{13}$$

On the other hand, for $x \in \mathcal{Z}_{\mathcal{V}}$ as in (6), we have:

$$(x|\xi) = x_{11}\xi_{11} + 2x_{11}{}^{\mathrm{t}}\tilde{u}v + x_{11}{}^{\mathrm{t}}\tilde{u}\psi(\xi')\tilde{u} + (\tilde{x}'|\xi). \tag{14}$$

Owing to Lemma 1 (iii), the element $\xi \in \mathcal{Z}_{\mathcal{V}}$ belongs to $\mathcal{P}^*_{\mathcal{V}}$ if and only if the right-hand side is strictly positive for all $x_{11} \geq 0$, $\tilde{U} \in \mathcal{W}$ and $\tilde{x}' \in \overline{\mathcal{P}_{\mathcal{V}'}}$ with $(x_{11}, \tilde{x}') \neq (0,0)$. Assume $\xi \in \mathcal{P}^*_{\mathcal{V}}$. Considering the case $x_{11} = 0$, we have $(\tilde{x}'|\xi') > 0$ for all $\tilde{x}' \in \overline{\mathcal{P}_{\mathcal{V}'}} \backslash \{0\}$, which means that $\xi' \in \mathcal{P}^*_{\mathcal{V}'}$. Then, the quantity in (13) is strictly positive for non-zero U because $U^{\mathrm{t}}U$ belongs to $\overline{\mathcal{P}_{\mathcal{V}}} \setminus \{0\}$. Thus, $\psi(\xi')$ is positive definite, and (14) is rewritten as:

$$(x|\xi) = x_{11}(\xi_{11} - {}^{\mathrm{t}}v\psi(\xi')^{-1}v) + x_{11}{}^{\mathrm{t}}(\tilde{u} + \psi(\xi')^{-1}v)\psi(\xi')(\tilde{u} + \psi(\xi')^{-1}v) + (\tilde{x}'|\xi'). \tag{15}$$

Therefore, we obtain:

$$\mathcal{P}^*_{\mathcal{V}} = \left\{ \xi \in \mathcal{Z}_{\mathcal{V}} \,;\, \xi' \in \mathcal{P}^*_{\mathcal{V}'} \text{ and } \xi_{11} - {}^{\mathrm{t}}v\psi(\xi')^{-1}v > 0 \right\}. \tag{16}$$

On the other hand, we see from (12) that:

$$\phi_1(\xi) = \begin{pmatrix} 1 & {}^{\mathrm{t}}v\psi(\xi')^{-1} \\ & I_{q_1} \end{pmatrix} \begin{pmatrix} \xi_{11} - {}^{\mathrm{t}}v\psi(\xi')^{-1}v & \\ & \psi(\xi') \end{pmatrix} \begin{pmatrix} 1 & \\ \psi(\xi')^{-1}v & I_{q_1} \end{pmatrix}. \tag{17}$$

Hence, we deduce (11) from (16) and (17). □

We note that, if $q_1 > 0$, the $(1,1)$-component of the inverse matrix $\phi_1(\xi)^{-1}$ is given by:

$$(\phi_1(\xi)^{-1})_{11} = (\xi_{11} - {}^{\mathrm{t}}v\psi(\xi')^{-1}v)^{-1} \tag{18}$$

thanks to (17).

3. Koszul–Vinberg Characteristic Function of $\mathcal{P}^*_{\mathcal{V}}$

We denote by $\varphi_{\mathcal{V}}$ the Koszul–Vinberg characteristic function of $\mathcal{P}^*_{\mathcal{V}}$. In this section, we give an explicit formula of $\varphi_{\mathcal{V}}$.

Recall that the linear map $\psi : \mathcal{Z}_{\mathcal{V}'} \to \mathrm{Sym}(q_1, \mathbb{R})$ plays an important role in the proof of Theorem 1. We shall introduce similar linear maps $\psi_k : \mathcal{Z}_{\mathcal{V}} \to \mathrm{Sym}(q_k, \mathbb{R})$ for k such that $q_k > 0$. Let $\mathcal{W}_k$ be the subspace of $\widetilde{\mathcal{W}}_k$ consisting of $W \in \widetilde{\mathcal{W}}_k$ for which $w_1 = x_{kk} = 0$. Then, clearly, $\mathcal{W}_k \simeq \sum^{\oplus}_{l>k} \mathcal{V}_{lk}$ and $\dim \mathcal{W}_k = q_k$. If $q_k > 0$, using the same orthogonal basis of $\mathcal{V}_{lk}$ as in the previous section, we have the isomorphism $\mathcal{W}_k \ni W \mapsto w = \mathrm{vect}(W) \in \mathbb{R}^{q_k}$. Similarly to (8), we define ψ_k by:

$$(W^{\mathrm{t}}W|\xi) = {}^{\mathrm{t}}w\psi_k(\xi)w \qquad (\xi \in \mathcal{Z}_{\mathcal{V}},\ W \in \mathcal{W}_k,\ w = \mathrm{vect}(W) \in \mathbb{R}^{q_k}). \tag{19}$$

Then, we have:

$$\phi_k(\xi) = \begin{pmatrix} \xi_{kk} & {}^{\mathrm{t}}v_k \\ v_k & \psi_k(\xi) \end{pmatrix} \qquad (\xi \in \mathcal{Z}_{\mathcal{V}}), \tag{20}$$

where $v_k \in \mathbb{R}^{q_k}$ is a vector corresponding to the $\mathcal{W}_k$-component of ξ. If $\xi \in \mathcal{P}_\mathcal{V}^*$, we see from (19) that $\psi_k(\xi)$ is positive definite. In this case, we have:

$$\phi_k(\xi) = \begin{pmatrix} 1 & {}^{\mathrm{t}}v_k\psi_k(\xi)^{-1} \\ & I_{q_k} \end{pmatrix} \begin{pmatrix} \xi_{kk} - {}^{\mathrm{t}}v_k\psi_k(\xi)^{-1}v_k & \\ & \psi_k(\xi) \end{pmatrix} \begin{pmatrix} 1 & \\ \psi_k(\xi)^{-1}v_k & I_{q_k} \end{pmatrix}, \tag{21}$$

so that we get a generalization of (18), that is,

$$(\phi_k(\xi)^{-1})_{11} = (\xi_{kk} - {}^{\mathrm{t}}v_k\psi_k(\xi)^{-1}v_k)^{-1}. \tag{22}$$

On the other hand, if $q_k = 0$, then $\phi_k(\xi)^{-1} = \xi_{kk}^{-1}$.

We remark that $\psi_1(\xi) = \psi(\xi')$, and that some part of the argument above is parallel to the proof of Theorem 1.

Theorem 2. *The Koszul–Vinberg characteristic function $\varphi_\mathcal{V}$ of $\mathcal{P}_\mathcal{V}^*$ is given by the following formula:*

$$\varphi_\mathcal{V}(\xi) = C_\mathcal{V} \prod_{k=1}^{r} (\phi_k(\xi)^{-1})_{11}^{1+q_k/2} \prod_{q_k>0} (\det \psi_k(\xi))^{-1/2} \qquad (\xi \in \mathcal{P}_\mathcal{V}^*), \tag{23}$$

where $C_\mathcal{V} := (2\pi)^{(N-r)/2} \prod_{k=1}^{r} \Gamma(1 + \frac{q_k}{2})$ and $N := \dim \mathcal{Z}_\mathcal{V}$.

Proof. We shall show the statement by induction on the rank as in the proof of Theorem 1. Then, it suffices to show that:

$$\varphi_\mathcal{V}(\xi) = (2\pi)^{q_1/2}\Gamma(1 + \frac{q_1}{2})(\phi_1(\xi)^{-1})_{11}^{1+q_1/2}(\det \psi_1(\xi))^{-\mathrm{sgn}(q_1)/2}\varphi_{\mathcal{V}'}(\xi') \tag{24}$$

for $\xi \in \mathcal{P}_\mathcal{V}^*$ as in (10), where $(\det \psi_1(\xi))^{-\mathrm{sgn}(q_1)/2}$ is interpreted as:

$$(\det \psi_1(\xi))^{-\mathrm{sgn}(q_1)/2} := \begin{cases} 1 & (q_1 = 0), \\ (\det \psi_1(\xi))^{-1/2} & (q_1 > 0). \end{cases}$$

When $q_1 = 0$, we have:

$$\begin{aligned} \varphi_\mathcal{V}(\xi) &= \int_0^\infty \int_{\mathcal{P}_{\mathcal{V}'}} e^{-x_{11}\xi_{11}} e^{-(x'|\xi')} dx_{11}\, dx' \\ &= \xi_{11}^{-1} \varphi_{\mathcal{V}'}(\xi'), \end{aligned}$$

which means (24).

When $q_1 > 0$, the Euclidean measure dx equals $2^{q_1/2}x_{11}^{q_1}\, dx_{11} d\tilde{u} d\tilde{x}'$ by the change of variables in (6). Indeed, the coefficient $2^{q_1/2}$ comes from the normalization of the inner product on $\mathcal{W} \simeq \mathbb{R}^{q_1}$ regarded as a subspace of $\mathcal{Z}_\mathcal{V}$. Then, we have by (15):

$$\begin{aligned} \varphi_\mathcal{V}(\xi) = \int_0^\infty \int_{\mathbb{R}^{q_1}} \int_{\mathcal{P}_{\mathcal{V}'}} & e^{-x_{11}(\xi_{11} - {}^{\mathrm{t}}v\psi(\xi')^{-1}v)} e^{-x_{11}{}^{\mathrm{t}}(\tilde{u}+\psi(\xi')^{-1}v)\psi(\xi')(\tilde{u}+\psi(\xi')^{-1}v)} e^{-(\tilde{x}'|\xi')} \\ & \times 2^{q_1/2} x_{11}^{q_1}\, dx_{11} d\tilde{u} d\tilde{x}'. \end{aligned}$$

By the Gaussian integral formula, we have:

$$\int_{\mathbb{R}^{q_1}} e^{-x_{11}{}^{\mathrm{t}}(\tilde{u}+\psi(\xi')^{-1}v)\psi(\xi')(\tilde{u}+\psi(\xi')^{-1}v)} d\tilde{u} = \pi^{q_1/2} x_{11}^{-q_1/2} (\det \psi(\xi'))^{-1/2}.$$

Therefore, we get:

$$\begin{aligned}\varphi_{\mathcal{V}}(\xi) &= (2\pi)^{q_1/2}(\det\psi(\xi'))^{-1/2}\int_0^\infty e^{-x_{11}(\xi_{11}-{}^{t}v\psi(\xi')^{-1}v)}x_{11}^{q_1/2}\,dx_{11}\int_{\mathcal{P}_{\mathcal{V}'}} e^{-(\tilde{x}'|\xi')}\,d\tilde{x}' \\ &= (2\pi)^{q_1/2}(\det\psi_1(\xi))^{-1/2}\Gamma(1+\frac{q_1}{2})(\xi_{11}-{}^{t}v\psi(\xi')^{-1}v)^{-1-q_k/2}\varphi_{\mathcal{V}'}(\xi'),\end{aligned}$$

which together with (18) leads us to (24). □

Example 2. *Let $\mathcal{V} = \{\mathcal{V}_{lk}\}_{1\le k<l\le 3}$ be as in Example 1. For:*

$$\xi = \begin{pmatrix} \xi_1 & 0 & \xi_4 & 0 \\ 0 & \xi_1 & 0 & \xi_5 \\ \xi_4 & 0 & \xi_2 & \xi_6 \\ 0 & \xi_5 & \xi_6 & \xi_3 \end{pmatrix} \in \mathcal{Z}_{\mathcal{V}}, \tag{25}$$

we have:

$$\phi_1(\xi) = \begin{pmatrix} \xi_1 & \xi_4 & \xi_5 \\ \xi_4 & \xi_2 & 0 \\ \xi_5 & 0 & \xi_3 \end{pmatrix}, \quad \phi_2(\xi) = \begin{pmatrix} \xi_2 & \xi_6 \\ \xi_6 & \xi_3 \end{pmatrix}, \quad \phi_3(\xi) = \xi_3,$$

$$\psi_1(\xi) = \begin{pmatrix} \xi_2 & 0 \\ 0 & \xi_3 \end{pmatrix}, \quad \psi_2(\xi) = \xi_3.$$

The cone $\mathcal{P}_{\mathcal{V}}^*$ is described as:

$$\mathcal{P}_{\mathcal{V}}^* = \left\{ \xi \in \mathcal{Z}_{\mathcal{V}};\ \begin{vmatrix} \xi_1 & \xi_4 & \xi_5 \\ \xi_4 & \xi_2 & 0 \\ \xi_5 & 0 & \xi_3 \end{vmatrix} > 0,\ \begin{vmatrix} \xi_2 & \xi_6 \\ \xi_6 & \xi_3 \end{vmatrix} > 0,\ \xi_3 > 0 \right\},$$

and its Koszul–Vinberg characteristic function $\varphi_{\mathcal{V}}$ is expressed as:

$$\begin{aligned}\varphi_{\mathcal{V}}(\xi) &= C_{\mathcal{V}} \left\{ \begin{vmatrix} \xi_1 & \xi_4 & \xi_5 \\ \xi_4 & \xi_2 & 0 \\ \xi_5 & 0 & \xi_3 \end{vmatrix} / (\xi_2\xi_3) \right\}^{-2} \left\{ \begin{vmatrix} \xi_2 & \xi_6 \\ \xi_6 & \xi_3 \end{vmatrix} / \xi_3 \right\}^{-3/2} \xi_3^{-1} \cdot (\xi_2\xi_3)^{-1/2}(\xi_3)^{-1/2} \\ &= C_{\mathcal{V}} \begin{vmatrix} \xi_1 & \xi_4 & \xi_5 \\ \xi_4 & \xi_2 & 0 \\ \xi_5 & 0 & \xi_3 \end{vmatrix}^{-2} \begin{vmatrix} \xi_2 & \xi_6 \\ \xi_6 & \xi_3 \end{vmatrix}^{-3/2} \xi_2^{3/2}\xi_3^{3/2},\end{aligned}$$

where $C_{\mathcal{V}} = (2\pi)^{3/2}\Gamma(2)\Gamma(3/2)\Gamma(1) = \sqrt{2}\pi^2$.

Suppose that the cone $\mathcal{P}_{\mathcal{V}}$ is homogeneous. Then, $\mathcal{P}_{\mathcal{V}}^*$, as well as $\mathcal{P}_{\mathcal{V}}$, is a homogeneous cone of rank 3, so that the Koszul–Vinberg characteristic function of $\mathcal{P}_{\mathcal{V}}^*$ has at most three irreducible factors (see [8]). However, we have seen that there are four irreducible factors in the function $\varphi_{\mathcal{V}}$. Therefore, we conclude that neither $\mathcal{P}_{\mathcal{V}}$, nor $\mathcal{P}_{\mathcal{V}}^*$ is homogeneous.

4. Γ-Type Integral Formulas

For an $n \times n$ matrix $A = (A_{ij})$ and $1 \leq m \leq n$, we denote by $A^{[m]}$ the upper-left $m \times m$ submatrix $(A_{ij})_{i,j \leq m}$ of A. Put $M_k := \sum_{i=1}^{k} n_k$ $(k = 1, \ldots, r)$. For $\underline{s} = (s_1, \ldots, s_r) \in \mathbb{C}^r$, we define functions $\Delta_{\underline{s}}^{\mathcal{V}}$ on $\mathcal{P}_{\mathcal{V}}$ and $\delta_{\underline{s}}^{\mathcal{V}}$ on $\mathcal{P}_{\mathcal{V}}^*$ respectively by:

$$
\begin{aligned}
\Delta_{\underline{s}}^{\mathcal{V}}(x) &:= (\det x^{[M_1]})^{s_1/n_1} \prod_{k=2}^{r} \Big(\frac{\det x^{[M_k]}}{\det x^{[M_{k-1}]}} \Big)^{s_k/n_k} && (26) \\
&= (\det x)^{s_r/n_r} \prod_{k=1}^{r-1} (\det x^{[M_k]})^{s_k/n_k - s_{k-1}/n_{k-1}} \qquad (x \in \mathcal{P}_{\mathcal{V}}), \\
\delta_{\underline{s}}^{\mathcal{V}}(\xi) &:= \prod_{k=1}^{r} (\phi_k(\xi)^{-1})_{11}^{-s_k} && (27) \\
&= \prod_{q_k=0} \xi_{kk}^{s_k} \prod_{q_k>0} (\xi_{kk} - {}^t v_k \psi_k(\xi)^{-1} v_k)^{s_k} \qquad (\xi \in \mathcal{P}_{\mathcal{V}}^*).
\end{aligned}
$$

Recall (22) for the second equality of (27).

For $\underline{a} = (a_1, \ldots, a_r) \in \mathbb{R}_{>0}^r$, let $D_{\underline{a}}$ denote the diagonal matrix defined by:

$$
D_{\underline{a}} := \begin{pmatrix} a_1 I_{n_1} & & & \\ & a_2 I_{n_2} & & \\ & & \ddots & \\ & & & a_r I_{n_r} \end{pmatrix} \in GL(n, \mathbb{R}).
$$

Then, the linear map $\mathcal{Z}_{\mathcal{V}} \ni x \mapsto D_a x D_a \in \mathcal{Z}_{\mathcal{V}}$ preserves both $\mathcal{P}_{\mathcal{V}}$ and $\mathcal{P}_{\mathcal{V}}^*$, and we have:

$$
\Delta_{\underline{s}}^{\mathcal{V}}(D_a x D_a) = \Big(\prod_{k=1}^{r} a_k^{2s_k}\Big) \Delta_{\underline{s}}^{\mathcal{V}}(x) \qquad (x \in \mathcal{P}_{\mathcal{V}}), \tag{28}
$$

$$
\delta_{\underline{s}}^{\mathcal{V}}(D_a \xi D_a) = \Big(\prod_{k=1}^{r} a_k^{2s_k}\Big) \delta_{\underline{s}}^{\mathcal{V}}(\xi) \qquad (\xi \in \mathcal{P}_{\mathcal{V}}). \tag{29}
$$

Assume $q_1 > 0$. For $B \in \mathcal{W}$, we denote by τ_B the linear transform on $\mathcal{Z}_{\mathcal{V}}$ given by:

$$
\begin{aligned}
\tau_B x &:= \begin{pmatrix} I_{n_1} & \\ B & I_{n-n_1} \end{pmatrix} \begin{pmatrix} x_{11} I_{n_1} & {}^t U \\ U & x' \end{pmatrix} \begin{pmatrix} I_{n_1} & {}^t B \\ & I_{n-n_1} \end{pmatrix} \\
&= \begin{pmatrix} x_{11} I_{n_1} & {}^t U + x_{11} {}^t B \\ U + x_{11} B & x' + U {}^t B + B {}^t U + x_{11} B {}^t B \end{pmatrix},
\end{aligned}
$$

where $x \in \mathcal{Z}_{\mathcal{V}}$ is as in (3). Indeed, since:

$$
U^t B + B^t U = (U + B)^t (U + B) - U^t U - B^t B \in \mathcal{Z}_{\mathcal{V}'},
$$

the matrix $\tau_B x$ belongs to $\mathcal{Z}_{\mathcal{V}}$. Clearly, τ_B preserves $\mathcal{P}_{\mathcal{V}}$, and we have:

$$
\Delta_{\underline{s}}^{\mathcal{V}}(\tau_B x) = \Delta_{\underline{s}}^{\mathcal{V}}(x) \qquad (x \in \mathcal{P}_{\mathcal{V}}). \tag{30}
$$

The formula (5) is rewritten as:

$$
\tau_{-x_{11}^{-1} U}(x) = \begin{pmatrix} x_{11} I_{n_1} & \\ & x' - x_{11}^{-1} U^t U \end{pmatrix},
$$

which together with (30) tells us that:

$$\Delta_{\underline{s}}^{\mathcal{V}}(x) = x_{11}^{s_1}\Delta_{\underline{s}'}^{\mathcal{V}'}(x' - x_{11}^{-1}U^{\mathrm{t}}U), \tag{31}$$

where $\underline{s}' := (s_2, \ldots, s_r) \in \mathbb{C}^{r-1}$.

Let us consider the adjoint map $\tau_B^* : \mathcal{Z}_{\mathcal{V}} \to \mathcal{Z}_{\mathcal{V}}$ of τ_B with respect to the standard inner product. Let $b \in \mathbb{R}^{q_1}$ be the vector corresponding to $B \in \mathcal{W}$. For $x \in \mathcal{Z}_{\mathcal{V}}$ and $\xi \in \mathcal{Z}_{\mathcal{V}}$ as in (3) and (10), respectively, we observe that:

$$\begin{aligned}(\tau_B x|\xi) &= x_{11}\xi_{11} + 2{}^{\mathrm{t}}(u + x_{11}b)v + (x' + U^{\mathrm{t}}B + B^{\mathrm{t}}U + x_{11}B^{\mathrm{t}}B|\xi') \\ &= x_{11}(\xi_{11} + 2{}^{\mathrm{t}}bv + {}^{\mathrm{t}}b\psi(\xi')b) + 2{}^{\mathrm{t}}u(v + \psi(\xi')b) + (x'|\xi').\end{aligned}$$

Thus, if we write:

$$\iota(\xi_{11}, v, \xi') := \begin{pmatrix} \xi_{11}I_{n_1} & {}^{\mathrm{t}}V \\ V & \xi' \end{pmatrix},$$

we have:

$$\tau_B^*\iota(\xi_{11}, v, \xi') = \iota(\xi_{11} + 2{}^{\mathrm{t}}bv + {}^{\mathrm{t}}b\psi(\xi')b, v + \psi(\xi')b, \xi'). \tag{32}$$

Furthermore, we see from (12) that $\phi_1(\tau_B^*\iota(\xi_{11}, v, \xi'))$ equals:

$$\begin{pmatrix} \xi_{11} + 2{}^{\mathrm{t}}bv + {}^{\mathrm{t}}b\psi(\xi')b & {}^{\mathrm{t}}v + {}^{\mathrm{t}}b\psi(\xi') \\ v + \psi(\xi')b & \psi(\xi') \end{pmatrix} = \begin{pmatrix} 1 & {}^{\mathrm{t}}b \\ & I_{q_1} \end{pmatrix}\begin{pmatrix} \xi_{11} & {}^{\mathrm{t}}v \\ v & \psi(\xi') \end{pmatrix}\begin{pmatrix} 1 & \\ b & I_{q_1} \end{pmatrix},$$

so that we get for $\xi = \iota(\xi_{11}, v, \xi')$:

$$\phi_1(\tau_B^*\xi) = \begin{pmatrix} 1 & {}^{\mathrm{t}}b \\ & I_{q_1} \end{pmatrix}\phi_1(\xi)\begin{pmatrix} 1 & \\ b & I_{q_1} \end{pmatrix}.$$

Therefore:

$$(\phi_1(\tau_B^*\xi)^{-1})_{11} = (\phi_1(\xi)^{-1})_{11}.$$

On the other hand, we have for $\xi = \iota(\xi_{11}, v, \xi') \in \mathcal{P}_{\mathcal{V}}^*$:

$$\delta_{\underline{s}}^{\mathcal{V}}(\xi) = (\phi_1(\xi)^{-1})_{11}^{-s_1}\,\delta_{\underline{s}'}^{\mathcal{V}'}(\xi'). \tag{33}$$

Thus, we conclude that:

$$\delta_{\underline{s}}^{\mathcal{V}}(\tau_B^*\xi) = \delta_{\underline{s}}^{\mathcal{V}}(\xi). \tag{34}$$

Theorem 3. *When* $\Re s_k > -1 - q_k/2$ *for* $k = 1, \ldots, r$, *one has:*

$$\int_{\mathcal{P}_{\mathcal{V}}} e^{-(x|\xi)}\Delta_{\underline{s}}^{\mathcal{V}}(x)\,dx = C_{\mathcal{V}}^{-1}\gamma_{\mathcal{V}}(\underline{s})\,\delta_{-\underline{s}}^{\mathcal{V}}(\xi)\varphi_{\mathcal{V}}(\xi), \tag{35}$$

where $\gamma_{\mathcal{V}}(\underline{s}) := (2\pi)^{(N-r)/2}\prod_{k=1}^{r}\Gamma(s_k + 1 + \frac{q_k}{2})$.

Proof. Recalling Theorem 2, we rewrite the right-hand side of (35) as:

$$(2\pi)^{(N-r)/2}\prod_{k=1}^{r}\Gamma(s_k + 1 + \frac{q_k}{2})\prod_{k=1}^{r}(\phi_k(\xi)^{-1})_{11}^{s_k+1+q_k/2}\prod_{q_k>0}(\det\psi_k(\xi))^{-1/2},$$

which is similar to the right-hand side of (23). Thus, the proof is parallel to Theorem 2. Namely, by induction on the rank, it suffices to show that:

$$\begin{aligned} &\int_{\mathcal{P}_{\mathcal{V}}} e^{-(x|\xi)} \Delta_{\underline{s}}^{\mathcal{V}}(x)\, dx \\ &= (2\pi)^{q_1/2} \Gamma(s_1 + 1 + \frac{q_1}{2}) (\phi_1(\xi)^{-1})_{11}^{s_1+1+q_1/2} (\det \psi_1(\xi))^{-\mathrm{sgn}(q_1)/2} \\ &\times \int_{\mathcal{P}_{\mathcal{V}'}} e^{-(x'|\xi)} \Delta_{\underline{s}'}^{\mathcal{V}'}(x')\, dx' \end{aligned} \tag{36}$$

thanks to (33).

When $q_1 = 0$, we have $(x|\xi) = x_{11}\xi_{11} + (x'|\xi')$ and $\Delta_{\underline{s}}^{\mathcal{V}}(x) = x_{11}^{s_1} \Delta_{\underline{s}}^{\mathcal{V}'}(x')$. Thus:

$$\int_{\mathcal{P}_{\mathcal{V}}} e^{-(x|\xi)} \Delta_{\underline{s}}^{\mathcal{V}}(x)\, dx = \int_0^\infty e^{-x_{11}\xi_{11}} x_{11}^{s_1}\, dx_{11} \times \int_{\mathcal{P}_{\mathcal{V}'}} e^{-(x'|\xi)} \Delta_{\underline{s}'}^{\mathcal{V}'}(x')\, dx'.$$

Since $\int_0^\infty e^{-x_{11}\xi_{11}} x_{11}^{s_1}\, dx_{11} = \Gamma(s_1+1)\xi_{11}^{-s_1-1}$, we get (36).

When $q_1 > 1$, we use the change of variable (6). Since $\tilde{x}' = x' - x_{11}^{-1} U^{\mathrm{t}} U$, we have $\Delta_{\underline{s}}^{\mathcal{V}}(x) = x_{11}^{s_1} \Delta_{\underline{s}'}^{\mathcal{V}'}(\tilde{x}')$ by (31). Therefore, by the same Gaussian integral formula as in the proof of Theorem 2, the integral $\int_{\mathcal{P}_{\mathcal{V}}} e^{-(x|\xi)} \Delta_{\underline{s}}^{\mathcal{V}}(x)\, dx$ equals:

$$\begin{aligned} &\int_0^\infty \int_W \int_{\mathcal{P}_{\mathcal{V}'}} e^{-x_{11}(\xi_{11} - {}^{\mathrm{t}}v\psi(\xi')^{-1}v)} e^{-x_{11}{}^{\mathrm{t}}(\tilde{u}+\psi(\xi')^{-1}v)\psi(\xi')(\tilde{u}+\psi(\xi')^{-1}v)} e^{-(\tilde{x}'|\xi')} x_{11}^{s_1} \Delta_{\underline{s}'}^{\mathcal{V}'}(\tilde{x}') \\ &\qquad \times 2^{q_1/2} x_{11}^{q_1}\, dx_{11} d\tilde{u} d\tilde{x}' \\ &= (2\pi)^{q_1/2} (\det \psi(\xi))^{-1/2} \int_0^\infty e^{-x_{11}(\xi_{11} - {}^{\mathrm{t}}v\psi(\xi')^{-1}v)} x_{11}^{s_1+q_1/2}\, dx_{11} \\ &\times \int_{\mathcal{P}_{\mathcal{V}'}} e^{-(\tilde{x}'|\xi')} \Delta_{\underline{s}'}^{\mathcal{V}'}(\tilde{x}')\, d\tilde{x}' \\ &= (2\pi)^{q_1/2} (\det \psi(\xi))^{-1/2} \Gamma(s_k + 1 + \frac{q_1}{2}) (\xi_{11} - {}^{\mathrm{t}}v\psi(\xi')^{-1}v)^{-s_k-1-q_k/2} \\ &\times \int_{\mathcal{P}_{\mathcal{V}'}} e^{-(\tilde{x}'|\xi')} \Delta_{\underline{s}'}^{\mathcal{V}'}(\tilde{x}')\, d\tilde{x}'. \end{aligned}$$

Hence, we get (36) by (18). □

We shall obtain an integral formula over $\mathcal{P}_{\mathcal{V}}^*$ as follows.

Theorem 4. *When $\Re s_k > q_k/2$ for $k = 1, \ldots, r$, one has:*

$$\int_{\mathcal{P}_{\mathcal{V}}^*} e^{-(x|\xi)} \delta_{\underline{s}}^{\mathcal{V}}(\xi)\, \varphi_{\mathcal{V}}(\xi)\, d\xi = C_{\mathcal{V}} \Gamma_{\mathcal{V}}(\underline{s}) \Delta_{-\underline{s}}^{\mathcal{V}}(x) \qquad (x \in \mathcal{P}_{\mathcal{V}}), \tag{37}$$

where $\Gamma_{\mathcal{V}}(\underline{s}) := (2\pi)^{(N-r)/2} \prod_{k=1}^r \Gamma(s_k - q_k/2)$.

Proof. Using (24), (31) and (33), we rewrite (37) as:

$$\begin{aligned} &\int_{\mathcal{P}_{\mathcal{V}}^*} e^{-(x|\xi)} (\phi_1(\xi)^{-1})_{11}^{-s_1+1+q_1/2} (\det \psi_1(\xi))^{-\mathrm{sgn}(q_1)/2} \delta_{\underline{s}'}^{\mathcal{V}'}(\xi')\, \varphi_{\mathcal{V}'}(\xi')\, d\xi \\ &= C_{\mathcal{V}'} (2\pi)^{q_1/2} \Gamma(s_1 - q_1/2) \Gamma_{\mathcal{V}'}(\underline{s}')\, x_{11}^{-s_1} \Delta_{-\underline{s}'}^{\mathcal{V}'}(\tilde{x}'), \end{aligned} \tag{38}$$

where:

$$\tilde{x}' := \begin{cases} x' & (q_1 = 0), \\ x' - x_{11}^{-1} U^{\mathrm{t}} U & (q_1 > 0). \end{cases}$$

Therefore, by induction on the rank, it suffices to show that the left-hand side of (38) equals:

$$(2\pi)^{q_1/2}\Gamma(s_1 - q_1/2)x_{11}^{-s_1} \int_{\mathcal{P}^*_{\mathcal{V}'}} e^{-(\tilde{x}'|\xi')}\delta^{\mathcal{V}'}_{\underline{s}'}(\xi')\,\varphi_{\mathcal{V}'}(\xi')\,d\xi'. \tag{39}$$

When $q_1 = 0$, since $d\xi = d\xi_{11}d\xi'$, the left-hand side of (38) equals:

$$\int_0^\infty e^{-x_{11}\xi_{11}}\xi_{11}^{s_1-1}\,d\xi_{11} \int_{\mathcal{P}^*_{\mathcal{V}'}} e^{-(x'|\xi')}\delta^{\mathcal{V}'}_{\underline{s}'}(\xi')\,\varphi_{\mathcal{V}'}(\xi')\,d\xi',$$

which coincides with (39) in this case.

Assume $q_1 > 0$. Keeping (16) and (18) in mind, we put $\tilde{\xi}_{11} := \xi_{11} - {}^{\mathrm{t}}v\psi(\xi')^{-1}v = (\phi_1(\xi)^{-1})_{11}^{-1} > 0$. By the change of variables $\xi = \iota(\tilde{\xi}_{11} + {}^{\mathrm{t}}v\psi(\xi')^{-1}v,\ v,\ \xi')$, we have $d\xi = 2^{q_1/2}\,d\tilde{\xi}_{11}dvd\xi'$. On the other hand, we observe:

$$\begin{aligned}(x|\xi) &= x_{11}(\tilde{\xi}_{11} + {}^{\mathrm{t}}v\psi(\xi')^{-1}v) + 2{}^{\mathrm{t}}uv + (x'|\xi')\\ &= x_{11}\tilde{\xi}_{11} + x_{11}{}^{\mathrm{t}}(v + x_{11}^{-1}\psi(\xi')u)\psi(\xi')^{-1}(v + x_{11}^{-1}\psi(\xi')u) + (x - x_{11}^{-1}U{}^{\mathrm{t}}U|\xi').\end{aligned}$$

Thus, the left-hand side of (39) equals:

$$\begin{aligned}&\int_0^\infty \int_{\mathbb{R}^{q_1}} \int_{\mathcal{P}^*_{\mathcal{V}'}} e^{-x_{11}\tilde{\xi}_{11}} e^{-x_{11}{}^{\mathrm{t}}(v+x_{11}^{-1}\psi(\xi')u)\psi(\xi')^{-1}(v+x_{11}^{-1}\psi(\xi')u)} e^{-(x-x_{11}^{-1}U{}^{\mathrm{t}}U|\xi')}\\ &\qquad\times\tilde{\xi}_{11}^{s_1-1-q_1/2}(\det\psi(\xi'))^{-1/2}\delta^{\mathcal{V}'}_{\underline{s}'}(\xi')\,\varphi_{\mathcal{V}'}(\xi')\,2^{q_1/2}\,d\tilde{\xi}_{11}dvd\xi'.\end{aligned} \tag{40}$$

By the Gaussian integral formula, we have:

$$\int_{\mathbb{R}^{q_1}} e^{-x_{11}{}^{\mathrm{t}}(v+x_{11}^{-1}\psi(\xi')u)\psi(\xi')^{-1}(v+x_{11}^{-1}\psi(\xi')u)}\,dv = \pi^{q_1/2}x_{11}^{-q_1/2}(\det\psi(\xi'))^{1/2},$$

so that (40) equals:

$$(2\pi)^{q_1/2}x_{11}^{-q_1/2}\int_0^\infty e^{-x_{11}\tilde{\xi}_{11}}\tilde{\xi}_{11}^{s_1-1-q_1/2}\,d\tilde{\xi}_{11}\int_{\mathcal{P}^*_{\mathcal{V}'}} e^{-(x-x_{11}^{-1}U{}^{\mathrm{t}}U|\xi')}\delta^{\mathcal{V}'}_{\underline{s}'}(\xi')\,\varphi_{\mathcal{V}'}(\xi')d\xi',$$

which coincides with (39) because: $\int_0^\infty e^{-x_{11}\tilde{\xi}_{11}}\tilde{\xi}_{11}^{s_1-1-q_1/2}\,d\tilde{\xi}_{11} = \Gamma(s_1 - q_1/2)x_{11}^{-s_1+q_1/2}$. □

Example 3. *Let $\mathcal{Z}_{\mathcal{V}}$ be as in Example 1, and let $x \in \mathcal{P}_{\mathcal{V}}$ and $\xi \in \mathcal{P}^*_{\mathcal{V}}$ be as in (2) and (25), respectively. Then, we have for $\underline{s} = (s_1, s_2, s_3) \in \mathbb{C}^3$,*

$$\begin{aligned}\Delta^{\mathcal{V}}_{\underline{s}}(x) &= (x_{11}^2)^{s_1/2-s_2}\begin{vmatrix}x_1 & 0 & x_4\\ 0 & x_1 & 0\\ x_4 & 0 & x_2\end{vmatrix}^{s_2-s_3}(\det x)^{s_3}\\ &= x_{11}^{s_1-s_2-s_3}\begin{vmatrix}x_1 & x_4\\ x_4 & x_2\end{vmatrix}^{s_2-s_3}(\det x)^{s_3},\end{aligned}$$

and:

$$\begin{aligned}\delta^{\mathcal{V}}_{\underline{s}}(\xi) &= (\xi_1 - \frac{\xi_4^2}{\xi_2} - \frac{\xi_5^2}{\xi_3})^{s_1}(\xi_2 - \frac{\xi_6^2}{\xi_3})^{s_2}\xi_3^{s_3}\\ &= \begin{vmatrix}\xi_1 & \xi_4 & \xi_5\\ \xi_4 & \xi_2 & 0\\ \xi_5 & 0 & \xi_3\end{vmatrix}^{s_1}\begin{vmatrix}\xi_2 & \xi_6\\ \xi_6 & \xi_3\end{vmatrix}^{s_2}\xi_2^{-s_1}\xi_3^{s_3-s_1-s_2}.\end{aligned}$$

When $\Re s_1 > -2, \Re s_2 > -3/2$ *and* $\Re s_3 > -1$, *the integral formula (35) holds with:*

$$\gamma_{\mathcal{V}}(\underline{s}) = (2\pi)^{3/2}\Gamma(s_1+2)\Gamma(s_2+3/2)\Gamma(s_3+1).$$

Furthermore, when $\Re s_1 > 1$, $\Re s_2 > 1/2$ *and* $\Re s_3 > 0$, *the integral formula (37) holds with:*

$$\Gamma_{\mathcal{V}}(\underline{s}) = (2\pi)^{3/2}\Gamma(s_1-1)\Gamma(s_2-1/2)\Gamma(s_3).$$

5. Multiplicative Legendre Transform of Generalized Power Functions

For $\underline{s} \in \mathbb{R}^r_{>0}$, we see that $\log \Delta_{-\underline{s}}$ is a strictly convex function on the cone $\mathcal{P}_{\mathcal{V}}$. In fact, $\Delta_{-\underline{s}}$ is defined naturally on $\mathcal{P}_n$ as a product of powers of principal minors, and it is well known that such $\log \Delta_{-\underline{s}}$ is strictly convex on the whole $\mathcal{P}_n$. In this section, we shall show that $\log \Delta^{\mathcal{V}}_{-\underline{s}}$ and $\log \delta^{\mathcal{V}}_{-\underline{s}}$ are related by the Fenchel–Legendre transform.

For $x \in \mathcal{P}_{\mathcal{V}}$, we denote by $\mathcal{I}^{\mathcal{V}}_{\underline{s}}(x)$ the minus gradient $-\nabla \log \Delta_{-\underline{s}}(x)$ at x with respect to the inner product. Namely, $\mathcal{I}^{\mathcal{V}}_{\underline{s}}(x)$ is an element of $\mathcal{Z}_{\mathcal{V}}$ for which:

$$(\mathcal{I}^{\mathcal{V}}_{\underline{s}}(x)|y) = -\left(\frac{d}{dt}\right)_{t=0} \log \Delta_{-\underline{s}}(x+ty) \qquad (y \in \mathcal{Z}_{\mathcal{V}}).$$

Similarly, $\mathcal{J}^{\mathcal{V}}_{\underline{s}}(\xi) := -\nabla \log \delta_{-\underline{s}}(\xi)$ is defined for $\xi \in \mathcal{P}^*_{\mathcal{V}}$. If $q_1 > 0$, then for any $B \in \mathcal{W}$, we have:

$$\mathcal{I}^{\mathcal{V}}_{\underline{s}} \circ \tau_B = \tau_B^* \circ \mathcal{I}^{\mathcal{V}}_{\underline{s}}, \tag{41}$$

$$\mathcal{J}^{\mathcal{V}}_{\underline{s}} \circ \tau_B^* = \tau_B \circ \mathcal{J}^{\mathcal{V}}_{\underline{s}} \tag{42}$$

owing to (30) and (34), respectively.

Theorem 5. *For any* $\underline{s} \in \mathbb{R}^r_{>0}$, *the map* $\mathcal{I}^{\mathcal{V}}_{\underline{s}} : \mathcal{P}_{\mathcal{V}} \to \mathcal{Z}_{\mathcal{V}}$ *gives a diffeomorphism from* $\mathcal{P}_{\mathcal{V}}$ *onto* $\mathcal{P}^*_{\mathcal{V}}$, *and* $\mathcal{J}^{\mathcal{V}}_{\underline{s}}$ *gives the inverse map.*

Proof. We shall prove the statement by induction on the rank. When $r = 1$, we have $\mathcal{I}^{\mathcal{V}}_{\underline{s}}(x_{11}I_{n_1}) = \frac{s_1}{x_{11}}I_{n_1} = \mathcal{J}^{\mathcal{V}}_{\underline{s}}(x_{11}I_{n_1})$ for $x_{11} > 0$. Thus, the statement is true in this case.

When $r > 1$, assume that the statement holds for the system of rank $r-1$. Let $\mathcal{Z}^0_{\mathcal{V}}$ be the subspace of $\mathcal{Z}_{\mathcal{V}}$ defined by:

$$\mathcal{Z}^0_{\mathcal{V}} := \left\{ \begin{pmatrix} x_{11}I_{n_1} & 0 \\ 0 & x' \end{pmatrix} ; x_{11} \in \mathbb{R},\ x' \in \mathcal{Z}_{\mathcal{V}'} \right\}.$$

By direct computation with (31) and (33), we have:

$$\mathcal{I}^{\mathcal{V}}_{\underline{s}} \begin{pmatrix} x_{11}I_{n_1} & 0 \\ 0 & x' \end{pmatrix} = \begin{pmatrix} \frac{s_1}{x_{11}}I_{n_1} & 0 \\ 0 & \mathcal{I}^{\mathcal{V}'}_{\underline{s}'}(x') \end{pmatrix}, \tag{43}$$

$$\mathcal{J}^{\mathcal{V}}_{\underline{s}} \begin{pmatrix} \xi_{11}I_{n_1} & 0 \\ 0 & \xi' \end{pmatrix} = \begin{pmatrix} \frac{s_1}{\xi_{11}}I_{n_1} & 0 \\ 0 & \mathcal{J}^{\mathcal{V}'}_{\underline{s}'}(\xi') \end{pmatrix} \tag{44}$$

for $x_{11}, \xi_{11} > 0$, $x' \in \mathcal{P}_{\mathcal{V}'}$ and $\xi' \in \mathcal{P}^*_{\mathcal{V}'}$. By the induction hypothesis, we see that $\mathcal{I}^{\mathcal{V}}_{\underline{s}} : \mathcal{P}_{\mathcal{V}} \cap \mathcal{Z}^0_{\mathcal{V}} \to \mathcal{P}^*_{\mathcal{V}} \cap \mathcal{Z}^0_{\mathcal{V}}$ is bijective with the inverse map $\mathcal{J}^{\mathcal{V}}_{\underline{s}} : \mathcal{P}^*_{\mathcal{V}} \cap \mathcal{Z}^0_{\mathcal{V}} \to \mathcal{P}_{\mathcal{V}} \cap \mathcal{Z}^0_{\mathcal{V}}$.

If $q_1 = 0$, the statement holds because $\mathcal{Z}_{\mathcal{V}} = \mathcal{Z}^0_{\mathcal{V}}$. Assume $q_1 > 0$. Lemma 1 (ii) tells us that, for $x \in \mathcal{P}_{\mathcal{V}}$, there exist unique $x^0 \in \mathcal{Z}^0_{\mathcal{V}} \cap \mathcal{P}_{\mathcal{V}}$ and $B \in \mathcal{W}$ for which $x = \tau_B x^0$. Similarly, we see from (32) that, for $\xi \in \mathcal{P}^*_{\mathcal{V}}$, there exist unique $\xi^0 \in \mathcal{Z}^0_{\mathcal{V}} \cap \mathcal{P}^*_{\mathcal{V}}$ and $C \in \mathcal{W}$ for which $\xi = \tau^*_C \xi^0$. Therefore, we deduce from (41) and (42) that $\mathcal{I}^{\mathcal{V}}_{\underline{s}} : \mathcal{P}_{\mathcal{V}} \mapsto \mathcal{P}^*_{\mathcal{V}}$ is a bijection with $\mathcal{J}^{\mathcal{V}}_{\underline{s}}$ the inverse map. □

Proposition 1. *Let* $\underline{s} \in \mathbb{R}^r_{>0}$. *For* $\xi \in \mathcal{P}^*_{\mathcal{V}}$, *one has:*

$$\Delta_{-\underline{s}}(\mathcal{J}_{\underline{s}}(\xi))^{-1} = (\prod_{k=1}^{r} s_k^{s_k})\delta_{-\underline{s}}(\xi). \tag{45}$$

Proof. We prove the statement by induction on the rank. When $r = 1$, the equality (45) is verified directly. Indeed, the left-hand side of (45) is computed as $(\frac{s_1}{\xi_{11}})^{s_1} = s_1^{s_1}\xi_{11}^{-s_1}$.

When $r > 1$, assume that (45) holds for a system of rank $r - 1$. We deduce from (31), (33), (43), (44) and the induction hypothesis that (45) holds for $\xi \in \mathcal{P}^*_{\mathcal{V}} \cap \mathcal{Z}^0_{\mathcal{V}}$. Therefore, (45) holds for all $\xi \in \mathcal{P}^*_{\mathcal{V}}$ by (30), (34) and (42). □

In general, for a non-zero function f, the function $\frac{1}{f \circ (\nabla \log f)^{-1}}$ is called the multiplicative Legendre transform of f. Thanks to Theorem 5 and Proposition 1, we see that the multiplicative Legendre transform of $\Delta_{-\underline{s}}(x)$ is equal to $\delta_{-\underline{s}}(-\xi)$ on $-\mathcal{P}^*_{\mathcal{V}}$ up to constant multiple. As a corollary, we arrive at the following result.

Theorem 6. *The Fenchel–Legendre transform of the convex function* $\log \Delta_{-\underline{s}}$ *on* $\mathcal{P}_{\mathcal{V}}$ *is equal to the function* $\log \delta_{-\underline{s}}(-\xi)$ *of* $\xi \in -\mathcal{P}^*$ *up to constant addition.*

6. Application to Statistics and Optimization

Take $\underline{s} \in \mathbb{R}^r$ for which $s_k > q_k/2$ $(k = 1, \ldots, r)$. We define a measure $\rho^{\mathcal{V}}_{\underline{s}}$ on $\mathcal{P}^*_{\mathcal{V}}$ by:

$$\rho^{\mathcal{V}}_{\underline{s}}(d\xi) := C_{\mathcal{V}}^{-1}\Gamma_{\mathcal{V}}(\underline{s})^{-1}\delta^{\mathcal{V}}_{\underline{s}}(\xi)\varphi_{\mathcal{V}}(\xi)\,d\xi \qquad (\xi \in \mathcal{P}^*_{\mathcal{V}}). \tag{46}$$

Theorem 4 states that:

$$\int_{\mathcal{P}^*_{\mathcal{V}}} e^{-(x|\xi)}\rho^{\mathcal{V}}_{\underline{s}}(d\xi) = \Delta^{\mathcal{V}}_{-\underline{s}}(x) \qquad (x \in \mathcal{P}_{\mathcal{V}}).$$

Then, we obtain the natural exponential family generated by $\rho^{\mathcal{V}}_{\underline{s}}$, that is a family $\{\mu^{\mathcal{V}}_{\underline{s},x}\}_{x \in \mathcal{P}_{\mathcal{V}}}$ of probability measures on $\mathcal{P}^*_{\mathcal{V}}$ given by:

$$\mu^{\mathcal{V}}_{\underline{s},x}(d\xi) := \Delta^{\mathcal{V}}_{\underline{s}}(x)e^{-(x|\xi)}\rho^{\mathcal{V}}_{\underline{s}}(d\xi).$$

In particular, when $\underline{s} = (n_1\alpha, n_2\alpha, \ldots, n_r\alpha)$ for sufficiently large α, we have $\mu^{\mathcal{V}}_{\underline{s},x}(d\xi) = (\det x)^{\alpha}e^{-(x|\xi)}\rho^{\mathcal{V}}_{\underline{s}}(d\xi)$. We call $\mu^{\mathcal{V}}_{\underline{s},x}$ the Wishart distributions on $\mathcal{P}^*_{\mathcal{V}}$ in general.

From a sample $\xi_0 \in \mathcal{P}^*_{\mathcal{V}}$, let us estimate the parameter $x \in \mathcal{P}_{\mathcal{V}}$ in such a way that the likelihood function $\Delta^{\mathcal{V}}_{\underline{s}}(x)e^{-(x|\xi)}$ attains its maximum at the estimator x_0. Then, we have the likelihood equation $\xi_0 = \mathcal{I}^{\mathcal{V}}_{\underline{s}}(x_0)$, whereas Theorem 5 gives a unique solution by $x_0 = \mathcal{J}^{\mathcal{V}}_{\underline{s}}(\xi_0)$.

The same argument leads us to the following result in semidefinite programming. For a fixed $\xi_0 \in \mathcal{P}^*_{\mathcal{V}}$ and $\alpha > 0$, a unique solution x_0 of the minimization problem of $(x|\xi_0) - \alpha \log \det x$ subject to $x \in \mathcal{P}_{\mathcal{V}} = \mathcal{Z}_{\mathcal{V}} \cap \mathcal{P}_n$ is given by $x_0 = \mathcal{J}^{\mathcal{V}}_{\underline{s}}(\xi_0)$, where $\underline{s} = (n_1\alpha, \ldots, n_r\alpha)$. Note that $\mathcal{J}^{\mathcal{V}}_{\underline{s}}$ is a rational map because $\delta^{\mathcal{V}}_{\underline{s}}$ is a product of powers of rational functions.

7. Special Cases

7.1. Matrix Realization of Homogeneous Cones

Let us assume that the system $\mathcal{V} = \{\mathcal{V}_{lk}\}_{1 \le k < l \le r}$ satisfies not only the conditions (V1) and (V2), but also the following:

(V3) $A \in \mathcal{V}_{lk}$, $B \in \mathcal{V}_{kj} \Rightarrow AB \in \mathcal{V}_{lj}$ $\quad (1 \le j < k < l \le r)$.

Then, the set $H_{\mathcal{V}}$ of lower triangular matrices T of the form:

$$T = \begin{pmatrix} T_{11} & & & \\ T_{21} & T_{22} & & \\ \vdots & & \ddots & \\ T_{r1} & T_{r2} & \dots & T_{rr} \end{pmatrix}$$

becomes a linear Lie group, and $H_{\mathcal{V}}$ acts on the space $\mathcal{Z}_{\mathcal{V}}$ by $\rho(T)x := Tx\,{}^{t}T$ $(T \in H_{\mathcal{V}}, x \in \mathcal{Z}_{\mathcal{V}})$. The group $H_{\mathcal{V}}$ acts on the cone $\mathcal{P}_{\mathcal{V}}$ simply transitively by this action ρ, so that $\mathcal{P}_{\mathcal{V}}$ is a homogeneous cone. Moreover, it is shown in [15] that every homogeneous cone is linearly isomorphic to such $\mathcal{P}_{\mathcal{V}}$ (see also [18]).

Let $\mathcal{V}^0 = \{\mathcal{V}^0_{lk}\}_{1 \le k < l \le 3}$ be the system given by $\mathcal{V}^0_{21} = \{0\}$ and $\mathcal{V}^0_{lk} = \mathbb{R}$ $((l,k) \neq (2,1))$. Then:

$$\mathcal{Z}_{\mathcal{V}^0} = \left\{ \begin{pmatrix} x_1 & 0 & x_4 \\ 0 & x_2 & x_5 \\ x_4 & x_5 & x_3 \end{pmatrix} ; x_1, \dots, x_5 \in \mathbb{R} \right\},$$

and $\mathcal{P}_{\mathcal{V}^0} := \mathcal{Z}_{\mathcal{V}^0} \cap \mathcal{P}_3$ is homogeneous because (V1)–(V3) are satisfied in this case. On the other hand, let $\mathcal{V}^1 = \{\mathcal{V}^1_{lk}\}_{1 \le k < l \le 3}$ be the system given by $\mathcal{V}^1_{31} = \{0\}$ and $\mathcal{V}^1_{lk} = \mathbb{R}$ $((l,k) \neq (3,1))$. Then:

$$\mathcal{Z}_{\mathcal{V}^1} = \left\{ \begin{pmatrix} x_1 & x_4 & 0 \\ x_4 & x_2 & x_5 \\ 0 & x_5 & x_3 \end{pmatrix} ; x_1, \dots, x_5 \in \mathbb{R} \right\}.$$

Note that $\mathcal{V}^1$ satisfies only (V1) and (V2), but $\mathcal{P}_{\mathcal{V}^1}$ is homogeneous because $\mathcal{P}_{\mathcal{V}^1}$ is isomorphic to the homogeneous cone $\mathcal{P}_{\mathcal{V}^0}$ via the map:

$$\mathcal{P}_{\mathcal{V}^1} \ni \begin{pmatrix} x_1 & x_4 & 0 \\ x_4 & x_2 & x_5 \\ 0 & x_5 & x_3 \end{pmatrix} \mapsto \begin{pmatrix} 1 & 0 & 0 \\ 0 & 0 & 1 \\ 0 & 1 & 0 \end{pmatrix} \begin{pmatrix} x_1 & x_4 & 0 \\ x_4 & x_2 & x_5 \\ 0 & x_5 & x_3 \end{pmatrix} \begin{pmatrix} 1 & 0 & 0 \\ 0 & 0 & 1 \\ 0 & 1 & 0 \end{pmatrix} = \begin{pmatrix} x_1 & 0 & x_4 \\ 0 & x_3 & x_5 \\ x_4 & x_5 & x_2 \end{pmatrix} \in \mathcal{P}_{\mathcal{V}^0}.$$

This example tells us that our matrix realization of a convex cone is not unique and that the condition (V3) is merely a sufficient condition for the homogeneity of the cone.

Many ideas in this work are inspired by the theory of homogeneous cones. The notion of generalized power functions, as well as the Γ-type integral formulas are due to Gindikin [8] (see also [23]). The Wishart distributions for homogeneous cones are studied in [17,21,24,25].

7.2. Cones Associated with Chordal Graphs

If $n_1 = n_2 = \cdots = n_r = 1$, then $\mathcal{V}_{lk}$ equals either $\mathbb{R}$ or $\{0\}$. In this case, $\mathcal{Z}_{\mathcal{V}}$ is the space of symmetric matrices with prescribed zero components. Such a space is described by using an undirected graph in the graphical model theory.

Let us recall some notion in the graph theory. Let G be a graph and V_G the set of vertices of G. We assume that G has no multiple edge, that is, for any two vertices $i, j \in V_G$, either there is one edge connecting them or there is no edge between them. These relations of the vertices i and j are denoted by $i \sim j$ and $i \not\sim j$, respectively. Assume further that G has no loop, which means that $i \not\sim i$ for $i \in V_G$. We define the edge set $E_G \subset V_G \times V_G$ by:

$$E_G := \{ (i,j) \in V_G \times V_G ; i \sim j \}.$$

Since V_G and E_G have all of the information of G, the graph G is often identified with the pair (V_G, E_G). For a non-empty subset V' of V_G, put $E' := E_G \cap (V' \times V')$. The graph $G' := (V', E')$ is called an induced subgraph of G. The graph G is said to be chordal or decomposable if G contains no cycle of length greater than three as an induced subgraph, and said to be A_4-free if G contains no A_4 graph $\bullet - \bullet - \bullet - \bullet$ as an induced subgraph. Let $\preceq$ be a total order on the vertex set V_G, and for $i \in V_G$, put $V_G^{[i]} := \{ j \in V_G \,;\, i \sim j \text{ and } i \preceq j \} \subset V_G$. Then, $\preceq$ is said to be an eliminating order on the graph G if the induced subgraph with the vertex set $V_G^{[i]}$ is complete for each $i \in V_G$. It is known that there exists an eliminating order on G if and only if the graph G is chordal.

Let us identify the vertex set V_G with $\{1, 2, \dots, r\}$. Let $\mathcal{Z}_G$ be the space of symmetric matrices $x = (x_{ij}) \in \mathrm{Sym}(r, \mathbb{R})$, such that, if $i \neq j$ and $i \not\sim j$, then $x_{ij} = 0$. Define $\mathcal{P}_G := \mathcal{Z}_G \cap \mathcal{P}_r$. We can show ([11] (Theorem 2.2), [26]) that the cone $\mathcal{P}_G$ is homogeneous if and only if the graph G is chordal and A_4-free. On the other hand, it is known in the graphical model theory as well as the sparse matrix linear algebra that even though $\mathcal{P}_G$ is not homogeneous, various formulas still hold for $\mathcal{P}_G$ if G is chordal.

The cone $\mathcal{P}_G$ is expressed as $\mathcal{P}_{\mathcal{V}}$ with $n_1 = n_2 = \cdots = n_r = 1$ and:

$$\mathcal{V}_{ji} = \begin{cases} \mathbb{R} & (j \sim i), \\ \{0\} & (j \not\sim i). \end{cases}$$

Then, the condition (V2) means exactly that the order $\leq$ is an eliminating order on G. Therefore, any cone $\mathcal{P}_G$ with chordal G can be treated as $\mathcal{P}_{\mathcal{V}}$ in our framework. Most of the integral formulas for $\mathcal{P}_G$ in [11,27] can be deduced from Theorems 3 and 4, while the Wishart distribution is a central object in the theory of graphical model. In [28], the analysis for generalized power functions associated with all eliminating orders is discussed for a specific graph $A_n : \bullet - \bullet - \cdots - \bullet$ by direct computations.

Acknowledgments: The author would like to express sincere gratitude to Piotr Graczyk and Yoshihiko Konno for stimulating discussions about this subject. He is also grateful to Frédéric Barbaresco for his interest in and encouragement of this work. He thanks to anonymous referees for valuable comments, which were helpful for the improvement of the present paper. This work was supported by JSPS KAKENHI Grant Number 16K05174.

Conflicts of Interest: The authors declare no conflict of interest.

References

1. Koszul, J.L. Ouverts convexes homogènes des espaces affines. *Math. Z.* **1962**, *79*, 254–259.
2. Vinberg, E.B. The theory of convex homogeneous cones. *Trans. Moscow Math. Soc.* **1963**, *12*, 340–403.
3. Vey, J. Sur les automorphismes affines des ouverts convexes saillants. *Annali della Scuola Normale Superiore di Pisa* **1970**, *24*, 641–665.
4. Shima, H. *The Geometry of Hessian Structures*; World Scientific: Hackensack, NJ, USA, 2007.
5. Nesterov, Y.; Nemirovskii, A. *Interior-Point Polynomial Algorithms in Convex Programming*; Society for Industrial and Applied Mathematics: Philadelphia, PA, USA, 1994.
6. Barbaresco, F. Koszul information geometry and Souriau geometric temperature/capacity of Lie group thermodynamics. *Entropy* **2014**, *16*, 4521–4565.
7. Barbaresco, F. Symplectic structure of information geometry: Fisher etric and Euler-Poincaré equation of Souriau Lie group thermodynamics. In *Geometric Science of Information*; Nielsen, F., Barbaresco, F., Eds.; (Lecture Notes in Computer Science); Springer International Publishing: Basel, Switzerland, 2015; Volume 9389, pp. 529–540.
8. Gindikin, S.G. Analysis in homogeneous domains. *Russ. Math. Surv.* **1964**, *19*, 1–89.
9. Truong, V.A.; Tunçel, L. Geometry of homogeneous convex cones, duality mapping, and optimal self-concordant barriers. *Math. Program.* **2004**, *100*, 295–316.
10. Tunçel, L.; Xu, S. On homogeneous convex cones, the Caratheodory number, and the duality mapping. *Math. Oper. Res.* **2001**, *26*, 234–247.
11. Letac, G.; Massam, H. Wishart distributions for decomposable graphs. *Ann. Stat.* **2007**, *35*, 1278–1323.
12. Rothaus, O.S. The construction of homogeneous convex cones. *Ann. Math.* **1966**, *83*, 358–376.
13. Xu, Y.-C. *Theory of Complex Homogeneous Bounded Domains*; Kluwer: Dordrecht, The Netherlands, 2005.

14. Chua, C.B. Relating homogeneous cones and positive definite cones via T-algebras. *SIAM J. Optim.* **2003**, *14*, 500–506.
15. Ishi, H. On symplectic representations of normal j-algebras and their application to Xu's realizations of Siegel domains. *Differ. Geom. Appl.* **2006**, *24*, 588–612.
16. Yamasaki, T.; Nomura, T. Realization of homogeneous cones through oriented graphs. *Kyushu J. Math.* **2015**, *69*, 11–48.
17. Ishi, H. Homogeneous cones and their applications to statistics. In *Modern Methods of Multivariate Statistics;* Graczyk, P., Hassairi, A, Eds.; Hermann: Paris, France, 2014; pp. 135–154.
18. Ishi, H. Matrix realization of homogeneous cones. In *Geometric Science of Information;* Nielsen, F., Barbaresco, F., Eds.; (Lecture Notes in Computer Science); Springer International Publishing: Basel, Switzerland, 2015; Volume 9389, pp. 248–256.
19. Lauritzen, S.L. *Graphical Models;* Clarendon Press: Oxford, UK, 1996.
20. Faraut, J.; Korányi, A. *Analysis on Symmetric Cones;* Clarendon Press: Oxford, UK; 1994.
21. Graczyk, P.; Ishi, H. Riesz measures and Wishart laws associated with quadratic maps. *J. Math. Soc. Jpn.* **2014**, *66*, 317–348.
22. Shima, H. Homogeneous Hessian manifolds. *Ann. Inst. Fourier* **1980**, *30*, 91–128.
23. Güler, O.; Tunçel, L. Characterization of the barrier parameter of homogeneous convex cones. *Math. Program. A* **1988**, *81*, 55–76.
24. Andersson, S.A.; Wojnar, G.G. Wishart distributions on homogeneous cones. *J. Theor. Probab.* **2004**, *17*, 781–818.
25. Graczyk, P.; Ishi, H.; Kołodziejek, B. Wishart exponential families and variance function on homogeneous cones. *HAL* **2016**, submitted for publication.
26. Ishi, H. On a class of homogeneous cones consisting of real symmetric matrices. *Josai Math. Monogr.* **2013**, *6*, 71–80.
27. Roverato, A. Cholesky decomposition of a hyper inverse Wishart matrix. *Biometrika* **2000**, *87*, 99–112.
28. Graczyk, P.; Ishi, H.; Mamane, S. Wishart exponential families on cones related to A_n graphs. *HAL* **2016**, submitted for publication.

Chapter 3:
Divergence Geometry and Information Geometry

entropy

MDPI

Article

A Proximal Point Algorithm for Minimum Divergence Estimators with Application to Mixture Models †

Diaa Al Mohamad * and Michel Broniatowski

Laboratoire de Statistique Théorique et Appliquée, Université Pierre et Marie CURIE, 4 place Jussieu, 75005 Paris, France; michel.broniatowski@upmc.fr

* Correspondence: diaa.almohamad@gmail.com; Tel.: +33-7-62-59-17-73

† This paper is an extended version of our paper published in the 2nd Conference on Geometric Science of Information, Palaiseau, France, 28–30 October 2015.

Academic Editors: Frédéric Barbaresco and Frank Nielsen

Received: 11 June 2016; Accepted: 21 July 2016; Published: 27 July 2016

Abstract: Estimators derived from a divergence criterion such as $\varphi-$divergences are generally more robust than the maximum likelihood ones. We are interested in particular in the so-called minimum dual φ–divergence estimator (MDφDE), an estimator built using a dual representation of φ–divergences. We present in this paper an iterative proximal point algorithm that permits the calculation of such an estimator. The algorithm contains by construction the well-known Expectation Maximization (EM) algorithm. Our work is based on the paper of Tseng on the likelihood function. We provide some convergence properties by adapting the ideas of Tseng. We improve Tseng's results by relaxing the identifiability condition on the proximal term, a condition which is not verified for most mixture models and is hard to be verified for "non mixture" ones. Convergence of the EM algorithm in a two-component Gaussian mixture is discussed in the spirit of our approach. Several experimental results on mixture models are provided to confirm the validity of the approach.

Keywords: φ–divergences; robust estimation; EM algorithm; proximal-point algorithms; mixture models

1. Introduction

The Expectation Maximization (EM) algorithm is a well-known method for calculating the maximum likelihood estimator of a model where incomplete data is considered. For example, when working with mixture models in the context of clustering, the labels or classes of observations are unknown during the training phase. Several variants of the EM algorithm were proposed (see [1]). Another way to look at the EM algorithm is as a proximal point problem (see [2,3]). Indeed, one may rewrite the conditional expectation of the complete log-likelihood as a sum of the log-likelihood function and a distance-like function over the conditional densities of the labels provided an observation. Generally, the proximal term has a regularization effect in the sense that a proximal point algorithm is more stable and frequently outperforms classical optimization algorithms (see [4]). Chrétien and Hero [5] prove superlinear convergence of a proximal point algorithm derived from the EM algorithm. Notice that EM-type algorithms usually enjoy no more than linear convergence.

Taking into consideration the need for robust estimators, and the fact that the maximum likelihood estimator (MLE) is the least robust estimator among the class of divergence-type estimators that we present below, we generalize the EM algorithm (and the version of Tseng [2]) by replacing the log-likelihood function by an estimator of a $\varphi-$divergence between the *true distribution* of the data and the model. A φ–divergence in the sense of Csiszár [6] is defined in the same way as [7] by:

$$D_{\varphi}(Q,P) = \int \varphi\left(\frac{dQ}{dP}(y)\right) dP(y),$$

where φ is a nonnegative strictly convex function. Examples of such divergences are: the Kullback–Leibler (KL) divergence , the modified KL divergence, the Hellinger distanceamong others. All these well-known divergences belong to the class of Cressie-Read functions [8] defined by

$$\varphi_{\gamma}(x) = \frac{x^{\gamma} - \gamma x + \gamma - 1}{\gamma(\gamma-1)} \text{ for } \gamma \in \mathbb{R} \setminus \{0,1\}. \tag{1}$$

for $\gamma = \frac{1}{2}, 0, 1$ respectively. For $\gamma \in \{0,1\}$, the limit is calculated, and we denote $\varphi_0(x) = -\log x + x - 1$ for the case of the modified KL and $\varphi_1(x) = x \log x - x + 1$ for the KL.

Since the $\varphi-$divergence calculus uses the unknown true distribution, we need to estimate it. We consider the dual estimator of the divergence introduced independently by [9,10]. The use of this estimator is motivated by many reasons. Its minimum coincides with the MLE for $\varphi(t) = -\log(t) + t - 1$. In addition, it has the same form for discrete and continuous models, and does not consider any partitioning or smoothing.

Let $(P_{\phi})_{\phi \in \Phi}$ be a parametric model with $\Phi \subset \mathbb{R}^d$, and denote ϕ^T as the *true* set of parameters. Let dy be the Lebesgue measure defined on $\mathbb{R}$. Suppose that $\forall \phi \in \Phi$, the probability measure P_{ϕ} is absolutely continuous with respect to dy and denote p_{ϕ} the corresponding probability density. The dual estimator of the $\varphi-$divergence given an $n-$sample $y_1, \cdots, y_n$ is given by:

$$\hat{D}_{\varphi}(p_{\phi}, p_{\phi_T}) = \sup_{\alpha \in \Phi} \int \varphi'\left(\frac{p_{\phi}}{p_{\alpha}}\right)(x) p_{\phi}(x) dx - \frac{1}{n}\sum_{i=1}^{n} \varphi^{\#}\left(\frac{p_{\phi}}{p_{\alpha}}\right)(y_i), \tag{2}$$

with $\varphi^{\#}(t) = t\varphi'(t) - \varphi(t)$. Al Mohamad [11] argues that this formula works well under the model; however, when we are not, this quantity largely underestimates the divergence between the true distribution and the model, and proposes the following modification:

$$\tilde{D}_{\varphi}(p_{\phi}, p_{\phi_T}) = \int \varphi'\left(\frac{p_{\phi}}{K_{n,w}}\right)(x) p_{\phi}(x) dx - \frac{1}{n}\sum_{i=1}^{n} \varphi^{\#}\left(\frac{p_{\phi}}{K_{n,w}}\right)(y_i), \tag{3}$$

where $K_{n,w}$ is the Rosenblatt–Parzen kernel estimate with window parameter w. Whether it is $\hat{D}_{\varphi}$, or $\tilde{D}_{\varphi}$, the minimum dual $\varphi-$divergence estimator (MDφDE) is defined as the argument of the infimum of the dual approximation:

$$\hat{\phi}_n = \arg\inf_{\phi \in \Phi} \hat{D}_{\varphi}(p_{\phi}, p_{\phi_T}), \tag{4}$$

$$\tilde{\phi}_n = \arg\inf_{\phi \in \Phi} \tilde{D}_{\varphi}(p_{\phi}, p_{\phi_T}). \tag{5}$$

Asymptotic properties and consistency of these two estimators can be found in [7,11]. Robustness properties were also studied using the influence function approach in [11,12]. The kernel-based MDφDE (5) seems to be a *better* estimator than the classical MDφDE (4) in the sense that the former is robust whereas the later is generally not. Under the model, the estimator given by (4) is, however, more efficient, especially when the true density of the data is unbounded. More investigation is needed in the context of unbounded densities, since we may use asymmetric kernels in order to improve the efficiency of the kernel-based MDφDE, see [11] for more details.

In this paper, we propose calculation of the MDφDE using an iterative procedure based on the work of Tseng [2] on the log-likelihood function. This procedure has the form of a proximal point algorithm, and extends the EM algorithm. Our convergence proof demands some regularity (continuity and differentiability) of the estimated divergence with respect to the parameter vector ϕ) which is not simply checked using (2). Recent results in the book of Rockafellar and Wets [13] provide sufficient conditions to prove continuity and differentiability of supremal functions of the form of (2) with respect to ϕ. Differentiability with respect to ϕ still remains a very hard task; therefore, our results cover cases when the objective function is not differentiable.

The paper is organized as follows: in Section 2, we present the general context. We also present the derivation of our algorithm from the EM algorithm and passing by Tseng's generalization. In Section 3, we present some convergence properties. We discuss in Section 4 a variant of the algorithm with a theoretical global infimum, and an example of the two-Gaussian mixture model and a convergence proof of the EM algorithm in the spirit of our approach. Finally, Section 5 contains simulations confirming our claim about the efficiency and the robustness of our approach in comparison with the MLE. The algorithm is also applied to the so-called minimum density power divergence (MDPD) introduced by [14].

2. A Description of the Algorithm

2.1. General Context and Notations

Let (X, Y) be a couple of random variables with joint probability density function $f(x, y|\phi)$ parametrized by a vector of parameters $\phi \in \Phi \subset \mathbb{R}^d$. Let $(X_1, Y_1), \cdots, (X_n, Y_n)$ be n copies of (X, Y) independently and identically distributed. Finally, let $(x_1, y_1), \cdots, (x_n, y_n)$ be n realizations of the n copies of (X, Y). The x_is are the unobserved data (labels) and the y_is are the observations. The vector of parameters ϕ is unknown and needs to be estimated. The observed data y_i are supposed to be real numbers, and the labels x_i belong to a space $\mathcal{X}$ not necessarily finite unless mentioned otherwise. The marginal density of the observed data is given by $p_\phi(y) = \int f(x, y|\phi)dx$, where dx is a measure defined on the label space (for example, the counting measure if we work with mixture models).

For a parametrized function f with a parameter a, we write $f(x|a)$. We use the notation ϕ^k for sequences with the index above. The derivatives of a real valued function ψ defined on $\mathbb{R}$ are denoted ψ', ψ'', etc. We denote ∇f the gradient of a real function f defined on $\mathbb{R}^d$. For a generic function of two (vectorial) arguments $D(\phi|\theta)$, then $\nabla_1 D(\phi|\theta)$ denotes the gradient with respect to the first (vectorial) variable. Finally, for any set A, we use $int(A)$ to denote the interior of A.

2.2. EM Algorithm and Tseng's Generalization

The EM algorithm estimates the unknown parameter vector by (see [15]):

$$\phi^{k+1} = \arg\max_{\Phi} \mathbb{E}\left[\log(f(\mathbf{X}, \mathbf{Y}|\phi)) \,\middle|\, \mathbf{Y} = \mathbf{y}, \phi^k\right],$$

where $\mathbf{X} = (X_1, \cdots, X_n)$, $\mathbf{Y} = (Y_1, \cdots, Y_n)$ and $\mathbf{y} = (y_1, \cdots, y_n)$. By independence between the couples (X_i, Y_i)'s, the previous iteration may be written as:

$$\begin{aligned}\phi^{k+1} &= \arg\max_{\Phi} \sum_{i=1}^{n} \mathbb{E}\left[\log(f(X_i, Y_i|\phi)) \,\middle|\, Y_i = y_i, \phi^k\right] \\ &= \arg\max_{\Phi} \sum_{i=1}^{n} \int_{\mathcal{X}} \log(f(x, y_i|\phi)) h_i(x|\phi^k) dx, \end{aligned} \tag{6}$$

where $h_i(x|\phi^k) = \frac{f(x, y_i|\phi^k)}{p_{\phi^k}(y_i)}$ is the conditional density of the labels (at step k) provided y_i which we suppose to be positive $dx-$almost everywhere. It is well-known that the EM iterations can be rewritten as a difference between the log-likelihood and a *Kullback–Liebler* distance-like function. Indeed,

$$\begin{aligned}\phi^{k+1} &= \arg\max_{\Phi} \sum_{i=1}^{n} \int_{\mathcal{X}} \log\left(h_i(x|\phi) \times p_\phi(y_i)\right) h_i(x|\phi^k)dx \\ &= \arg\max_{\Phi} \sum_{i=1}^{n} \int_{\mathcal{X}} \log\left(p_\phi(y_i)\right) h_i(x|\phi^k)dx + \sum_{i=1}^{n} \int_{\mathcal{X}} \log\left(h_i(x|\phi)\right) h_i(x|\phi^k)dx \\ &= \arg\max_{\Phi} \sum_{i=1}^{n} \log\left(p_\phi(y_i)\right) + \sum_{i=1}^{n} \int_{\mathcal{X}} \log\left(\frac{h_i(x|\phi)}{h_i(x|\phi^k)}\right) h_i(x|\phi^k)dx \\ &\quad + \sum_{i=1}^{n} \int_{\mathcal{X}} \log\left(h_i(x|\phi^k)\right) h_i(x|\phi^k)dx.\end{aligned}$$

The final line is justified by the fact that $h_i(x|\phi)$ is a density, therefore it integrates to 1. The additional term does not depend on ϕ and, hence, can be omitted. We now have the following iterative procedure:

$$\phi^{k+1} = \arg\max_{\Phi} \sum_{i=1}^{n} \log\left(p_\phi(y_i|\phi)\right) + \sum_{i=1}^{n} \int_{\mathcal{X}} \log\left(\frac{h_i(x|\phi)}{h_i(x|\phi^k)}\right) h_i(x|\phi^k)dx.$$

The previous iteration has the form of a proximal point maximization of the log-likelihood, i.e., a perturbation of the log-likelihood by a distance-like function defined on the conditional densities of the labels. Tseng [2] generalizes this iteration by allowing any nonnegative convex function ψ to replace the $t \mapsto -\log(t)$ function. Tseng's recurrence is defined by:

$$\phi^{k+1} = \arg\sup_{\phi} J(\phi) - D_\psi(\phi, \phi^k), \tag{7}$$

where J is the log-likelihood function and D_ψ is given by:

$$D_\psi(\phi, \phi^k) = \sum_{i=1}^{n} \int_{\mathcal{X}} \psi\left(\frac{h_i(x|\phi)}{h_i(x|\phi^k)}\right) h_i(x|\phi^k)dx, \tag{8}$$

for any real nonnegative convex function ψ such that $\psi(1) = \psi'(1) = 0$. $D_\psi(\phi_1, \phi_2)$ is nonnegative, and $D_\psi(\phi_1, \phi_2) = 0$ if and only if $\forall i, h_i(x|\phi_1) = h_i(x|\phi_2)$ dx almost everywhere.

2.3. Generalization of Tseng's Algorithm

We use the relationship between maximizing the log-likelihood and minimizing the Kullback–Liebler divergence to generalize the previous algorithm. We, therefore, replace the log-likelihood function by an estimate of a $\varphi-$divergence D_φ between the true distribution and the model. We use the dual estimators of the divergence presented earlier in the introduction (2) or (3), which we denote in the same manner $\hat{D}_\varphi$, unless mentioned otherwise. Our new algorithm is defined by:

$$\phi^{k+1} = \arg\inf_{\phi} \hat{D}_\varphi(p_\phi, p_{\phi_T}) + \frac{1}{n} D_\psi(\phi, \phi^k), \tag{9}$$

where $D_\psi(\phi, \phi^k)$ is defined by (8). When $\varphi(t) = -\log(t) + t - 1$, it is easy to see that we get recurrence (7). Indeed, for the case of (2) we have:

$$\hat{D}_\varphi(p_\phi, p_{\phi_T}) = \sup_{\alpha} \frac{1}{n} \sum_{i=1}^{n} \log(p_\alpha(y_i)) - \frac{1}{n} \sum_{i=1}^{n} \log(p_\phi(y_i)).$$

Using the fact that the first term in $\hat{D}_\varphi(p_\phi, p_{\phi_T})$ does not depend on ϕ, so it does not count in the arg inf defining ϕ^{k+1}, we easily get (7). The same applies for the case of (3). For notational simplicity, from now on, we redefine D_ψ with a normalization by n, i.e.,

$$D_\psi(\phi, \phi^k) = \frac{1}{n}\sum_{i=1}^{n}\int_{\mathcal{X}} \psi\left(\frac{h_i(x|\phi)}{h_i(x|\phi^k)}\right) h_i(x|\phi^k)dx. \tag{10}$$

Hence, our set of algorithms is redefined by:

$$\phi^{k+1} = \arg\inf_{\phi} \hat{D}_\varphi(p_\phi, p_{\phi_T}) + D_\psi(\phi, \phi^k). \tag{11}$$

We will see later that this iteration forces the divergence to decrease and that, under suitable conditions, it converges to a (local) minimum of $\hat{D}_\varphi(p_\phi, p_{\phi_T})$. It results that algorithm (11) being a way to calculate both the MDφDE (4) and the kernel-based MDφDE (5).

3. Some Convergence Properties of ϕ^k

We show here how, according to some possible situations, one may prove convergence of the algorithm defined by (11). Let ϕ^0 be a given initialization, and define

$$\Phi^0 := \{\phi \in \Phi : \hat{D}_\varphi(p_\phi, p_{\phi_T}) \leq \hat{D}_\varphi(p_{\phi^0}, p_{\phi_T})\},$$

which we suppose to be a subset of *int*(Φ). The idea of defining this set in this context is inherited from the paper Wu [16], which provided the first *correct proof* of convergence for the EM algorithm. Before going any further, we recall the following definition of a (generalized) stationary point.

Definition 1. *Let $f : \mathbb{R}^d \to \mathbb{R}$ be a real valued function. If f is differentiable at a point ϕ^* such that $\nabla f(\phi^*) = 0$, we then say that ϕ^* is a stationary point of f. If f is not differentiable at ϕ^* but the subgradient of f at ϕ^*, say $\partial f(\phi^*)$, exists such that $0 \in \partial f(\phi^*)$, then ϕ^* is called a generalized stationary point of f.*

Remark 1. *In the whole paper, the subgradient is defined for any function not necessarily convex (see Definition 8.3) in [13] for more details.*

We will be using the following assumptions:

A0. Functions $\phi \mapsto \hat{D}_\varphi(p_\phi|p_{\phi_T}), D_\psi$ are lower semicontinuous;
A1. Functions $\phi \mapsto \hat{D}_\varphi(p_\phi|p_{\phi_T}), D_\psi$ and $\nabla_1 D_\psi$ are defined and continuous on, respectively, $\Phi, \Phi \times \Phi$ and $\Phi \times \Phi$;
AC. Function $\phi \mapsto \nabla\hat{D}_\varphi(p_\phi|p_{\phi_T})$ is defined and continuous on Φ;
A2. Φ^0 is a compact subset of int(Φ);
A3. $D_\psi(\phi, \bar{\phi}) > 0$ for all $\bar{\phi} \neq \phi \in \Phi$.

Recall also that we suppose that $h_i(x|\phi) > 0, dx - a.e.$ We relax the convexity assumption of function ψ. We only suppose that ψ is nonnegative and $\psi(t) = 0$ iff $t = 1$. In addition, $\psi'(t) = 0$ if $t = 1$.

Continuity and differentiability assumptions of function $\phi \mapsto \hat{D}_\varphi(p_\phi|p_{\phi_T})$ for the case of (3) can be easily checked using Lebesgue theorems. The continuity assumption for the case of (2) can be checked using Theorem 1.17 or Corollary 10.14 in [13]. Differentiability can also be checked using Corollary 10.14 or Theorem 10.31 in the same book. In what concerns D_ψ, continuity and differentiability can be obtained merely by fulfilling Lebesgue theorems conditions. When working with mixture models, we only need the continuity and differentiability of ψ and functions h_i. The later is easily deduced from regularity assumptions on the model. For assumption A2, there is no universal method, see Section 4.2 for an Example. Assumption A3 can be checked using Lemma 2 in [2].

We start the convergence properties by proving that the objective function $\hat{D}_\varphi(p_\phi|p_{\phi_T})$ decreases alongside the the sequence $(\phi^k)_k$, and give a possible set of conditions for the existence of the sequence $(\phi^k)_k$.

Proposition 1. *(a) Assume that the sequence $(\phi^k)_k$ is well defined in Φ, then $\hat{D}_\varphi(p_{\phi^{k+1}}|p_{\phi_T}) \leq \hat{D}_\varphi(p_{\phi^k}|p_{\phi_T})$, and (b) $\forall k, \phi^k \in \Phi^0$. (c) Assume A0 and A2 are verified, then the sequence $(\phi^k)_k$ is defined and bounded. Moreover, the sequence $(\hat{D}_\varphi(p_{\phi^k}|p_{\phi^T}))_k$ converges.*

Proof. We prove (a). We have by definition of the arginf:

$$\hat{D}_\varphi(p_{\phi^{k+1}}, p_{\phi_T}) + D_\psi(\phi^{k+1}, \phi^k) \leq \hat{D}_\varphi(p_{\phi^k}, p_{\phi_T}) + D_\psi(\phi^k, \phi^k).$$

We use the fact that $D_\psi(\phi^k, \phi^k) = 0$ for the right-hand side and that $D_\psi(\phi^{k+1}, \phi^k) \geq 0$ for the left-hand side of the previous inequality. Hence, $\hat{D}_\varphi(p_{\phi^{k+1}}, p_{\phi_T}) \leq \hat{D}_\varphi(p_{\phi^k}, p_{\phi_T})$.

We prove (b) using the decreasing property previously proved in (a). We have by recurrence $\forall k, \hat{D}_\varphi(p_{\phi^{k+1}}, p_{\phi_T}) \leq \hat{D}_\varphi(p_{\phi^k}, p_{\phi_T}) \leq \cdots \leq \hat{D}_\varphi(p_{\phi^0}, p_{\phi_T})$. The result follows directly by definition of Φ^0.

We prove (c) by induction on k. For $k = 0$, clearly ϕ^0 is well defined since we choose it. The choice of the initial point ϕ^0 of the sequence may influence the convergence of the sequence. See the Example of the Gaussian mixture in Section 4.2. Suppose, for some $k \geq 0$, that ϕ^k exists. We prove that the infimum is attained in Φ^0. Let $\phi \in \Phi$ be any vector at which the value of the optimized function has a value less than its value at ϕ^k, i.e., $\hat{D}_\varphi(p_\phi, p_{\phi_T}) + D_\psi(\phi, \phi^k) \leq \hat{D}_\varphi(p_{\phi^k}, p_{\phi_T}) + D_\psi(\phi^k, \phi^k)$. We have:

$$\begin{aligned}
\hat{D}_\varphi(p_\phi, p_{\phi_T}) &\leq \hat{D}_\varphi(p_\phi, p_{\phi_T}) + D_\psi(\phi, \phi^k) \\
&\leq \hat{D}_\varphi(p_{\phi^k}, p_{\phi_T}) + D_\psi(\phi^k, \phi^k) \\
&\leq \hat{D}_\varphi(p_{\phi^k}, p_{\phi_T}) \\
&\leq \hat{D}_\varphi(p_{\phi^0}, p_{\phi_T}).
\end{aligned}$$

The first line follows from the non negativity of D_ψ. As $\hat{D}_\varphi(p_\phi, p_{\phi_T}) \leq \hat{D}_\varphi(p_{\phi^0}, p_{\phi_T})$, then $\phi \in \Phi^0$. Thus, the infimum can be calculated for vectors in Φ^0 instead of Φ. Since Φ^0 is compact and the optimized function is lower semicontinuous (the sum of two lower semicontinuous functions), then the infimum exists and is attained in Φ^0. We may now define ϕ^{k+1} to be a vector whose corresponding value is equal to the infimum.

Convergence of the sequence $(\hat{D}_\varphi(p_{\phi^k}, p_{\phi_T}))_k$ comes from the fact that it is non increasing and bounded. It is non increasing by virtue of (a). Boundedness comes from the lower semicontinuity of $\phi \mapsto \hat{D}_\varphi(p_\phi, p_{\phi_T})$. Indeed, $\forall k, \hat{D}_\varphi(p_{\phi^k}, p_{\phi_T}) \geq \inf_{\phi \in \Phi^0} \hat{D}_\varphi(p_\phi, p_{\phi_T})$. The infimum of a proper lower semicontinuous function on a compact set exists and is attained on this set. Hence, the quantity $\inf_{\phi \in \Phi^0} \hat{D}_\varphi(p_\phi, p_{\phi_T})$ exists and is finite. This ends the proof. □

Compactness in part (c) can be replaced by inf-compactness of function $\phi \mapsto \hat{D}_\varphi(p_\phi|p_{\phi_T})$ and continuity of D_ψ with respect to its first argument. The convergence of the sequence $(\hat{D}_\varphi(\phi^k|\phi_T))_k$ is an interesting property, since, in general, there is no theoretical guarantee, or it is difficult to prove that the whole sequence $(\phi^k)_k$ converges. It may also continue to fluctuate around a minimum. The decrease of the error criterion $\hat{D}_\varphi(\phi^k|\phi_T)$ between two iterations helps us decide when to stop the iterative procedure.

Proposition 2. *Suppose A1 verified, Φ^0 is closed and $\{\phi^{k+1} - \phi^k\} \to 0$.*

(a) If AC is verified, then any limit point of $(\phi^k)_k$ is a stationary point of $\phi \mapsto \hat{D}_\varphi(p_\phi|p_{\phi^T})$;
(b) If AC is dropped, then any limit point of $(\phi^k)_k$ is a "generalized" stationary point of $\phi \mapsto \hat{D}_\varphi(p_\phi|p_{\phi^T})$, i.e., zero belongs to the subgradient of $\phi \mapsto \hat{D}_\varphi(p_\phi|p_{\phi^T})$ calculated at the limit point.

Proof. We prove (a). Let $(\phi^{n_k})_k$ be a convergent subsequence of $(\phi^k)_k$ which converges to ϕ^∞. First, $\phi^\infty \in \Phi^0$, because Φ^0 is closed and the subsequence (ϕ^{n_k}) is a sequence of elements of Φ^0 (proved in Proposition 1b).

Let us now show that the subsequence (ϕ^{n_k+1}) also converges to ϕ^∞. We simply have:

$$\|\phi^{n_k+1} - \phi^\infty\| \quad \leq \quad \|\phi^{n_k} - \phi^\infty\| + \|\phi^{n_k+1} - \phi^{n_k}\|.$$

Since $\phi^{k+1} - \phi^k \to 0$ and $\phi^{n_k} \to \phi^\infty$, we conclude that $\phi^{n_k+1} \to \phi^\infty$.

By definition of ϕ^{n_k+1}, it verifies the infimum in recurrence (11), so that the gradient of the optimized function is zero:

$$\nabla \hat{D}_\varphi(p_{\phi^{n_k+1}}, p_{\phi_T}) + \nabla D_\psi(\phi^{n_k+1}, \phi^{n_k}) = 0.$$

Using the continuity assumptions A1 and AC of the gradients, one can pass to the limit with no problem:

$$\nabla \hat{D}_\varphi(p_{\phi^\infty}, p_{\phi_T}) + \nabla D_\psi(\phi^\infty, \phi^\infty) = 0.$$

However, the gradient $\nabla D_\psi(\phi^\infty, \phi^\infty) = 0$ because (recall that $\psi'(1) = 0$) for any $\phi \in \Phi$

$$\nabla D_\psi(\phi, \phi) = \sum_{i=1}^{n} \int_{\mathcal{X}} \frac{\nabla h_i(x|\phi)}{h_i(x|\phi)} \psi' \left(\frac{h_i(x|\phi)}{h_i(x|\phi)} \right) h_i(x|\phi) dx = \sum_{i=1}^{n} \int_{\mathcal{X}} \nabla h_i(x|\phi) \psi'(1) dx,$$

which is equal to zero since $\psi'(1) = 0$. This implies that $\nabla \hat{D}_\varphi(p_{\phi^\infty}, p_{\phi_T}) = 0$.

We prove (b). We use again the definition of the arginf. As the optimized function is not necessarily differentiable at the points of the sequence $(\phi^k)_k$, a necessary condition for ϕ^{k+1} to be an infimum is that 0 belongs to the subgradient of the function on ϕ^{k+1}. Since $D_\psi(\phi, \phi^k)$ is assumed to be differentiable, the optimality condition is translated into:

$$-\nabla D_\psi(\phi^{k+1}, \phi^k) \in \partial \hat{D}_\varphi(p_{\phi^{k+1}}, p_{\phi_T}) \quad \forall k.$$

Since $\hat{D}_\varphi(p_\phi, p_{\phi_T})$ is continuous, then its subgradient is outer semicontinuous (see [13] Chapter 8, Proposition 7). We use the same arguments presented in (a) to conclude the existence of two subsequences $(\phi^{n_k})_k$ and $(\phi^{n_k+1})_k$ which converge to the same limit ϕ^∞. By definition of outer semicontinuity, and since $\phi^{n_k+1} \to \phi^\infty$, we have:

$$\limsup_{\phi^{n_k+1} \to \phi^\infty} \partial \hat{D}_\varphi(p_{\phi^{n_k+1}}, p_{\phi_T}) \subset \partial \hat{D}_\varphi(p_{\phi^\infty}, p_{\phi_T}). \tag{12}$$

We want to prove that $0 \in \limsup_{\phi^{n_k+1} \to \phi^\infty} \partial \hat{D}_\varphi(p_{\phi^{n_k+1}}, p_{\phi_T})$. By definition of the (outer) limsup (see [13] Chapter 4, Definition 1 or Chapter 5B):

$$\limsup_{\phi \to \phi^\infty} \partial \hat{D}_\varphi(p_\phi, p_{\phi_T}) = \left\{ u | \exists \phi^k \to \phi^\infty, \exists u^k \to u \text{ with } u^k \in \partial \hat{D}_\varphi(p_{\phi^k}, p_{\phi_T}) \right\}.$$

In our scenario, $\phi = \phi^{n_k+1}$, $\phi^k = \phi^{n_k+1}$, $u = 0$ and $u^k = \nabla_1 D_\psi(\phi^{n_k+1}, \phi^{n_k})$. The continuity of $\nabla_1 D_\psi$ with respect to both arguments and the fact that the two subsequences ϕ^{n_k+1} and ϕ^{n_k} converge to the same limit, imply that $u^k \to \nabla_1 D_\psi(\phi^\infty, \phi^\infty) = 0$. Hence, $u = 0 \in \limsup_{\phi^{n_k+1} \to \phi^\infty} \partial \hat{D}_\varphi(p_{\phi^{n_k+1}}, p_{\phi_T})$. By inclusion (12), we get our result:

$$0 \in \partial \hat{D}_\varphi(p_{\phi^\infty}, p_{\phi_T}).$$

This ends the proof. □

The assumption $\{\phi^{k+1} - \phi^k\} \to 0$ used in Proposition 2 is not easy to be checked unless one has a close formula of ϕ^k. The following proposition gives a method to prove such assumption. This method seems simpler, but it is not verified in many mixture models (see Section 4.2 for a counter Example).

Proposition 3. *Assume that A1, A2 and A3 are verified, then* $\{\phi^{k+1} - \phi^k\} \to 0$. *Thus, by Proposition 2 (according to whether AC is verified or not), any limit point of the sequence* ϕ^k *is a (generalized) stationary point of* $\hat{D}_\varphi(.|\phi_T)$.

Proof. By contradiction, let us suppose that $\phi^{k+1} - \phi^k$ does not converge to 0. There exists a subsequence such that $\|\phi^{N_0(k)+1} - \phi^{N_0(k)}\| > \varepsilon$, $\forall k \geq k_0$. Since $(\phi^k)_k$ belongs to the compact set Φ^0, there exists a convergent subsequence $(\phi^{N_1 \circ N_0(k)})_k$ such that $\phi^{N_1 \circ N_0(k)} \to \bar{\phi}$. The sequence $(\phi^{N_1 \circ N_0(k)+1})_k$ belongs to the compact set Φ^0; therefore, we can extract a further subsequence $(\phi^{N_2 \circ N_1 \circ N_0(k)+1})_k$ such that $\phi^{N_2 \circ N_1 \circ N_0(k)+1} \to \tilde{\phi}$. Besides $\hat{\phi} \neq \tilde{\phi}$. Finally since the sequence $(\phi^{N_1 \circ N_0(k)})_k$ is convergent, a further subsequence also converges to the same limit $\bar{\phi}$. We have proved the existence of a subsequence of $(\phi^k)_k$ such that $\phi^{N(k)+1} - \phi^{N(k)}$ does not converge to 0 and such that $\phi^{N(k)+1} \to \tilde{\phi}$, $\phi^{N(k)} \to \bar{\phi}$ with $\bar{\phi} \neq \tilde{\phi}$.

The real sequence $(\hat{D}_\varphi(p_{\phi^k}, p_{\phi_T}))_k$ converges as proved in Proposition 1c. As a result, both sequences $\hat{D}_\varphi(p_{\phi^{N(k)+1}}, p_{\phi_T})$ and $\hat{D}_\varphi(p_{\phi^{N(k)}}, p_{\phi_T})$ converge to the same limit being subsequences of the same convergent sequence. In the proof of Proposition 1, we can deduce the following inequality:

$$\hat{D}(p_{\phi^{k+1}}, p_{\phi_T}) + D_\psi(\phi^{k+1}, \phi^k) \leq \hat{D}(p_{\phi^k}, p_{\phi_T}), \tag{13}$$

which is also verified for any substitution of k by $N(k)$. By passing to the limit on k, we get $D_\psi(\tilde{\phi}, \bar{\phi}) \leq 0$. However, the distance-like function D_ψ is nonnegative, so that it becomes zero. Using assumption A3, $D_\psi(\tilde{\phi}, \bar{\phi}) = 0$ implies that $\tilde{\phi} = \bar{\phi}$. This contradicts the hypothesis that $\phi^{k+1} - \phi^k$ does not converge to 0.

The second part of the Proposition is a direct result of Proposition 2. □

Corollary 1. *Under assumptions of Proposition 3, the set of accumulation points of* $(\phi^k)_k$ *is a connected compact set. Moreover, if* $\phi \mapsto \hat{D}(p_\phi, p_{\phi_T})$ *is strictly convex in the neighborhood of a limit point of the sequence* $(\phi^k)_k$*, then the whole sequence* $(\phi^k)_k$ *converges to a local minimum of* $\hat{D}(p_\phi, p_{\phi_T})$.

Proof. Since the sequence $(\phi)_k$ is bounded and verifies $\phi^{k+1} - \phi^k \to 0$, then Theorem 28.1 in [17] implies that the set of accumulation points of $(\phi^k)_k$ is a connected compact set. It is not empty since Φ^0 is compact. The remaining of the proof is a direct result of Theorem 3.3.1 from [18]. The strict concavity of the objective function around an accumulation point is replaced here by the strict convexity of the estimated divergence. □

Proposition 3 and Corollary 1 describe what we may hope to get of the sequence ϕ^k. Convergence of the whole sequence is bound by a local convexity assumption in the neighborhood of a limit point. Although simple, this assumption remains difficult to be checked since we do not know where might be the limit points. In addition, assumption A3 is very restrictive, and is not verified in mixture models.

Propositions 2 and 3 were developed for the likelihood function in the paper of Tseng [2]. Similar results for a general class of functions replacing $\hat{D}_\varphi$ and D_ψ which may not be differentiable (but still continuous) are presented in [3]. In these results, assumption A3 is essential. Although in [18] this problem is avoided, their approach demands that the log-likelihood has $-\infty$ limit as $\|\phi\| \to \infty$. This is simply not verified for mixture models. We present a similar method to the one in [18] based on the idea of Tseng [2] of using the set Φ^0 which is valid for mixtures. We lose, however, the guarantee of consecutive decrease of the sequence $(\phi^k)_k$.

Proposition 4. *Assume A1, AC and A2 verified. Any limit point of the sequence* $(\phi^k)_k$ *is a stationary point of* $\phi \to \hat{D}(p_\phi, p_{\phi_T})$. *If AC is dropped, then 0 belongs to the subgradient of* $\phi \mapsto \hat{D}(p_\phi, p_{\phi_T})$ *calculated at the limit point.*

Proof. If $(\phi^k)_k$ converges to, say, ϕ^∞, then the result falls simply from Proposition 2.

If $(\phi^k)_k$ does not converge. Since Φ^0 is compact and $\forall k, \phi^k \in \Phi^0$ (proved in Proposition 1), there exists a subsequence $(\phi^{N_0(k)})_k$ such that $\phi^{N_0(k)} \to \tilde{\phi}$. Let us take the subsequence $(\phi^{N_0(k)-1})_k$. This subsequence does not necessarily converge; it is still contained in the compact Φ^0, so that we can extract a further subsequence $(\phi^{N_1 \circ N_0(k)-1})_k$ which converges to, say, $\bar{\phi}$. Now, the subsequence $(\phi^{N_1 \circ N_0(k)})_k$ converges to $\tilde{\phi}$, because it is a subsequence of $(\phi^{N_0(k)})_k$. We have proved until now the existence of two convergent subsequences $\phi^{N(k)-1}$ and $\phi^{N(k)}$ with a priori different limits. For simplicity and without any loss of generality, we will consider these subsequences to be ϕ^k and ϕ^{k+1}, respectively.

Conserving previous notations, suppose that $\phi^{k+1} \to \tilde{\phi}$ and $\phi^k \to \bar{\phi}$. We use again inequality (13):

$$\hat{D}(p_{\phi^{k+1}}, p_{\phi_T}) + D_\psi(\phi^{k+1}, \phi^k) \leq \hat{D}(p_{\phi^k}, p_{\phi_T}).$$

By taking the limits of the two parts of the inequality as k tends to infinity, and using the continuity of the two functions, we have

$$\hat{D}(p_{\tilde{\phi}}, p_{\phi_T}) + D_\psi(\tilde{\phi}, \bar{\phi}) \leq \hat{D}(p_{\bar{\phi}}, p_{\phi_T}).$$

Recall that under A1-2, the sequence $\left(\hat{D}_\varphi(p_{\phi^k}, p_{\phi_T})\right)_k$ converges, so that it has the same limit for any subsequence, i.e., $\hat{D}(p_{\tilde{\phi}}, p_{\phi_T}) = \hat{D}(p_{\bar{\phi}}, p_{\phi_T})$. We also use the fact that the distance-like function D_ψ is non negative to deduce that $D_\psi(\tilde{\phi}, \bar{\phi}) = 0$. Looking closely at the definition of this divergence (10), we get that if the sum is zero, then each term is also zero since all terms are nonnegative. This means that:

$$\forall i \in \{1, \cdots, n\}, \quad \int_{\mathcal{X}} \psi\left(\frac{h_i(x|\tilde{\phi})}{h_i(x|\bar{\phi})}\right) h_i(x|\bar{\phi}) dx = 0.$$

The integrands are nonnegative functions, so they vanish almost everywhere with respect to the measure dx defined on the space of labels.

$$\forall i \in \{1, \cdots, n\}, \quad \psi\left(\frac{h_i(x|\tilde{\phi})}{h_i(x|\bar{\phi})}\right) h_i(x|\bar{\phi}) = 0 \quad dx - a.e.$$

The conditional densities h_i are supposed to be positive (which can be ensured by a suitable choice of the initial point ϕ^0), i.e., $h_i(x|\bar{\phi}) > 0, dx - a.e.$ Hence, $\psi\left(\frac{h_i(x|\tilde{\phi})}{h_i(x|\bar{\phi})}\right) = 0, dx - a.e.$ On the other hand, ψ is chosen in a way that $\psi(z) = 0$ iff $z = 1$. Therefore:

$$\forall i \in \{1, \cdots, n\}, \quad h_i(x|\tilde{\phi}) = h_i(x|\bar{\phi}) \quad dx - a.e. \tag{14}$$

Since ϕ^{k+1} is, by definition, an infimum of $\phi \mapsto \hat{D}(p_\phi, p_{\phi_T}) + D_\psi(\phi, \phi^k)$, then the gradient of this function is zero on ϕ^{k+1}. It results that:

$$\nabla \hat{D}(p_{\phi^{k+1}}, p_{\phi_T}) + \nabla D_\psi(\phi^{k+1}, \phi^k) = 0, \quad \forall k.$$

Taking the limit on k, and using the continuity of the derivatives, we get that:

$$\nabla \hat{D}(p_{\tilde{\phi}}, p_{\phi_T}) + \nabla D_\psi(\tilde{\phi}, \bar{\phi}) = 0. \tag{15}$$

Let us write explicitly the gradient of the second divergence:

$$\nabla D_\psi(\tilde{\phi}, \bar{\phi}) = \sum_{i=1}^{n} \int_{\mathcal{X}} \frac{\nabla h_i(x|\tilde{\phi})}{h_i(x|\bar{\phi})} \psi'\left(\frac{h_i(x|\tilde{\phi})}{h_i(x|\bar{\phi})}\right) h_i(x|\bar{\phi}).$$

We use now the identities (14), and the fact that $\psi'(1) = 0$, to deduce that:

$$\nabla D_\psi(\tilde{\phi}, \bar{\phi}) = 0.$$

This entails using (15) that $\nabla\hat{D}(p_{\tilde{\phi}}, p_{\phi_T}) = 0$.

Comparing the proved result with the notation considered at the beginning of the proof, we have proved that the limit of the subsequence $(\phi^{N_1 \circ N_0(k)})_k$ is a stationary point of the objective function. Therefore, the final step is to deduce the same result on the original convergent subsequence $(\phi^{N_0(k)})_k$. This is simply due to the fact that $(\phi^{N_1 \circ N_0(k)})_k$ is a subsequence of the convergent sequence $(\phi^{N_0(k)})_k$, hence they have the same limit.

When assumption AC is dropped, similar arguments to those used in the proof of Proposition 2b. are employed. The optimality condition in (11) implies :

$$-\nabla D_{\psi}(\phi^{k+1}, \phi^k) \in \partial \hat{D}_{\varphi}(p_{\phi^{k+1}}, p_{\phi_T}) \quad \forall k.$$

Function $\phi \mapsto \hat{D}_{\varphi}(p_{\phi}, p_{\phi_T})$ is continuous, hence its subgradient is outer semicontinuous and:

$$\limsup_{\phi^{k+1} \to \phi^{\infty}} \partial \hat{D}_{\varphi}(p_{\phi^{k+1}}, p_{\phi_T}) \subset \partial \hat{D}_{\varphi}(p_{\tilde{\phi}}, p_{\phi_T}). \tag{16}$$

By definition of the limsup:

$$\limsup_{\phi \to \phi^{\infty}} \partial \hat{D}_{\varphi}(p_{\phi}, p_{\phi_T}) = \left\{ u | \exists \phi^k \to \phi^{\infty}, \exists u^k \to u \text{ with } u^k \in \partial \hat{D}_{\varphi}(p_{\phi^k}, p_{\phi_T}) \right\}.$$

In our scenario, $\phi = \phi^{k+1}$, $\phi^k = \phi^{k+1}$, $u = 0$ and $u^k = \nabla_1 D_{\psi}(\phi^{k+1}, \phi^k)$. We have proved above in this proof that $\nabla_1 D_{\psi}(\tilde{\phi}, \tilde{\phi}) = 0$ using only the convergence of $(\hat{D}_{\varphi}(p_{\phi^k}, p_{\phi_T}))_k$, inequality (13) and the properties of D_{ψ}. Assumption AC was not needed. Hence, $u^k \to 0$. This proves that $u = 0 \in \limsup_{\phi^{k+1} \to \phi^{\infty}} \partial \hat{D}_{\varphi}(p_{\phi^{n_k+1}}, p_{\phi_T})$. Finally, using the inclusion (16), we get our result:

$$0 \in \partial \hat{D}_{\varphi}(p_{\tilde{\phi}}, p_{\phi_T}),$$

which ends the proof. □

The proof of the previous proposition is very similar to the proof of Proposition 2. The key idea is to use the sequence of conditional densities $h_i(x|\phi^k)$ instead of the sequence ϕ^k. According to the application, one may be interested only in Proposition 1 or in Propositions 2–4. If one is interested in the parameters, Propositions 2 to 4 should be used, since we need a stable limit of $(\phi^k)_k$. If we are only interested in minimizing an error criterion $\hat{D}_{\varphi}(p_{\phi}, p_{\phi_T})$ between the estimated distribution and the true one, Proposition 1 should be sufficient.

4. Case Studies

4.1. An Algorithm With Theoretically Global Infimum Attainment

We present a variant of algorithm (11) which ensures *theoretically* the convergence to a global infimum of the objective function $\hat{D}_{\varphi}(p_{\phi}, p_{\phi^T})$ as soon as there exists a convergent subsequence of $(\phi^k)_k$. The idea is the same as Theorem 3.2.4 in [18]. Define ϕ^{k+1} by:

$$\phi^{k+1} = \arg\inf_{\phi} \hat{D}_{\varphi}(p_{\phi}, p_{\phi^T}) + \beta_k D_{\psi}(\phi, \phi^k).$$

The proof of convergence is very simple and does not depend on the differentiability of any of the two functions $\hat{D}_{\varphi}$ or D_{ψ}. We only assume A1 and A2 to be verified. Let $(\phi^{N(k)})_k$ be a convergent subsequence. Let ϕ^{∞} be its limit. This is guaranteed by the compactness of Φ^0 and the fact that the whole sequence $(\phi^k)_k$ resides in Φ^0 (see Proposition 1b). Suppose also that the sequence $(\beta_k)_k$ converges to 0 as k goes to infinity.

Now assumptions of Theorem 3.2.4. from [18] are verified. Thus, using the same lines from the proof of this theorem (inverting all inequalities since we are minimizing instead of maximizing), we may prove that ϕ^{∞} is a global infimum of the estimated divergence, that is

$$\hat{D}_{\varphi}(p_{\phi^{\infty}}, p_{\phi^T}) \leq \hat{D}_{\varphi}(p_{\phi}, p_{\phi^T}), \qquad \forall \phi \in \Phi.$$

The problem with this approach is that it depends heavily on the fact that the supremum on each step of the algorithm is calculated exactly. This does not happen in general unless function $\hat{D}_{\varphi}(p_{\phi}, p_{\phi^T}) + \beta_k D_{\psi}(\phi, \phi^k)$ is convex or that we dispose of an algorithm that can perfectly solve non convex optimization problems (In this case, there is no meaning in applying an iterative proximal algorithm. We would have used the optimization algorithm directly on the objective function $\hat{D}_{\varphi}(p_{\phi}, p_{\phi^T})$). Although in our approach, we use a similar assumption to prove the consecutive decreasing of $\hat{D}_{\varphi}(p_{\phi}, p_{\phi^T})$, we can replace the infimum calculus in (11) by two things. We require at each step that we find a local infimum of $\hat{D}_{\varphi}(p_{\phi}, p_{\phi^T}) + D_{\psi}(\phi, \phi^k)$ whose evaluation with $\phi \mapsto \hat{D}_{\varphi}(p_{\phi}, p_{\phi^T})$ is less than the previous term of the sequence ϕ^k. If we can no longer find any local minima verifying the claim, the procedure stops with $\phi^{k+1} = \phi^k$. This ensures the availability of all the proofs presented in this paper with no change.

4.2. The Two-Component Gaussian Mixture

We suppose that the model $(p_{\phi})_{\phi \in \Phi}$ is a mixture of two gaussian densities, and that we are only interested in estimating the means $\mu = (\mu_1, \mu_2) \in \mathbb{R}^2$ and the proportion $\lambda \in [\eta, 1-\eta]$. The use of η is to avoid cancellation of any of the two components, and to keep the hypothesis $h_i(x|\phi) > 0$ for $x = 1, 2$ verified. We also suppose that the components variances are reduced ($\sigma_i = 1$). The model takes the form

$$p_{\lambda,\mu}(x) = \frac{\lambda}{\sqrt{2\pi}} e^{-\frac{1}{2}(x-\mu_1)^2} + \frac{1-\lambda}{\sqrt{2\pi}} e^{-\frac{1}{2}(x-\mu_2)^2}. \tag{17}$$

Here, $\Phi = [\eta, 1-\eta] \times \mathbb{R}^2$. The regularization term D_{ψ} is defined by (8) where:

$$h_i(1|\phi) = \frac{\lambda e^{-\frac{1}{2}(y_i-\mu_1)^2}}{\lambda e^{-\frac{1}{2}(y_i-\mu_1)^2} + (1-\lambda) e^{-\frac{1}{2}(y_i-\mu_2)^2}}, \quad h_i(2|\phi) = 1 - h_i(1|\phi).$$

Functions h_i are clearly of class $\mathcal{C}^1(\text{int}(\Phi))$, and so does D_{ψ}. We prove that Φ^0 is closed and bounded, which is sufficient to conclude its compactness, since the space $[\eta, 1-\eta] \times \mathbb{R}^2$ provided with the euclidean distance is complete.

If we are using the dual estimator of the $\varphi-$divergence given by (2), then assumption A0 can be verified using the maximum theorem of Berge [19]. There is still a great difficulty in studying the properties (closedness or compactness) of the set Φ^0. Moreover, all convergence properties of the sequence ϕ^k require the continuity of the estimated $\varphi-$divergence $\hat{D}_{\varphi}(p_{\phi}, p_{\phi^T})$ with respect to ϕ. In order to prove the continuity of the estimated divergence, we need to assume that Φ is compact, i.e., assume that the means are included in an interval of the form $[\mu_{\min}, \mu_{\max}]$. Now, using Theorem 10.31 from [13], $\phi \mapsto \hat{D}_{\varphi}(p_{\phi}, p_{\phi^T})$ is continuous and differentiable almost everywhere with respect to ϕ.

The compactness assumption of Φ implies directly the compactness of Φ^0. Indeed,

$$\begin{aligned} \Phi^0 &= \left\{ \phi \in \Phi, \hat{D}_{\varphi}(p_{\phi}, p_{\phi^T}) \leq \hat{D}_{\varphi}(p_{\phi^0}, p_{\phi^T}) \right\} \\ &= \hat{D}_{\varphi}(p_{\phi}, p_{\phi^T})^{-1} \left((-\infty, \hat{D}_{\varphi}(p_{\phi^0}, p_{\phi^T})] \right). \end{aligned}$$

Φ^0 is then the inverse image by a continuous function of a closed set, so it is closed in Φ. Hence, it is compact.

Conclusion 1. *Using Propositions 4 and 1, if* $\Phi = [\eta, 1-\eta] \times [\mu_{\min}, \mu_{\max}]^2$*, the sequence* $(\hat{D}_\varphi(p_{\phi^k}, p_{\phi^T}))_k$ *defined through Formula (2) converges and there exists a subsequence* $(\phi^{N(k)})$ *which converges to a stationary point of the estimated divergence. Moreover, every limit point of the sequence* $(\phi^k)_k$ *is a stationary point of the estimated divergence.*

If we are using the kernel-based dual estimator given by (3) with a Gaussian kernel density estimator, then function $\phi \mapsto \hat{D}_\varphi(p_\phi, p_{\phi^T})$ is continuously differentiable over Φ even if the means μ_1 and μ_2 are not bounded. For example, take $\varphi = \varphi_\gamma$ defined by (1). There is one condition which relates the window of the kernel, say w, with the value of γ. Indeed, using Formula (3), we can write

$$\hat{D}_\varphi(p_\phi, p_{\phi^T}) = \frac{1}{\gamma-1}\int \frac{p_\phi^\gamma}{K_{n,w}^{\gamma-1}}(y)dy - \frac{1}{\gamma n}\sum_{i=1}^{n}\frac{p_\phi^\gamma}{K_{n,w}^\gamma}(y_i) - \frac{1}{\gamma(\gamma-1)}.$$

In order to study the continuity and the differentiability of the estimated divergence with respect to ϕ, it suffices to study the integral term. We have

$$\frac{p_\phi^\gamma}{K_{n,w}^{\gamma-1}}(y) = \frac{\left(\frac{\lambda}{\sqrt{2\pi}}\exp\left[-\frac{1}{2}(y-\mu_1)^2\right] + \frac{1-\lambda}{\sqrt{2\pi}}\exp\left[-\frac{1}{2}(y-\mu_2)^2\right]\right)^\gamma}{\left(\frac{1}{nw}\sum_{i=1}^n \exp\left[-\frac{(y-y_i)^2}{2w^2}\right]\right)^{\gamma-1}}.$$

The dominating term at infinity in the nominator is $\exp(-\gamma y^2/2)$, whereas it is $\exp(-(\gamma-1)y^2/(2w^2))$ in the denominator. It suffices now in order that the integrand to be bounded by an integrable function independently of $\phi = (\lambda, \mu)$ that we have $-\gamma + (\gamma-1)/w^2 < 0$. That is $-\gamma w^2 + \gamma - 1 < 0$, which is equivalent to $\gamma(w^2-1) < -1$. This argument also holds if we differentiate the integrand with respect to λ or either of the means μ_1 or μ_2. For $\gamma = 2$ (the Pearson's χ^2), we need $w^2 > 1/2$. For $\gamma = 1/2$ (the Hellinger), there is no condition on w.

Closedness of Φ^0 is proved similarly to the previous case. Boundedness, however, must be treated differently since Φ is not necessarily compact and is supposed to be $\Phi = [\eta, 1-\eta] \times \mathbb{R}^2$. For simplicity, take $\varphi = \varphi_\gamma$. The idea is to choose ϕ^0 an initialization for the proximal algorithm in a way that Φ^0 does not include unbounded values of the means. Continuity of $\phi \mapsto \hat{D}_\varphi(p_\phi, p_{\phi^T})$ permits calculation of the limits when either (or both) of the means tends to infinity. If both the means go to infinity, then $p_\phi(x) \to 0, \forall x$. Thus, for $\gamma \in (0,\infty) \setminus \{1\}$, we have $\hat{D}_\varphi(p_\phi, p_{\phi^T}) \to \frac{1}{\gamma(\gamma-1)}$. For $\gamma < 0$, the limit is infinity. If only one of the means tends to ∞, then the corresponding component vanishes from the mixture. Thus, if we choose ϕ^0 such that:

$$\hat{D}_\varphi(p_{\phi^0}, p_{\phi^T}) < \min\left(\frac{1}{\gamma(\gamma-1)}, \inf_{\lambda,\mu}\hat{D}_\varphi(p_{(\lambda,\infty,\mu)}, p_{\phi^T})\right) \text{ if } \gamma \in (0,\infty)\setminus\{1\}, \tag{18}$$

$$\hat{D}_\varphi(p_{\phi^0}, p_{\phi^T}) < \inf_{\lambda,\mu}\hat{D}_\varphi(p_{(\lambda,\infty,\mu)}, p_{\phi^T}) \quad \text{if } \gamma < 0, \tag{19}$$

then the algorithm starts at a point of Φ whose function value is inferior to the limits of $\hat{D}_\varphi(p_\phi, p_{\phi^T})$ at infinity. By Proposition 1, the algorithm will continue to decrease the value of $\hat{D}_\varphi(p_\phi, p_{\phi^T})$ and never goes back to the limits at infinity. In addition, the definition of Φ^0 permits to conclude that if ϕ^0 is chosen according to conditions (18) and (19), then Φ^0 is bounded. Thus, Φ^0 becomes compact. Unfortunately the value of $\inf_{\lambda,\mu}\hat{D}_\varphi(p_{(\lambda,\infty,\mu)}, p_{\phi^T})$ can be calculated but numerically. We will see next that in the case of the likelihood function, a similar condition will be imposed for the compactness of Φ^0, and there will be no need for any numerical calculus.

Conclusion 2. *Using Propositions 4 and 1, under conditions (18) and (19) the sequence* $(\hat{D}_\varphi(p_{\phi^k}, p_{\phi^T}))_k$ *defined through Formula (3) converges and there exists a subsequence* $(\phi^{N(k)})$ *that converges to a stationary* *point of the estimated divergence. Moreover, every limit point of the sequence $(\phi^k)_k$ is a stationary point of the estimated divergence.*

In the case of the likelihood $\varphi(t) = -\log(t) + t - 1$, the set Φ^0 can be written as:

$$\begin{aligned}\Phi^0 &= \left\{\phi \in \Phi, J_{\mathcal{N}}(\phi) \geq J_{\mathcal{N}}(\phi^0)\right\} \\ &= J_{\mathcal{N}}^{-1}\left([J_{\mathcal{N}}(\phi^0), +\infty)\right),\end{aligned}$$

where $J_{\mathcal{N}}$ is the log-likelihood function of the Gaussian mixture model. The log-likelihood function $J_{\mathcal{N}}$ is clearly of class $\mathcal{C}^1(\text{int}(\Phi))$. We prove that Φ^0 is closed and bounded which is sufficient to conclude its compactness, since the space $[\eta, 1-\eta] \times \mathbb{R}^2$ provided with the euclidean distance is complete.

Closedness. The set Φ^0 is the inverse image by a continuous function (the log-likelihood) of a closed set. Therefore it is closed in $[\eta, 1-\eta] \times \mathbb{R}^2$.

Boundedness. By contradiction, suppose that Φ^0 is unbounded, then there exists a sequence $(\phi^l)_l$ which tends to infinity. Since $\lambda^l \in [\eta, 1-\eta]$, then either of μ_1^l or μ_2^l tends to infinity. Suppose that both μ_1^l and μ_2^l tend to infinity, we then have $J_{\mathcal{N}}(\phi^l) \to -\infty$. Any finite initialization ϕ^0 will imply that $J_{\mathcal{N}}(\phi^0) > -\infty$ so that $\forall \phi \in \Phi^0, J_{\mathcal{N}}(\phi) \geq J_{\mathcal{N}}(\phi^0) > -\infty$. Thus, it is impossible for both μ_1^l and μ_2^l to go to infinity.

Suppose that $\mu_1^l \to \infty$, and that μ_2^l converges (or that μ_2^l is bounded; in such case we extract a convergent subsequence) to $\mu 2$. The limit of the likelihood has the form:

$$L(\lambda, \infty, \phi_2) = \prod_{i=1}^{n} \frac{(1-\lambda)}{\sqrt{2\pi}} e^{-\frac{1}{2}(y_i - \mu_2)^2},$$

which is bounded by its value for $\lambda = 0$ and $\mu_2 = \frac{1}{n}\sum_{i=1}^{n} y_i$. Indeed, since $1 - \lambda \leq 1$, we have:

$$L(\lambda, \infty, \phi_2) \leq \prod_{i=1}^{n} \frac{1}{\sqrt{2\pi}} e^{-\frac{1}{2}(y_i - \mu_2)^2}.$$

The right-hand side of this inequality is the likelihood of a Gaussian model $\mathcal{N}(\mu_2, 0)$, so that it is maximized when $\mu_2 = \frac{1}{n}\sum_{i=1}^{n} y_i$. Thus, if ϕ^0 is chosen in a way that $J_{\mathcal{N}}(\phi^0) > J_{\mathcal{N}}\left(0, \infty, \frac{1}{n}\sum_{i=1}^{n} y_i\right)$, the case when μ_1 tends to infinity and μ_2 is bounded would never be allowed. For the other case where $\mu_2 \to \infty$ and μ_1 is bounded, we choose ϕ^0 in a way that $J_{\mathcal{N}}(\phi^0) > J_{\mathcal{N}}\left(1, \frac{1}{n}\sum_{i=1}^{n} y_i, \infty\right)$. In conclusion, with a choice of ϕ^0 such that:

$$J_{\mathcal{N}}(\phi^0) > \max\left[J_{\mathcal{N}}\left(0, \infty, \frac{1}{n}\sum_{i=1}^{n} y_i\right), J_{\mathcal{N}}\left(1, \frac{1}{n}\sum_{i=1}^{n} y_i, \infty\right)\right], \tag{20}$$

the set Φ^0 is bounded.

This condition on ϕ^0 is very natural and means that we need to begin at a point at least better than the extreme cases where we only have one component in the mixture. This can be easily verified by choosing a random vector ϕ^0, and calculating the corresponding log-likelihood value. If $J_{\mathcal{N}}(\phi^0)$ does not verify the previous condition, we draw again another random vector until satisfaction.

Conclusion 3. *Using Propositions 4 and 1, under condition (20) the sequence $(J_{\mathcal{N}}(\phi^k))_k$ converges and there exists a subsequence $(\phi^{N(k)})$ which converges to a stationary point of the likelihood function. Moreover, every limit point of the sequence $(\phi^k)_k$ is a stationary point of the likelihood.*

Assumption A3 is not fulfilled (this part applies for all aforementioned situations). As mentioned in the paper of Tseng [2], for the two Gaussian mixture example, by changing μ_1 and μ_2 by the same amount and suitably adjusting λ, the value of $h_i(x|\phi)$ would be unchanged. We explore this more thoroughly by writing the corresponding equations. Let us suppose, absurdly, that for distinct ϕ and ϕ', we have $D_\psi(\phi|\phi') = 0$. By definition of D_ψ, it is given by a sum of nonnegative terms, which implies that all terms need to be equal to zero. The following lines are equivalent $\forall i \in \{1, \cdots, n\}$:

$$\begin{aligned} h_i(0|\lambda, \mu_1, \mu_2) &= h_i(0|\lambda', \mu_1', \mu_2'), \\ \frac{\lambda e^{-\frac{1}{2}(y_i-\mu_1)^2}}{\lambda e^{-\frac{1}{2}(y_i-\mu_1)^2} + (1-\lambda)e^{-\frac{1}{2}(y_i-\mu_2)^2}} &= \frac{\lambda' e^{-\frac{1}{2}(y_i-\mu_1')^2}}{\lambda' e^{-\frac{1}{2}(y_i-\mu_1')^2} + (1-\lambda')e^{-\frac{1}{2}(y_i-\mu_2')^2}}, \\ \log\left(\frac{1-\lambda}{\lambda}\right) - \frac{1}{2}(y_i-\mu_2)^2 + \frac{1}{2}(y_i-\mu_1)^2 &= \log\left(\frac{1-\lambda'}{\lambda'}\right) - \frac{1}{2}(y_i-\mu_2')^2 + \frac{1}{2}(y_i-\mu_1')^2. \end{aligned}$$

Looking at this set of n equations as an equality of two polynomials on y of degree 1 at n points, we deduce that as we have two distinct observations, say, y_1 and y_2, the two polynomials need to have the same coefficients. Thus, the set of n equations is equivalent to the following two equations:

$$\left\{ \begin{aligned} \mu_1 - \mu_2 &= \mu_1' - \mu_2' \\ \log\left(\frac{1-\lambda}{\lambda}\right) + \frac{1}{2}\mu_1^2 - \frac{1}{2}\mu_2^2 &= \log\left(\frac{1-\lambda'}{\lambda'}\right) + \frac{1}{2}\mu_1'^2 - \frac{1}{2}\mu_2'^2. \end{aligned} \right. \tag{21}$$

These two equations with three variables have an infinite number of solutions. Take, for example, $\mu_1 = 0$, $\mu_2 = 1$, $\lambda = \frac{2}{3}$, $\mu_1' = \frac{1}{2}$, $\mu_2' = \frac{3}{2}$, $\lambda' = \frac{1}{2}$.

Remark 2. *The previous conclusion can be extended to any two-component mixture of exponential families having the form:*

$$p_\phi(y) = \lambda e^{\sum_{i=1}^{m_1} \theta_{1,i} y^i - F(\theta_1)} + (1-\lambda) e^{\sum_{i=1}^{m_2} \theta_{2,i} y^i - F(\theta_2)}.$$

One may write the corresponding n *equations. The polynomial of y_i has a degree of at most* $\max(m_1, m_2)$. *Thus, if one disposes of* $\max(m_1, m_2) + 1$ *distinct observations, the two polynomials will have the same set of coefficients. Finally, if $(\theta_1, \theta_2) \in \mathbb{R}^{d-1}$ with $d >$* $\max(m_1, m_2)$, *then assumption A3 does not hold.*

Unfortunately, we have no an information about the difference between consecutive terms $\|\phi^{k+1} - \phi^k\|$ except for the case of $\psi(t) = \varphi(t) = -\log(t) + t - 1$ which corresponds to the classical EM recurrence:

$$\lambda^{k+1} = \frac{1}{n}\sum_{i=1}^n h_i(0|\phi^k), \quad \mu_1^{k+1} = \frac{\sum_{i=1}^n y_i h_i(0|\phi^k)}{\sum_{i=1}^n h_i(0|\phi^k)} \quad \mu_1^{k+1} = \frac{\sum_{i=1}^n y_i h_i(1|\phi^k)}{\sum_{i=1}^n h_i(1|\phi^k)}.$$

Tseng [2] has shown that we can prove directly that $\phi^{k+1} - \phi^k$ converges to 0.

5. Simulation Study

We summarize the results of 100 experiments on 100 samples by giving the average of the estimates and the error committed, and the corresponding standard deviation. The criterion error is the total variation distance (TVD), which is calculated using the $L1$ distance. Indeed, the Scheffé Lemma (see [20] (Page 129)) states that:

$$\sup_{A \in \mathcal{B}_n(\mathbb{R})} \left|P_\phi(A) - P_{\phi^T}(A)\right| = \frac{1}{2}\int_{\mathbb{R}} \left|p_\phi(y) - p_{\phi^T}(y)\right| dy.$$

The TVD gives a measure of the maximum error we may commit when we use the estimated model in lieu of the true distribution. We consider the Hellinger divergence for estimators based on φ−divergences, which corresponds to $\varphi(t) = \frac{1}{2}(\sqrt{t} - 1)^2$. Our preference of the Hellinger divergence is that we hope to obtain robust estimators without loss of efficiency (see [21]). D_ψ is calculated with

$\psi(t) = \frac{1}{2}(\sqrt{t}-1)^2$. The kernel-based MD$\varphi$DE is calculated using the Gaussian kernel, and the window is calculated using Silverman's rule. We included in the comparison the minimum density power divergence (MDPD) of [14]. The estimator is defined by:

$$\begin{aligned} \hat{\phi}_n &= \arg\inf_{\phi\in\Phi} \int p_\phi^{1+a}(z)dz - \frac{a+1}{a}\frac{1}{n}\sum_i^n p_\phi^a(y_i) \\ &= \arg\inf_{\phi\in\Phi} \mathbb{E}_{P_\phi}\left[p_\phi^a\right] - \frac{a+1}{a}\mathbb{E}_{P_n}\left[p_\phi^a\right], \end{aligned} \quad (22)$$

where $a \in (0,1]$. This is a Bregman divergence and is known to have good efficiency and robustness for a good choice of the tradeoff parameter. According to the simulation results in [11], the value of $a = 0.5$ seems to give a good tradeoff between robustness against outliers and a good performance under the model. Notice that the MDPD coincides with MLE when a tends to zero. Thus, our methodology presented here in this article, is applicable on this estimator and the proximal point algorithm can be used to calculate the MDPD. The proximal term will be kept the same, i.e., $\psi(t) = \frac{1}{2}(\sqrt{t}-1)^2$.

Remark 3 (Note on the robustness of the used estimators). *In Section 3, we have proved under mild conditions that the proximal point algorithm (11) ensures the decrease of the estimated divergence. This means that when we use the dual Formulas (2) and (3), then the proximal point algorithm (11) returns at convergence the estimators defined by (4) and (5), respectively. Similarly, if we use the density power divergence of Basu et al. [14], then the proximal-point algorithm returns at convergence the MDPD defined by (22). The robustness properties of the dual estimators (4) and (5) are studied in [12] and [11] respectively using the influence function (IF) approach. On the other hand, the robustness properties of the MDPD are studied using the IF approach in [14]. The MDφDE (4) has generally an unbounded IF (see [12] Section 3.1), whereas the kernel-based MDφDE's IF may be bounded for example in a Gaussian model and for any $\varphi-$divergence with $\varphi = \varphi_\gamma$ with $\gamma \in (0,1)$, see [11] Example 2. On the other hand, the MDPD has generally a bounded IF if the tradeoff parameter a is positive, and, in particular, in the Gaussian model. The MDPD becomes more robust as the tradeoff parameter a increases (see Section 3.3 in [14]). Therefore, we should expect that the proximal point algorithm produces robust estimators in the case of the kernel-based MDφDE and the MDPD, and thus obtain better results than the MLE calculated using the EM algorithm.*

Simulations from two mixture models are given below—a Gaussian mixture and a Weibull mixture. The MLE for both mixtures was calculated using the EM algorithm.

Optimizations were carried out using the Nelder–Mead algorithm [22] under the statistical tool R [23]. Numerical integrations in the Gaussian mixture were calculated using the `distrExIntegrate` function of package `distrEx`. It is a slight modification of the standard function `integrate`. It performs a Gauss–Legendre quadrature when function `integrate` returns an error. In the Weibull mixture, we used the `integral` function from package `pracma`. Function `integral` includes a variety of adaptive numerical integration methods such as Kronrod–Gauss quadrature, Romberg's method, Gauss–Richardson quadrature, Clenshaw–Curtis (not adaptive) and (adaptive) Simpson's method. Although function `integral` is slow, it performs better than other functions even if the integrand has a relatively bad behavior.

5.1. The Two-Component Gaussian Mixture Revisited

We consider the Gaussian mixture (17) presented earlier with true parameters $\lambda = 0.35$, $\mu_1 = -2, \mu_2 = 1.5$ and known variances equal to 1. Contamination was done by adding in the original sample to the five lowest values random observations from the uniform distribution $\mathcal{U}[-5,-2]$. We also added to the five largest values random observations from the uniform distribution $\mathcal{U}[2,5]$. Results are summarized in Table 1. The EM algorithm was initialized according to condition (20). This condition gave good results when we are under the model, whereas it did not always result in good estimates (the proportion converged towards 0 or 1) when outliers were added, and thus the EM algorithm was reinitialized manually.

Table 1. The mean and the standard deviation of the estimates and the errors committed in a 100 run experiment of a two-component Gaussian mixture. The true set of parameters is $\lambda = 0.35$, $\mu_1 = -2$, $\mu_2 = 1.5$.

Estimation Method	λ	sd (λ)	μ_1	sd (μ_1)	μ_2	sd (μ_2)	TVD	sd (TVD)
Without Outliers								
Classical MDφDE	0.349	0.049	−1.989	0.207	1.511	0.151	0.061	0.029
New MDφDE–Silverman	0.349	0.049	−1.987	0.208	1.520	0.155	0.062	0.029
MDPD $a = 0.5$	0.360	0.053	−1.997	0.226	1.489	0.135	0.065	0.025
EM (MLE)	0.360	0.054	−1.989	0.204	1.493	0.136	0.064	0.025
With 10% Outliers								
Classical MDφDE	0.357	0.022	−2.629	0.094	1.734	0.111	0.146	0.034
New MDφDE–Silverman	0.352	0.057	−1.756	0.224	1.358	0.132	0.087	0.033
MDPD $a = 0.5$	0.364	0.056	−1.819	0.218	1.404	0.132	0.078	0.030
EM (MLE)	0.342	0.064	−2.617	0.288	1.713	0.172	0.150	0.034

Figure 1 shows the values of the estimated divergence for both Formulas (2) and (3) on a logarithmic scale at each iteration of the algorithm.

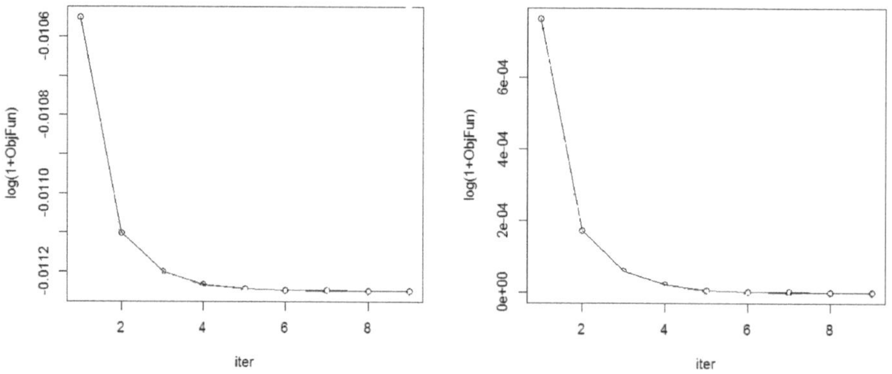

Figure 1. Decrease of the (estimated) Hellinger divergence between the true density and the estimated model at each iteration in the Gaussian mixture. The figure to the left is the curve of the values of the kernel-based dual Formula (3). The figure to the right is the curve of values of the classical dual Formula (2). Values are taken at a logarithmic scale $\log(1 + x)$.

Concerning our simulation results, the total variation of all four estimation methods is very close when we are under the model. When we added outliers, the classical MDφDE was as sensitive as the maximum likelihood estimator. The error was doubled. Both the kernel-based MDφDE and the MDPD are clearly robust since the total variation of these estimators under contamination has slightly increased.

5.2. The Two-Component Weibull Mixture Model

We consider a two-component Weibull mixture with unknown shapes $\nu_1 = 1.2, \nu_2 = 2$ and a proportion $\lambda = 0.35$. The scales are known an equal to $\sigma_1 = 0.5, \sigma_2 = 2$. The desity function is given by:

$$p_\phi(x) = 2\lambda\alpha_1(2x)^{\alpha_1-1}e^{-(2x)^{\alpha_1}} + (1-\lambda)\frac{\alpha_2}{2}\left(\frac{x}{2}\right)^{\alpha_2-1}e^{-\left(\frac{x}{2}\right)^{\alpha_2}}. \tag{23}$$

Contamination was done by replacing 10 observations of each sample chosen randomly by 10 i.i.d. observations drawn from a Weibull distribution with shape $\nu = 0.9$ and scale $\sigma = 3$. Results are summarized in Table 2. Notice that it would have been better to use asymmetric kernels in order to build the kernel-based MDφDE since their use in the context of positive-supported distributions is advised in order to reduce the bias at zero, see [11] for a detailed comparison with symmetric kernels. This is not, however, the goal of this paper. In addition, the use of symmetric kernels in this mixture model gave satisfactory results.

Simulations results in Table 2 confirm once more the validity of our proximal point algorithm and the clear robustness of both the kernel-based MDφDE and the MDPD.

Table 2. The mean and the standard deviation of the estimates and the errors committed in a 100-run experiment of a two-component Weibull mixture. The true set of parameter is $\lambda = 0.35, \nu_1 = 1.2, \nu_2 = 2$.

Estimation Method	λ	sd (λ)	μ_1	sd (μ_1)	μ_2	sd (μ_2)	TVD	sd (TVD)
Without Outliers								
Classical MDφDE	0.356	0.066	1.245	0.228	2.055	0.237	0.052	0.025
New MDφDE–Silverman	0.387	0.067	1.229	0.241	2.145	0.289	0.058	0.029
MDPD $a = 0.5$	0.354	0.068	1.238	0.230	2.071	0.345	0.056	0.029
EM (MLE)	0.355	0.066	1.245	0.228	2.054	0.237	0.052	0.025
With 10% Outliers								
Classical MDφDE	0.250	0.085	1.089	0.300	1.470	0.335	0.092	0.037
New MDφDE–Silverman	0.349	0.076	1.122	0.252	1.824	0.324	0.067	0.034
MDPD $a = 0.5$	0.322	0.077	1.158	0.236	1.858	0.344	0.060	0.029
EM (MLE)	0.259	0.095	0.941	0.368	1.565	0.325	0.095	0.035

6. Conclusions

We introduced in this paper a proximal-point algorithm that permits calculation of divergence-based estimators. We studied the theoretical convergence of the algorithm and verified it in a two-component Gaussian mixture. We performed several simulations which confirmed that the algorithm works and is a way to calculate divergence-based estimators. We also applied our proximal algorithm on a Bregman divergence estimator (the MDPD), and the algorithm succeeded to produce the MDPD. Further investigations about the role of the proximal term and a comparison with direct optimization methods in order to show the practical use of the algorithm may be considered in a future work.

Acknowledgments: The authors are grateful to Laboratoire de Statistique Théorique et Appliquée, Université Pierre et Marie Curie, for financial support.

Author Contributions: Michel Broniatowski proposed use of a proximal-point algorithm in order to calculate the MDφDE. Michel Broniatowski proposed building a work based on the paper of [2]. Diaa Al Mohamad proposed the generalization in Section 2.3 and provided all of the convergence results in Section 3. Diaa Al Mohamad also conceived the simulations. Finally, Michel Broniatowski contributed to improving the text written by Diaa Al Mohamad. Both authors have read and approved the final manuscript.

Conflicts of Interest: The authors declare no conflict of interest.

References

1. McLachlan, G.J.; Krishnan, T. *The EM Algorithm and Extensions*; Wiley: Hoboken, NJ, USA, 2007.
2. Tseng, P. An Analysis of the EM Algorithm and Entropy-Like Proximal Point Methods. *Math. Oper. Res.* **2004**, *29*, 27–44.
3. Chrétien, S.; Hero, A.O. Generalized Proximal Point Algorithms and Bundle Implementations. Available online: http://www.eecs.umich.edu/techreports/systems/cspl/cspl-316.pdf (acceesed on 25 July 2016).
4. Goldstein, A.; Russak, I. How good are the proximal point algorithms? *Numer. Funct. Anal. Optim.* **1987**, *9*, 709–724.

5. Chrétien, S.; Hero, A.O. Acceleration of the EM algorithm via proximal point iterations. In Proceedings of the IEEE International Symposium on Information Theory, Cambridge, MA, USA, 16–21 August 1998.
6. Csiszár, I. Eine informationstheoretische Ungleichung und ihre anwendung auf den Beweis der ergodizität von Markoffschen Ketten. *Publ. Math. Inst. Hung. Acad. Sci.* **1963**, *8*, 95–108. (In German)
7. Broniatowski, M.; Keziou, A. Parametric estimation and tests through divergences and the duality technique. *J. Multivar. Anal.* **2009**, *100*, 16–36.
8. Cressie, N.; Read, T.R.C. Multinomial goodness-of-fit tests. *J. R. Stat. Soc. Ser. B* **1984**, *46*, 440–464.
9. Broniatowski, M.; Keziou, A. Minimization of divergences on sets of signed measures. *Stud. Sci. Math. Hung.* **2006**, *43*, 403–442.
10. Liese, F.; Vajda, I. On Divergences and Informations in Statistics and Information Theory. *IEEE Trans. Inf. Theory* **2006**, *52*, 4394–4412.
11. Al Mohamad, D. Towards a better understanding of the dual representation of phi divergences. **2016**, arXiv:1506.02166.
12. Toma, A.; Broniatowski, M. Dual divergence estimators and tests: Robustness results. *J. Multivar. Anal.* **2011**, *102*, 20–36.
13. Rockafellar, R.T.; Wets, R.J.B. *Variational Analysis*, 3rd ed.; Springer: Berlin/Heidelberg, Germany, 1998.
14. Basu, A.; Harris, I.R.; Hjort, N.L.; Jones, M.C. Robust and Efficient Estimation by Minimising a Density Power Divergence. *Biometrika* **1998**, *85*, 549–559.
15. Dempster, A.P.; Laird, N.M.; Rubin, D.B. Maximum likelihood from incomplete data via the EM algorithm. *J. R. Stat. Soc. B* **1977**, *39*, 1–38.
16. Wu, C.F.J. On the Convergence Properties of the EM Algorithm. *Ann. Stat.* **1983**, *11*, 95–103.
17. Ostrowski, A. *Solution of Equations and Systems of Equations*; Academic Press: Cambridge, MA, USA, 1966.
18. Chrétien, S.; Hero, A.O. On EM algorithms and their proximal generalizations. *ESAIM Probabil. Stat.* **2008**, *12*, 308–326.
19. Berge, C. *Topological Spaces: Including a Treatment of Multi-valued Functions, Vector Spaces, and Convexity*; Dover Publications: Mineola, NY, USA, 1963.
20. Meister, A. *Deconvolution Problems in Nonparametric Statistics*; Springer: Berlin/Heidelberg, Germany, 2009.
21. Jiménz, R.; Shao, Y. On robustness and efficiency of minimum divergence estimators. *Test* **2001**, *10*, 241–248.
22. Nelder, J.A.; Mead, R. A Simplex Method for Function Minimization. *Comput. J.* **1965**, *7*, 308–313.
23. The R Core Team. *R: A Language and Environment for Statistical Computing;* R Foundation for Statistical Computing: Vienna, Austria, 2013.

entropy

MDPI

Article

Geometry Induced by a Generalization of Rényi Divergence

David C. de Souza [1,†], Rui F. Vigelis [2,*,†] and Charles C. Cavalcante [3,†]

1 Instituto Federal do Ceará, Campus Maracanaú, Fortaleza 61939-140, Brazil; davidcs@ifce.edu.br
2 Computer Engineering School, Campus Sobral, Federal University of Ceará, Sobral 62010-560, Brazil
3 Department of Teleinformatics Engineering, Federal University of Ceará, Fortaleza 60455-900, Brazil; charles@ufc.br
* Correspondence: rfvigelis@ufc.br
† These authors contributed equally to this work.

Academic Editors: Frédéric Barbaresco and Frank Nielsen
Received: 6 September 2016; Accepted: 11 November 2016; Published: 17 November 2016

Abstract: In this paper, we propose a generalization of Rényi divergence, and then we investigate its induced geometry. This generalization is given in terms of a φ-function, the same function that is used in the definition of non-parametric φ-families. The properties of φ-functions proved to be crucial in the generalization of Rényi divergence. Assuming appropriate conditions, we verify that the generalized Rényi divergence reduces, in a limiting case, to the φ-divergence. In generalized statistical manifold, the φ-divergence induces a pair of dual connections $D^{(-1)}$ and $D^{(1)}$. We show that the family of connections $D^{(\alpha)}$ induced by the generalization of Rényi divergence satisfies the relation $D^{(\alpha)} = \frac{1-\alpha}{2} D^{(-1)} + \frac{1+\alpha}{2} D^{(1)}$, with $\alpha \in [-1, 1]$.

Keywords: Rényi divergence; φ-function; φ-divergence; φ-family; statistical manifold; information geometry

1. Introduction

Information geometry, the study of statistical models equipped with a differentiable structure, was pioneered by the work of Rao [1], and gained maturity with the work of Amari and many others [2–4]. It has been successfully applied in many different areas, such as statistical inference, machine learning, signal processing or optimization [4,5]. In appropriate statistical models, the differentiable structure is induced by a (statistical) divergence. The Kullback–Leibler divergence induces a Riemannian metric, called the Fisher–Rao metric, and a pair of dual connections, the exponential and mixture connections. A statistical model endowed with the Fisher–Rao metric is called a (classical) statistical manifold. Amari also considered a family of α-divergences that induce a family of α-connections.

Much research in recent years has focused on the geometry of non-standard statistical models [6–8]. These models are defined in terms of a deformed exponential (also called ϕ-exponential). In particular, κ-exponential models and q-exponential families are investigated in [9,10]. Non-parametric (or infinite-dimensional) φ-families were introduced by the authors in [11,12], which generalize exponential families in the non-parametric setting [13–16]. Based on the similarity between exponential and φ-families, we defined the so-called φ-divergence, with respect to which the Kullback–Leibler divergence is a particular case. Statistical models equipped with a geometric structure induced by φ-divergences, which are called generalized statistical manifolds, are investigated in [17,18]. With respect to these connections, parametric φ-families are dually flat.

The φ-divergence is intrinsically related to the (ρ, τ)-model of Zhang, which was proposed in [19,20], extended to the infinite-dimension setting in [21], and explained in more details in [22,23].

For instance, the metric induced by φ-divergence and the (ρ, τ)-generalization of the Fisher–Rao metric, for the choices $\rho = \varphi^{-1}$ and $f = \rho^{-1}$, differ by a conformal factor.

Among many attempts to generalize Kullback–Leibler divergence, Rényi divergence [24] is one of the most successful, having found many applications [25]. In the present paper, we propose a generalization of Rényi divergence, which we use to define a family of α-connections. This generalization is based on an interpretation of Rényi divergence as a kind of normalizing function. To generalize Rényi divergence, we considered functions satisfying some suitable conditions. To a function for which these conditions hold, we give the name of φ-function. In a limiting case, the generalized Rényi divergence reduces to the φ-divergence. In [17,18], the φ-divergence gives rise to a pair of dual connections $D^{(-1)}$ and $D^{(1)}$. We show that the connection $D^{(\alpha)}$ induced by the generalization of Rényi divergence satisfies the convex combination $D^{(\alpha)} = \frac{1-\alpha}{2} D^{(-1)} + \frac{1+\alpha}{2} D^{(1)}$.

Eguchi in [26] investigated a geometry based on a normalizing function similar to the one used in the generalization of Rényi divergence. In [26], results were derived supposing that this normalizing function exists; conditions for its existence were not given. In the present paper, the existence of the normalizing function is ensured by conditions involved in the definition of φ-functions.

The rest of the paper is organized as follows. In Section 2, φ-functions are introduced and some properties are discussed. The Rényi divergence is generalized in Section 3. We investigate in Section 4 the geometry induced by the generalization of Rényi divergence. Section 4.2 provides evidence of the role of the generalized Rényi divergence in φ-families.

2. φ-Functions

Rényi divergence is defined in terms of the exponential function (to be more precise, the logarithm). A way of generalizing Rényi divergence is to replace the exponential function by another function, which satisfies some suitable conditions. To a function for which these conditions hold, we give the name φ-function. In this section, we define and investigate some properties of φ-functions.

Let (T, Σ, μ) be a measure space. Although we do not restrict our analysis to a particular measure space, the reader can think of T as the set of real numbers $\mathbb{R}$, Σ as the Borel σ-algebra on $\mathbb{R}$, and μ as the Lebesgue measure. We can also consider T to be a discrete set, a case in which μ is the counting measure.

We say that $\varphi\colon \mathbb{R} \to (0, \infty)$ is a *φ-function* if the following conditions are satisfied:

(a1) $\varphi(\cdot)$ is convex;
(a2) $\lim_{u\to-\infty} \varphi(u) = 0$ and $\lim_{u\to\infty} \varphi(u) = \infty$;
(a3) there exists a measurable function $u_0\colon T \to (0, \infty)$ such that

$$\int_T \varphi(c + \lambda u_0) d\mu < \infty, \qquad \text{for all } \lambda > 0, \tag{1}$$

for each measurable function $c\colon T \to \mathbb{R}$ satisfying $\int_T \varphi(c) d\mu = 1$.

Thanks to condition (a3), we can generalize Rényi divergence using φ-functions. These conditions appeared first at [12] where the authors constructed non-parametric φ-families of probability distributions. We remark that if T is finite, condition (a3) is always satisfied.

Examples of functions $\varphi\colon \mathbb{R} \to (0, \infty)$ satisfying (a1)–(a3) abound. An example of great relevance is the exponential function $\varphi(u) = \exp(u)$, which satisfies conditions (a1)–(a3) with $u_0 = \mathbf{1}_T$. Another example of φ-function is the Kaniadakis' κ-exponential [12,27,28].

Example 1. *The Kaniadakis' κ-exponential* $\exp_\kappa\colon \mathbb{R} \to (0, \infty)$ *for* $\kappa \in [-1, 1]$ *is defined as*

$$\exp_\kappa(u) = \begin{cases} (\kappa u + \sqrt{1 + \kappa^2 u^2})^{1/\kappa}, & \text{if } \kappa \neq 0, \\ \exp(u), & \text{if } \kappa = 0, \end{cases}$$

whose inverse is the so called the Kaniadakis' κ-logarithm $\log_\kappa\colon (0,\infty) \to \mathbb{R}$, *which is given by*

$$\log_\kappa(u) = \begin{cases} \frac{u^\kappa - u^{-\kappa}}{2\kappa}, & \text{if } \kappa \neq 0, \\ \ln(u), & \text{if } \kappa = 0. \end{cases}$$

It is clear that $\exp_\kappa(\cdot)$ *satisfies (a1) and (a2). Let* $u_0\colon T \to (0,\infty)$ *be any measurable function for which* $\int_T \exp_\kappa(u_0)d\mu < \infty$. *We will show that* u_0 *satisfies expression (1). For any* $u \in \mathbb{R}$ *and* $\alpha \geq 1$, *we can write*

$$\begin{aligned} \exp_\kappa(\alpha u) &= \alpha^{1/|\kappa|}(|\kappa|u + \sqrt{1/\alpha^2 + |\kappa|^2u^2})^{1/|\kappa|} \\ &\leq \alpha^{1/|\kappa|}(|\kappa|u + \sqrt{1 + |\kappa|^2u^2})^{1/|\kappa|} \\ &= \alpha^{1/|\kappa|}\exp_\kappa(u), \end{aligned}$$

where we used that $\exp_\kappa(\cdot) = \exp_{-\kappa}(\cdot)$. *Then, we conclude that* $\int_T \exp_\kappa(\alpha u_0)d\mu < \infty$ *for all* $\alpha \geq 0$. *Fix any measurable function* $c\colon T \to \mathbb{R}$ *such that* $\int_T \varphi(c)d\mu = 1$. *For each* $\lambda > 0$, *we have*

$$\begin{aligned} \int_T \exp_\kappa(c + \lambda u_0)d\mu &\leq \frac{1}{2}\int_T \exp_\kappa(2c)d\mu + \frac{1}{2}\int_T \exp_\kappa(2\lambda u_0)d\mu \\ &\leq 2^{1/|\kappa|-1}\int_T \exp_\kappa(c)d\mu + 2^{1/|\kappa|-1}\int_T \exp_\kappa(\lambda u_0)d\mu \\ &< \infty, \end{aligned}$$

which shows that $\exp_\kappa(\cdot)$ *satisfies (a3). Therefore, the Kaniadakis' κ-exponential* $\exp_\kappa(\cdot)$ *is an example of* φ*-function.*

The restriction that $\int_T \varphi(c)d\mu = 1$ can be weakened, as asserted in the next result.

Lemma 1. *Let* $\widetilde{c}\colon T \to \mathbb{R}$ *be any measurable function such that* $\int_T \varphi(\widetilde{c})d\mu < \infty$. *Then,* $\int_T \varphi(\widetilde{c} + \lambda u_0)d\mu < \infty$ *for all* $\lambda > 0$.

Proof. Notice that if $\int_T \varphi(\widetilde{c})d\mu \geq 1$, then $\int_T \varphi(\widetilde{c} - \alpha u_0)d\mu = 1$ for some $\alpha > 1$. From the definition of u_0, it follows that $\int_T \varphi(\widetilde{c} + \lambda u_0)d\mu = \int_T \varphi(c + (\alpha + \lambda)u_0)d\mu < \infty$, where $c = \widetilde{c} - \alpha u_0$. Now assume that $\int_T \varphi(\widetilde{c})d\mu < 1$. Consider any measurable set $A \subseteq T$ with measure $0 < \mu(A) < \mu(T)$. Let $u\colon T \to [0,\infty)$ be a measurable function supported on A satisfying $\varphi(\widetilde{c} + u)\mathbf{1}_A = [\varphi(\widetilde{c}) + \alpha]\mathbf{1}_A$, where $\alpha = (1 - \int_T \varphi(\widetilde{c})d\mu)/\mu(A)$. Defining $c = (\widetilde{c} + u)\mathbf{1}_A + \widetilde{c}\mathbf{1}_{T\setminus A}$, we see that $\int_T \varphi(c)d\mu = 1$. By the definition of u_0, we can write

$$\int_T \varphi(\widetilde{c} + \lambda u_0)d\mu \leq \int_T \varphi(c + \lambda u_0)d\mu < \infty, \qquad \text{for any } \lambda > 0,$$

which is the desired result. □

As a consequence of Lemma 1, condition (a3) can be replaced by the following one:

(a3′) There exists a measurable function $u_0\colon T \to (0,\infty)$ such that

$$\int_T \varphi(c + \lambda u_0)d\mu < \infty, \qquad \text{for all } \lambda > 0, \tag{2}$$

for each measurable function $c\colon T \to \mathbb{R}$ for which $\int_T \varphi(c)d\mu < \infty$.

Without the equivalence between conditions (a3) and (a3′), we could not generalize Rényi divergence in the manner we propose. In fact, φ-functions could be defined directly in terms of (a3′), without mentioning (a3). We chose to begin with (a3) because this condition appeared initially in [12].

Not all functions $\varphi\colon \mathbb{R} \to (0,\infty)$, for which conditions (a1) and (a2) hold, satisfy condition (a3). Such a function is given below.

Example 2. *Assume that the underlying measure μ is σ-finite and non-atomic. This is the case of the Lebesgue measure. Let us consider the function*

$$\varphi(u) = \begin{cases} e^{(u+1)^2/2}, & u \geq 0, \\ e^{(u+1/2)}, & u \leq 0, \end{cases} \tag{3}$$

which clearly is convex, and satisfies the limits $\lim_{u\to-\infty}\varphi(u) = 0$ and $\lim_{u\to\infty}\varphi(u) = \infty$. Given any measurable function $u_0\colon T \to (0,\infty)$, we will find a measurable function $c\colon T \to \mathbb{R}$ with $\int_T \varphi(c)d\mu < \infty$, for which expression (2) is not satisfied.

For each $m \geq 1$, we define

$$v_m(t) := \Big(m\frac{\log(2)}{u_0(t)} - \frac{u_0(t)}{2} - 1\Big)\mathbf{1}_{E_m}(t),$$

where $E_m = \{t \in T : m\frac{\log(2)}{u_0(t)} - \frac{u_0(t)}{2} - 1 > 0\}$. Because $v_m \uparrow \infty$, we can find a sub-sequence $\{v_{m_n}\}$ such that

$$\int_{E_{m_n}} e^{(v_{m_n}+u_0+1)^2/2} d\mu \geq 2^n.$$

According to (Lemma 8.3 in [29]) , there exists a sub-sequence $w_k = v_{m_{n_k}}$ and pairwise disjoint sets $A_k \subseteq E_{m_{n_k}}$ for which

$$\int_{A_k} e^{(w_k+u_0+1)^2/2} d\mu = 1.$$

Let us define $c = \overline{c}\mathbf{1}_{T\setminus A} + \sum_{k=1}^\infty w_k \mathbf{1}_{A_k}$, where $A = \bigcup_{k=1}^\infty A_k$ and $\overline{c}$ is any measurable function such that $\varphi(\overline{c}(t)) > 0$ for $t \in T \setminus A$ and $\int_{T\setminus A}\varphi(\overline{c})d\mu < \infty$. Observing that

$$e^{(w_k(t)+u_0(t)+1)^2/2} = 2^{m_{n_k}} e^{(w_k(t)+1)^2/2}, \qquad \text{for } t \in A_k,$$

we get

$$\int_{A_k} e^{(w_k+1)^2/2} d\mu = \frac{1}{2^{m_{n_k}}}, \qquad \text{for every } m \geq 1.$$

Then, we can write

$$\begin{aligned} \int_T \varphi(c)d\mu &= \int_{T\setminus A}\varphi(\overline{c})d\mu + \sum_{k=1}^\infty \int_{A_k} e^{(w_k+1)^2/2} d\mu \\ &= \int_{T\setminus A}\varphi(\overline{c})d\mu + \sum_{k=1}^\infty \frac{1}{2^{m_{n_k}}} < \infty. \end{aligned}$$

On the other hand,

$$\begin{aligned} \int_T \varphi(c+u_0)d\mu &= \int_{T\setminus A}\varphi(\overline{c})d\mu + \sum_{k=1}^\infty \int_{A_k} e^{(u_0+w_k+1)^2/2} d\mu \\ &= \int_{T\setminus A}\varphi(\overline{c})d\mu + \sum_{k=1}^\infty 1 = \infty, \end{aligned}$$

which shows that (2) is not satisfied.

3. Generalization of Rényi Divergence

In this section, we provide a generalization of Rényi divergence, which is given in terms of a φ-function. This generalization also depends on a parameter $\alpha \in [-1,1]$; for $\alpha = \pm 1$, it is defined as a limit. Supposing that the underlying φ-function is continuously differentiable, we show that this limit exists and results in the φ-divergence [12]. In what follows, all probability distributions are assumed to have positive density. In other words, they belong to the collection

$$\mathcal{P}_\mu = \left\{ p \in L^0 : \int_T p d\mu = 1 \text{ and } p > 0 \right\},$$

where L^0 is the space of all real-valued, measurable functions on T, with equality μ-a.e. (μ-almost everywhere).

The Rényi divergence of order $\alpha \in (-1,1)$ between two probability distributions p and q in $\mathcal{P}_\mu$ is defined as

$$\mathcal{D}_{\mathrm{R}}^{(\alpha)}(p \parallel q) = \frac{4}{\alpha^2 - 1} \log\left(\int_T p^{\frac{1-\alpha}{2}} q^{\frac{1+\alpha}{2}} d\mu \right). \tag{4}$$

For $\alpha = \pm 1$, the Rényi divergence is defined by taking a limit:

$$\mathcal{D}_{\mathrm{R}}^{(-1)}(p \parallel q) = \lim_{\alpha \downarrow -1} \mathcal{D}_{\mathrm{R}}^{(\alpha)}(p \parallel q), \tag{5}$$

$$\mathcal{D}_{\mathrm{R}}^{(1)}(p \parallel q) = \lim_{\alpha \uparrow 1} \mathcal{D}_{\mathrm{R}}^{(\alpha)}(p \parallel q). \tag{6}$$

Under some conditions, the limits in (5) and (6) are finite-valued, and converge to the Kullback–Leibler divergence. In other words,

$$\mathcal{D}_{\mathrm{R}}^{(-1)}(p \parallel q) = \mathcal{D}_{\mathrm{R}}^{(1)}(q \parallel p) = \mathcal{D}_{\mathrm{KL}}(p \parallel q) < \infty,$$

where $\mathcal{D}_{\mathrm{KL}}(p \parallel q)$ denotes the Kullback–Leibler divergence between p and q, which is given by

$$\mathcal{D}_{\mathrm{KL}}(p \parallel q) = \int_T p \log\left(\frac{p}{q}\right) d\mu.$$

These conditions are stated in Proposition 1, given in the end of this section, for the case involving the generalized Rényi divergence.

The Rényi divergence in its standard form is given by

$$\mathcal{D}^{(\alpha)}(p \parallel q) = \frac{1}{1-\alpha} \log\left(\int_T p^\alpha q^{1-\alpha} d\mu \right), \qquad \text{for } \alpha \in (0,1). \tag{7}$$

Expression (4) is related to this form by

$$\mathcal{D}_{\mathrm{R}}^{(\alpha)}(p \parallel q) = \frac{2}{1-\alpha} \mathcal{D}^{((1-\alpha)/2)}(p \parallel q).$$

Beyond the change of variables, which results in α ranging in $[-1,1]$, expressions (4) and (7) differ by the factor $2/(1-\alpha)$. We opted to insert the term $2/(1-\alpha)$ so that some kind of symmetry could be maintained when the limits $\alpha \downarrow -1$ and $\alpha \uparrow 1$ are considered. In addition, the geometry induced by the version (4) conforms with Amari's notation [5].

The Rényi divergence $\mathcal{D}_{\mathrm{R}}^{(\alpha)}(\cdot \parallel \cdot)$ can be defined for every $\alpha \in \mathbb{R}$. However, for $\alpha \notin (-1,1)$, the expression (4) may not be finite-valued for every p and q in $\mathcal{P}_\mu$. To avoid some technicalities, we just consider $\alpha \in [-1,1]$.

Given p and q in $\mathcal{P}_\mu$, let us define

$$\kappa(\alpha) = -\log\Big(\int_T p^{\frac{1-\alpha}{2}} q^{\frac{1+\alpha}{2}} d\mu\Big), \qquad \text{for } \alpha \in [-1,1],$$

which can be used to express the Rényi divergence as

$$\mathcal{D}_{\mathrm{R}}^{(\alpha)}(p \parallel q) = \frac{4}{1-\alpha^2}\kappa(\alpha), \qquad \text{for } \alpha \in (-1,1).$$

The function $\kappa(\alpha)$, which depends on p and q, can be defined as the unique non-negative real number for which

$$\int_T \exp\Big(\frac{1-\alpha}{2}\ln(p) + \frac{1+\alpha}{2}\ln(q) + \kappa(\alpha)\Big) d\mu = 1. \tag{8}$$

The function $\kappa(\alpha)$ makes the role of a normalizing term. The generalization of Rényi divergence, which we propose, is based on the interpretation of $\kappa(\alpha)$ given in (8). Instead of the exponential function, we consider a φ-function in (8).

Fix any φ-function $\varphi\colon \mathbb{R} \to (0,\infty)$. Given any p and q in $\mathcal{P}_\mu$, we take $\kappa(\alpha) = \kappa(\alpha; p, q) \geq 0$ so that

$$\int_T \varphi\Big(\frac{1-\alpha}{2}\varphi^{-1}(p) + \frac{1+\alpha}{2}\varphi^{-1}(q) + \kappa(\alpha)u_0\Big) d\mu = 1, \tag{9}$$

or, in other words, the term inside the integral is a probability distribution in $\mathcal{P}_\mu$. The existence and uniqueness of $\kappa(\alpha)$ as defined in (9) is guaranteed by condition (a3').

We define a generalization of the Rényi divergence of order $\alpha \in (-1,1)$ as

$$\mathcal{D}_\varphi^{(\alpha)}(p \parallel q) = \frac{4}{1-\alpha^2}\kappa(\alpha). \tag{10}$$

For $\alpha = \pm 1$, this generalization is defined as a limit:

$$\mathcal{D}_\varphi^{(-1)}(p \parallel q) = \lim_{\alpha \downarrow -1} \mathcal{D}_\varphi^{(\alpha)}(p \parallel q), \tag{11}$$

$$\mathcal{D}_\varphi^{(1)}(p \parallel q) = \lim_{\alpha \uparrow 1} \mathcal{D}_\varphi^{(\alpha)}(p \parallel q). \tag{12}$$

The cases $\alpha = \pm 1$ are related to a generalization of the Kullback–Leibler divergence, the so-called φ-divergence, which was introduced by the authors in [12]. The φ-divergence is given by (It was pointed out to us by an anonymous referee that this form of divergence is a special case of the (ρ,τ)-divergence for $\rho = \varphi^{-1}$ and $f = \rho^{-1}$ (see Section 3.5 in [19]) apart from a conformal factor, which is the denominator of (13)):

$$\mathcal{D}_\varphi(p \parallel q) = \frac{\int_T \frac{\varphi^{-1}(p) - \varphi^{-1}(q)}{(\varphi^{-1})'(p)} d\mu}{\int_T \frac{u_0}{(\varphi^{-1})'(p)} d\mu}. \tag{13}$$

Under some conditions, the limit in (11) or (12) is finite-valued and converges to the φ-divergence:

$$\mathcal{D}_\varphi^{(-1)}(p \parallel q) = \mathcal{D}_\varphi^{(1)}(q \parallel p) = \mathcal{D}_\varphi(p \parallel q) < \infty. \tag{14}$$

To show (14), we make use of the following result.

Lemma 2. *Assume that $\varphi(\cdot)$ is continuously differentiable. If for $\alpha_0, \alpha_1 \in \mathbb{R}$, the expression*

$$\int_T \varphi\Big(\frac{1-\alpha}{2}\varphi^{-1}(p) + \frac{1+\alpha}{2}\varphi^{-1}(q)\Big) d\mu < \infty \tag{15}$$

is satisfied for all $\alpha \in [\alpha_0, \alpha_1]$*, then the derivative of* $\kappa(\alpha)$ *exists at any* $\alpha \in (\alpha_0, \alpha_1)$*, and is given by*

$$\frac{\partial \kappa}{\partial \alpha}(\alpha) = -\frac{1}{2}\frac{\int_T [\varphi^{-1}(q) - \varphi^{-1}(p)]\varphi'(c_\alpha)d\mu}{\int_T \varphi'(c_\alpha)u_0 d\mu}, \tag{16}$$

where $c_\alpha = \frac{1-\alpha}{2}\varphi^{-1}(p) + \frac{1+\alpha}{2}\varphi^{-1}(q) + \kappa(\alpha)u_0$.

Proof. For $\alpha \in (\alpha_0, \alpha_1)$ and $\kappa > 0$, define

$$g(\alpha, \kappa) = \int_T \varphi\Big(\frac{1-\alpha}{2}\varphi^{-1}(p) + \frac{1+\alpha}{2}\varphi^{-1}(q) + \kappa u_0\Big)d\mu.$$

The function $\kappa(\alpha)$ is defined implicitly by $g(\alpha, \kappa(\alpha)) = 1$. If we show that

(i) the function $g(\alpha, \kappa)$ is continuous in a neighborhood of $(\alpha, \kappa(\alpha))$,
(ii) the partial derivatives $\frac{\partial g}{\partial \alpha}$ and $\frac{\partial g}{\partial \kappa}$ exist and are continuous at $(\alpha, \kappa(\alpha))$,
(iii) and $\frac{\partial g}{\partial \kappa}(\alpha, \kappa(\alpha)) > 0$,

then by the Implicit Function Theorem $\kappa(\alpha)$ is differentiable at $\alpha \in (\alpha_0, \alpha_1)$, and

$$\frac{\partial \kappa}{\partial \alpha}(\alpha) = -\frac{(\partial g/\partial \alpha)(\alpha, \kappa(\alpha))}{(\partial g/\partial \kappa)(\alpha, \kappa(\alpha))}. \tag{17}$$

We begin by verifying that $g(\alpha, \kappa)$ is continuous. For fixed $\alpha \in (\alpha_0, \alpha_1)$ and $\kappa > 0$, set $\kappa_0 = 2\kappa$. Denoting $A = \{t \in T : \varphi^{-1}(q(t)) > \varphi^{-1}(p(t))\}$, we can write

$$\begin{aligned}\varphi\Big(\frac{1-\beta}{2}\varphi^{-1}(p) + \frac{1+\beta}{2}\varphi^{-1}(q) + \lambda u_0\Big) &\le \varphi\Big(\varphi^{-1}(p) + \frac{1+\beta}{2}[\varphi^{-1}(q) - \varphi^{-1}(p)] + \kappa_0 u_0\Big)\\ &\le \varphi\Big(\varphi^{-1}(p) + \frac{1+\alpha_1}{2}[\varphi^{-1}(q) - \varphi^{-1}(p)] + \kappa_0 u_0\Big)\mathbf{1}_A\\ &+ \varphi\Big(\varphi^{-1}(p) + \frac{1+\alpha_0}{2}[\varphi^{-1}(q) - \varphi^{-1}(p)] + \kappa_0 u_0\Big)\mathbf{1}_{T\setminus A},\end{aligned} \tag{18}$$

for every $\beta \in (\alpha_0, \alpha_1)$ and $\lambda \in (0, \kappa_0)$. Because the function on the right-hand side of (18) is integrable, we can apply the Dominated Convergence Theorem to conclude that

$$\lim_{(\beta,\lambda)\to(\alpha,\kappa)} g(\beta, \lambda) = g(\alpha, \kappa).$$

Now, we will show that the derivative of $g(\alpha, \kappa)$ with respect to α exists and is continuous. Consider the difference

$$\frac{g(\gamma, \lambda) - g(\beta, \lambda)}{\gamma - \beta} = \int_T \frac{1}{\gamma - \beta}\Big[\varphi\Big(c_\beta + \frac{\gamma - \beta}{2}[\varphi^{-1}(q) - \varphi^{-1}(p)] + \lambda u_0\Big) - \varphi(c_\beta + \lambda u_0)\Big]d\mu, \tag{19}$$

where $c_\beta = \frac{1-\beta}{2}\varphi^{-1}(p) + \frac{1+\beta}{2}\varphi^{-1}(q)$. Represent by $f_{\beta,\gamma,\lambda}$ the function inside the integral sign in (19). For fixed $\alpha \in (\alpha_0, \alpha_1)$ and $\kappa > 0$, denote $\overline{\alpha}_0 = (\alpha_0 + \alpha)/2$, $\overline{\alpha}_1 = (\alpha + \alpha_1)/2$, and $\kappa_0 = 2\kappa$. Because $\varphi(\cdot)$ is convex and increasing, it follows that

$$|f_{\beta,\gamma,\lambda}| \le f_{\overline{\alpha}_1,\alpha_1,\kappa_0}\mathbf{1}_A - f_{\overline{\alpha}_0,\alpha_0,\kappa_0}\mathbf{1}_{T\setminus A} =: f, \qquad \text{for all } \beta, \gamma \in (\overline{\alpha}_0, \overline{\alpha}_1) \text{ and } \lambda \in (0, \kappa_0),$$

where $A = \{t \in T : \varphi^{-1}(q(t)) > \varphi^{-1}(p(t))\}$. Observing that f is integrable, we can use the Dominated Convergence Theorem to get

$$\lim_{\gamma\to\beta}\int_T f_{\beta,\gamma,\lambda}d\mu = \int_T \big(\lim_{\gamma\to\beta} f_{\beta,\gamma,\lambda}\big)d\mu,$$

and then

$$\frac{\partial g}{\partial \alpha}(\beta,\lambda) = \frac{1}{2}\int_T [\varphi^{-1}(q) - \varphi^{-1}(p)]\varphi'(c_\beta + \lambda u_0)d\mu. \tag{20}$$

For $\beta \in (\overline{\alpha}_0, \overline{\alpha}_1)$ and $\lambda \in (0, \kappa_0)$, the function inside the integral sign in (20) is dominated by f. As a result, a second use of the Dominated Convergence Theorem shows that $\frac{\partial g}{\partial \alpha}$ is continuous at (α, κ):

$$\lim_{(\beta,\lambda)\to(\alpha,\kappa)} \frac{\partial g}{\partial \alpha}(\beta,\lambda) = \frac{\partial g}{\partial \alpha}(\alpha,\kappa).$$

Using similar arguments, one can show that $\frac{\partial g}{\partial \kappa}(\alpha,\kappa)$ exists and is continuous at any $\alpha \in (\alpha_0, \alpha_1)$ and $\kappa > 0$, and is given by

$$\frac{\partial g}{\partial \kappa}(\alpha,\kappa) = \int_T u_0 \varphi'(c_\alpha + \kappa u_0)d\mu. \tag{21}$$

Clearly, expression (21) implies that $\frac{\partial g}{\partial \kappa}(\alpha,\kappa) > 0$ for all $\alpha \in (0, \alpha_0)$ and $\kappa > 0$.

We proved that items (i)–(iii) are satisfied. As consequence, the derivative of $\kappa(\alpha)$ exists at any $\alpha \in (\alpha_0, \alpha_1)$. Expression (16) for the derivative of $\kappa(\alpha)$ follows from (17), (20) and (21). □

As an immediate consequence of Lemma 2, we get the proposition below.

Proposition 1. *Assume that $\varphi(\cdot)$ is continuously differentiable.*

(a) *If, for some $\alpha_0 < -1$, expression (15) is satisfied for all $\alpha \in [\alpha_0, -1)$, then*

$$\mathcal{D}_\varphi^{(-1)}(p \parallel q) = \lim_{\alpha \downarrow -1} \mathcal{D}_\varphi^{(\alpha)}(p \parallel q) = 2\frac{\partial \kappa}{\partial \alpha}(-1) = \mathcal{D}_\varphi(p \parallel q) < \infty.$$

(b) *If, for some $\alpha_1 > 1$, expression (15) is satisfied for all $\alpha \in (1, \alpha_1]$, then*

$$\mathcal{D}_\varphi^{(1)}(p \parallel q) = \lim_{\alpha \uparrow 1} \mathcal{D}_\varphi^{(\alpha)}(p \parallel q) = -2\frac{\partial \kappa}{\partial \alpha}(1) = \mathcal{D}_\varphi(q \parallel p) < \infty.$$

4. Generalized Statistical Manifolds

Statistical manifolds consist of a collection of probability distributions endowed with a metric and α-connections, which are defined in terms of the derivative of $l(t;\theta) = \log p(t;\theta)$. In a generalized statistical manifold, the metric and connection are defined in terms of $f(t;\theta) = \varphi^{-1}(p(t;\theta))$. Instead of the logarithm, we consider the inverse $\varphi^{-1}(\cdot)$ of a φ-function. Generalized statistical manifolds were introduced by the authors in [17,18]. Among examples of the generalized statistical manifold, (parametric) φ-families of probability distributions are of greatest importance. The non-parametric counterpart was investigated in [11,12]. The metric in φ-families can be defined as the Hessian of a function; i.e., φ-families are Hessian manifolds [30]. In [17,18], the φ-divergence gives rise to a pair of dual connections $D^{(-1)}$ and $D^{(1)}$; and then for $\alpha \in (-1, 1)$ the α-connection $D^{(\alpha)}$ is defined as the convex combination $D^{(\alpha)} = \frac{1-\alpha}{2}D^{(-1)} + \frac{1+\alpha}{2}D^{(1)}$. In the present paper, we show that the connection induced by $\mathcal{D}_\varphi^{(\alpha)}(\cdot \parallel \cdot)$, the generalization of Rényi divergence, corresponds to $D^{(\alpha)}$.

4.1. Definitions

Let $\varphi\colon \mathbb{R} \to (0,\infty)$ be a φ-function. A *generalized statistical manifold* $\mathcal{P} = \{p(t;\theta) : \theta \in \Theta\}$ is a collection of probability distributions $p_\theta(t) := p(t;\theta)$, indexed by parameters $\theta = (\theta^1, \dots, \theta^n) \in \Theta$ in a one-to-one relation, such that

(m1) Θ is a domain (open and connected set) in $\mathbb{R}^n$;
(m2) $p(t;\theta)$ is differentiable with respect to θ;

(m3) the matrix $g = (g_{ij})$ defined by

$$g_{ij} = -E'_\theta\Big[\frac{\partial^2\varphi^{-1}(p_\theta)}{\partial\theta^i\partial\theta^j}\Big], \tag{22}$$

is positive definite at each $\theta \in \Theta$, where

$$E'_\theta[\cdot] = \frac{\int_T (\cdot)\varphi'(\varphi^{-1}(p_\theta))d\mu}{\int_T u_0\varphi'(\varphi^{-1}(p_\theta))d\mu}; \tag{23}$$

(m4) the operations of integration with respect to μ and differentiation with respect to θ^i commute in all calculations found below, which are related to the metric and connections.

The matrix $g = (g_{ij})$ equips $\mathcal{P}$ with a metric. By the chain rule, the tensor related to $g = (g_{ij})$ is invariant under change of coordinates. The (classical) statistical manifold is a particular case in which $\varphi(u) = \exp(u)$ and $u_0 = \mathbf{1}_T$.

We introduce a notation similar to Equation (23) that involves higher order derivatives of $\varphi(\cdot)$. For each $n \geq 1$, we define

$$E^{(n)}_\theta[\cdot] = \frac{\int_T (\cdot)\varphi^{(n)}(\varphi^{-1}(p_\theta))d\mu}{\int_T u_0\varphi'(\varphi^{-1}(p_\theta))d\mu}. \tag{24}$$

We also use $E'_\theta[\cdot]$, $E''_\theta[\cdot]$ and $E'''_\theta[\cdot]$ to denote $E^{(n)}_\theta[\cdot]$ for $n = 1, 2, 3$, respectively. The notation (24) appears in expressions related to the metric and connections.

Using property (m4), we can find an alternate expression for g_{ij} as well as an identification involving tangent spaces. The matrix $g = (g_{ij})$ can be equivalently defined by

$$g_{ij} = E''_\theta\Big[\frac{\partial\varphi^{-1}(p_\theta)}{\partial\theta^i}\frac{\partial\varphi^{-1}(p_\theta)}{\partial\theta^j}\Big]. \tag{25}$$

As a consequence of this equivalence, the tangent space $T_{p_\theta}\mathcal{P}$ can be identified with $\widetilde{T}_{p_\theta}\mathcal{P}$, the vector space spanned by $\frac{\partial\varphi^{-1}(p_\theta)}{\partial\theta^i}$, and endowed with the inner product $\langle\widetilde{X},\widetilde{Y}\rangle_\theta := E''_\theta[\widetilde{X}\widetilde{Y}]$. The mapping

$$\sum_i a_i\frac{\partial}{\partial\theta^i} \mapsto \sum_i a_i\frac{\partial\varphi^{-1}(p_\theta)}{\partial\theta^i}$$

defines an isometry between $T_{p_\theta}\mathcal{P}$ and $\widetilde{T}_{p_\theta}\mathcal{P}$.

To verify (25), we differentiate $\int_T p_\theta d\mu = 1$, with respect to θ^i, to get

$$0 = \frac{\partial}{\partial\theta^i}\int_T p_\theta d\mu = \int_T \frac{\partial}{\partial\theta^i}\varphi(\varphi^{-1}(p_\theta))d\mu = \int_T \frac{\partial\varphi^{-1}(p_\theta)}{\partial\theta^i}\varphi'(\varphi^{-1}(p_\theta))d\mu. \tag{26}$$

Now, differentiating with respect to θ^j, we obtain

$$0 = \int_T \frac{\partial^2\varphi^{-1}(p_\theta)}{\partial\theta^i\partial\theta^j}\varphi'(\varphi^{-1}(p_\theta))d\mu + \int_T \frac{\partial\varphi^{-1}(p_\theta)}{\partial\theta^i}\frac{\partial\varphi^{-1}(p_\theta)}{\partial\theta^j}\varphi''(\varphi^{-1}(p_\theta))d\mu,$$

and then (25) follows. In view of (26), we notice that every vector $\widetilde{X}$ belonging to $\widetilde{T}_{p_\theta}\mathcal{P}$ satisfies $E'_\theta[\widetilde{X}] = 0$.

The metric $g = (g_{ij})$ gives rise to a Levi–Civita connection ∇ (i.e., a torsion-free, metric connection), whose corresponding Christoffel symbols Γ_{ijk} are given by

$$\Gamma_{ijk} := \frac{1}{2}\Big(\frac{\partial g_{ki}}{\partial\theta^j} + \frac{\partial g_{kj}}{\partial\theta^i} - \frac{\partial g_{ij}}{\partial\theta^k}\Big). \tag{27}$$

Using expression (25) to calculate the derivatives in (27), we can express

$$\begin{aligned}\Gamma_{ijk} = E''_\theta\Big[\frac{\partial^2\varphi^{-1}(p_\theta)}{\partial\theta^i\partial\theta^j}\frac{\partial\varphi^{-1}(p_\theta)}{\partial\theta^k}\Big] + \frac{1}{2}E'''_\theta\Big[\frac{\partial\varphi^{-1}(p_\theta)}{\partial\theta^i}\frac{\partial\varphi^{-1}(p_\theta)}{\partial\theta^j}\frac{\partial\varphi^{-1}(p_\theta)}{\partial\theta^k}\Big] \\ - \frac{1}{2}E''_\theta\Big[\frac{\partial\varphi^{-1}(p_\theta)}{\partial\theta^i}\frac{\partial\varphi^{-1}(p_\theta)}{\partial\theta^k}\Big]E''_\theta\Big[u_0\frac{\partial\varphi^{-1}(p_\theta)}{\partial\theta^j}\Big] - \frac{1}{2}E''_\theta\Big[\frac{\partial\varphi^{-1}(p_\theta)}{\partial\theta^j}\frac{\partial\varphi^{-1}(p_\theta)}{\partial\theta^k}\Big]E''_\theta\Big[u_0\frac{\partial\varphi^{-1}(p_\theta)}{\partial\theta^i}\Big] \\ + \frac{1}{2}E''_\theta\Big[\frac{\partial\varphi^{-1}(p_\theta)}{\partial\theta^i}\frac{\partial\varphi^{-1}(p_\theta)}{\partial\theta^j}\Big]E''_\theta\Big[u_0\frac{\partial\varphi^{-1}(p_\theta)}{\partial\theta^k}\Big].\end{aligned}$$

As we will show later, the Levi–Civita connection ∇ corresponds to the connection derived from the divergence $\mathcal{D}^{(\alpha)}_\varphi(\cdot \,\|\, \cdot)$ with $\alpha = 0$.

4.2. φ-Families

Let $c\colon T \to \mathbb{R}$ be a measurable function for which $p = \varphi(c)$ is a probability density in $\mathcal{P}_\mu$. Fix measurable functions $u_1, \dots, u_n\colon T \to \mathbb{R}$. A *(parametric) φ-family* $\mathcal{F}_p = \{p_\theta : \theta \in \Theta\}$, centered at $p = \varphi(c)$, is a set of probability distributions in $\mathcal{P}_\mu$, whose members can be written in the form

$$p_\theta := \varphi\Big(c + \sum_{i=1}^n \theta^i u_i - \psi(\theta)u_0\Big), \qquad \text{for each } \theta = (\theta^i) \in \Theta, \tag{28}$$

where $\psi\colon \Theta \to [0, \infty)$ is a *normalizing function*, which is introduced so that expression (28) defines a probability distribution belonging to $\mathcal{P}_\mu$.

The functions $u_1, \dots, u_n$ are not arbitrary. They are chosen to satisfy the following assumptions:

(i) $u_0, u_1, \dots, u_n$ are linearly independent,
(ii) $\int_T u_i \varphi'(c) d\mu = 0$, and
(iii) there exists $\varepsilon > 0$ such that $\int_T \varphi(c + \lambda u_i) d\mu < \infty$, for all $\lambda \in (-\varepsilon, \varepsilon)$.

Moreover, the domain $\Theta \subseteq \mathbb{R}^n$ is defined as the set of all vectors $\theta = (\theta^i)$ for which

$$\int_T \varphi\Big(c + \lambda \sum_{i=1}^n \theta^i u_i\Big) d\mu < \infty, \qquad \text{for some } \lambda > 1.$$

Condition (i) implies that the mapping defined by (28) is one-to-one. Assumption (ii) makes of ψ a non-negative function. Indeed, by the convexity of $\varphi(\cdot)$, along with (ii), we can write

$$\int_T \varphi(c) d\mu = \int_T \Big[\varphi(c) + \Big(\sum_{i=1}^n \theta^i u_i\Big)\varphi'(c)\Big] d\mu \leq \int_T \varphi\Big(c + \sum_{i=1}^n \theta^i u_i\Big) d\mu,$$

which implies $\psi(\theta) \geq 0$. By condition (iii), the domain Θ is an open neighborhood of the origin. If the set T is finite, condition (iii) is always satisfied. One can show that the domain Θ is open and convex. Moreover, the normalizing function ψ is also convex (or strictly convex if $\varphi(\cdot)$ is strictly convex). Conditions (ii) and (iii) also appears in the definition of non-parametric φ-families. For further details, we refer to [11,12].

In a φ-family $\mathcal{F}_p$, the matrix (g_{ij}) given by (22) or (25) can be expressed as the Hessian of ψ. If $\varphi(\cdot)$ is strictly convex, then(g_{ij}) is positive definite. From

$$\frac{\partial\varphi^{-1}(p_\theta)}{\partial\theta^i} = u_i - \frac{\partial\psi}{\partial\theta^i}, \qquad -\frac{\partial^2\varphi^{-1}(p_\theta)}{\partial\theta^i\partial\theta^j} = -\frac{\partial^2\psi}{\partial\theta^i\partial\theta^j},$$

it follows that $g_{ij} = \frac{\partial^2\psi}{\partial\theta^i\partial\theta^j}$.

The next two results show how the generalization of Rényi divergence and the φ-divergence are related to the normalizing function in φ-families.

Proposition 2. *In a φ-family $\mathcal{F}_p$, the generalization of Rényi divergence for $\alpha \in (-1,1)$ can be expressed in terms of the normalizing function ψ as follows:*

$$\mathcal{D}_\varphi^{(\alpha)}(p_\theta \parallel p_\vartheta) = \frac{2}{1+\alpha}\psi(\theta) + \frac{2}{1-\alpha}\psi(\vartheta) - \frac{4}{1-\alpha^2}\psi\Big(\frac{1-\alpha}{2}\theta + \frac{1+\alpha}{2}\vartheta\Big), \tag{29}$$

for all $\theta, \vartheta \in \Theta$.

Proof. Recall the definition of $\kappa(\alpha)$ as the real number for which

$$\int_T \varphi\Big(\frac{1-\alpha}{2}\varphi^{-1}(p_\theta) + \frac{1+\alpha}{2}\varphi^{-1}(p_\vartheta) + \kappa(\alpha)u_0\Big)d\mu = 1.$$

Using expression (28) for probability distributions in $\mathcal{F}_p$, we can write

$$\begin{aligned}\frac{1-\alpha}{2}\varphi^{-1}(p_\theta) &+ \frac{1+\alpha}{2}\varphi^{-1}(p_\vartheta) + \kappa(\alpha)u_0 \\ &= c + \sum_{i=1}^n\Big(\frac{1-\alpha}{2}\theta^i + \frac{1+\alpha}{2}\vartheta^i\Big)u_i - \Big(\frac{1-\alpha}{2}\psi(\theta) + \frac{1+\alpha}{2}\psi(\vartheta) - \kappa(\alpha)\Big)u_0 \\ &= c + \sum_{i=1}^n\Big(\frac{1-\alpha}{2}\theta^i + \frac{1+\alpha}{2}\vartheta^i\Big)u_i - \psi\Big(\frac{1-\alpha}{2}\theta + \frac{1+\alpha}{2}\vartheta\Big)u_0.\end{aligned}$$

The last equality is a consequence of the domain Θ being convex. Thus, it follows that

$$\kappa(\alpha) = \frac{1-\alpha}{2}\psi(\theta) + \frac{1+\alpha}{2}\psi(\vartheta) - \psi\Big(\frac{1-\alpha}{2}\theta + \frac{1+\alpha}{2}\vartheta\Big).$$

By the definition of $\mathcal{D}_\varphi^{(\alpha)}(\cdot \parallel \cdot)$, we get (29). □

Proposition 3. *In a φ-family $\mathcal{F}_p$, the φ-divergence is related to the normalizing function ψ by the equality*

$$\mathcal{D}_\varphi(p_\theta \parallel p_\vartheta) = \psi(\vartheta) - \psi(\theta) - \nabla\psi(\theta)\cdot(\vartheta - \theta), \tag{30}$$

for all $\theta, \vartheta \in \Theta$.

Proof. To show (30), we use

$$\frac{\partial\psi}{\partial\theta^i}(\theta) = \frac{\int_T u_i\varphi'(\varphi^{-1}(p_\theta))d\mu}{\int_T u_0\varphi'(\varphi^{-1}(p_\theta))d\mu},$$

which is a consequence of (Lemma 10 in [12]). In view of $(\varphi^{-1})'(u) = 1/\varphi'(\varphi^{-1}(u))$, expression (13) with $p = p_\theta$ and $q = p_\vartheta$ results in

$$\mathcal{D}_\varphi(p_\theta \parallel p_\vartheta) = \frac{\int_T[\varphi^{-1}(p_\theta) - \varphi^{-1}(p_\vartheta)]\varphi'(\varphi^{-1}(p_\theta))d\mu}{\int_T u_0\varphi'(\varphi^{-1}(p_\theta))d\mu}. \tag{31}$$

Inserting into (31) the difference

$$\begin{aligned}\varphi^{-1}(p_\theta) - \varphi^{-1}(p_\vartheta) &= \Big(c + \sum_{i=1}^n\theta^i u_i - \psi(\theta)u_0\Big) - \Big(c + \sum_{i=1}^n\vartheta^i u_i - \psi(\vartheta)u_0\Big) \\ &= \psi(\vartheta)u_0 - \psi(\theta)u_0 - \sum_{i=1}^n(\vartheta^i - \theta^i)u_i,\end{aligned}$$

we get expression (30). □

In Proposition 2, the expression on the right-hand side of Equation (29) defines a divergence on its own, which was investigated by Jun Zhang in [19]. Proposition 3 asserts that the φ-divergence $\mathcal{D}_\varphi(p_\vartheta \parallel p_\theta)$ coincides with the Bregman divergence [31,32] associated with the normalizing function ψ for points ϑ and θ in Θ. Because ψ is convex and attains a minimum at $\theta = 0$, it follows that $\frac{\partial\psi}{\partial\theta^i}(\theta) = 0$ at $\theta = 0$. As a result, equality (30) reduces to $\mathcal{D}_\varphi(p \parallel p_\theta) = \psi(\theta)$.

4.3. Geometry Induced by $\mathcal{D}^{(\alpha)}_\varphi(\cdot \parallel \cdot)$

In this section, we assume that $\varphi(\cdot)$ is continuously differentiable and strictly convex. The latter assumption guarantees that

$$\mathcal{D}^{(\alpha)}_\varphi(p \parallel q) = 0 \quad \text{if and only if} \quad p = q. \tag{32}$$

The generalized Rényi divergence induces a metric $g = (g_{ij})$ in generalized statistical manifolds $\mathcal{P}$. This metric is given by

$$g_{ij} = -\Big[\Big(\frac{\partial}{\partial\theta^i}\Big)_p \Big(\frac{\partial}{\partial\theta^j}\Big)_q \mathcal{D}^\alpha_\varphi(p \parallel q)\Big]_{q=p}. \tag{33}$$

To show that this expression defines a metric, we have to verify that g_{ij} is invariant under change of coordinates, and (g_{ij}) is positive definite. The first claim follows from the chain rule. The positive definiteness of (g_{ij}) is a consequence of Proposition 4, which is given below.

Proposition 4. *The metric induced by $\mathcal{D}^{(\alpha)}_\varphi(\cdot \parallel \cdot)$ coincides with the metric given by (22) or (25).*

Proof. Fix any $\alpha \in (-1,1)$. Applying the operator $(\frac{\partial}{\partial\theta^j})_{p_\theta}$ to

$$\int_T \varphi(c_\alpha)d\mu = 1,$$

where $c_\alpha = \frac{1-\alpha}{2}\varphi^{-1}(p_\theta) + \frac{1+\alpha}{2}\varphi^{-1}(p_\vartheta) + \kappa(\alpha)u_0$, we obtain

$$\int_T \Big(\frac{1+\alpha}{2}\frac{\partial\varphi^{-1}(p_\vartheta)}{\partial\theta^j} + \Big(\frac{\partial}{\partial\theta^j}\Big)_{p_\vartheta}\kappa(\alpha)u_0\Big)\varphi'(c_\alpha)d\mu = 0,$$

which results in

$$\Big(\frac{\partial}{\partial\theta^j}\Big)_{p_\vartheta}\kappa(\alpha) = -\frac{1+\alpha}{2}\frac{\int_T \frac{\partial\varphi^{-1}(p_\vartheta)}{\partial\theta^j}\varphi'(c_\alpha)d\mu}{\int_T u_0\varphi'(c_\alpha)d\mu}.$$

By the standard differentiation rules, we can write

$$\begin{aligned}\Big(\frac{\partial}{\partial\theta^i}\Big)_{p_\theta}\Big(\frac{\partial}{\partial\theta^j}\Big)_{p_\vartheta}\kappa(\alpha) = &-\frac{1+\alpha}{2}\frac{\int_T[\frac{1-\alpha}{2}\frac{\partial\varphi^{-1}(p_\theta)}{\partial\theta^i} + (\frac{\partial}{\partial\theta^i})_{p_\theta}\kappa(\alpha)u_0]\frac{\partial\varphi^{-1}(p_\vartheta)}{\partial\theta^j}\varphi''(c_\alpha)d\mu}{\int_T u_0\varphi'(c_\alpha)d\mu} \\ &+\frac{1+\alpha}{2}\frac{\int_T \frac{\partial\varphi^{-1}(p_\vartheta)}{\partial\theta^j}\varphi'(c_\alpha)d\mu}{\int_T u_0\varphi'(c_\alpha)d\mu}\frac{\int_T u_0[\frac{1-\alpha}{2}\frac{\partial\varphi^{-1}(p_\theta)}{\partial\theta^i} + (\frac{\partial}{\partial\theta^i})_{p_\theta}\kappa(\alpha)u_0]\varphi''(c_\alpha)d\mu}{\int_T u_0\varphi'(c_\alpha)d\mu}.\end{aligned} \tag{34}$$

Noticing that $\int_T \frac{\partial\varphi^{-1}(p_\vartheta)}{\partial\theta^j}\varphi'(c_\alpha)d\mu = 0$ for $p_\vartheta = p_\theta$, the second term on the right-hand side of Equation (34) vanishes, and then

$$\Big[\Big(\frac{\partial}{\partial\theta^i}\Big)_{p_\theta}\Big(\frac{\partial}{\partial\theta^j}\Big)_{p_\vartheta}\kappa(\alpha)\Big]_{p_\vartheta=p_\theta} = -\frac{1-\alpha^2}{4}\frac{\int_T \frac{\partial\varphi^{-1}(p_\theta)}{\partial\theta^i}\frac{\partial\varphi^{-1}(p_\theta)}{\partial\theta^j}\varphi''(\varphi^{-1}(p_\theta))d\mu}{\int_T u_0\varphi'(\varphi^{-1}(p_\theta))d\mu}.$$

If we use the notation introduced in (24), we can write

$$g_{ij} = -\Big[\Big(\frac{\partial}{\partial\theta^i}\Big)_{p_\theta}\Big(\frac{\partial}{\partial\theta^j}\Big)_{p_\vartheta}\mathcal{D}_\varphi^{(\alpha)}(p_\theta \parallel p_\vartheta)\Big]_{p_\vartheta=p_\theta} = E''_\theta\Big[\frac{\partial\varphi^{-1}(p_\theta)}{\partial\theta^i}\frac{\partial\varphi^{-1}(p_\theta)}{\partial\theta^j}\Big].$$

It remains to show the case $\alpha = \pm1$. Comparing (13) and (23), we can write

$$\mathcal{D}_\varphi(p_\theta \parallel p_\vartheta) = E'_\theta[\varphi^{-1}(p_\theta) - \varphi^{-1}(p_\vartheta)]. \tag{35}$$

We use the equivalent expressions

$$g_{ij} = \Big[\Big(\frac{\partial^2}{\partial\theta^i\partial\theta^j}\Big)_p\mathcal{D}_\varphi^\alpha(p \parallel q)\Big]_{q=p} = \Big[\Big(\frac{\partial^2}{\partial\theta^i\partial\theta^j}\Big)_q\mathcal{D}_\varphi^\alpha(p \parallel q)\Big]_{q=p},$$

which follows from condition (32), to infer that

$$g_{ij} = \Big[\Big(\frac{\partial^2}{\partial\theta^i\partial\theta^j}\Big)_{p_\theta}\mathcal{D}_\varphi(p_\theta \parallel p_\vartheta)\Big]_{p_\vartheta=p_\theta} = -E'_\theta\Big[\frac{\partial^2\varphi^{-1}(p_\theta)}{\partial\theta^i\partial\theta^j}\Big]. \tag{36}$$

Because $\mathcal{D}_\varphi^{(-1)}(p \parallel q) = \mathcal{D}_\varphi^{(1)}(q \parallel p) = \mathcal{D}_\varphi(p \parallel q)$, we conclude that the metric defined by (22) coincides with the metric induced by $\mathcal{D}_\varphi^{(-1)}(\cdot \parallel \cdot)$ and $\mathcal{D}_\varphi^{(1)}(\cdot \parallel \cdot)$. □

In generalized statistical manifolds, the generalized Rényi divergence $\mathcal{D}_\varphi^{(\alpha)}(\cdot \parallel \cdot)$ induces a connection $D^{(\alpha)}$, whose Christoffel symbols $\Gamma_{ijk}^{(\alpha)}$ are given by

$$\Gamma_{ijk}^{(\alpha)} = -\Big[\Big(\frac{\partial^2}{\partial\theta^i\partial\theta^j}\Big)_p\Big(\frac{\partial}{\partial\theta^k}\Big)_q\mathcal{D}_\varphi^{(\alpha)}(p \parallel q)\Big]_{q=p}.$$

Because $\mathcal{D}_\varphi^{(\alpha)}(p \parallel q) = \mathcal{D}_\varphi^{(-\alpha)}(q \parallel p)$, it follows that $D^{(\alpha)}$ and $D^{(-\alpha)}$ are mutually dual for any $\alpha \in [-1,1]$. In other words, $\Gamma_{ijk}^{(\alpha)}$ and $\Gamma_{ijk}^{(-\alpha)}$ satisfy the relation $\frac{\partial g_{jk}}{\partial\theta^i} = \Gamma_{ijk}^{(\alpha)} + \Gamma_{ikj}^{(-\alpha)}$. A development involving expression (35) results in

$$\Gamma_{ijk}^{(1)} = E''_\theta\Big[\frac{\partial^2\varphi^{-1}(p_\theta)}{\partial\theta^i\partial\theta^j}\frac{\partial\varphi^{-1}(p_\theta)}{\partial\theta^k}\Big] - E'_\theta\Big[\frac{\partial^2\varphi^{-1}(p_\theta)}{\partial\theta^i\partial\theta^j}\Big]E''_\theta\Big[u_0\frac{\partial\varphi^{-1}(p_\theta)}{\partial\theta^k}\Big], \tag{37}$$

and

$$\begin{aligned}\Gamma_{ijk}^{(-1)} = {} & E''_\theta\Big[\frac{\partial^2\varphi^{-1}(p_\theta)}{\partial\theta^i\partial\theta^j}\frac{\partial\varphi^{-1}(p_\theta)}{\partial\theta^k}\Big] + E'''_\theta\Big[\frac{\partial\varphi^{-1}(p_\theta)}{\partial\theta^i}\frac{\partial\varphi^{-1}(p_\theta)}{\partial\theta^j}\frac{\partial\varphi^{-1}(p_\theta)}{\partial\theta^k}\Big] \\ & - E''_\theta\Big[\frac{\partial\varphi^{-1}(p_\theta)}{\partial\theta^j}\frac{\partial\varphi^{-1}(p_\theta)}{\partial\theta^k}\Big]E''_\theta\Big[u_0\frac{\partial\varphi^{-1}(p_\theta)}{\partial\theta^i}\Big] \\ & - E''_\theta\Big[\frac{\partial\varphi^{-1}(p_\theta)}{\partial\theta^i}\frac{\partial\varphi^{-1}(p_\theta)}{\partial\theta^k}\Big]E''_\theta\Big[u_0\frac{\partial\varphi^{-1}(p_\theta)}{\partial\theta^j}\Big].\end{aligned} \tag{38}$$

For $\alpha \in (-1,1)$, the Christoffel symbols $\Gamma_{ijk}^{(\alpha)}$ can be written as a convex combination of $\Gamma_{ijk}^{(-1)}$ and $\Gamma_{ijk}^{(-1)}$, as asserted in the next result.

Proposition 5. *The Christoffel symbols $\Gamma_{ijk}^{(\alpha)}$ induced by the divergence $\mathcal{D}_\varphi^{(\alpha)}(\cdot \parallel \cdot)$ satisfy the relation*

$$\Gamma_{ijk}^{(\alpha)} = \frac{1-\alpha}{2}\Gamma_{ijk}^{(-1)} + \frac{1+\alpha}{2}\Gamma_{ijk}^{(1)}, \qquad \text{for } \alpha \in [-1,1]. \tag{39}$$

Proof. For $\alpha = \pm 1$, equality (39) follows trivially. Thus, we assume $\alpha \in (-1,1)$. By (34), we can write

$$\Big(\frac{\partial}{\partial\theta^i}\Big)_{p_\theta}\Big(\frac{\partial}{\partial\theta^k}\Big)_{p_\theta}\kappa(\alpha) = -\frac{1+\alpha}{2}\frac{\int_T[\frac{1-\alpha}{2}\frac{\partial\varphi^{-1}(p_\theta)}{\partial\theta^i} + (\frac{\partial}{\partial\theta^i})_{p_\theta}\kappa(\alpha)u_0]\frac{\partial\varphi^{-1}(p_\theta)}{\partial\theta^k}\varphi''(c_\alpha)d\mu}{\int_T u_0\varphi'(c_\alpha)d\mu} + \frac{1+\alpha}{2}\frac{\int_T\frac{\partial\varphi^{-1}(p_\theta)}{\partial\theta^k}\varphi'(c_\alpha)d\mu}{\int_T u_0\varphi'(c_\alpha)d\mu}\frac{\int_T u_0[\frac{1-\alpha}{2}\frac{\partial\varphi^{-1}(p_\theta)}{\partial\theta^i} + (\frac{\partial}{\partial\theta^i})_{p_\theta}\kappa(\alpha)u_0]\varphi''(c_\alpha)d\mu}{\int_T u_0\varphi'(c_\alpha)d\mu}. \tag{40}$$

Applying $(\frac{\partial}{\partial\theta^j})_{p_\theta}$ to the first term on the right-hand side of (40), and then equating $p_\theta = p_\theta$, we obtain

$$-\frac{1-\alpha^2}{4}E''_\theta\Big[\frac{\partial^2\varphi^{-1}(p_\theta)}{\partial\theta^i\partial\theta^j}\frac{\partial\varphi^{-1}(p_\theta)}{\partial\theta^k}\Big] - \frac{1+\alpha}{2}\Big(\frac{\partial^2}{\partial\theta^i\partial\theta^j}\Big)_{p_\theta}\kappa(\alpha)E''_\theta\Big[u_0\frac{\partial\varphi^{-1}(p_\theta)}{\partial\theta^k}\Big] - \frac{1-\alpha^2}{4}\frac{1-\alpha}{2}E'''_\theta\Big[\frac{\partial\varphi^{-1}(p_\theta)}{\partial\theta^i}\frac{\partial\varphi^{-1}(p_\theta)}{\partial\theta^j}\frac{\partial\varphi^{-1}(p_\theta)}{\partial\theta^k}\Big] + \frac{1-\alpha^2}{4}\frac{1-\alpha}{2}E''_\theta\Big[\frac{\partial\varphi^{-1}(p_\theta)}{\partial\theta^i}\frac{\partial\varphi^{-1}(p_\theta)}{\partial\theta^k}\Big]E''_\theta\Big[u_0\frac{\partial\varphi^{-1}(p_\theta)}{\partial\theta^j}\Big]. \tag{41}$$

Similarly, if we apply $(\frac{\partial}{\partial\theta^j})_{p_\theta}$ to the second term on the right-hand side of (40), and make $p_\theta = p_\theta$, we get

$$\frac{1-\alpha^2}{4}\frac{1-\alpha}{2}E''_\theta\Big[\frac{\partial\varphi^{-1}(p_\theta)}{\partial\theta^j}\frac{\partial\varphi^{-1}(p_\theta)}{\partial\theta^k}\Big]E''_\theta\Big[u_0\frac{\partial\varphi^{-1}(p_\theta)}{\partial\theta^i}\Big]. \tag{42}$$

Collecting (41) and (42), we can write

$$\begin{aligned}\Gamma^{(\alpha)}_{ijk} &= -\frac{4}{1-\alpha^2}\Big[\Big(\frac{\partial^2}{\partial\theta^i\partial\theta^j}\Big)_{p_\theta}\Big(\frac{\partial}{\partial\theta^k}\Big)_{p_\theta}\kappa(\alpha)\Big]_{p_\theta=p_\theta} \\ &= E''_\theta\Big[\frac{\partial^2\varphi^{-1}(p_\theta)}{\partial\theta^i\partial\theta^j}\frac{\partial\varphi^{-1}(p_\theta)}{\partial\theta^k}\Big] + \frac{1-\alpha}{2}E'''_\theta\Big[\frac{\partial\varphi^{-1}(p_\theta)}{\partial\theta^i}\frac{\partial\varphi^{-1}(p_\theta)}{\partial\theta^j}\frac{\partial\varphi^{-1}(p_\theta)}{\partial\theta^k}\Big] \\ &\quad - \frac{1-\alpha}{2}E''_\theta\Big[\frac{\partial\varphi^{-1}(p_\theta)}{\partial\theta^j}\frac{\partial\varphi^{-1}(p_\theta)}{\partial\theta^k}\Big]E''_\theta\Big[u_0\frac{\partial\varphi^{-1}(p_\theta)}{\partial\theta^i}\Big] \\ &\quad - \frac{1-\alpha}{2}E''_\theta\Big[\frac{\partial\varphi^{-1}(p_\theta)}{\partial\theta^i}\frac{\partial\varphi^{-1}(p_\theta)}{\partial\theta^k}\Big]E''_\theta\Big[u_0\frac{\partial\varphi^{-1}(p_\theta)}{\partial\theta^j}\Big] \\ &\quad - \frac{1+\alpha}{2}E'_\theta\Big[\frac{\partial^2\varphi^{-1}(p_\theta)}{\partial\theta^i\partial\theta^j}\Big]E''_\theta\Big[u_0\frac{\partial\varphi^{-1}(p_\theta)}{\partial\theta^k}\Big],\end{aligned} \tag{43}$$

where we used

$$\begin{aligned}\Big(\frac{\partial^2}{\partial\theta^i\partial\theta^j}\Big)_{p_\theta}\kappa(\alpha) &= \frac{1-\alpha^2}{4}\Big[\Big(\frac{\partial^2}{\partial\theta^i\partial\theta^j}\Big)_{p_\theta}\mathcal{D}^{(\alpha)}_\varphi(p_\theta \parallel p_\theta)\Big]_{p_\theta=p_\theta} \\ &= \frac{1-\alpha^2}{4}g_{ij} = -\frac{1-\alpha^2}{4}E'_\theta\Big[\frac{\partial^2\varphi^{-1}(p_\theta)}{\partial\theta^i\partial\theta^i}\Big].\end{aligned}$$

Expression (39) follows from (37), (38) and (43). □

5. Conclusions

In [17,18], the authors introduced a pair of dual connections $D^{(-1)}$ and $D^{(1)}$ induced by φ-divergence. The main motivation of the present work was to find a (non-trivial) family of α-divergences, whose induced α-connections are convex combinations of $D^{(-1)}$ and $D^{(1)}$. As a result of our efforts, we proposed a generalization of Rényi divergence. The connection $D^{(\alpha)}$ induced by the generalization of Rényi divergence satisfies the relation $D^{(\alpha)} = \frac{1-\alpha}{2}D^{(-1)} + \frac{1+\alpha}{2}D^{(1)}$. To generalize Rényi divergence, we made use of properties of φ-functions. This makes evident the importance of φ-functions in the geometry of non-standard models. In standard statistical manifolds, even though Amari's α-divergence and Rényi divergence (with $\alpha \in [-1,1]$) do not coincide, they induce the same family of α-connections. This striking result requires further investigation. Future work should focus on how the generalization of Rényi divergence is related to Zhang's (ρ, τ)-divergence, and also how the present proposal is related to the model presented in [33].

Acknowledgments: The authors are indebted to the anonymous reviewers for their valuable comments and corrections, which led to a great improvement of this paper. Charles C. Cavalcante also thanks the CNPq (Proc. 309055/2014-8) for partial funding.

Author Contributions: All authors contributed equally to the design of the research. The research was carried out by all authors. Rui F. Vigelis and Charles C. Cavalcante gave the central idea of the paper and managed the organization of it. Rui F. Vigelis wrote the paper. All the authors read and approved the final manuscript.

Conflicts of Interest: The authors declare no conflict of interest.

References

1. Rao, C.R. Information and the accuracy attainable in the estimation of statistical parameters. *Bull. Calcutta Math. Soc.* **1945**, *37*, 81–91.
2. Amari, S.-I. Differential geometry of curved exponential families—Curvatures and information loss. *Ann. Stat.* **1982**, *10*, 357–385.
3. Amari, S.-I. *Differential-Geometrical Methods in Statistics*; Springer: Berlin/Heidelberg, Germany, 1985; Volume 28.
4. Amari, S.-I.; Nagaoka, H. *Methods of Information Geometry (Translations of Mathematical Monographs)*; American Mathematical Society: Providence, RI, USA, 2000; Volume 191.
5. Amari, S.-I. *Information Geometry and Its Applications*; Applied Mathematical Sciences Series; Springer: Berlin/Heidelberg, Germany, 2016; Volume 194.
6. Amari, S.-I.; Ohara, A.; Matsuzoe, H. Geometry of deformed exponential families: Invariant, dually-flat and conformal geometries. *Physica A* **2012**, *391*, 4308–4319.
7. Matsuzoe, H. Hessian structures on deformed exponential families and their conformal structures. *Differ. Geom. Appl.* **2014**, *35* (Suppl.), 323–333.
8. Naudts, J. Estimators, escort probabilities, and ϕ-exponential families in statistical physics. *J. Inequal. Pure Appl. Math.* **2004**, *5*, 102.
9. Pistone, G. κ-exponential models from the geometrical viewpoint. *Eur. Phys. J. B* **2009**, *70*, 29–37.
10. Amari, S.-I.; Ohara, A. Geometry of q-exponential family of probability distributions. *Entropy* **2011**, *13*, 1170–1185.
11. Vigelis, R.F.; Cavalcante, C.C. The Δ_2-Condition and φ-Families of Probability Distributions. In *Geometric Science of Information*; Springer: Berlin/Heidelberg, Germany, 2013; Volume 8085, pp. 729–736.
12. Vigelis, R.F.; Cavalcante, C.C. On φ-families of probability distributions. *J. Theor. Probab.* **2013**, *26*, 870–884.
13. Cena, A.; Pistone, G. Exponential statistical manifold. *Ann. Inst. Stat. Math.* **2007**, *59*, 27–56.
14. Grasselli, M.R. Dual connections in nonparametric classical information geometry. *Ann. Inst. Stat. Math.* **2010**, *62*, 873–896.
15. Pistone, G.; Sempi, C. An infinite-dimensional geometric structure on the space of all the probability measures equivalent to a given one. *Ann. Stat.* **1995**, *23*, 1543–1561.
16. Santacroce, M.; Siri, P.; Trivellato, B. New results on mixture and exponential models by Orlicz spaces. *Bernoulli* **2016**, *22*, 1431–1447.
17. Vigelis, R.F.; Cavalcante, C.C. Information Geometry: An Introduction to New Models for Signal Processing. In *Signals and Images*; CRC Press: Boca Raton, FL, USA, 2015; pp. 455–491.
18. Vigelis, R.F.; de Souza, D.C.; Cavalcante, C.C. New Metric and Connections in Statistical Manifolds. In *Geometric Science of Information*; Springer: Berlin/Heidelberg, Germany, 2015; Volume 9389, pp. 222–229.
19. Zhang, J. Divergence function, duality, and convex analysis. *Neural Comput.* **2004**, *16*, 159–195.
20. Zhang, J. Referential Duality and Representational Duality on Statistical Manifolds. In Proceedings of the 2nd International Symposium on Information Geometry and Its Applications, Pescara, Italy, 12–16 December 2005; pp. 58–67.
21. Zhang, J. Nonparametric information geometry: From divergence function to referential-representational biduality on statistical manifolds. *Entropy* **2013**, *15*, 5384–5418.
22. Zhang, J. Divergence Functions and Geometric Structures They Induce on a Manifold. In *Geometric Theory of Information*; Springer: Berlin/Heidelberg, Germany, 2014; pp. 1–30.

23. Zhang, J. On monotone embedding in information geometry. *Entropy* **2015**, *17*, 4485–4489.
24. Rényi, A. On measures of entropy and information. In *Proceedings of 4th Berkeley Symposium on Mathematical Statistics and Probability*; University California Press: Berkeley, CA, USA, 1961; Volume I, pp. 547–561.
25. Van Erven, T.; Harremoës, P. Rényi divergence and Kullback-Leibler divergence. *IEEE Trans. Inform. Theory* **2014**, *60*, 3797–3820.
26. Eguchi, S.; Komori, O. Path Connectedness on a Space of Probability Density Functions. In *Geometric Science of Information*; Springer: Berlin/Heidelberg, Germany, 2015; Volume 9389, pp. 615–624.
27. Kaniadakis, G.; Lissia, M.; Scarfone, A.M. Deformed logarithms and entropies. *Physica A* **2004**, *340*, 41–49.
28. Kaniadakis, G. Theoretical foundations and mathematical formalism of the power-law tailed statistical distributions. *Entropy* **2013**, *15*, 3983–4010.
29. Musielak, J. *Orlicz Spaces and Modular Spaces*; Springer: Berlin/Heidelberg, Germany, 1983; Volume 1034.
30. Shima, H. *The Geometry of Hessian Structures*; World Scientific: Singapore, 2007.
31. Banerjee, A.; Merugu, S.; Dhillon, I.S.; Ghosh, J. Clustering with Bregman Divergences. *J. Mach. Learn. Res.* **2005**, *6*, 1705–1749.
32. Bregman, L.M. The relaxation method of finding the common point of convex sets and its application to the solution of problems in convex programming. *USSR Comput. Math. Math. Phys.* **1967**, *7*, 200–217.
33. Zhanga, J.; Hästö, P. Statistical manifold as an affine space: A functional equation approach. *J. Math. Psychol.* **2006**, *50*, 60–65.

entropy

MDPI

Article

Guaranteed Bounds on Information-Theoretic Measures of Univariate Mixtures Using Piecewise Log-Sum-Exp Inequalities

Frank Nielsen [1,2,*] and Ke Sun [3]

1 Computer Science Department LIX, École Polytechnique, 91128 Palaiseau Cedex, France
2 Sony Computer Science Laboratories Inc, Tokyo 141-0022, Japan
3 King Abdullah University of Science and Technology, Thuwal 23955, Saudi Arabia; sunk.edu@gmail.com
* Correspondence: Frank.Nielsen@acm.org; Tel.: +33-1-7757-8070

Academic Editor: Antonio M. Scarfone
Received: 20 October 2016; Accepted: 5 December 2016; Published: 9 December 2016

Abstract: Information-theoretic measures, such as the entropy, the cross-entropy and the Kullback–Leibler divergence between two mixture models, are core primitives in many signal processing tasks. Since the Kullback–Leibler divergence of mixtures provably does not admit a closed-form formula, it is in practice either estimated using costly Monte Carlo stochastic integration, approximated or bounded using various techniques. We present a fast and generic method that builds algorithmically closed-form lower and upper bounds on the entropy, the cross-entropy, the Kullback–Leibler and the α-divergences of mixtures. We illustrate the versatile method by reporting our experiments for approximating the Kullback–Leibler and the α-divergences between univariate exponential mixtures, Gaussian mixtures, Rayleigh mixtures and Gamma mixtures.

Keywords: information geometry; mixture models; α-divergences; log-sum-exp bounds

1. Introduction

Mixture models are commonly used in signal processing. A typical scenario is to use mixture models [1–3] to smoothly model histograms. For example, Gaussian Mixture Models (GMMs) can be used to convert grey-valued images into binary images by building a GMM fitting the image intensity histogram and then choosing the binarization threshold as the average of the Gaussian means [1]. Similarly, Rayleigh Mixture Models (RMMs) are often used in ultrasound imagery [2] to model histograms, and perform segmentation by classification. When using mixtures, a fundamental primitive is to define a proper statistical distance between them. The Kullback–Leibler (KL) divergence [4], also called relative entropy or information discrimination, is the most commonly used distance. Hence the main target of this paper is to faithfully measure the KL divergence. Let $m(x) = \sum_{i=1}^{k} w_i p_i(x)$ and $m'(x) = \sum_{i=1}^{k'} w'_i p'_i(x)$ be two finite statistical density mixtures of k and k' components, respectively. Notice that the Cumulative Density Function (CDF) of a mixture is like its density also a convex combinations of the component CDFs. However, beware that a mixture is not a sum of random variables (RVs). Indeed, sums of RVs have convolutional densities. In statistics, the mixture components $p_i(x)$ are often parametric: $p_i(x) = p(x;\theta_i)$, where θ_i is a vector of parameters. For example, a mixture of Gaussians (MoG also used as a shortcut instead of GMM) has each component distribution parameterized by its mean μ_i and its covariance matrix Σ_i (so that the parameter vector is $\theta_i = (\mu_i, \Sigma_i)$). Let $\mathcal{X} = \{x \in \mathbb{R} : p(x;\theta) > 0\}$ be the support of the component distributions. Denote by $H_\times(m, m') = -\int_{\mathcal{X}} m(x) \log m'(x) \mathrm{d}x$ the cross-entropy [4] between two continuous mixtures of densities m and m', and denote by $H(m) = H_\times(m,m) = \int_{\mathcal{X}} m(x)\log\frac{1}{m(x)}\mathrm{d}x = -\int_{\mathcal{X}} m(x)\log m(x)\mathrm{d}x$ the Shannon entropy [4]. Then the Kullback–Leibler divergence between m and m' is given by:

$$\mathrm{KL}(m:m') = H_\times(m,m') - H(m) = \int_{\mathcal{X}} m(x)\log\frac{m(x)}{m'(x)}\mathrm{d}x \geq 0. \quad (1)$$

The notation ":" is used instead of the usual comma "," notation to emphasize that the distance is not a metric distance since neither is it symmetric ($\mathrm{KL}(m:m') \neq \mathrm{KL}(m':m)$), nor does it satisfy the triangular inequality [4] of metric distances ($\mathrm{KL}(m:m') + \mathrm{KL}(m':m'') \not\geq \mathrm{KL}(m:m'')$). When the natural base of the logarithm is chosen, we get a differential entropy measure expressed in nat units. Alternatively, we can also use the base-2 logarithm ($\log_2 x = \frac{\log x}{\log 2}$) and get the entropy expressed in bit units. Although the KL divergence is available in closed-form for many distributions (in particular as equivalent Bregman divergences for exponential families [5], see Appendix C), it was proven that the Kullback–Leibler divergence between two (univariate) GMMs is not analytic [6] (see also the particular case of a GMM of two components with the same variance that was analyzed in [7]). See Appendix A for an analysis. Note that the differential entropy may be negative. For example, the differential entropy of a univariate Gaussian distribution is $\log(\sigma\sqrt{2\pi e})$, and is therefore negative when the standard variance $\sigma < \frac{1}{\sqrt{2\pi e}} \approx 0.242$. We consider continuous distributions with entropies well-defined (entropy may be undefined for singular distributions like Cantor's distribution [8]).

1.1. Prior Work

Many approximation techniques have been designed to beat the computationally intensive Monte Carlo (MC) stochastic estimation: $\widehat{\mathrm{KL}}_s(m:m') = \frac{1}{s}\sum_{i=1}^{s}\log\frac{m(x_i)}{m'(x_i)}$ with $x_1,\ldots,x_s \sim m(x)$ (s independently and identically distributed (i.i.d.) samples $x_1,\ldots,x_s$). The MC estimator is asymptotically consistent, $\lim_{s\to\infty}\widehat{\mathrm{KL}}_s(m:m') = \mathrm{KL}(m:m')$, so that the "true value" of the KL of mixtures is estimated in practice by taking a very large sample (say, $s = 10^9$). However, we point out that the MC estimator gives as output a stochastic *approximation*, and therefore does not guarantee deterministic bounds (confidence intervals may be used). Deterministic lower and upper bounds of the integral can be obtained by various numerical integration techniques using quadrature rules. We refer to [9–12] for the current state-of-the-art approximation techniques and bounds on the KL of GMMs. The latest work for computing the entropy of GMMs is [13]. It considers arbitrary finely tuned bounds of the entropy of isotropic Gaussian mixtures (a case encountered when dealing with KDEs, kernel density estimators). However, there is a catch in the technique of [13]: It relies on solving the unique roots of some log-sum-exp equations (See Theorem 1 of [13], p. 3342) that do not admit a closed-form solution. Thus it is a hybrid method that contrasts with our combinatorial approach. Bounds of the KL divergence between mixture models can be generalized to bounds of the likelihood function of mixture models [14], because log-likelihood is just the KL between the empirical distribution and the mixture model up to a constant shift.

In information geometry [15], a mixture family of linearly independent probability distributions $p_1(x),\ldots,p_k(x)$ is defined by the convex combination of those non-parametric component distributions: $m(x;\eta) = \sum_{i=1}^{k}\eta_i p_i(x)$ with $\eta_i > 0$ and $\sum_{i=1}^{k}\eta_i = 1$. A mixture family induces a dually flat space where the Kullback–Leibler divergence is equivalent to a Bregman divergence [5,15] defined on the η-parameters. However, in that case, the Bregman convex generator $F(\eta) = \int m(x;\eta)\log m(x;\eta)\mathrm{d}x$ (the Shannon information) is not available in closed-form. Except for the family of multinomial distributions that is both a mixture family (with closed-form $\mathrm{KL}(m:m') = \sum_{i=1}^{k} m_i\log\frac{m_i}{m'_i}$, the discrete KL [4]) and an exponential family [15].

1.2. Contributions

In this work, we present a simple and efficient method that builds algorithmically a closed-form formula that guarantees both deterministic lower and upper bounds on the KL divergence within an additive factor of $\log k + \log k'$. We then further refine our technique to get improved adaptive bounds. For univariate GMMs, we get the non-adaptive bounds in $O(k \log k + k' \log k')$ time, and the adaptive bounds in $O(k^2 + k'^2)$ time. To illustrate our generic technique, we demonstrate it based on Exponential Mixture Models (EMMs), Gamma mixtures, RMMs and GMMs. We extend our preliminary results on KL divergence [16] to other information theoretical measures such as the differential entropy and α-divergences.

1.3. Paper Outline

The paper is organized as follows. Section 2 describes the algorithmic construction of the formula using piecewise log-sum-exp inequalities for the cross-entropy and the Kullback–Leibler divergence. Section 3 instantiates this algorithmic principle to the entropy and discusses related works. Section 4 extends the proposed bounds to the family of alpha divergences. Section 5 discusses an extension of the lower bound to f-divergences. Section 6 reports our experimental results on several mixture families. Finally, Section 7 concludes this work by discussing extensions to other statistical distances. Appendix A proves that the Kullback–Leibler divergence of mixture models is not analytic [6]. Appendix B reports the closed-form formula for the KL divergence between scaled and truncated distributions of the same exponential family [17] (that include Rayleigh, Gaussian and Gamma distributions among others). Appendix C shows that the KL divergence between two mixtures can be approximated by a Bregman divergence.

2. A Generic Combinatorial Bounding Algorithm Based on Density Envelopes

Let us bound the cross-entropy $H_\times(m : m')$ by deterministic lower and upper bounds, $L_\times(m : m') \leq H_\times(m : m') \leq U_\times(m : m')$, so that the bounds on the Kullback–Leibler divergence $\mathrm{KL}(m : m') = H_\times(m : m') - H_\times(m : m)$ follows as:

$$L_\times(m : m') - U_\times(m : m) \leq \mathrm{KL}(m : m') \leq U_\times(m : m') - L_\times(m : m). \tag{2}$$

Since the cross-entropy of two mixtures $\sum_{i=1}^{k} w_i p_i(x)$ and $\sum_{j=1}^{k'} w'_j p'_j(x)$:

$$H_\times(m : m') = -\int_{\mathcal{X}} \left(\sum_{i=1}^{k} w_i p_i(x)\right) \log \left(\sum_{j=1}^{k'} w'_j p'_j(x)\right) \mathrm{d}x \tag{3}$$

has a log-sum term of positive arguments, we shall use bounds on the log-sum-exp (lse) function [18,19]:

$$\mathrm{lse}\left(\{x_i\}_{i=1}^{l}\right) = \log\left(\sum_{i=1}^{l} e^{x_i}\right).$$

We have the following basic inequalities:

$$\max\{x_i\}_{i=1}^{l} < \mathrm{lse}\left(\{x_i\}_{i=1}^{l}\right) \leq \log l + \max\{x_i\}_{i=1}^{l}. \tag{4}$$

The left-hand-side (LHS) strict inequality holds because $\sum_{i=1}^{l} e^{x_i} > \max\{e^{x_i}\}_{i=1}^{l} = \exp\left(\max\{x_i\}_{i=1}^{l}\right)$ since $e^x > 0, \forall x \in \mathbb{R}$. The right-hand-side (RHS) inequality follows from the fact that $\sum_{i=1}^{l} e^{x_i} \leq l \max\{e^{x_i}\}_{i=1}^{l} = l \exp(\max\{x_i\}_{i=1}^{l})$, and equality holds if and only if $x_1 = \cdots = x_l$. The lse function is convex but not strictly convex, see exercise 7.9 [20]. It is known [21] that the conjugate of the lse function is the negative entropy restricted to the probability simplex. The lse function enjoys the following translation identity property: $\mathrm{lse}\left(\{x_i\}_{i=1}^{l}\right) = c + \mathrm{lse}\left(\{x_i - c\}_{i=1}^{l}\right), \forall c \in \mathbb{R}$. Similarly, we can also lower bound the lse function by $\log l + \min\{x_i\}_{i=1}^{l}$. We write equivalently that for l positive numbers $x_1, \ldots, x_l$,

$$\max\left\{\log\max\{x_i\}_{i=1}^{l}, \log l + \log\min\{x_i\}_{i=1}^{l}\right\} \leq \log\sum_{i=1}^{l} x_i \leq \log l + \log\max\{x_i\}_{i=1}^{l}. \tag{5}$$

In practice, we seek matching lower and upper bounds that minimize the bound gap. The gap of that ham-sandwich inequality in Equation (5) is $\min\{\log\frac{\max_i x_i}{\min_i x_i}, \log l\}$, which is upper bounded by $\log l$.

A mixture model $\sum_{j=1}^{k'} w'_j p'_j(x)$ must satisfy

$$\begin{aligned}\max\left\{\max\{\log w'_j p'_j(x)\}_{j=1}^{k'},\ \log k' + \min\{\log w'_j p'_j(x)\}_{j=1}^{k'}\right\}\\ \leq \log\left(\sum_{j=1}^{k'} w'_j p'_j(x)\right) \leq \log k' + \max\{\log w'_j p'_j(x)\}_{j=1}^{k'}\end{aligned} \tag{6}$$

point-wisely for any $x \in \mathcal{X}$. Therefore we shall bound the integral term $\int_{\mathcal{X}} m(x)\log\left(\sum_{j=1}^{k'} w'_j p'_j(x)\right)\mathrm{d}x$ in Equation (3) using piecewise lse inequalities where the min and max are kept unchanged. We get

$$L_\times(m:m') = -\int_{\mathcal{X}} m(x)\max\{\log w'_j p'_j(x)\}_{j=1}^{k'}\mathrm{d}x - \log k', \tag{7}$$

$$U_\times(m:m') = -\int_{\mathcal{X}} m(x)\max\left\{\min\{\log w'_j p'_j(x)\}_{j=1}^{k'} + \log k',\ \max\{\log w'_j p'_j(x)\}_{j=1}^{k'}\right\}\mathrm{d}x. \tag{8}$$

In order to calculate $L_\times(m:m')$ and $U_\times(m:m')$ efficiently using closed-form formula, let us compute the upper and lower envelopes of the k' real-valued functions $\{w'_j p'_j(x)\}_{j=1}^{k'}$ defined on the support $\mathcal{X}$, that is, $\mathcal{E}_U(x) = \max\{w'_j p'_j(x)\}_{j=1}^{k'}$ and $\mathcal{E}_L(x) = \min\{w'_j p'_j(x)\}_{j=1}^{k'}$. These envelopes can be computed exactly using techniques of computational geometry [22,23] provided that we can calculate the roots of the equation $w'_r p'_r(x) = w'_s p'_s(x)$, where $w'_r p'_r(x)$ and $w'_s p'_s(x)$ are a pair of weighted components. (Although this amounts to solve quadratic equations for Gaussian or Rayleigh distributions, the roots may not always be available in closed form, e.g. in the case of Weibull distributions.)

Let the envelopes be combinatorially described by ℓ elementary interval pieces in the form $I_r = (a_r, a_{r+1})$ partitioning the support $\mathcal{X} = \uplus_{r=1}^{\ell} I_r$ (with $a_1 = \min\mathcal{X}$ and $a_{\ell+1} = \max\mathcal{X}$). Observe that on each interval I_r, the maximum of the functions $\{w'_j p'_j(x)\}_{j=1}^{k'}$ is given by $w'_{\delta(r)} p'_{\delta(r)}(x)$, where $\delta(r)$ indicates the weighted component dominating all the others, i.e., the arg max of $\{w'_j p'_j(x)\}_{j=1}^{k'}$ for any $x \in I_r$, and the minimum of $\{w'_j p'_j(x)\}_{j=1}^{k'}$ is given by $w'_{\epsilon(r)} p'_{\epsilon(r)}(x)$.

To fix ideas, when mixture components are univariate Gaussians, the upper envelope $\mathcal{E}_U(x)$ amounts to find equivalently the lower envelope of k' parabolas (see Figure 1) which has linear complexity, and can be computed in $O(k'\log k')$-time [24], or in output-sensitive time $O(k'\log\ell)$ [25], where ℓ denotes the number of parabola segments in the envelope. When the Gaussian mixture components have all the same weight and variance (e.g., kernel density estimators), the upper envelope amounts to find a lower envelope of cones: $\min_j |x - \mu'_j|$ (a Voronoi diagram in arbitrary dimension).

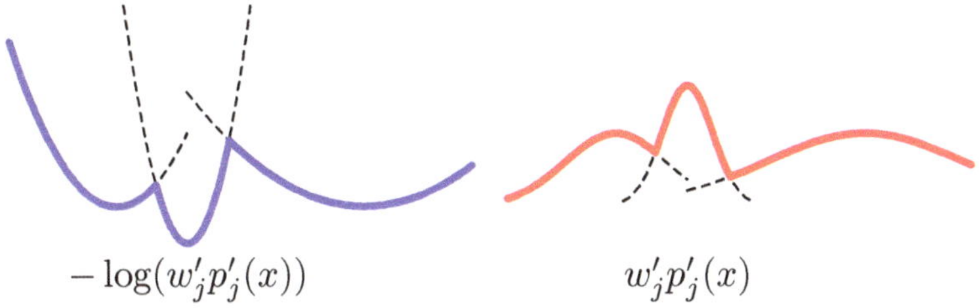

Figure 1. Lower envelope of parabolas corresponding to the upper envelope of weighted components of a Gaussian mixture with $k' = 3$ components.

To proceed once the envelopes have been built, we need to calculate two types of definite integrals on those elementary intervals: (i) the probability mass in an interval $\int_a^b p(x)\mathrm{d}x = \Phi(b) - \Phi(a)$ where Φ denotes the Cumulative Distribution Function (CDF); and (ii) the partial cross-entropy $-\int_a^b p(x)\log p'(x)\mathrm{d}x$ [26]. Thus let us define these two quantities:

$$C_{i,j}(a,b) = -\int_a^b w_i p_i(x)\log(w'_j p'_j(x))\mathrm{d}x, \quad (9)$$

$$M_i(a,b) = -\int_a^b w_i p_i(x)\mathrm{d}x. \quad (10)$$

By Equations (7) and (8), we get the bounds of $H_\times(m:m')$ as

$$L_\times(m:m') = \sum_{r=1}^{\ell}\sum_{s=1}^{k} C_{s,\delta(r)}(a_r,a_{r+1}) - \log k',$$

$$U_\times(m:m') = \sum_{r=1}^{\ell}\sum_{s=1}^{k} \min\left\{C_{s,\delta(r)}(a_r,a_{r+1}),\ C_{s,\epsilon(r)}(a_r,a_{r+1}) - M_s(a_r,a_{r+1})\log k'\right\}. \quad (11)$$

The size of the lower/upper bound formula depends on the envelope complexity ℓ, the number k of mixture components, and the closed-form expressions of the integral terms $C_{i,j}(a,b)$ and $M_i(a,b)$. In general, when a pair of weighted component densities intersect in at most p points, the envelope complexity is related to the Davenport–Schinzel sequences [27]. It is quasi-linear for bounded $p = O(1)$, see [27].

Note that in symbolic computing, the Risch semi-algorithm [28] solves the problem of computing indefinite integration in terms of elementary functions provided that there exists an oracle (hence the term "semi-algorithm") for checking whether an expression is equivalent to zero or not (however it is unknown whether there exists an algorithm implementing the oracle or not).

We presented the technique by bounding the cross-entropy (and entropy) to deliver lower/upper bounds on the KL divergence. When only the KL divergence needs to be bounded, we rather consider the ratio term $\frac{m(x)}{m'(x)}$. This requires to partition the support $\mathcal{X}$ into elementary intervals by overlaying the critical points of both the lower and upper envelopes of $m(x)$ and $m'(x)$, which can be done in linear time. In a given elementary interval, since $\max\{k\min_i\{w_ip_i(x)\}, \max_i\{w_ip_i(x)\}\} \leq m(x) \leq k\max_i\{w_ip_i(x)\}$, we then consider the inequalities:

$$\frac{\max\{k\min_i\{w_ip_i(x)\}, \max_i\{w_ip_i(x)\}\}}{k'\max_j\{w'_jp'_j(x)\}} \leq \frac{m(x)}{m'(x)} \leq \frac{k\max_i\{w_ip_i(x)\}}{\max\{k'\min_j\{w'_jp'_j(x)\}, \max_j\{w'_jp'_j(x)\}\}}. \quad (12)$$

We now need to compute definite integrals of the form $\int_a^b w_1 p(x;\theta_1)\log\frac{w_2p(x;\theta_2)}{w_3p(x;\theta_3)}\mathrm{d}x$ (see Appendix B for explicit formulas when considering scaled and truncated exponential families [17]). (Thus for exponential families, the ratio of densities removes the auxiliary carrier measure term.)

We call these bounds CELB and CEUB for Combinatorial Envelope Lower and Upper Bounds, respectively.

2.1. Tighter Adaptive Bounds

We shall now consider shape-dependent bounds improving over the additive $\log k + \log k'$ non-adaptive bounds. This is made possible by a decomposition of the lse function explained as follows. Let $t_i(x_1, \ldots, x_k) = \log\left(\sum_{j=1}^{k} e^{x_j - x_i}\right)$. By translation identity of the lse function,

$$\mathrm{lse}(x_1, \ldots, x_k) = x_i + t_i(x_1, \ldots, x_k) \tag{13}$$

for all $i \in [k]$. Since $e^{x_j - x_i} = 1$ if $j = i$, and $e^{x_j - x_i} > 0$, we have necessarily $t_i(x_1, \ldots, x_k) > 0$ for any $i \in [k]$. Since Equation (13) is an identity for all $i \in [k]$, we minimize the residual $t_i(x_1, \ldots, x_k)$ by maximizing x_i. Denoting by $x_{(1)}, \ldots, x_{(k)}$ the sequence of numbers sorted in non-decreasing order, the decomposition

$$\mathrm{lse}(x_1, \ldots, x_k) = x_{(k)} + t_{(k)}(x_1, \ldots, x_k) \tag{14}$$

yields the smallest residual. Since $x_{(j)} - x_{(k)} \leq 0$ for all $j \in [k]$, we have

$$t_{(k)}(x_1, \ldots, x_k) = \log\left(1 + \sum_{j=1}^{k-1} e^{x_{(j)} - x_{(k)}}\right) \leq \log k.$$

This shows the bounds introduced earlier can indeed be improved by a more accurate computation of the residual term $t_{(k)}(x_1, \ldots, x_k)$.

When considering 1D GMMs, let us now bound $t_{(k)}(x_1, \ldots, x_k)$ in a combinatorial range $I_r = (a_r, a_{r+1})$. Let $\delta = \delta(r)$ denote the index of the dominating weighted component in this range. Then,

$$\forall x \in I_r, \forall i, \quad \exp\left(-\log\sigma_i - \frac{(x-\mu_i)^2}{2\sigma_i^2} + \log w_i\right) \leq \exp\left(-\log\sigma_\delta - \frac{(x-\mu_\delta)^2}{2\sigma_\delta^2} + \log w_\delta\right).$$

Thus we have:

$$\log m(x) = \log\frac{w_\delta}{\sigma_\delta\sqrt{2\pi}} - \frac{(x-\mu_\delta)^2}{2\sigma_\delta^2} + \log\left(1 + \sum_{i\neq\delta}\exp\left(-\frac{(x-\mu_i)^2}{2\sigma_i^2} + \log\frac{w_i}{\sigma_i} + \frac{(x-\mu_\delta)^2}{2\sigma_\delta^2} - \log\frac{w_\delta}{\sigma_\delta}\right)\right).$$

Now consider the ratio term:

$$\rho_{i,\delta}(x) = \exp\left(-\frac{(x-\mu_i)^2}{2\sigma_i^2} + \log\frac{w_i\sigma_\delta}{w_\delta\sigma_i} + \frac{(x-\mu_\delta)^2}{2\sigma_\delta^2}\right).$$

It is maximized in $I_r = (a_r, a_{r+1})$ by maximizing equivalently the following quadratic equation:

$$l_{i,\delta}(x) = -\frac{(x-\mu_i)^2}{2\sigma_i^2} + \log\frac{w_i\sigma_\delta}{w_\delta\sigma_i} + \frac{(x-\mu_\delta)^2}{2\sigma_\delta^2}.$$

Setting the derivative to zero ($l'_{i,\delta}(x) = 0$), we get the root (when $\sigma_i \neq \sigma_\delta$)

$$x_{i,\delta} = \left(\frac{\mu_\delta}{\sigma_\delta^2} - \frac{\mu_i}{\sigma_i^2}\right) \Big/ \left(\frac{1}{\sigma_\delta^2} - \frac{1}{\sigma_i^2}\right).$$

If $x_{i,\delta} \in I_r$, the ratio $\rho_{i,\delta}(x)$ can be bounded in the slab I_r by considering the extreme values of the three element set $\{\rho_{i,\delta}(a_r), \rho_{i,\delta}(x_{i,\delta}), \rho_{i,\delta}(a_{r+1})\}$. Otherwise $\rho_{i,\delta}(x)$ is monotonic in I_r, its bounds in I_r

are given by $\{\rho_{i,\delta}(a_r), \rho_{i,\delta}(a_{r+1})\}$. In any case, let $\rho_{i,\delta}^{\min}(r)$ and $\rho_{i,\delta}^{\max}(r)$ represent the resulting lower and upper bounds of $\rho_{i,\delta}(x)$ in I_r. Then t_δ is bounded in the range I_r by:

$$0 < \log\left(1+\sum_{i\neq\delta}\rho_{i,\delta}^{\min}(r)\right) \leq t_\delta \leq \log\left(1+\sum_{i\neq\delta}\rho_{i,\delta}^{\max}(r)\right) \leq \log k.$$

In practice, we always get better bounds using the shape-dependent technique at the expense of computing overall $O(k^2)$ intersection points of the pairwise densities. We call those bounds CEALB and CEAUB for Combinatorial Envelope Adaptive Lower Bound and Combinatorial Envelope Adaptive Upper Bound.

Let us illustrate one scenario where this adaptive technique yields very good approximations. Consider a GMM with all variance σ^2 tending to zero (a mixture of k Diracs). Then in a combinatorial slab I_r, we have $\rho_{i,\delta}^{\max}(r) \to 0$ for all $i \neq \delta$, and therefore we get tight bounds.

As a related technique, we could also upper bound $\int_{a_r}^{a_{r+1}} \log m(x)\mathrm{d}x$ by $(a_{r+1}-a_r)\log m(a_r, a_{r+1})$ where $m(x, x')$ denotes the maximal value of the mixture density in the range (x, x'). This maximal value is either found at the slab extremities, or is a mode of the GMM. It then requires to find the modes of a GMM [29,30], for which no analytical solution is known in general.

2.2. Another Derivation Using the Arithmetic-Geometric Mean Inequality

Let us start by considering the inequality of arithmetic and geometric weighted means (AGI, Arithmetic-Geometric Inequality) applied to the mixture component distributions:

$$m(x) = \sum_{i=1}^{k} w_i p(x;\theta_i) \geq \prod_{i=1}^{k} p(x;\theta_i)^{w_i}$$

with equality holds iff. $\theta_1 = \ldots = \theta_k$.

To get a tractable formula with a positive remainder of the log-sum term $\log m(x)$, we need to have the log argument greater or equal to 1, and thus we shall write the positive remainder:

$$R(x) = \log\left(\frac{m(x)}{\prod_{i=1}^{k} p(x;\theta_i)^{w_i}}\right) \geq 0.$$

Therefore, we can decompose the log-sum into a tractable part and a remainder as:

$$\log m(x) = \sum_{i=1}^{k} w_i \log p(x;\theta_i) + \log\left(\frac{m(x)}{\prod_{i=1}^{k} p(x;\theta_i)^{w_i}}\right). \tag{15}$$

For exponential families, the first term can be integrated accurately. For the second term, we notice that $\prod_{i=1}^{k} p(x;\theta_i)^{w_i}$ is a distribution in the same exponential family. We denote $p(x;\theta_0) = \prod_{i=1}^{k} p(x;\theta_i)^{w_i}$. Then

$$R(x) = \log\left(\sum_{i=1}^{k} w_i \frac{p(x;\theta_i)}{p(x;\theta_0)}\right)$$

As the ratio $p(x;\theta_i)/p(x;\theta_0)$ can be bounded above and below using techniques in Section 2.1, $R(x)$ can be correspondingly bounded. Notice the similarity between Equations (14) and (15). The key difference with the adaptive bounds is that, here we choose $p(x;\theta_0)$ instead of the dominating component in $m(x)$ as the "reference distribution" in the decomposition. This subtle difference is not presented in detail in our experimental studies but discussed here for completeness. Essentially, the gap of the bounds is up to the difference between the geometric average and the arithmetic average. In the extreme case that all mixture components are identical, this gap will reach zero. Therefore we expect good quality bounds with a small gap when the mixture components are similar as measured by KL divergence.

2.3. Case Studies

In the following, we instantiate the proposed method for several prominent cases on the mixture of exponential family distributions.

2.3.1. The Case of Exponential Mixture Models

An exponential distribution has density $p(x;\lambda) = \lambda\exp(-\lambda x)$ defined on $\mathcal{X} = [0,\infty)$ for $\lambda > 0$. Its CDF is $\Phi(x;\lambda) = 1-\exp(-\lambda x)$. Any two components $w_1 p(x;\lambda_1)$ and $w_2 p(x;\lambda_2)$ (with $\lambda_1 \neq \lambda_2$) have a unique intersection point

$$x^\star = \frac{\log(w_1\lambda_1)-\log(w_2\lambda_2)}{\lambda_1-\lambda_2} \tag{16}$$

if $x^\star \geq 0$; otherwise they do not intersect. The basic formulas to evaluate the bounds are

$$C_{i,j}(a,b) = \log\left(\lambda'_j w'_j\right) M_i(a,b) + w_i\lambda'_j\left[\left(a+\frac{1}{\lambda_i}\right)e^{-\lambda_i a} - \left(b+\frac{1}{\lambda_i}\right)e^{-\lambda_i b}\right], \tag{17}$$

$$M_i(a,b) = -w_i\left(e^{-\lambda_i a}-e^{-\lambda_i b}\right). \tag{18}$$

2.3.2. The Case of Rayleigh Mixture Models

A Rayleigh distribution has density $p(x;\sigma) = \frac{x}{\sigma^2}\exp\left(-\frac{x^2}{2\sigma^2}\right)$, defined on $\mathcal{X} = [0,\infty)$ for $\sigma > 0$. Its CDF is $\Phi(x;\sigma) = 1-\exp\left(-\frac{x^2}{2\sigma^2}\right)$. Any two components $w_1 p(x;\sigma_1)$ and $w_2 p(x;\sigma_2)$ (with $\sigma_1 \neq \sigma_2$) must intersect at $x_0 = 0$ and can have at most one other intersection point

$$x^\star = \sqrt{\log\frac{w_1\sigma_2^2}{w_2\sigma_1^2} \Big/ \left(\frac{1}{2\sigma_1^2}-\frac{1}{2\sigma_2^2}\right)} \tag{19}$$

if the square root is well defined and $x^\star > 0$. We have

$$\begin{aligned} C_{i,j}(a,b) = {} & \log\frac{w'_j}{(\sigma'_j)^2}M_i(a,b) + \frac{w_i}{2(\sigma'_j)^2}\left[(a^2+2\sigma_i^2)e^{-\frac{a^2}{2\sigma_i^2}} - (b^2+2\sigma_i^2)e^{-\frac{b^2}{2\sigma_i^2}}\right] \\ & - w_i\int_a^b \frac{x}{\sigma_i^2}\exp\left(-\frac{x^2}{2\sigma_i^2}\right)\log x\,dx, \end{aligned} \tag{20}$$

$$M_i(a,b) = -w_i\left(e^{-\frac{a^2}{2\sigma_i^2}} - e^{-\frac{b^2}{2\sigma_i^2}}\right). \tag{21}$$

The last term in Equation (20) does not have a simple closed form (it requires the exponential integral, Ei). One need a numerical integrator to compute it.

2.3.3. The Case of Gaussian Mixture Models

The Gaussian density $p(x;\mu,\sigma) = \frac{1}{\sqrt{2\pi}\sigma}e^{-(x-\mu)^2/(2\sigma^2)}$ has support $\mathcal{X} = \mathbb{R}$ and parameters $\mu \in \mathbb{R}$ and $\sigma > 0$. Its CDF is $\Phi(x;\mu,\sigma) = \frac{1}{2}\left[1+\mathrm{erf}(\frac{x-\mu}{\sqrt{2}\sigma})\right]$, where erf is the Gauss error function. The intersection point $x^\star$ of two components $w_1 p(x;\mu_1,\sigma_1)$ and $w_2 p(x;\mu_2,\sigma_2)$ can be obtained by solving the quadratic equation $\log(w_1 p(x;\mu_1,\sigma_1)) = \log(w_2 p(x;\mu_2,\sigma_2))$, which gives at most two solutions. As shown in Figure 1, the upper envelope of Gaussian densities corresponds to the lower envelope of parabolas. We have

$$C_{i,j}(a,b) = M_i(a,b)\left(\log w_j' - \log\sigma_j' - \frac{1}{2}\log(2\pi) - \frac{1}{2(\sigma_j')^2}\left((\mu_j'-\mu_i)^2+\sigma_i^2\right)\right)$$
$$+\frac{w_i\sigma_i}{2\sqrt{2\pi}(\sigma_j')^2}\left[(a+\mu_i-2\mu_j')e^{-\frac{(a-\mu_i)^2}{2\sigma_i^2}} - (b+\mu_i-2\mu_j')e^{-\frac{(b-\mu_i)^2}{2\sigma_i^2}}\right], \tag{22}$$

$$M_i(a,b) = -\frac{w_i}{2}\left(\operatorname{erf}\left(\frac{b-\mu_i}{\sqrt{2}\sigma_i}\right) - \operatorname{erf}\left(\frac{a-\mu_i}{\sqrt{2}\sigma_i}\right)\right). \tag{23}$$

2.3.4. The Case of Gamma Distributions

For simplicity, we only consider gamma distributions with the shape parameter $k > 0$ fixed and the scale $\lambda > 0$ varying. The density is defined on $(0,\infty)$ as $p(x;k,\lambda) = \frac{x^{k-1}e^{-\frac{x}{\lambda}}}{\lambda^k\Gamma(k)}$, where $\Gamma(\cdot)$ is the gamma function. Its CDF is $\Phi(x;k,\lambda) = \gamma(k, x/\lambda)/\Gamma(k)$, where $\gamma(\cdot,\cdot)$ is the lower incomplete gamma function. Two weighted gamma densities $w_1 p(x;k,\lambda_1)$ and $w_2 p(x;k,\lambda_2)$ (with $\lambda_1 \neq \lambda_2$) intersect at a unique point

$$x^\star = \left(\log\frac{w_1}{\lambda_1^k} - \log\frac{w_2}{\lambda_2^k}\right) \Big/ \left(\frac{1}{\lambda_1} - \frac{1}{\lambda_2}\right) \tag{24}$$

if $x^\star > 0$; otherwise they do not intersect. From straightforward derivations,

$$C_{i,j}(a,b) = \log\frac{w_j'}{(\lambda_j')^k\Gamma(k)}M_i(a,b) + w_i\int_a^b \frac{x^{k-1}e^{-\frac{x}{\lambda_i}}}{\lambda_i^k\Gamma(k)}\left(\frac{x}{\lambda_j'} - (k-1)\log x\right)dx, \tag{25}$$

$$M_i(a,b) = -\frac{w_i}{\Gamma(k)}\left(\gamma\left(k,\frac{b}{\lambda_i}\right) - \gamma\left(k,\frac{a}{\lambda_i}\right)\right). \tag{26}$$

Similar to the case of Rayleigh mixtures, the last term in Equation (25) relies on numerical integration.

3. Upper-Bounding the Differential Entropy of a Mixture

First, consider a finite parametric mixture model $m(x) = \sum_{i=1}^k w_i p(x;\theta_i)$. Using the chain rule of the entropy, we end up with the well-known lemma:

Lemma 1. *The entropy of a d-variate mixture is upper bounded by the sum of the entropy of its marginal mixtures:* $H(m) \leq \sum_{i=1}^d H(m_i)$*, where* m_i *is the 1D marginal mixture with respect to variable* x_i*.*

Since the 1D marginals of a multivariate GMM are univariate GMMs, we thus get a loose upper bound. A generic sample-based probabilistic bound is reported for the entropies of distributions with given support [31]: The method builds probabilistic upper and lower piecewisely linear CDFs based on an i.i.d. finite sample set of size n and a given deviation probability threshold. It then builds algorithmically between those two bounds the maximum entropy distribution [31] with a so-called string-tightening algorithm.

Instead, we proceed as follows: Consider finite mixtures of component distributions defined on the full support $\mathbb{R}^d$ that have finite component means and variances (like exponential families). Then we shall use the fact that the maximum entropy distribution with prescribed mean and variance is a Gaussian distribution, and conclude the upper bound by plugging the mixture mean and variance in the differential entropy formula of the Gaussian distribution. In general, the maximum entropy with moment constraints yields as a solution an exponential family.

Without loss of generality, consider GMMs in the form $m(x) = \sum_{i=1}^{k} w_i p(x;\mu_i,\Sigma_i)$ ($\Sigma_i = \sigma_i^2$ for univariate Gaussians). The mean $\bar{\mu}$ of the mixture is $\bar{\mu} = \sum_{i=1}^{k} w_i\mu_i$ and the variance is $\bar{\sigma}^2 = E[m^2] - E[m]^2$. Since $E[m^2] = \sum_{i=1}^{k} w_i \int x^2 p(x;\mu_i,\Sigma_i)\mathrm{d}x = \sum_{i=1}^{k} w_i \left(\mu_i^2 + \sigma_i^2\right)$, we deduce that

$$\bar{\sigma}^2 = \sum_{i=1}^{k} w_i(\mu_i^2 + \sigma_i^2) - \left(\sum_{i=1}^{k} w_i\mu_i\right)^2 = \sum_{i=1}^{k} w_i \left[(\mu_i - \bar{\mu})^2 + \sigma_i^2\right].$$

The entropy of a random variable with a prescribed variance $\bar{\sigma}^2$ is maximal for the Gaussian distribution with the same variance $\bar{\sigma}^2$, see [4]. Since the differential entropy of a Gaussian is $\log(\bar{\sigma}\sqrt{2\pi e})$, we deduce that the entropy of the GMM is upper bounded by

$$H(m) \leq \frac{1}{2}\log(2\pi e) + \frac{1}{2}\log\sum_{i=1}^{k} w_i \left[(\mu_i - \bar{\mu})^2 + \sigma_i^2\right].$$

This upper bound can be easily generalized to arbitrary dimensionality. We get the following lemma:

Lemma 2. *The entropy of a d-variate GMM* $m(x) = \sum_{i=1}^{k} w_i p(x;\mu_i,\Sigma_i)$ *is upper bounded by* $\frac{d}{2}\log(2\pi e) + \frac{1}{2}\log\det\Sigma$, *where* $\Sigma = \sum_{i=1}^{k} w_i(\mu_i\mu_i^\top + \Sigma_i) - \left(\sum_{i=1}^{k} w_i\mu_i\right)\left(\sum_{i=1}^{k} w_i\mu_i^\top\right)$.

In general, exponential families have finite moments of any order [17]: In particular, we have $E[t(X)] = \nabla F(\theta)$ and $V[t(X)] = \nabla^2 F(\theta)$. For Gaussian distribution, we have the sufficient statistics $t(x) = (x, x^2)$ so that $E[t(X)] = \nabla F(\theta)$ yields the mean and variance from the log-normalizer. It is easy to generalize Lemma 2 to mixtures of exponential family distributions.

Note that this bound (called the Maximum Entropy Upper Bound in [13], MEUB) is tight when the GMM approximates a single Gaussian. It is fast to compute compared to the bound reported in [9] that uses Taylor' s expansion of the log-sum of the mixture density.

A similar argument cannot be applied for a lower bound since a GMM with a given variance may have entropy tending to $-\infty$. For example, assume the 2-component mixture's mean is zero, and that the variance approximates 1 by taking $m(x) = \frac{1}{2}G(x;-1,\epsilon) + \frac{1}{2}G(x;1,\epsilon)$ where G denotes the Gaussian density. Letting $\epsilon \to 0$, we get the entropy tending to $-\infty$.

We remark that our log-sum-exp inequality technique yields a $\log 2$ additive approximation range in the case of a Gaussian mixture with two components. It thus generalizes the bounds reported in [7] to GMMs with arbitrary variances that are not necessarily equal.

To see the bound gap, we have

$$\begin{aligned} -\sum_r \int_{I_r} m(x)\left(\log k + \log\max_i w_i p_i(x)\right)\mathrm{d}x &\leq H(m) \\ &\leq -\sum_r \int_{I_r} m(x)\max\left\{\log\max_i w_i p_i(x), \log k + \log\min_i w_i p_i(x)\right\}\mathrm{d}x. \end{aligned} \tag{27}$$

Therefore the gap is at most

$$\begin{aligned} \Delta &= \min\left\{\sum_r \int_{I_r} m(x)\log\frac{\max_i w_i p_i(x)}{\min_i w_i p_i(x)}\mathrm{d}x,\ \log k\right\} \\ &= \min\left\{\sum_s\sum_r \int_{I_r} w_s p_s(x)\log\frac{\max_i w_i p_i(x)}{\min_i w_i p_i(x)}\mathrm{d}x,\ \log k\right\}. \end{aligned} \tag{28}$$

Thus to compute the gap error bound of the differential entropy, we need to integrate terms in the form

$$\int w_a p_a(x)\log\frac{w_b p_b(x)}{w_c p_c(x)}\mathrm{d}x.$$

See Appendix B for a closed-form formula when dealing with exponential family components.

4. Bounding the α-Divergence

The α-divergence [15,32–34] between $m(x) = \sum_{i=1}^{k} w_i p_i(x)$ and $m'(x) = \sum_{i=1}^{k'} w'_i p'_i(x)$ is defined as

$$D_\alpha(m : m') = \frac{1}{\alpha(1-\alpha)}\left(1 - \int_{\mathcal{X}} m(x)^\alpha m'(x)^{1-\alpha} \mathrm{d}x\right), \tag{29}$$

which clearly satisfies $D_\alpha(m : m') = D_{1-\alpha}(m' : m)$. The α-divergence is *a family* of information divergences parametrized by $\alpha \in \mathbb{R} \setminus \{0,1\}$. Let $\alpha \to 1$, we get the KL divergence (see [35] for a proof):

$$\lim_{\alpha \to 1} D_\alpha(m : m') = \mathrm{KL}(m : m') = \int_{\mathcal{X}} m(x) \log \frac{m(x)}{m'(x)} \mathrm{d}x, \tag{30}$$

and $\alpha \to 0$ gives the reverse KL divergence:

$$\lim_{\alpha \to 0} D_\alpha(m : m') = \mathrm{KL}(m' : m).$$

Other interesting values [33] include $\alpha = 1/2$ (squared Hellinger distance), $\alpha = 2$ (Pearson Chi-square distance), $\alpha = -1$ (Neyman Chi-square distance), etc. Notably, the Hellinger distance is a valid distance metric which satisfies non-negativity, symmetry, and the triangle inequality. In general, $D_\alpha(m : m')$ only satisfies non-negativity so that $D_\alpha(m : m') \geq 0$ for any $m(x)$ and $m'(x)$. It is neither symmetric nor admitting the triangle inequality. Minimization of α-divergences allows one to choose a trade-off between mode fitting and support fitting of the minimizer [36]. The minimizer of α-divergences including MLE as a special case has interesting connections with transcendental number theory [37].

To compute $D_\alpha(m : m')$ for given $m(x)$ and $m'(x)$ reduces to evaluate the Hellinger integral [38,39]:

$$H_\alpha(m : m') = \int_{\mathcal{X}} m(x)^\alpha m'(x)^{1-\alpha} \mathrm{d}x, \tag{31}$$

which in general does not have a closed form, as it was known that the α-divergence of mixture models is not analytic [6]. Moreover, $H_\alpha(m : m')$ may diverge making the α-divergence unbounded. Once $H_\alpha(m : m')$ can be solved, the Rényi and Tsallis divergences [35] and in general Sharma–Mittal divergences [40] can be easily computed. Therefore the results presented here directly extend to those divergence families.

Similar to the case of KL divergence, the Monte Carlo stochastic estimation of $H_\alpha(m : m')$ can be computed either as

$$\hat{H}_\alpha^n(m : m') = \frac{1}{n} \sum_{i=1}^{n} \left(\frac{m'(x_i)}{m(x_i)}\right)^{1-\alpha},$$

where $x_1, \ldots, x_n \sim m(x)$ are i.i.d. samples, or as

$$\hat{H}_\alpha^n(m : m') = \frac{1}{n} \sum_{i=1}^{n} \left(\frac{m(x_i)}{m'(x_i)}\right)^{\alpha},$$

where $x_1, \ldots, x_n \sim m'(x)$ are i.i.d. In either case, it is consistent so that $\lim_{n\to\infty} \hat{H}_\alpha^n(m : m') = H_\alpha(m : m')$. However, MC estimation requires a large sample and does not guarantee deterministic bounds. The techniques described in [41] work in practice for very close distributions, and do not apply between mixture models. We will therefore derive combinatorial bounds for $H_\alpha(m : m')$. The structure of this Section is parallel with Section 2 with necessary reformulations for a clear presentation.

4.1. Basic Bounds

For a pair of given $m(x)$ and $m'(x)$, we only need to derive bounds of $H_\alpha(m:m')$ in Equation (31) so that $L_\alpha(m:m') \le H_\alpha(m:m') \le U_\alpha(m:m')$. Then the α-divergence $D_\alpha(m:m')$ can be bounded by a linear transformation of the range $[L_\alpha(m:m'),\ U_\alpha(m:m')]$. In the following we always assume without loss of generality $\alpha \ge 1/2$. Otherwise we can bound $D_\alpha(m:m')$ by considering equivalently the bounds of $D_{1-\alpha}(m':m)$.

Recall that in each elementary slab I_r, we have

$$\max\left\{k w_{\epsilon(r)} p_{\epsilon(r)}(x), w_{\delta(r)} p_{\delta(r)}(x)\right\} \le m(x) \le k w_{\delta(r)} p_{\delta(r)}(x). \tag{32}$$

Notice that $k w_{\epsilon(r)} p_{\epsilon(r)}(x)$, $w_{\delta(r)} p_{\delta(r)}(x)$, and $k w_{\delta(r)} p_{\delta(r)}(x)$ are all single component distributions up to a scaling coefficient. The general thinking is to bound the multi-component mixture $m(x)$ by single component distributions in each elementary interval, so that the integral in Equation (31) can be computed in a piecewise manner.

For the convenience of notation, we rewrite Equation (32) as

$$c_{\nu(r)} p_{\nu(r)}(x) \le m(x) \le c_{\delta(r)} p_{\delta(r)}(x), \tag{33}$$

where

$$c_{\nu(r)} p_{\nu(r)}(x) := k w_{\epsilon(r)} p_{\epsilon(r)}(x) \quad \text{or} \quad w_{\delta(r)} p_{\delta(r)}(x), \tag{34}$$

$$c_{\delta(r)} p_{\delta(r)}(x) := k w_{\delta(r)} p_{\delta(r)}(x). \tag{35}$$

If $1/2 \le \alpha < 1$, then both x^α and $x^{1-\alpha}$ are monotonically increasing on $\mathbb{R}^+$. Therefore we have

$$A^\alpha_{\nu(r),\nu'(r)}(I_r) \le \int_{I_r} m(x)^\alpha m'(x)^{1-\alpha} \mathrm{d}x \le A^\alpha_{\delta(r),\delta'(r)}(I_r), \tag{36}$$

where

$$A^\alpha_{i,j}(I) = \int_I (c_i p_i(x))^\alpha \left(c'_j p'_j(x)\right)^{1-\alpha} \mathrm{d}x, \tag{37}$$

and I denotes an interval $I = (a,b) \subset \mathbb{R}$. The other case $\alpha > 1$ is similar by noting that x^α and $x^{1-\alpha}$ are monotonically increasing and decreasing on $\mathbb{R}^+$, respectively. In conclusion, we obtain the following bounds of $H_\alpha(m:m')$:

$$\text{If } 1/2 \le \alpha < 1,\ L_\alpha(m:m') = \sum_{r=1}^{\ell} A^\alpha_{\nu(r),\nu'(r)}(I_r), \quad U_\alpha(m:m') = \sum_{r=1}^{\ell} A^\alpha_{\delta(r),\delta'(r)}(I_r); \tag{38}$$

$$\text{if } \alpha > 1,\ L_\alpha(m:m') = \sum_{r=1}^{\ell} A^\alpha_{\nu(r),\delta'(r)}(I_r), \quad U_\alpha(m:m') = \sum_{r=1}^{\ell} A^\alpha_{\delta(r),\nu'(r)}(I_r). \tag{39}$$

The remaining problem is to compute the definite integral $A^\alpha_{i,j}(I)$ in the above equations. Here we assume all mixture components are in the same exponential family so that $p_i(x) = p(x;\theta_i) = h(x)\exp\left(\theta_i^\top t(x) - F(\theta_i)\right)$, where $h(x)$ is a base measure, $t(x)$ is a vector of sufficient statistics, and the function F is known as the cumulant generating function. Then it is straightforward from Equation (37) that

$$A^\alpha_{i,j}(I) = c_i^\alpha (c'_j)^{1-\alpha} \int_I h(x) \exp\left(\left(\alpha\theta_i + (1-\alpha)\theta'_j\right)^\top t(x) - \alpha F(\theta_i) - (1-\alpha) F(\theta'_j)\right) \mathrm{d}x. \tag{40}$$

If $1/2 \leq \alpha < 1$, then $\bar{\theta} = \alpha\theta_i + (1-\alpha)\theta'_j$ belongs to the natural parameter space $\mathcal{M}_\theta$. Therefore $A^{\alpha}_{i,j}(I)$ is bounded and can be computed from the CDF of $p(x;\bar{\theta})$ as

$$A^{\alpha}_{i,j}(I) = c_i^{\alpha}(c'_j)^{1-\alpha}\exp(F(\bar{\theta}) - \alpha F(\theta_i) - (1-\alpha)F(\theta'_j))\int_I p(x;\bar{\theta})\,\mathrm{d}x. \tag{41}$$

The other case $\alpha > 1$ is more difficult: if $\bar{\theta} = \alpha\theta_i + (1-\alpha)\theta'_j$ still lies in $\mathcal{M}_\theta$, then $A^{\alpha}_{i,j}(I)$ can be computed by Equation (41). Otherwise we try to solve it by a numerical integrator. This is not ideal as the integral may diverge, or our approximation may be too loose to conclude. We point the reader to [42] and Equations (61)–(69) in [35] for related analysis with more details. As computing $A^{\alpha}_{i,j}(I)$ only requires $O(1)$ time, the overall computational complexity (without considering the envelope computation) is $O(\ell)$.

4.2. Adaptive Bounds

This section derives the shape-dependent bounds which improve the basic bounds in Section 4.1. We can rewrite a mixture model $m(x)$ in a slab I_r as

$$m(x) = w_{\zeta(r)}p_{\zeta(r)}(x)\left(1 + \sum_{i\neq\zeta(r)}\frac{w_i p_i(x)}{w_{\zeta(r)}p_{\zeta(r)}(x)}\right), \tag{42}$$

where $w_{\zeta(r)}p_{\zeta(r)}(x)$ is a weighted component in $m(x)$ serving as a *reference*. We only discuss the case that the reference is chosen as the dominating component, i.e., $\zeta(r) = \delta(r)$. However it is worth to note that the proposed bounds do not depend on this particular choice. Therefore the ratio

$$\frac{w_i p_i(x)}{w_{\zeta(r)}p_{\zeta(r)}(x)} = \frac{w_i}{w_{\zeta(r)}}\exp\left(\left(\theta_i - \theta_{\zeta(r)}\right)^{\top} t(x) - F(\theta_i) + F(\theta_{\zeta(r)})\right) \tag{43}$$

can be bounded in a sub-range of $[0,1]$ by analyzing the extreme values of $t(x)$ in the slab I_r. This can be done because $t(x)$ usually consists of polynomial functions with finite critical points which can be solved easily. Correspondingly the function $\left(1 + \sum_{i\neq\zeta(r)}\frac{w_i p_i(x)}{w_{\zeta(r)}p_{\zeta(r)}(x)}\right)$ in I_r can be bounded in a subrange of $[1,k]$, denoted as $[\omega_{\zeta(r)}(I_r), \Omega_{\zeta(r)}(I_r)]$. Hence

$$\omega_{\zeta(r)}(I_r)w_{\zeta(r)}p_{\zeta(r)}(x) \leq m(x) \leq \Omega_{\zeta(r)}(I_r)w_{\zeta(r)}p_{\zeta(r)}(x). \tag{44}$$

This forms better bounds of $m(x)$ than Equation (32) because each component in the slab I_r is analyzed more accurately. Therefore, we refine the fundamental bounds of $m(x)$ by replacing the Equations (34) and (35) with

$$c_{\nu(r)}p_{\nu(r)}(x) := \omega_{\zeta(r)}(I_r)w_{\zeta(r)}p_{\zeta(r)}(x), \tag{45}$$

$$c_{\delta(r)}p_{\delta(r)}(x) := \Omega_{\zeta(r)}(I_r)w_{\zeta(r)}p_{\zeta(r)}(x). \tag{46}$$

Then, the improved bounds of H_α are given by Equations (38) and (39) according to the above replaced definition of $c_{\nu(r)}p_{\nu(r)}(x)$ and $c_{\delta(r)}p_{\delta(r)}(x)$.

To evaluate $\omega_{\zeta(r)}(I_r)$ and $\Omega_{\zeta(r)}(I_r)$ requires iterating through all components in each slab. Therefore the computational complexity is increased to $O(\ell(k+k'))$.

4.3. Variance-Reduced Bounds

This section further improves the proposed bounds based on variance reduction [43]. By assumption, $\alpha \geq 1/2$, then $m(x)^{\alpha}m'(x)^{1-\alpha}$ is more similar to $m(x)$ rather than $m'(x)$. The ratio

$m(x)^\alpha m'(x)^{1-\alpha}/m(x)$ is likely to have a small variance when x varies inside a slab I_r, especially when α is close to 1. We will therefore bound this ratio term in

$$\int_{I_r} m(x)^\alpha m'(x)^{1-\alpha}\mathrm{d}x = \int_{I_r} m(x)\left(\frac{m(x)^\alpha m'(x)^{1-\alpha}}{m(x)}\right)\mathrm{d}x = \sum_{i=1}^{k}\int_{I_r} w_i p_i(x)\left(\frac{m'(x)}{m(x)}\right)^{1-\alpha}\mathrm{d}x. \tag{47}$$

No matter $\alpha < 1$ or $\alpha > 1$, the function $x^{1-\alpha}$ must be monotonic on $\mathbb{R}^+$. In each slab I_r, $(m'(x)/m(x))^{1-\alpha}$ ranges between these two functions:

$$\left(\frac{c'_{\nu'(r)}p'_{\nu'(r)}(x)}{c_{\delta(r)}p_{\delta(r)}(x)}\right)^{1-\alpha} \quad \text{and} \quad \left(\frac{c'_{\delta'(r)}p'_{\delta'(r)}(x)}{c_{\nu(r)}p_{\nu(r)}(x)}\right)^{1-\alpha}, \tag{48}$$

where $c_{\nu(r)}p_{\nu(r)}(x)$, $c_{\delta(r)}p_{\delta(r)}(x)$, $c'_{\nu'(r)}p'_{\nu'(r)}(x)$ and $c'_{\delta'(r)}p'_{\delta'(r)}(x)$ are defined in Equations (45) and (46). Similar to the definition of $A^\alpha_{i,j}(I)$ in Equation (37), we define

$$B^\alpha_{i,j,l}(I) = \int_I w_i p_i(x)\left(\frac{c'_l p'_l(x)}{c_j p_j(x)}\right)^{1-\alpha}\mathrm{d}x. \tag{49}$$

Therefore we have,

$$L_\alpha(m:m') = \min\mathcal{S},\quad U_\alpha(m:m') = \max\mathcal{S},$$
$$\mathcal{S} = \left\{\sum_{r=1}^{\ell}\sum_{i=1}^{k} B^\alpha_{i,\delta(r),\nu'(r)}(I_r),\ \sum_{r=1}^{\ell}\sum_{i=1}^{k} B^\alpha_{i,\nu(r),\delta'(r)}(I_r)\right\}. \tag{50}$$

The remaining problem is to evaluate $B^\alpha_{i,j,l}(I)$ in Equation (49). Similar to Section 4.1, assuming the components are in the same exponential family with respect to the natural parameters θ, we get

$$B^\alpha_{i,j,l}(I) = w_i\frac{c_l'^{1-\alpha}}{c_j^{1-\alpha}}\exp\left(F(\bar\theta) - F(\theta_i) - (1-\alpha)F(\theta'_l) + (1-\alpha)F(\theta_j)\right)\int_I p(x;\bar\theta)\mathrm{d}x. \tag{51}$$

If $\bar\theta = \theta_i + (1-\alpha)\theta'_l - (1-\alpha)\theta_j$ is in the natural parameter space, $B^\alpha_{i,j,l}(I)$ can be computed from the CDF of $p(x;\bar\theta)$; otherwise $B^\alpha_{i,j,l}(I)$ can be numerically integrated by its definition in Equation (49). The computational complexity is the same as the bounds in Section 4.2, i.e., $O(\ell(k+k'))$.

We have introduced three pairs of deterministic lower and upper bounds that enclose the true value of α-divergence between univariate mixture models. Thus the gap between the upper and lower bounds provides the additive approximation factor of the bounds. We conclude by emphasizing that the presented methodology can be easily generalized to other divergences [35,40] relying on Hellinger-type integrals $H_{\alpha,\beta}(p:q) = \int p(x)^\alpha q(x)^\beta \mathrm{d}x$ like the γ-divergence [44] as well as entropy measures [45].

5. Lower Bounds of the f-Divergence

The f-divergence between two distributions $m(x)$ and $m'(x)$ (not necessarily mixtures) is defined for a *convex generator* f by:

$$D_f(m:m') = \int m(x) f\left(\frac{m'(x)}{m(x)}\right)\mathrm{d}x.$$

If $f(x) = -\log x$, then $D_f(m:m') = \mathrm{KL}(m:m')$.

Let us partition the support $\mathcal{X} = \uplus_{r=1}^{\ell} I_r$ arbitrarily into elementary ranges, which *do not necessarily correspond to the envelopes*. Denote by M_I the probability mass of a mixture $m(x)$ in the range I: $M_I = \int_I m(x)dx$. Then

$$D_f(m : m') = \sum_{r=1}^{\ell} M_{I_r} \int_{I_r} \frac{m(x)}{M_{I_r}} f\left(\frac{m'(x)}{m(x)}\right) \mathrm{d}x.$$

Note that in range I_r, $\frac{m(x)}{M_{I_r}}$ is a unit weight distribution. Thus by Jensen's inequality $f(E[X]) \leq E[f(X)]$, we get

$$D_f(m : m') \geq \sum_{r=1}^{\ell} M_{I_r} f\left(\int_{I_r} \frac{m(x)}{M_{I_r}} \frac{m'(x)}{m(x)} \mathrm{d}x\right) = \sum_{r=1}^{\ell} M_{I_r} f\left(\frac{M'_{I_r}}{M_{I_r}}\right). \tag{52}$$

Notice that the RHS of Equation (52) is the f-divergence between $(M_{I_1}, \cdots, M_{I_\ell})$ and $(M'_{I_1}, \cdots, M'_{I_\ell})$, denoted by $D_f^{\mathcal{I}}(m : m')$. In the special case that $\ell = 1$ and $I_1 = \mathcal{X}$, the above Equation (52) turns out to be the usual Gibbs' inequality: $D_f(m : m') \geq f(1)$, and Csiszár generator is chosen so that $f(1) = 0$. In conclusion, for a fixed (coarse-grained) countable partition of $\mathcal{X}$, we recover the well-know information monotonicity [46] of the f-divergences:

$$D_f(m : m') \geq D_f^{\mathcal{I}}(m : m') \geq 0.$$

In practice, we get closed-form lower bounds when $M_I = \int_a^b m(x)\mathrm{d}x = \Phi(b) - \Phi(a)$ is available in closed-form, where $\Phi(\cdot)$ denote the CDF. In particular, if $m(x)$ is a mixture model, then its CDF can be computed by linearly combining the CDFs of its components.

To wrap up, we have proved that coarse-graining by making a finite partition of the support $\mathcal{X}$ yields a lower bound on the f-divergence by virtue of the information monotonicity. Therefore, instead of doing Monte Carlo stochastic integration:

$$\hat{D}_f^n(m : m') = \frac{1}{n} \sum_{i=1}^{n} f\left(\frac{m'(x_i)}{m(x_i)}\right),$$

with $x_1, \ldots, x_n \sim_{\text{i.i.d.}} m(x)$, it could be better to sort those n samples and consider the coarse-grained partition:

$$\mathcal{I} = (-\infty, x_{(1)}] \cup \left(\uplus_{i=1}^{n-1} (x_{(i)}, x_{(i+1)}]\right) \cup (x_{(n)}, +\infty)$$

to get a *guaranteed lower bound* on the f-divergence. We will call this bound CGQLB for Coarse Graining Quantization Lower Bound.

Given a budget of n splitting points on the range $\mathcal{X}$, it would be interesting to find the best n points that maximize the lower bound $D_f^{\mathcal{I}}(m : m')$. This is ongoing research.

6. Experiments

We perform an empirical study to verify our theoretical bounds. We simulate four pairs of mixture models $\{(\mathtt{EMM}_1, \mathtt{EMM}_2), (\mathtt{RMM}_1, \mathtt{RMM}_2), (\mathtt{GMM}_1, \mathtt{GMM}_2), (\mathtt{GaMM}_1, \mathtt{GaMM}_2)\}$ as the test subjects. The component type is implied by the model name, where `GaMM` stands for Gamma mixtures. The components of each mixture model are given as follows.

1. $\mathtt{EMM}_1$'s components, in the form (λ_i, w_i), are given by $(0.1, 1/3)$, $(0.5, 1/3)$, $(1, 1/3)$; $\mathtt{EMM}_2$'s components are $(2, 0.2)$, $(10, 0.4)$, $(20, 0.4)$.
2. $\mathtt{RMM}_1$'s components, in the form (σ_i, w_i), are given by $(0.5, 1/3)$, $(2, 1/3)$, $(10, 1/3)$; $\mathtt{RMM}_2$ consists of $(5, 0.25)$, $(60, 0.25)$, $(100, 0.5)$.

3. $\texttt{GMM}_1$'s components, in the form (μ_i, σ_i, w_i), are $(-5, 1, 0.05)$, $(-2, 0.5, 0.1)$, $(5, 0.3, 0.2)$, $(10, 0.5, 0.2)$, $(15, 0.4, 0.05)$, $(25, 0.5, 0.3)$, $(30, 2, 0.1)$; $\texttt{GMM}_2$ consists of $(-16, 0.5, 0.1)$, $(-12, 0.2, 0.1)$, $(-8, 0.5, 0.1)$, $(-4, 0.2, 0.1)$, $(0, 0.5, 0.2)$, $(4, 0.2, 0.1)$, $(8, 0.5, 0.1)$, $(12, 0.2, 0.1)$, $(16, 0.5, 0.1)$.
4. $\texttt{GaMM}_1$'s components, in the form (k_i, λ_i, w_i), are $(2, 0.5, 1/3)$, $(2, 2, 1/3)$, $(2, 4, 1/3)$; $\texttt{GaMM}_2$ consists of $(2, 5, 1/3)$, $(2, 8, 1/3)$, $(2, 10, 1/3)$.

We compare the proposed bounds with Monte Carlo estimation with different sample sizes in the range $\{10^2, 10^3, 10^4, 10^5\}$. For each sample size configuration, we report the 0.95 confidence interval by Monte Carlo estimation using the corresponding number of samples. Figure 2a–d shows the input signals as well as the estimation results, where the proposed bounds CELB, CEUB, CEALB, CEAUB, CGQLB are presented as horizontal lines, and the Monte Carlo estimations over different sample sizes are presented as error bars. We can loosely consider the average Monte Carlo output with the largest sample size (10^5) as the underlying truth, which is clearly inside our bounds. This serves as an empirical justification on the correctness of the bounds.

A key observation is that the bounds can be *very tight*, especially when the underlying KL divergence has a large magnitude, e.g., $\mathrm{KL}(\texttt{RMM}_2 : \texttt{RMM}_1)$. This is because the gap between the lower and upper bounds is always guaranteed to be within $\log k + \log k'$. Because KL is unbounded [4], in the general case two mixture models may have a large KL. Then our approximation gap is relatively very small. On the other hand, we also observed that the bounds in certain cases, e.g., $\mathrm{KL}(\texttt{EMM}_2 : \texttt{EMM}_1)$, are not as tight as the other cases. When the underlying KL is small, the bounds are not as informative as the general case.

Comparatively, there is a significant improvement of the shape-dependent bounds (CEALB and CEAUB) over the combinatorial bounds (CELB and CEUB). In all investigated cases, the adaptive bounds can roughly shrink the gap by half of its original size at the cost of additional computation.

Note that, the bounds are accurate and must contain the true value. Monte Carlo estimation gives no guarantee on where the true value is. For example, in estimating $\mathrm{KL}(\texttt{GMM}_1 : \texttt{GMM}_2)$, Monte Carlo estimation based on 10^4 samples can go beyond our bounds! It therefore suffers from a larger estimation error.

CGQLB as a simple-to-implement technique shows surprising good performance in several cases, e.g., $\mathrm{KL}(\texttt{RMM}_1, \texttt{RMM}_2)$. Although it requires a large number of samples, we can observe that increasing sample size has limited effect on improving this bound. Therefore, in practice, one may intersect the range defined by CEALB and CEAUB with the range defined by CGQLB with a small sample size (e.g., 100) to get better bounds.

We simulates a set of Gaussian mixture models besides the above $\texttt{GMM}_1$ and $\texttt{GMM}_2$. Figure 3 shows the GMM densities as well as their differential entropy. A detailed explanation of the components of each GMM model is omitted for brevity.

The key observation is that CEUB (CEAUB) is *very tight* in most of the investigated cases. This is because that the upper envelope that is used to compute CEUB (CEAUB) gives a very good estimation of the input signal.

Notice that MEUB only gives an upper bound of the differential entropy as discussed in Section 3. In general the proposed bounds are tighter than MEUB. However, this is not the case when the mixture components are merged together and approximate one single Gaussian (and therefore its entropy can be well approximated by the Gaussian entropy), as shown in the last line of Figure 3.

For α-divergence, the bounds introduced in Sections 4.1–4.3 are denoted as "Basic", "Adaptive" and "VR", respectively. Figure 4 visualizes these GMMs and plots the estimations of their α-divergences against α. The red lines mean the upper envelope. The dashed vertical lines mean the elementary intervals. The components of $\texttt{GMM}_1$ and $\texttt{GMM}_2$ are more separated than $\texttt{GMM}_3$ and $\texttt{GMM}_4$. Therefore these two pairs present different cases. For a clear presentation, only VR (which is expected to be better than Basic and Adaptive) is shown. We can see that, visually in the big scale, VR tightly surrounds the true value.

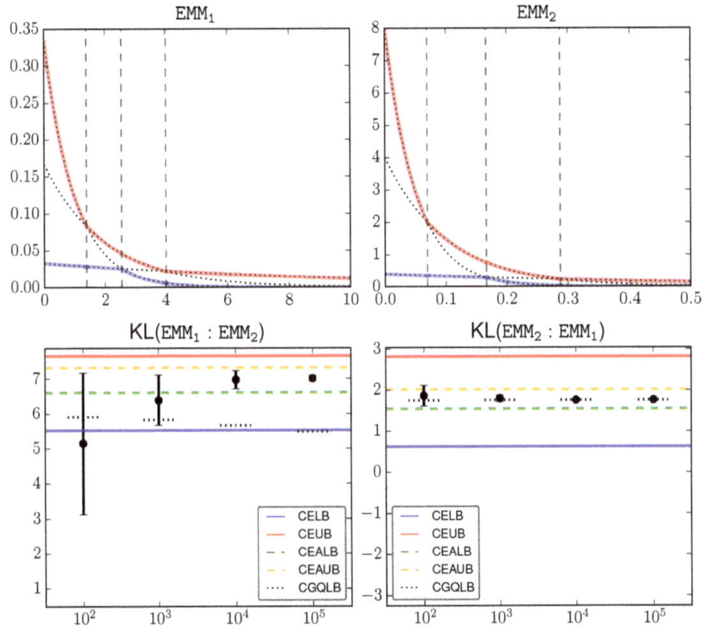

(**a**) KL divergence between two exponential mixture models

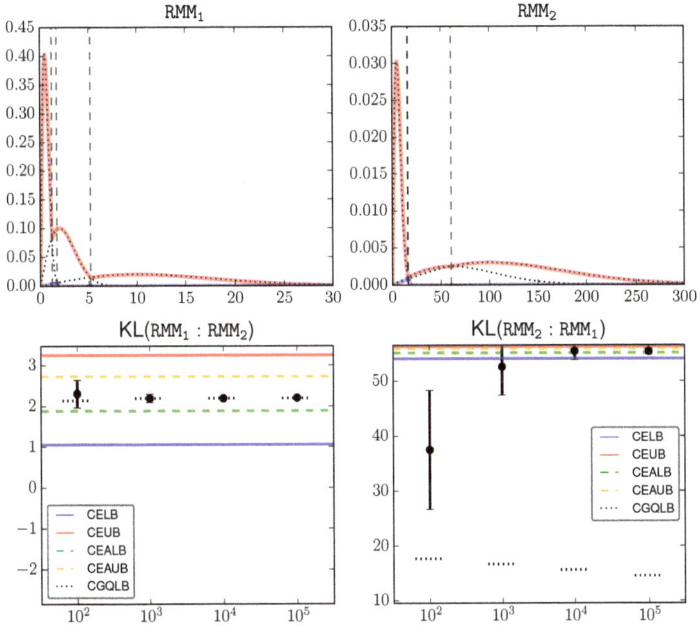

(**b**) KL divergence between two Rayleigh mixture models

Figure 2. *Cont.*

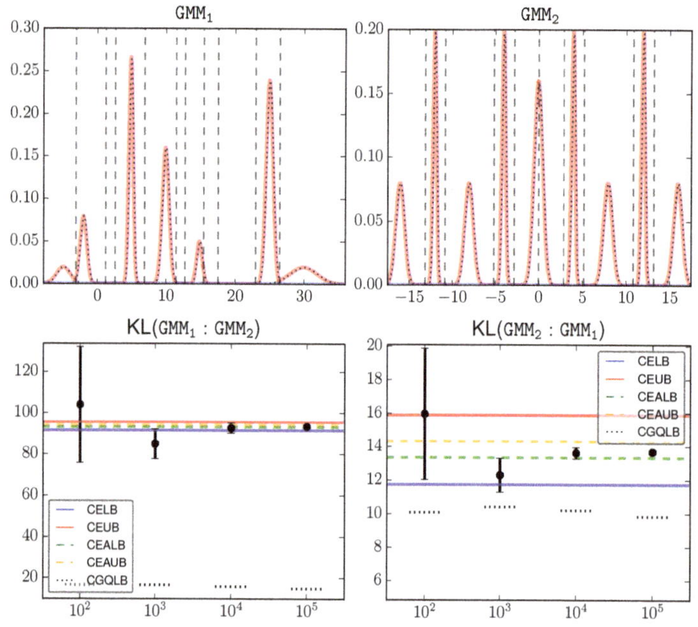

(**c**) KL divergence between two Gaussian mixture models

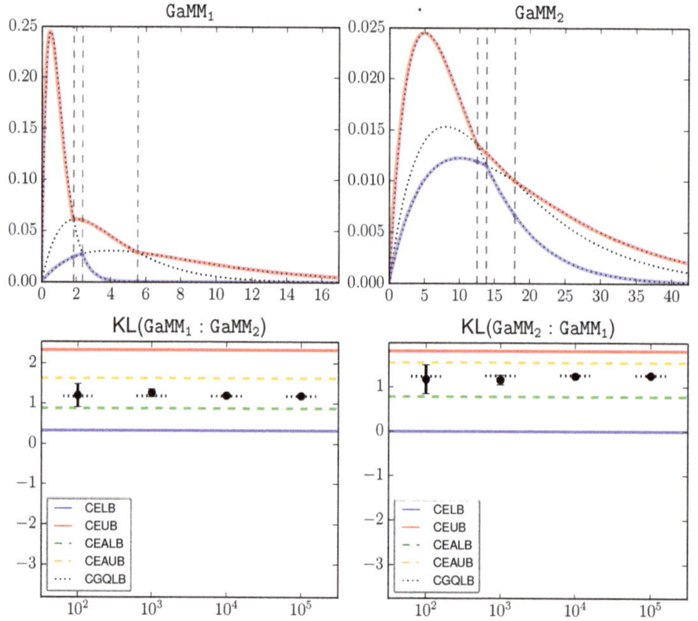

(**d**) KL divergence between two Gamma mixture models

Figure 2. Lower and upper bounds on the KL divergence between mixture models. The y-axis means KL divergence. Solid/dashed lines represent the combinatorial/adaptive bounds, respectively. The error-bars show the 0.95 confidence interval by Monte Carlo estimation using the corresponding sample size (x-axis). The narrow dotted bars show the CGQLB estimation w.r.t. the sample size.

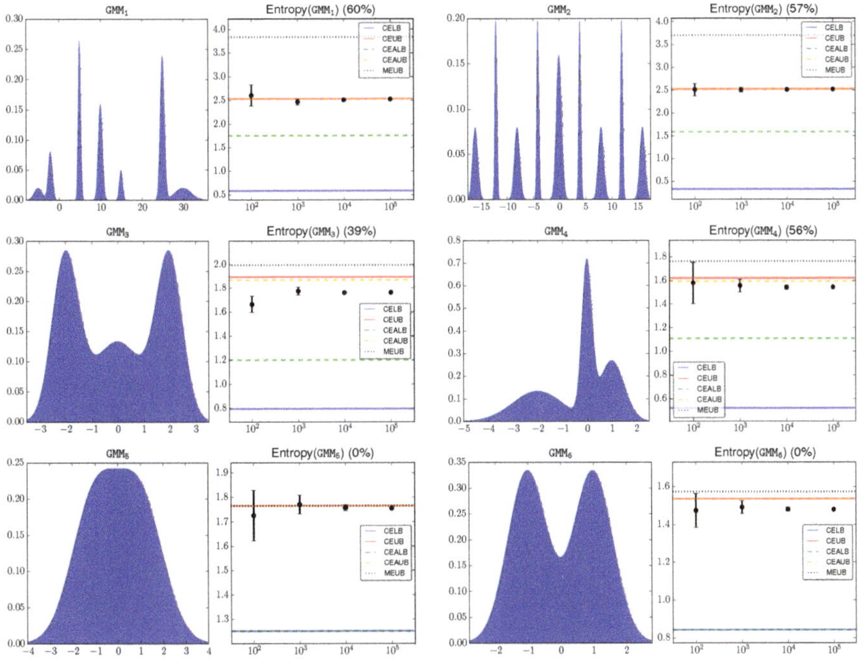

Figure 3. Lower and upper bounds on the differential entropy of Gaussian mixture models. On the left of each subfigure is the simulated GMM signal. On the right of each subfigure is the estimation of its differential entropy. Note that a subset of the bounds coincide with each other in several cases.

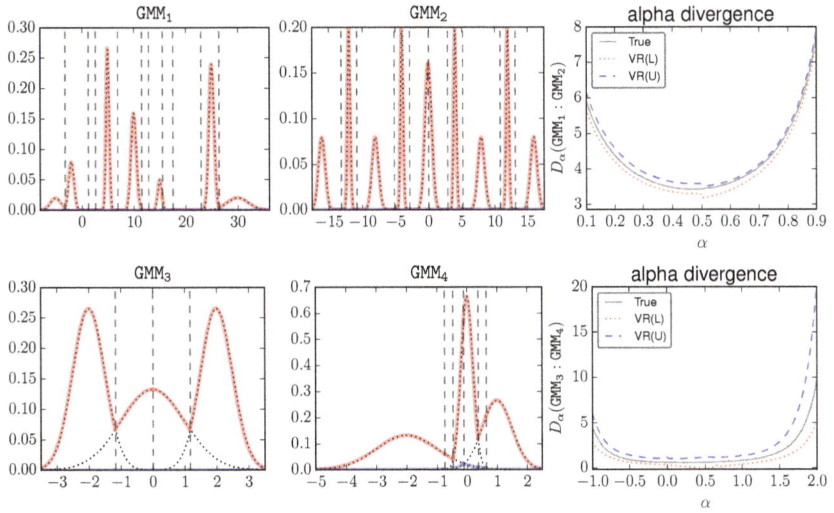

Figure 4. Two pairs of Gaussian Mixture Models and their α-divergences against different values of α. The "true" value of D_α is estimated by MC using 10^4 random samples. VR(L) and VR(U) denote the variation reduced lower and upper bounds, respectively. The range of α is selected for each pair for a clear visualization.

For a more quantitative comparison, Table 1 shows the estimated α-divergence by MC, Basic, Adaptive, and VR. As D_α is defined on $\mathbb{R} \setminus \{0,1\}$, the KL bounds CE(A)LB and CE(A)UB are presented for $\alpha = 0$ or 1. Overall, we have the following order of gap size: Basic > Adaptive > VR, and VR is recommended in general for bounding α-divergences. There are certain cases that the upper VR bound is looser than Adaptive. In practice one can compute the intersection of these bounds as well as the trivial bound $D_\alpha(m : m') \geq 0$ to get the best estimation.

Table 1. The estimated D_α and its bounds. The 95% confidence interval is shown for MC.

	α	MC(10^2)	MC(10^3)	MC(10^4)	Basic		Adaptive		VR	
					L	U	L	U	L	U
	0	15.96 ± 3.9	12.30 ± 1.0	13.63 ± 0.3	11.75	15.89	12.96	14.63		
	0.01	13.36 ± 2.9	10.63 ± 0.8	11.66 ± 0.3	-700.50	11.73	-77.33	11.73	11.40	12.27
GMM_1 & GMM_2	0.5	3.57 ± 0.3	3.47 ± 0.1	3.47 ± 0.07	-0.60	3.42	3.01	3.42	3.17	3.51
	0.99	40.04 ± 7.7	37.22 ± 2.3	38.58 ± 0.8	-333.90	39.04	5.36	38.98	38.28	38.96
	1	104.01 ± 28	84.96 ± 7.2	92.57 ± 2.5	91.44	95.59	92.76	94.41		
	0	0.71 ± 0.2	0.63 ± 0.07	0.62 ± 0.02	0.00	1.76	0.00	1.16		
	0.01	0.71 ± 0.2	0.63 ± 0.07	0.62 ± 0.02	-179.13	7.63	-38.74	4.96	0.29	1.00
GMM_3 & GMM_4	0.5	0.82 ± 0.3	0.57 ± 0.1	0.62 ± 0.04	-5.23	0.93	-0.71	0.85	-0.18	1.19
	0.99	0.79 ± 0.3	0.76 ± 0.1	0.80 ± 0.03	-165.72	12.10	-59.76	9.11	0.37	1.28
	1	0.80 ± 0.3	0.77 ± 0.1	0.81 ± 0.03	0.00	1.82	0.31	1.40		

Note the similarity between KL in Equation (30) and the expression in Equation (47). We give without a formal analysis that: CEAL(U)B is equivalent to VR at the limit $\alpha \to 0$ or $\alpha \to 1$. Experimentally as we slowly set $\alpha \to 1$, we can see that VR is consistent with CEAL(U)B.

7. Concluding Remarks and Perspectives

We have presented a fast versatile method to compute bounds on the Kullback–Leibler divergence between mixtures by building algorithmic formulae. We reported on our experiments for various mixture models in the exponential family. For univariate GMMs, we get a guaranteed bound of the KL divergence of two mixtures m and m' with k and k' components within an additive approximation factor of $\log k + \log k'$ in $O\left((k+k')\log(k+k')\right)$-time. Therefore, the larger the KL divergence, the better the bound when considering a multiplicative $(1+\alpha)$-approximation factor, since $\alpha = \frac{\log k + \log k'}{\mathrm{KL}(m:m')}$. The adaptive bounds are guaranteed to yield better bounds at the expense of computing potentially $O\left(k^2 + (k')^2\right)$ intersection points of pairwise weighted components.

Our technique also yields the bound for the Jeffreys divergence (the symmetrized KL divergence: $J(m,m') = \mathrm{KL}(m:m') + \mathrm{KL}(m':m)$) and the Jensen–Shannon divergence [47] (JS):

$$\mathrm{JS}(m,m') = \frac{1}{2}\left(\mathrm{KL}\left(m : \frac{m+m'}{2}\right) + \mathrm{KL}\left(m' : \frac{m+m'}{2}\right)\right),$$

since $\frac{m+m'}{2}$ is a mixture model with $k+k'$ components. One advantage of this statistical distance is that it is symmetric, always bounded by $\log 2$, and its square root yields a metric distance [48]. The log-sum-exp inequalities may also be used to compute some Rényi divergences [35]:

$$R_\alpha(m,p) = \frac{1}{\alpha-1}\log\left(\int m(x)^\alpha p(x)^{1-\alpha}\right)\mathrm{d}x,$$

when α is an integer, $m(x)$ a mixture and $p(x)$ a single (component) distribution. Getting fast guaranteed tight bounds on statistical distances between mixtures opens many avenues. For example, we may consider building hierarchical mixture models by merging iteratively two mixture components so that those pairs of components are chosen so that the KL distance between the full mixture and the simplified mixture is minimized.

In order to be useful, our technique is unfortunately limited to univariate mixtures: indeed, in higher dimensions, we can still compute the maximization diagram of weighted components (an additively weighted Bregman–Voronoi diagram [49,50] for components belonging to the same exponential family). However, it becomes more complex to compute in the elementary Voronoi cells V, the functions $C_{i,j}(V)$ and $M_i(V)$ (in 1D, the Voronoi cells are segments). We may obtain hybrid algorithms by approximating or estimating these functions. In 2D, it is thus possible to obtain lower and upper bounds on the Mutual Information [51] (MI) when the joint distribution $m(x,y)$ is a 2D mixture of Gaussians:

$$I(M;M') = \int m(x,y)\log\frac{m(x,y)}{m(x)m'(y)}\mathrm{d}x\mathrm{d}y.$$

Indeed, the marginal distributions $m(x)$ and $m'(y)$ are univariate Gaussian mixtures.

A Python code implementing those computational-geometric methods for reproducible research is available online [52].

Acknowledgments: The authors gratefully thank the referees for their comments. This work was carried out while Ke Sun was visiting Frank Nielsen at Ecole Polytechnique, Palaiseau, France.

Author Contributions: Frank Nielsen and Ke Sun contributed to the theoretical results as well as to the writing of the article. Ke Sun implemented the methods and performed the numerical experiments. All authors have read and approved the final manuscript.

Conflicts of Interest: The authors declare no conflict of interest.

Appendix A. The Kullback–Leibler Divergence of Mixture Models Is Not Analytic [6]

Ideally, we aim at getting a finite length closed-form formula to compute the KL divergence of finite mixture models. However, this is provably mathematically intractable [6] because of the log-sum term in the integral, as we shall prove below.

Analytic expressions encompass closed-form formula and may include special functions (e.g., Gamma function) but do not allow to use limits or integrals. An analytic function $f(x)$ is a C^∞ function (infinitely differentiable) such that around any point x_0 the k-order Taylor series $T_k(x) = \sum_{i=0}^{k}\frac{f^{(i)}(x_0)}{i!}(x-x_0)^i$ converges to $f(x)$: $\lim_{k\to\infty} T_k(x) = f(x)$ for x belonging to a neighborhood $N_r(x_0) = \{x : |x-x_0| \leq r\}$ of x_0, where r is called the radius of convergence. The analytic property of a function is equivalent to the condition that for each $k \in \mathbb{N}$, there exists a constant c such that $\left|\frac{\mathrm{d}^k f}{\mathrm{d}x^k}(x)\right| \leq c^{k+1}k!$.

To prove that the KL of mixtures is not analytic (hence does not admit a closed-form formula), we shall adapt the proof reported in [6] (in Japanese, we thank Professor Aoyagi for sending us his paper [6]). We shall prove that $\mathrm{KL}(p : q)$ is not analytic for $p(x) = G(x;0,1)$ and $q(x;w) = (1-w)G(x;0,1) + wG(x;1,1)$, where $w \in (0,1)$, and $G(x;\mu,\sigma) = \frac{1}{\sqrt{2\pi}\sigma}\exp(-\frac{(x-\mu)^2}{2\sigma^2})$ is the density of a univariate Gaussian of mean μ and standard deviation σ. Let $D(w) = \mathrm{KL}(p(x) : q(x;w))$ denote the KL divergence between these two mixtures (p has a single component and q has two components).

We have

$$\log\frac{p(x)}{q(x;w)} = \log\frac{\exp\left(-\frac{x^2}{2}\right)}{(1-w)\exp\left(-\frac{x^2}{2}\right) + w\exp\left(-\frac{(x-1)^2}{2}\right)} = -\log(1 + w(e^{x-\frac{1}{2}} - 1)). \tag{A1}$$

Therefore

$$\frac{\mathrm{d}^k D}{\mathrm{d}w^k} = \frac{(-1)^k}{k}\int p(x)(e^{x-\frac{1}{2}} - 1)\mathrm{d}x.$$

Let x_0 be the root of the equation $e^{x-\frac{1}{2}} - 1 = e^{\frac{x}{2}}$ so that for $x \geq x_0$, we have $e^{x-\frac{1}{2}} - 1 \geq e^{\frac{x}{2}}$. It follows that:

$$\left|\frac{\mathrm{d}^k D}{\mathrm{d}w^k}\right| \geq \frac{1}{k}\int_{x_0}^{\infty} p(x)e^{\frac{kx}{2}}\mathrm{d}x = \frac{1}{k}e^{\frac{k^2}{8}}A_k$$

with $A_k = \int_{x_0}^{\infty} \frac{1}{\sqrt{2\pi}} \exp(-\frac{x-\frac{k}{2}}{2})\mathrm{d}x$. When $k \to \infty$, we have $A_k \to 1$. Consider $k_0 \in \mathbb{N}$ such that $A_{k_0} > 0.9$. Then the radius of convergence r is such that:

$$\frac{1}{r} \geq \lim_{k\to\infty} \left(\frac{1}{kk!} 0.9 \exp\left(\frac{k^2}{8}\right) \right)^{\frac{1}{k}} = \infty.$$

Thus the convergence radius is $r = 0$, and therefore the KL divergence is not an analytic function of the parameter w. The KL of mixtures is an example of a non-analytic smooth function. (Notice that the absolute value is not analytic at 0.)

Appendix B. Closed-Form Formula for the Kullback–Leibler Divergence between Scaled and Truncated Exponential Families

When computing approximation bounds for the KL divergence between two mixtures $m(x)$ and $m'(x)$, we end up with the task of computing $\int_{\mathcal{D}} w_a p_a(x) \log \frac{w'_b p'_b(x)}{w'_c p'_c(x)} \mathrm{d}x$ where $\mathcal{D} \subseteq \mathcal{X}$ is a subset of the full support $\mathcal{X}$. We report a generic formula for computing these formula when the mixture (scaled and truncated) components belong to the same exponential family [17]. An exponential family has canonical log-density written as $l(x;\theta) = \log p(x;\theta) = \theta^\top t(x) - F(\theta) + k(x)$, where $t(x)$ denotes the sufficient statistics, $F(\theta)$ the log-normalizer (also called cumulant function or partition function), and $k(x)$ an auxiliary carrier term.

Let $\mathrm{KL}(w_1p_1 : w_2p_2 : w_3p_3) = \int_{\mathcal{X}} w_1 p_1(x) \log \frac{w_2 p_2(x)}{w_3 p_3(x)} \mathrm{d}x = H_\times(w_1p_1 : w_3p_3) - H_\times(w_1p_1 : w_2p_2)$. Since it is a difference of two cross-entropies, we get for three distributions belonging to the same exponential family [26] the following formula:

$$\mathrm{KL}(w_1p_1 : w_2p_2 : w_3p_3) = w_1 \log \frac{w_2}{w_3} + w_1(F(\theta_3) - F(\theta_2) - (\theta_3 - \theta_2)^\top \nabla F(\theta_1)).$$

Furthermore, when the support is restricted, say to support range $\mathcal{D} \subseteq \mathcal{X}$, let $m_{\mathcal{D}}(\theta) = \int_{\mathcal{D}} p(x;\theta)\mathrm{d}x$ denote the mass and $\tilde{p}(x;\theta) = \frac{p(x;\theta)}{m_{\mathcal{D}}(\theta)}$ the normalized distribution. Then we have:

$$\int_{\mathcal{D}} w_1 p_1(x) \log \frac{w_2 p_2(x)}{w_3 p_3(x)} \mathrm{d}x = m_{\mathcal{D}}(\theta_1)(\mathrm{KL}(w_1\tilde{p}_1 : w_2\tilde{p}_2 : w_3\tilde{p}_3)) - \log \frac{w_2 m_{\mathcal{D}}(\theta_3)}{w_3 m_{\mathcal{D}}(\theta_2)}.$$

When $F_{\mathcal{D}}(\theta) = F(\theta) - \log m_{\mathcal{D}}(\theta)$ is strictly convex and differentiable then $\tilde{p}(x;\theta)$ is an exponential family and the closed-form formula follows straightforwardly. Otherwise, we still get a closed-form but need more derivations. For univariate distributions, we write $\mathcal{D} = (a,b)$ and $m_{\mathcal{D}}(\theta) = \int_a^b p(x;\theta)\mathrm{d}x = P_\theta(b) - P_\theta(a)$ where $P_\theta(a) = \int^a p(x;\theta)\mathrm{d}x$ denotes the cumulative distribution function.

The usual formula for truncated and scaled Kullback–Leibler divergence is:

$$\mathrm{KL}_{\mathcal{D}}(wp(x;\theta) : w'p(x;\theta')) = w m_{\mathcal{D}}(\theta) \left(\log \frac{w}{w'} + B_F(\theta' : \theta)\right) + w(\theta' - \theta)^\top \nabla m_{\mathcal{D}}(\theta), \tag{B1}$$

where $B_F(\theta' : \theta)$ is a Bregman divergence [5]:

$$B_F(\theta' : \theta) = F(\theta') - F(\theta) - (\theta' - \theta)^\top \nabla F(\theta).$$

This formula extends the classic formula [5] for full regular exponential families (by setting $w = w' = 1$ and $m_{\mathcal{D}}(\theta) = 1$ with $\nabla m_{\mathcal{D}}(\theta) = 0$).

Similar formulæ are available for the cross-entropy and entropy of exponential families [26].

Appendix C. On the Approximation of KL between Smooth Mixtures by a Bregman Divergence [5]

Clearly, since Bregman divergences are always finite while KL divergences may diverge, we need extra conditions to assert that the KL between mixtures can be approximated by Bregman divergences.

We require that the Jeffreys divergence between mixtures be finite in order to approximate the KL between mixtures by a Bregman divergence. We loosely derive this observation (Careful derivations will be reported elsewhere) using two different approaches:

- First, continuous mixture distributions have smooth densities that can be arbitrarily closely approximated using a single distribution (potentially multi-modal) belonging to the Polynomial Exponential Families [53,54] (PEFs). A polynomial exponential family of order D has log-likelihood $l(x;\theta) \propto \sum_{i=1}^{D} \theta_i x^i$: Therefore, a PEF is an exponential family with polynomial sufficient statistics $t(x) = (x, x^2, \ldots, x^D)$. However, the log-normalizer $F_D(\theta) = \log \int \exp(\theta^\top t(x))\mathrm{d}x$ of a D-order PEF is not available in closed-form: It is computationally intractable. Nevertheless, the KL between two mixtures $m(x)$ and $m'(x)$ can be theoretically approximated closely by a Bregman divergence between the two corresponding PEFs: $\mathrm{KL}(m(x) : m'(x)) \simeq \mathrm{KL}(p(x;\theta) : p(x;\theta')) = B_{F_D}(\theta':\theta)$, where θ and θ' are the natural parameters of the PEF family $\{p(x;\theta)\}$ approximating $m(x)$ and $m'(x)$, respectively (i.e., $m(x) \simeq p(x;\theta)$ and $m'(x) \simeq p(x;\theta')$). Notice that the Bregman divergence of PEFs has necessarily finite value but the KL of two smooth mixtures can potentially diverge (infinite value), hence the conditions on Jeffreys divergence to be finite.

- Second, consider two finite mixtures $m(x) = \sum_{i=1}^{k} w_i p_i(x)$ and $m'(x) = \sum_{j=1}^{k'} w'_j p'_j(x)$ of k and k' components (possibly with heterogeneous components $p_i(x)$'s and $p'_j(x)$'s), respectively. In information geometry, a mixture family is the set of convex combination of fixed component densities. Thus in statistics, a mixture is understood as a convex combination of parametric components while in information geometry a mixture family is the set of convex combination of fixed components. Let us consider the mixture families $\{g(x;(w,w'))\}$ generated by the $D = k + k'$ fixed components $p_1(x), \ldots, p_k(x), p'_1(x), \ldots, p'_{k'}(x)$:

$$\left\{ g(x;(w,w')) = \sum_{i=1}^{k} w_i p_i(x) + \sum_{j=1}^{k'} w'_j p'_j(x) \; : \; \sum_{i=1}^{k} w_i + \sum_{j=1}^{k'} w'_j = 1 \right\}$$

  We can approximate arbitrarily finely (with respect to total variation) mixture $m(x)$ for any $\epsilon > 0$ by $g(x;\alpha) \simeq (1-\epsilon)m(x) + \epsilon m'(x)$ with $\alpha = ((1-\epsilon)w, \epsilon w')$ (so that $\sum_{i=1}^{k+k'} \alpha_i = 1$) and $m'(x) \simeq g(x;\alpha') = \epsilon m(x) + (1-\epsilon)m'(x)$ with $\alpha' = (\epsilon w, (1-\epsilon)w')$ (and $\sum_{i=1}^{k+k'} \alpha'_i = 1$). Therefore $\mathrm{KL}(m(x) : m'(x)) \simeq \mathrm{KL}(g(x;\alpha) : g(x;\alpha')) = B_{F^*}(\alpha : \alpha')$, where $F^*(\alpha) = \int g(x;\alpha) \log g(x;\alpha) \mathrm{d}x$ is the Shannon information (negative Shannon entropy) for the composite mixture family. Again, the Bregman divergence $B_{F^*}(\alpha : \alpha')$ is necessarily finite but $\mathrm{KL}(m(x) : m'(x))$ between mixtures may be potentially infinite when the KL integral diverges (hence, the condition on Jeffreys divergence finiteness). Interestingly, this Shannon information can be arbitrarily closely approximated when considering isotropic Gaussians [13]. Notice that the convex conjugate $F(\theta)$ of the continuous Shannon neg-entropy $F^*(\eta)$ is the log-sum-exp function on the inverse soft map.

References

1. Huang, Z.K.; Chau, K.W. A new image thresholding method based on Gaussian mixture model. *Appl. Math. Comput.* **2008**, *205*, 899–907.
2. Seabra, J.; Ciompi, F.; Pujol, O.; Mauri, J.; Radeva, P.; Sanches, J. Rayleigh mixture model for plaque characterization in intravascular ultrasound. *IEEE Trans. Biomed. Eng.* **2011**, *58*, 1314–1324.
3. Julier, S.J.; Bailey, T.; Uhlmann, J.K. Using Exponential Mixture Models for Suboptimal Distributed Data Fusion. In Proceedings of the 2006 IEEE Nonlinear Statistical Signal Processing Workshop, Cambridge, UK, 13–15 September 2006; IEEE: New York, NY, USA, 2006; pp. 160–163.
4. Cover, T.M.; Thomas, J.A. *Elements of Information Theory*; John Wiley & Sons: Hoboken, NJ, USA, 2012.
5. Banerjee, A.; Merugu, S.; Dhillon, I.S.; Ghosh, J. Clustering with Bregman divergences. *J. Mach. Learn. Res.* **2005**, *6*, 1705–1749.

6. Watanabe, S.; Yamazaki, K.; Aoyagi, M. *Kullback Information of Normal Mixture is Not an Analytic Function*; Technical Report of IEICE Neurocomputing; The Institute of Electronics, Information and Communication Engineers, Tokyo, Japan, 2004; pp. 41–46. (In Japanese)
7. Michalowicz, J.V.; Nichols, J.M.; Bucholtz, F. Calculation of differential entropy for a mixed Gaussian distribution. *Entropy* **2008**, *10*, 200–206.
8. Pichler, G.; Koliander, G.; Riegler, E.; Hlawatsch, F. Entropy for singular distributions. In Proceedings of the IEEE International Symposium on Information Theory (ISIT), Honolulu, HI, USA, 29 June–4 July 2014; pp. 2484–2488.
9. Huber, M.F.; Bailey, T.; Durrant-Whyte, H.; Hanebeck, U.D. On entropy approximation for Gaussian mixture random vectors. In Proceedings of the IEEE International Conference on Multisensor Fusion and Integration for Intelligent Systems, Seoul, Korea, 20–22 August 2008; IEEE: New York, NY, USA, 2008; pp. 181–188.
10. Yamada, M.; Sugiyama, M. Direct importance estimation with Gaussian mixture models. *IEICE Trans. Inf. Syst.* **2009**, *92*, 2159–2162.
11. Durrieu, J.L.; Thiran, J.P.; Kelly, F. Lower and upper bounds for approximation of the Kullback-Leibler divergence between Gaussian Mixture Models. In Proceedings of the IEEE International Conference on Acoustics, Speech and Signal Processing (ICASSP), Kyoto, Japan, 25–30 March 2012; IEEE: New York, NY, USA, 2012; pp. 4833–4836.
12. Schwander, O.; Marchand-Maillet, S.; Nielsen, F. Comix: Joint estimation and lightspeed comparison of mixture models. In Proceedings of the 2016 IEEE International Conference on Acoustics, Speech and Signal Processing, ICASSP 2016, Shanghai, China, 20–25 March 2016; pp. 2449–2453.
13. Moshksar, K.; Khandani, A.K. Arbitrarily Tight Bounds on Differential Entropy of Gaussian Mixtures. *IEEE Trans. Inf. Theory* **2016**, *62*, 3340–3354.
14. Mezuman, E.; Weiss, Y. A Tight Convex Upper Bound on the Likelihood of a Finite Mixture. *arXiv* **2016**, arXiv:1608.05275.
15. Amari, S.-I. *Information Geometry and Its Applications*; Springer: Tokyo, Japan, 2016; Volume 194.
16. Nielsen, F.; Sun, K. Guaranteed Bounds on the Kullback–Leibler Divergence of Univariate Mixtures. *IEEE Signal Process. Lett.* **2016**, *23*, 1543–1546.
17. Nielsen, F.; Garcia, V. Statistical exponential families: A digest with flash cards. *arXiv* **2009**, arXiv:0911.4863.
18. Calafiore, G.C.; El Ghaoui, L. *Optimization Models*; Cambridge University Press: Cambridge, UK, 2014.
19. Shen, C.; Li, H. On the dual formulation of boosting algorithms. *IEEE Trans. Pattern Anal. Mach. Intell.* **2010**, *32*, 2216–2231.
20. Beck, A. *Introduction to Nonlinear Optimization: Theory, Algorithms, and Applications with MATLAB*; Society for Industrial and Applied Mathematics: Philadelphia, PA, USA, 2014.
21. Boyd, S.; Vandenberghe, L. *Convex Optimization*; Cambridge University Press: Cambridge, UK, 2004.
22. De Berg, M.; van Kreveld, M.; Overmars, M.; Schwarzkopf, O.C. *Computational Geometry*; Springer: Heidelberg, Germany, 2000.
23. Setter, O.; Sharir, M.; Halperin, D. *Constructing Two-Dimensional Voronoi Diagrams via Divide-and-Conquer of Envelopes in Space*; Springer: Heidelberg, Germany, 2010.
24. Devillers, O.; Golin, M.J. Incremental algorithms for finding the convex hulls of circles and the lower envelopes of parabolas. *Inf. Process. Lett.* **1995**, *56*, 157–164.
25. Nielsen, F.; Yvinec, M. An output-sensitive convex hull algorithm for planar objects. *Int. J. Comput. Geom. Appl.* **1998**, *8*, 39–65.
26. Nielsen, F.; Nock, R. Entropies and cross-entropies of exponential families. In Proceedings of the 17th IEEE International Conference on Image Processing (ICIP), Hong Kong, China, 26–29 September 2010; IEEE: New York, NY, USA, 2010; pp. 3621–3624.
27. Sharir, M.; Agarwal, P.K. *Davenport-Schinzel Sequences and Their Geometric Applications*; Cambridge University Press: Cambridge, UK, 1995.
28. Bronstein, M. Algorithms and computation in mathematics. In *Symbolic Integration. I. Transcendental Functions*; Springer: Berlin, Germany, 2005.
29. Carreira-Perpinan, M.A. Mode-finding for mixtures of Gaussian distributions. *IEEE Trans. Pattern Anal. Mach. Intell.* **2000**, *22*, 1318–1323.

30. Aprausheva, N.N.; Sorokin, S.V. Exact equation of the boundary of unimodal and bimodal domains of a two-component Gaussian mixture. *Pattern Recognit. Image Anal.* **2013**, *23*, 341–347.
31. Learned-Miller, E.; DeStefano, J. A probabilistic upper bound on differential entropy. *IEEE Trans. Inf. Theory* **2008**, *54*, 5223–5230.
32. Amari, S.-I. α-Divergence Is Unique, Belonging to Both f-Divergence and Bregman Divergence Classes. *IEEE Trans. Inf. Theory* **2009**, *55*, 4925–4931.
33. Cichocki, A.; Amari, S.I. Families of Alpha- Beta- and Gamma-Divergences: Flexible and Robust Measures of Similarities. *Entropy* **2010**, *12*, 1532–1568.
34. Póczos, B.; Schneider, J. On the Estimation of α-Divergences. In Proceedings of the 14th International Conference on Artificial Intelligence and Statistics, Ft. Lauderdale, FL, USA, 11–13 April 2011; pp. 609–617.
35. Nielsen, F.; Nock, R. On Rényi and Tsallis entropies and divergences for exponential families. *arXiv* **2011**, arXiv:1105.3259.
36. Minka, T. *Divergence Measures and Message Passing*; Technical Report MSR-TR-2005-173; Microsoft Research: Cambridge, UK, 2005.
37. Améndola, C.; Drton, M.; Sturmfels, B. Maximum Likelihood Estimates for Gaussian Mixtures Are Transcendental. *arXiv* **2015**, arXiv:1508.06958.
38. Hellinger, E. Neue Begründung der Theorie quadratischer Formen von unendlichvielen Veränderlichen. *J. Reine Angew. Math.* **1909**, *136*, 210–271. (In German)
39. Van Erven, T.; Harremos, P. Rényi divergence and Kullback-Leibler divergence. *IEEE Trans. Inf. Theory* **2014**, *60*, 3797–3820.
40. Nielsen, F.; Nock, R. A closed-form expression for the Sharma-Mittal entropy of exponential families. *J. Phys. A Math. Theor.* **2012**, *45*, 032003.
41. Nielsen, F.; Nock, R. On the Chi Square and Higher-Order Chi Distances for Approximating f-Divergences. *IEEE Signal Process. Lett.* **2014**, *21*, 10–13.
42. Nielsen, F.; Boltz, S. The Burbea-Rao and Bhattacharyya centroids. *IEEE Trans. Inf. Theory* **2011**, *57*, 5455–5466.
43. Jarosz, W. Efficient Monte Carlo Methods for Light Transport in Scattering Media. Ph.D. Thesis, University of California, San Diego, CA, USA, 2008.
44. Fujisawa, H.; Eguchi, S. Robust parameter estimation with a small bias against heavy contamination. *J. Multivar. Anal.* **2008**, *99*, 2053–2081.
45. Havrda, J.; Charvát, F. Quantification method of classification processes. Concept of structural α-entropy. *Kybernetika* **1967**, *3*, 30–35.
46. Liang, X. A Note on Divergences. *Neural Comput.* **2016**, *28*, 2045–2062.
47. Lin, J. Divergence measures based on the Shannon entropy. *IEEE Trans. Inf. Theory* **1991**, *37*, 145–151.
48. Endres, D.M.; Schindelin, J.E. A new metric for probability distributions. *IEEE Trans. Inf. Theory* **2003**, *49*, 1858–1860.
49. Nielsen, F.; Boissonnat, J.D.; Nock, R. On Bregman Voronoi diagrams. In Proceedings of the Eighteenth Annual ACM-SIAM Symposium on Discrete Algorithms, New Orleans, LA, USA, 7–9 January 2007; Society for Industrial and Applied Mathematics: Philadelphia, PA, USA, 2007; pp. 746–755.
50. Boissonnat, J.D.; Nielsen, F.; Nock, R. Bregman Voronoi diagrams. *Discret. Comput. Geom.* **2010**, *44*, 281–307.
51. Foster, D.V.; Grassberger, P. Lower bounds on mutual information. *Phys. Rev. E* **2011**, *83*, 010101.
52. Nielsen, F.; Sun, K. PyKLGMM: Python Software for Computing Bounds on the Kullback-Leibler Divergence between Mixture Models. 2016. Available online: https://www.lix.polytechnique.fr/~nielsen/KLGMM/ (accessed on 6 December 2016).
53. Cobb, L.; Koppstein, P.; Chen, N.H. Estimation and moment recursion relations for multimodal distributions of the exponential family. *J. Am. Stat. Assoc.* **1983**, *78*, 124–130.
54. Nielsen, F.; Nock, R. Patch matching with polynomial exponential families and projective divergences. In Proceedings of the 9th International Conference Similarity Search and Applications (SISAP), Tokyo, Japan, 24–26 October 2016.

entropy

MDPI

Article

A Sequence of Escort Distributions and Generalizations of Expectations on q-Exponential Family

Hiroshi Matsuzoe

Department of Computer Science and Engineering, Nagoya Institute of Technology, Nagoya 466-8555, Japan; matsuzoe1@nitech.ac.jp; Tel.: +81-52-735-5143

Academic Editors: Frédéric Barbaresco and Frank Nielsen
Received: 26 October 2016; Accepted: 19 December 2016; Published: 25 December 2016

Abstract: In the theory of complex systems, long tailed probability distributions are often discussed. For such a probability distribution, a deformed expectation with respect to an escort distribution is more useful than the standard expectation. In this paper, by generalizing such escort distributions, a sequence of escort distributions is introduced. As a consequence, it is shown that deformed expectations with respect to sequential escort distributions effectively work for anomalous statistics. In particular, it is shown that a Fisher metric on a q-exponential family can be obtained from the escort expectation with respect to the second escort distribution, and a cubic form (or an Amari–Chentsov tensor field, equivalently) is obtained from the escort expectation with respect to the third escort distribution.

Keywords: escort distribution; escort expectation; statistical manifold; deformed exponential family; Tsallis statistics; information geometry

MSC: 53A15; 53B50; 62F99; 94A14

---

1. Introduction

Long tailed probability distributions and their related probability distributions are important objects in anomalous statistical physics (cf. [1–3]). For such long tailed probability distributions, the standard expectation does not exist in general. Therefore, the notion of escort distribution has been introduced [4]. Since an escort distribution gives a suitable weight for tail probability, the escort expectation which is the expectation with respect to an escort distribution is more useful than the standard one.

In anomalous statistics, a deformed exponential function and a deformed logarithm function play essential roles. In fact, a deformed exponential family is an important statistical model in anomalous statistics. Such a statistical model is described by such a deformed exponential function. In particular, the set of all q-normal distributions (or Student's t-distributions, equivalently) is a q-exponential family, which is described by a q-deformed exponential function [5] (see also [6,7]).

On the other hand, a generalized score function is defined from a deformed logarithm function. In the previous works, the author showed that a deformed score function is unbiased with respect to the escort expectation [8,9]. This implies that a deformed score function is regarded as an estimating function on a deformed exponential family. In addition, in information geometry, it is known that a deformed exponential family has a statistical manifold structure. Then a deformed score function is regarded as a tangent vector on this statistical manifold [6,10]. Therefore, properties of escort expectations are closely related to geometric structures on a deformed exponential family.

In this paper, we introduce a sequence of escort distributions, then we consider a sequential structure of escort expectations. It is known that a deformed exponential family naturally has at least three kind of different statistical manifold structures [6,11]. Then we show that such statistical manifold structures can be obtained from a sequential structure of escort expectations. In particular, we show that a Fisher metric on a q-exponential family can be obtained from the deformed expectations with respect to the second escort distribution, and a cubic form (or an Amari–Chentsov tensor field, equivalently) is obtained from the deformed expectations with respect to the third escort distribution.

This paper is written based on the proceeding paper [7]. However, this paper focuses on deformed expectations of a q-exponential family, whereas the previous paper focused on deformed independences. We remark that several authors have been studying deformed expectations recently. See [12,13], for example.

2. Deformed Exponential Families

In this paper, we assume that all objects are smooth for simplicity. Let us review preliminary facts about deformed exponential functions and deformed exponential families. For more details, see [2,6], for example. Historically, Tsallis [14] introduced the notion of q-exponential function and Naudts [5] introduced the notion of q-exponential family together with a further generalization. Such a historical note is provided in [2].

Let $\mathbf{R}_{++}$ be the set of all positive real numbers, $\mathbf{R}_{++} := \{x \in \mathbf{R} | x > 0\}$. Let χ be a strictly increasing function from $\mathbf{R}_{++}$ to $\mathbf{R}_{++}$. We define a *χ-logarithm function* or a *deformed logarithm function* by

$$\ln_\chi s := \int_1^s \frac{1}{\chi(t)} dt.$$

The inverse of $\ln_\chi s$ is called a *χ-exponential function* or a *deformed exponential function*, which is defined by

$$\exp_\chi t := 1 + \int_0^t u(s) ds,$$

where the function $u(s)$ is given by $u(\ln_\chi s) = \chi(s)$.

From now on, we suppose that χ is a power function, that is, $\chi(t) = t^q$. Then the deformed logarithm and the deformed exponential are defined by

$$\ln_q s := \frac{s^{1-q} - 1}{1-q}, \qquad (s > 0),$$

$$\exp_q t := (1 + (1-q)t)^{\frac{1}{1-q}}, \qquad (1 + (1-q)t > 0).$$

We say that $\ln_q s$ is a *q-logarithm function* and $\exp_q t$ is a *q-exponential function*. In this case, the function $u(s)$ is given by

$$u(s) = (1 + (1-q)s)^{\frac{q}{1-q}} = \{\exp_q s\}^q.$$

By taking a limit $q \to 1$, these functions coincide with the standard logarithm $\ln s$ and the standard exponential $\exp t$, respectively.

A statistical model S_q is called a *q-exponential family* if

$$S_q := \left\{ p(x,\theta) \middle| p(x;\theta) = \exp_q \left[\sum_{i=1}^n \theta^i F_i(x) - \psi(\theta) \right], \theta \in \Theta \subset \mathbf{R}^n \right\}, \qquad (1)$$

where $F_1(x), \dots, F_n(x)$ are functions on a sample space Ω, $\theta = {}^t(\theta^1, \dots, \theta^n)$ is a parameter, and $\psi(\theta)$ is the normalization with respect to the parameter θ. Under suitable conditions, S_q is regarded as a manifold with a local coordinate system $\{\theta^1, \dots, \theta^n\}$. In this case, we call $\{\theta^i\}$ a *natural coordinate system*.

In this paper, we focus on the q-exponential case. However, many results for the q-exponential family can be generalized for the χ-exponential family (cf. [6,8]). We remark that a q-exponential family and a χ-exponential family have further generalizations. See [15], for example.

Example 1 (Student's t-distribution (cf. [2,6,7])). *Fix a number q $(1 < q < 1 + 2/d,\ d \in \mathbf{N})$, and set $\nu = -d - 2/(1-q)$. We define a d-dimensional Student's t-distribution with degree of freedom ν or a q-Gaussian distribution by*

$$p_q(x;\mu,\Sigma) := \frac{\Gamma\left(\frac{1}{q-1}\right)}{(\pi\nu)^{\frac{d}{2}}\Gamma\left(\frac{\nu}{2}\right)\sqrt{\det(\Sigma)}}\left[1 + \frac{1}{\nu}{}^t(x-\mu)\Sigma^{-1}(x-\mu)\right]^{\frac{1}{1-q}},$$

where $X = {}^t(X_1, \ldots, X_d)$ is a random vector on $\mathbf{R}^d$, $\mu = {}^t(\mu^1, \ldots, \mu^d)$ is a location vector on $\mathbf{R}^d$ and Σ is a scale matrix on $\mathrm{Sym}^+(d)$. For simplicity, we assume that Σ is invertible. Otherwise, we should choose a suitable basis $\{v^\alpha\}$ on $\mathrm{Sym}^+(d)$ such that $\Sigma = \sum_\alpha w_\alpha v^\alpha$. Then, the set of all Student's t-distributions is a q-exponential family. In fact, setting parameters by

$$z_q = \frac{(\pi\nu)^{\frac{d}{2}}\Gamma\left(\frac{\nu}{2}\right)\sqrt{\det(\Sigma)}}{\Gamma\left(\frac{1}{q-1}\right)}, \quad \tilde{R} = \frac{z_q^{q-1}}{(1-q)d+2}\Sigma^{-1}, \quad \text{and} \quad \theta = 2\tilde{R}\mu, \tag{2}$$

we have

$$\begin{aligned} p_q(x;\mu,\Sigma) &= \frac{1}{z_q}\left[1 + \frac{1}{\nu}{}^t(x-\mu)\Sigma^{-1}(x-\mu)\right]^{\frac{1}{1-q}} \\ &= \left[\left(\frac{1}{z_q}\right)^{1-q} - \frac{1-q}{(1-q)d+2}\left(\frac{1}{z_q}\right)^{1-q}{}^t(x-\mu)\Sigma^{-1}(x-\mu)\right]^{\frac{1}{1-q}} \\ &= \exp_q\left[-{}^t(x-\mu)\tilde{R}(x-\mu) + \ln_q \frac{1}{z_q}\right] \\ &= \exp_q\left[\sum_{i=1}^d \theta^i x_i - \sum_{i=1}^d \tilde{R}_{ii}x_i^2 - 2\sum_{i<j}\tilde{R}_{ij}x_i x_j - \frac{1}{4}{}^t\theta\tilde{R}^{-1}\theta + \ln_q \frac{1}{z_q}\right]. \end{aligned}$$

Since $\theta \in \mathbf{R}^d$ and $\tilde{R} \in \mathrm{Sym}^+(d)$, the set of all Student's t-distributions is a $d(d+3)/2$-dimensional q-exponential family. The normalization $\psi(\theta)$ is given by

$$\psi(\theta) = \frac{1}{4}{}^t\theta\tilde{R}^{-1}\theta - \ln_q \frac{1}{z_q}.$$

A univariate Student's t-distribution is a well-known probability distribution in elementary statistics. We denote it by

$$t_\nu(x;\mu,\sigma) := \frac{1}{Z_q}\exp_q\left[-\frac{(x-\mu)^2}{(3-q)\sigma^2}\right], \tag{3}$$

where $\mu \in \mathbf{R}$ is a location parameter, $\sigma \in \mathbf{R}_{++}$ is a scale parameter, and Z_q is the normalization defined by

$$Z_q = \sqrt{\frac{3-q}{q-1}}\mathrm{Beta}\left(\frac{3-q}{2(q-1)}, \frac{1}{2}\right)\sigma.$$

In this case, the degree of freedom is $\nu = (3-q)/(q-1)$. Conversely, the parameter q is give by

$$q = \frac{\nu+3}{\nu+1}. \tag{4}$$

3. Escort Distributions and Generalizations of Expectations

In anomalous statistics, a generalized expectation, called an escort expectation, is often discussed since the standard expectation does not exist in general (cf. [2,5,6]). In this section, we recall generalizations of expectations and introduce a sequential structure of escort distributions.

Let S_q be a q-exponential family. For a given $p(x;\theta) \in S_q$ we define the *q-escort distribution* $P_q(x;\theta)$ of $p(x;\theta)$ and the *normalized q-escort distribution* $P_q^{esc}(x;\theta)$ by

$$\begin{aligned} P_q(x;\theta) &:= P_{q,(1)}(x;\theta) := \{p(x;\theta)\}^q, \\ P_q^{esc}(x;\theta) &:= \frac{1}{Z_q(p)}\{p(x;\theta)\}^q, \quad \text{where} \quad Z_q(p) = \int_\Omega \{p(x;\theta)\}^q dx, \end{aligned}$$

respectively. For a q-exponential family $S_q = \{p_q(x;\theta)\}$, the set of normalized escort distributions $S_{q'} = \{P_q^{esc}(x;\theta)\}$ is a q'-exponential family with $q' = (2q-1)/q$.

Example 2. *Let $t_\nu(x;\mu,\sigma)$ be a univariate Student's t-distribution with degree of freedom ν. Then its normalized escort distribution is also a univariate Student's t-distribution with degree of freedom $\nu+2$. In fact, from Equation (4), a direct calculation shows that*

$$q' = \frac{2q-1}{q} = \frac{\nu+5}{\nu+3}.$$

This implies that the degree of freedom $\nu' = \nu+2$. Therefore, we obtain a sequence of escort distributions from a given Student's t-distribution t_ν:

$$t_\nu \to t_{\nu+2} \to t_{\nu+4} \to \cdots.$$

This sequence is called a τ-sequence, and the procedure to obtain from a given t-distribution to another t-distribution through an escort distribution is called the τ-transformation [16].

For a given $p_q(x;\theta) \in S_q$, we can define the escort of an escort distribution

$$\widetilde{P}_q(x;\theta) := P_{q,(2)}(x;\theta) := q\{P_q(x;\theta)\}^{q'} = q\{p_q(x;\theta)\}^{2q-1}.$$

We call $\widetilde{P}_q(x;\theta)$ the *second escort distribution* of $p_q(x;\theta)$. The coefficient q before $\{p_q(x;\theta)\}^{2q-1}$ comes from considerations of U-information geometry [17]. We will discuss in the latter part of Section 5.

Similarly, we can define the *n-th escort distribution* $P_{q,(n)}(x;\theta)$ from the sequence of escort distributions:

$$P_{q,(n)}(x;\theta) := \{q(2q-1)\cdots((n-1)q-(n-2))\}\{p_q(x;\theta)\}^{nq-(n-1)}. \tag{5}$$

Let $f(x)$ be a function on Ω. The *q-expectation* $E_{q,p}[f(x)]$ and the *normalized q-expectation* $E_{q,p}^{esc}[f(x)]$ with respect to $p(x;\theta) \in S_q$ are defined by

$$\begin{aligned} E_{q,p}[f(x)] &:= \int_\Omega f(x) P_q(x;\theta) dx, \\ E_{q,p}^{esc}[f(x)] &:= \int_\Omega f(x) P_q^{esc}(x;\theta) dx, \end{aligned}$$

respectively. We denote by $\widetilde{E_{q,p}}[f(x)]$ the expectation with respect to the second escort distribution $\widetilde{P}_q(x;\theta)$, that is,

$$\widetilde{E_{q,p}}[f(x)] := \int_\Omega f(x)\widetilde{P}_q(x;\theta)dx = q\int_\Omega f(x)\{p_q(x;\theta)\}^{2q-1}dx.$$

Since a differential of a power function is also a power function, we can give a characterization for escort distributions.

Proposition 1. *Suppose that S_q is a q-exponential family defined by (1). Then the n-th escort distribution is given by the n-th differential of q-exponential function. That is, by setting $u(t) = (\exp_q t)'$, we have the following formula:*

$$\begin{aligned}
p_q(x;\theta) &= \exp_q\left(\sum_{i=1}^{n} \theta^i F_i(x) - \psi(\theta)\right), \\
P_q(x;\theta) = P_{q,(1)}(x;\theta) &= u\left(\sum_{i=1}^{n} \theta^i F_i(x) - \psi(\theta)\right), \\
\widetilde{P}_q(x;\theta) = P_{q,(2)}(x;\theta) &= u'\left(\sum_{i=1}^{n} \theta^i F_i(x) - \psi(\theta)\right), \\
&\vdots \\
P_{q,(n)}(x;\theta) &= u^{(n-1)}\left(\sum_{i=1}^{n} \theta^i F_i(x) - \psi(\theta)\right), \\
&\vdots
\end{aligned}$$

Proof. Since a q-exponential function is $\exp_q(x) = (1 + (1-q))^{1/(1-q)}$, its differential is given by

$$u(x) = \frac{1-q}{1-q}(1+(1-q)x)^{\frac{1}{1-q}-1} = (1+(1-q)x)^{\frac{q}{1-q}} = \{\exp_q x\}^q.$$

Therefore, we obtain $P_q(x;\theta) = u\left(\sum_{i=1}^{n} \theta^i F_i(x) - \psi(\theta)\right)$.

By induction, the n-th differential of $u(x)$ coincides with the n-th escort distribution $P_{q,(n)}$, which is given by Equation (5). □

4. Statistical Manifolds and Their Generalized Conformal Structures

In this section, we us review the geometry of statistical manifolds. For more details about the geometry of statistical manifolds, see [18,19].

Let (S, g) be a Riemannian manifold and ∇ be a torsion-free affine connection on S. We say that the triplet (S, ∇, g) is a *statistical manifold* if ∇g is totally symmetric. In this case, we can define a totally symmetric $(0,3)$-tensor field by

$$C(X,Y,Z) := (\nabla_X g)(Y,Z) = Xg(Y,Z) - g(\nabla_X Y, Z) - g(Y, \nabla_X Z),$$

where X, Y and Z are arbitrary vector fields on S. The tensor field C is called a *cubic form* or an *Amari–Chentsov tensor field*.

The notion of statistical manifold was introduced by Lauritzen [20]. He called the triplet (S, g, C) a statistical manifold. In this paper, the definition is followed to Kurose [18]. Though these two definitions are different, the other statistical manifold structure can be obtained from a given one, However, the motivation for the notion of conformal equivalence using (S, g, C) is different from that one using (S, ∇, g), which we will discuss in the latter part of this section.

For a given statistical manifold (S, ∇, g), we can define another torsion-free affine connection ∇^* on S by

$$Xg(Y,Z) = g(\nabla_X Y, Z) + g(Y, \nabla^*_X Z).$$

The connection ∇^* is called the *dual connection* of ∇ with respect to g. The triplet (S, ∇^*, g) is also a statistical manifold, which is called the *dual statistical manifold* of (S, ∇, g). The cubic form is given by the difference of two affine connections ∇^* and ∇:

$$C(X,Y,Z) = g(\nabla^*_X Y - \nabla_X Y, Z).$$

We define generalized conformal structures for statistical manifolds followed to Kurose [18]. Two statistical manifolds (S, ∇, g) and $(S, \bar{\nabla}, \bar{g})$ are said to be 1-*conformally equivalent* if there exists a function $\lambda : S \to \mathbf{R}_{++}$ such that

$$\bar{g}(X,Y) = \lambda g(X,Y), \quad (6)$$

$$\bar{\nabla}_X Y = \nabla_X Y - g(X,Y)\mathrm{grad}_g(\ln \lambda), \quad (7)$$

where $\mathrm{grad}_g(\ln \lambda)$ is the gradient vector field of $\ln \lambda$ with respect to g, that is, $g(X, \ln \lambda) = X(\ln \lambda)$. We say that (S, ∇, g) is 1-*conformally flat* if (S, ∇, g) is locally 1-conformally equivalent to a flat statistical manifold.

Two statistical manifolds (S, ∇, g) and $(S, \bar{\nabla}, \bar{g})$ are said to be (-1)-*conformally equivalent* if there exists a function $\lambda : S \to \mathbf{R}_{++}$ such that

$$\bar{g}(X,Y) = \lambda g(X,Y),$$

$$\bar{\nabla}_X Y = \nabla_X Y + d(\ln \lambda)(Y)X + d(\ln \lambda)(X)Y, \quad (8)$$

where $d(\ln \lambda)(X) = X(\ln \lambda)$. If two statistical manifolds (S, ∇, g) and $(S, \bar{\nabla}, \bar{g})$ are 1-conformally equivalent, then their dual statistical manifolds (S, ∇^*, g) and $(S, \bar{\nabla}^*, \bar{g})$ are (-1)-conformally equivalent.

Proposition 2. *If two statistical manifolds (S, ∇, g) and $(S, \bar{\nabla}, \bar{g})$ are* 1*-conformally equivalent, then their cubic forms have the following relation:*

$$\frac{1}{\lambda}\bar{C}(X,Y,Z) = C(X,Y,Z) + g(Y,Z)d(\ln \lambda)(X) + g(Z,X)d(\ln \lambda)(Y) + g(X,Y)d(\ln \lambda)(Z).$$

Proof. From Equations (7) and (8), we obtain

$$\bar{\nabla}_X Y = \nabla_X Y + d(\ln \lambda)(Y)X + d(\ln \lambda)(X)Y + g(X,Y)\mathrm{grad}_g(\ln \lambda).$$

By taking an inner product with respect to g, we obtain the result. □

5. Statistical Manifold Structures on q-Exponential Families

In this section, we consider statistical manifold structures on a q-exponential family. It is known that a q-exponential family naturally has at least three kinds of statistical manifold structures (cf. [6,8]). We reformulate these structures from the viewpoint of the sequence of escort distributions. In this paper, we omit the details about information geometry. See [21,22] for further details.

Firstly, we review basic facts about q-exponential family. Let S_q be a q-exponential family. The normalization $\psi(\theta)$ on S_q is convex, but may not be strictly convex. In fact, we obtain the following proposition.

Proposition 3. *Let $S_q = \{p(x;\theta)\}$ be a q-exponential family. Then the normalization function $\psi(\theta)$ is convex.*

Proof. Set $u(x) = (\exp_q x)'$ and $\partial_i = \partial/\partial\theta^i$. Then we have

$$\begin{aligned}
\partial_i p(x;\theta) &= u\left(\sum \theta^k F_k(x) - \psi(\theta)\right)(F_i(x) - \partial_i\psi(\theta)), \\
\partial_i\partial_j p(x;\theta) &= u'\left(\sum \theta^k F_k(x) - \psi(\theta)\right)(F_i(x) - \partial_i\psi(\theta))(F_j(x) - \partial_j\psi(\theta)) \\
&- u\left(\sum \theta^k F_k(x) - \psi(\theta)\right)\partial_i\partial_j\psi(\theta).
\end{aligned} \tag{9}$$

Since $\partial_i \int_\Omega p(x;\theta)dx = \int_\Omega \partial_i p(x;\theta)dx = 0$ and $\int_\Omega \partial_i\partial_j p(x;\theta)dx = 0$, we have

$$\begin{aligned}
Z_q(p) &= \int_\Omega \{(p(x;\theta)\}^q dx = \int_\Omega u\left(\sum \theta^k F_k(x) - \psi(\theta)\right)dx, \\
\partial_i\partial_j\psi(\theta) &= \frac{1}{Z_q(p)}\int_\Omega u'\left(\sum \theta^k F_k(x) - \psi(\theta)\right)(F_i(x) - \partial_i\psi(\theta))(F_j(x) - \partial_j\psi(\theta))dx.
\end{aligned} \tag{10}$$

For an arbitrary vector $c = {}^t(c^1, c^2, \ldots, c^n) \in \mathbf{R}^n$, since $Z_q(p) > 0$ and $u''(x) > 0$, we have

$$\sum_{i,j=1}^{n} c^i c^j(\partial_i\partial_j\psi(\theta)) = \frac{1}{Z_q(p)}\int_\Omega u''\left(\sum_{k=1}^{n}\theta^k F_k(x) - \psi(\theta)\right)\left\{\sum_{i=1}^{n} c^i(F_i(x) - \partial_i\psi(\theta))\right\}^2 dx \geq 0.$$

This implies that the Hessian matrix $(\partial_i\partial_j\psi(\theta))$ is semi-positive definite. □

We assume that ψ is strictly convex in this paper. Under this assumption, we can induce many geometric structures for a q-exponential family.

Since ψ is strictly convex, we can define a Riemannian metric and a cubic form by

$$\begin{aligned}
g^q_{ij}(\theta) &:= \partial_i\partial_j\psi(\theta), \\
C^q_{ijk}(\theta) &:= \partial_i\partial_j\partial_k\psi(\theta).
\end{aligned}$$

We call g^q and C^q a *q-Fisher metric* and a *q-cubic form*, respectively [23,24]. Since g^q is a Hessian of a function ψ, g^q is a *Hessian metric*, and ψ is the *potential* of g^q with respect to the natural coordinate $\{\theta^i\}$ [25].

For a fixed real number α, set

$$g^q\left(\nabla^{q(\alpha)}_X Y, Z\right) := g^q\left(\nabla^{q(0)}_X Y, Z\right) - \frac{\alpha}{2}C^q(X,Y,Z), \tag{11}$$

where $\nabla^{q(0)}$ is the Levi-Civita connection with respect to g^q. Since g^q is a Hessian metric, from standard arguments in Hessian geometry [25], $\nabla^{q(e)} := \nabla^{q(1)}$ and $\nabla^{q(m)} := \nabla^{q(-1)}$ are flat affine connections and mutually dual with respect to g^q. Therefore, the triplets $(S_q, \nabla^{q(e)}, g^q)$ and $(S_q, \nabla^{q(m)}, g^q)$ are flat statistical manifolds, and the quadruplet $(S_q, g^q, \nabla^{q(e)}, \nabla^{q(m)})$ is a dually flat space.

Under q-expectations, we have the following proposition (cf. [10]).

Proposition 4. *For S_q a q-exponential family,* (1) *Set $\eta_i = E^{esc}_{q,p}[F_i(x)]$. Then $\{\eta_i\}$ is a $\nabla^{q(m)}$-affine coordinate system such that*

$$g^q\left(\frac{\partial}{\partial\theta^i}, \frac{\partial}{\partial\eta_j}\right) = \delta^j_i.$$

(2) *Set $\phi(\eta) = E^{esc}_{q,p}[\log_q p(x;\theta)]$, then $\phi(\eta)$ is the potential of g^q with respect to $\{\eta_i\}$.*

Next, let us consider the standard Fisher metric and the standard cubic form. Suppose that $S := \{p(x;\theta)\}$ is a statistical model. Set $p_\theta := p(x;\theta)$, for simplicity. We define the *(standard) Fisher metric* g^F on S_q by

$$g^F_{ij}(\theta) := \int_\Omega (\partial_i \ln p_\theta)(\partial_j \ln p_\theta) p_\theta dx,$$

and the *(standard) cubic form* or the *Amari–Chentsov vector field* C^F by

$$C^F_{ijk}(\theta) := \int_\Omega (\partial_i \ln p_\theta)(\partial_j \ln p_\theta)(\partial_j \ln p_\theta) p_\theta dx.$$

From similar arguments of (11), we can define an *α-connection* $\nabla^{(\alpha)}$ on S_q, and we can obtain a statistical manifold structure $(S_q, \nabla^{(\alpha)}, g^F)$. In this case, $(S_q, \nabla^{(\alpha)}, g^F)$ is called an *invariant statistical manifold* [21,22].

A Fisher metric and a cubic form have the following representation using a sequence of escort distributions,

Theorem 1. *Let S_q be a q-exponential family. For $p(x;\theta) \in S_q$, suppose that $P_{q,(2)}(x;\theta)$ and $P_{q,(3)}(x;\theta)$ are the second and the third escort distribution of $p(x;\theta)$, respectively. Then the Fisher metric g^F and the cubic form C^F are given as follows:*

$$g^F_{ij}(\theta) = \frac{1}{q}\int_\Omega (\partial_i \ln_q p_\theta)(\partial_j \ln_q p_\theta) P_{q,(2)}(x;\theta) dx, \tag{12}$$

$$C^F_{ijk}(\theta) = \frac{1}{q(2q-1)}\int_\Omega (\partial_i \ln_q p_\theta)(\partial_j \ln_q p_\theta)(\partial_k \ln_q p_\theta) P_{q,(3)}(x;\theta) dx. \tag{13}$$

Proof. Differentiating the q-logarithm, we have

$$\partial_i \ln_q p_\theta = \partial_i \left(\frac{p_\theta^{1-q} - 1}{1-q}\right) = p_\theta^{-q} \partial_i p(\theta) = p_\theta^{1-q} \partial_i \ln p(\theta).$$

Therefore, we obtain

$$\begin{aligned}
\frac{1}{q}\int_\Omega (\partial_i \ln_q p_\theta)(\partial_j \ln_q p_\theta) P_{q,(2)}(x;\theta) dx &= \int_\Omega p_\theta^{1-q}(\partial_i \ln p_\theta) p_\theta^{1-q}(\partial_j \ln p_\theta) p_\theta^{2q-1}(x;\theta) dx \\
&= \int_\Omega (\partial_i \ln p_\theta)(\partial_j \ln p_\theta) p_\theta(x;\theta) dx \\
&= g^F_{ij}(\theta).
\end{aligned}$$

By a similar argument, we obtain the representation for C^F. □

We define an *α-divergence* $D^{(\alpha)}$ with $\alpha = 1 - 2q$ and a *q-relative entropy* (or a *normalized Tsallis relative entropy*) D^T_q by

$$D^{(1-2q)}(p(x), r(x)) = \frac{1}{q} E_{q,p}[\ln_q p(x) - \ln_q r(x)] = \frac{1 - \int_\Omega p(x)^q r(x)^{1-q} dx}{q(1-q)}, \tag{14}$$

$$D^T_q(p(x), r(x)) = E^{esc}_{q,p}[\ln_q p(x) - \ln_q r(x)] = \frac{1 - \int_\Omega p(x)^q r(x)^{1-q} dx}{(1-q) Z_q(p)}, \tag{15}$$

respectively. It is known that the α-divergence $D^{(1-2q)}(r, p)$ induces a statistical manifold structure $(S_q, g^F, \nabla^{(2q-1)})$, where g^F is the Fisher metric on S_q and $\nabla^{(2q-1)}$ is the α-connection with $\alpha = 2q - 1$, and the q-relative entropy $D^T_q(r, p)$ induces $(S_q, g, \nabla^{q(e)})$.

Theorem 2 (cf. [10,24]). *For a q-exponential family S_q, two statistical manifolds $(S_q, g^F, \nabla^{(2q-1)})$ and $(S_q, g, \nabla^{q(e)})$ are 1-conformally equivalent. In particular, an invariant statistical manifold $(S_q, g^F, \nabla^{(2q-1)})$ is 1-conformally flat. Riemannian metrics and cubic forms have the following relations:*

$$g^q_{ij}(\theta) = \frac{q}{Z_q(p)} g^F_{ij}(\theta), \tag{16}$$

$$\begin{aligned} C^q_{ijk}(\theta) &= \frac{q}{Z_q(p)}(2q-1)C^F_{ijk}(\theta) \\ &- \frac{q}{Z_q(p)}\left\{ g^F_{ij}\partial_k \ln Z_q(p) + g^F_{jk}(\theta)\partial_i \ln Z_q(p) + g^F_{ki}(\theta)\partial_j \ln Z_q(p) \right\}. \end{aligned} \tag{17}$$

Proof. The results were essentially obtained in [10]. However, we give a simpler proof for Equations (16) and (17). The key idea is a sequence of escort distributions and the escort representations of g^F and C^F in Theorem 1.

From Equation (10), we directly obtain the conformal equivalence relation (16) using the escort representation of g^F in (12).

By differentiating (9) and taking an integration, we obtain

$$\begin{aligned} 0 &= \int_\Omega u''\left(\sum \theta^l F_l(x) - \psi(\theta)\right)(F_i(x) - \partial_i\psi(\theta))(F_j(x) - \partial_j\psi(\theta))(F_k(x) - \partial_k\psi(\theta))dx \\ &- \int_\Omega u'\left(\sum \theta^l F_l(x) - \psi(\theta)\right)(F_k(x) - \partial_k\psi(\theta))\partial_i\partial_j\psi(\theta)dx \\ &- \int_\Omega u'\left(\sum \theta^l F_l(x) - \psi(\theta)\right)(F_i(x) - \partial_i\psi(\theta))\partial_j\partial_k\psi(\theta)dx \\ &- \int_\Omega u'\left(\sum \theta^l F_l(x) - \psi(\theta)\right)(F_j(x) - \partial_j\psi(\theta))\partial_k\partial_i\psi(\theta)dx \\ &- \int_\Omega u\left(\sum \theta^l F_l(x) - \psi(\theta)\right)\partial_i\partial_j\partial_k\psi(\theta)dx. \end{aligned}$$

Since $Z_q(p) = \int_\Omega P_q(x;\theta)dx$, we have

$$\partial_i Z_q(p) = \partial_i \int_\Omega P_q(x;\theta)dx = \int_\Omega \partial_i P_q(x;\theta)dx = \int_\Omega \tilde{P}_q(x;\theta)(F_i(x) - \partial_i\psi(\theta))dx.$$

From the escort representation of C^F in (13), and Proposition 1, we obtain Equation (17) since $g^q_{ij}(\theta) = \partial_i\partial_j\psi(\theta)$ and $C^q_{ijk}(\theta) = \partial_i\partial_j\partial_k\psi(\theta)$. □

We remark that the cubic form of $(S_q, g^F, \nabla^{(2q-1)})$ is not C^F but $(2q-1)C^F$.

The difference of a α-divergence and a q-relative entropy is only the normalization $q/Z_q(p)$. This implies that a normalization for probability density imposes a generalized conformal change for a statistical model.

In the next part of this section, let us consider another statistical manifold on S_q (cf. [6,17,26]).

Recall that a Fisher metric g^F has the following representation:

$$g^F_{ij}(\theta) = \int_\Omega (\partial_i \ln p_\theta)(\partial_j p_\theta)dx.$$

In information geometry, $\partial_i \ln p_\theta$ is called an *e-representation (exponential representation)* of p_θ, and $\partial_j p_\theta$ is called a *m-representation (mixture representation)*. Intuitively, $\partial_i \ln p_\theta$ and $\partial_j p_\theta$ are regarded as tangent vectors on a statistical model. Hence a Fisher metric is regarded as a L^2-inner product of e- and m-representations.

Let us generalize e- and m-representations for a q-exponential family. For $p_\theta \in S_q$, we call $\partial_i \ln_q p_\theta$ a *q-score function*. Then we define a Riemannian metric g^M by

$$g^M_{ij}(\theta) = \int_\Omega (\partial_i \ln_q p_\theta)(\partial_j p_\theta)dx. \tag{18}$$

By differentiating the above equation, we can define mutually dual torsion-free affine connections $\nabla^{M(e)}$ and $\nabla^{M(m)}$:

$$\begin{aligned}\Gamma^{M(e)}_{ij,k}(\theta) &:= \int_\Omega (\partial_i\partial_j \ln_q p_\theta)(\partial_k p_\theta)dx,\\ \Gamma^{M(m)}_{ij,k}(\theta) &:= \int_\Omega (\partial_k \ln_q p_\theta)(\partial_i\partial_j p_\theta)dx,\end{aligned}$$

where $\Gamma^{M(e)}_{ij,k}$ and $\Gamma^{M(m)}_{ij,k}$ are the Christoffel symbols of $\nabla^{M(e)}$ and $\nabla^{M(m)}$ of the first kind, respectively. It is known that g^M is a Hessian metric, and the quadruplet $(S_q, g^M, \nabla^{M(e)}, \nabla^{M(m)})$ is a dually flat space. In addition, a natural parameter $\{\theta^i\}$ is a $\nabla^{M(e)}$-affine coordinate sysem. Therefore, the cubic form for $(S_q, \nabla^{M(e)}, g^M)$ is

$$C^M_{ijk}(\theta) = \Gamma^{M(m)}_{ij,k}(\theta). \tag{19}$$

We remark that the statistical manifold structure $(S_q, \nabla^{M(e)}, g^M)$ is induced from a *β-divergence* [17,26] (or a *density power divergence* [27]):

$$D_{1-q}(p,r) := \int_\Omega \left\{ p(x)\frac{p(x)^{1-q} - r(x)^{1-q}}{1-q} - \frac{p(x)^{2-q} - r(x)^{2-q}}{2-q} \right\} dx. \tag{20}$$

Theorem 3. *For the statistical manifold structure $(S_q, \nabla^{M(e)}, g^M)$, the escort representations of the Riemannian metric g^M and the cubic form C^M are given as follows:*

$$g^M_{ij}(\theta) = \int_\Omega (\partial_i \ln_q p_\theta)(\partial_j \ln_q p_\theta)P_q(x;\theta)dx, \tag{21}$$

$$C^M_{ijk}(\theta) = \int_\Omega (\partial_i \ln_q p_\theta)(\partial_j \ln_q p_\theta)(\partial_k \ln_q p_\theta)\tilde{P}_q(x;\theta)dx. \tag{22}$$

Proof. For the Riemannian metric g^M, since $\partial_i p_\theta = (\partial_i \ln_q p_\theta)P_q(x;\theta)$, we immediately obtain Equation (21) from the definition of g^M.

Let us consider the expression for cubic form (22). The q-score function $\partial_i \ln_q p_\theta$ is unbiased under the q-expectation. In fact,

$$E_{q,p}[\partial_i \ln_q p_\theta] = \int_\Omega (\partial_j \ln_q p_\theta)P_q(x;\theta)dx = \int_\Omega \partial_j p_\theta dx = 0.$$

From Equation (19), we obtain

$$\begin{aligned}C^M_{ijk}(\theta) &= \Gamma^{M(m)}_{ij,k}(x;\theta)\\ &= \int_\Omega (\partial_k \ln_q p_\theta)(\partial_i\partial_j p_\theta)dx\\ &= \int_\Omega (\partial_k \ln_q p_\theta)\partial_i \left\{ (\partial_j \ln_q p_\theta)\, P_q(x;\theta) \right\} dx\\ &= -\partial_{ij}\psi(\theta)\int_\Omega (\partial_k \ln_q p_\theta)P_q(x;\theta)dx + \int_\Omega (\partial_k \ln_q p_\theta)(\partial_j \ln_q p_\theta)\{\partial_i P_q(x;\theta)\}dx\\ &= \int_\Omega (\partial_k \ln_q p_\theta)(\partial_j \ln_q p_\theta)(\partial_i \ln_q p_\theta)\tilde{P}_q(x;\theta)dx.\end{aligned}$$

□

We remark that Naudts [5] gave another generalization of Fisher metric g^N, which is defined by

$$g^N_{ij}(\theta) := \int_\Omega \frac{1}{P^{esc}_q(x;\theta)}(\partial_i p_\theta)(\partial_j p_\theta)dx,$$

The metric g^N is conformally equivalent to g^M with conformal factor $Z_q(p_\theta) = \int_\Omega \{p(x;\theta)\}^q dx$. That is, $g^N(\theta) = Z_q(p_\theta) g^M(\theta)$. (See also [6]). Naudts gave a further generalization of Fisher metric and he showed a Cramér–Rao type bound theorem [5].

6. Concluding Remarks

In this paper, we introduced a sequence of escort distributions. Then we gave representations of Riemannian metrics and cubic forms from a viewpoint of the sequence of escort distributions.

In particular, we can define the following $(0,2)$-tensor fields on a q-exponential family. For $p_\theta \in S_q$, set $\eta_i = \partial_i \psi(\theta)$.

(1) From the standard expectation, we obtain

$$\begin{aligned} g_{ij}^{(0)}(\theta) := G_{ij}(\theta) &:= \int_\Omega (\partial_i \ln_q p_\theta)(\partial_j \ln_q p_\theta) p_\theta dx \\ &= E_p[(F_i(x) - \eta_i)(F_j(x) - \eta_j)]. \end{aligned}$$

The tensor G is a covariance matrix. However, G may not be important in anomalous statistics.

(2) From the q-expectation, we obtain

$$\begin{aligned} g_{ij}^{(1)}(\theta) := g_{ij}^M(\theta) &= \int_\Omega (\partial_i \ln_q p_\theta)(\partial_j \ln_q p_\theta) \{p_\theta\}^q dx \\ &= \int_\Omega (\partial_i \ln_q p_\theta)(\partial_j \ln_q p_\theta) P_q(x;\theta) dx \\ &= E_{q,p}[(F_i(x) - \eta_i)(F_j(x) - \eta_j)]. \end{aligned}$$

The Riemannian metric g^M is a Hessian metric, and it is induced from the β-divergence (20).

(3) From the expectation with respect to the second escort distribution, we obtain

$$\begin{aligned} g_{ij}^{(2)}(\theta) := g_{ij}^F(\theta) &= \int_\Omega (\partial_i \ln_q p_\theta)(\partial_j \ln_q p_\theta) \{p_\theta\}^{2q-1} dx \\ &= \frac{1}{q} \int_\Omega (\partial_i \ln_q p_\theta)(\partial_j \ln_q p_\theta) \tilde{P}_q(x;\theta) dx \\ &= \frac{1}{q} E_{q,(2),p}[(F_i(x) - \eta_i)(F_j(x) - \eta_j)]. \\ g_{ij}^q(\theta) &= \frac{Z_q(p)}{q} g_{ij}^F. \end{aligned}$$

The Riemannian metric g^F is a Fisher metric. Hence g^F is invariant to the choice of reference measure on Ω, but it is not a Hessian metric. In addition, g^F is induced from the α-divergence (14). The conformal Riemannian metric g^q is a q-Fisher metric. It is a Hessian metric, and it is induced from a normalized Tsallis relative entropy (15).

We may define a Riemannian metric and a cubic form from higher order escort expectations:

$$\begin{aligned} g_{ij}^{(n)}(\theta) &:= \int_\Omega (\partial_i \ln_q p_\theta)(\partial_j \ln_q p_\theta) P_{q,(n)}(x;\theta) dx, \\ C_{ij}^{(n)}(\theta) &:= \int_\Omega (\partial_i \ln_q p_\theta)(\partial_j \ln_q p_\theta)(\partial_k \ln_q p_\theta) P_{q,(n+1)}(x;\theta) dx. \end{aligned}$$

Then we obtain a sequence of statistical manifold structures.

$$(S_q, g^{(1)}, C^{(1)}) \to (S_q, g^{(2)}, C^{(2)}) \to \cdots \to (S_q, g^{(n)}, C^{(n)}) \to \cdots$$

However, the geometric meaning of this sequence is not clear at this moment. Elucidating geometric properties of this sequence is a future problem.

Acknowledgments: This research was partially supported by JSPS (Japan Society for the Promotion of Science), KAKENHI (Grants-in-Aid for Scientific Research) Grant Numbers JP26108003 and JP15K04842.

Conflicts of Interest: The author declares no conflict of interest.

References

1. Tsallis, C. *Introduction to Nonextensive Statistical Mechanics: Approaching a Complex World*; Springer: New York, NY, USA, 2009.
2. Naudts, J. *Generalised Thermostatistics*; Springer: London, UK, 2011.
3. Kaniadakis, G. Theoretical foundations and mathematical formalism of the power-law tailed statistical distributions. *Entropy* **2013**, *15*, 3983–4010.
4. Beck, C.; Schlögl, F. *Thermodynamics of Chaotic Systems: An Introduction*; Cambridge University Press: Cambridge, UK, 1993.
5. Naudts, J. Estimators, escort probabilities, and ϕ-exponential families in statistical physics. *J. Inequal. Pure Appl. Math.* **2004**, *5*, 102.
6. Matsuzoe, H.; Henmi, M. Hessian structures and divergence functions on deformed exponential families. In *Geometric Theory of Information, Signals and Communication Technology*; Nielsen, F., Ed.; Springer: Basel, Switzerland, 2014; pp. 57–80.
7. Sakamoto, M.; Matsuzoe, H. A generalization of independence and multivariate Student's t-distributions. In *Geometric Science of Information*, Proceedings of Second International Conference on Geometric Science of Information (GSI 2015), Palaiseau, France, 28–30 October 2015; Volume 9389, pp. 740–749.
8. Matsuzoe, H.; Wada, T. Deformed algebras and generalizations of independence on deformed exponential families. *Entropy* **2015**, *17*, 5729–5751.
9. Wada, T.; Matsuzoe, H.; Scarfone, A.M. Dualistic hessian structures among the thermodynamic potentials in the κ-Thermostatistics. *Entropy* **2015**, *17*, 7213–7229.
10. Matsuzoe, H. Statistical manifolds and geometry of estimating functions. In *Prospects of Differential Geometry and Its Related Fields, Proceedings of the 3rd International Colloquium on Differential Geomentry and Its Related Fields, Veliko Tarnovo, Bulgaria, 3–7 September 2012*; Adachi, T., Hashimoto, H., Hristov, M.J., Eds.; World Scientific: Hackensack, NJ, USA, 2013; pp. 187–202.
11. Matsuzoe, H.; Henmi, M. Hessian structures on deformed exponential families. In *Geometric Science of Information, Proceedings of First International Conference on Geometric Science of Information (GSI 2013), Paris, France, 28–30 August 2013*; Springer: Berlin/Heidelberg, Germany, 2015; Volume 8085, pp. 275–282.
12. Eguchi, S.; Komori, O. Path connectedness on a space of probability density functions. In *Geometric Science of Information, Proceedings of Second International Conference on Geometric Science of Information (GSI 2015), Palaiseau, France, 28–30 October 2015*; Springer: Berlin/Heidelberg, Germany, 2015; Volume 9389, pp. 615–624.
13. Scarfone, A.M.; Matsuzoe, H.; Wada, T. Consistency of the structure of Legendre transform in thermodynamics with the Kolmogorov-Nagumo average. *Phys. Lett. A* **2016**, *380*, 3022–3028.
14. Tsallis, C. What are the numbers that experiments provide? *Quim. Nova* **1994**, *17*, 468–471.
15. Zhang, J. On monotone embedding in information geometry. *Entropy* **2015**, *17*, 4485–4499.
16. Tanaka M. Meaning of an escort distribution and τ-transformation. *J. Phys. Conf. Ser.* **2010**, *201*, 012007.
17. Murata, N.; Takenouchi, T.; Kanamori, T.; Eguchi, S. Information geometry of U-boost and Bregman divergence. *Neural Comput.* **2004**, *16*, 1437–1481.
18. Kurose, T. On the divergences of 1-conformally flat statistical manifolds. *Tôhoku Math. J.* **1994**, *46*, 427–433.
19. Matsuzoe, H. Statistical manifolds and affine differential geometry. *Adv. Stud. Pure Math.* **2010**, *57*, 303–321.
20. Lauritzen, S.L. Statistical manifolds. In *Differential Geometry in Statistical Inferences*; Gupta, S.S., Ed.; IMS Lecture Notes Monograph Series 10; Institute of Mathematical Statistics: Hayward, CA, USA, 1987; pp. 96–163.
21. Amari, S.; Nagaoka, H. *Method of Information Geometry*; Translations of Mathematical Monographs; American Mathematical Society: Providence, RI, USA; Oxford University Press: Oxford, UK, 2000.
22. Amari, S. *Information Geometry and Its Applications*; Springer: Tokyo, Japan, 2016.
23. Amari, S.; Ohara, A.; Matsuzoe, H. Geometry of deformed exponential families: Invariant, dually-flat and conformal geometry. *Physica A* **2012**, *391*, 4308–4319.

24. Matsuzoe, H.; Ohara, A. Geometry for q-exponential families. In *Recent Progress in Differential Geometry and Its Related Fields, Proceedings of the 2nd International Colloquium on Differential Geomentry and Its Related Fields, Veliko Tarnovo, Bulgaria, 6–10 September 2010*; Adachi, T., Hashimoto, H., Hristov, M.J., Eds.; World Scientific: Hackensack, NJ, USA, 2011; pp. 55–71.
25. Shima, H. *The Geometry of Hessian Structures*; World Scientific: Hackensack, NJ, USA, 2007.
26. Ohara, A.; Wada, T. Information geometry of q-Gaussian densities and behaviors of solutions to related diffusion equations. *J. Phys. A* **2010**, *43*, 035002.
27. Basu, A.; Harris, I.R.; Hjort, N.L.; Jones, M.C. Robust and efficient estimation by minimising a density power divergence. *Biometrika* **1998**, *85*, 549–559.

entropy

MDPI

Article

The Information Geometry of Sparse Goodness-of-Fit Testing

Paul Marriott [1,*], Radka Sabolová [2], Germain Van Bever [3] and Frank Critchley [2]

1 Department of Statistics and Actuarial Science, University of Waterloo, 200 University Avenue West, Waterloo, ON N2L 3G1, Canada

2 School of Mathematics and Statistics, The Open University, Walton Hall, Milton Keynes, Buckinghamshire MK7 6AA, UK; radka.sabolova@open.ac.uk (R.S.); f.critchley@open.ac.uk (F.C.)

3 Department of Mathematics & ECARES, Université libre de Bruxelles, Avenue F.D. Roosevelt 42, 1050 Brussels, Belgium; gvbever@ulb.ac.be

* Correspondence: pmarriot@uwaterloo.ca; Tel.: +1-519-888-4567

Academic Editors: Frédéric Barbaresco and Frank Nielsen

Received: 31 August 2016; Accepted: 19 November 2016; Published: 24 November 2016

Abstract: This paper takes an information-geometric approach to the challenging issue of goodness-of-fit testing in the high dimensional, low sample size context where—potentially—boundary effects dominate. The main contributions of this paper are threefold: first, we present and prove two new theorems on the behaviour of commonly used test statistics in this context; second, we investigate—in the novel environment of the extended multinomial model—the links between information geometry-based divergences and standard goodness-of-fit statistics, allowing us to formalise relationships which have been missing in the literature; finally, we use simulation studies to validate and illustrate our theoretical results and to explore currently open research questions about the way that discretisation effects can dominate sampling distributions near the boundary. Novelly accommodating these discretisation effects contrasts sharply with the essentially continuous approach of skewness and other corrections flowing from standard higher-order asymptotic analysis.

Keywords: extended multinomial models; goodness-of-fit testing; information geometry

1. Introduction

We start by emphasising the threefold achievements of this paper, spelled out in detail in terms of the paper's section structure below. First, we present and prove two new theorems on the behaviour of some standard goodness-of-fit statistics in the high dimensional, low sample size context, focusing on behaviour "near the boundary" of the extended multinomial family. We also comment on the methods of proof which allow explicit calculations of higher order moments in this context. Second, working again explicitly in the extended multinomial context, we fill a hole in the literature by linking information-geometric-based divergences and standard goodness-of-fit statistics. Finally, we use simulation studies to explore discretisation effects that can dominate sampling distributions "near the boundary". Indeed, we illustrate and explore how—in the high dimensional, low sample size context—all distributions are affected by boundary effects. We also use these simulation results to explore currently open research questions. As can be seen, the overarching theme is the importance of working in the geometry of the extended exponential family [1], rather than the traditional manifold-based structure of information geometry.

In more detail, the paper extends and builds on the results of [2], and we use notation and definitions consistently across these two papers. Both papers investigate the issue of goodness-of-fit testing in the high dimensional sparse extended multinomial context, using the tools of Computational Information Geometry (CIG) [1].

Section 2 gives formal proofs of two results, Theorems 1 and 2, which were announced in [2]. These results explore the sampling performance of standard goodness-of-fit statistics—Wald, Pearson's χ^2, score and deviance—in the sparse setting. In particular, they look at the case where the data generation process is "close to the boundary" of the parameter space where one or more cell probabilities vanish. This complements results in much of the literature, where the centre of the parameter space—i.e., the uniform distribution—is often the focus of attention.

Section 3 starts with a review of the links between Information Geometry (IG) [3] and goodness-of-fit testing. In particular, it looks at the power family of Cressie and Read [4,5] in terms of the geometric theory of divergences. In the case of regular exponential families, these links have been well-explored in the literature [6], as has the corresponding sampling behaviour [7]. What is novel here is the exploration of the geometry with respect to the closure of the exponential family; i.e., the extended multinomial model—a key tool in CIG. We illustrate how the boundary can dominate the statistical properties in ways that are surprising compared to standard—and even high-order—analyses, which are asymptotic in sample size.

Through simulation experiments, Section 4 explores the consequences of working in the sparse multinomial setting, with the design of the numerical experiments being inspired by the information geometry.

2. Sampling Distributions in the Sparse Case

One of the first major impacts that information geometry had on statistical practice was through the geometric analysis of higher order asymptotic theory (e.g., [8,9]). Geometric interpretations and invariant expressions of terms in the higher order corrections to approximations of sampling distributions are a good example, [8] (Chapter 4). Geometric terms are used to correct for skewness and other higher order moment (cumulant) issues in the sampling distributions. However, these correction terms grow very large near the boundary [1,10]. Since this region plays a key role in modelling in the sparse setting—the maximum likelihood estimator (MLE) often being on the boundary—extensions to the classical theory are needed. This paper, together with [2], start such a development. This work is related to similar ideas in categorical, (hierarchical) log–linear, and graphical models [1,11–13]. As stated in [13], "their statistical properties under sparse settings are still very poorly understood. As a result, analysis of such data remains exceptionally difficult".

In this section we show why the Wald—equivalently, the Pearson χ^2 and score statistics—are unworkable when near the boundary of the extended multinomial model, but that the deviance has a simple, accurate, and tractable sampling distribution—even for moderate sample sizes. We also show how the higher moments of the deviance are easily computable, in principle allowing for higher order adjustments. However, we also make some observations about the appropriateness of these classical adjustments in Section 4.

First, we define some notation, consistent with that of [2]. With i ranging over $\{0,1,...,k\}$, let $n=(n_i)\sim$ Multinomial $(N,(\pi_i))$, where here each $\pi_i>0$. In this context, the Wald, Pearson's χ^2, and score statistics all coincide, their common value, W, being

$$W:=\sum_{i=0}^{k}\frac{(\pi_i-n_i/N)^2}{\pi_i}\equiv\frac{1}{N^2}\sum_{i=0}^{k}\frac{n_i^2}{\pi_i}-1.$$

Defining $\pi^{(\alpha)}:=\sum_i\pi_i^\alpha$, we note the inequality, for each $m\geq1$,

$$\pi^{(-m)}-(k+1)^{m+1}\geq0,$$

in which equality holds if and only if $\pi_i\equiv1/(k+1)$—i.e., iff (π_i) is uniform. We then have the following theorem, which establishes that the statistic W is unworkable as $\pi_{\min}:=\min(\pi_i)\to0$ for fixed k and N.

Theorem 1. *For $k > 1$ and $N \geq 6$, the first three moments of W are:*

$$E(W) = \frac{k}{N}, \quad Var(W) = \frac{\left\{\pi^{(-1)} - (k+1)^2\right\} + 2k(N-1)}{N^3}$$

and $E[\{W - E(W)\}^3]$ *given by*

$$\frac{\left\{\pi^{(-2)} - (k+1)^3\right\} - (3k + 25 - 22N)\left\{\pi^{(-1)} - (k+1)^2\right\} + g(k,N)}{N^5},$$

where $g(k, N) = 4(N-1)k(k + 2N - 5) > 0$.

In particular, for fixed k and N, as $\pi_{\min} \to 0$

$$Var(W) \to \infty \text{ and } \gamma(W) \to +\infty,$$

where $\gamma(W) := E[\{W - E(W)\}^3]/\{Var(W)\}^{3/2}$.

A detailed proof is found in Appendix A, and we give here an outline of its important features. The machinery developed is capable of delivering much more than a proof of Theorem 1. As indicated there, it provides a generic way to explicitly compute arbitrary moments or mixed moments of multinomial counts, and could in principle be implemented by computer algebra. Overall, there are four stages. First, a key recurrence relation is established; secondly, it is exploited to deliver moments of a single cell count. Third, mixed moments of any order are derived from those of lower order, exploiting a certain functional dependence. Finally, results are combined to find the first three moments of W, higher moments being similarly obtainable.

The practical implication of Theorem 1 is that standard first (and higher-order) asymptotic approximations to the sampling distribution of the Wald, χ^2, and score statistics break down when the data generation process is "close to" the boundary, where at least one cell probability is zero. This result is qualitatively similar to results in [10], which shows how asymptotic approximations to the distribution of the maximum likelihood estimate fail; for example, in the case of logistic regression, when the boundary is close in terms of distances as defined by the Fisher information.

Unlike statistics considered in Theorem 1, the deviance has a workable distribution in the same limit: that is, for fixed N and k as we approach the boundary of the probability simplex. In sharp contrast to that theorem, we see the very stable and workable behaviour of the k-asymptotic approximation to the distribution of the deviance, in which the number of cells increases without limit.

Define the deviance D via

$$\begin{aligned} D/2 &= \sum\nolimits_{\{0 \leq i \leq k: n_i > 0\}} n_i \log(n_i/N) - \sum_{i=0}^{k} n_i \log(\pi_i) \\ &= \sum\nolimits_{\{0 \leq i \leq k: n_i > 0\}} n_i \log(n_i/\mu_i), \end{aligned}$$

where $\mu_i := E(n_i) = N\pi_i$. We will exploit the characterisation that the multinomial random vector (n_i) has the same distribution as a vector of independent Poisson random variables conditioned on their sum. Specifically, let the elements of (n_i^*) be *independently* distributed as Poisson $Po(\mu_i)$. Then, $N^* := \sum_{i=0}^{k} n_i^* \sim Po(N)$, while $(n_i) := (n_i^* | N^* = N) \sim$ Multinomial$(N, (\pi_i))$. Define the vector

$$S^* := \begin{pmatrix} N^* \\ D^*/2 \end{pmatrix} = \sum_{i=0}^{k} \begin{pmatrix} n_i^* \\ n_i^* \log(n_i^*/\mu_i) \end{pmatrix},$$

where D^* is defined implicitly and $0 \log 0 := 0$. The terms ν, τ, and ρ are defined by the first two moments of S^* via the vectors

$$\begin{pmatrix} N \\ \nu \end{pmatrix} := E(S^*) = \begin{pmatrix} N \\ \sum_{i=0}^{k} E(n_i^* \log(n_i^*/\mu_i)) \end{pmatrix}, \tag{1}$$

$$\begin{pmatrix} N & \rho\tau\sqrt{N} \\ \cdot & \tau^2 \end{pmatrix} := Cov(S^*) = \begin{pmatrix} N & \sum_{i=0}^{k} C_i \\ \cdot & \sum_{i=0}^{k} V_i \end{pmatrix}, \tag{2}$$

where $C_i := Cov(n_i^*, n_i^* \log(n_i^*/\mu_i))$ and $V_i := Var(n_i^* \log(n_i^*/\mu_i))$.

Theorem 2. *Each of the terms ν, τ, and ρ remains bounded as $\pi_{\min} \to 0$.*

We start with some preliminary remarks. We use the following notation: $\mathcal{N} := \{1, 2, ...\}$ denotes the natural numbers, while $\mathcal{N}_0 := \{0\} \cup \mathcal{N}$. Throughout, $X \sim Po(\mu)$ denotes a Poisson random variable having positive mean μ—that is, X is discrete with support $\mathcal{N}_0$ and probability mass function $p : \mathcal{N}_0 \to (0, 1)$ given by:

$$p(x) := e^{-\mu}\mu^x / x! \ (\mu > 0). \tag{3}$$

Putting:

$$\forall m \in \mathcal{N}_0, \ F^{[m]}(\mu) := \Pr(X \leq m) = \textstyle\sum_{x=0}^{m} p(x) \in (0, 1), \tag{4}$$

for given μ, $\{1 - F^{[m]}(\mu)\}$ is strictly decreasing with m, vanishing as $m \to \infty$. For all $(x, m) \in \mathcal{N}_0^2$, we define $x_{(m)}$ by:

$$x_{(0)} := 1; \ x_{(m)} := x(x-1)...(x-(m-1)) \ (m \in \mathcal{N}) \tag{5}$$

so that, if $x \geq m$, $x_{(m)} = x!/(x-m)!$.

The set $\mathcal{A}_0$ comprises all functions $a_0 : (0, \infty) \to R$ such that, as $\xi \to 0_+$:

(i) $a_0(\xi)$ tends to an infinite limit $a_0(0_+) \in \{-\infty, +\infty\}$, while: (ii) $\xi a_0(\xi) \to 0$.

Of particular interest here, by l'Hôspital's rule,

$$\forall m \in \mathcal{N}, \ (\log)^m \in \mathcal{A}_0, \tag{6}$$

where $(\log)^m : \xi \to (\log \xi)^m$ $(\xi > 0)$. For each $a_0 \in \mathcal{A}_0$, $\overline{a_0}$ denotes its continuous extension from $(0, \infty)$ to $[0, \infty)$—that is: $\overline{a_0}(0) := a_0(0_+)$; $\overline{a_0}(\xi) := a_0(\xi)$ $(\xi > 0)$—while, appealing to continuity, we also define $0\overline{a_0}(0) := 0$. Overall, denoting the extended reals by $\overline{R} := R \cup \{-\infty\} \cup \{+\infty\}$, and putting

$$\mathcal{A} := \{a : \mathcal{N}_0 \to \overline{R} \text{ such that } 0a(0) = 0\}$$

we have that $\mathcal{A}$ contains the disjoint union:

$$\{\text{all functions } a : \mathcal{N}_0 \to R\} \cup \{\overline{a_0}|_{\mathcal{N}_0} : a_0 \in \mathcal{A}_0\}.$$

We refer to $\overline{a_0}|_{\mathcal{N}_0}$ as *the member of $\mathcal{A}$ based on $a_0 \in \mathcal{A}_0$*.

We make repeated use of two simple facts. First:

$$\forall x \in \mathcal{N}_0, \ 0 \leq \log(x+1) \leq x, \tag{7}$$

equality holding in both places if, and only if, $x = 0$. Second, (3) and (5) give:

$$\forall (x, m) \in \mathcal{N}_0^2 \text{ with } x \geq m, \ x_{(m)} p(x) = \mu^m p(x-m) \tag{8}$$

so that, by definition of $\mathcal{A}$:

$$\forall m \in \mathcal{N}_0, \forall a \in \mathcal{A},\ E(X_{(m)}a(X)) = \mu^m E(a(X+m)), \tag{9}$$

equality holding trivially when $m = 0$. In particular, taking $a = 1 \in \mathcal{A}$—that is, $a(x) = 1$ $(x \in \mathcal{N}_0)$—(9) recovers, at once, the Poisson factorial moments:

$$\forall m \in \mathcal{N}_0,\ E(X_{(m)}) = \mu^m$$

whence, in further particular, we also recover:

$$E(X) = \mu,\ E(X^2) = \mu^2 + \mu \text{ and } E(X^3) = \mu^3 + 3\mu^2 + \mu. \tag{10}$$

We are ready now to prove Theorem 2.

Proof of Theorem 2. In view of (1) and (2), it suffices to show that the first two moments of S^* remain bounded as $\pi_{\min} \to 0$. By the Cauchy–Schwarz inequality, this in turn is a direct consequence of the following result. □

Lemma 1. *Let $X \sim Po(\mu)$ $(\mu > 0)$, and put $X_\mu := X\log(X/\mu)$, with $0\log 0 := 0$. Then, there exist $b^{(1)}, b^{(2)} : (0,\infty) \to (0,\infty)$ such that:*

(a) $0 \leq E(X_\mu) \leq b^{(1)}(\mu)$ and $0 \leq E(X_\mu^2) \leq b^{(2)}(\mu)$, while:

(b) for $i = 1, 2 :\ b^{(i)}(\mu) \to 0$ as $\mu \to 0_+$.

Proof. By (6), $a_0^{(1)}(\xi) := \log(\xi/\mu) \in \mathcal{A}_0$. Taking $m = 1$ and $a \in \mathcal{A}$ based on $a_0^{(1)}$ in (9), and using (7), gives at once the stated bounds on $E(X_\mu)$ with $b^{(1)}(\mu) = \mu(\mu - \log\mu)$, which does indeed tend to 0 as $\mu \to 0_+$.

Further, let $a_0^{(2)}(\xi) := \xi(\log(\xi/\mu))^2$. Taking $m = 1$ and a as the restriction of $a_0^{(2)}$ to $\mathcal{N}_0$ in (9) gives $E(X_\mu^2) = \mu E(a^{(2)}(X+1))$. Noting that

$$\{x \in \mathcal{N}_0 : \log((x+1)/\mu) < 0\} = \begin{cases} \emptyset & (\mu \leq 1) \\ \{0, ..., \overline{\mu} - 2\} & (\mu > 1) \end{cases},$$

in which $\overline{\mu}$ denotes the smallest integer greater than or equal to μ, and putting

$$B(\mu) := \begin{cases} 0 & (\mu \leq 1) \\ \mu\sum_{x=0}^{\overline{\mu}-2} a^{(2)}(x+1)p(x) & (\mu > 1) \end{cases},$$

(7), (10), and l'Hôpital's rule give the stated bounds on $E(X_\mu^2)$, with

$$\begin{aligned} b^{(2)}(\mu) &= B(\mu) + \mu\textstyle\sum_{x=0}^{\infty}(x+1)(x - \log\mu)^2 p(x) \\ &= B(\mu) + \mu E\{X^3 + X^2(1 - 2\log\mu) + X((\log\mu)^2 - 2\log\mu) + (\log\mu)^2\} \\ &= B(\mu) + \mu^4 + 4\mu^3 + 2\mu^2 + \mu(\log\mu)^2 + (\mu\log\mu)^2 - 2\mu(\mu+2)(\mu\log\mu) \end{aligned}$$

which, indeed, tends to 0 as $\mu \to 0_+$. □

As a result of Theorem 2, the distribution of the deviance is stable in this limit. Further, as noted in [2], each of ν, τ, and ρ can be easily and accurately approximated by standard truncate and bound methods in the limit as $\pi_{\min} \to 0$. These are detailed in Appendix B.

3. Divergences and Goodness-of-Fit

The emphasis of this section is the importance of the boundary of the extended multinomial when understanding the links between information geometric divergences and families of goodness-of-fit statistics. For completeness, a set of well-known results linking the Power-Divergence family and information geometry in the manifold sense are surveyed in Sections 3.1–3.3. The extension to the extended multinomial family is discussed in Section 3.4, where we make clear how the global behaviour of divergences is dominated by boundary effects. This complements the usual local analysis, which links divergences with the Fisher information, [8]. Perhaps the key point is that, since counts in the data can be zero, information geometric structures should also allow probabilities to be zero. Hence, closures of exponential families seem to be the correct geometric object to work on.

3.1. The Power-Divergence Family

The results of Section 2 concern the boundary behaviour of two important members of a rich class of goodness-of-fit statistics. An important unifying framework which encompasses these and other important statistics can be found in [5] (page 16) with the so-called Power-Divergence statistics. These are defined, for $-\infty < \lambda < \infty$, by

$$2NI^{\lambda}\left(\frac{n}{N} : \pi\right) := \frac{2}{\lambda(\lambda+1)} \sum_{i=0}^{k} n_i \left[\left(\frac{n_i}{N\pi_i}\right)^{\lambda} - 1\right], \tag{11}$$

with the cases $\lambda = -1, 0$ being defined by taking the appropriate limit to give

$$\lim_{\lambda \to -1} 2NI^{\lambda}\left(\frac{n}{N} : \pi\right) = 2\sum_{i=0}^{k} N\pi_i \log\left(N\pi_i/n_i\right), \ \lim_{\lambda \to 0} 2NI^{\lambda}\left(\frac{n}{N} : \pi\right) = 2\sum_{i=0}^{k} n_i \log\left(n_i/N\pi_i\right).$$

Important special cases are shown in Table 1 (whose first column is described below in Section 3.3), and we also note the case $\lambda = 2/3$, which Read and Cressie recommend [5] (page 79) as a reasonably robust statistic with an easily calculable critical value for small N. In a sense, it lies "between" the Pearson χ^2 and deviance statistics, which we compared in Section 2.

Table 1. Special cases of the Power-Divergence statistics.

$\alpha := 1 + 2\lambda$	λ	Formula	Name
3	1	$\sum_{i=0}^{k} \frac{(n_i - N\pi_i)^2}{N\pi_i}$	Pearson χ^2
7/3	2/3	$\frac{9}{5}\sum_{i=0}^{k} n_i \left[\left(\frac{n_i}{N\pi_i}\right)^{\frac{2}{3}} - 1\right]$	Read–Cressie
1	0	$2\sum_{i=0}^{k} n_i \log\left(n_i/N\pi_i\right)$	Twice log-likelihood (deviance)
0	$-\frac{1}{2}$	$4\sum_{i=0}^{k} \left(\sqrt{n_i} - \sqrt{N\pi_i}\right)^2$	Freeman–Tukey or Hellinger
−1	−1	$2\sum_{i=0}^{k} N\pi_i \log\left(N\pi_i/n_i\right)$	Twice modified log-likelihood
−3	−2	$\sum_{i=0}^{k} \frac{(n_i - N\pi_i)^2}{n_i}$	Neyman χ^2

This paper is primarily concerned with the sparse case where many of the n_i counts are zero, and we are also interested in letting probabilities, π_i, becoming arbitrarily small, or even zero.

3.2. Literature Review

Before we look at this, we briefly review the literature on the geometry of goodness-of-fit statistics. A good source for the historical developments (in the discrete context) can be found in [5] (pages 131–153) and [7]. Important examples include the analysis of contingency tables, log-linear, and discrete graphical models. Testing is often used to check the consistency of a parametric model with given data, and to check dependency assumptions such as independence between categorical variables. However, we note an important caveat: as pointed out by [14,15], the fact that a parametric model "passes" a goodness-of-fit test only weakly constrains the resulting inference. The essential point here is that goodness-of-fit is a necessary, but not sufficient, condition for model choice, since—in general—many models will be empirically supported. This issue has recently been explored geometrically in [16] using CIG.

There have been many possible test statistics proposed for goodness-of-fit testing, and one of the attractions of the Power-Divergence family, defined in (11), is that the most important ones are included in the family and indexed by a single scalar λ. Of course, when there is a choice of test statistic, different inferences can result from different choices. One of the main themes of [5] is to give the analyst insight about selecting a particular λ. Key considerations for making the selection of λ include the tractability of the sampling distribution, its power against important alternatives, and interpretation when hypotheses are rejected.

The first order, asymptotic in N, χ^2-sampling distribution for all members of the Power-Divergence family, which is appropriate when all observed counts are "large enough", is the most commonly used tool, and a very attractive feature of the family. However, this can fail badly in the "sparse" case and when the model is close to the boundary. Elementary, moment based corrections, to improve small sample performance, are discussed in [5] (Chapter 5). More formal asymptotic approaches to these issues include the doubly asymptotic, in N and k, approach of [17], discussed in Section 2 and similar normal approximation ideas in [18]. See also [19]. Extensive simulation experiments have been undertaken to learn in practice what 'large enough' means, see [5,20,21].

When there are nuisance parameters to be estimated (as is common), [22] points out that it is the sampling distribution *conditional* upon these estimates which needs to be approximated, and proposes higher order methods based on the Edgeworth expansion. Simulation approaches are often used in the conditional context due to the common intractability of the conditional distribution [23,24], and importance sampling methods play an important role—see [25–27]. Other approaches used to investigate the sampling distribution include jackknifing [28], the Chen–Stein method [29], and detailed asymptotic analysis in [30–32].

In very high dimensional model spaces, considerations of the power of tests rarely generates uniformly best procedures but, we feel, geometry can be an important tool in understanding the choices that need to be made. Further, [5], states the situation is "complicated", showing this through simulation experiments. One of the reasons for Read and Cressie's preferred choice of $\lambda = 2/3$ is its good power against some important types of alternative–the so-called bump or dip cases–as well as the relative tractability of its sampling distribution under the null. Other considerations about power can be found in [33] which looks specifically at mixture model based alternatives.

3.3. Links with Information Geometry

At the time that the Power-Divergence family was being examined, there was a parallel development in Information Geometry; oddly, however, it seemed to have taken some time before the links between the two areas were fully recognised. A good treatment of these links can be found in [6] (Chapter 9). Since it is important to understand the extreme values of divergence functions, considerations of convexity can clearly play an important role. The general class of Bregman divergences, [6,34] (page 240), and [35] (page 13) is very useful here. For each Bregman divergence, there will exist affine parameters of the exponential family in which the divergence function is convex. In the class of product Poisson models—which are the key building blocks of log–linear models—all members of the Power-Divergence family have the Bregman property. These are then α-divergences, capable of generating the complete Information Geometry of the model [35], with the link between α and λ given in Table 1. The α-representation highlights the duality properties, which are a cornerstone of Information Geometry, but which is rather hidden in the λ representation. The Bregman divergence representation for the Poisson is given in Table 2. The divergence parameter—in which we have convexity—is shown for each λ, as is the so-called potential function, which generates the complete information geometry for these models.

Table 2. Power-Divergence in the Poisson model with mean μ, where $\lambda^* = 1 - \lambda$.

λ	α	**Divergence** $D_\lambda(\mu_1, \mu_2)$	**Divergence Parameter** ξ	**Potential**
-1	-1	$\mu_1 - \mu_2 - \mu_2(\log(\mu_1) - \log(\mu_2))$	$\xi = \log(\mu)$	$\exp(\xi)$
0	1	$\mu_2 - \mu_1 - \mu_1(\log(\mu_2) - \log(\mu_1))$	$\xi = \mu$	$\xi\log(\xi) - \xi$
$\lambda \neq 0, -1$	$\alpha \neq \pm 1$	$\frac{\left(\lambda^*\mu_1 - \lambda^*\mu_2 - \mu_2\left(\left(\frac{\mu_1}{\mu_2}\right)^{\lambda^*} - 1\right)\right)}{\lambda^*(1-\lambda^*)}$	$\xi = \frac{1}{\lambda^*}\mu^{\lambda^*}$	$\frac{(\lambda^*\xi)^{1/\lambda^*}}{1-\lambda^*}$

3.4. Extended Multinomial Case

In this paper, we are focusing on the class of log–linear models where the multinomial is the underlying class of distributions; that is, we condition on the sample size, N, being fixed in the product Poisson space. In particular, we focus on extended multinomials, which includes the closure of the multinomials, so we have a boundary. Due to the conditioning (which induces curvature), only the cases where $\lambda = 0, -1$ remain Bregman divergences, but all are still divergences in the sense of being Csiszár f-divergences [36,37].

The closure of an exponential family (e.g., [11,38–40]), and its application in the theory of log–linear models has been explored in [12,13,41,42]. The key here is understanding the limiting behaviour in the natural—$\alpha = 1$ in the sense of [8]—parameter space. This can be done by considering the polar dual [43], or, alternatively, the directions of recession—[12] or [42]. The boundary polytope determines key statistical properties of the model, including the behaviour of the sampling distribution of (functions of) the MLE and the shape of level sets of divergence functions.

Figures 1 and 2 show level sets of the $\alpha = \pm 1$ Power-Divergences in the $(+1)$-affine and (-1)-affine parameters (Panels (a) and (b), respectively) for the $k = 2$ extended multinomial model. The boundary polytope in this case is a simple triangle "at infinity", and the shape of this is strongly reflected in the behaviour of the level sets. In Figure 1, we show—in the simplex $\left\{(\pi_0, \pi_1, \pi_2) | \sum_{i=0}^{2} \pi_i = 1, \pi_i \geq 0\right\}$—the level sets of the $\alpha = -1$ divergence, which, in the Csiszár f-divergence form, is

$$K(\pi^0, \pi) := \sum_{i=0}^{2} \log\left(\frac{\pi_i^0}{\pi_i}\right) \pi_i^0.$$

The figures show how in Panel (a), the directions of recession dominate the shape of level sets, and in Panel (b) the duals of these directions (i.e., the vertices of the simplex) each have different maximal behaviour. The lack of convexity of the level sets in Panel (a) corresponds to the fact that the natural parameters are not the affine divergence parameters for this divergence, so we do not expect convex behaviour. In Panel (b), we do get non-convex level sets, as expected.

Figure 2 shows the same story, but this time for the dual divergence,

$$K^*(\pi, \pi^0) := K(\pi^0, \pi).$$

Now, the affine divergence parameters are shown in Panel (a), the natural parameters. We see that in the limit the shape of the divergence is converging to that of the polar of the boundary polytope. In general, local behaviour is quadratic, but boundary behaviour is polygonal.

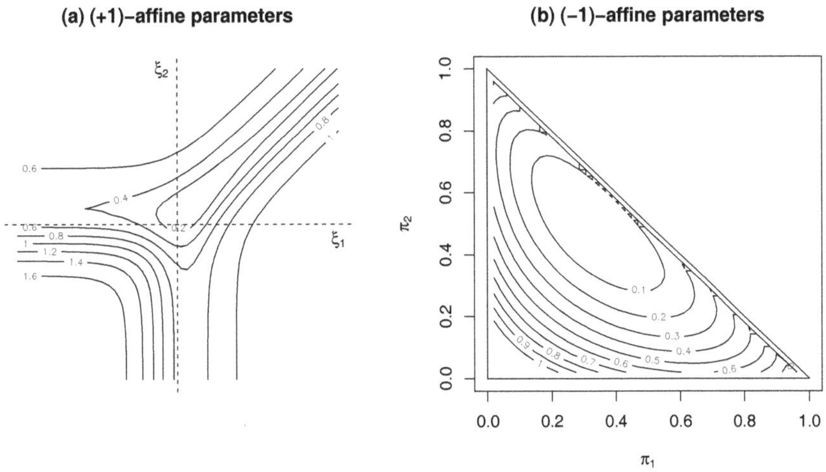

Figure 1. Level sets of $K(\pi^0, \pi)$, for fixed $\pi^0 = (\frac{1}{6}, \frac{2}{6}, \frac{3}{6})$ in: (**a**) the natural parameters, and (**b**) the mean parameters.

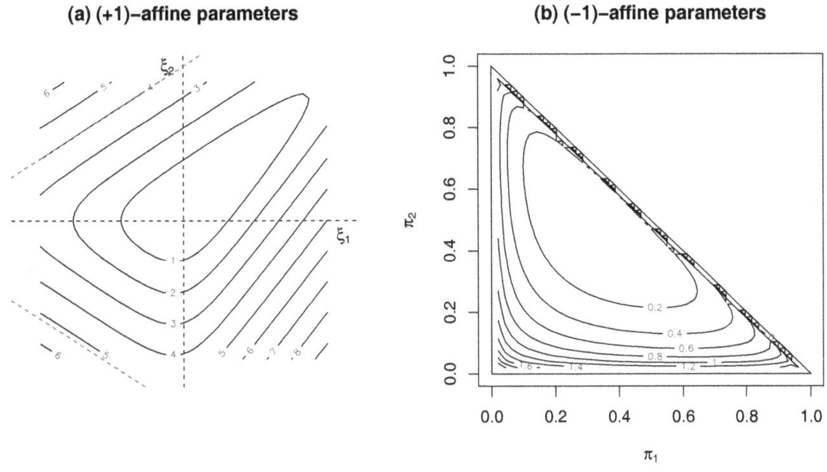

Figure 2. Level sets of $K^*(\pi^0, \pi)$, for fixed $\pi^0 = (\frac{1}{6}, \frac{2}{6}, \frac{3}{6})$ in: (**a**) the natural parameters, and (**b**) the mean parameters.

4. Simulation Studies

In this section, we undertake simulation studies to numerically explore what has been discussed above. Separate sub-sections address three general topics—focusing on one particular instance of each, as follows:

1. The transition as (N, k) varies between discrete and continuous features of the sampling distributions of goodness-of-fit statistics—focusing on the behaviour of the deviance at the uniform discrete distribution;
2. The comparative behaviour of a range of Power-Divergence statistics—focusing on the relative stability of their sampling distributions near the boundary;
3. The lack of uniformity—across the parameter space—of the finite sample adequacy of standard asymptotic sampling distributions, focusing on testing independence in 2×2 contingency tables.

For each topic, the results presented invite further investigation.

4.1. Transition Between Discrete and Continuous Features of Sampling Distributions

Earlier work [2] used the decomposition:

$$D^*/2 = \sum_{\{0 \le i \le k : n_i^* > 0\}} n_i^* \log(n_i^*/\mu_i) = \Gamma^* + \Delta^*,$$

$$\Gamma^* := \sum_{i=0}^{k} \alpha_i n_i^* \text{ and } \Delta^* := \sum_{\{0 \le i \le k : n_i^* > 1\}} n_i^* \log n_i^* \ge 0, \text{ where } \alpha_i := -\log \mu_i,$$

to show that a particularly bad case for the adequacy of any continuous approximation to the sampling distribution of the deviance $D := D^*|(N^* = N)$ is the uniform discrete distribution: $\pi_i = 1/(k+1)$. In this case, the Γ^* term contributes a constant to the deviance, while the Δ^* term has no contributions from cells with 0 or 1 observations—these being in the vast majority in the $N << k$ situation considered here. In other words, *all* of the variability in D comes from that between the $n_i \log n_i$ values for the (relatively rare) cell counts above 1. This gives rise to a discreteness phenomenon termed "granularity" in [2], whose meaning was conveyed graphically there in the case $N = 30$ and $k = 200$. Work by Holst [19] predicts that continuous (indeed, normal) approximations will improve with larger values of N/k, as is intuitive. Remarkably, simply doubling the sample size to $N = 60$ was shown in [2] to be sufficient to give a good enough approximation for most goodness-of-fit testing purposes. In other words, N being 30% of $k = 200$ was found to be good enough for practical purposes.

Here, we illustrate the role of k-asymptotics (Section 2) in this transition between discrete and continuous features by repeating the above analyses for different values of k. Figures 3 and 4 (where $k = 100$ while $N = 20$ and 40, respectively) are qualitatively the same as those presented in [2]. The difference here is that the smaller value of k means that a higher value of N/k (40%) is needed in Figure 4 to adequately remove the granularity evident in Figure 3. For $k = 400$, the figures with $N = 50$ and $N = 100$ (omitted here for brevity) are, again, qualitatively the same as in [2]—the larger value of k needing only a smaller value of N/k (25%) for practical purposes. Note the QQ-plots used in these two figures are relative to normal quantiles.

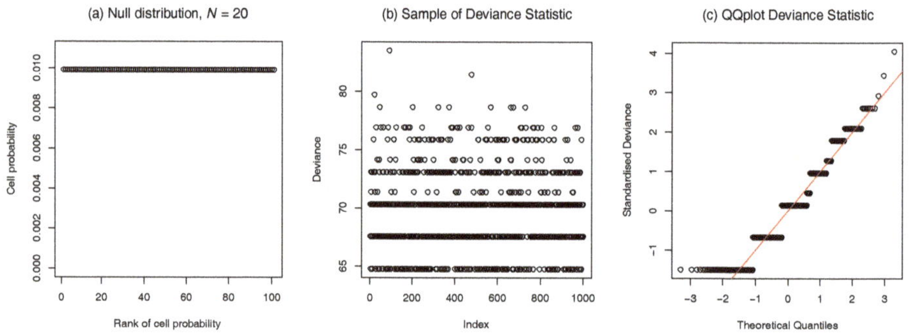

Figure 3. $k = 100$, $N = 20$.

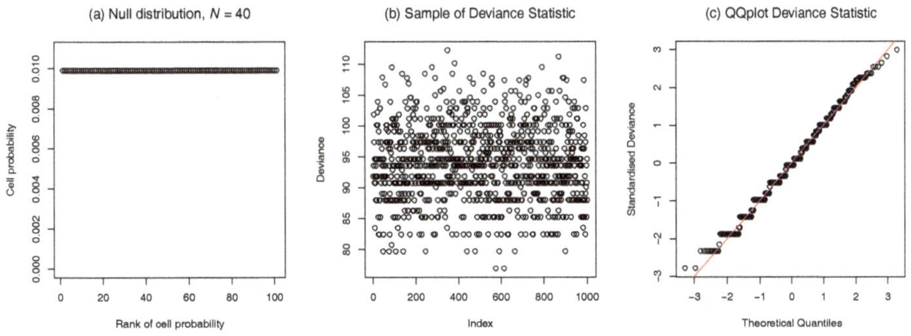

Figure 4. $k = 100$, $N = 40$.

The results of this section show the universality of boundary effects. The simulations of Figures 3 and 4 are undertaken under the uniform model, which might be felt to be far from the boundary. In fact, the results show that in the high dimensional, low sample size case, all distributions are "close to" the boundary, and that discretisation effects can dominate.

4.2. Comparative Behaviour of Power-Divergence Statistics near the Boundary

Here we study the relative stability—near the boundary of the simplex—of the sampling distributions of a range of Power-Divergence statistics indexed by Amari's parameter α. Figure 5 shows histograms for six different values of α, $N = 50$, $k = 200$, and exponentially decreasing values of $\{\pi_i\}$, as plotted in Figure 6. In it, red lines depict kernel density estimates using the bandwidth suggested in [44].

These sampling distributions differ markedly. The instability for $\alpha = 3$ expected from Theorem 1 is clearly visible: very large values contribute to high variance and skewness. Analogous instability features (albeit at a lower level) remain with the Cressie–Read recommended value $\alpha = 7/3$. In contrast (as expected from the discussion around Theorem 2), the distribution of the deviance ($\alpha = 1$) is stable and roughly normal. Lower values of α retain these same features.

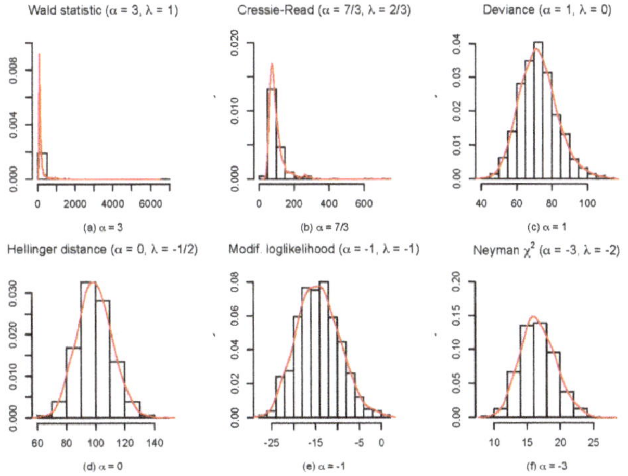

Figure 5. Sampling distributions for six members of the Power-Divergence family.

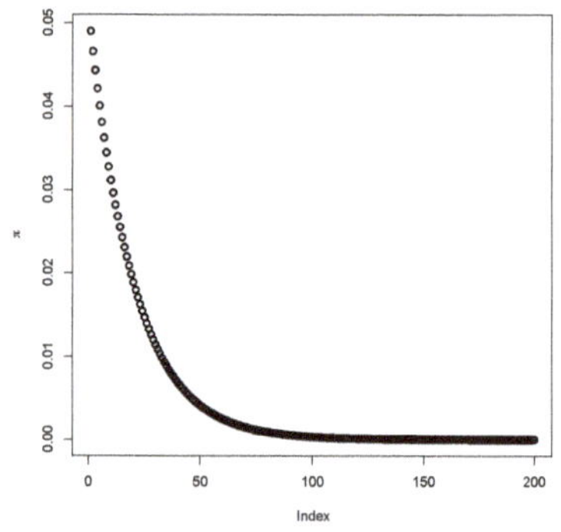

Figure 6. Exponentially decreasing values of π_i.

4.3. Variation in Finite Sample Adequacy of Asymptotic Distributions across the Parameter Space

Pearson's χ^2 statistic ($\alpha = 3$) is widely used to test independence in contingency tables, a standard rule-of-thumb for its validity being that each expected cell frequency should be at least 5. For illustrative purposes, we consider 2×2 contingency tables, the relevant N-asymptotic null distribution being χ_1^2. We assess the adequacy of this asymptotic approximation by comparing nominal and actual significance levels of this test, based on 10,000 replications. Particular interest lies in how these actual levels vary across different data generation processes within the same null hypothesis of independence.

Figures 7 and 8 show the actual level of the Pearson χ^2 test for nominal levels 0.1 and 0.05 for sample sizes $N = 20$ and $N = 50$, with π_r and π_c denoting row and column probabilities, respectively. The above general rule applies only at the central black dot in Figure 7, and inside the closed black curved region in Figure 8. The actual level was computed for all pairs of values of π_r and π_c, averaged using the symmetry of the parameter space, and smoothed using the kernel smoother for irregular 2D data (implemented in the package *fields* in R). In each case, the white tone contains the nominal level, while red tones correspond to liberal and blue tones to conservative actual levels.

The finite sample adequacy of this standard asymptotic test clearly varies across the parameter space. In particular, its nominal and actual levels agree well at some parameter values outside the standard rule-of-thumb region; and, conversely, disagree somewhat at other parameter values inside it. Intriguingly, the agreement between nominal and actual levels does not improve everywhere with sample size. Overall, the clear patterns evident in this lack of uniformity invite further theoretical investigation.

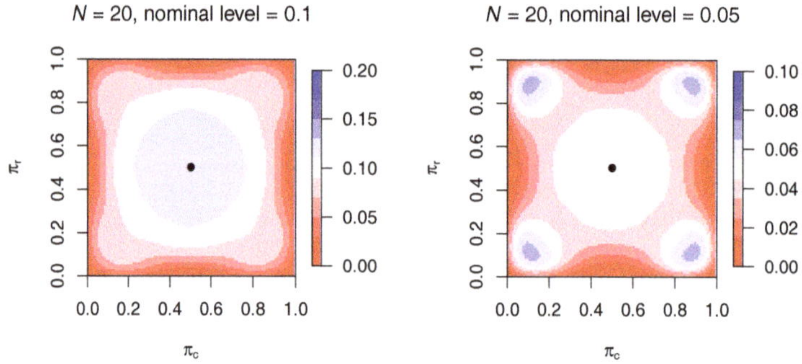

Figure 7. Heatmap of the actual level of the test for $N = 20$ at nominal levels 0.1 and 0.05; the standard rule-of-thumb (where expected counts are greater than 5) applies only at the black dot.

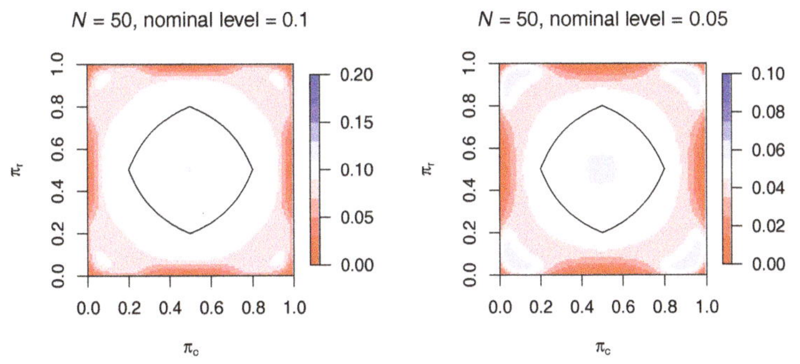

Figure 8. Heatmap of the actual level of the test for $N = 50$ at nominal levels 0.1 and 0.05; the standard rule-of-thumb (where expected counts are greater than 5) applies inside the closed black curved region.

5. Discussion

This paper has illustrated the key importance of working with the boundary of the closure of exponential families when studying goodness-of-fit testing in the high dimensional, low sample size context. Some of this work is new (Section 2), while some uses the structure of extended exponential families to add insight to standard results in the literature (Section 3). The last section, Section 4, uses simulation studies to start to explore open questions in this area.

One open question—related to the results of Theorems 1 and 2—is to see if a unified theory, for all values of α, and over large classes of extended exponential families, can be developed.

Acknowledgments: The authors would like to thank the EPSRC for the support of grant number EP/L010429/1. Germain Van Bever would also like to thank FRS-FNRS for its support through the grant FC84444. We would also like to thank the referees for very helpful comments.

Author Contributions: All four authors made critical contributions to the paper. R.S. made key contribution to, especially, Section 4. P.M. and F.C. provided the overall structure and key content details of the paper. G.V.B. provided invaluable suggestions throughout.

Conflicts of Interest: The authors declare no conflicts of interest.

Appendix A. Proof of Theorem 1

We start by noting an important recurrence relation which will be exploited in the computations below. By definition, for any $t := (t_i) \in R^{k+1}$, $n = (n_i)$ has moment generating function

$$M(t;N) := E\{\exp(t^T n)\} = [m(t)]^N$$

with $m(t) = \sum_{i=0}^{k} a_i$ and $a_i = a_i(t_i) = \pi_i e^{t_i}$. Putting

$$f_{N,i}(t;r) := N_{(r)}\,[m(t)]^{N-r}\,a_i^r \quad (0 \le r \le N),$$

where

$$N_{(r)} := {}^{N}P_r = \begin{cases} 1 & \text{if } r = 0 \\ N(N-1)...(N-(r-1)) & \text{if } r \in \{1,...,N\} \end{cases},$$

we have

$$M(t;N) = f_{N,i}(t;0) \quad (0 \le i \le k) \tag{A1}$$

and the recurrence relation:

$$\frac{\partial f_{N,i}(t;r)}{\partial t_i} = f_{N,i}(t;r+1) + r f_{N,i}(t;r) \quad (0 \le i \le k;\ 0 \le r < N). \tag{A2}$$

When there is no risk of confusion, we may abbreviate $M(t;N)$ to M and $f_{N,i}(t;r)$ to $f_N(r)$, or even to $f(r)$—so that (A1) becomes $M = f(0)$. Again, we may write $\partial^r M(t;N)/\partial t_i^r$ as M_r, $\partial^{r+s} M(t;N)/\partial t_i^r \partial t_j^s$ as $M_{r,s}$ and $\partial^{r+s+u} M(t;N)/\partial t_i^r \partial t_j^s \partial t_l^u$ as $M_{r,s,u}$, with similar conventions for higher order mixed derivatives.

We can now use this to explicitly calculate low order moments of the count vectors. Using $E(n_i^r) = \partial^r M(t;N)/\partial t_i^r|_{t=0}$, the first N moments of n_i now follow from (A1) and repeated use of (A2), noting that $m(0) = 1$ and $a_i(0) = \pi_i$.

In particular, the first 6 moments of each n_i can be obtained as follows, where $N \geq 6$ is assumed. Using (A1) and (A2), we have

$$\begin{aligned}
M_1 &= f(1) \\
M_2 &= f(2) + f(1) \\
M_3 &= f(3) + 2f(2) + f(2) + f(1) = f(3) + 3f(2) + f(1) \\
M_4 &= f(4) + \ \ 6f(3) + \ \ 7f(2) + \ \ f(1) \\
M_5 &= f(5) + 10f(4) + 25f(3) + 15f(2) + \ \ f(1) \\
M_6 &= f(6) + 15f(5) + 65f(4) + 90f(3) + 31f(2) + f(1).
\end{aligned}$$

Substituting in, we have

$$\begin{aligned}
E(n_i) &= N\pi_i \\
E(n_i^2) &= N_{(2)}\pi_i^2 + N\pi_i \\
E(n_i^3) &= N_{(3)}\pi_i^3 + 3N_{(2)}\pi_i^2 + N\pi_i \\
E(n_i^4) &= N_{(4)}\pi_i^4 + 6N_{(3)}\pi_i^3 + 7N_{(2)}\pi_i^2 + N\pi_i \\
E(n_i^5) &= N_{(5)}\pi_i^5 + 10N_{(4)}\pi_i^4 + 25N_{(3)}\pi_i^3 + 15N_{(2)}\pi_i^2 + N\pi_i \\
E(n_i^6) &= N_{(6)}\pi_i^6 + 15N_{(5)}\pi_i^5 + 65N_{(4)}\pi_i^4 + 90N_{(3)}\pi_i^3 + 31N_{(2)}\pi_i^2 + N\pi_i.
\end{aligned}$$

This can be formalised in the following Lemma

Lemma A1. *The integer coefficients in any expansion*

$$M_r = \sum_{s=1}^{r} c_r(s) f(s) \quad (1 \leq r \leq N)$$

can be computed using $c_r(1) = c_r(r) = 1$ *together, for* $r \geq 3$, *with the update:*

$$c_r(s) = c_{r-1}(s-1) + s c_{r-1}(s) \quad (1 < s < r).$$

We note that if M_r is required for $r > N$, we may repeatedly differentiate

$$M_N = \sum_{s=1}^{N} c_N(s) f(s)$$

w.r.t. t_i, noting that $f(N) = N! a_i^N$ no longer depends on $m(t)$ so that, for all $h > 0$, $\partial^h f(N)/\partial t_i^h = N^h f(N)$.

Mixed moments of any order can be derived from those of lower order, exploiting the fact that a_i depends on t only via t_i. We illustrate this by deriving those required for the second and third moments of W.

First consider the mixed moments required for the second moment of W. Of course, $Var(W) = 0$ if $k = 0$. Otherwise, $k > 0$, and computing $Var(W)$ requires $E(n_i^2 n_j^2)$ for $i \neq j$. We find this as follows, assuming $N \geq 4$.

The relation $M_2 = f(2) + f(1)$ established above gives

$$\partial^2 M/\partial t_j^2 = N_{(2)} a_j^2 f_{N-2}(0) + N a_j f_{N-1}(0). \tag{A3}$$

Repeated use of (A3) now gives

$$M_{2,2} = N_{(4)} a_i^2 a_j^2 f_{N-4}(0) + N_{(3)} a_i a_j (a_i + a_j) f_{N-3}(0) + N_{(2)} a_i a_j f_{N-2}(0) \tag{A4}$$

so that

$$E(n_i^2 n_j^2) = N_{(4)} \pi_i^2 \pi_j^2 + N_{(3)} \pi_i \pi_j (\pi_i + \pi_j) + N_{(2)} \pi_i \pi_j.$$

We further look at the mixed moments needed for the third moment of W. For the skewness of W, we need $E(n_i^2 n_j^4)$ for $i \neq j$ and, when $k > 1$, $E(n_i^2 n_j^2 n_l^2)$ for i, j, l distinct. We find these similarly, as follows, assuming $k > 1$ and $N \geq 6$.

Equation (A4) above gives

$$\partial^2 M/\partial t_j^2 \partial t_l^2 = N_{(4)} a_j^2 a_l^2 f_{N-4}(0) + N_{(3)} a_j a_l (a_j + a_l) f_{N-3}(0) + N_{(2)} a_j a_l f_{N-2}(0)$$

from which, using (A3) repeatedly, we have

$$\begin{aligned} M_{2,2,2} &= a_j^2 a_l^2 \{N_{(6)} a_i^2 f_{N-6}(0) + N_{(5)} a_i f_{N-5}(0)\} + a_j a_l (a_j + a_l) \{N_{(5)} a_i^2 f_{N-5}(0) + N_{(4)} a_i f_{N-4}(0)\} + \\ &\quad a_j a_l \{N_{(4)} a_i^2 f_{N-4}(0) + N_{(3)} a_i f_{N-3}(0)\} \\ &= N_{(6)} a_i^2 a_j^2 a_l^2 f_{N-6}(0) + N_{(5)} a_i a_j a_l \{a_i a_j + a_j a_l + a_l a_i\} f_{N-5}(0) + N_{(4)} a_i a_j a_l \{a_i + a_j + a_l\} f_{N-4}(0) + \\ &\quad N_{(3)} a_i a_j a_l f_{N-3}(0) \end{aligned}$$

so that $E(n_i^2 n_j^2 n_l^2)$ equals

$$N_{(6)} \pi_i^2 \pi_j^2 \pi_l^2 + N_{(5)} \pi_i \pi_j \pi_l \{\pi_i \pi_j + \pi_j \pi_l + \pi_l \pi_i\} + N_{(4)} \pi_i \pi_j \pi_l \{\pi_i + \pi_j + \pi_l\} + N_{(3)} \pi_i \pi_j \pi_l.$$

Finally, the relation $M_4 = f(4) + 6f(3) + 7f(2) + f(1)$ established above gives

$$\partial^4 M/\partial t_j^4 = N_{(4)} a_j^4 f_{N-4}(0) + 6N_{(3)} a_j^3 f_{N-3}(0) + 7N_{(2)} a_j^2 f_{N-2}(0) + N a_j f_{N-1}(0)$$

so that, again using (A3) repeatedly, yields

$$E(n_i^2 n_j^4) = N_{(6)} \pi_i^2 \pi_j^4 + N_{(5)} \pi_i \pi_j^3 (6\pi_i + \pi_j) + N_{(4)} \pi_i \pi_j^2 (7\pi_i + 6\pi_j) + N_{(3)} \pi_i \pi_j (\pi_i + 7\pi_j) + N_{(2)} \pi_i \pi_j.$$

Combining above results, we obtain here the first three moments of W. Higher moments may be found similarly.

We first look at $E(W)$. We have $W = \frac{1}{N^2} \sum_{i=0}^{k} \frac{n_i^2}{\pi_i} - 1$ and $E(n_i^2) = N_{(2)} \pi_i^2 + N\pi_i$, so that

$$E(W) = \frac{N_{(2)}}{N^2} + \frac{(k+1)}{N} - 1 = \frac{k}{N}.$$

The variance is computed by recalling that $N^2(W+1) = \sum_i \frac{n_i^2}{\pi_i}$, while $E(W) = \frac{k}{N}$,

$$Var(W) = Var(W+1) = \frac{A^{(2)}}{N^4} - \left(\frac{k}{N} + 1\right)^2,$$

where

$$A^{(2)} := N^4 E\{(W+1)^2\} = \sum_i \frac{E(n_i^4)}{\pi_i^2} + \sum\sum_{i \neq j} \frac{E(n_i^2 n_j^2)}{\pi_i \pi_j}.$$

Using expressions for $E(n_i^4)$ and $E(n_i^2 n_j^2)$ established above, and putting

$$\pi^{(\alpha)} := \sum_i \pi_i^\alpha,$$

we have

$$\begin{aligned}\sum_i \frac{E(n_i^4)}{\pi_i^2} &= \sum_i \{N_{(4)}\pi_i^2 + 6N_{(3)}\pi_i + 7N_{(2)} + N\pi_i^{-1}\} \\ &= N_{(4)}\pi^{(2)} + 6N_{(3)} + 7N_{(2)}(k+1) + N\pi^{(-1)}\end{aligned}$$

and

$$\begin{aligned}\sum\sum_{i\neq j} \frac{E(n_i^2 n_j^2)}{\pi_i \pi_j} &= \sum_{i\neq j}\{N_{(4)}\pi_i\pi_j + N_{(3)}(\pi_i + \pi_j) + N_{(2)}\} \\ &= N_{(4)}(1-\pi^{(2)}) + 2N_{(3)}k + N_{(2)}k(k+1),\end{aligned}$$

so that

$$A^{(2)} = N_{(4)} + 2N_{(3)}(k+3) + N_{(2)}(k+1)(k+7) + N\pi^{(-1)},$$

whence

$$\begin{aligned}Var(W) &= \frac{N_{(4)} + 2N_{(3)}(k+3) + N_{(2)}(k+1)(k+7) + N\pi^{(-1)}}{N^4} - \left(1 + \frac{k}{N}\right)^2 \\ &= \frac{\left\{\pi^{(-1)} - (k+1)^2\right\} + 2k(N-1)}{N^3}, \text{ after some simplification.}\end{aligned}$$

Note that $Var(W)$ depends on (π_i) *only* via $\pi^{(-1)}$ while, by strict convexity of $x \to 1/x$ $(x > 0)$,

$$\pi^{(-1)} \geq (k+1)^2, \text{equality holding iff } \pi_i \overset{i}{\equiv} 1/(k+1).$$

Thus, for given k and N, $Var(W)$ is strictly increasing as (π_i) departs from uniformity, tending to ∞ as one or more $\pi_i \to 0_+$.

Finally, for these calculations, we look at $E[\{W - E(W)\}^3]$. Recalling again that $N^2(W+1) = \sum_i \frac{n_i^2}{\pi_i}$,

$$\begin{aligned}E[\{W - E(W)\}^3] &= E[\{(W+1) - E(W+1)\}^3] \\ &= N^{-6}A^{(3)} - 3\,Var(W)(E(W)+1) - (E(W)+1)^3,\end{aligned}$$

where $A^{(3)} := N^6 E\{(W+1)^3\}$ is given by

$$A^{(3)} = \sum_i \frac{E(n_i^6)}{\pi_i^3} + 3\sum\sum_{i\neq j} \frac{E(n_i^2 n_j^4)}{\pi_i \pi_j^2} + \sum\sum\sum_{i,j,l \text{ distinct}} \frac{E(n_i^2 n_j^2 n_l^2)}{\pi_i \pi_j \pi_l}.$$

Given that

$$E(W) = k/N \text{ and } Var(W) = \frac{\left\{\pi^{(-1)} - (k+1)^2\right\} + 2k(N-1)}{N^3},$$

it suffices to find $A^{(3)}$.

Using expressions for $E(n_i^6)$, $E(n_i^2 n_j^2 n_l^2)$, and $E(n_i^2 n_j^4)$ established above, we have

$$\sum_i \frac{E(n_i^6)}{\pi_i^3} = N_{(6)}\pi^{(3)} + 15N_{(5)}\pi^{(2)} + 65N_{(4)} + 90N_{(3)}(k+1) + 31N_{(2)}\pi^{(-1)} + N\pi^{(-2)}$$

$$\sum\sum_{i \neq j} \frac{E(n_i^2 n_j^4)}{\pi_i \pi_j^2} = N_{(6)} \pi_i \pi_j^2 + N_{(5)} \pi_j (6\pi_i + \pi_j) + N_{(4)} (7\pi_i + 6\pi_j) + N_{(3)} (\pi_i / \pi_j + 7) + N_{(2)} \pi_j^{-1}$$
$$= N_{(6)} \{\pi^{(2)} - \pi^{(3)}\} + N_{(5)} \{6 + (k-6)\pi^{(2)}\} +$$
$$13N_{(4)} k + N_{(3)} \{\pi^{(-1)} + (7k-1)(k+1)\} + N_{(2)} k \pi^{(-1)}$$

and

$$\sum\sum\sum_{i,j,l \text{ distinct}} \frac{E(n_i^2 n_j^2 n_l^2)}{\pi_i \pi_j \pi_l} = N_{(6)} \{1 + 2\pi^{(3)} - 3\pi^{(2)}\} + 3N_{(5)} (k-1)\{1 - \pi^{(2)}\} +$$
$$3N_{(4)} k(k-1) + N_{(3)} k(k^2 - 1)$$

so that, after some simplification,

$$A^{(3)} = N_{(6)} + 3N_{(5)} (k+5) + N_{(4)} \{3k(k+12) + 65\} +$$
$$N_{(3)} \{k^3 + 21k^2 + 107k + 87\} + 3N_{(3)} \pi^{(-1)} + N_{(2)} (31 + 3k) \pi^{(-1)} + N\pi^{(-2)}.$$

Substituting in and simplifying, we find $E[\{W - E(W)\}^3]$ to be:

$$\frac{\left\{\pi^{(-2)} - (k+1)^3\right\} - (3k + 25 - 22N) \left\{\pi^{(-1)} - (k+1)^2\right\} + g(k,N)}{N^5},$$

where

$$g(k,N) = 4(N-1)k(k + 2N - 5) > 0.$$

Note that $E[\{W - E(W)\}^3]$ depends on (π_i) *only* via $\pi^{(-1)}$ and the *larger* quantity $\pi^{(-2)}$. In particular, for given k and N, the skewness of W tends to $+\infty$ as one or more $\pi_i \to 0_+$.

Appendix B. Truncate and Bound Approximations

In the notation of Lemma 1, it suffices to find truncate and bound approximations for each of $E(X_\mu)$, $E(X.X_\mu)$, and $E(X_\mu^2)$.

For all r, s in $\mathcal{N}$, define $h_{r,s}(\mu) := E\{(\log(X + r))^s\}$. Appropriate choices of $m \in \mathcal{N}_0$ and $a \in \mathcal{A}$ in (9), together with (10), give:

$$E(X_\mu) = \mu h_{1,1}(\mu) - \mu \log \mu,$$
$$E(X.X_\mu) = \{\mu^2 h_{2,1}(\mu) + \mu h_{1,1}(\mu)\} - (\mu^2 + \mu) \log \mu, \text{ and:}$$
$$E(X_\mu^2) = \mu^2 h_{2,2}(\mu) + \mu h_{1,2}(\mu) + (\mu^2 + \mu)(\log \mu)^2 - 2 \log \mu \{\mu^2 h_{2,1}(\mu) + \mu h_{1,1}(\mu)\},$$

so that it suffices to truncate and bound $h_{r,s}(\mu)$ for $r, s \in \{1, 2\}$.

For all r, s in $\mathcal{N}$, and for all $m \in \mathcal{N}_0$, we write:

$$h_{r,s}(\mu) = h_{r,s}^{[m]}(\mu) + \varepsilon_{r,s}^{[m]}(\mu)$$

in which:

$$h_{r,s}^{[m]}(\mu) := \sum_{x=0}^{m} \{(\log(x+r))^s\} p(x) \text{ and } \varepsilon_{r,s}^{[m]}(\mu) := \sum_{x=m+1}^{\infty} \{(\log(x+r))^s\} p(x).$$

Using again (7), the "error term" $\varepsilon_{r,s}^{[m]}(\mu)$ has lower and upper bounds:

$$0 < \varepsilon_{r,s}^{[m]}(\mu) < \bar{\varepsilon}_{r,s}^{[m]}(\mu) := \sum_{x=m+1}^{\infty} (x + (r-1))^s p(x).$$

Restricting attention now to $r, s \in \{1,2\}$, as we may, and requiring $m \geq s$ so that $F^{[m-s]}(\mu)$ given by (4) is defined, (8) gives:

$$\bar{\varepsilon}_{1,1}^{[m]}(\mu) = \Sigma_{x=m+1}^{\infty} x p(x) = \mu \Sigma_{x=m}^{\infty} p(x) = \mu\{1 - F^{[m-1]}(\mu)\},$$

$$\bar{\varepsilon}_{2,1}^{[m]}(\mu) = \Sigma_{x=m+1}^{\infty}(x+1)p(x) = \bar{\varepsilon}_{1,1}^{[m]}(\mu) + \{1 - F^{[m]}(\mu)\},$$

$$\begin{aligned} \bar{\varepsilon}_{1,2}^{[m]}(\mu) &= \Sigma_{x=m+1}^{\infty} x^2 p(x) = \Sigma_{x=m+1}^{\infty}\{x(x-1) + x\}p(x) \\ &= \mu^2\{1 - F^{[m-2]}(\mu)\} + \bar{\varepsilon}_{1,1}^{[m]}(\mu) \end{aligned}$$

and:

$$\begin{aligned} \bar{\varepsilon}_{2,2}^{[m]}(\mu) &= \Sigma_{x=m+1}^{\infty}(x+1)^2 p(x) = \Sigma_{x=m+1}^{\infty}\{x^2 + (x+1) + x\}p(x) \\ &= \bar{\varepsilon}_{1,2}^{[m]}(\mu) + \bar{\varepsilon}_{2,1}^{[m]}(\mu) + \bar{\varepsilon}_{1,1}^{[m]}(\mu). \end{aligned}$$

Accordingly, for given μ, each $\bar{\varepsilon}_{r,s}^{[m]}(\mu)$ decreases strictly to zero with m providing—to any desired accuracy—truncate and bound approximations for each of ν, τ, and ρ. In this connection, we note that the upper tail probabilities involved here can be bounded by standard Chernoff arguments.

References

1. Critchley, F.; Marriott, P. Computational Information Geometry in Statistics: Theory and practice. *Entropy* **2014**, *16*, 2454–2471.
2. Marriott, P.; Sabolova, R.; Van Bever, G.; Critchley, F. Geometry of goodness-of-fit testing in high dimensional low sample size modelling. In *Geometric Science of Information: Second International Conference, GSI 2015, Palaiseau, France, October 28–30, 2015, Proceedings*; Nielsen, F., Barbaresco, F., Eds.; Springer: Berlin, Germany, 2015; pp. 569–576.
3. Amari, S.-I.; Nagaoka, H. *Methods of Information Geometry*; Translations of Mathematical Monographs; American Mathematical Society: Providence, RI, USA, 2000.
4. Cressie, N.; Read, T.R.C. Multinomial goodness-of-fit tests. *J. R. Stat. Soc. B* **1984**, *46*, 440–464.
5. Read, T.R.C.; Cressie, N.A.C. *Goodness-of-Fit Statistics for Discrete Multivariate Data*; Springer: New York, NY, USA, 1988.
6. Kass, R.E.; Vos, P.W. *Geometrical Foundations of Asymptotic Inference*; John Wiley & Sons, Inc.: Hoboken, NJ, USA, 1997.
7. Agresti, A. *Categorical Data Analysis*, 3rd ed.; Wiley: Hoboken, NJ, USA, 2013.
8. Amari, S.-I. Differential-geometrical methods in statistics. In *Lecture Notes in Statistics*; Springer: New York, NY, USA, 1985; Volume 28.
9. Barndorff-Nielsen, O.E.; Cox, D.R. *Asymptotic Techniques for Use in Statistics*; Chapman & Hall: London, UK, 1989.
10. Anaya-Izquierdo, K.; Critchley, F.; Marriott, P. When are first-order asymptotics adequate? A diagnostic. *STAT* **2014**, *3*, 17–22.
11. Lauritzen, S.L. *Graphical Models*; Clarendon Press: Oxford, UK, 1996.
12. Geyer, C.J. Likelihood inference in exponential families and directions of recession. *Electron. J. Stat.* **2009**, *3*, 259–289.
13. Fienberg, S.E.; Rinaldo, A. Maximum likelihood estimation in log-linear models. *Ann. Stat.* **2012**, *40*, 996–1023.
14. Eguchi, S.; Copas, J. Local model uncertainty and incomplete-data bias. *J. R. Stat. Soc. B* **2005**, *67*, 1–37.
15. Copas, J.; Eguchi, S. Likelihood for statistically equivalent models. *J. R. Stat. Soc. B* **2010**, *72*, 193–217.

16. Anaya-Izquierdo, K.; Critchley, F.; Marriott, P.; Vos, P. On the geometric interplay between goodness-of-fit and estimation: Illustrative examples. In *Computational Information Geometry: For Image and Signal Processing*; Lecture Notes in Computer Science (LNCS); Nielsen, F., Dodson, K., Critchley, F., Eds.; Springer: Berlin, Germany, 2016.
17. Morris, C. Central limit theorems for multinomial sums. *Ann. Stat.* **1975**, *3*, 165–188.
18. Osius, G.; Rojek, D. Normal goodness-of-fit tests for multinomial models with large degrees of freedom. *JASA* **1992**, *87*, 1145–1152.
19. Holst, L. Asymptotic normality and efficiency for certain goodness-of-fit tests. *Biometrika* **1972**, *59*, 137–145.
20. Koehler, K.J.; Larntz, K. An empirical investigation of goodness-of-fit statistics for sparse multinomials. *JASA* **1980**, *75*, 336–344.
21. Koehler, K.J. Goodness-of-fit tests for log-linear models in sparse contingency tables. *JASA* **1986**, 81, 483–493.
22. McCullagh, P. The conditional distribution of goodness-of-fit statistics for discrete data. *JASA* **1986**, *81*, 104–107.
23. Forster, J.J.; McDonald, J.W.; Smith, P.W.F. Monte Carlo exact conditional tests for log-linear and logistic models. *J. R. Stat. Soc. B* **1996**, *58*, 445–453.
24. Kim, D.; Agresti, A. Nearly exact tests of conditional independence and marginal homogeneity for sparse contingency tables. *Comput. Stat. Data Anal.* **1997**, *24*, 89–104.
25. Booth, J.G.; Butler, R.W. An importance sampling algorithm for exact conditional tests in log-linear models. *Biometrika* **1999**, *86*, 321–332.
26. Caffo, B.S.; Booth, J.G. Monte Carlo conditional inference for log-linear and logistic models: A survey of current methodology. *Stat. Methods Med. Res.* **2003**, *12*, 109–123.
27. Lloyd, C.J. Computing highly accurate or exact P-values using importance sampling. *Comput. Stat. Data Anal.* **2012**, *56*, 1784–1794.
28. Simonoff, J.S. Jackknifing and bootstrapping goodness-of-fit statistics in sparse multinomials. *JASA* **1986**, *81*, 1005–1011.
29. Gaunt, R.E.; Pickett, A.; Reinert, G. Chi-square approximation by Stein's method with application to Pearson's statistic. *arXiv* **2015**, arXiv:1507.01707.
30. Fan, J.; Hung, H.-N.; Wong, W.-H. Geometric understanding of likelihood ratio statistics. *JASA* **2000**, *95*, 836–841.
31. Ulyanov, V.V.; Zubov, V.N. Refinement on the convergence of one family of goodness-of-fit statistics to chi-squared distribution. *Hiroshima Math. J.* **2009**, *39*, 133–161.
32. Asylbekov, Z.A.; Zubov, V.N.; Ulyanov, V.V. On approximating some statistics of goodness-of-fit tests in the case of three-dimensional discrete data. *Sib. Math. J.* **2011**, *52*, 571–584.
33. Zelterman, D. Goodness-of-fit tests for large sparse multinomial distributions. *JASA* **1987**, *82*, 624–629.
34. Bregman, L.M. The relaxation method of finding the common point of convex sets and its application to the solution of problems in convex programming. *USSR Comp. Math. Math.* **1967**, *7*, 200–217.
35. Amari, S.-I. *Information Geometry and Its Applications*; Springer: Tokyo, Japan, 2015.
36. Csiszár, I. On topological properties of f-divergences. *Stud. Sci. Math. Hung.* **1967**, *2*, 329–339.
37. Csiszár, I. Information measures: A critical survey. In *Transactions of the Seventh Prague Conference on Information Theory, Statistical Decision Functions, Random Processes and of the 1974 European Meeting of Statisticians*; Kozesnik, J., Ed.; Springer: Houten, The Netherlands, 1977; Volume B, pp. 73–86.
38. Barndorff-Nielsen, O. *Information and Exponential Families in Statistical Theory*; John Wiley & Sons, Ltd.: Chichester, UK, 1978.
39. Brown, L.D. *Fundamentals of Statistical Exponential Families with Applications in Statistical Decision Theory*; Lecture Notes-Monograph Series; Integrated Media Systems (IMS): Hayward, CA, USA, 1986; Volume 9.
40. Csiszár, I.; Matúš, F. Closures of exponential families. *Ann. Probab.* **2005**, *33*, 582–600.
41. Eriksson, N.; Fienberg, S.E.; Rinaldo, A.; Sullivant, S. Polyhedral conditions for the nonexistence of the MLE for hierarchical log-linear models. *J. Symb. Comput.* **2006**, *41*, 222–233.

42. Rinaldo, A.; Feinberg, S.; Zhou, Y. On the geometry of discrete exponential families with applications to exponential random graph models. *Electron. J. Stat.* **2009**, *3*, 446–484.
43. Critchley, F.; Marriott, P. Computing with Fisher geodesics and extended exponential families. *Stat. Comput.* **2016**, *26*, 325–332.
44. Sheather, S.J.; Jones, M.C. A reliable data-based bandwidth selection method for kernel density estimation. *J. R. Stat. Soc. B* **1991**, *53*, 683–690.

Chapter 4:
Density of Probability on Manifold and Metric Space

entropy

MDPI

Article

Kernel Density Estimation on the Siegel Space with an Application to Radar Processing †

Emmanuel Chevallier [1,*], Thibault Forget [2,3], Frédéric Barbaresco [2] and Jesus Angulo [3]

1 Department of Computer Science and Applied Mathematics, Weizmann Institute of Science, Rehovot 7610001, Israel
2 Thales Air Systems, Surface Radar Business Line, Advanced Radar Concepts Business Unit, Voie Pierre-Gilles de Gennes, Limours 91470, France; thibault.forget@mines-paristech.fr (T.F.); frederic.barbaresco@thalesgroup.com (F.B.)
3 CMM-Centre de Morphologie Mathématique, MINES ParisTech, PSL-Research University, Paris 75006, France; jesus.angulo@mines-paristech.fr
* Correspondence: emmanuelchevallier1@gmail.com; Tel.: +972-58-693-7744

† This paper is an extended version of our paper published in the 2nd conference on Geometric Science of Information, Paris, France, 28–30 October 2015.

Academic Editors: Arye Nehorai, Satyabrata Sen and Murat Akcakaya
Received: 13 August 2016; Accepted: 31 October 2016; Published: 11 November 2016

Abstract: This paper studies probability density estimation on the Siegel space. The Siegel space is a generalization of the hyperbolic space. Its Riemannian metric provides an interesting structure to the Toeplitz block Toeplitz matrices that appear in the covariance estimation of radar signals. The main techniques of probability density estimation on Riemannian manifolds are reviewed. For computational reasons, we chose to focus on the kernel density estimation. The main result of the paper is the expression of Pelletier's kernel density estimator. The computation of the kernels is made possible by the symmetric structure of the Siegel space. The method is applied to density estimation of reflection coefficients from radar observations.

Keywords: kernel density estimation; Siegel space; symmetric spaces; radar signals

1. Introduction

Various techniques can be used to estimate the density of probability measure in the Euclidean spaces, such as histograms, kernel methods, or orthogonal series. These methods can sometimes be adapted to densities in Riemannian manifolds. The computational cost of the density estimation depends on the isometry group of the manifold. In this paper, we study the special case of the Siegel space. The Siegel space is a generalization of the hyperbolic space. It has a structure of symmetric Riemannian manifold, which enables the adaptation of different density estimation methods at a reasonable cost. Convergence rates of the density estimation using kernels and orthogonal series were gradually generalized to Riemannian manifolds (see [1–3]).

The Siegel space appears in radar processing in the study of Toeplitz block Toeplitz matrices, whose blocks represent covariance matrices of a radar signal (see [4–6]). The Siegel also appears in statistical mechanics, see [7] and was recently used in image processing (see [8]). Information geometry is now a standard framework in radar processing (see [4–6,9–13]). The information geometry on positive definite Teoplitz block Teoplitz matrices is directly related to the metric on the Siegel space (see [14]). Indeed, Toeplitz block Toeplitz matrices can be represented by a symmetric positive definite matrix and a point laying in a product of Siegel disks. The metric considered on Toeplitz block Toeplitz matrices is induced by the product metric between a metric on the symmetric positive definite matrices and the Siegel disks metrics (see [4–6,9,14]).

One already encounters the problem of density estimation in the hyperbolic space for electrical impedance [15], networks [16] and radar signals [17]. In [18], a generalization of the Gaussian law on the hyperbolic space was proposed. Apart from [19], where authors propose a generalization of the Gaussian law, probability density estimation on the Siegel space has not yet been addressed.

The contributions of the paper are the following. We review the main non parametric density estimation techniques on the Siegel disk. We provide some rather simple explicit expressions of the kernels defined by Pelletier in [1]. These expressions make the kernel density estimation the most adapted method. We present visual results of estimated densities in the simple case where the Siegel disk reduces to the Poincaré disk.

The paper begins with an introduction to the Siegel space in Section 2. Section 3 reviews the main non-parametric density estimation techniques on the Siegel space. Section 3.3 contains the original results of the paper. Section 4 presents an application to radar data estimation.

2. The Siegel Space

This section presents facts about the Siegel space. The interested reader can find more details in [20,21]. The necessary background on Lie groups and symmetric space can be found in [22].

2.1. The Siegel Upper Half Space

The Siegel upper half space is a generalization of the Poincaré upper half space (see [23]) for a description of the hyperbolic space. Let $Sym(n)$ be the space of real symmetric matrices of size $n \times n$ and $Sym_+(n)$ the set of real symmetric positive definite matrices of size $n \times n$. The Siegel upper half space is defined by

$$\mathcal{H}_n = \{Z = X + iY | X \in Sym(n), Y \in Sym_+(n)\}.$$

$\mathcal{H}_n$ is equipped with the following metric:

$$ds = 2tr(Y^{-1}dZY^{-1}d\overline{Z}).$$

The set of real symplectic matrices $Sp(n, \mathbb{R})$ is defined by

$$g \in Sp(n, \mathbb{R}) \Leftrightarrow g^t J g = J,$$

where

$$J = \begin{pmatrix} 0 & I_n \\ -I_n & 0 \end{pmatrix},$$

and I_n is the $n \times n$ identity matrix. $Sp(n, \mathbb{R})$ is a subgroup of $SL_{2n}(\mathbb{R})$, the set of $2n \times 2n$ invertible matrices of determinant 1. Let $g = \begin{pmatrix} A & B \\ C & D \end{pmatrix} \in Sp(n, \mathbb{R})$. The metric ds is invariant under the following action of $Sp(n, \mathbb{R})$,

$$g.Z = (AZ + B)(CZ + D)^{-1}.$$

This action is transitive, i.e.,

$$\forall Z \in \mathcal{H}_n, \exists g \in Sp(n, \mathbb{R}), g.iI = Z.$$

The stabilizer K of iI is the set of elements g of $Sp(n, \mathbb{R})$ whose action leaves iI fixed. K is a subgroup of $Sp(n, \mathbb{R})$ called the isotropy group. We can verify that

$$K = \left\{ \begin{pmatrix} A & B \\ -B & A \end{pmatrix}, A + iB \in SU(n) \right\}.$$

A symmetric space is a Riemannian manifold, where the reversal of the geodesics is well defined and is an isometry. Formally, $exp_p(u) \mapsto exp_p(-u)$ is an isometry for each p on the manifold, where u is a vector in the tangent space at p, and exp_p the Riemannian exponential application at p. In other words, the symmetry around each point is an isometry. $\mathcal{H}_n$ is a symmetric space (see [20]). The structure of a symmetric space can be studied through its isometry group and the Lie algebra of its isometry group. The present work will make use of the Cartan and Iwasawa decompositions of the Lie algebra of $Sp(n,\mathbb{R})$ (see [22]). Let $\mathfrak{sp}(n,\mathbb{R})$ be the Lie algebra of $Sp(n,\mathbb{R})$. Given A, B and C three real $n \times n$ matrices, let denote $\begin{pmatrix} A & B \\ C & -A^t \end{pmatrix} = (A,B,C)$. We have

$$\mathfrak{sp}(n,\mathbb{R}) = \{(A,B,C) | B \text{ and } C \text{ symmetric}\}.$$

The Cartan decomposition of $\mathfrak{sp}(n,\mathbb{R})$ is given by

$$\mathfrak{sp}(n,\mathbb{R}) = \mathfrak{t} \oplus \mathfrak{p},$$

where

$$\mathfrak{t} = \{(A,B,-B) | B \text{ symmetric and } A \text{ skew-symmetric}\},$$

$$\mathfrak{p} = \{(A,B,B) | A,B, \text{ symmetric}\}. \quad (1)$$

The Iwasawa decomposition is given by

$$\mathfrak{sp}(n,\mathbb{R}) = \mathfrak{t} \oplus \mathfrak{a} \oplus \mathfrak{n},$$

where

$$\mathfrak{a} = \{(H,0,0) | H \text{ diagonal}\},$$

$$\mathfrak{n} = \{(A,B,0) | A \text{ upper triangular with 0 on the diagonal}, B \text{ symmetric}\}.$$

It can be shown that

$$\mathfrak{p} = \cup_{k \in K} Ad_k(\mathfrak{a}), \quad (2)$$

where Ad is the adjoint representation of $Sp(n,\mathbb{R})$.

2.2. The Siegel Disk

The Siegel disk $\mathcal{D}_n$ is the set of complex matrices $\{Z | I - Z^*Z \geq 0\}$, where $\geq$ stands for the Loewner order (see [24] for details on the Loewner order). Recall that for A and B two Hermitian matrices, $A \geq B$ with respect to the Loewner order means that $A - B$ is positive definite. The transformation

$$Z \in \mathcal{H}_n \mapsto (Z - iI)(Z + iI)^{-1} \in \mathcal{D}_n$$

is an isometry between the Siegel upper half space and the Siegel disk. Let $C = \begin{pmatrix} I & -iI \\ I & iI \end{pmatrix}$. The application $g \in Sp(n,\mathbb{R}) \mapsto CgC^{-1}$ identifies the set of isometries of $\mathcal{H}_n$ and of $\mathcal{D}_n$. Thus, it can be shown that a matrix $g = \begin{pmatrix} A & B \\ \overline{A} & \overline{B} \end{pmatrix} \in Sp(n,\mathbb{C})$ acts isometrically on $\mathcal{D}_n$ by

$$g.Z = (AZ + B)(\overline{A}Z + \overline{B})^{-1},$$

where $\overline{A}$ stands for the conjugate of A. The point iI in $\mathcal{H}_n$ is identified with the null matrix noted 0 in $\mathcal{D}_n$. Let $Z \in \mathcal{D}_n$. There exists P a diagonal matrix with decreasing positive real entries and U a unitary matrix such that $Z = UPU^t$. Let $\tau_1 \geq ... \geq \tau_n$ be such that

$$P = \begin{pmatrix} th(\tau_1) & & \\ & \ddots & \\ & & th(\tau_n) \end{pmatrix}.$$

Let

$$A_0 = \begin{pmatrix} ch(\tau_1) & & \\ & \ddots & \\ & & ch(\tau_n) \end{pmatrix}, B_0 = \begin{pmatrix} sh(\tau_1) & & \\ & \ddots & \\ & & sh(\tau_n) \end{pmatrix}$$

and

$$g_Z = \begin{pmatrix} U & 0 \\ 0 & \overline{U} \end{pmatrix} \cdot \begin{pmatrix} A_0 & B_0 \\ A_0 & B_0 \end{pmatrix}.$$

It can be shown that

$$g_Z \in Sp(n, \mathbb{C}) \text{ and } g_Z.0 = Z. \tag{3}$$

We provide now a correspondence between the elements of $\mathcal{D}_n$ and the elements of $\mathfrak{p}$ defined in Equation (1). Let

$$H_Z = \begin{pmatrix} \tau_1 & & & & & \\ & \ddots & & & & \\ & & \tau_n & & & \\ & & & -\tau_1 & & \\ & & & & \ddots & \\ & & & & & -\tau_n \end{pmatrix} \in \mathfrak{a}, \tag{4}$$

and

$$a_Z = \begin{pmatrix} e^{\tau_1} & & & & & \\ & \ddots & & & & \\ & & e^{\tau_n} & & & \\ & & & e^{-\tau_1} & & \\ & & & & \ddots & \\ & & & & & e^{-\tau_n} \end{pmatrix} \in A = exp(\mathfrak{a}).$$

It can be shown that there exists $k \in K$ such that

$$Cexp(Ad_k(H_Z))C^{-1}.0 = Z,$$

or equivalently

$$Cka_ZkC^{-1}.0 = Z.$$

Recall that Equation (2) gives $Ad_k(H) \in \mathfrak{p}$ and $kak \in exp(\mathfrak{p})$. The distance between Z and 0 in $\mathcal{D}_n$ is given by

$$d(0, Z) = \left(2\sum \tau_i^2\right)^{1/2} \tag{5}$$

(see p. 292 in [20]).

3. Non Parametric Density Estimation on the Siegel Space

Let Ω be a space, endowed with a σ-algebra and a probability measure p. Let X be a random variable $\Omega \to \mathcal{D}_n$. The Riemannian measure of $\mathcal{D}_n$ is called vol and the measure on $\mathcal{D}_n$ induced by X is noted μ_X. We assume that μ_X has a density, noted f, with respect to vol, and that the support of X is a compact set noted $Supp$. Let $(x_1, ..., x_k) \in \mathcal{D}_n^k$ be a set of draws of X.

The Dirac measure at a point $a \in \mathcal{D}_n$ is denoted δ_a. Let $\mu_k = \frac{1}{k}\sum_{i=1}^k \delta_{x_i}$ denotes the empirical measure of the set of draws. This section presents four non-parametric techniques of estimation of the density f from the set of draws $(x_1, ..., x_k)$. The estimated density at x in $\mathcal{D}_n$ is noted $\hat{f}_k(x) = \hat{f}(x, x_1, ..., x_k)$. The relevance of a density estimation technique depends on several aspects. When the space allows it, the estimation technique should equally consider each direction and location. This leads to an isotropy and a homogeneity condition. In the kernel method, for instance, a kernel density function K_{x_i} is placed at each observation x_i. Firstly, in order to treat directions equally, the function K_{x_i} should be invariant under the isotropy group of x_i; Secondly, for another observation x_j, functions K_{x_i} and K_{x_j} should be similar up to the isometries that send x_i on x_j. These considerations strongly depend on the geometry of the space: if the space is not homogeneous and the isotropy group is empty, these indifference principles have no meaning. Since the Siegel space is symmetric, it is homogeneous and has a non empty isotropy group. Thus, the density estimation technique should be chosen accordingly.

The convergence of the different estimation techniques is widely studied. Results were first obtained in the Euclidean case, and are gradually extended to the probability densities on manifold (see [1,2,15,25]).

The last relevant aspect is computational. Each estimation technique has its own computational framework that presents pros and cons given the different applications. For instance, the estimation by orthogonal series needs an initial pre-processing, but provides a fast evaluation of the estimated density in compact manifolds.

3.1. Histograms

The histogram is the simplest density estimation method. Given a partition of the space $\mathcal{D}_n = \cup_i A_i$, the estimated density is given by

$$\hat{f}(x \in A_i) = \frac{1}{vol(A_i)} \sum_{j=1}^{k} 1_{A_i}(x_j),$$

where 1_{A_i} stands for the indicator function of A_i. Following the considerations of the previous sections, the elements of the partition should firstly be as isotropic as possible, and secondly as similar as possible to each other. Regarding the problem of histograms, the case of the Siegel space is similar to the case of the hyperbolic space. There exist various uniform polygonal tilings on the Siegel space that could be used to compute histograms. However, there are ratio $\lambda \in \mathbb{R}$ for which there is no homothety. Thus, it is not always possible to adapt the size of the bins to a given set of draws of the random variable. Modifying the size of the bins can require a change of the structure of the tiling. This is why the study of histograms has not been deepened.

3.2. Orthogonal Series

The estimation of the density f can be made out of the estimation of the scalar product between f and a set of "orthonormal" functions $\{e_j\}$. The most standard choice for $\{e_j\}$ is the eigenfunctions of the Laplacian. When the variable X takes its values in $\mathbb{R}^n$, this estimation technique becomes the characteristic function method. When the underlying space is compact, the spectrum of the Laplacian operator is countable, while when the space is non-compact, the spectrum is uncountable. In the first case, the estimation of the density f is made through the estimation of a sum, while in the second case is made through the estimation of an integral. In practice, the second situation presents a larger computational complexity. Unfortunately, eigenfunctions of the Laplacian operator are known on $\mathcal{D}_n$ but not on compact sub-domains. This is why the study of this method has not been deepened.

3.3. Kernels

Let $\mathcal{K} : \mathbb{R}_+ \rightarrow \mathbb{R}_+$ be a map which verifies the following properties:

(i) $\int_{\mathbb{R}^d} \mathcal{K}(||x||)dx = 1$;
(ii) $\int_{\mathbb{R}^d} x\mathcal{K}(||x||)dx = 0$;
(iii) $\mathcal{K}(x > 1) = 0$;
(iv) $sup(\mathcal{K}(x)) = \mathcal{K}(0)$.

Let $p \in \mathcal{D}_n$. Generally, given a point p on a Riemannian manifold, exp_p defines an injective application only on a neighborhood of 0. On the Siegel space, exp_p is injective on the whole space. When the tangent space $T_p\mathcal{D}_n$ is endowed with the local scalar product,

$$||u|| = d(p, exp_p(u)),$$

where $||.||$ is the Euclidean distance associated with the local scalar product and $d(.,.)$ is the Riemannian distance. The corresponding Lebesgue measure on $T_p\mathcal{D}_n$ is noted Leb_p. Let $exp_p^*(Leb_p)$ denote the push-forward measure of Lep_p by exp_p. The function θ_p defined by:

$$\theta_p : q \mapsto \theta_p(q) = \frac{dvol}{dexp_p^*(Leb_p)}(q) \tag{6}$$

is the density of the Riemannian measure on $\mathcal{D}_n$ with respect to the Lebesgue measure Leb_p after the identification of $\mathcal{D}_n$ and $T_p\mathcal{D}_n$ induced by exp_p (see Figure 1).

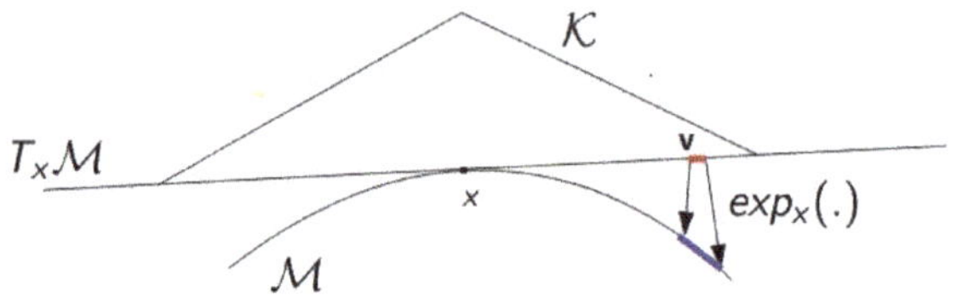

Figure 1. $\mathcal{M}$ is a Riemannian manifold, and $T_x\mathcal{M}$ is its tangent space at x. The exponential application induces a volume change θ_x between $T_x\mathcal{M}$ and $\mathcal{M}$.

Given $\mathcal{K}$ and a positive radius r, the estimator of f proposed by [1] is defined by:

$$\hat{f}_k = \frac{1}{k}\sum_i \frac{1}{r^n}\frac{1}{\theta_{x_i}(x)}\mathcal{K}\left(\frac{d(x,x_i)}{r}\right). \tag{7}$$

The corrective factor $\theta_{x_i}(x)^{-1}$ is necessary since the kernel $\mathcal{K}$ originally integrates to one with respect to the Lebesgue measure and not with respect to the Riemannian measure. It can be noticed that this estimator is the usual kernel estimator in the case of Euclidean space. When the curvature of the space is negative, which is the case of the Siegel space, the distribution placed over each sample x_i has x_i as intrinsic mean. The following theorem provides convergence rate of the estimator. It is a direct adaptation of Theorem 3.1 of [1].

Theorem 1. *Let $(\mathcal{M}, g)$ be a Riemannian manifold of dimension n and μ its Riemannian volume measure. Let X be a random variable taking its values in a compact subset C of $(\mathcal{M}, g)$. Let $0 < r \leq r_{inj}$, where r_{inj} is the infimum of the injectivity radius on C. Assume the law of X has a twice differentiable density f with respect to* *the Riemannian volume measure. Let $\hat{f}_k$ be the estimator defined in Equation (7). There exists a constant C_f such that*

$$\int_{x\in\mathcal{M}} E_{x_1,\ldots,x_k}[(f(x)-\hat{f}_k(x))^2]d\mu \leq C_f(\frac{1}{kr^n}+r^4). \tag{8}$$

If $r \sim k^{\frac{-1}{n+4}}$,

$$\int_{x\in\mathcal{M}} E_{x_1,\ldots,x_k}[(f(x)-\hat{f}_k(x))^2]d\mu = O(k^{-\frac{4}{n+4}}). \tag{9}$$

Proof. See Appendix A. □

It can be checked that on the Siegel space $r_{inj} = +\infty$ and that, for an isometry α, we have:

$$\hat{f}_k(x, x_1, ..., x_k) = \hat{f}_k(\alpha(x), \alpha(x_1), ..., \alpha(x_k)).$$

Each location and direction are processed as similarly as possible. This density estimator can be used for data classification on Riemannian manifolds, see [26].

In order to obtain the explicit expression of the estimator, one must have the explicit expression of the Riemannian exponential, of its inverse, and of the function θ_p (see Equations (6) and (7)). These expressions are difficult and sometimes impossible to obtain for general Riemannian manifolds. In the case of the Siegel space, the symmetric structure makes the computation possible. Since the space is homogeneous, the computation can be made at the origin $iI \in \mathcal{H}_n$ or $0 \in \mathcal{D}_n$ and transported to the whole space. In the present work, the random variable lays in $\mathcal{D}_n$. However, in the literature, the Cartan and Iwasawa decompositions are usually given for the isometry group of $\mathcal{H}_n$. Thus, our computation starts in $\mathcal{H}_n$ before moving to $\mathcal{D}_n$.

The Killing form on the Lie algebra $\mathfrak{sp}(n,\mathbb{R})$ of the isometry group of $\mathcal{H}_n$ induces a scalar product on $\mathfrak{p}$. This scalar product can be transported on $exp(\mathfrak{p})$ by left multiplication. This operation gives $exp(\mathfrak{p})$ a Riemannian structure. It can be shown that on this Riemannian manifold, the Riemannian exponential at the identity coincides with the group exponential. Furthermore,

$$\begin{array}{rcccl} \phi & : & exp(\mathfrak{p}) & \rightarrow & \mathcal{H}_n \\ & & g & \mapsto & g.iI \end{array} \tag{10}$$

is a bijective isometry, up to a scaling factor. Since the volume change θ_p is invariant under rescaling of the metric, this scaling factor has no impact. Thus, $\mathcal{H}_n$ can be identified with $exp(\mathfrak{p})$ and $exp_{iI\in\mathcal{H}_n}$ with $exp_{|\mathfrak{p}}$. The expression of the Riemannian exponential is difficult to obtain in general; however, it boils down to the group exponential in the case of symmetric spaces. This is the main element of the computation of θ_p. The Riemannian volume measure on $exp(\mathfrak{p})$ is noted vol'. Let

$$\begin{array}{rcccl} \psi & : & K\times\mathfrak{a} & \rightarrow & \mathfrak{p} \\ & & (k,H) & \mapsto & Ad_k(H). \end{array}$$

Let $\mathfrak{a}^+$ be the diagonal matrices with strictly decreasing positive eigenvalues. Let Λ^+ be the set of positive roots of $\mathfrak{sp}(n,\mathbb{R})$ as described in p. 282 in [20] ,

$$\Lambda^+ = \{e_i + e_j, i \leq j\} \cup \{e_i - e_j, i < j\},$$

where $e_i(H)$ is the i-th diagonal term of the diagonal'matrix H. Let $C_c(E)$ be the set of continuous compactly supported functions on the space E. In [27], at page 73, it is given that for all $t \in C_c(\mathfrak{p})$, there exists $c_1 > 0$ such that

$$\int_{\mathfrak{p}} t(Y)dY = c_1 \int_K \int_{\mathfrak{a}^+} t(\psi(k,H)) \prod_{\lambda\in\Lambda^+} \lambda(H)dkdH, \tag{11}$$

where dY is a Lebesgue measure on the coefficients of Y. Let $\tilde{\mathfrak{p}} = \psi(K \times \mathfrak{a}^+)$. $\lambda \in \Lambda^+$ never vanishes on $\mathfrak{a}^+$ and $\mathfrak{p} \setminus \tilde{\mathfrak{p}}$ has a null measure. Thus,

$$\int_{\tilde{\mathfrak{p}}} t(Y) \prod_{\lambda \in \Lambda^+} \frac{1}{\lambda(H_Y)} dY = c_1 \int_{K'} \int_{\mathfrak{a}^+} t(Ad_k(H)) dk dH, \tag{12}$$

where H_Y is the point in $\mathfrak{a}^+$ such that there exists k in K such that $\psi(k, H_Y) = Y$. Calculation in p. 73 in [27] also gives that for all $t \in C_c(\mathfrak{p})$, there exists $c_2 > 0$, such that

$$\int_{Sp(n,\mathbb{R})} t(g) dg = c_2 \int_K \int_{\mathfrak{a}^+} \int_K t(k_2.exp(Ad_{k_1}(H))) J(H) dk_1 dH dk_2, \tag{13}$$

where dg is the Haar measure on $Sp(n, \mathbb{R})$ and

$$\begin{aligned} J(H) &= \prod_{\lambda \in \Lambda^+} e^{\lambda(H)} - e^{-\lambda(H)} \\ &= 2^{|\Lambda^+|} \prod_{\lambda \in \Lambda^+} \sinh(\lambda(H)). \end{aligned}$$

Thus, for all $t \in C_c(Sp(n, \mathbb{R})/K)$,

$$\int_{Sp(n,\mathbb{R})/K} t(x) dx = c_2 \int_K \int_{\mathfrak{a}^+} t(exp(Ad_k(H))) J(H) dk dH, \tag{14}$$

where dx is the invariant measure on $Sp(n, \mathbb{R})/K$. After identifying $Sp(n, \mathbb{R})/K$ and $exp(\mathfrak{p})$, the Riemannian measure on $exp(\mathfrak{p})$ coincides with the invariant measure on $Sp(n, \mathbb{R})/K$. Thus, for all $t \in C_c(exp(\mathfrak{p}))$,

$$\int_{exp(\mathfrak{p})} t(x) dvol' = c_2 \int_K \int_{\mathfrak{a}^+} t(exp(Ad_k(H))) J(H) dk dH. \tag{15}$$

Using the notation H_Y of Equation (12),

$$\int_{\tilde{\mathfrak{p}}} t(exp(Y)) J(H_Y) \prod_{\lambda \in \Lambda^+} \frac{1}{\lambda(H_Y)} dY = c_1 \int_K \int_{\mathfrak{a}^+} t(exp(Ad_k(H))) J(H) dk dH. \tag{16}$$

Combining Equations (15) and (16), we obtain that there exists c_3 such that

$$\int_{\tilde{\mathfrak{p}}} t(exp(Y)) \prod_{\lambda \in \Lambda^+} \frac{\sinh(\lambda(H_Y))}{\lambda(H_Y)} dY = c_3 \int_{exp(\mathfrak{p})} t(x) dvol'. \tag{17}$$

The term $\frac{\sinh(\lambda(H))}{\lambda(H)}$ can be extended by continuity on $\mathfrak{a}$; thus,

$$\int_{\mathfrak{p}} t(exp(Y)) \prod_{\lambda \in \Lambda^+} \frac{\sinh(\lambda(H_Y))}{\lambda(H_Y)} dY = c_3 \int_{exp(\mathfrak{p})} t(x) dvol'. \tag{18}$$

Let dY be the Lebesgue measure corresponding to the metric. Then, the exponential application does not introduce a volume change at $0 \in \mathfrak{p}$. Since $H_0 = 0$ and $\frac{\sinh(\lambda(H))}{\lambda(H)} \underset{H \to 0}{\longrightarrow} 1$, we have $c_3 = 1$. Let log denote the inverse of the exponential application. We have

$$\frac{dlog^*(vol')}{dY} = \prod_{\lambda \in \Lambda^+} \frac{\sinh(\lambda(H_Y))}{\lambda(H_Y)}.$$

Since ϕ from Equation (10) is an isometry up to a scaling factor, if $Y \in \mathfrak{p}$ and $C\phi(exp(Y))C^{-1} = exp_0(u \in T_0\mathcal{D}_n)$, then

$$\frac{dlog^*(vol)}{dLeb_0}(u) = \frac{dlog^*(vol')}{dY}(Y),$$

where Leb_0 refers to the Lebesgue measure on the tangent space $T_0\mathcal{D}_n$ as in Equation (6). Given $Z \in \mathcal{D}_n$, H_Z from Equation (4) verifies $C\phi(exp(Ad_k(H_Z)))C^{-1} = Z$ for some k in K. Thus,

$$\theta_0(Z) = \frac{dlog^*(vol')}{dY}(Ad_k(H_Z)) = \prod_{\lambda \in \Lambda^+} \frac{\sinh(\lambda(H_Z))}{\lambda(H_Z)}.$$

We have then

$$\theta_0(Z) = \prod_{i<j} \frac{\sinh(\tau_i - \tau_j)}{\tau_i - \tau_j} \prod_{i \leq j} \frac{\sinh(\tau_i + \tau_j)}{\tau_i + \tau_j},$$

where the (τ_i) are described in Section 2.2. Given $Z_1, Z_2 \in \mathcal{D}_n$,

$$\theta_{Z_1}(Z_2) = \theta_0(g_{Z_1}^{-1}.Z_2),$$

where $g_{Z_1}^{-1}$ is defined in Equation (3). It is thus possible to use the density estimator defined in Equation (7). Indeed,

$$\frac{1}{\theta_{Z_1}(Z_2)} \mathcal{K}\left(\frac{d(Z_1, Z_2)}{r}\right) = \prod_{i<j} \frac{\tau_i - \tau_j}{\sinh(\tau_i - \tau_j)} \prod_{i \leq j} \frac{\tau_i + \tau_j}{\sinh(\tau_i + \tau_j)} \mathcal{K}\left(\frac{(2\sum \tau_i^2)^{1/2}}{r}\right), \tag{19}$$

where the (τ_i) are the diagonal elements of $H_{g_{Z_1}^{-1}.Z_2}$. Recall that when $n = 1$, the Siegel disk corresponds to the Poincaré disk. Thus, we retrieve the expression of the kernel for the hyperbolic space,

$$\frac{1}{\theta_{Z_1}(Z_2)} \mathcal{K}\left(\frac{d(Z_1, Z_2)}{r}\right) = \frac{2\tau}{\sinh(2\tau)} \mathcal{K}\left(\frac{(2\tau^2)^{1/2}}{r}\right). \tag{20}$$

4. Application to Radar Processing

4.1. Radar Data

In space time adaptative radar processing (STAP), the signal is formed by a succession of matrices X representing the realization of a temporal and spatial process. Let $\mathcal{B}_{n,m}^+$ be the set of positive definite block Teoplitz matrices composed of $n \times n$ blocks of $m \times m$ matrices (PD BT). For a stationary signal, the autocorrelation matrix R is PD BT (see [5,6,14]). Authors of [5,6,14] proposed a generalization of Verblunsky coefficients and defined a parametrization of PD BT matrices,

$$\begin{array}{ccc} \mathcal{B}_{n,m}^+ & \to & Sym^+ \times \mathbb{D}_n^{m-1} \\ \mathcal{R} & \mapsto & (P_0, \Omega_1, ..., \Omega_{m-1}), \end{array} \tag{21}$$

in which the metric induced by the Kähler potential is the product metric of an affine invariant metric on Sym^+ and $m-1$ times the metric of the Siegel disk, up to a scaling factor. When the signal is not Gaussian, reflection/Verblunsky coefficients in Poincaré or Siegel Disks should be normalized as described in [28] by a normalized Burg algorithm. Among other references, positive definite block Teoplitz matrices have been studied in the context of STAP-radar processing in [4–6].

4.2. Marginal Densities of Reflection Coefficients

In this section, we show density estimation results of the marginal parameters Ω_k. For the sake of visualization, only the Siegel disk $\mathbb{D}_1$ is considered. Recall that $\mathbb{D}_1$ coincides with the Poincaré disk.

The results are partly extracted from the conference paper [17]. Data used in the experimental tests are radar observations from THALES X-band Radar, recorded during 2014 field trials campaign at Toulouse Blagnac Airport for European FP7 UFO study (Ultra-Fast wind sensOrs for wake-vortex hazards mitigation) (see [29,30]). Data are representative of Turbulent atmosphere monitored by radar. Figure 2 illustrates the density estimation of six coefficients on the Poincaré unit disk under a rainy environment. The densities are individually re-scaled for visualization purposes. For each environment, the dataset is composed of 120 draws. The densities of the coefficients Ω_k are representative of the background. This information on the background is expected to ease the detection of interesting targets.

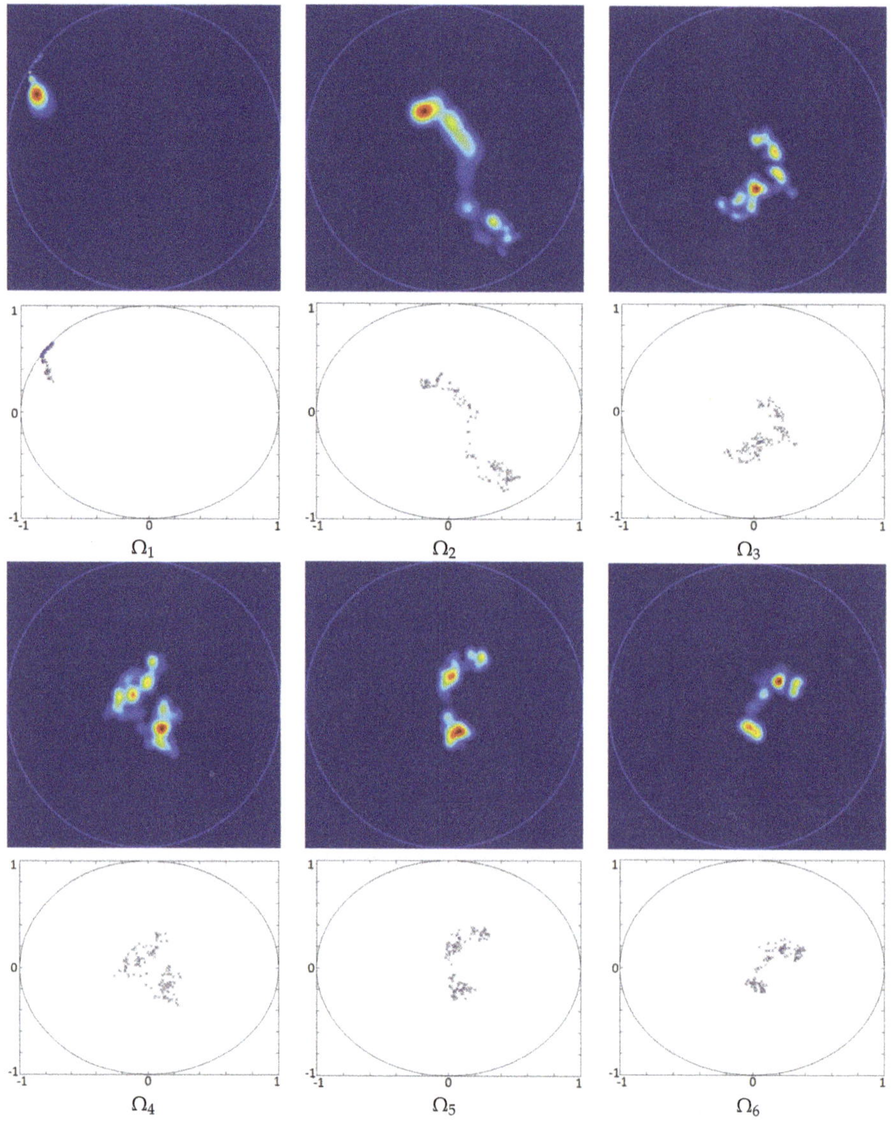

Figure 2. Estimation of the density of six coefficients Ω_k under rainy conditions. The expression of the used kernel is $K(x) = \frac{3}{\pi}(1 - x^2)^2 \mathbf{1}_{x<1}$. Densities are rescaled for visual purposes.

4.3. Radar Clutter Segmentation

Clutter refers to background Doppler signal related to meteorological conditions (e.g., wind in wooded areas, currents and breaking waves on water), which hinders detection of small and slow targets. At each range, a set of reflection coefficients are computed from the Doppler spectrum (see [31]). This set of coefficients is a point in the Poincaré poly-disk. From this set of points in the poly-disk, it is possible to estimate the underlying density. Segmenting clutter, i.e., determining zones of homogeneous Doppler characteristics (see Figure 3), enables the improvement of detection algorithms on each zone. The mean-shift algorithm enables segmentation of the space according to the kernel density estimation of a set of points. It was introduced by Fukunaga and Hostetler in 1975 (see [32]). It corresponds to a gradient ascent of the density estimator (see [33]) for a study of the statistical consistency of the gradient lines estimation. Each data point moves to a local mode of the density estimator, which yields as many clusters as modes. This algorithm has been generalized on manifolds in [34], and applied to radar images in [35]. It can thus be used to segment the set of points in the Poincaré poly-disk. Unfortunately, the mean-shift algorithm requires working with a kernel depending only on the distance to its barycenter, which is not the case of the kernel defined in Equation (19). Thus, the computations are performed without the use of the corrective term θ_p. It is possible to solve this problem by replacing the corrective term by its average at a given radius, which leads to a kernel depending only on the distance to its barycenter. Our future work will focus on the computation of these averages. Let

$$\hat{f}_r^{\mathcal{K}}(x) = \frac{c_d}{k} \sum_{i=1}^{k} \frac{1}{r^n} \mathcal{K}\left(\frac{d(x_i, x)^2}{r^2}\right),$$

where c_n is a normalization constant. Let $g = -k'$.

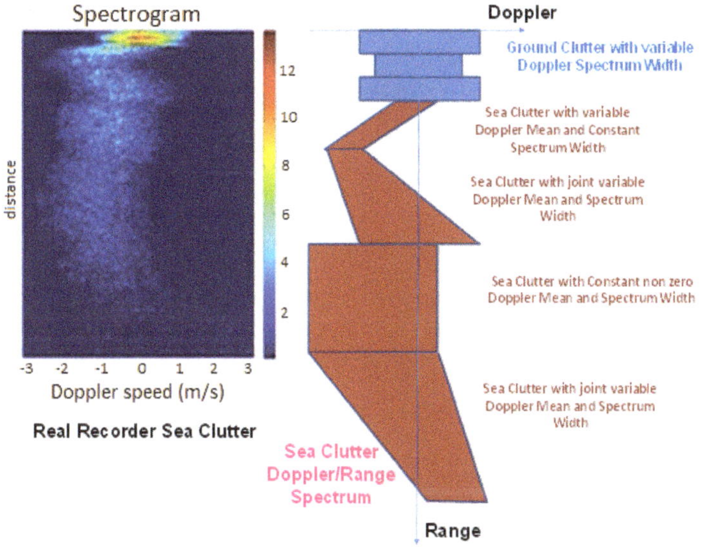

Figure 3. Mean and width variability of sea clutter Doppler spectrum.

The mean-shift is defined by

$$m(x) = \sum_{i=1}^{k} \frac{\frac{1}{r^{n+2}} g\left(\frac{d(x,x_i)^2}{r^2}\right)}{\sum_{i=1}^{k} \frac{1}{r^{n+2}} g\left(\frac{d(x,x_i)^2}{r^2}\right)} log_x(x_i) \propto \frac{\nabla f_r^{\mathcal{K}}}{f_r^{g}},$$

where $m(x)$ is in the tangent space at x. The algorithm moves from x to $exp_x(m(x))$ until convergence to a local maximum. The points of the space are segmented according to the local maxima to which they converge.

In order to assess the quality of unsupervised classification, we use the notion of Silhouette, see [36], which computes for each point a proximity criterion with respect to other points of the same cluster and other points of different clusters (see Figure 4). Let x be in the cluster A. We respectively define $a(x) = min_{y \in A} d(x,y)$ and $b(x) = min_{y \neq A} d(x,y)$, the minimum distance to points of the same (resp. other) class(es). The Silhouette of x is

$$\frac{a(x) - b(x)}{max\{a(x), b(x)\}},$$

which takes values between -1 and 1, respectively, when the data point is considered "badly" and "well" clustered. The average of all the silhouettes provides an indication of the relevance of the classification. One can represent graphically the silhouette profile by plotting for each class horizontal segments of the length of the silhouette value (see Figure 5).

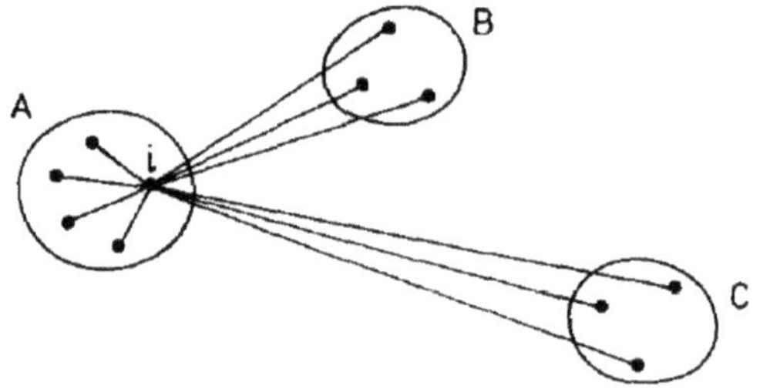

Figure 4. Intra and inter cluster distances.

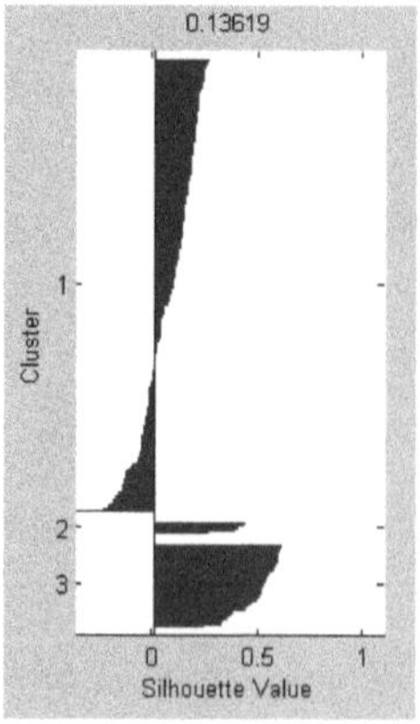

Figure 5. Example of silhouette.

In order to test the Riemannian Mean Shift performance, we generate simple synthetic radar clutter data. Given 250 range cells, we generate 125 cells of ground clutter (wind) centered at 0 $m \cdot s^{-1}$, of spectral width 5 $m \cdot s^{-1}$, to which we add 125 cells of rain clutter, centered at 5 $m \cdot s^{-1}$, of spectral width 10 $m \cdot s^{-1}$. This clutter is sampled 10 times and the segmentation is performed on each simulation (see Figures 6–8).

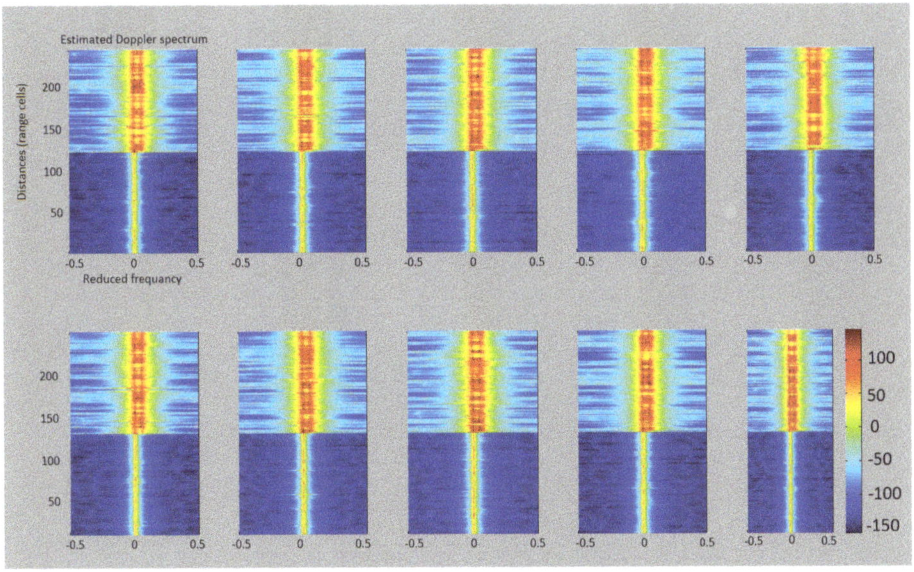

Figure 6. Autoregressive spectra.

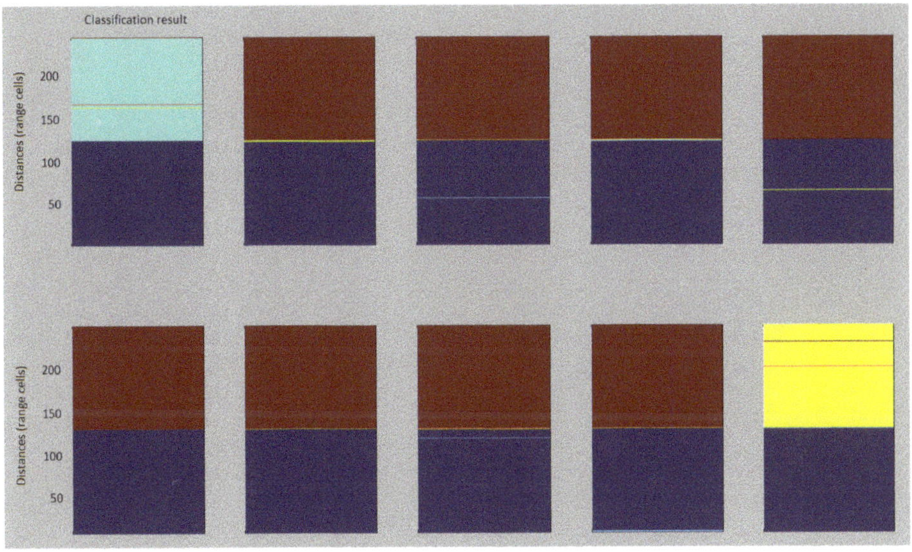

Figure 7. Classification results (one color per cluster).

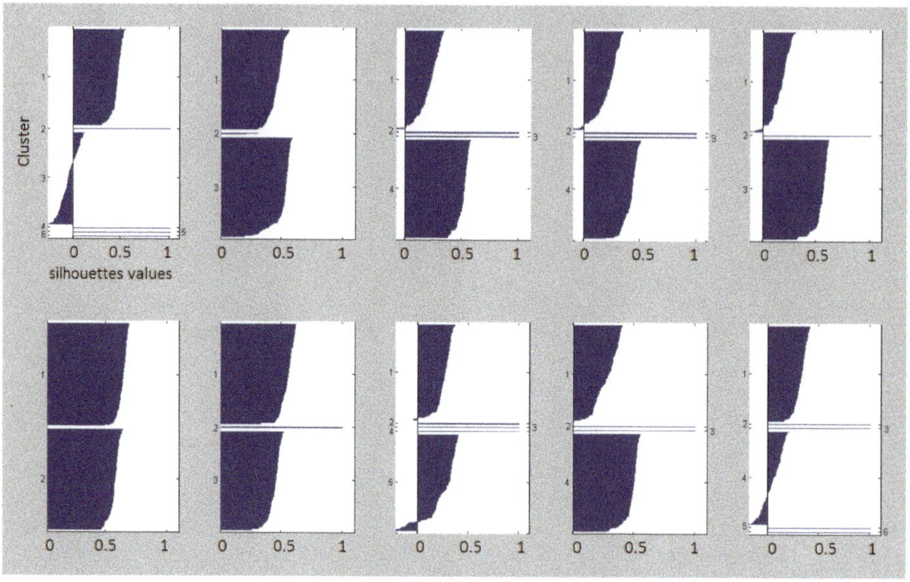

Figure 8. Silhouettes.

It can be seen that, apart from a few outliers, the two clutters are well classified and that the algorithm was able to distinguish between two zones of different Doppler characteristics.

We then test our algorithm on real sea clutter data (see Figures 9–11).

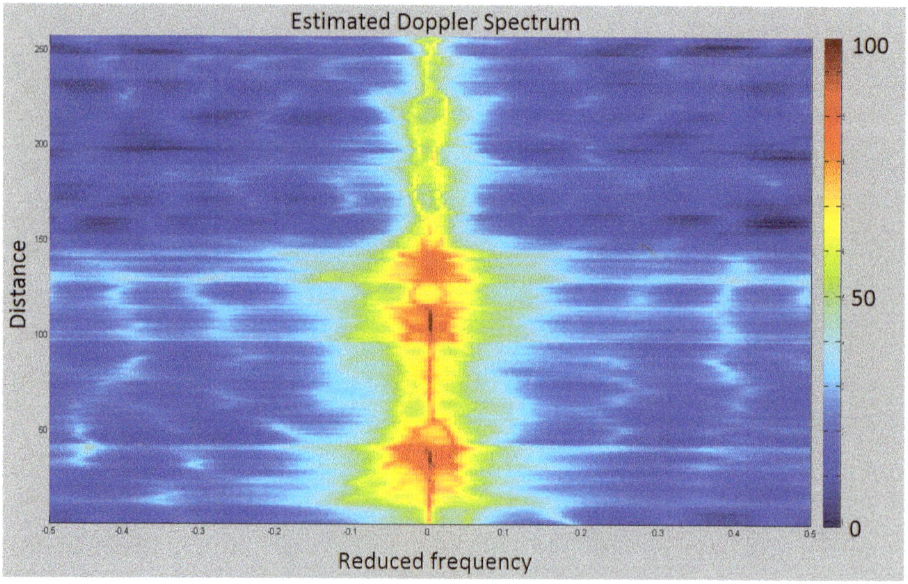

Figure 9. Autoregressive spectrum.

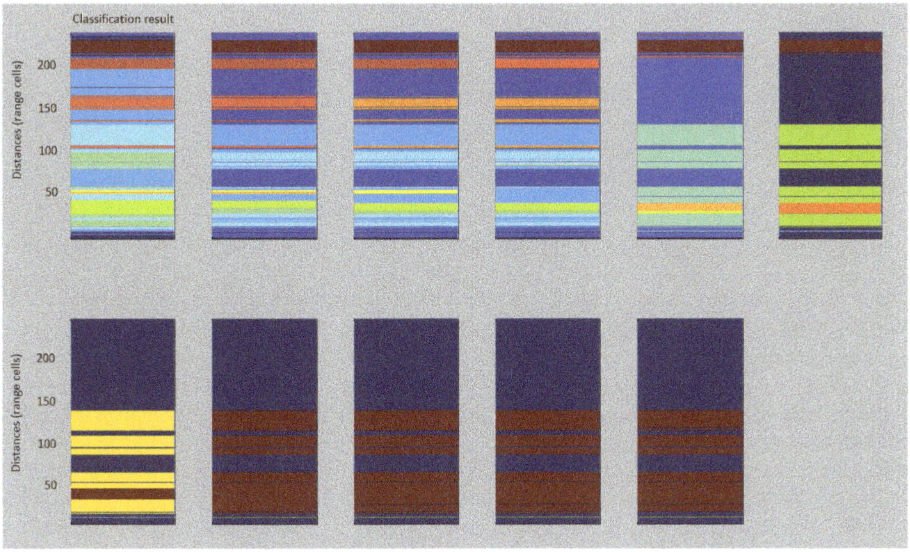

Figure 10. Classification results for varying radii size in the density estimator (10 to 20 closest neighbours).

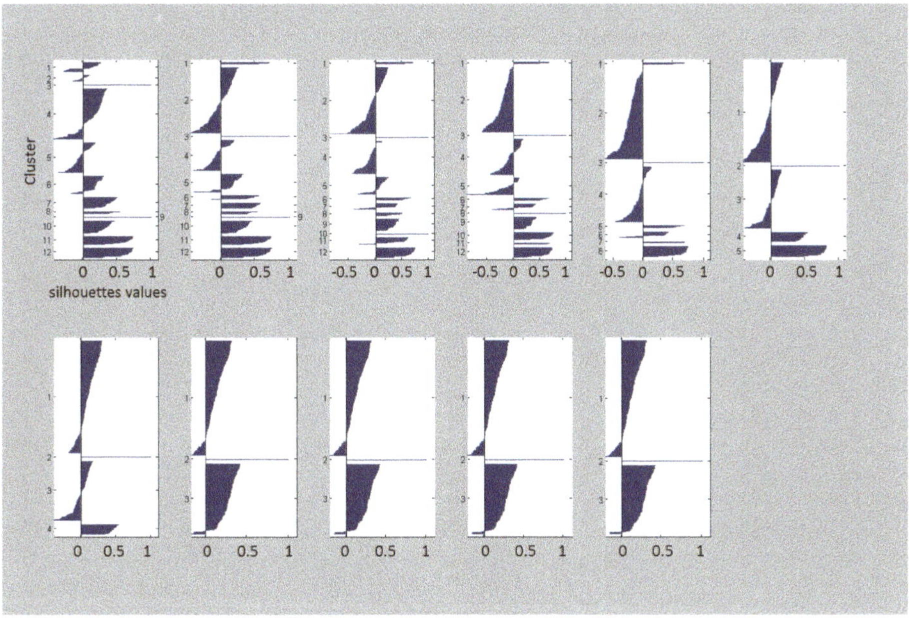

Figure 11. Silhouettes.

The results are more difficult to interpret in that case. The Doppler spectra are varying quite a lot along the range axis. Even though it looks over-segmented, the first classification (kernel size defined by the distance to the 10th closest neighbor point) displays the highest average silhouette value.

5. Conclusions

Three non parametric density estimation techniques have been considered. The main advantage of histograms in the Euclidean context is their simplicity of use. This makes histograms an interesting tool despite the fact that they do not present optimal convergence rates. On the Siegel space, histograms lose their simplicity advantage. They were thus not deeply studied. The orthogonal series density estimation also presents technical disadvantages on the Siegel space. Indeed, the series become integrals, which make the numerical computation of the estimator more difficult than in the Euclidean case. On the other hand, the use of the kernel density estimator does not present major differences with the Euclidean case. The convergence rate obtained in [1] can be extended to compactly supported random variables on non compact Riemannian manifolds. Furthermore, the corrective term whose computation is required to use Euclidean kernels on Riemannian manifolds turns out to have a reasonably simple expression. Our future efforts will concentrate on the use of kernel density estimation on the Siegel space in radar signal processing. As the experimental section suggests, we strongly believe that the estimation of the densities of the Ω_k will provide an interesting description of the different backgrounds. This non-parametric method of density estimation should be compared with parametric ones, as "Maximum Entropy Density" (Gibbs density) on homogenesous manifold as proposed in [37] based on the works of Jean-Marie Souriau. As proposed in [38], a median-shift approach might also be investigated.

Acknowledgments: The authors would like to thank Salem Said, Michal Zidor and Dmitry Gourevitch for the help they provided in the understanding of symmetric spaces and the Siegel space.

Author Contributions: Emmanuel Chevallier carried out the mathematical development. Thibault Forget has set up the Radar clutter segmentation. Frédéric Barbaresco has introduced Poincaré/Siegel Half space and Poincaré/Siegel Disk parameterization for Radar Doppler and Space-Time Adaptive Processing based on Metric spaces deduced from Information Geometry. This parameterization has been re-used in this paper. Jesus Angulo was the Ph.D. supervisor of Emmanuel Chevallier and participates in the supervision of master thesis of Thibault Forget, both thesis are at the origin of this study. All authors have read and approved the final manuscript.

Conflicts of Interest: The authors declare no conflict of interest.

Appendix Demonstration of Theorem 1

Lemma A1. *Let (M, g) be a Riemannian manifold, let C be a compact subset of M and let U be a relatively compact open subset of M containing C. Then, there is a compact Riemannian manifold (M', g') such that U is an open subset of M', the inclusion $i : U \hookrightarrow M'$ is a diffeomorphism onto its image and $g' = g$ on $\overline{U}$.*

Proof. We can assume that M is not compact. Let $f : M \to \mathbb{R}$ be a smooth function on M which tends to $+\infty$ at infinity. Since $\overline{U}$ is compact, $f^{-1}(] - \infty, a[)$ contains $\overline{U}$ for a large enough. By Sard Theorem, there exists a value $a \in \mathbb{R}$ such that $f^{-1}(a)$ contains no critical point of f and such that $f^{-1}(] - \infty, a[)$ contains $\overline{U}$. It follows that $N = f^{-1}(] - \infty, a])$ is a submanifold with boundary of M. Since f tends to $+\infty$ at infinity, N is compact as well as its boundary $\partial N = f^{-1}(\{a\})$.

Call M' the double of N. It is a compact manifold which contains N such that the inclusion $i : N \hookrightarrow M'$ is a diffeomorphism onto its image (see [39], Theorem 5.9 and Definition 5.10). Choose any metric g_0 on M'. Consider two open subsets W_1 and W_2 in M' and two smooth functions $f_1, f_2 : M' \to [0, 1]$ such that

$$\overline{U} \subset W_1 \subset \overline{W_1} \subset W_2 \subset \overline{W_2} \subset \operatorname{int} N,$$

the interior of N,

$$f_1(x) = 1$$

on $\overline{W_1}$, vanishes outside of W_2, and

$$f_2(x) = 1$$

outside W_1, and vanishes in $\overline{U}$. Define g' on M' by

$$g' = f_1 g + f_2 g_0$$

on N and

$$g' = f_2 g_0$$

outside of N. Since $f_1 + f_2 > 0$, g' is positive definite everywhere on M'. Since f_1 vanishes outside of W_2, g' is smooth on M'. Finally, since $f_1 = 1$ and $f_2 = 0$ on $\overline{U}$, $g' = g$ on $\overline{U}$. □

We can now prove Theorem 1. Let X be a random variable as in Theorem 1. Following the notations of the theorem and the lemma, let $U = \{x \in \mathcal{M}, d(x, C) < r_{inj}\}$. U is open, relatively compact and contains C. Let $(\mathcal{M}', g')$ be as in the lemma. Let $\hat{f}$ and $\hat{f}'$ be the kernel density estimators defined on M and M', respectively. Theorem 3.1 of [1] provides the desired results for $\hat{f}'$. For $r \leq r_{inj}$, the support and the values on the support of $\hat{f}'$ and $\hat{f}$ coincide. Thus, the desired result also holds for $\hat{f}$.

References

1. Pelletier, B. Kernel density estimation on Riemannian manifolds. *Stat. Probab. Lett.* **2005**, *73*, 297–304.
2. Hendriks, H. Nonparametric estimation of a probability density on a Riemannian manifold using Fourier expansions. *Ann. Stat.* **1990**, *18*, 832–849.
3. Asta, D.M. Kernel Density Estimation on Symmetric Spaces. In *Geometric Science of Information*; Springer: Berlin/Heidelberg, Germany, 2015; Volume 9389, pp. 779–787.
4. Barbaresco, F. Robust statistical radar processing in Fréchet metric space: OS-HDR-CFAR and OS-STAP processing in siegel homogeneous bounded domains. In Proceedings of the 2011 12th International Radar Symposium (IRS), Leipzig, Germany, 7–9 Septerber 2011.
5. Barbaresco, F. Information Geometry of Covariance Matrix: Cartan-Siegel Homogeneous Bounded Domains, Mostow/Berger Fibration and Fréchet Median. In *Matrix Information Geometry*; Bhatia, R., Nielsen, F., Eds.; Springer: Berlin/Heidelberg, Germany, 2012; pp. 199–256.
6. Barbaresco, F. Information geometry manifold of Toeplitz Hermitian positive definite covariance matrices: Mostow/Berger fibration and Berezin quantization of Cartan-Siegel domains. *Int. J. Emerg. Trends Signal Process.* **2013**, *1*, 1–87.
7. Berezin, F.A. Quantization in complex symmetric spaces. *Izv. Math.* **1975**, *9*, 341–379.
8. Lenz, R. Siegel Descriptors for Image Processing. *IEEE Signal Process. Lett.* **2016**, *25*, 625–628.
9. Barbaresco, F. Robust Median-Based STAP in Inhomogeneous Secondary Data: Frechet Information Geometry of Covariance Matrices. In Proceedings of the 2nd French-Singaporian SONDRA Workshop on EM Modeling, New Concepts and Signal Processing For Radar Detection and Remote Sensing, Cargese, France, 25–28 May 2010.
10. Degurse, J.F.; Savy, L.; Molinie, J.P.; Marcos, S. A Riemannian Approach for Training Data Selection in Space-Time Adaptive Processing Applications. In Proceedings of the 2013 14th International Radar Symposium (IRS), Dresden, Germany, 19–21 June 2013; Volume 1, pp. 319–324.
11. Degurse, J.F.; Savy, L.; Marcos, S. Information Geometry for radar detection in heterogeneous environments. In Proceedings of the 33rd International Workshop on Bayesian Inference and Maximum Entropy Methods in Science and Engineering, Amboise, France, 21–26 September 2014.
12. Barbaresco, F. Koszul Information Geometry and Souriau Geometric Temperature/Capacity of Lie Group Thermodynamics. *Entropy* **2014**, *16*, 4521–4565.
13. Barbaresco, F. New Generation of Statistical Radar Processing based on Geometric Science of Information: Information Geometry, Metric Spaces and Lie Groups Models of Radar Signal Manifolds. In Proceedings of the 4th French-Singaporian Radar Workshop SONDRA, Lacanau, France, 23 May 2016.
14. Jeuris, B.; Vandebril, R. The Kahler mean of Block-Toeplitz matrices with Toeplitz structured block. *SIAM J. Matrix Anal. Appl.* **2015**, *37*, 1151–1175.
15. Huckemann, S.; Kim, P.; Koo, J.; Munk, A. Mobius deconvolution on the hyperbolic plan with application to impedance density estimation. *Ann. Stat.* **2010**, *38*, 2465–2498.
16. Asta, D.; Shalizi, C. Geometric network comparison. 2014, arXiv:1411.1350.

17. Chevallier, E.; Barbaresco, F.; Angulo, J. Probability density estimation on the hyperbolic space applied to radar processing. In *Geometric Science of Information*; Springer: Berlin/Heidelberg, Germany, 2015; pp. 753–761.
18. Said, S.; Bombrun, L.; Berthoumieu, Y. New Riemannian Priors on the Univariate Normal Model. *Entropy* **2014**, *16*, 4015–4031.
19. Said, S.; Hatem, H.; Bombrun, L.; Baba, C.;Vemuri, B.C. Gaussian distributions on Riemannian symmetric spaces: Statistical learning with structured covariance matrices. 2016, arXiv:1607.06929.
20. Terras, A. *Harmonic Analysis on Symmetric Spaces and Applications II*; Springer: Berlin/Heidelberg, Germany, 2012.
21. Siegel, C.L. Symplectic geometry. *Am. J. Math.* **1943**, *65*, doi:10.2307/2371774.
22. Helgason, S. *Differential Geometry, Lie Groups, and Symmetric Spaces*; Academic Press: Cambridge, MA, USA, 1979.
23. Cannon, J.W.; Floyd, W.J.; Kenyon, R.; Parry, W.R. Hyperbolic geometry. In *Flavors of Geometry*; Cambridge University Press: Cambridge, UK, 1997; Volume 31, pp. 59–115.
24. Bhatia, R. Matrix Analysis. In *Graduate Texts in Mathematics-169*; Springer: Berlin/Heidelberg, Germany, 1997.
25. Kim, P.; Richards, D. Deconvolution density estimation on the space of positive definite symmetric matrices. In *Nonparametric Statistics and Mixture Models: A Festschrift in Honor of Thomas P. Hettmansperger*; World Scientific Publishing: Singapore, 2008; pp. 147–168.
26. Loubes, J.-M.; Pelletier, B. A kernel-based classifier on a Riemannian manifold. *Stat. Decis.* **2016**, *26*, 35–51.
27. Gangolli, R.; Varadarajan, V.S. *Harmonic Analysis of Spherical Functions on Real Reductive Groups*; Springer: Berlin/Heidelberg, Germany, 1988.
28. Decurninge, A.; Barbaresco, F. Robust Burg Estimation of Radar Scatter Matrix for Mixtures of Gaussian Stationary Autoregressive Vectors. 2016, arxiv:1601.02804.
29. Barbaresco, F. Eddy Dissipation Rate (EDR) retrieval with ultra-fast high range resolution electronic-scanning X-band airport radar: Results of European FP7 UFO Toulouse Airport trials. In Proceedings of the 2015 16th International Radar Symposium, Dresden, Germany, 24–26 June 2015.
30. Oude Nijhuis, A.C.P.; Thobois, L.P.; Barbaresco, F. *Monitoring of Wind Hazards and Turbulence at Airports with Lidar and Radar Sensors and Mode-S Downlinks: The UFO Project*; Bulletin of the American Meteorological Society, 2016, submitted for publication.
31. Barbaresco, F.; Forget, T.; Chevallier, E.; Angulo, J. Doppler spectrum segmentation of radar sea clutter by mean-shift and information geometry metric. In Proceedings of the 17th International Radar Symposium (IRS), Krakow, Poland, 10–12 May 2016; pp. 1–6.
32. Fukunaga, K.; Hostetler, L.D. The Estimation of the Gradient of a Density Function, with Applications in Pattern Recognition. *Proc. IEEE Trans. Inf. Theory* **1975**, *21*, 32–40.
33. Arias-Castro, E.; Mason, D.; Pelletier, B. On the estimation of the gradient lines of a density and the consistency of the mean-shift algorithm. *J. Mach. Learn. Res.* **2000**, *17*, 1–28.
34. Subbarao, R.; Meer, P. Nonlinear Mean Shift over Riemannian Manifolds. *Int. J. Comput. Vis.* **2009**, *84*, doi:10.1007/s11263-008-0195-8.
35. Wang, Y.H.; Han, C.Z. PolSAR Image Segmentation by Mean Shift Clustering in the Tensor Space. *Acta Autom. Sin.* **2010**, *36*, 798–806.
36. Rousseeuw, P. Silhouettes: A graphical aid to the interpretation and validation of cluster analysis. *J. Comput. Appl. Math.* **1987**, *1*, 53–65.
37. Barbaresco, F. Geometric Theory of Heat from Souriau Lie Groups Thermodynamics and Koszul Hessian Geometry: Applications in Information Geometry for Exponential Families. *Entropy* **2016**, *18*, 386.
38. Wanga, Y.; Huang, X.; Wua, L. Clustering via geometric median shift over Riemannian manifolds. *Inf. Sci.* **2013**, *220*, 292–305.
39. Munkres, J.R. Elementary Differential Topology. In *Annals of Mathematics Studies-54*; Princeton University Press: Princeton, NJ, USA, 1967.

entropy

MDPI

Article

Riemannian Laplace Distribution on the Space of Symmetric Positive Definite Matrices

Hatem Hajri [1,*,†], Ioana Ilea [1,2,†], Salem Said [1,†], Lionel Bombrun [1,†] and Yannick Berthoumieu [1,†]

1 Groupe Signal et Image, CNRS Laboratoire IMS, Institut Polytechnique de Bordeaux, Université de Bordeaux, UMR 5218, Talence 33405, France; ioana.ilea@u-bordeaux.fr (I.I.); salem.said@u-bordeaux.fr (S.S.); lionel.bombrun@u-bordeaux.fr (L.B.); Yannick.Berthoumieu@ims-bordeaux.fr (Y.B.)
2 Communications Department, Technical University of Cluj-Napoca, 71-73 Dorobantilor street, Cluj-Napoca 3400, Romania
* Correspondence: hatem.hajri@ims-bordeaux.fr; Tel.: +33-5-4000-6540
† These authors contributed equally to this work.

Academic Editors: Frédéric Barbaresco and Frank Nielsen
Received: 19 December 2015; Accepted: 8 March 2016; Published: 16 March 2016

Abstract: The Riemannian geometry of the space $\mathcal{P}_m$, of $m \times m$ symmetric positive definite matrices, has provided effective tools to the fields of medical imaging, computer vision and radar signal processing. Still, an open challenge remains, which consists of extending these tools to correctly handle the presence of outliers (or abnormal data), arising from excessive noise or faulty measurements. The present paper tackles this challenge by introducing new probability distributions, called Riemannian Laplace distributions on the space $\mathcal{P}_m$. First, it shows that these distributions provide a statistical foundation for the concept of the Riemannian median, which offers improved robustness in dealing with outliers (in comparison to the more popular concept of the Riemannian center of mass). Second, it describes an original expectation-maximization algorithm, for estimating mixtures of Riemannian Laplace distributions. This algorithm is applied to the problem of texture classification, in computer vision, which is considered in the presence of outliers. It is shown to give significantly better performance with respect to other recently-proposed approaches.

Keywords: symmetric positive definite matrices; Laplace distribution; expectation-maximization; Bayesian information criterion; texture classification

1. Introduction

Data with values in the space $\mathcal{P}_m$, of $m \times m$ symmetric positive definite matrices, play an essential role in many applications, including medical imaging [1,2], computer vision [3–7] and radar signal processing [8,9]. In these applications, the location where a dataset is centered has a special interest. While several definitions of this location are possible, its meaning as a representative of the set should be clear. Perhaps, the most known and well-used quantity to represent a center of a dataset is the Fréchet mean. Given a set of points $Y_1, \cdots, Y_n$ in $\mathcal{P}_m$, their Fréchet mean is defined to be:

$$\text{Mean}(Y_1, \cdots, Y_n) = \text{argmin}_{Y \in \mathcal{P}_m} \sum_{i=1}^{n} d^2(Y, Y_i) \tag{1}$$

where d is Rao's Riemannian distance on $\mathcal{P}_m$ [10,11].

Statistics on general Riemannian manifolds have been powered by the development of different tools for geometric measurements and new probability distributions on manifolds [12,13]. On the manifold $(\mathcal{P}_m, d)$, the major advances in this field have been achieved by the recent papers [14,15], which introduce the Riemannian Gaussian distribution on $(\mathcal{P}_m, d)$. This distribution depends on two parameters $\bar{Y} \in \mathcal{P}_m$ and $\sigma > 0$, and its density with respect to the Riemannian volume form $dv(Y)$ of $\mathcal{P}_m$ (see Formula (13) in Section 2) is:

$$\frac{1}{Z_m(\sigma)} \exp\left[-\frac{d^2(Y,\bar{Y})}{2\sigma^2}\right] \quad (2)$$

where $Z_m(\sigma)$ is a normalizing factor depending only on σ (and not on $\bar{Y}$).

For the Gaussian distribution Equation (2), the maximum likelihood estimate (MLE) for the parameter $\bar{Y}$ based on observations $Y_1, \cdots, Y_n$ corresponds to the mean Equation (1). In [15], a detailed study of statistical inference for this distribution was given and then applied to the classification of data in $\mathcal{P}_m$, showing that it yields better performance, in comparison to recent approaches [2].

When a dataset contains extreme values (or outliers), because of the impact of these values on d^2, the mean becomes less useful. It is usually replaced with the Riemannian median:

$$\text{Median}(Y_1, \cdots, Y_n) = \text{argmin}_{Y \in \mathcal{P}_m} \sum_{i=1}^{n} d(Y, Y_i) \quad (3)$$

Definition Equation (3) corresponds to that of the median in statistics based on ordering of the values of a sequence. However, this interpretation does not continue to hold on $\mathcal{P}_m$. In fact, the Riemannian distance on $\mathcal{P}_m$ is not associated with any norm, and it is therefore only possible to compare distances of a set of matrices to a reference matrix.

In the presence of outliers, the Gaussian distribution on $\mathcal{P}_m$ also loses its robustness properties. The main contribution of the present paper is to remedy this problem by introducing the Riemannian Laplace distribution while maintaining the same one-to-one relation between MLE and the Riemannian median. This will be shown to offer considerable improvement in dealing with outliers.

This paper is organized as follows.

Section 2 reviews the Riemannian geometry of $\mathcal{P}_m$, when this manifold is equipped with the Riemannian metric known as the Rao–Fisher or affine invariant metric [10,11]. In particular, it gives analytic expressions for geodesic curves, Riemannian distance and recalls the invariance of Rao's distance under affine transformations.

Section 3 introduces the Laplace distribution $\mathcal{L}(\bar{Y}, \sigma)$ through its probability density function with respect to the volume form $dv(Y)$:

$$p(Y|\bar{Y}, \sigma) = \frac{1}{\zeta_m(\sigma)} \exp\left[-\frac{d(Y,\bar{Y})}{\sigma}\right]$$

here, σ lies in an interval $]0, \sigma_{\max}[$ with $\sigma_{\max} < \infty$. This is because the normalizing constant $\zeta_m(\sigma)$ becomes infinite for $\sigma \geq \sigma_{\max}$. It will be shown that $\zeta_m(\sigma)$ depend only on σ (and not on $\bar{Y}$) for all $\sigma < \sigma_{\max}$. This important fact leads to simple expressions of MLEs of $\bar{Y}$ and σ. In particular, the MLE of $\bar{Y}$ based on a family of observations $Y_1, \cdots, Y_N$ sampled from $\mathcal{L}(\bar{Y}, \sigma)$ is given by the median of $Y_1, \cdots, Y_N$ defined by Equation (3) where d is Rao's distance.

Section 4 focuses on mixtures of Riemannian Laplace distributions on $\mathcal{P}_m$. A distribution of this kind has a density:

$$p(Y|(\varpi_\mu, \bar{Y}_\mu, \sigma_\mu)_{1 \leq \mu \leq M}) = \sum_{\mu=1}^{M} \varpi_\mu p(Y|\bar{Y}_\mu, \sigma_\mu) \quad (4)$$

with respect to the volume form $dv(Y)$. Here, M is the number of mixture components, $\varpi_\mu > 0$, $\bar{Y}_\mu \in \mathcal{P}_m, \sigma_\mu > 0$ for all $1 \leq \mu \leq M$ and $\sum_{\mu=1}^{M} \varpi_\mu = 1$. A new EM (expectation-maximization) algorithm that computes maximum likelihood estimates of the mixture parameters $(\varpi_\mu, \bar{Y}_\mu, \sigma_\mu)_{1 \leq \mu \leq M}$ is provided. The problem of the order selection of the number M in Equation (4) is also discussed and performed using the Bayesian information criterion (BIC) [16].

Section 5 is an application of the previous material to the classification of data with values in $\mathcal{P}_m$, which contain outliers (abnormal data points). Assume to be given a training sequence $Y_1, \cdots, Y_n \in \mathcal{P}_m$. Using the EM algorithm developed in Section 4, it is possible to subdivide this sequence into disjoint classes. To classify new data points, a classification rule is proposed. The robustness of this rule lies in the fact that it is based on the distances between new observations and the respective medians of classes instead of the means [15]. This rule will be illustrated by an application to the problem of texture classification in computer vision. The obtained results show improved performance with respect to recent approaches which use the Riemannian Gaussian distribution [15] and the Wishart distribution [17].

2. Riemannian Geometry of $\mathcal{P}_m$

The geometry of Siegel homogeneous bounded domains, such as Kähler homogeneous manifolds, have been studied by Felix A. Berezin [18] and P. Malliavin [19]. The structure of Kähler homogeneous manifolds has been used in [20,21] to parameterize (Toeplitz–) Block–Toeplitz matrices. This led to a Hessian metric from information geometry theory with a Kähler potential given by entropy and to an algorithm to compute medians of (Toeplitz–)Block–Toeplitz matrices by Karcher flow on Mostow/Berger fibration of a Siegel disk. Optimal numerical schemes of this algorithm in a Siegel disk have been studied, developed and validated in [22–24].

This section introduces the necessary background on the Riemannian geometry of $\mathcal{P}_m$, the space of symmetric positive definite matrices of size $m \times m$. Precisely, $\mathcal{P}_m$ is equipped with the Riemannian metric known as the affine-invariant metric. First, analytic expressions are recalled for geodesic curves and Riemannian distance. Then, two properties are stated, which are fundamental to the following. These are affine-invariance of the Riemannian distance and the existence and uniqueness of Riemannian medians.

The affine-invariant metric, called the Rao–Fisher metric in information geometry, has the following expression:

$$g_Y(A, B) = \operatorname{tr}(Y^{-1}AY^{-1}B) \tag{5}$$

where $Y \in \mathcal{P}_m$ and $A, B \in T_Y\mathcal{P}_m$, the tangent space to $\mathcal{P}_m$ at Y, which is identified with the vector space of $m \times m$ symmetric matrices. The Riemannian metric Equation (5) induces a Riemannian distance on $\mathcal{P}_m$ as follows. The length of a smooth curve $c : [0,1] \to \mathcal{P}_m$ is given by:

$$L(c) = \int_0^1 \sqrt{g_{c(t)}(\dot{c}(t), \dot{c}(t))}\, dt \tag{6}$$

where $\dot{c}(t) = \frac{dc}{dt}$. For $Y, Z \in \mathcal{P}_m$, the Riemannian distance $d(Y, Z)$, called Rao's distance in information geometry, is defined to be:

$$d(Y, Z) = \inf\{\, L(c), c : [0,1] \to \mathcal{P}_m \text{ is a smooth curve with } c(0) = Y, c(1) = Z\} .$$

This infimum is achieved by a unique curve $c = \gamma$, called the geodesic connecting Y and Z, which has the following equation [10,25]:

$$\gamma(t) = Y^{1/2}\,(Y^{-1/2}ZY^{-1/2})^t\, Y^{1/2} \tag{7}$$

Here, and throughout the following, all matrix functions (for example, square root, logarithm or power) are understood as symmetric matrix functions [26]. By definition, $d(Y, Z)$ coincides with $L(\gamma)$, which turns out to be:

$$d^2(Y, Z) = \operatorname{tr}\,[\log(Y^{-1/2}ZY^{-1/2})]^2 \tag{8}$$

Equipped with the affine-invariant metric Equation (5), the space $\mathcal{P}_m$ enjoys two useful properties, which are the following. The first property is invariance under affine transformations [10,25]. Recall that an affine transformation of $\mathcal{P}_m$ is a mapping $Y \mapsto Y \cdot A$, where A is an invertible real matrix of size $m \times m$,

$$Y \cdot A = A^{\dagger} Y A \tag{9}$$

and † denotes the transpose. Denote by $\mathrm{GL}(m)$ the group of $m \times m$ invertible real matrices on $\mathcal{P}_m$. Then, the action of $\mathrm{GL}(m)$ on $\mathcal{P}_m$ is transitive. This means that for any $Y, Z \in \mathcal{P}_m$, there exists $A \in \mathrm{GL}(m)$, such that $Y.A = Z$. Moreover, the Riemannian distance Equation (8) is invariant by affine transformations in the sense that for all $Y, Z \in \mathcal{P}_m$:

$$d(Y, Z) = d(Y \cdot A, Z \cdot A) \tag{10}$$

where $Y \cdot A$ and $Z \cdot A$ are defined by Equation (9). The transitivity of the action Equation (9) and the isometry property Equation (10) make $\mathcal{P}_m$ a Riemannian homogeneous space.

The affine-invariant metric Equation (5) turns $\mathcal{P}_m$ into a Riemannian manifold of negative sectional curvature [10,27]. As a result, $\mathcal{P}_m$ enjoys the property of the existence and uniqueness of Riemannian medians. The Riemannian median of N points $Y_1, \cdots, Y_N \in \mathcal{P}_m$ is defined to be:

$$\hat{Y}_N = \operatorname{argmin}_Y \sum_{n=1}^{N} d(Y, Y_n) \tag{11}$$

where $d(Y, Y_n)$ is the Riemannian distance Equation (8). If $Y_1, \cdots, Y_N$ do not belong to the same geodesic, then $\hat{Y}_N$ exists and is unique [28]. More generally, for any probability measure π on $\mathcal{P}_m$, the median of π is defined to be:

$$\hat{Y}_\pi = \operatorname{argmin}_Y \int_{\mathcal{P}_m} d(Y, Z) d\pi(Z) \tag{12}$$

Note that Equation (12) reduces to Equation (11) for $\pi = \frac{1}{N} \sum_{n=1}^{N} \delta_{Y_n}$. If the support of π is not carried by a single geodesic, then again, $\hat{Y}_\pi$ exists and is unique by the main result of [28].

To end this section, consider the Riemannian volume associated with the affine-invariant Riemannian metric [10]:

$$dv(Y) = \det(Y)^{-\frac{m+1}{2}} \prod_{i \le j} dY_{ij} \tag{13}$$

where the indices denote matrix elements. The Riemannian volume is used to define the integral of a function $f : \mathcal{P}_m \to \mathbb{R}$ as:

$$\int_{\mathcal{P}_m} f(Y) dv(Y) = \int \ldots \int f(Y) \det(Y)^{-\frac{m+1}{2}} \prod_{i \le j} dY_{ij} \tag{14}$$

where the integral on the right-hand side is a multiple integral over the $m(m+1)/2$ variables, Y_{ij} with $i \le j$. The integral Equation (14) is invariant under affine transformations. Precisely:

$$\int_{\mathcal{P}_m} f(Y \cdot A) dv(Y) = \int_{\mathcal{P}_m} f(Y) dv(Y) \tag{15}$$

where $Y \cdot A$ is the affine transformation given by Equation (9). It takes on a simplified form when $f(Y)$ only depends on the eigenvalues of Y. Precisely, let the spectral decomposition of Y be given by $Y = U^{\dagger} \operatorname{diag}(e^{r_1}, \cdots, e^{r_m})\, U$, where U is an orthogonal matrix and $e^{r_1}, \cdots, e^{r_m}$ are the eigenvalues of Y. Assume that $f(Y) = f(r_1, \ldots, r_m)$, then the invariant integral Equation (14) reduces to:

$$\int_{\mathcal{P}_m} f(Y) dv(Y) = c_m \times \int_{\mathbb{R}^m} f(r_1, \cdots, r_m) \prod_{i<j} \sinh\left(\frac{|r_i - r_j|}{2}\right) dr_1 \cdots dr_m \tag{16}$$

where the constant c_m is given by $c_m = \frac{1}{m!} \times \omega_m \times 8^{\frac{m(m-1)}{4}}, \omega_m = \frac{\pi^{m^2/2}}{\Gamma_m(m/2)}$ and Γ_m is the multivariate gamma function given in [29]. See Appendix A for the derivation of Equation (16) from Equation (14).

3. Riemannian Laplace Distribution on $\mathcal{P}_m$

3.1. Definition of $\mathcal{L}(\bar{Y}, \sigma)$

The Riemannian Laplace distribution on $\mathcal{P}_m$ is defined by analogy with the well-known Laplace distribution on $\mathbb{R}$. Recall the density of the Laplace distribution on $\mathbb{R}$,

$$p(x|\bar{x}, \sigma) = \frac{1}{2\sigma} e^{-|x-\bar{x}|/\sigma}$$

where $\bar{x} \in \mathbb{R}$ and $\sigma > 0$. This is a density with respect to the length element dx on $\mathbb{R}$. The density of the Riemannian Laplace distribution on $\mathcal{P}_m$ will be given by:

$$p(Y|\bar{Y}, \sigma) = \frac{1}{\zeta_m(\sigma)} \exp\left[-\frac{d(Y, \bar{Y})}{\sigma}\right] \tag{17}$$

here, $\bar{Y} \in \mathcal{P}_m$, $\sigma > 0$, and the density is with respect to the Riemannian volume element Equation (13) on $\mathcal{P}_m$. The normalizing factor $\zeta_m(\sigma)$ appearing in Equation (17) is given by the integral:

$$\int_{\mathcal{P}_m} \exp\left[-\frac{d(Y, \bar{Y})}{\sigma}\right] dv(Y)$$

Assume for now that this integral is finite for some choice of $\bar{Y}$ and σ. It is possible to show that its value does not depend on $\bar{Y}$. To do so, recall that the action of $\mathrm{GL}(m)$ on $\mathcal{P}_m$ is transitive. As a consequence, there exists $A \in \mathcal{P}_m$, such that $\bar{Y} = I.A$, where $I.A$ is defined as in Equation (9). From Equation (10), it follows that $d(Y, \bar{Y}) = d(Y, I.A) = d(Y.A^{-1}, I)$. From the invariance property Equation (15):

$$\int_{\mathcal{P}_m} \exp\left[-\frac{d(Y, \bar{Y})}{\sigma}\right] dv(Y) = \int_{\mathcal{P}_m} \exp\left[-\frac{d(Y, I)}{\sigma}\right] dv(Y) \tag{18}$$

The integral on the right does not depend on $\bar{Y}$, which proves the above claim.
The last integral representation and formula Equation (16) lead to the following explicit expression:

$$\zeta_m(\sigma) = c_m \times \int_{\mathbb{R}^m} e^{-\frac{|r|}{\sigma}} \prod_{i<j} \sinh\left(\frac{|r_i - r_j|}{2}\right) dr_1 \cdots dr_m \tag{19}$$

where $|r| = (r_1^2 + \cdots + r_2^m)^{\frac{1}{2}}$ and c_m is the same constant as in Equation (16) (see Appendix B for more details on the derivation of Equation (19)).

A distinctive feature of the Riemannian Laplace distribution on $\mathcal{P}_m$, in comparison to the Laplace distribution on $\mathbb{R}$ is that there exist certain values of σ for which it cannot be defined. This is because the integral Equation (19) diverges for certain values of this parameter. This leads to the following definition.

Definition 1. *Set $\sigma_m = \sup\{\sigma > 0 : \zeta_m(\sigma) < \infty\}$. Then, for $\bar{Y} \in \mathcal{P}_m$ and $\sigma \in (0, \sigma_m)$, the Riemannian Laplace distribution on $\mathcal{P}_m$, denoted by $\mathcal{L}(\bar{Y}, \sigma)$, is defined as the probability distribution on $\mathcal{P}_m$, whose density with respect to $dv(Y)$ is given by Equation (17), where $\zeta_m(\sigma)$ is defined by Equation (19).*

The constant σ_m in this definition satisfies $0 < \sigma_m < \infty$ for all m and takes the value $\sqrt{2}$ for $m = 2$ (see Appendix C for proofs).

3.2. Sampling from $\mathcal{L}(\bar{Y},\sigma)$

The current section presents a general method for sampling from the Laplace distribution $\mathcal{L}(\bar{Y},\sigma)$. This method relies in part on the following transformation property.

Proposition 1. *Let Y be a random variable in $\mathcal{P}_m$. For all $A \in \mathrm{GL}(m)$,*

$$Y \sim \mathcal{L}(\bar{Y},\sigma) \implies Y \cdot A \sim \mathcal{L}(\bar{Y} \cdot A,\sigma)$$

where $Y \cdot A$ is given by Equation (9).

Proof. Let $\varphi : \mathcal{P}_m \to \mathbb{R}$ be a test function. If $Y \sim \mathcal{L}(\bar{Y},\sigma)$ and $Z = Y \cdot A$, then the expectation of $\varphi(Z)$ is given by:

$$E[\varphi(Z)] = \int_{\mathcal{P}_m} \varphi(X \cdot A)\, p(X|\,\bar{Y},\sigma)\, dv(X) \;=\; \int_{\mathcal{P}_m} \varphi(X)\, p(X \cdot A^{-1}|\,\bar{Y},\sigma)\, dv(X)$$

where the equality is a result of Equation (15). However, $p(X \cdot A^{-1}|\,\bar{Y},\sigma) = p(X|\,\bar{Y} \cdot A,\sigma)$ by Equation (10), which proves the proposition.
□

The following algorithm describes how to sample from $\mathcal{L}(\bar{Y},\sigma)$ where $0 < \sigma < \sigma_m$. For this, it is first required to sample from the density p on $\mathbb{R}^m$ defined by:

$$p(r) = \frac{c_m}{\zeta_m(\sigma)} e^{-\frac{|r|}{\sigma}} \prod_{i<j} \sinh\left(\frac{|r_i - r_j|}{2}\right), \quad r = (r_1, \cdots, r_m).$$

This can be done by a usual Metropolis algorithm [30].

It is also required to sample from the uniform distribution on $\mathrm{O}(m)$, the group of real orthogonal $m \times m$ matrices. This can be done by generating A, an $m \times m$ matrix, whose entries are i.i.d. with normal distribution $\mathcal{N}(0,1)$, then the orthogonal matrix U, in the decomposition $A = UT$ with T upper triangular, is uniformly distributed on $\mathrm{O}(m)$ [29] (p. 70). Sampling from $\mathcal{L}(\bar{Y},\sigma)$ can now be described as follows.

Algorithm 1 Sampling from $\mathcal{L}(\bar{Y},\sigma)$.

1: Generate i.i.d. samples $(r_1, \cdots, r_m) \in \mathbb{R}^m$ with density p
2: Generate U from a uniform distribution on $\mathrm{O}(m)$
3: $X \leftarrow U^{\dagger}\mathrm{diag}(e^{r_1}, \cdots, e^{r_m})U$
4: $Y \leftarrow X.\bar{Y}^{\frac{1}{2}}$

Note that the law of X in Step 3 is $\mathcal{L}(I,\sigma)$; the proof of this fact is given in Appendix D. Finally, the law of Y in Step 4 is $\mathcal{L}(I.\bar{Y}^{\frac{1}{2}} = \bar{Y},\sigma)$ by proposition Equation (1).

3.3. Estimation of $\bar{Y}$ and σ

The current section considers maximum likelihood estimation of the parameters $\bar{Y}$ and σ, based on independent observations $Y_1, \ldots, Y_N$ from the Riemannian Laplace distribution $\mathcal{L}(\bar{Y},\sigma)$. The main results are contained in Propositions 2 and 3 below.

Proposition 2 states the existence and uniqueness of the maximum likelihood estimates $\hat{Y}_N$ and $\hat{\sigma}_N$ of $\bar{Y}$ and σ. In particular, the maximum likelihood estimate $\hat{Y}_N$ of $\bar{Y}$ is the Riemannian median of $Y_1, \ldots, Y_N$, defined by Equation (11). Numerical computation of $\hat{Y}_N$ will be considered and carried out using a Riemannian sub-gradient descent algorithm [8].

Proposition 3 states the convergence of the maximum likelihood estimate $\hat{Y}_N$ to the true value of the parameter $\bar{Y}$. It is based on Lemma 1, which states that the parameter $\bar{Y}$ is the Riemannian median of the distribution $\mathcal{L}(\bar{Y}, \sigma)$ in the sense of definition Equation (12).

Proposition 2 (MLE and median)**.** *The maximum likelihood estimate of the parameter $\bar{Y}$ is the Riemannian median $\hat{Y}_N$ of $Y_1, \ldots, Y_N$. Moreover, the maximum likelihood estimate of the parameter σ is the solution $\hat{\sigma}_N$ of:*

$$\sigma^2 \times \frac{d}{d\sigma} \log \zeta_m(\sigma) = \frac{1}{N} \sum_{n=1}^{N} d(\bar{Y}, Y_n) \tag{20}$$

Both $\hat{Y}_N$ and $\hat{\sigma}_N$ exist and are unique for any realization of the samples $Y_1, \ldots, Y_N$.

Proof of Proposition 2. The log-likelihood function, of the parameters $\bar{Y}$ and σ, can be written as:

$$\begin{aligned} \sum_{n=1}^{N} \log p(Y_n | \bar{Y}, \sigma) &= \sum_{n=1}^{N} \log \left(\frac{1}{\zeta_m(\sigma)} e^{-\frac{d(\bar{Y}, Y_n)}{\sigma}} \right) \\ &= -N \log \zeta_m(\sigma) - \frac{1}{\sigma} \sum_{n=1}^{N} d(\bar{Y}, Y_n) \end{aligned}$$

As the first term in the last expression does not contain $\bar{Y}$,

$$\text{argmax}_{\bar{Y}} \sum_{n=1}^{N} \log p(Y_n | \bar{Y}, \sigma) = \text{argmin}_{\bar{Y}} \sum_{n=1}^{N} d(\bar{Y}, Y_n)$$

The quantity on the right is exactly $\hat{Y}_N$ by Equation (11). This proves the first claim. Now, consider the function:

$$F(\eta) = -N \log(\zeta_m(\frac{-1}{\eta})) + \eta \sum_{n=1}^{N} d(\hat{Y}_N, Y_n), \quad \eta < \frac{-1}{\sigma_m}$$

This function is strictly concave, since it is the logarithm of the moment generating function of a positive measure. Note that $\lim_{\eta \to \frac{-1}{\sigma_m}} F(\eta) = -\infty$, and admit for a moment that $\lim_{\eta \to -\infty} F(\eta) = -\infty$. By the strict concavity of F, there exists a unique $\hat{\eta}_N < \frac{-1}{\sigma_m}$ (which is the maximum of F), such that $F'(\hat{\eta}_N) = 0$. It follows that $\hat{\sigma}_N = \frac{-1}{\hat{\eta}_N}$ lies in $(0, \sigma_m)$ and satisfies Equation (20). The uniqueness of $\hat{\sigma}_N$ is a consequence of the uniqueness of $\hat{\eta}_N$. Thus, the proof is complete. Now, it remains to check that $\lim_{\eta \to -\infty} F(\eta) = -\infty$ or just $\lim_{\sigma \to +\infty} \frac{1}{\sigma} \log(\zeta_m(\frac{1}{\sigma})) = 0$. Clearly:

$$\prod_{i<j} \sinh \left(\frac{|r_i - r_j|}{2} \right) \leq A_m e^{B_m |r|}$$

where A_m and B_m are two constants only depending on m. Using this, it follows that:

$$\frac{1}{\sigma} \log(\zeta_m(\frac{1}{\sigma})) \leq \frac{1}{\sigma} \log(c_m A_m) + \frac{1}{\sigma} \log \left(\int_{\mathbb{R}^m} \exp((-\sigma + B_m)|r|) dr_1 \cdots dr_m \right) \tag{21}$$

However, for some constant C_m only depending on m,

$$\int_{\mathbb{R}^m} \exp((-\sigma + B_m)|r|) dr_1 \cdots dr_m = C_m \int_0^{\infty} \exp((-\sigma + B_m)u) u^{m-1} du$$

$$\leq (m-1)! C_m \int_0^{\infty} \exp((-\sigma + B_m + 1)u) du = \frac{(m-1)! C_m}{\sigma - B_m - 1}$$

Combining this bound and Equation (21) yields $\lim_{\sigma \to +\infty} \frac{1}{\sigma} \log(\zeta_m(\frac{1}{\sigma})) = 0$. □

Remark 1. *Replacing F in the previous proof with* $F(\eta) = -\log(\zeta_m(\frac{-1}{\eta})) + \eta c$ *where* $c > 0$ *shows that the equation:*

$$\sigma^2 \times \frac{d}{d\sigma} \log \zeta_m(\sigma) = c$$

has a unique solution $\sigma \in (0, \sigma_m)$. *This shows in particular that* $\sigma \mapsto \sigma^2 \times \frac{d}{d\sigma} \log \zeta_m(\sigma)$ *is a bijection from* $(0, \sigma_m)$ *to* $(0, \infty)$.

Consider now the numerical computation of the maximum likelihood estimates $\hat{Y}_N$ and $\hat{\sigma}_N$ given by Proposition 2. Computation of $\hat{Y}_N$ consists in finding the Riemannian median of $Y_1, \ldots, Y_N$, defined by Equation (11). This can be done using the Riemannian sub-gradient descent algorithm of [8]. The k-th iteration of this algorithm produces an approximation $\hat{Y}_N^k$ of $\hat{Y}_N$ in the following way.

For $k = 1, 2, \ldots$, let Δ_k be the symmetric matrix:

$$\Delta_k = \frac{1}{N} \sum_{n=1}^{N} \frac{\mathrm{Log}_{\hat{Y}_N^{k-1}}(Y_n)}{||\mathrm{Log}_{\hat{Y}_N^{k-1}}(Y_n)||} \tag{22}$$

Here, Log is the Riemannian logarithm mapping inverse to the the Riemannian exponential mapping:

$$\mathrm{Exp}_Y(\Delta) = Y^{1/2} \exp\left(Y^{-1/2}\, \Delta\, Y^{-1/2}\right) Y^{1/2} \tag{23}$$

and $||\mathrm{Log}_a(b)|| = \sqrt{g_a(b, b)}$. Then, $\hat{Y}_N^k$ is defined to be:

$$\hat{Y}_N^k = \mathrm{Exp}_{\hat{Y}_N^{k-1}}(\tau_k \Delta_k) \tag{24}$$

where $\tau_k > 0$ is a step size, which can be determined using a backtracking procedure.

Computation of $\hat{\sigma}_N$ requires solving a non-linear equation in one variable. This is readily done using Newton's method.

It is shown now that the empirical Riemannian median $\hat{Y}_N$ converges almost surely to the true median $\bar{Y}$. This means that $\hat{Y}_N$ is a consistent estimator of $\bar{Y}$. The proof of this fact requires few notations and a preparatory lemma.

For $\bar{Y} \in \mathcal{P}_m$ and $\sigma \in (0, \sigma_m)$, let:

$$\mathcal{E}(Y|\,\bar{Y}, \sigma) = \int_{\mathcal{P}_m} d(Y, Z)\, p(Z|\,\bar{Y}, \sigma) dv(Z)$$

The following lemma shows how to find $\bar{Y}$ and σ from the function $Y \mapsto \mathcal{E}(Y|\,\bar{Y}, \sigma)$.

Lemma 1. *For any* $\bar{Y} \in \mathcal{P}_m$ *and* $\sigma \in (0, \sigma_m)$, *the following properties hold*

(i) $\bar{Y}$ *is given by:*

$$\bar{Y} = \mathrm{argmin}_Y\, \mathcal{E}(Y|\,\bar{Y}, \sigma) \tag{25a}$$

That is, $\bar{Y}$ *is the Riemannian median of* $\mathcal{L}(\bar{Y}, \sigma)$.

(ii) σ *is given by:*

$$\sigma = \Phi\left(\mathcal{E}(\bar{Y}|\,\bar{Y}, \sigma)\right) \tag{25b}$$

where the function Φ *is the inverse function of* $\sigma \mapsto \sigma^2 \times d \log \zeta_m(\sigma)/d\sigma$.

Proof of Lemma 1. (i) Let $\mathcal{E}(Y) = \mathcal{E}(Y|\,\bar{Y}, \sigma)$. According to Theorem 2.1 in [28], this function has a unique global minimum, which is also a unique stationary point. Thus, to prove that $\bar{Y}$ is the minimum point of $\mathcal{E}$, it will suffice to check that for any geodesic γ starting from $\bar{Y}$, $\frac{d}{dt}|_{t=0}\mathcal{E}(\gamma(t)) = 0$ [31] (p. 76). Note that:

$$\frac{d}{dt}|_{t=0}\mathcal{E}(\gamma(t)) = \int_{\mathcal{P}_m} \frac{d}{dt}|_{t=0} d(\gamma(t), Z)\, p(Z|\, \bar{Y}, \sigma) dv(Z) \tag{26}$$

where for all $Z \neq \bar{Y}$ [32]:

$$\frac{d}{dt}|_{t=0} d(\gamma(t), Z) = -g_{\bar{Y}}(\log_{\bar{Y}}(Z), \gamma'(0)) d(\bar{Y}, Z)^{-1}$$

The integral in Equation (26) is, up to a constant,

$$\frac{d}{dt}|_{t=0} \int_{\mathcal{P}_m} p(Z|\, \gamma(t), \sigma) dv(Z) = 0$$

since $\int_{\mathcal{P}_m} p(Z|\, \gamma(t), \sigma) dv(Z) = 1$.

(ii) Differentiating $\int_{\mathcal{P}_m} \exp(-\frac{d(Z,\bar{Y})}{\sigma}) dv(Z) = \zeta_m(\sigma)$ with respect to σ, it comes that:

$$\sigma^2 \times d \log \zeta_m(\sigma)/d\sigma = \sigma^2 \frac{\zeta_m'(\sigma)}{\zeta_m(\sigma)} = \int_{\mathcal{P}_m} d(Z, \bar{Y}) p(Z|\bar{Y}, \sigma) dv(Z) = \mathcal{E}(\bar{Y}|\, \bar{Y}, \sigma)$$

which proves (ii). □

Proposition 3 (Consistency of $\hat{Y}_N$). *Let $Y_1, Y_2, \cdots$ be independent samples from a Laplace distribution $G(\bar{Y}, \sigma)$. The empirical median $\hat{Y}_N$ of $Y_1, \ldots, Y_N$ converges almost surely to $\bar{Y}$, as $N \to \infty$.*

Proof of Proposition 3. Corollary 3.5 in [33] (p. 49) states that if (Y_n) is a sequence of i.i.d. random variables on $\mathcal{P}_m$ with law π, then the Riemannian median $\hat{Y}_N$ of $Y_1, \cdots, Y_N$ converges almost surely as $N \to \infty$ to $\hat{Y}_\pi$, the Riemannian median of π defined by Equation (12). Applying this result to $\pi = \mathcal{L}(\bar{Y}, \sigma)$ and using $\hat{Y}_\pi = \bar{Y}$, which follows from item (i) of Lemma 1, shows that $\hat{Y}_N$ converges almost surely to $\bar{Y}$. □

4. Mixtures of Laplace Distributions

There are several motivations for considering mixtures of distributions in general. The most natural approach is to envisage a dataset as constituted of several subpopulations. Another approach is the fact that there is a support for the argument that mixtures of distributions provide a good approximation to most distributions in a spirit similar to wavelets.

The present section introduces the class of probability distributions that are finite mixtures of Riemannian Laplace distributions on $\mathcal{P}_m$. These constitute the main theoretical tool, to be used for the target application of the present paper, namely the problem of texture classification in computer vision, which will be treated in Section 5.

A mixture of Riemannian Laplace distributions is a probability distribution on $\mathcal{P}_m$, whose density with respect to the Riemannian volume element Equation (13) has the following expression:

$$p(Y|(\varpi_\mu, \bar{Y}_\mu, \sigma_\mu)_{1\leq\mu\leq M}) = \sum_{\mu=1}^{M} \varpi_\mu \times p(Y|\, \bar{Y}_\mu, \sigma_\mu) \tag{27}$$

where ϖ_μ are nonzero weights, whose sum is equal to one, $\bar{Y}_\mu \in \mathcal{P}_m$ and $\sigma_\mu \in (0, \sigma_m)$ for all $1 \leq \mu \leq M$, and the parameter M is called the number of mixture components.

Section 4.1 describes a new EM algorithm, which computes the maximum likelihood estimates of the mixture parameters $(\varpi_\mu, \bar{Y}_\mu, \sigma_\mu)_{1\leq\mu\leq M}$, based on independent observations $Y_1, \ldots, Y_N$ from the mixture distribution Equation (27).

Section 4.2 considers the problem of order selection for mixtures of Riemannian Laplace distributions. Precisely, this consists of finding the number M of mixture components in Equation (27) that realizes the best representation of a given set of data $Y_1, \ldots, Y_N$. This problem is solved by computing the BIC criterion, which is here found in explicit form for the case of mixtures of Riemannian Laplace distributions on $\mathcal{P}_m$.

4.1. Estimation of the Mixture Parameters

In this section, $Y_1, \ldots, Y_N$ are i.i.d. samples from Equation (27). Based on these observations, an EM algorithm is proposed to estimate $(\varpi_\mu, \bar{Y}_\mu, \sigma_\mu)_{1\leq\mu\leq M}$. The derivation of this algorithm can be carried out similarly to [15].

To explain how this algorithm works, define for all $\vartheta = \{(\varpi_\mu, \bar{Y}_\mu, \sigma_\mu)\}$,

$$\omega_\mu(Y_n, \vartheta) = \frac{\varpi_\mu \times p(Y_n | \bar{Y}_\mu, \sigma_\mu)}{\sum_{s=1}^{M} \varpi_s \times p(Y_n | \bar{Y}_s, \sigma_s)}, \qquad N_\mu(\vartheta) = \sum_{n=1}^{N} \omega_\mu(Y_n) \tag{28}$$

The algorithm iteratively updates $\hat{\vartheta} = \{(\hat{\varpi}_\mu, \hat{Y}_\mu, \hat{\sigma}_\mu)\}$, which is an approximation of the maximum likelihood estimate of the mixture parameters $\vartheta = (\varpi_\mu, \bar{Y}_\mu, \sigma_\mu)$ as follows.

- Update for $\hat{\varpi}_\mu$: Based on the current value of $\hat{\vartheta}$, assign to $\hat{\varpi}_\mu$ the new value $\hat{\varpi}_\mu = N_\mu(\hat{\vartheta})/N$.
- Update for $\hat{Y}_\mu$: Based on the current value of $\hat{\vartheta}$, assign to $\hat{Y}_\mu$ the value:

$$\hat{Y}_\mu = \text{argmin}_Y \sum_{n=1}^{N} \omega_\mu(Y_n, \hat{\vartheta})\, d(Y, Y_n) \tag{29}$$

- Update for $\hat{\sigma}_\mu$: Based on the current value of $\hat{\vartheta}$, assign to $\hat{\sigma}_\mu$ the new value:

$$\hat{\sigma}_\mu = \Phi\left(N_\mu^{-1}(\hat{\vartheta}) \times \textstyle\sum_{n=1}^{N} \omega_\mu(Y_n, \hat{\vartheta})\, d(\hat{Y}_\mu, Y_n)\right) \tag{30}$$

where the function Φ is defined in Proposition 1.

These three update rules should be performed in the above order. Realization of the update rules for $\hat{\varpi}_\mu$ and $\hat{\sigma}_\mu$ is straightforward. The update rule for $\hat{Y}_\mu$ is realized using a slight modification of the sub-gradient descent algorithm described in Section 3.2. More precisely, the factor $1/N$ appearing in Equation (22) is only replaced with $\omega_\mu(Y_n, \hat{\vartheta})$ at each iteration.

In practice, the initial conditions $(\hat{\varpi}_{\mu_0}, \hat{Y}_{\mu_0}, \hat{\sigma}_{\mu_0})$ in this algorithm were chosen in the following way. The weights (ϖ_{μ_0}) are uniform and equal to $1/M$; $(\hat{Y}_{\mu_0})$ are M different observations from the set $\{Y_1, .., Y_N\}$ chosen randomly; and $(\hat{\sigma}_{\mu_0})$ is computed from (ϖ_{μ_0}) and $(\hat{Y}_{\mu_0})$ according to the rule Equation (30). Since the convergence of the algorithm depends on the initial conditions, the EM algorithm is run several times, and the best result is retained, *i.e.*, the one maximizing the log-likelihood function.

4.2. The Bayesian Information Criterion

The BIC was introduced by Schwarz to find the appropriate dimension of a model that will fit a given set of observations [16]. Since then, BIC has been used in many Bayesian modeling problems where priors are hard to set precisely. In large sample settings, the fitted model favored by BIC ideally corresponds to the candidate model that is *a posteriori* most probable; *i.e.*, the model that is rendered most plausible by the data at hand. One of the main features of the BIC is its easy computation, since it is only based on the empirical log-likelihood function.

Given a set of observations $\{Y_1, \cdots, Y_N\}$ arising from Equation (27) where M is unknown, the BIC consists of choosing the parameter:

$$\bar{M} = \text{argmax}_M BIC(M)$$

where:

$$BIC(M) = LL - \frac{1}{2} \times DF \times \log(N) \tag{31}$$

Here, LL is the log-likelihood given by:

$$LL = \sum_{n=1}^{N} \log \left(\sum_{k=1}^{M} \hat{\varpi}_k p(Y_n | \hat{Y}_k, \hat{\sigma}_k) \right) \tag{32}$$

and DF is the number of degrees of freedom of the statistical model:

$$DF = M \times \frac{m(m+1)}{2} + M + M - 1 \tag{33}$$

In Formula (32), $(\hat{\varpi}_k, \hat{Y}_k, \hat{\sigma}_k)_{1 \le k \le M}$ are obtained from an EM algorithm as stated in Section 4.1 assuming the exact dimension is M. Finally, note that in Formula (33), $M \times \frac{m(m+1)}{2}$ (respectively M and $M-1$) corresponds to the number of degrees of freedom associated with $(\hat{Y}_k)_{1 \le k \le M}$ (respectively $(\hat{\sigma}_k)_{1 \le k \le M}$ and $(\hat{\varpi}_k)_{1 \le k \le M}$).

5. Application to Classification of Data on $\mathcal{P}_m$

Recently, several approaches have used the Riemannian distance in general as the main innovation in image or signal classification problems [2,15,34]. It turns out that the use of this distance leads to more accurate results (in comparison, for example, with the Euclidean distance). This section proposes an application that follows a similar approach, but in addition to the Riemannian distance, it also relies on a statistical approach. It considers the application of the Riemannian Laplace distribution (RLD) to the classification of data in $\mathcal{P}_m$ and gives an original Laplace classification rule, which can be used to carry out the task of classification, even in the presence of outliers. It also applies this classification rule to the problem of texture classification in computer vision, showing that it leads to improved results in comparison with recent literature.

Section 5.1 considers, from the point of view of statistical learning, the classification of data with values in $\mathcal{P}_m$. Given data points $Y_1, \cdots, Y_N \in \mathcal{P}_m$, this proceeds in two steps, called the learning phase and the classification phase, respectively. The learning phase uncovers the class structure of the data, by estimating a mixture model using the EM algorithm developed in Section 4.1. Once training is accomplished, data points are subdivided into disjoint classes. Classification consists of associating each new data point to the most suitable class. For this, a new classification rule will be established and shown to be optimal.

Section 5.2 is the implementation of the Laplace classification rule together with the BIC criterion to texture classification in computer vision. It highlights the advantage of the Laplace distribution in the presence of outliers and shows its better performance compared to recent approaches.

5.1. Classification Using Mixtures of Laplace Distributions

Assume to be given a set of training data $Y_1, \cdots, Y_N$. These are now modeled as a realization of a mixture of Laplace distributions:

$$p(Y) = \sum_{\mu=1}^{M} \varpi_\mu \times p(Y | \bar{Y}_\mu, \sigma_\mu) \tag{34}$$

In this section, the order M in Equation (34) is considered as known. The training phase of these data consists of learning its structure as a family of M disjoint classes $C_\mu, \mu = 1, \cdots, M$. To be more precise, depending on the family (ϖ_μ), some of these classes may be empty. Training is done by applying the EM algorithm described in Section 4.1. As a result, each class C_μ is represented by a triple $(\hat{\varpi}_\mu, \hat{Y}_\mu, \hat{\sigma}_\mu)$ corresponding to maximum likelihood estimates of $(\varpi_\mu, Y_\mu, \sigma_\mu)$. Each observation Y_n is now associated with the class C_{μ^*} where $\mu^* = \text{argmax}_\mu \omega(Y_n, \hat{\vartheta})$ (recall the definition from Equation (28)). In this way, $\{Y_1, \cdots, Y_N\}$ is subdivided into M disjoint classes.

The classification phase requires a classification rule. Following [15], the optimal rule (in the sense of a Bayesian risk criterion given in [35]) consists of associating any new data Y_t to the class C_{μ^*} where:

$$\mu^* = \text{argmax}_\mu \left\{ \hat{N}_\mu \times p(Y_t | \hat{Y}_\mu, \hat{\sigma}_\mu) \right\} \tag{35}$$

Here, $\hat{N}_\mu$ is the number of elements in C_μ. Replacing $\hat{N}_\mu$ with $N \times \hat{\varpi}_\mu$, Equation (35) becomes $\text{argmax}_\mu \hat{\varpi}_\mu \times p(Y_t | \hat{Y}_\mu, \hat{\sigma}_\mu)$. Note that when the weights ϖ_μ in Equation (34) are assumed to be equal, this rule reduces to a maximum likelihood classification rule $\max_\mu p(Y_t | \hat{Y}_\mu, \hat{\sigma}_\mu)$. A quick look at the expression Equation (17) shows that Equation (35) can also be expressed as:

$$\mu^* = \text{argmin}_\mu \left\{ -\log \hat{\varpi}_\mu + \log \zeta(\hat{\sigma}_\mu) + \frac{d(Y_t, \hat{Y}_\mu)}{\hat{\sigma}_\mu} \right\} \tag{36}$$

The rule Equation (36) will be called the Laplace classification rule. It favors clusters C_μ having a larger number of data points (the minimum contains $-\log \hat{\varpi}_\mu$) or a smaller dispersion away from the median (the minimum contains $\log \zeta(\hat{\sigma}_\mu)$). When choosing between two clusters with the same number of points and the same dispersion, this rule favors the one whose median is closer to Y_t. If the number of data points inside clusters and the respective dispersions are neglected, then Equation (36) reduces to the nearest neighbor rule involving only the Riemannian distance introduced in [2].

The analogous rules of Equation (36) for the Riemannian Gaussian distribution (RGD) [15] and the Wishart distribution (WD) [17] on $\mathcal{P}_m$ can be established by replacing $p(Y_t | \hat{Y}_\mu, \hat{\sigma}_\mu)$ in Equation (35) with the RGD and the WD and then following the same reasoning as before. Recall that a WD depends on an expectation $\Sigma \in \mathcal{P}_m$ and a number of degrees of freedom n [29]. For the WD, Equation (36) becomes:

$$\mu^* = \text{argmin}_\mu \left\{ -2 \log \hat{\varpi}(\mu) - \hat{n}(\mu) \left(\log \det \left(\hat{\Sigma}^{-1}(\mu) Y_t \right) - \text{tr}(\hat{\Sigma}^{-1}(\mu) Y_t) \right) \right\}$$

Here, $\hat{\varpi}(\mu)$, $\hat{\Sigma}(\mu)$ and $\hat{n}(\mu)$ denote maximum likelihood estimates of the true parameters $\varpi(\mu)$, $\Sigma(\mu)$ and $n(\mu)$, which define the mixture model (these estimates can be computed as in [36,37]).

5.2. Application to Texture Classification

This section presents an application of the mixture of Laplace distributions to the context of texture classification on the MIT Vision Texture (VisTex) database [38]. The purpose of this experiment is to classify the textures, by taking into consideration the within-class diversity. In addition, the influence of outliers on the classification performances is analyzed. The obtained results for the RLD are compared to those given by the RGD [15] and the WD [17].

The VisTex database contains 40 images, considered as being 40 different texture classes. The database used for the experiment is obtained after several steps. First of all, each texture is decomposed into 169 patches of 128×128 pixels, with an overlap of 32 pixels, giving a total number of 6760 textured patches. Next, some patches are corrupted, in order to introduce abnormal data into the dataset. Therefore, their intensity is modified by applying a gradient of luminosity. For each class, between zero and 60 patches are modified in order to become outliers. An example of a VisTex texture with one of its patches and an outlier patch are shown in Figure 1.

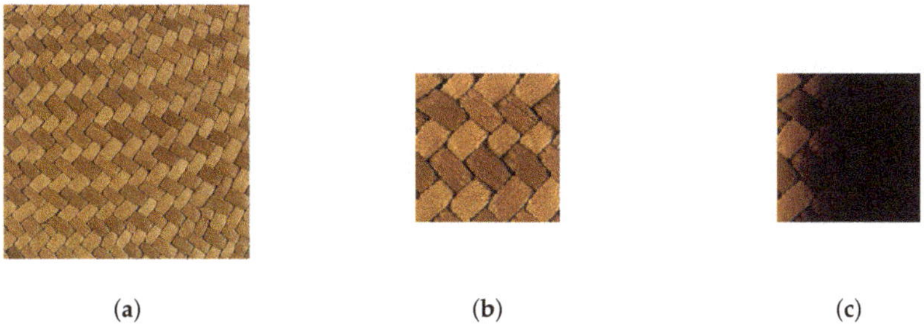

Figure 1. Example of a texture from the VisTex database (**a**), one of its patches (**b**) and the corresponding outlier (**c**).

Once the database is built, it is 15-times equally and randomly divided in order to obtain the training and the testing sets that are further used in the supervised classification algorithm. Then, for each patch in both databases, a feature vector has to be computed. The luminance channel is first extracted and then normalized in intensity. The grayscale patches are filtered using the stationary wavelet transform Daubechies db4 filter (see [39]), with two scales and three orientations. To model the wavelet sub-bands, various stochastic models have been proposed in the literature. Among them, the univariate generalized Gaussian distribution has been found to accurately model the empirical histogram of wavelet sub-bands [40]. Recently, it has been proposed to model the spatial dependency of wavelet coefficients. To this aim, the wavelet coefficients located in a $p \times q$ spatial neighborhood of the current spatial position are clustered in a random vector. The realizations of these vectors can be further modeled by elliptical distributions [41,42], copula-based models [43,44], *etc.* In this paper, the wavelet coefficients are considered as being realizations of zero-mean multivariate Gaussian distributions. In addition, for this experiment the spatial information is captured by using a vertical (2×1) and a horizontal (1×2) neighborhood. Next, the 2×2 sample covariance matrices are estimated for each wavelet sub-band and each neighborhood. Finally, each patch is represented by a set of $F = 12$ covariance matrices (2 scales $\times$ 3 orientations $\times$ 2 neighborhoods) denoted $Y = [Y_1, \cdots, Y_F]$.

The estimated covariance matrices are elements of $\mathcal{P}_m$, with $m = 2$, and therefore, they can be modeled by Riemannian Laplace distributions. More precisely, in order to take into consideration the within-class diversity, each class in the training set is viewed as a realization of a mixture of Riemannian Laplace distributions (Equation (27)) with M mixture components, characterized by $(\varpi_\mu, \bar{Y}_{\mu,f}, \sigma_{\mu,f})$, having $\bar{Y}_{\mu,f} \in \mathcal{P}_2$, with $\mu = 1, \cdots, M$ and $f = 1, \cdots, F$. Since the sub-bands are assumed to be independent, the probability density function is given by:

$$p(Y|(\varpi_\mu, \bar{Y}_{\mu,f}, \sigma_{\mu,f})_{1 \leq \mu \leq M, 1 \leq f \leq F}) = \sum_{\mu=1}^{M} \varpi_\mu \prod_{f=1}^{F} p(Y_f | \bar{Y}_{\mu,f}, \sigma_{\mu,f}) \tag{37}$$

The learning step of the classification is performed using the EM algorithm presented in Section 4, and the number of mixture components is determined using the BIC criterion recalled in Equation (31). Note that for the considered model given in Equation (37), the degree of freedom is expressed as:

$$DF = M - 1 + M \times F \times \left(\frac{m(m+1)}{2} + 1 \right) \tag{38}$$

since one centroid and one dispersion parameter should be estimated per feature and per component of the mixture model. In practice, the number of mixture components M varies between two and five, and the M yielding to the highest BIC criterion is retained. As mentioned earlier, the EM algorithm is sensitive to the initial conditions. In order to minimize this influence, for this experiment, the EM algorithm is repeated 10 times, and the result maximizing the log-likelihood function is retained. Finally, the classification is performed by assigning each element $Y_t \in \mathcal{P}_2$ in the testing set to the class of the closest cluster μ^*, given by:

$$\mu^* = \operatorname{argmin}_{\mu} \left\{ -\log \hat{\varpi}_\mu + \sum_{f=1}^{F} \log \zeta(\hat{\sigma}_{\mu,f}) + \sum_{f=1}^{F} \frac{d(Y_t, \hat{Y}_{\mu,f})}{\hat{\sigma}_{\mu,f}} \right\} \tag{39}$$

This expression is obtained starting from Equations (36) and (37), knowing that F features are extracted for each patch.

The classification results of the proposed model (solid red line), expressed in terms of overall accuracy, shown in Figure 2, are compared to those given by a fixed number of mixture components (that is, for $M = 3$, dashed red line) and with those given when the within-class diversity is not considered (that is, for $M = 1$, dotted red line). In addition, the classification performances given by the RGD model (displayed in black) proposed in [15] and the WD model (displayed in blue) proposed in [17] are also considered. For each of these models, the number of mixture components is first computed using the BIC, and next, it is fixed to $M = 3$ and $M = 1$. For all of the considered models, the classification rate is given as a function of the number of outliers, which varies between zero and 60 for each class.

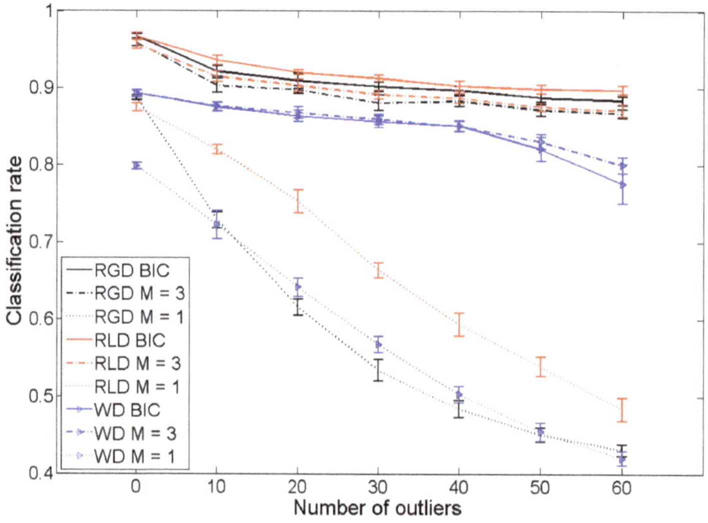

Figure 2. Classification results.

It is shown that, as the number of outliers increases, the RLD gives progressively better results than the RGD and the WD. The results are improved by using the BIC criterion for choosing the suitable number of clusters. In conclusion, the mixture of RLDs combined with the BIC criterion to estimate the best number of mixture components can minimize the influence of abnormal samples present in the dataset, illustrating the relevance of the proposed method.

6. Conclusions

Motivated by the problem of outliers in statistical data, this paper introduces a new distribution on the space $\mathcal{P}_m$ of $m \times m$ symmetric positive definite matrices, called the Riemannian Laplace distribution.

Denoted throughout the paper by $\mathcal{L}(\bar{Y},\sigma)$, where $\bar{Y} \in \mathcal{P}_m$ and $\sigma > 0$ are the indexing parameters, this distribution may be thought of as specifying the law of a family of observations on $\mathcal{P}_m$ concentrated around the location $\bar{Y}$ and having dispersion σ. If d denotes Rao's distance on $\mathcal{P}_m$ and $dv(Y)$ its associated volume form, the density of $\mathcal{L}(\bar{Y},\sigma)$ with respect to $dv(Y)$ is proportional to $\exp(-\frac{d(Y,\bar{Y})}{\sigma})$. Interestingly, the normalizing constant depends only on σ (and not on $\bar{Y}$). This allows us to deduce exact expressions for maximum likelihood estimates of $\bar{Y}$ and σ relying on the Riemannian median on $\mathcal{P}_m$. These estimates are also computed numerically by means of sub-gradient algorithms. The estimation of parameters in mixture models of Laplace distributions are also considered and performed using a new expectation-maximization algorithm. Finally, the main theoretical results are illustrated by an application to texture classification. The proposed experiment consists of introducing abnormal data (outliers) into a set of images from the VisTex database and analyzing their influences on the classification performances. Each image is characterized by a set of 2×2 covariance matrices modeled as mixtures of Riemannian Laplace distributions in the space $\mathcal{P}_2$. The number of mixtures is estimated using the BIC criterion. The obtained results are compared to those given by the Riemannian Gaussian distribution, showing the better performance of the proposed method.

Acknowledgments: This study has been carried out with financial support from the French State, managed by the French National Research Agency (ANR) in the frame of the "Investments for the future" Programme initiative d'excellence (IdEX) Bordeaux-CPU (ANR-10-IDEX-03-02).

Author Contributions: Hatem Hajri and Salem Said carried out the mathematical development and specified the algorithms. Ioana Ilea and Lionel Bombrun conceived and designed the experiments. Yannick Berthoumieu gave the central idea of the paper and managed the main tasks and experiments. Hatem Hajri wrote the paper. All the authors read and approved the final manuscript.

Conflicts of Interest: The authors declare no conflict of interest.

Appendix: Proofs of Some Technical Points

The subsections below provide proofs (using the same notations) of certain points in the paper.

A. Derivation of Equation (16) from Equation (14)

For $U \in \mathrm{O}(m)$ and $r = (r_1, \cdots, r_m) \in \mathbb{R}^m$, let $Y(r,U) = U^{\dagger}\,\mathrm{diag}(e^{r_1}, \cdots, e^{r_m})\,U$. On $\mathrm{O}(m)$, consider the exterior product $\det(\theta) = \bigwedge_{i<j} \theta_{ij}$, where $\theta_{ij} = \sum_k U_{jk} dU_{ik}$.

Proposition 4. *For all test functions $f : \mathcal{P}_m \to \mathbb{R}$,*

$$\int_{\mathcal{P}_m} f(Y)\,dv(Y) = (m!\,2^m)^{-1} \times 8^{\frac{m(m-1)}{4}} \int_{\mathrm{O}(m)} \int_{\mathbb{R}^m} f(Y(r,U))\,\det(\theta) \prod_{i<j} \sinh\left(\frac{|r_i - r_j|}{2}\right) \prod_{i=1}^{m} dr_i$$

This proposition allows one to deduce Equation (16) from Equation (14), since $\int_{\mathrm{O}(m)} \det(\theta) = \frac{2^m \pi^{m^2/2}}{\Gamma_m(m/2)}$ (see [29], p. 70).

Sketch of the proof of Proposition 4. In a differential form, the Rao–Fisher metric on $\mathcal{P}_m$ is:

$$ds^2(Y) = \mathrm{tr}[Y^{-1}dY]^2$$

For $U \in \mathrm{O}(m)$ and $(a_1, \cdots, a_m) \in (\mathbb{R}_+^*)^m$, let $Y = U^{\dagger}\,\mathrm{diag}(a_1, \cdots, a_m)\,U$. Then:

$$ds^2(Y) = \sum_{j=1}^{m} \frac{da_j^2}{a_j^2} + 2 \sum_{1 \le i < j \le m} \frac{(a_i - a_j)^2}{a_i a_j} \theta_{ij}^2$$

(see [10], p. 24). Let $a_i = e^{r_i}$, then simple calculations show that:

$$ds^2(Y) = \sum_{j=1}^{m} dr_j^2 + 8\sum_{i<j} \sinh^2\left(\frac{r_i - r_j}{2}\right)\theta_{ij}^2$$

As a consequence, the volume element $dv(Y)$ is written as:

$$dv(Y) = 8^{\frac{m(m-1)}{4}} \det(\theta) \prod_{i<j} \sinh\left(\frac{|r_i - r_j|}{2}\right) \prod_{i=1}^{m} dr_i$$

This proves the proposition (the factor $m!\,2^m$ comes from the fact that the correspondence between Y and (r, U) is not unique: $m!$ corresponds to all possible reorderings of $r_1, \ldots, r_m$, and 2^m corresponds to the orientation of the columns of U).

B. Derivation of Equation (19)

By Equations (16) and (18), to prove Equation (19), it is sufficient to prove that for all $Y \in \mathcal{P}_m$, $d(Y, I) = (\sum_{i=1}^{m} r_i^2)^{1/2}$ if the spectral decomposition of Y is $Y = U^{\dagger} \operatorname{diag}(e^{r_1}, \cdots, e^{r_m})\, U$, where U is an orthogonal matrix. Note that $d(Y, I) = d(\operatorname{diag}(e^{r_1}, \cdots, e^{r_m}).U, I) = d(\operatorname{diag}(e^{r_1}, \cdots, e^{r_m}).U, I.U)$, where . is the affine transformation given by Equation (9). By Equation (10), it comes that $d(Y, I) = d(\operatorname{diag}(e^{r_1}, \cdots, e^{r_m}), I)$, and so, $d(Y, I) = (\sum_{i=1}^{m} r_i^2)^{1/2}$ holds using the explicit expression Equation (8).

C. The Normalizing Factor $\zeta_m(\sigma)$

The subject of this section is to prove these two claims:

(i) $0 < \sigma_m < \infty$ for all $m \geq 2$;
(ii) $\sigma_2 = \sqrt{2}$.

To check (i), note that $\prod_{i<j} \sinh\left(\frac{|r_i - r_j|}{2}\right) \leq \exp(C|r|)$ for some constant C. Thus, for σ small enough, the integral $I_m(\sigma) = \int_{\mathbb{R}^m} e^{-\frac{|r|}{\sigma}} \prod_{i<j} \sinh\left(\frac{|r_i - r_j|}{2}\right) dr$ given in Equation (19) is finite, and consequently, $\sigma_m > 0$.

Fix $A > 0$, such that $\sinh(\frac{x}{2}) \geq \exp(\frac{x}{4})$ for all $x \geq A$. Then:

$$I_m(\sigma) \geq \int_{C} \exp\left(\frac{1}{4}\sum_{i<j}(r_j - r_i) - \frac{|r|}{\sigma}\right) dr$$

where C is the set of infinite Lebesgue measures:

$$C = \{r = (r_1, \cdots, r_m) \in \mathbb{R}^m : r_i \in [2(i-1)A, (2i-1)A], 1 \leq i \leq m-1, r_m \geq 2(m-1)A\}$$

Now:

$$\frac{1}{4}\sum_{i<j}(r_j - r_i) = \frac{1}{4}r_m + \frac{1}{4}(-r_1 + \sum_{i<j,(i,j)\neq(1,m)} (r_j - r_i))$$

Assume $m \geq 3$ (the case $m = 2$ is easy to deal with separately). Then, on C, $\frac{1}{4}\sum_{i<j}(r_j - r_i) \geq \frac{1}{4}r_m + C'$ and $\frac{|r|}{\sigma} \leq \frac{(C'' + r_m^2)^{\frac{1}{2}}}{\sigma}$, where C' and C'' are two positive constants (not depending on r). However, for σ large enough:

$$\frac{1}{4}\sum_{i<j}(r_j - r_i) - \frac{|r|}{\sigma} \geq \frac{1}{4}r_m + C' - \frac{(C'' + r_m^2)^{\frac{1}{2}}}{\sigma} \geq 0.$$

and so, the integral $I_m(\sigma)$ diverges. This shows that σ_m is finite.

(ii) Note the following easy inequalities $|r_1 - r_2| \leq |r_1| + |r_2| \leq \sqrt{2}|r|$, which yield $\sinh(\frac{|r_1-r_2|}{2}) \leq \frac{1}{2}e^{\frac{|r|}{\sqrt{2}}}$. This last inequality shows that $\zeta_2(\sigma)$ is finite for all $\sigma < \sqrt{2}$. In order to check $\zeta_2(\sqrt{2}) = \infty$, it is necessary to show:

$$\int_{\mathbb{R}^2} \exp(-\frac{|r|}{\sqrt{2}} + \frac{|r_1 - r_2|}{2})dr_1 dr_2 = \infty \tag{40}$$

The last integral is, up to a constant, greater than $\int_{\mathcal{C}} \exp\left(-|r| + \frac{|r_1-r_2|}{\sqrt{2}}\right) dr_1 dr_2$, where:

$$\mathcal{C} = \{(r_1, r_2) \in \mathbb{R}^2 : r_1 \geq -r_2, r_2 \leq 0\} = \{(r_1, r_2) \in \mathbb{R}^2 : r_1 \geq |r_2|, r_2 \leq 0\}.$$

On $\mathcal{C}$,

$$-|r| + \frac{|r_1 - r_2|}{\sqrt{2}} = -|r| + \frac{r_1 - r_2}{\sqrt{2}} \geq -\sqrt{2}r_1 + \frac{r_1 - r_2}{\sqrt{2}} = \frac{-r_1 - r_2}{\sqrt{2}}$$

However, $\int_{\mathcal{C}} \exp\left(\frac{-r_1-r_2}{\sqrt{2}}\right) dr_1 dr_2 = \infty$ by integrating with respect to r_1 and then r_2, which shows Equation (40).

D. The Law of X in Algorithm 1

As stated in Appendix A, the uniform distribution on $\mathrm{O}(m)$ is given by $\frac{1}{\omega'_m}\det(\theta)$, where $\omega'_m = \frac{2^m \pi^{m^2/2}}{\Gamma_m(m/2)}$. Let $Y(s, V) = V^\dagger \operatorname{diag}(e^{s_1}, \cdots, e^{s_m})\, V$, with $s = (s_1, \cdots, s_m)$. Since $X = Y(r, U)$, for any test function $\varphi : \mathcal{P}_m \to \mathbb{R}$,

$$E[\varphi(X)] = \frac{1}{\omega'_m} \int_{\mathrm{O}(m)\times\mathbb{R}^m} \varphi(Y(s, V)) p(s) \det(\theta) \prod_{i=1}^{m} ds_i \tag{41}$$

Here, $\det(\theta) = \bigwedge_{i<j} \theta_{ij}$ and $\theta_{ij} = \sum_k V_{jk} dV_{ik}$. On the other hand, by Proposition 4, $\int_{\mathcal{P}_m} \varphi(Y)\, p(Y|\, I, \sigma)\, dv(Y)$ can be expressed as:

$$(m!\, 2^m)^{-1} \times 8^{\frac{m(m-1)}{4}} \frac{1}{\zeta_m(\sigma)} \int_{\mathrm{O}(m)} \int_{\mathbb{R}^m} \varphi(Y(s, V)) e^{-\frac{|s|}{\sigma}} \det(\theta) \prod_{i<j} \sinh\left(\frac{|s_i - s_j|}{2}\right) \prod_{i=1}^{m} ds_i$$

which coincides with Equation (41).

References

1. Pennec, X.; Fillard, P.; Ayache, N. A Riemannian framework for tensor computing. *Int. J. Comput. Vis.* **2006**, *66*, 41–66.
2. Barachant, A.; Bonnet, S.; Congedo, M.; Jutten, C. Multiclass Brain–Computer Interface Classification by Riemannian Geometry. *IEEE Trans. Biomed. Eng.* **2012**, *59*, 920–928.
3. Jayasumana, S.; Hartley, R.; Salzmann, M.; Li, H.; Harandi, M. Kernel Methods on the Riemannian Manifold of Symmetric Positive Definite Matrices. In Proceedings of the IEEE Conference on Computer Vision and Pattern Recognition (CVPR), Portland, OR, USA, 23–28 June 2013; pp. 73–80.
4. Zheng, L.; Qiu, G.; Huang, J.; Duan, J. Fast and accurate Nearest Neighbor search in the manifolds of symmetric positive definite matrices. In Proceedings of the IEEE International Conference on Acoustics, Speech and Signal Processing (ICASSP), Florence, Italy, 4–9 May 2014; pp. 3804–3808.
5. Dong, G.; Kuang, G. Target recognition in SAR images via classification on Riemannian manifolds. *IEEE Geosci. Remote Sens. Lett.* **2015**, *21*, 199–203.
6. Tuzel, O.; Porikli, F.; Meer, P. Pedestrian detection via classification on Riemannian manifolds. *IEEE Trans. Pattern Anal. Mach. Intell.* **2008**, *30*, 1713–1727.
7. Caseiro, R.; Henriques, J.F.; Martins, P.; Batista, J. A nonparametric Riemannian framework on tensor field with application to foreground segmentation. *Pattern Recognit.* **2012**, *45*, 3997–4017.

8. Arnaudon, M.; Barbaresco, F.; Yang, L. Riemannian Medians and Means With Applications to Radar Signal Processing. *IEEE J. Sel. Top. Signal Process.* **2013**, *7*, 595–604.
9. Arnaudon, M.; Yang, L.; Barbaresco, F. Stochastic algorithms for computing p-means of probability measures, Geometry of Radar Toeplitz covariance matrices and applications to HR Doppler processing. In Proceedings of International International Radar Symposium (IRS), Leipzig, Germany, 7–9 September 2011; pp. 651–656.
10. Terras, A. *Harmonic Analysis on Symmetric Spaces and Applications*; Springer-Verlag: New York, NY, USA, 1988; Volume II.
11. Atkinson, C.; Mitchell, A. Rao's distance measure. *Sankhya Ser. A* **1981**, *43*, 345–365.
12. Pennec, X. Probabilities and statistics on Riemannian manifolds: Basic tools for geometric measurements. In Procedings of the IEEE Workshop on Nonlinear Signal and Image Processing, Antalya, Turkey, 20–23 June 1999; pp. 194–198.
13. Pennec, X. Intrinsic statistics on Riemannian manifolds: Basic tools for geometric measurements. *J. Math. Imaging Vis.* **2006**, *25*, 127–154.
14. Guang, C.; Baba C.V. A Novel Dynamic System in the Space of SPD Matrices with Applications to Appearance Tracking. *SIAM J. Imaging Sci.* **2013**, *6*, 592–615.
15. Said, S.; Bombrun, L.; Berthoumieu, Y.; Manton, J. Riemannian Gaussian distributions on the space of symmetric positive definite matrices. 2015, arXiv:1507.01760.
16. Schwarz, G. Estimating the Dimension of a Model. *Ann. Stat.* **1978**, *6*, 461–464.
17. Lee, J.S.; Grunes, M.R.; Ainsworth, T.L.; Du, L.J.; Schuler, D.L.; Cloude, S.R. Unsupervised classification using polarimetric decomposition and the complex Wishart classifier. *IEEE Trans. Geosci. Remote Sens.* **1999**, *37*, 2249–2258.
18. Berezin, F.A. Quantization in complex symmetric spaces. *Izv. Akad. Nauk SSSR Ser. Mat.* **1975**, *39*, 363–402.
19. Malliavin, P. Invariant or quasi-invariant probability measures for infinite dimensional groups, Part II: Unitarizing measures or Berezinian measures. *Jpn. J. Math.* **2008**, *3*, 19–47.
20. Barbaresco, F. *Information Geometry of Covariance Matrix: Cartan-Siegel Homogeneous Bounded Domains, Mostow/Berger Fibration and Fréchet Median, Matrix Information Geometry*; Bhatia, R., Nielsen, F., Eds.; Springer: New York, NY, USA, 2012; pp. 199–256.
21. Barbaresco, F. Information geometry manifold of Toeplitz Hermitian positive definite covariance matrices: Mostow/Berger fibration and Berezin quantization of Cartan-Siegel domains. *Int. J. Emerg. Trends Signal Process.* **2013**, *1*, 1–11.
22. Jeuris, B.; Vandebril, R. Averaging block-Toeplitz matrices with preservation of Toeplitz block structure. In Proceedings of the SIAM Conference on Applied Linear Algebra (ALA), Atlanta, GA, USA, 20–26 October 2015.
23. Jeuris, B.; Vandebril, R. The Kähler Mean of Block-Toeplitz Matrices with Toeplitz Structured Block. Available online: http://www.cs.kuleuven.be/publicaties/rapporten/tw/TW660.pdf (accessed on 10 March 2016).
24. Jeuris, B. Riemannian Optimization for Averaging Positive Definite Matrices. Ph.D. Thesis, University of Leuven, Leuven, Belgium, 2015.
25. Maass, H. Siegel's modular forms and Dirichlet series. In *Lecture Notes in Mathematics*; Springer-Verlag: New York, NY, USA, 1971; Volume 216.
26. Higham, N.J. *Functions of Matrices, Theory and Computation*; Society for Industrial and Applied Mathematics: Philadelphia, PA, USA, 2008.
27. Helgason, S. *Differential Geometry, Lie Groups, and Symmetric Spaces*; American Mathematical Society: Providence, RI, USA, 2001.
28. Afsari, B. Riemannian L^p center of mass: Existence, uniqueness and convexity. *Proc. Am. Math. Soc.* **2011**, *139*, 655–673.
29. Muirhead, R.J. *Aspects of Multivariate Statistical Theory*; John Wiley & Sons: New York, NY, USA, 1982.
30. Robert, C.P.; Casella, G. *Monte Carlo Statistical Methods*; Springer-Verlag: Berlin, Germany, 2004.
31. Udriste, C. *Convex Functions and Optimization Methods on Riemannian Manifolds*; Mathematics and Its Applications; Kluwer Academic Publishers: Dordrecht, The Netherlands, 1994.
32. Chavel, I. *Riemannian Geometry, a Modern Introduction*; Cambridge University Press: Cambridge, UK, 2006.
33. Yang, L. Médianes de Mesures de Probabilité dans les Variétés Riemanniennes et Applications à la Détection de Cibles Radar. Ph.D. Thesis, L'université de Poitiers, Poitiers, France, 2011. (In French)

34. Li, Y.; Wong, K.M. Riemannian distances for signal classification by power spectral density. *IEEE J. Sel. Top. Sig. Process.* **2013**, *7*, 655–669.
35. Hastie, T.; Tibshirani, R.; Friedman, J. *The Elements of Statistical Learning: Data Mining, Inference, and Prediction*; Springer: Berlin, Germany, 2009.
36. Saint-Jean, C.; Nielsen, F. A new implementation of k-MLE for mixture modeling of Wishart distributions. In *Geometric Science of Information (GSI)*; Springer-Verlag: Berlin/Heidelberg, Germany, 2013; pp. 249–256.
37. Hidot, S.; Saint-Jean, C. An expectation-maximization algorithm for the Wishart mixture model: Application to movement clustering. *Pattern Recognit. Lett.* **2010**, *31*, 2318–2324.
38. VisTex: Vision Texture Database. MIT Media Lab Vision and Modeling Group. Available online: http://vismod.media.mit.edu/pub/ (accessed on 9 March 2016).
39. Daubechies, I. *Ten Lectures on Wavelets*; Society for Industrial and Applied Mathematics: Philadelphia, PA, USA, 1992.
40. Do, M.N.; Vetterli, M. Wavelet-Based Texture Retrieval Using Generalized Gaussian Density and Kullback-Leibler Distance. *IEEE Trans. Image Process.* **2002**, *11*, 146–158.
41. Bombrun, L.; Berthoumieu, Y.; Lasmar, N.-E.; Verdoolaege, G. Mutlivariate Texture Retrieval Using the Geodesic Distance between Elliptically Distributed Random Variables. In Proceedings of 2011 18th IEEE International Conference on Image Processing (ICIP), Brussels, Belgium, 11–14 September 2011.
42. Verdoolaege, G.; Scheunders, P. On the Geometry of Multivariate Generalized Gaussian Models. *J. Math. Imaging Vis.* **2012**, *43*, 180–193.
43. Stitou, Y.; Lasmar, N.-E.; Berthoumieu, Y. Copulas based Multivariate Gamma Modeling for Texture Classification. In Proceedings of the IEEE International Conference on Acoustic Speech and Signal Processing, Taipei, Taiwan, 19–24 April 2009; pp. 1045–1048.
44. Kwitt, R.; Uhl, A. Lightweight Probabilistic Texture Retrieval. *IEEE Trans. Image Process.* **2010**, *19*, 241–253.

Chapter 5:
Statistics on Paths and on Riemannian Manifolds

entropy

MDPI

Article

Entropy Minimizing Curves with Application to Flight Path Design and Clustering †

Stéphane Puechmorel * and Florence Nicol

Laboratoire de Mathématiques Appliquées, Informatique et Automatique pour l'Aérien (MAIAA), Département Sciences et Ingénierie de la Navigation Aérienne (SINA), École Nationale de l'Aviation Civile (ENAC), 7 avenue Edouard Belin CS 54005, 31055 Toulouse, France; florence.nicol@enac.fr

* Correspondence: stephane.puechmorel@enac.fr; Tel.: +33-5-6217-9503

† This paper is an extended version of our paper published in the 2nd Conference on Geometric Science of Information, Paris, France, 28–30 October 2015.

Academic Editors: Frédéric Barbaresco and Frank Nielsen
Received: 26 July 2016; Accepted: 8 September 2016; Published: 15 September 2016

Abstract: Air traffic management (ATM) aims at providing companies with a safe and ideally optimal aircraft trajectory planning. Air traffic controllers act on flight paths in such a way that no pair of aircraft come closer than the regulatory separation norms. With the increase of traffic, it is expected that the system will reach its limits in the near future: a paradigm change in ATM is planned with the introduction of trajectory-based operations. In this context, sets of well-separated flight paths are computed in advance, tremendously reducing the number of unsafe situations that must be dealt with by controllers. Unfortunately, automated tools used to generate such planning generally issue trajectories not complying with operational practices or even flight dynamics. In this paper, a means of producing realistic air routes from the output of an automated trajectory design tool is investigated. For that purpose, the entropy of a system of curves is first defined, and a mean of iteratively minimizing it is presented. The resulting curves form a route network that is suitable for use in a semi-automated ATM system with human in the loop. The tool introduced in this work is quite versatile and may be applied also to unsupervised classification of curves: an example is given for French traffic.

Keywords: curve system entropy; curves manifold; curve clustering; probability distribution estimation; air traffic management

1. Introduction

Based on recent studies [1], traffic in Europe is expected to grow by an average yearly rate of 2.6%, yielding a net increase of two million flights per year at the 2020 horizon. The long-term forecast gives a two-fold increase in 2050 over the current traffic, pointing out the need for a paradigm change in the way flights are managed. Two major framework programs, SESAR (Single European Sky Air traffic management Research) in Europe and Nextgen in the U.S. have been launched in order to first investigate potential solutions and to deploy them in a second phase. One of the main changes that the air traffic management (ATM) system will undergo is a switch from airspace-based to trajectory-based operations with a delegation of the separation task to the crews. Within this framework, trajectories become the basic object of ATM, changing the way air traffic controllers will be working. In order to alleviate the workload of controllers, trajectories will be planned several weeks in advance in such a way that close encounters are minimized and ideally removed. For that purpose, several tools are currently being developed; most of them coming from the field of robotics [2]. Unfortunately, flight paths issued by these algorithms are not tractable for a human controller and need to be simplified. The purpose of the present work is to introduce an automated procedure that takes as input a set of trajectories and outputs a simplified one that can be used in an operational context. Please note that the separation norm constraints were not taken into account in this work. In our algorithm, we cannot enforce the regulatory separation norms, just construct clusters with low interactions. According to the applications, the results of our algorithm may be used as an initial solution of a post-processing algorithm based on optimal control in order to keep in line with the regulatory constraints. Using entropy associated with a curves system, a gradient descent is performed in order to reduce it so as to straighten trajectories while avoiding areas with low aircraft density, thus enforcing route-like behavior. This effect is related to the fact that entropy-minimizing distributions favor concentration.

2. Entropy-Minimizing Curves

2.1. Motivation

As previously mentioned, air traffic management of the future will make an intensive use of 4D trajectories as a basic object. Full automation is a far-reaching concept that will probably not be implemented before 2040–2050, and even in such a situation, it will be necessary to keep humans in the loop so as to gain a wide societal acceptance of the concept. Starting from SESAR or Nextgen initial deployment and aiming towards this ultimate objective, a transition phase with human-system cooperation will take place. Since ATC controllers are used to a well-structured network of routes, it is advisable to post-process the 4D trajectories issued by automated systems in order to make them as close as possible to line segments connecting beacons. To perform this task, in an automatic way, flight paths will be deformed so as to minimize an entropy criterion that enforces avoidance of low density areas and at the same time penalizes length. Compared to already available bundling algorithms [3] that tend to move curves to high density areas, this new procedure generates geometrically-correct curves, without excess curvature.

Let a set $\gamma_1, \ldots, \gamma_N$ of smooth curves be given that will be aircraft flight paths for the air traffic application. It will be assumed in the sequel that all curves are smooth mappings from $[0,1]$ to a domain Ω of $\mathbb{R}^q$ with everywhere non-vanishing derivatives in $]0,1[$. This last condition allows one to view them as smooth immersions with boundaries and is sound from the application point of view, as aircraft velocities are bounded below by the efficiency consideration and ultimately by the stall and, therefore, cannot vanish expect at the endpoints. In air traffic applications, the dimension of the state space is generally two and sometimes three when the evolution of the aircraft in the vertical plane is of interest.

The approach taken in this work is first to get a sound definition of spatial density associated with a curve system, then to derive from it an entropy that will be minimized.

2.2. Spatial Density of a System of Curves

Due to the fact that aircraft positions are acquired through radar measurements, a trajectory is known only at discrete sampling times. In the operational context, the sampling period ranges from 4 to 10 s, which corresponds roughly to a 100–250-m traveling distance. Derived from that, a classical performance indicator used in ATM is the aircraft density [4], obtained from the sampled positions $\gamma_i(t_j)$, $j = 1, \ldots, n_i$ on each flight path γ_i, $i = 1, \ldots, N$. It is constructed from a partition U_k, $k = 1, \ldots, P$ of Ω by counting the number of samples occurring in a given U_k, then dividing out by the total number of samples $n = \sum_{i=1}^{N} n_i$. More formally, the density d_k in the subset U_k of Ω is expressed as:

$$d_k = n^{-1} \sum_{i=1}^{N} \sum_{j=1}^{n_i} 1_{U_k} \left(\gamma_i(t_j)\right) \tag{1}$$

with 1_{U_k} the characteristic function of the set U_k. It seems natural to extend the density obtained from samples to another one based on the trajectories themselves using an integral form:

$$d_k = \lambda^{-1} \sum_{i=1}^{N} \int_0^1 1_{U_k} (\gamma_i(t))\, dt \tag{2}$$

where the normalizing constant λ is chosen so that d_k is a discrete probability distribution:

$$\lambda = \sum_{k=1}^{P} \sum_{i=1}^{N} \int_0^1 1_{U_k} (\gamma_i(t))\, dt = \sum_{i=1}^{N} \int_0^1 \sum_{k=1}^{P} 1_{U_k} (\gamma_i(t))\, dt$$

and since $U_k, k = 1, \ldots, P$ is a partition:

$$\sum_{k=1}^{P} 1_{U_k} (\gamma_i(t)) = 1 \tag{3}$$

so that $\lambda = N$.

Density can be viewed as an empirical probability distribution with the U_k considered as bins in an histogram. It is thus natural to extend the above computation so as to give rise to a continuous distribution on Ω. For that purpose, local weighting techniques, such as kernel density estimation methods, are well known in nonparametric statistics, because they are a useful data-driven way to yield continuous density estimation. Many references may be found in the literature as in [5,6]. Given the observations, the resulting estimation will be the sum of weights taking into account the distance between the observations and the location x at which the density has to be estimated; the more an observation is close to x, the greater is the weighting. The weights are defined by selecting a summable function centered on the observations, called a kernel, usually denoted by $K\colon \mathbb{R} \to \mathbb{R}^+$ in the univariate case, and a smoothed version of the Parzen–Rosenblatt density estimator [7,8] is used. Standard choices for the K function are the ones used for nonparametric kernel estimation, like the Epanechnikov function [9]:

$$K\colon x \mapsto \left(1 - x^2\right) 1_{[-1,1]}(x).$$

There exists a large variety of kernel functions, and any density function satisfying the normalization condition can be considered, so that the estimation is a probability density. Moreover, the kernel function is a symmetric positive function, with the first moment equal to zero and a finite second order moment. In the multivariate case, a multivariate kernel function $\mathcal{K}\colon \mathbb{R}^q \to \mathbb{R}^+$ is selected that can be expressed by means of a real kernel K associated with a norm, denoted by $\|.\|$, in $\mathbb{R}^q$ as follows:

$$\mathcal{K}(x) = K(\|x\|), \qquad x \in \mathbb{R}^q.$$

The normalization condition becomes:

$$\int_{\mathbb{R}^q} \mathcal{K}(x)\, dx = \int_{\mathbb{R}^q} K(\|x\|)\, dx = 1.$$

A kernel version of the density is then defined as a mapping d from Ω to $[0,1]$:

$$d\colon x \mapsto \frac{\sum_{i=1}^{N} \int_0^1 K(\|x - \gamma_i(t)\|)\, dt}{\sum_{i=1}^{N} \int_\Omega \int_0^1 K(\|x - \gamma_i(t)\|)\, dt dx}. \tag{4}$$

Normalizing the kernel is not mandatory, as the normalization occurs with the definition of d. It is nevertheless easier to consider these kinds of kernels, as is done in nonparametric density estimation.

Note that when K is compactly supported, which is the case of the Epanechnikov function and all of its relatives, it becomes:

$$\int_{\Omega} K\left(\|x-\gamma_i(t)\|\right) dx = \int_{\mathbb{R}^q} K\left(\|x\|\right) dx$$

provided that Ω contains the set:

$$\{x \in \mathbb{R}^q, \inf_{i=1\ldots N, t\in[0,1]} \|x-\gamma_i(t)\| \leq A\}$$

where the interval $[-A, A]$ contains the support of K. The case of kernels with unbounded support, like Gaussian functions, may be dealt with provided $\Omega = \mathbb{R}^q$. In the application considered, only compactly-supported kernels are used, mainly to allow fast machine implementation of the density computation.

Using the polar coordinates (ρ, θ) and the rotation invariance of the integrand, the relation becomes:

$$Vol\left(\mathbb{S}^{q-1}\right) \int_{\mathbb{R}^+} K(\rho)\rho^{q-1} d\rho = 1$$

which yields a normalizing constant of $2/\pi$ for the Epanechnikov function in dimension two, instead of the usual $3/4$ in the real case. When the normalization condition is fulfilled, the expression of the density simplifies to:

$$d\colon x \mapsto N^{-1} \sum_{i=1}^{N} \int_0^1 K\left(\|x-\gamma_i(t)\|\right) dt. \tag{5}$$

The normalizing constant is the same as in (2).

As an example, one day of traffic over France is considered and pictured in Figure 1 with the corresponding density map, computed on an evenly-spaced grid with a normalized Epanechnikov kernel.

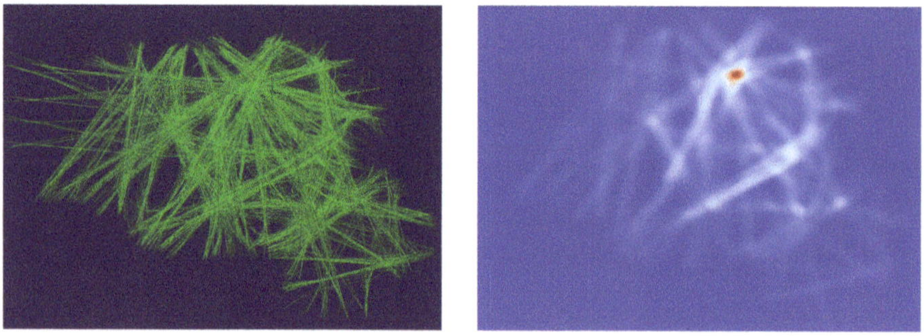

(a) **(b)**

Figure 1. (**a**) Traffic over France; (**b**) Associated density.

Unfortunately, density computed this way suffers a severe flaw for the ATM application: it is not related only to the shape of trajectories, but also to the time behavior. Formally, it is defined on the set $\mathrm{Imm}\left([0,1], \mathbb{R}^q\right)$ of smooth immersions from $[0,1]$ to $\mathbb{R}^q$ while the space of primary interest will be the quotient by smooth diffeomorphisms of the interval $[0,1]$, $\mathrm{Imm}\left([0,1], \mathbb{R}^q\right)/\mathrm{Diff}([0,1])$. Invariance of the density under the action of $\mathrm{Diff}([0,1])$ is obtained as in [10] by adding a term related to velocity in the integrals. The new definition of d becomes:

$$\tilde{d}\colon x \mapsto \frac{\sum_{i=1}^{N} \int_0^1 K\left(\|x-\gamma_i(t)\|\right) \|\gamma_i'(t)\| dt}{\sum_{i=1}^{N} \int_{\Omega} \int_0^1 K\left(\|x-\gamma_i(t)\|\right) \|\gamma_i'(t)\| dt dx}. \tag{6}$$

Assuming again a normalized kernel and letting l_i be the length of the curve γ_i, the expression of the density simplifies to:

$$\tilde{d}\colon x \mapsto \frac{\sum_{i=1}^{N} \int_0^1 K\left(\|x-\gamma_i(t)\|\right) \|\gamma_i'(t)\| dt}{\sum_{i=1}^{N} l_i}. \tag{7}$$

The new Diff-invariant density is pictured in Figure 2 along with the standard density. While the overall aspect of the plot is similar, one can observe that routes are more apparent in the right picture and that the density peak located above the Paris area is of less importance and less symmetric due to the fact that near airports, aircraft are slowing down, and this effect exaggerates the density with the non-invariant definition.

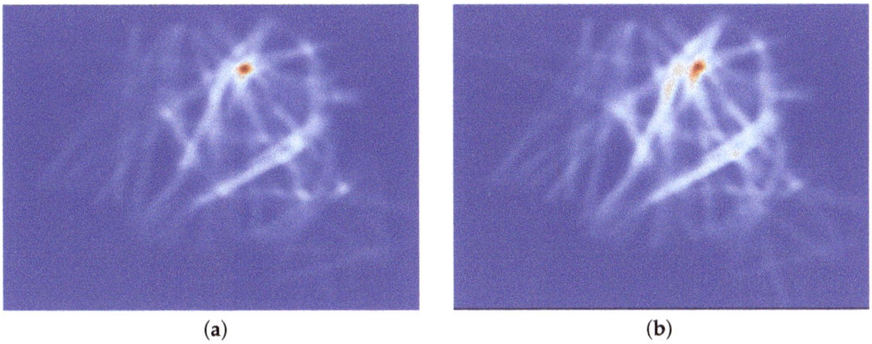

Figure 2. Density (**a**) and Diff invariant density (**b**) for 12 February 2013 traffic.

The extension of the two-dimensional defined that way to the general case of curves in an arbitrary space $\mathbb{R}^q$ is straightforward.

2.3. Further Properties of the Density

In this section, the curves considered are a smooth mapping from the closed interval $[0,1]$ to $\mathbb{R}^q$, with a non-vanishing derivative in $]0,1[$. All multivariate kernels $\mathcal{K}$ will be assumed smooth, positive, with a unit integral and of the form $x \mapsto K\left(\|x\|\right)$. However, it is not required that they are compactly supported unless explicitly stated. All results are presented for the whole space $\mathbb{R}^q$, but apply almost verbatim to an open subset.

Definition 1. *Let f be a smooth summable mapping from $\mathbb{R}$ to $\mathbb{R}$. The scaling f_ν of f is defined, for each $\nu > 0$, to be the mapping:*

$$f_\nu\colon x \in \mathbb{R} \mapsto \frac{1}{\nu} f\left(\frac{x}{\nu}\right).$$

It is clear that the L^1-norm of the original mapping is preserved by the scaling. Given a summable kernel function K from $\mathbb{R}$ to $\mathbb{R}^+$, it defines a multivariate kernel $\mathcal{K}$ on $\mathbb{R}^q$ that maps x to $K(\|x\|)$. One may derive from it a parametrized family of kernels in $\mathbb{R}$ by mapping each ν in $]0,1]$ to the scaled kernel K_ν. If the original K is of unit integral, so are all of the K_ν.

Proposition 1. *Let $\gamma\colon [0,1] \to \mathbb{R}^q$ be a smooth path with a non-vanishing derivative in $]0,1[$. Let $K_\nu, \nu > 0$ be a parametrized family of unit integral kernels. The family of Borel measures μ_ν defined for any Borel set A by:*

$$\mu_\nu(A) = \int_A \int_0^1 K_\nu\left(\|x-\gamma(t)\|\right) \|\gamma'(t)\| dt dx$$

is tight and converges narrowly to the Lebesgue measure on $\gamma\left([0,1]\right)$.

Proof. Let $\epsilon > 0$ be given. By the summability of K, there exists a positive real number r, such that:

$$\int_{\mathbb{R}^q - B(0,r)} K(\|x\|)dx < \epsilon$$

with $B(0,r)$ the open ball of radius r centered at the origin. Since $B(0,r) \subset B(0, r\nu^{-1})$ for $\nu > 0$, the same holds for all of the family K_ν. Let $B(0,M)$ be an open ball containing $\gamma([0,1])$. Then:

$$\begin{aligned} \mu_\nu\left(\mathbb{R}^q - \overline{B(0,M+r)}\right) &= \int_{\mathbb{R}^q - B(0,M+r)} \int_0^1 K_\nu\left(\|x-\gamma(t)\|\right)\|\gamma'(t)\|dtdx \\ &= \int_0^1 \int_{\mathbb{R}^q - B(0,M+r)} K_\nu\left(\|x-\gamma(t)\|\right)\|\gamma'(t)\|dxdt \\ &\leq \epsilon \int_0^1 \|\gamma'(t)\|dt = \epsilon\, l(\gamma) \end{aligned} \tag{8}$$

where $l(\gamma)$ denotes the length of γ. This proves the tightness of the family K_ν.

Let $f\colon \mathbb{R}^q \to \mathbb{R}$ be a bounded continuous mapping. It becomes:

$$\begin{aligned} I_\nu(f) = \int_{\mathbb{R}^q}\int_0^1 K_\nu\left(\|x-\gamma(t)\|\right) f(x)\|\gamma'(t)\|dtdx &= \int_0^1 \int_{\mathbb{R}^q} K_\nu\left(\|x-\gamma(t)\|\right) f(x)\|\gamma'(t)\|dxdt \\ &= \int_0^1 \int_{\mathbb{R}^q} K\left(\|x\|\right) f\left(x\nu+\gamma(t)\right)\|\gamma'(t)\|dxdt \end{aligned} \tag{9}$$

and since f is bounded, the dominated convergence theorem shows that:

$$\lim_{\nu\to 0} I_\nu(f) = \int_0^1 f\left(\gamma(t)\right)\|\gamma'(t)\|dt$$

proving the second part of the claim. □

The density in (7) is for a single curve of the form $d(x) = l(\gamma)^{-1}\int_0^1 K\left(\|x-\gamma(t)\|\right)\|\gamma'(t)\|dt$ with $l(\gamma)$ the length of the curve γ. It is invariant under the change of the parameter and can be written in a more concise way as:

$$\int_0^1 K\left(\|x-\gamma(\eta)\|\right)\|d\eta \tag{10}$$

where η is the arclength times $l(\gamma)^{-1}$.

This form allows a simple probabilistic interpretation of the density d: if a point u is drawn on the curve γ according to a uniform distribution and independently a vector v in $\mathbb{R}^q$ with a density $\mathcal{K}$ (the multivariate kernel corresponding to K), then the density of $x = u + v$ is given by Equation (10).

Proposition 2. *If the multivariate kernel $\mathcal{K}$ has a finite second moment, that is the univariate kernel K is such that:*

$$M = \int_{\mathbb{R}^+} r^{q+1} K(r)\, dr < +\infty$$

then the Wasserstein distance between the densities d_1, d_2 associated with smooth curves γ_1, γ_2 is bounded by:

$$2Vol(\mathbb{S}^{q-1})M + D(\gamma_1, \gamma_2)$$

with :

$$D(\gamma_1,\gamma_2) = \int_0^1 \|\gamma_1(\eta) - \gamma_2(\eta)\|^2 d\eta$$

where each curve is parametrized by the scaled arclength as in (10).

Proof. Let us consider the plan [11] given by the density:

$$d\colon (x,y) \mapsto \int_0^1 K(\|x-\gamma_1(\eta)\|)K(\|y-\gamma_2(\eta)\|)d\eta$$

where each curve is parametrized by the scaled arclength. The associated transport cost is given by:

$$C = \int_{\mathbb{R}^q \times \mathbb{R}^q} \|x-y\|^2 \int_0^1 K\left(\|x-\gamma_1(\eta)\|\right) K\left(\|y-\gamma_2(\eta)\|\right) d\eta dx dy$$

letting $u = y - x$ and using Fubini gives:

$$C = \int_0^1 \int_{\mathbb{R}^q} K\left(\|x-\gamma_1(\eta)\|\right) \int_{R^q} \|u\|^2 K\left(\|u+x-\gamma_2(\eta)\|\right) du dx d\eta.$$

The inner term can be written as:

$$\begin{aligned} \int_{R^q} \|u\|^2 K\left(\|u+x-\gamma_2(\eta)\|\right) du &= \int_{R^q} \|u+\gamma_2(\eta)-x\|^2 K\left(\|u\|\right) du \\ &= \int_{\mathbb{R}^q} \|u\|^2 K(\|u\|) du + 2\langle \gamma_2(\eta)-x, \int_{\mathbb{R}^q} uK(\|u\|)du\rangle + \|\gamma_2(\eta)-x\|^2. \end{aligned} \tag{11}$$

The integral:

$$\int_{\mathbb{R}^q} uK(\|u\|)du$$

is zero and, using spherical coordinates:

$$\int_{\mathbb{R}^q} \|u\|^2 K(\|u\|)du = \int_{\mathbb{R}^+} r^{q+1}K(r) \int_{\mathbb{S}^{q-1}} d\sigma\, dr = Vol(\mathbb{S}^{q-1})M$$

with $M = \int_{\mathbb{R}^+} r^{q+1}K(r)$. Putting back this value in the expression of the cost gives:

$$\begin{aligned} C &= Vol(\mathbb{S}^{q-1})M \int_0^1 \int_{\mathbb{R}^q} K\left(\|x-\gamma_1(\eta)\|\right) dx d\eta + \int_0^1 \int_{\mathbb{R}^q} K\left(\|x-\gamma_1(\eta)\|\right) \|\gamma_2(\eta)-x\|^2 dx d\eta \\ &= Vol(\mathbb{S}^{q-1})M + \int_0^1 \int_{\mathbb{R}^q} K\left(\|x-\gamma_1(\eta)\|\right) \|\gamma_2(\eta)-x\|^2 dx d\eta \\ &= Vol(\mathbb{S}^{q-1})M + \int_0^1 \int_{\mathbb{R}^q} K\left(\|x\|\right) \|\gamma_2(\eta)-\gamma_1(\eta)+x\|^2 dx d\eta. \end{aligned} \tag{12}$$

Finally:

$$\begin{aligned} \int_{\mathbb{R}^q} K\left(\|x\|\right) \|\gamma_2(\eta)-\gamma_1(\eta)+x\|^2 dx = &\int_{\mathbb{R}^q} K\left(\|x\|\right) \|\gamma_2(\eta)-\gamma_1(\eta)\|^2 dx d\eta \\ &+ 2\langle \gamma_2(\eta)-\gamma_1(\eta), \int_{\mathbb{R}^q} xK\left(\|x\|\right) dx \\ &+ Vol(\mathbb{S}^{q-1})M. \end{aligned} \tag{13}$$

As before, the middle term vanishes, and the first one integrates to:

$$\int_0^1 \|\gamma_2(\eta)-\gamma_1(\eta)\|^2 d\eta$$

so that:

$$C = 2Vol(\mathbb{S}^{q-1})M + \int_0^1 \|\gamma_2(\eta)-\gamma_1(\eta)\|^2 d\eta.$$

□

This result indicates that the densities associated with curves γ_1, γ_2 using the smoothing process described above cannot be too far (with respect to the Wasserstein distance) from each other if the geometric L^2 distance $D(\gamma_1, \gamma_2)$ is small. In fact, the upper bound in Proposition 2 can be interpreted as the cost of moving the smoothed density around γ_1 to the uniform distribution on the curve, then moving γ_1 to γ_2, keeping points with equal scaled arclength in correspondence, and finally, moving the uniform distribution on γ_2 to the smoothed density.

Having the density at hand, the entropy of the system of curves $\gamma_1, \ldots, \gamma_N$ is defined the usual way as:

$$E(\gamma_1, \ldots, \gamma_N) = -\int_\Omega \tilde{d}(x) \log\left(\tilde{d}(x)\right) dx.$$

The entropy is dependent on the particular choice of the kernel K. As mentioned before, it is a common practice in the field of non-parametric statistics to introduce a tuning parameter $\nu > 0$ in the kernel, called bandwidth, so that it is expressed as a scaled version $K = f_\nu$ of a given function $f\colon \mathbb{R}^+ \to \mathbb{R}^+$. The value of ν is the most influential parameter in the estimation of the density and must be selected carefully. For curve clustering applications, it is defined by the desired interaction length: if ν tends to zero, the curves will behave as independent objects, while on the other end of the scale, very high bandwidth will tend to remove the influence of the curves themselves. For the moment, no automated means of finding an optimal ν was used, although it will be part of a future work.

2.4. Minimizing the Entropy

In order to fulfill the initial requirement of finding bundles of curve segments as straight as possible, one seeks after the system of curves minimizing the entropy $E(\gamma_1, \ldots, \gamma_N)$, or equivalently maximizing:

$$\int_\Omega \tilde{d}(x) \log\left(\tilde{d}(x)\right) dx.$$

The reason why this criterion gives the expected behavior will become more apparent after derivation of its gradient at the end of this part. Nevertheless, when considering a single trajectory, it is intuitive that the most concentrated density distribution is obtained with a straight segment connecting the endpoints: this point will be made rigorous later.

Letting ϵ be a perturbation of the curve γ_j, such that $\epsilon(0) = \epsilon(1) = 0$, the first order expansion of $-E(\gamma_1, \ldots, \gamma_N)$ will be computed in order to get a maximizing displacement field, analogous to a gradient ascent (the choice has been made to maximize the opposite of the entropy, so that the algorithm will be a gradient ascent one) in the finite dimensional setting. The notation:

$$\frac{\partial F}{\partial \gamma_j}$$

will be used in the sequel to denote the derivative of a function F of the curve γ_j in the sense that for a perturbation ϵ:

$$F(\gamma_j + \epsilon) = F(\gamma_j) + \frac{\partial F}{\partial \gamma_j}(\epsilon) + o(\|\epsilon\|_2).$$

First of all, please note that since $\tilde{d}$ has integral one over the domain Ω:

$$\int_\Omega \frac{\partial \tilde{d}(x)}{\partial \gamma_j}(\epsilon) dx = 0$$

so that:

$$-\frac{\partial}{\partial \gamma_j} E(\gamma_1, \ldots, \gamma_N)(\epsilon) = \int_\Omega \frac{\partial \tilde{d}(x)}{\partial \gamma_j}(\epsilon) \log\left(\tilde{d}(x)\right) dx. \tag{14}$$

Starting from the expression of $\tilde{d}$ given in Equation (7), the first order expansion of $\tilde{d}$ with respect to the perturbation ϵ of γ_j is obtained as a sum of a term coming from the numerator:

$$\int_0^1 K\left(\|x - \gamma_j(t)\|\right) \|\gamma_j'(t)\| dt. \tag{15}$$

and a second one coming from the length of γ_j in the denominator. This last term is obtained from the usual first order variation formula of a curve length:

$$\int_{[0,1]} \left\| \gamma_j'(t) + \epsilon'(t) \right\| dt =$$
$$\int_{[0,1]} \left\| \gamma_j'(t) \right\| dt + \int_{[0,1]} \left\langle \frac{\gamma_j'(t)}{\|\gamma_j'(t)\|}, \epsilon'(t) \right\rangle dt + o(\|\epsilon\|_2).$$

Using an integration by parts, the first order term can be written as:

$$\int_{[0,1]} \left\langle \frac{\gamma_j'(t)}{\|\gamma_j'(t)\|}, \epsilon'(t) \right\rangle dt = - \int_{[0,1]} \left\langle \left(\frac{\gamma_j''(t)}{\|\gamma_j'(t)\|} \right)_{\mathcal{N}}, \epsilon(t) \right\rangle dt \tag{16}$$

with:
$$\left(\frac{\gamma_j''(t)}{\|\gamma_j'(t)\|} \right)_{\mathcal{N}} = \frac{\gamma_j''(t)}{\|\gamma_j'(t)\|} - \frac{\gamma_j'(t)}{\|\gamma_j'(t)\|} \left\langle \frac{\gamma_j'(t)}{\|\gamma_j'(t)\|}, \frac{\gamma_j''(t)}{\|\gamma_j'(t)\|} \right\rangle$$

the normal component of:
$$\frac{\gamma_j''(t)}{\|\gamma_j'(t)\|}.$$

Please note that when dealing with planar curves (i.e., with values in $\mathbb{R}^2$), it is $\kappa_j(t)N_j(t)$ with κ_j (resp. N_j) the curvature (resp. the unit normal vector) of γ_j.

The integral in (15) can be expanded in a similar fashion. Using as above the notation $()_{\mathcal{N}}$ for normal components, the first order term is obtained as:

$$\int_{[0,1]} \left\langle \left(\frac{\gamma_j(t) - x}{\|\gamma_j(t) - x\|} \right)_{\mathcal{N}}, \epsilon(t) \right\rangle K'\left(\|\gamma_j(t) - x\|\right) \|\gamma_j'(t)\| dt$$
$$- \int_{[0,1]} \left\langle \left(\frac{\gamma_j''(t)}{\|\gamma_j'(t)\|} \right)_{\mathcal{N}}, \epsilon(t) \right\rangle K\left(\|\gamma_j(t) - x\|\right) dt. \tag{17}$$

From the expressions in (16) and (17), the first order variation of the entropy is:

$$\frac{1}{\sum_{i=1}^N l_i} \left(\int_{[0,1]} \left\langle \int_\Omega \left(\frac{\gamma_j(t) - x}{\|\gamma_j(t) - x\|} \right)_{\mathcal{N}} K'\left(\|\gamma_j(t) - x\|\right) \log(\tilde{d}(x)) dx, \epsilon(t) \right\rangle \|\gamma_j'(t)\| dt \right.$$
$$- \int_{[0,1]} \left(\int_\Omega K\left(\|\gamma_j(t) - x\|\right) \log(\tilde{d}(x)) dx \right) \left\langle \left(\frac{\gamma_j''(t)}{\|\gamma_j'(t)\|} \right)_{\mathcal{N}}, \epsilon(t) \right\rangle dt \tag{18}$$
$$\left. + \left(\int_\Omega \tilde{d}(x) \log(\tilde{d}(x)) dx \right) \int_{[0,1]} \left\langle \left(\frac{\gamma_j''(t)}{\|\gamma_j'(t)\|} \right)_{\mathcal{N}}, \epsilon(t) \right\rangle dt \right).$$

As expected, only moves normal to the trajectory will change at first order the value of the criterion: the displacement of the curve γ_j will thus be performed at t in the normal bundle to γ_j and is given, up to the $(\sum_{i=1}^{N} l_i)^{-1}$ term, by:

$$\begin{aligned}
&\int_{\Omega} \left(\frac{\gamma_j(t) - x}{\|\gamma_j(t) - x\|} \right)_{\mathcal{N}} K'\left(\|\gamma_j(t) - x\|\right) \log(\tilde{d}(x))dx \|\gamma_j'(t)\| \\
&- \left(\int_{\Omega} K\left(\|\gamma_j(t) - x\|\right) \log(\tilde{d}(x))dx \right) \left(\frac{\gamma_j''(t)}{\|\gamma_j'(t)\|} \right)_{\mathcal{N}} \\
&+ \left(\int_{\Omega} \tilde{d}(x) \log(\tilde{d}(x))dx \right) \left(\frac{\gamma_j''(t)}{\|\gamma_j'(t)\|} \right)_{\mathcal{N}}.
\end{aligned} \tag{19}$$

The first term in the expression will favor moves towards areas of high density, while the second and third ones are moving along normal vector and will straighten the trajectory. This last point can be enlightened by considering the case of a single planar curve with fixed endpoints.

Proposition 3. *Let a, b be fixed points in $\mathbb{R}^2$ and K be a kernel as in (7). The segment $[a, b]$ is a critical point for the entropy associated with the curve system in $\mathbb{R}^2$ consisting of single smooth paths with endpoints a, b.*

Proof. Let the segment $[a, b]$ be parametrized as $\gamma\colon t \in [0,1] \mapsto a + tv$ with v the vector $(b - a)$. Starting with the expression (19), it is clear that the second and third terms occurring in the formula will vanish as the second derivative of γ is zero. Let u be the unit normal vector to γ. Any point x in $\mathbb{R}^2$ can be written as $x = a + \theta v + \xi u$, $\theta, \xi \in \mathbb{R}$, so that $\gamma(t) - x = (t - \theta)v - \xi u$ and $\|\gamma(t) - x\| = \sqrt{(t-\theta)^2\|b - a\|^2 + \xi^2}$. The change of variables $x \to (\theta, \xi)$ has Jacobian $\|v\| = \|b - a\|$. For a fixed $t \in [0, 1]$, it becomes:

$$\begin{aligned}
&\int_{\mathbb{R}^2} \left(\frac{\gamma(t) - x}{\|\gamma(t) - x\|} \right)_{\mathcal{N}} K'\left(\|\gamma(t) - x\|\right) \log(\tilde{d}(x))dx \|\gamma'(t)\| = \\
&\|b - a\|^2 \int_{\mathbb{R}} \int_{\mathbb{R}} \frac{-\xi}{\sqrt{(t-\theta)^2\|b - a\|^2 + \xi^2}} K'\left(\sqrt{(t-\theta)^2\|b - a\|^2 + \xi^2} \right) \log(\tilde{d}(\theta, \xi))d\xi d\theta.
\end{aligned} \tag{20}$$

The density $\tilde{d}$ for the γ curve is expressed in ξ, θ coordinates as:

$$\int_{[0,1]} K\left(\sqrt{(t-\theta)^2\|b - a\|^2 + \xi^2} \right) dt$$

and is an even function in ξ. The same is true for $K'\left(\|\gamma(t) - x\|\right)$. Finally, the mapping:

$$\xi \mapsto \frac{-\xi}{\sqrt{(t-\theta)^2\|b - a\|^2 + \xi^2}}$$

is odd for a fixed θ, so that the whole integrand is odd as a function of ξ. By the Fubini theorem, integrating first in ξ will therefore yield a vanishing integral, proving the assertion. □

The result still holds in $\mathbb{R}^q$, the only different aspect being that x is now expanded as $x = a + \theta v + \sum_{i=1}^{q-1} \xi_i u_i$ with $u_i, i = 1, \ldots, q-1$ an orthonormal basis of the orthogonal complement of $\mathbb{R}v$ in $\mathbb{R}^q$. Rewriting $\gamma(t) - x = (t-\theta)v - \sum_{i=1}^{q-1} \xi_i u_i$ and $\|\gamma(t) - x\| = \sqrt{(t-\theta)^2\|b - a\|^2 + \sum_{i=1}^{q-1} \xi_i^2}$, the same parity argument can be applied on any of the components $\xi_i, i = 1, \ldots, q-1$, showing that the integral is vanishing.

The effect of curve straightening is present when minimizing the entropy of a whole curve system, but is counterbalanced by the gathering effect. Depending on the choice of the kernel bandwidth, one or the other effect is dominant: straightening is preeminent for low values, being the only remaining effect in the limit, while gathering dominates at high bandwidths. For the air traffic application, a rule of the thumb is to take 2–3-times the separation norm as an effective support for the kernel. Using an adaptive bandwidth may be of some interest also: starting with medium to high values favors curve gathering; then, gradually reducing it will straighten the trajectories.

Using the scaled arclength in the entropy gives an equivalent, but somewhat easier to interpret result. Starting with the expression (7) that takes in this case the form:

$$\tilde{d}\colon x \mapsto \frac{\sum_{i=1}^{N} l_i \int_0^1 K\left(\|x-\gamma_i(\eta)\|\right) d\eta}{\sum_{i=1}^{N} l_i}. \tag{21}$$

Let $i \in \{1,\ldots,N\}$ be fixed. An admissible variation of the curve γ_i is a smooth mapping from $]-a,a[\times[0,1]$ to $\mathbb{R}^q$, with $a>0$ satisfying the following properties:

(a) $\forall \eta \in [0,1], \phi(0,\eta) = \gamma_i(\eta)$.
(b) $\forall (t,\eta) \in]-a,a[\times]0,1[, \|\partial_\eta \phi(t,\eta)\| = l_\phi(t)$ with $l_\phi(t)$ the length of the curve $\eta \mapsto \phi(t,\eta)$.
(c) $\forall t \in]-a,a[, \phi(t,0) = \gamma_i(0),\ \phi(t,1) = \gamma_i(1)$.

Taking the derivative with respect to tat zero of Equation (b) yields:

$$\left\langle \partial_t \partial_\eta \phi(0,\eta), \partial_\eta \phi(0,\eta) \right\rangle = \partial_t l_\phi(0) l_i.$$

Letting $T(\eta)$ be the unit tangent vector to γ_i at η and noting that $\partial_\eta \phi(0,\eta) = l_i T(\eta)$, it becomes:

$$\left\langle \partial_t \partial_\eta \phi(0,\eta), T(\eta) \right\rangle = \partial_t l_\phi(t). \tag{22}$$

This relation puts a constraint on the variation of the tangential component of the curve derivative and shows that it has to be constant in η.

Proposition 4. *Let D be the mapping from $]-a,a[\times\mathbb{R}^q$ to $\mathbb{R}^+$ defined by:*

$$D\colon (t,x) \mapsto \frac{\sum_{j=1, j\neq i}^{N} l_j \int_0^1 K\left(\|x-\gamma_j(\eta)\|\right) d\eta + \int_0^1 K\left(\|x-\phi(t,\eta)\|\right) d\eta}{\sum_{j=1}^{N} l_j}.$$

where η refers collectively to the scaled arclength parameter for each curve. The partial derivative $\partial_t D(0,x)$ is given by:

$$\partial_t D(0,x) = \frac{l_i}{\sum_{j=1}^{N} l_j} \int_0^1 \left\langle \frac{\gamma_i(\eta)-x}{\|\gamma_i(\eta)-x\|}, \partial_t \phi(0,\eta) \right\rangle K'\left(\|\gamma_i(\eta)-x\|\right) d\eta.$$

The proof is straightforward and is omitted. From Proposition 4, the variation of the entropy is derived:

$$\partial_t E = -\int_{\mathbb{R}^q} \frac{l_i}{\sum_{j=1}^{N} l_j} \int_0^1 \left\langle \frac{\gamma_i(\eta)-x}{\|\gamma_i(\eta)-x\|}, \partial_t \phi(0,\eta) \right\rangle K'\left(\|\gamma_i(\eta)-x\|\right) d\eta dx. \tag{23}$$

This relation is equivalent to (18): it can be seen by splitting the terms into a normal and a tangential component. The first one yields:

$$-\int_{\mathbb{R}^q} \frac{l_i}{\sum_{j=1}^{N} l_j} \int_0^1 \left\langle \left(\frac{\gamma_i(\eta)-x}{\|\gamma_i(\eta)-x\|}\right)_{\mathcal{N}}, \left(\partial_t \phi(0,\eta)\right)_{\mathcal{N}} \right\rangle K'\left(\|\gamma_i(\eta)-x\|\right) d\eta dx.$$

For the tangential part, the starting point is the relation:

$$\begin{aligned}\partial_\eta \left(K\left(\|\phi(0,\eta)-x\|\right)T(\eta)\right) =& l_i K'\left(\|\phi(0,\eta)-x\|\right)\left\langle \frac{\phi(0,\eta)-x}{\|\phi(0,\eta)-x\|}, T(\eta)\right\rangle T(\eta) \\ &+ K\left(\|\phi(0,\eta)-x\|\right)\partial_\eta T(\eta).\end{aligned} \tag{24}$$

where the subscript $\mathcal{T}$ stands for tangential component. It becomes:

$$\begin{aligned}&\frac{l_i}{\sum_{j=1}^N l_j}\int_0^1 \left\langle \left(\frac{\gamma_i(\eta)-x}{\|\gamma_i(\eta)-x\|}\right)_{\mathcal{T}}, (\partial_t\phi(0,\eta))_{\mathcal{T}}\right\rangle K'\left(\|\gamma_i(\eta)-x\|\right)d\eta dx = \\ &\frac{l_i}{\sum_{j=1}^N l_j}\int_0^1 \left\langle \partial_\eta\left(K\left(\|\phi(0,\eta)-x\|\right)T(\eta)\right), (\phi(0,\eta))_{\mathcal{T}}\right\rangle d\eta dx \\ &-\frac{l_i}{\sum_{j=1}^N l_j}\int_0^1 \left\langle K\left(\|\phi(0,\eta)-x\|\right)\partial_\eta T(\eta), (\phi(0,\eta))_{\mathcal{T}}\right\rangle d\eta dx.\end{aligned} \tag{25}$$

With an integration by parts, the first integral in the right-hand side becomes:

$$\begin{aligned}&-\frac{l_i}{\sum_{j=1}^N l_j}\int_0^1 \left\langle K\left(\|\phi(0,\eta)-x\|\right)T(\eta), \partial_\eta\left(\partial_t\phi(0,\eta)\right)_{\mathcal{T}}\right\rangle d\eta dx = \\ &-\frac{l_i}{\sum_{j=1}^N l_j}\int_0^1 K\left(\|\phi(0,\eta)-x\|\right)\partial_t l_\phi(0) d\eta dx.\end{aligned} \tag{26}$$

Gathering terms, the expression (18) is recovered. As expected, only the normal components enter the relation, but it has to be noted that the tangential component of $\partial_t\phi(0,\eta)$ is not arbitrary and can be deduced from (22). The gradient with respect to the i-th curve is obtained from the expression of the entropy variation and can be written in its simplest form as:

$$\frac{l_i}{\sum_{j=1}^N l_j}\int_{\mathbb{R}^q}\int_0^1 \frac{\gamma_i(\eta)-x}{\|\gamma_i(\eta)-x\|}K'\left(\|\gamma_i(\eta)-x\|\right)d\eta \log\tilde{d}(x)dx. \tag{27}$$

where $\tilde{d}$ is the estimated spatial density. One must keep in mind the constraint on $\partial_t\phi(0,\eta)$ that is hidden within the apparent simplicity of the expression.

3. Numerical Implementation

The two formulations (19) and (27) of the gradient may be used. The first one is more complicated, but does not require any additional constraint to be taken into account. The second one cannot be applied readily as the tangential component must comply with Relation (22). In both cases, it is needed to evaluate a spatial integral, which may yield to prohibitive computational time, especially in high dimensions. In the air traffic application, only planar 3D curves are considered, greatly simplifying the problem. Nevertheless, the performance of the algorithms is still a concern, and the choice made was to replace the spatial integral by a discrete sum over an evenly-spaced grid. From now, it is assumed that all curves are planar, so that the ambient space for the spatial density $\tilde{d}$ is $\mathbb{R}^2$. Going back to the expression of $\tilde{d}$ given by (7), a first step is to replace the integral over t by a discrete sum. In practice, curves are described by a sequence of sampled points $\gamma_i(t_{ij})$ where the sampling times t_{ij} will be assumed to be identical for all curves. This assumption is not satisfied in the air traffic application, so that a pre-processing step must be taken before the actual entropy minimization stage. It will not be described here, as any standard interpolation procedure can be applied with negligible differences on the final result. To obtain the results presented here, a cubic spline smoother was used. Since the sampling times are assumed to be the same for all trajectories, the double subscript will be dropped, so that the samples on each trajectory will be denoted as $\gamma_{ij} = \gamma_i(t_j)$. It is further assumed that the derivative $\gamma'_{ij} = \gamma'_i(t_i)$ is available, most of the time through a numerical approximation. Given a quadrature formula on $[0,1]$ with points $t_j, j = 1, \ldots, m$ and weights $w_j, j = 1, \ldots, m$, the density may be approximated at $x \in \mathbb{R}^2$ by:

$$\tilde{d}(x) = \frac{1}{\sum_{i=1}^{N} l_i} \sum_{i=1}^{N} \sum_{j=1}^{m} w_j K\left(\|x - \gamma_{ij}\|\right) \|\gamma'_{ij}\|. \tag{28}$$

where the lengths $l_i, i = 1, \ldots, N$ are also obtained with the same quadrature rule:

$$l_i = \sum_{j=1}^{m} w_j \|\gamma'_{ij}\|.$$

When γ'_{ij} is computed in a numerical way, it may be expressed as:

$$\gamma'_{ij} = \sum_{k=1}^{m} \tilde{w}_{jk} \gamma_{i,k}.$$

where the weights $\tilde{w}_{jk}$ are often obtained through the application of the Lagrange interpolation formula to ensure exactness on polynomials up to a given degree. In a more compact form, it can be written in matrix form as:

$$\begin{pmatrix} \gamma'_{i1} \\ \vdots \\ \gamma'_{iq} \end{pmatrix} = \tilde{W} \begin{pmatrix} \gamma_{i1} \\ \vdots \\ \gamma_{iq} \end{pmatrix}$$

where the matrix $\tilde{W}$ has as entries the weights $\tilde{w}_{jk}$. The cost of evaluating $\tilde{d}$ at a single point is in $o(Nm)$, with the kernel evaluation being dominant. When dealing with points in $\mathbb{R}^2$ or $\mathbb{R}^3$ and compactly-supported kernels, a simple trick greatly reduces the time needed to compute $\tilde{d}$. First of all, the domain of interest is discretized on an evenly-spaced grid, so that the points of evaluation of the density $\tilde{d}$ are its vertices $x_{ij}, i = 1, \ldots, n_x, j = 1, \ldots, n_y$. The grid step δ_x (resp. δ_y) in the first (resp. second) coordinate is the difference between any two adjacent vertices $\delta_x = x_{i+1,j} - x_{i,j}$ (resp. $\delta_y = x_{i,j+1} - x_{i,j}$ (most of the time, $\delta_x = \delta_y$). Since the expression (28) is linear, the computation can be performed by accumulating values $K(\|x_{kl} - \gamma_{ij}\|)\|\gamma'_{ij}\|$ for a fixed couple (i,j), where only the points x_{kl} close enough to γ_{ij} are considered. In fact, the evaluation can be written as a 2D discrete convolution:

$$\tilde{d}(x_{kl}) = \sum_{i=1,\ldots,N, j=1,\ldots,m} w_j K(\|x_{kl} - \gamma_{ij}\|)\|\gamma'_{ij}\|. \tag{29}$$

When the support of K is small compared to the overall spatial domain, much computation is saved using this procedure. Furthermore, it can be thought of as 2D filtering, so that highly efficient algorithms coming from the field of image processing can be applied: in particular, computing the density on a graphics processing unit (GPU) is straightforward and allows one to decrease the computational time by at least a factor of ten. When dealing with the scaled arclength, the derivative term is not present, and a factor of l_i appears in from of the integral. The discrete version becomes:

$$\tilde{d}(x_{kl}) = \sum_{i=1,\ldots,N, j=1,\ldots,m} l_i w_j K(\|x_{kl} - \gamma_{ij}\|) \tag{30}$$

where $\gamma_{ij} = \gamma_i(\eta_j)$, η_j being in correspondence with t_j. Please note that the quadrature weights must be adapted to the abscissa $\eta_j, j = 1, \ldots, m$ and not to the $t_j, j = 1, \ldots, m$. Therefore, it is advisable to resample the curves so that the points $\eta_j, j = 1, \ldots, m$ are, for example, evenly spaced or of the Gauss–Lobatto form. The former was chosen for the experiments due to its ease of implementation, although the second form is probably more efficient from a numerical point of view and will be investigated in a second stage.

Having the density at hand, the gradient of the entropy with respect to the points $\gamma_{ij}, i = 1, \ldots, N, j = 1, \ldots, m$ can be easily computed using a straightforward application of the formula (19). When dealing with planar curves, a simplification occurs for the second derivative term since for a smooth curve γ_j:

$$\left(\frac{\gamma_j''(t)}{\|\gamma_j'(t)\|} \right)_{\mathcal{N}} = \kappa(t) N(t).$$

where κ is the curvature and N the unit normal vector. These quantities may be computed using numerical differentiation, but a coarse approximation based on the rotation rate of the vectors $\gamma_{i,j+1} - \gamma_{i,j}, \gamma_{i,j+2} - \gamma_{i,j+1}$ works well in many cases.

The case of scaled arclength parametrization needs some extra attention, due to the condition on the tangential component. The simplest approach is to move the points γ_{ij} according to an unconstrained gradient, then to re-sample the obtained curve so as to get adjusted γ_{ij} that correspond to the abscissa $\eta_j, j = 1, \ldots, m$.

In a numerical implementation, the scaling factor in front of the whole expression may be dropped due to the fact that all gradient-based algorithms will use an automatically-tuned step length. As usual with gradient algorithms, one must carefully select the step taking in the maximizing direction in order to avoid divergence. A simple fixed step strategy was first applied and gives satisfactory results on small datasets. A safer approach is to adapt the step size so as to ensure a sufficient decrease of the entropy. Due to the potentially huge dimension of the search space, this procedure has to be simple enough. An approximate quadratic search [12] was used in the final implementation.

The procedure applied to one day of traffic over France yields the picture of Figure 3. As expected, a route-like network emerges. In such a case, since the traffic comes from an already organized situation, the recovered network is indeed a subset of the route network in the french airspace. Please note that there is a trade-off between the density concentration and the minimal curvature of the recovered trajectories, as already mentioned. The kernel bandwidth was chosen empirically in the example presented, with the aid of visual interaction.

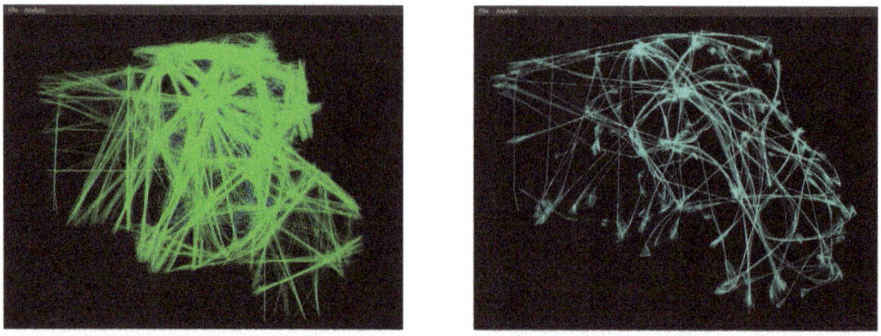

(a) (b)

Figure 3. Traffic of 24 February 2013: (**a**) Initial traffic; (**b**) Bundled traffic.

In the second example of Figure 4, the problem of automatic conflict solving is addressed. In the initial situation, aircraft are converging to a single point, which is unsafe. Air traffic controllers will proceed in such a case by diverting aircraft from their initial flight path so as to avoid each other, but only using very simple maneuvers. An automated tool will make full use of the available airspace, and the resulting set of trajectories may fail to be manageable by a human: in the event of a system failure, no backup can be provided by controllers. The entropy minimization procedure was added to an automated conflict solver in order to end up with flight paths still tractable by humans. The final result is shown in the right part of Figure 4, where encounters no longer exists, but aircraft are bound to simple trajectories, with a merging and a splitting point. Note that since the automated planner acts on velocity, all aircraft are separated in time on the inner part.

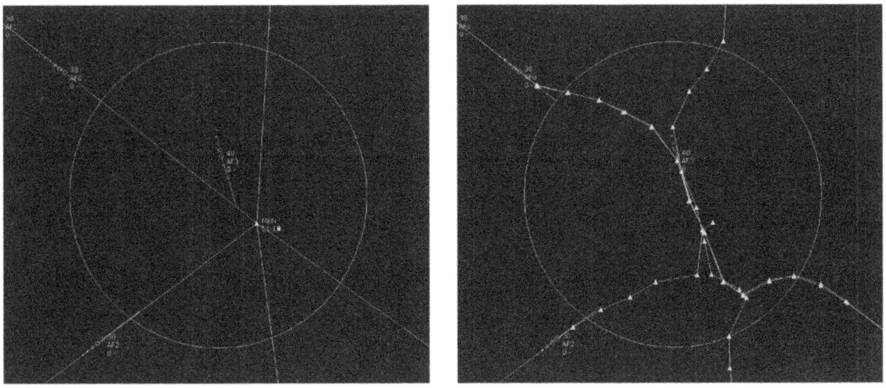

(a) (b)

Figure 4. (**a**) Initial flight plans; (**b**)Final flight plans.

4. Conclusions and Future Work

Algorithms coming from the field of shape spaces emerge as a valuable tool for applications in ATM. In this work, the foundations of a post-processing procedure that may be applied after an automated flight path planner are presented. Entropy minimization makes straight segment bundles emerge, which fulfills the operational requirements. Computational efficiency has to be improved in order to release a usable building block for future ATM systems. One way to address this issue is to compute kernel density estimators using GPUs, which excel in this kind of task, very similar to texture manipulations. Furthermore, statistical properties, such as the optimal choice of the bandwidth parameter in the kernel estimation, should be explored in more detail in the next step of this work.

Another important point that must be addressed in future works deals with the flight paths that are very similar in shape, but are oriented in opposite directions. As the spatial density is not sensitive to the directional information, the entropy-based procedure presented in this paper will tend to aggregate flight paths that should be sufficiently separated in order to prevent hazardous encounters. In [13], a notion of density based on position and velocity is developed. This work relies on Lie group modeling as a unifying state representation that takes into account the direction and the position of the curves. The curve system entropy has been extended to this setting.

Author Contributions: Stéphane Puechmorel has conceived the theoretical aspects of this work, as well as to the conception and design of the experiments; Florence Nicol has contributed to review the theoretical tools. Both authors have contributed to analyze the data and to write the paper. Both authors have read and approved the final manuscript.

Conflicts of Interest: The authors declare no conflict of interest.

References

1. De Bondt, A.; Leleu, C. *7-Year IFR Flight Movements and Service Units Forecast Update: 2014–2020*; EUROCONTROL: Brussels, Belgium, 2014.
2. Roussos, G.P.; Dimarogonas, D.V.; Kyriakopoulos, K.J. Distributed 3D navigation and collision avoidance for nonholonomic aircraft-like vehicles. In Proceedings of the 2009 European Control Conference, Budapest, Hungary, 23–26 August 2009.

3. Hurter, C.; Ersoy, O.; Telea, A. Smooth bundling of large streaming and sequence graphs. In Proceedings of the 6th PacificVis, Sydney, Australia, 26 February–1 March, 2013; pp. 41–48.
4. Harman, W.H. *Air Traffic Density and Distribution Measurements*; No. ATC-80; Lincoln Laboratory: Lexington, MA, USA, 3 May 1979.
5. Scott, D.W. *Multivariate Density Estimation: Theory, Practice, and Visualization*; Wiley: New York, NY, USA, 1992.
6. Silverman, B.W. *Density Estimation for Statistics and Data Analysis*; CRC: Boca Raton, FL, USA, 1986.
7. Parzen, E. On estimation of a probability density function and mode. *Ann. Math. Stat.* **1962**, *33*, 1065–1076.
8. Rosenblatt, M. Remarks on some nonparametric estimates of a density function. *Ann. Math. Statist.* **1956**, *27*, 832–837.
9. Epanechnikov, V.A. Non-parametric estimation of a multivariate probability density. *Theory Probab. Appl.* **1969**, *14*, 153–158.
10. Michor, P.W.; Mumford, D. Riemannian geometries on spaces of plane curves. *J. Eur. Math. Soc.* **2006**, *8*, 1–48, arXiv:math/0312384.
11. Ambrosio, L.; Gigli, N.; Savaré, G. *Gradient Flows: In Metric Spaces and in the Space of Probability Measures*; Springer: Basel, Switzerland, 2005.
12. Sun, W.; Yuan, Y.X. *Optimization Theory and Methods: Nonlinear Programming*; Springer: New York, NY, USA, 2006.
13. Nicol, F.; Puechmorel, S. Unsupervised aircraft trajectories clustering: A minimum entropy approach. In Proceedings of the Second International Conference on Big Data, Small Data, Linked Data and Open Data, Lisbon, Portugal, 21–25 February 2016.

entropy

MDPI

Article

Anisotropically Weighted and Nonholonomically Constrained Evolutions on Manifolds †

Stefan Sommer

Department of Computer Science, University of Copenhagen, DK-2100 Copenhagen E, Denmark; sommer@di.ku.dk

† This paper is an extended version of our paper published in the 2nd Conference on Geometric Science of Information, Paris, France, 28–30 October 2015.

Academic Editors: Frédéric Barbaresco and Frank Nielsen
Received: 1 September 2016; Accepted: 23 November 2016; Published: 26 November 2016

Abstract: We present evolution equations for a family of paths that results from anisotropically weighting curve energies in non-linear statistics of manifold valued data. This situation arises when performing inference on data that have non-trivial covariance and are anisotropic distributed. The family can be interpreted as most probable paths for a driving semi-martingale that through stochastic development is mapped to the manifold. We discuss how the paths are projections of geodesics for a sub-Riemannian metric on the frame bundle of the manifold, and how the curvature of the underlying connection appears in the sub-Riemannian Hamilton–Jacobi equations. Evolution equations for both metric and cometric formulations of the sub-Riemannian metric are derived. We furthermore show how rank-deficient metrics can be mixed with an underlying Riemannian metric, and we relate the paths to geodesics and polynomials in Riemannian geometry. Examples from the family of paths are visualized on embedded surfaces, and we explore computational representations on finite dimensional landmark manifolds with geometry induced from right-invariant metrics on diffeomorphism groups.

Keywords: sub-Riemannian geometry; geodesics; most probable paths; stochastic development; non-linear data analysis; statistics

---

1. Introduction

When manifold valued data have non-trivial covariance (i.e., when *anisotropy* asserts higher variance in some directions than others), non-zero curvature necessitates special care when generalizing Euclidean space normal distributions to manifold valued distributions: in the Euclidean situation, normal distributions can be seen as transition distributions of diffusion processes, but on the manifold, holonomy makes transport of covariance path-dependent in the presence of curvature, preventing a global notion of a spatially constant covariance matrix. To handle this, in the diffusion principal component analysis (PCA) framework [1], and with the class of anisotropic normal distributions on manifolds defined in [2,3], data on non-linear manifolds are modelled as being distributed according to transition distributions of anisotropic diffusion processes that are mapped from Euclidean space to the manifold by stochastic development (see [4]). The construction is connected to a non-bracket-generating sub-Riemannian metric on the bundle of linear frames of the manifold, the frame bundle, and the requirement that covariance stays covariantly constant gives a nonholonomically constrained system.

Velocity vectors and geodesic distances are conventionally used for estimation and statistics in Riemannian manifolds; for example, for estimation of the Frechét mean [5], for Principal Geodesic Analysis [6], and for tangent space statistics [7]. In contrast to this, anisotropy as modelled with anisotropic normal distributions makes a distance for a sub-Riemannian metric the natural vehicle for estimation and statistics. This metric naturally accounts for anisotropy in a similar way as the precision matrix weights the inner product in the negative log-likelihood of a Euclidean normal distribution. The connection between the weighted distance and statistics of manifold valued data was presented in [2], and the underlying sub-Riemannian and fiber-bundle geometry, together with properties of the generated densities, was further explored in [3]. The fundamental idea is to perform statistics on manifolds by maximum likelihood (ML) instead of parametric constructions that use, for example, approximating geodesic subspaces; by defining natural families of probability distributions (in this case using diffusion processes), ML parameter estimates give a coherent way to statistically model non-linear data. The anisotropically weighted distance and the resulting family of extremal paths arises in this situation when the diffusion processes have non-isotropic covariance (i.e., when the distribution is not generated from a standard Brownian motion).

In this paper, we focus on the family of *most probable paths* for the semi-martingales that drives the stochastic development, and in turn the manifold valued anisotropic stochastic processes. Such paths, as exemplified in Figure 1, extremize the anisotropically weighted action functional. We present derivations of evolution equations for the paths from different viewpoints, and we discuss the role of frames as representing either metrics or cometrics. In the derivation, we explicitly see the influence of the connection and its curvature. We then turn to the relation between the sub-Riemannian metric and the Sasaki–Mok metric on the frame bundle, and we develop a construction that allows the sub-Riemannian metric to be defined as a sum of a rank-deficient generator and an underlying Riemannian metric. Finally, we relate the paths to geodesics and polynomials in Riemannian geometry, and we explore computational representations on different manifolds including a specific case: the finite dimensional manifolds arising in the Large Deformation Diffeomorphic Metric Mapping (LDDMM) [8] landmark matching problem. The paper ends with a discussion concerning statistical implications, open questions, and concluding remarks.

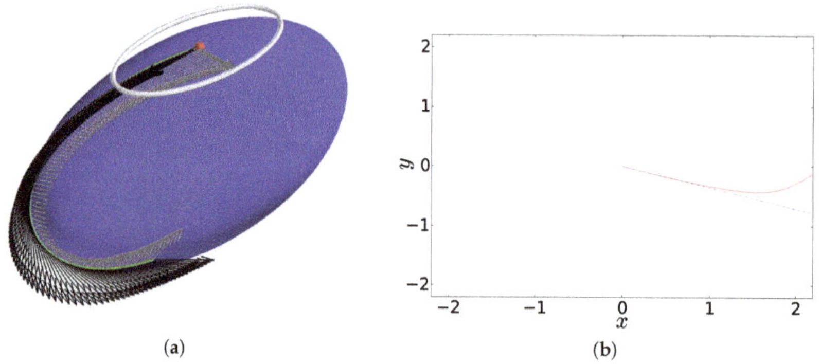

Figure 1. (**a**) A *most probable path* (MPP) for a driving Euclidean Brownian motion on an ellipsoid. The gray ellipsis over the starting point (red dot) indicates the covariance of the anisotropic diffusion. A frame u_t (black/gray vectors) representing the square root covariance is parallel transported along the curve, enabling the anisotropic weighting with the precision matrix in the action functional. With isotropic covariance, normal MPPs are Riemannian geodesics. In general situations, such as the displayed anisotropic case, the family of MPPs is much larger; (**b**) The corresponding anti-development in $\mathbb{R}^2$ (red line) of the MPP. Compare with the anti-development of a Riemannian geodesic with same initial velocity (blue dotted line). The frames $u_t \in \mathrm{GL}(\mathbb{R}^2, T_{x_t}M)$ provide local frame coordinates for each time t.

Background

Generalizing common statistical tools for performing inference on Euclidean space data to manifold valued data has been the subject of extensive work (e.g., [9]). Perhaps most fundamental is the notion of Frechét or Karcher means [5,10], defined as minimizers of the square Riemannian distance. Generalizations of the Euclidean principal component analysis procedure to manifolds are particularly relevant for data exhibiting anisotropy. Approaches include principal geodesic analysis (PGA, [6]), geodesic PCA (GPCA, [11]), principal nested spheres (PNS, [12]), barycentric subspace analysis (BSA, [13]), and horizontal component analysis (HCA, [14]). Common to these constructions are explicit representations of approximating low-dimensional subspaces. The fundamental challenge here is that the notion of Euclidean linear subspace on which PCA relies has no direct analogue in non-linear spaces.

A different approach taken by diffusion PCA (DPCA, [1,2]) and probabilistic PGA [15] is to base the PCA problem on a maximum likelihood fit of normal distributions to data. In Euclidean space, this approach was first introduced with probabilistic PCA [16]. In DPCA, the process of stochastic development [4] is used to define a class of anisotropic distributions that generalizes the family of Euclidean space normal distributions to the manifold context. DPCA is then a simple maximum likelihood fit in this family of distributions mimicking the Euclidean probabilistic PCA. The approach transfers the geometric complexities of defining subspaces common in the approaches listed above to the problem of defining a geometrically natural notion of normal distributions.

In Euclidean space, squared distances $\|x - x_0\|^2$ between observations x and the mean x_0 are affinely related to the negative log-likelihood of a normal distribution $\mathcal{N}(x_0, \mathrm{Id})$. This makes an ML fit of the mean such as performed in probabilistic PCA equivalent to minimizing squared distances. On a manifold, distances $d_g(x, x_0)^2$ coming from a Riemannian metric g are equivalent to tangent space distances $\|\mathrm{Log}_{x_0} x\|^2$ when mapping data from M to $T_{x_0}M$ using the inverse exponential map Log_{x_0}. Assuming $\mathrm{Log}_{x_0} x$ are distributed according to a normal distribution in the linear space $T_{x_0}M$, this restores the equivalence with a maximum likelihood fit. Let $\{e_1, \ldots, e_d\}$ be the standard basis for $\mathbb{R}^d$. If $u : \mathbb{R}^d \to T_{x_0}M$ is a linear invertible map with $ue_1, \ldots, ue_d$ orthonormal with respect to g, the normal distribution in $T_{x_0}M$ can be defined as $u\mathcal{N}(0, \mathrm{Id})$ (see Figure 2).

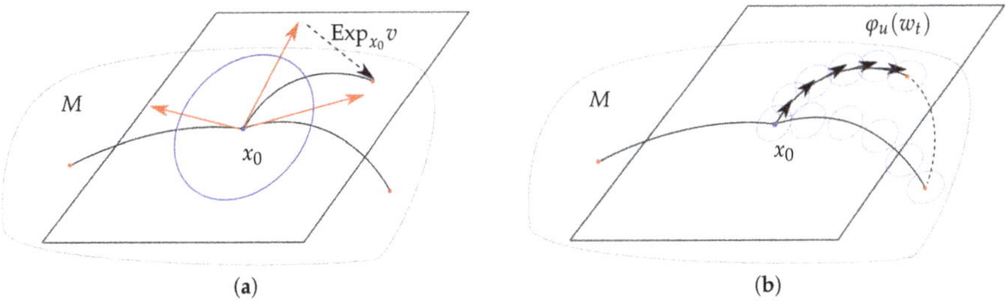

Figure 2. (**a**) Normal distributions $u\mathcal{N}(0, \mathrm{Id})$ in the tangent space $T_{x_0}M$ with covariance uu^T (blue ellipsis) can be mapped to the manifold by applying the exponential map Exp_{x_0} to sampled vectors $v \in T_{x_0}M$ (red vectors). This effectively linearises the geometry around x_0; (**b**) The stochastic development map φ_u maps $\mathbb{R}^d$ valued paths w_t to M by transporting the covariance in each step (blue ellipses) giving a covariance u_t along the entire sample path. The approach does not linearise around a single point. Holonomy of the connection implies that the covariance "rotates" around closed loops—an effect which can be illustrated by continuing the transport along the loop created by the dashed path. The anisotropic metric g_{FM} weights step lengths by the transported covariance at each time t.

The map u can be represented as a point in the frame bundle FM of M. When the orthonormal requirement on u is relaxed so that $u\mathcal{N}(0, \mathrm{Id})$ is a normal distribution in $T_{x_0}M$ with anisotropic covariance, the negative log-likelihood in $T_{x_0}M$ is related to $(u^{-1}\mathrm{Log}_{x_0} x)^T (u^{-1}\mathrm{Log}_{x_0} x)$ in the same way as the precision matrix Σ^{-1} is related to the negative log-likelihood $(x - x_0)^T \Sigma^{-1} (x - x_0)$ in

Euclidean space. The distance is thus weighted by the anisotropy of u, and u can be interpreted as a square root covariance matrix $\Sigma^{1/2}$.

However, the above approach does not specify how u changes when moving away from the base point x_0. The use of $\text{Log}_{x_0} x$ effectively linearises the geometry around x_0, but a geometrically natural way to relate u at points nearby to x_0 will be to parallel transport it, equivalently specifying that u when transported does not change as measured from the curved geometry. This constraint is *nonholonomic*, and it implies that any path from x_0 to x carries with it a parallel transport of u lifting paths from M to paths in the frame bundle FM. It therefore becomes natural to equip FM with a form of metric that encodes the anisotropy represented by u. The result is the sub-Riemannian metric on FM defined below that weights infinitesimal movements on M using the parallel transport of the frame u. Optimal paths for this metric are sub-Riemannian geodesics giving the family of *most probable paths* for the driving process that this paper concerns. Figure 1 shows one such path for an anisotropic normal distribution with M an ellipsoid embedded in $\mathbb{R}^3$.

2. Frame Bundles, Stochastic Development, and Anisotropic Diffusions

Let M be a finite dimensional manifold of dimension d with connection $\mathcal{C}$, and let x_0 be a fixed point in M. When a Riemannian metric is present, and $\mathcal{C}$ is its Levi–Civita connection, we denote the metric g_R. For a given interval $[0,T]$, we let $W(M)$ denote the Wiener space of continuous paths in M starting at x_0. Similarly, $W(\mathbb{R}^d)$ is the Wiener space of paths in $\mathbb{R}^d$. We let $H(\mathbb{R}^d)$ denote the subspace of $W(\mathbb{R}^d)$ of finite energy paths.

Let now $u = (u_1, \ldots, u_d)$ be a frame for T_xM, $x \in M$; i.e., $u_1, \ldots, u_d$ is an ordered set of linearly independent vectors in T_xM with $\text{span}\{u_1, \ldots, u_d\} = T_xM$. We can regard the frame as an isomorphism $u : \mathbb{R}^d \to T_xM$ with $u(e_i) = u_i$, where $e_1, \ldots, e_d$ denotes the standard basis in $\mathbb{R}^d$. Stochastic development (e.g., [4]) provides an invertible map φ_u from $W(\mathbb{R}^d)$ to $W(M)$. Through φ_u, Euclidean semi-martingales map to stochastic processes on M. When M is Riemannian and u orthonormal, the result is the Eells–Elworthy–Malliavin construction of Brownian motion [17]. We here outline the geometry behind development, stochastic development, the connection, and curvature, focusing in particular on frame bundle geometry.

2.1. The Frame Bundle

For each point $x \in M$, let F_xM be the set of frames for T_xM (i.e., the set of ordered bases for T_xM). The set $\{F_xM\}_{x \in M}$ can be given a natural differential structure as a fiber bundle on M called the frame bundle FM. It can equivalently be defined as the principal bundle $\text{GL}(\mathbb{R}^d, TM)$. We let the map $\pi : FM \to M$ denote the canonical projection. The kernel of $\pi_* : TFM \to TM$ is the sub-bundle of TFM that consists of vectors tangent to the fibers $\pi^{-1}(x)$. It is denoted the vertical subspace VFM. We will often work in a local trivialization $u = (x, u_1, \ldots, u_d) \in FM$, where $x = \pi(u) \in M$ denotes the base point, and for each $i = 1, \ldots, d$, $u_i \in T_xM$ is the ith frame vector. For $v \in T_xM$ and $u \in FM$ with $\pi(u) = x$, the vector $u^{-1}v \in \mathbb{R}^d$ expresses v in components in terms of the frame u. We will denote the vector $u^{-1}v$ frame coordinates of v.

For a differentiable curve x_t in M with $x = x_0$, a frame u for $T_{x_0}M$ can be parallel transported along x_t by parallel transporting each vector in the frame, thus giving a path $u_t \in FM$. Such paths are called horizontal, and have zero acceleration in the sense $\mathcal{C}(\dot{u}_{i,t}) = 0$. For each $x \in M$, their derivatives form a d-dimensional subspace of the $d + d^2$-dimensional tangent space T_uFM. This horizontal subspace HFM and the vertical subspace VFM together split the tangent bundle of FM (i.e., $TFM = HFM \oplus VFM$). The split induces a map $\pi_* : HFM \to TM$, see Figure 3. For fixed $u \in FM$, the restriction $\pi_*|_{H_uFM} : H_uFM \to T_xM$ is an isomorphism. Its inverse is called the horizontal lift and is denoted h_u in the following. Using h_u, horizontal vector fields H_e on FM are defined for vectors $e \in \mathbb{R}^d$ by $H_e(u) = h_u(ue)$. In particular, the standard basis $(e_1, \ldots, e_d)$ on $\mathbb{R}^d$ gives d globally defined horizontal vector fields $H_i \in HFM$, $i = 1, \ldots, d$ by $H_i = H_{e_i}$. Intuitively, the fields $H_i(u)$ model infinitesimal transformations in M of x_0 in direction $u_i = ue_i$ with corresponding infinitesimal parallel transport of the vectors $u_1, \ldots, u_d$ of the frame along the direction u_i. A *horizontal lift* of a differentiable curve $x_t \in M$ is a curve in FM tangent to HFM that projects to x_t. Horizontal lifts are unique up to the choice of initial frame u_0.

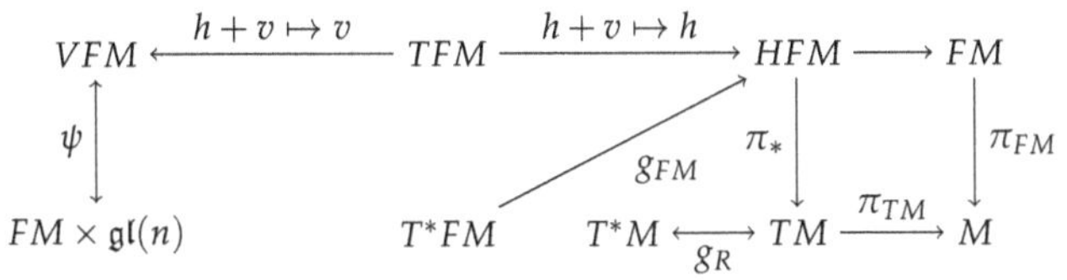

Figure 3. Relations between the manifold, frame bundle, the horizontal distribution HFM, the vertical bundle VFM, a Riemannian metric g_R, and the sub-Riemannian metric g_{FM}, defined below. The connection $\mathcal{C}$ provides the splitting $TFM = HFM \oplus VFM$. The restrictions $\pi_*|_{H_u M}$ are invertible maps $H_u M \to T_{\pi(u)} M$ with inverse h_u, the horizontal lift. Correspondingly, the vertical bundle VFM is isomorphic to the trivial bundle $FM \times \mathfrak{gl}(n)$. The metric $g_{FM} : T^*FM \to TFM$ has an image in the subspace HFM.

2.2. Development and Stochastic Development

Let x_t be a differentiable curve on M and u_t a horizontal lift. If s_t is a curve in $\mathbb{R}^d$ with components s_t^i such that $\dot{x}_t = H_i(u) s_t^i$, x_t is said to be a development of s_t. Correspondingly, s_t is the anti-development of x_t. For each t, the vector s_t contains frame coordinates of $\dot{x}_t$ as defined above. Similarly, let W_t be an $\mathbb{R}^d$ valued Brownian motion so that sample paths $W_t(\omega) \in W(\mathbb{R}^d)$. A solution to the stochastic differential equation $dU_t = \sum_{i=1}^d H_i(U_t) \circ dW_t^i$ in FM is called a stochastic development of W_t in FM. The solution projects to a stochastic development $X_t = \pi(U_t)$ in M. We call the process W_t in $\mathbb{R}^d$, that through φ maps to X_t, the driving process of X_t. Let $\varphi_u : W(\mathbb{R}^d) \to W(M)$ be the map that for fixed u sends a path in $\mathbb{R}^d$ to its development on M. Its inverse φ_u^{-1} is the anti-development in $\mathbb{R}^d$ of paths on M given u.

Equivalent to the fact that normal distributions $\mathcal{N}(0, \Sigma)$ in $\mathbb{R}^d$ can be obtained as the transition distributions of diffusion processes $\Sigma^{1/2} W_t$ stopped at time $t = 1$, a general class of distributions on the manifold M can be defined by stochastic development of processes W_t, resulting in M-valued random variables $X = X_1$. This family of distributions on M introduced in [2] is denoted *anisotropic normal distributions*. The stochastic development by construction ensures that U_t is horizontal, and the frames are thus parallel transported along the stochastic displacements. The effect is that the frames stay covariantly constant, thus resembling the Euclidean situation where $\Sigma^{1/2}$ is spatially constant and therefore does not change as W_t evolves. Thus, as further discussed in Section 3.2, the covariance is kept constant at each of the infinitesimal stochastic displacements. The existence of a smooth density for the target process X_t and small time asymptotics are discussed in [3].

Stochastic development gives a map $\int_{\text{Diff}} : FM \to \text{Prob}(M)$ to the space of probability distributions on M. For each point $u \in FM$, the map sends a Brownian motion in $\mathbb{R}^d$ to a distribution μ_u by stochastic development of the process U_t in FM, starting at u and letting μ_u be the distribution of $X = \pi(U_1)$. The pair (x, u), $x = \pi(u)$ is analogous to the parameters (μ, Σ) for a Euclidean normal distribution: the point $x \in M$ represents the starting point of the diffusion, and the frame u represents a square root $\Sigma^{1/2}$ of the covariance Σ. In the general situation where μ_u has smooth density, the construction can be used to fit the parameters u to data by maximum likelihood. As an example, diffusion PCA fits distributions obtained through $\int_{\text{Diff}}$ by maximum likelihood to observed samples in M; i.e., it optimizes for the most likely parameters $u = (x, u_1, \ldots, u_d)$ for the anisotropic diffusion process, giving a fit to the data of the manifold generalization of the Euclidean normal distribution.

2.3. Adapted Coordinates

For concrete expressions of the geometric constructions related to frame bundles, and for computational purposes, it is useful to apply coordinates that are adapted to the horizontal bundle HFM and the vertical bundle VFM together with their duals H^*FM and V^*FM. The notation below follows the notation used in, for example, [18]. Let $z = (u, \xi)$ be a local trivialization of T^*FM, and let (x^i, u^i_α) be coordinates on FM with u^i_α satisfying $u_\alpha = u^i_\alpha \partial_{x^i}$ for each $\alpha = 1, \ldots, d$.

To find a basis that is adapted to the horizontal distribution, define the d linearly independent vector fields $D_j = \partial_{x^j} - \Gamma^{h_\gamma}_j \partial_{u^h_\gamma}$ where $\Gamma^{h_\gamma}_j = \Gamma^h{}_{ji} u^i_\gamma$ is the contraction of the Christoffel symbols $\Gamma^h{}_{ij}$ for the connection C with u^i_α. We denote this adapted frame D. The vertical distribution is correspondingly spanned by $D_{j_\beta} = \partial_{u^j_\beta}$. The vectors $D^h = dx^h$, and $D^{h_\gamma} = \Gamma^{h_\gamma}_j dx^j + du^h_\gamma$ constitutes a dual coframe D^*. The map $\pi_* : HFM \to TM$ is in coordinates of the adapted frame $\pi_*(w^j D_j) = w^j \partial_{x^j}$. Correspondingly, the horizontal lift h_u is $h_u(w^j \partial_{x^j}) = w^j D_j$. The map $u : \mathbb{R}^d \to T_x M$ is given by the matrix $[u^i_\alpha]$ so that $uv = u^i_\alpha v^\alpha \partial_{x^i} = u_\alpha v^\alpha$.

Switching between standard coordinates and the adapted frame and coframes can be expressed in terms of the component matrices A below the frame and coframe induced by the coordinates (x^i, u^i_α) and the adapted frame D and coframe D^*. We have

$$
{}_{(\partial_{x^i}, \partial_{u^i_\alpha})} A_D = \begin{bmatrix} I & 0 \\ -\Gamma & I \end{bmatrix} \text{ with inverse } {}_D A_{(\partial_{x^i}, \partial_{u^i_\alpha})} = \begin{bmatrix} I & 0 \\ \Gamma & I \end{bmatrix}
$$

writing Γ for the matrix $[\Gamma^{h_\gamma}_j]$. Similarly, the component matrices of the dual frame D^* are

$$
{}_{(\partial_{x^i}, \partial_{u^i_\alpha})^*} A_{D^*} = \begin{bmatrix} I & \Gamma^T \\ 0 & I \end{bmatrix} \text{ and } {}_{D^*} A_{(\partial_{x^i}, \partial_{u^i_\alpha})^*} = \begin{bmatrix} I & -\Gamma^T \\ 0 & I \end{bmatrix} .
$$

2.4. Connection and Curvature

The TM valued connection $C : TM \times TM \to TM$ lifts to a principal connection $TFM \times TFM \to VFM$ on the principal bundle FM. C can then be identified with the $\mathfrak{gl}(n)$-valued connection form ω on TFM. The identification occurs by the isomorphism ψ between $FM \times \mathfrak{gl}(n)$ and VFM given by $\psi(u, v) = \frac{d}{dt} u \exp(tv)|_{t=0}$ (e.g., [19,20]).

The map ψ is equivariant with respect to the $\mathrm{GL}(n)$ action $g \mapsto ug^{-1}$ on FM. In order to explicitly see the connection between the usual covariant derivative $\nabla : \Gamma(TM) \times \Gamma(TM) \to \Gamma(TM)$ on M determined by C and C regarded as a connection on the principal bundle FM, following [19], we let $s : M \to TM$ be a local vector field on M; equivalently, $s \in \Gamma(TM)$ is a local section of TM. s determines a map $s^{FM} : FM \to \mathbb{R}^d$ by $s^{FM}(u) = u^{-1} s(\pi(u))$; i.e., it gives the coordinates of $s(x)$ in the frame u at x. The pushforward $(s^{FM})_* : TFM \to \mathbb{R}^d$ has in its ith component the exterior derivative $d(s^{FM})^i$. Let now $w(x)$ be a local section of FM. The composition $w \circ (s^{FM})_* \circ h_w : TM \to TM$ is identical to the covariant derivative $\nabla . s : TM \to TM$. The construction is independent of the choice of w because of the $\mathrm{GL}(n)$-equivariance of s^{FM}. The connection form ω can be expressed as the matrix $(s^{FM}_1 \circ h_w, \ldots, s^{FM}_d \circ h_w)$ when letting $s^{FM}_i(u) = e_i$.

The identification becomes particularly simple if the covariant derivative is taken along a curve x_t on which w_t is the horizontal lift. In this case, we can let $s_t = w_{t,i} s^i_t$. Then, $s^{FM}(w_t) = (s^1_t, \ldots, s^d_t)^T$, and

$$
w_t^{-1} \nabla_{\dot{x}_t} s = (s^F M)_* (h_{w_t}(\dot{x}_t)) = \tfrac{d}{dt} (s^1_t, \ldots, s^d_t)^T ; \tag{1}
$$

i.e., the covariant derivative takes the form of the standard derivative applied to the frame coordinates s^i_t.

The curvature tensor $R \in \mathcal{T}_1^3(M)$ gives the $\mathfrak{gl}(n)$-valued curvature form $\Omega : TFM \times TFM \to \mathfrak{gl}(n)$ on TFM by

$$\Omega(v_u, w_u) = u^{-1}R(\pi_*(v_u), \pi_*(w_u))u \ , \ v_u, w_v \in TFM \, .$$

Note that $\Omega(v_u, w_u) = \Omega(h_u(\pi_*(v_u)), h_u(\pi_*(w_u)))$, which we can use to write the curvature R as the $\mathfrak{gl}(n)$-valued map $R_u : T^2(T_{\pi(u)}M) \to \mathfrak{gl}(n)$, $(v, w) \mapsto \Omega(h_u(\pi_*(v_u)), h_u(\pi_*(w_u)))$ for fixed u. In coordinates, the curvature is

$$R_{ijk}{}^s = \Gamma^l{}_{ik}\Gamma^s{}_{jl} - \Gamma^l{}_{jk}\Gamma^s{}_{il} + \Gamma^s{}_{ik;j} - \Gamma^s{}_{jk;i}$$

where $\Gamma^s{}_{ik;j} = \partial_{x^j}\Gamma^s{}_{ik}$.

Let $x_{t,s}$ be a family of paths in M, and let $u_{t,s} \in \pi^{-1}(x_{t,s})$ be horizontal lifts of $x_{t,s}$ for each fixed s. Write $\dot{x}_{t,s} = \partial_t x_{t,s}$ and $\dot{u}_{t,s} = \partial_t u_{t,s}$. The s-derivative of $u_{t,s}$ can be regarded a pushforward of the horizontal lift and is the curve in TFM

$$\begin{aligned} \partial_s u_{t,s} &= \psi(u_{t,s}, \psi_{u_{0,s}}^{-1}(\mathcal{C}(\partial_s u_{0,s})) + \int_0^s \Omega(\dot{u}_{r,s}, \partial_s u_{r,s})dr) + h_{u_{t,s}}(\partial_s x_{t,s}) \\ &= \psi(u_{t,s}, \psi_{0,s}^{-1}(\mathcal{C}(\partial_s u_{0,s})) + \int_0^s R_{u_{r,s}}(\dot{x}_{r,s}, \partial_s x_{r,s})dr) + h_{u_{t,s}}(\partial_s x_{t,s}) \, . \end{aligned} \tag{2}$$

This follows from the structure equation $d\omega = -\omega \wedge \omega + \Omega$ (e.g., [21]). Note that the curve depends on the vertical variation $\mathcal{C}(\partial_s u_{0,s})$ at only one point along the curve. The remaining terms depend on the horizontal variation or, equivalently, $\partial_s x_{t,s}$. The t-derivative of $\partial_s u_{t,s}$ is the curve in $TTFM$ satisfying

$$\begin{aligned} \partial_s h_{u_{t,s}}(\dot{x}_{t,s}) &= \psi(u_{t,s}, R_{u_{t,s}}(\dot{x}_{t,s}, \partial_s x_{t,s})) + \partial_t \psi(u_{t,s}, \psi_{0,s}^{-1}(\mathcal{C}(\partial_s u_{0,s}))) + \partial_t(h_{u_{t,s}}(\partial_s x_{t,s})) \\ &= \psi(u_{t,s}, R_{u_{t,s}}(\dot{x}_{t,s}, \partial_s x_{t,s})) + \partial_t \psi(u_{t,s}, \psi_{0,s}^{-1}(\mathcal{C}(\partial_s u_{0,s}))) \\ &\quad + h_{u_{t,s}}(\partial_t \partial_s x_{t,s}) + (\partial_t h_{u_{t,s}})(\partial_s x_{t,s}). \end{aligned} \tag{3}$$

Here, the first and third term in the last expression are identified with elements of $T_{\partial_s u_{t,s}}TFM$ by the natural mapping $T_{u_{t,s}}FM \to T_{\partial_s u_{t,s}}TFM$. When $\mathcal{C}(\partial_s u_{0,s})$ is zero, the relation reflects the property that the curvature arises when computing brackets between horizontal vector fields. Note that the first term of (3) has values in VFM, while the third term has values in HFM.

3. The Anisotropically Weighted Metric

For a Euclidean driftless diffusion process with spatially constant stochastic generator Σ, the log-probability of a sample path can formally be written

$$\ln \tilde{p}_\Sigma(x_t) \propto -\int_0^1 \|\dot{x}_t\|_\Sigma^2 dt + c_\Sigma \tag{4}$$

with the norm $\|\cdot\|_\Sigma$ given by the inner product $\langle v, w\rangle_\Sigma = \left\langle \Sigma^{-1/2}v, \Sigma^{-1/2}w \right\rangle = v\Sigma^{-1}w$; i.e., the inner product weighted by the precision matrix Σ^{-1}. Though only formal, as the sample paths are almost surely nowhere differentiable, the interpretation can be given a precise meaning by taking limits of piecewise linear curves [21]. Turning to the manifold situation with the processes mapped to M by stochastic development, the probability of observing a differentiable path can either be given a precise meaning in the manifold by taking limits of small tubes around the curve, or in $\mathbb{R}^d$ by considering infinitesimal tubes around the anti-development of the curves. With the former formulation, a scalar curvature correction term must be added to (4), giving the Onsager–Machlup function [22]. The latter formulation corresponds to defining a notion of path density for the driving $\mathbb{R}^d$-valued process W_t. When M is Riemannian and Σ unitary, taking the maximum of (4) gives geodesics as most probable paths for the driving process.

Let now u_t be a path in FM, and choose a local trivialization $u_t = (x_t, u_{1,t}, \ldots, u_{d,t})$ such that the matrix $[u^i_{\alpha,t}]$ represents the square root covariance matrix $\Sigma^{1/2}$ at x_t. Since u_t being a frame defines an invertible map $\mathbb{R}^d \to T_{x_t}M$, the norm $\|\cdot\|_\Sigma$ above has a direct analogue in the norm $\|\cdot\|_{u_t}$ defined by the inner product

$$\langle v, w\rangle_{u_t} = \left\langle u_t^{-1}v, u_t^{-1}w\right\rangle_{\mathbb{R}^d} \tag{5}$$

for vectors $v, w \in T_{x_t}M$. The transport of the frame along paths in effect defines a transport of inner product along sample paths: the paths carry with them the inner product weighted by the precision matrix, which in turn is a transport of the square root covariance u_0 at x_0.

The inner product can equivalently be defined as a metric $g_u : T^*_xM \to T_xM$. Again using that u can be considered a map $\mathbb{R}^d \to T_xM$, g_u is defined by $\xi \mapsto u(\xi \circ u)^\sharp$, where $\sharp$ is the standard identification $(\mathbb{R}^d)^* \to \mathbb{R}^d$. The sequence of mappings defining g_u is illustrated below:

$$\begin{array}{ccccccc} T^*_xM & \to & (\mathbb{R}^d)^* & \to & \mathbb{R}^d & \to & T_xM \\ \xi & \mapsto & \xi \circ u & \mapsto & (\xi \circ u)^\sharp & \mapsto & u(\xi \circ u)^\sharp. \end{array} \tag{6}$$

This definition uses the $\mathbb{R}^d$ inner product in the definition of $\sharp$. Its inverse gives the cometric $g_u^{-1} : T_xM \to T^*_xM$; i.e., $v \mapsto (u^{-1}v)^\flat \circ u^{-1}$.

$$\begin{array}{ccccccc} T_xM & \to & \mathbb{R}^d & \to & (\mathbb{R}^d)^* & \to & T^*_xM \\ v & \mapsto & u^{-1}v & \mapsto & (u^{-1})^\flat & \mapsto & (u^{-1})^\flat \circ u^{-1}. \end{array} \tag{7}$$

3.1. Sub-Riemannian Metric on the Horizontal Distribution

We now lift the path-dependent metric defined above to a sub-Riemannian metric on HFM. For any $w, \tilde{w} \in H_uFM$, the lift of (5) by π_* is the inner product

$$\langle w, \tilde{w}\rangle = \left\langle u^{-1}\pi_* w, u^{-1}\pi_*\tilde{w}\right\rangle_{\mathbb{R}^d}.$$

The inner product induces a sub-Riemannian metric $g_{FM} : TFM^* \to HFM \subset TFM$ by

$$\langle w, g_{FM}(\xi)\rangle = (\xi|w)\,,\ \forall w \in H_uFM \tag{8}$$

with $(\xi|w)$ denoting the evaluation $\xi(w)$ for the covector $\xi \in T^*FM$. The metric g_{FM} gives FM a non-bracket-generating sub-Riemannian structure [23] on FM (see also Figure 3). It is equivalent to the lift

$$\xi \mapsto h_u(g_u(\xi \circ h_u))\,,\ \xi \in T_uFM \tag{9}$$

of the metric g_u above. In frame coordinates, the metric takes the form

$$u^{-1}\pi_* g_{FM}(\xi) = \begin{pmatrix} \xi(H_1(u)) \\ \vdots \\ \xi(H_d(u)) \end{pmatrix}. \tag{10}$$

In terms of the adapted coordinates for TFM described in Section 2.3, with $w = w^jD_j$ and $\tilde{w} = \tilde{w}^jD_j$, we have

$$\begin{aligned}\langle w, \tilde{w}\rangle &= \left\langle w^iD_i, \tilde{w}^jD_j\right\rangle = \left\langle u^{-1}w^i\partial_{x^i}, u^{-1}\tilde{w}^j\partial_{x^j}\right\rangle \\ &= \left\langle w^iu^\alpha_i, \tilde{w}^ju^\alpha_j\right\rangle_{\mathbb{R}^d} = \delta_{\alpha\beta}w^iu^\alpha_i\tilde{w}^ju^\beta_j = W_{ij}w^i\tilde{w}^j\end{aligned}$$

where $[u_i^\alpha]$ is the inverse of $[u_\alpha^i]$ and $W_{ij} = \delta_{\alpha\beta} u_i^\alpha u_j^\beta$. Define now $W^{kl} = \delta^{\alpha\beta} u_\alpha^k u_\beta^l$, so that $W^{ir} W_{rj} = \delta_j^i$ and $W_{ir} W^{rj} = \delta_i^j$. We can then write the metric g_{FM} directly as

$$g_{FM}(\xi_h D^h + \xi_{h_\gamma} D^{h_\gamma}) = W^{ih} \xi_h D_i, \tag{11}$$

because $\langle w, g_{FM}(\xi) \rangle = \left\langle w, W^{jh} \xi_h D_j \right\rangle = W_{ij} w^i W^{jh} \xi_h = w^i \xi_i = \xi_h D^h (w^j D_j) = \xi(w)$. One clearly recognizes the dependence on the horizontal H^*FM part of T^*FM only, and the fact that g_{FM} has image in HFM. The sub-Riemannian energy of an almost everywhere horizontal path u_t is

$$l_{FM}(u_t) = \int g_{FM}(\dot{u}_t, \dot{u}_t) dt;$$

i.e., the line element is $ds^2 = W_{ij} D^i D^j$ in adapted coordinates. The corresponding distance is given by

$$d_{FM}(u_1, u_2) = \inf\{ l_{FM}(\gamma) \mid \gamma(0) = u_1, \gamma(1) = u_2 \}.$$

If we wish to express g_{FM} in canonical coordinates on T^*FM, we can switch between the adapted frame and the coordinates $(x^i, u_\alpha^i, \xi^i, \xi_\alpha^i)$. From (11), g_{FM} has D, D^* components

$$D g_{FM,D^*} = \begin{bmatrix} W^{-1} & 0 \\ 0 & 0 \end{bmatrix}.$$

Therefore, g_{FM} has the following components in the coordinates $(x^i, u_\alpha^i, \xi_h, \xi_{h_\gamma})$

$$(\partial_{x^i}, \partial_{u_\alpha^i}) g_{FM, (\partial_x, \partial_{u_\alpha^i})^*} = (\partial_{x^i}, \partial_{u_\alpha^i}) A_D \, D g_{FM,D^*} \, D^* A_{(\partial_{x^i}, \partial_{u_\alpha^i})^*} = \begin{bmatrix} W^{-1} & -W^{-1}\Gamma^T \\ -\Gamma W^{-1} & \Gamma W^{-1} \Gamma^T \end{bmatrix}$$

or $g_{FM}^{ij} = W^{ij}$, $g_{FM}^{ij_\beta} = -W^{ih} \Gamma_h^{j_\beta}$, $g_{FM}^{i_\alpha j} = -\Gamma_h^{i_\alpha} W^{hj}$, and $g_{FM}^{i_\alpha j_\beta} = \Gamma_k^{i_\alpha} W^{kh} \Gamma_h^{j_\beta}$.

3.2. Covariance and Nonholonomicity

The metric g_{FM} encodes the anisotropic weighting given the frame u, thus up to an affine transformation measuring the energy of horizontal paths equivalently to the negative log-probability of sample paths of Euclidean anisotropic diffusions as formally given in (4). In addition, the requirement that paths must stay horizontal almost everywhere enforces that $C(\dot{u}_t) = 0$ a.e., i.e., that *no change of the covariance is measured by the connection*. The intuitive effect is that covariance is covariantly constant as seen by the connection. Globally, *curvature* of C will imply that the covariance changes when transported along closed loops, and *torsion* will imply that the base point "slips" when travelling along covariantly closed loops on M. However, the zero acceleration requirement implies that the covariance is as close to spatially constant as possible with the given connection. This is enabled by the parallel transport of the frame, and it ensures that the model closely resembles the Euclidean case with spatially constant stochastic generator.

With non-zero curvature of C, the horizontal distribution is non-integrable (i.e., the brackets $[H_i, H_j]$ are non-zero for some i, j). This prevents integrability of the horizontal distribution HFM in the sense of the Frobenius theorem. In this case, the horizontal constraint is *nonholonomic* similarly to nonholonomic constraints appearing in geometric mechanics (e.g., [24]). The requirement of covariantly constant covariance thus results in a nonholonomic system.

3.3. Riemannian Metrics on FM

If the horizontality constraint is relaxed, a related Riemannian metric on FM can be defined by pulling back a metric on $\mathfrak{gl}(n)$ to each fiber using the isomorphism $\psi(u, \cdot)^{-1} : V_u FM \to \mathfrak{gl}(n)$.

Therefore, the metric on HFM can be extended to a Riemannian metric on FM. Such metrics incorporate the anisotropically weighted metric on HFM, however, allowing vertical variations and thus that covariances can change unrestricted.

When M is Riemannian, the metric g_{FM} is in addition related to the Sasaki–Mok metric on FM [18] that extends the Sasaki metric on TM. As for the above Riemannian metric on FM, the Sasaki–Mok metric allows paths in FM to have derivatives in the vertical space VFM. On HFM, the Riemannian metric g_R is here lifted to the metric $g_{SM} = (v_u, w_u) = g_R(\pi_*(v_u), \pi_*(w_u))$ (i.e., the metric is not anisotropically weighted). The line element is in this case $ds^2 = g_{ij}dx^i dx^j + X_{\beta\alpha} g_{ij} D^{\alpha_i} D^{\beta_j}$.

Geodesics for g_{SM} are lifts of Riemannian geodesics for g_R on M, in contrast to the sub-Riemannian normal geodesics for g_{FM} which we will characterize below. The family of curves arising as projections to M of normal geodesics for g_{FM} includes Riemannian geodesics for g_R (and thus projections of geodesics for g_{SM}), but the family is in general larger than geodesics for g_R.

4. Constrained Evolutions

Extremal paths for (5) can be interpreted as most probable paths for the driving process W_t when u_0 defines an anisotropic diffusion. This is captured in the following definition [3]:

Definition 1. *A most probable path for the driving process (MPP) from $x = \pi(u_0) \in M$ to $y \in M$ is a smooth path $x_t : [0,1] \to M$ with $x_0 = x$ and $x_1 = y$ such that its anti-development $\varphi_{u_0}^{-1}(x_t)$ is a most probable path for W_t; i.e.,*

$$x_t \in argmin_{\sigma,\sigma_0=x,\sigma_1=y} \int_0^1 -L_{\mathbb{R}^d}(\varphi_{u_0}^{-1}(\sigma_t), \tfrac{d}{dt}\varphi_{u_0}^{-1}(\sigma_t))\, dt$$

with $L_{\mathbb{R}^d}$ being the Onsager–Machlup function for the process W_t on $\mathbb{R}^d$ [22].

The definition uses the one-to-one relation between $W(\mathbb{R}^d)$ and $W(M)$ provided by φ_{u_0} to characterize the paths using the $\mathbb{R}^d$ Onsager–Machlup function $L_{\mathbb{R}^d}$. When M is Riemannian with metric g_R, the Onsager–Machlup function for a g-Brownian motion on M is $L(x_t, \dot{x}_t) = -\frac{1}{2}\|\dot{x}_t\|_{g_R}^2 + \frac{1}{12}S_{g_R}(x_t)$ with S_{g_R} denoting the scalar curvature. This curvature term vanishes on $\mathbb{R}^d$, and therefore $L_{\mathbb{R}^d}(\gamma_t, \dot{\gamma}_t) = -\frac{1}{2}\|\dot{\gamma}_t\|^2$ for a curve $\gamma_t \in \mathbb{R}^d$.

By pulling $x_t \in M$ back to $\mathbb{R}^d$ using $\varphi_{u_0}^{-1}$, the construction removes the $\frac{1}{12}S_{g_R}(x_t)$ scalar curvature correction term present in the non-Euclidean Onsager–Machlup function. It thereby provides a relation between geodesic energy and most probable paths for the driving process. This is contained in the following characterization of most probable paths for the driving process as extremal paths of the sub-Riemannian distance [3] that follows from the Euclidean space Onsager–Machlup theorem [22].

Theorem 1 ([3]). *Let $Q(u_0)$ denote the principal sub-bundle of FM of points $z \in FM$ reachable from $u_0 \in FM$ by horizontal paths. Suppose the Hörmander condition is satisfied on $Q(u_0)$, and that $Q(u_0)$ has compact fibers. Then, most probable paths from x_0 to $y \in M$ for the driving process of X_t exist, and they are projections of sub-Riemannian geodesics in FM minimizing the sub-Riemannian distance from u_0 to $\pi^{-1}(y)$.*

Below, we will derive evolution equations for the set of such extremal paths that correspond to normal sub-Riemannian geodesics.

4.1. Normal Geodesics for g_{FM}

Connected to the metric g_{FM} is the Hamiltonian

$$H(z) = \frac{1}{2}(z|g_{FM}(z)) \tag{12}$$

on the symplectic space T^*FM. Letting $\hat{\pi}$ denote the projection on the bundle $T^*FM \to FM$, (8) gives

$$H(z) = \frac{1}{2}\langle g_{FM}(z)|g_{FM}(z)\rangle = \frac{1}{2}\|z \circ h_{\hat{\pi}(z)} \circ \hat{\pi}(z)\|^2_{(\mathbb{R}^d)^*} = \frac{1}{2}\sum_{i=1}^{d} \xi(H_i(u))^2.$$

Normal geodesics in sub-Riemannian manifolds satisfy the Hamilton–Jacobi equations [23] with Hamiltonian flow

$$\dot{z}_t = X_H = \Omega^\# dH(z) \tag{13}$$

where Ω here is the canonical symplectic form on T^*FM (e.g., [25]). We denote (13) the MPP equations, and we let projections $x_t = \pi_{T^*FM}(z_t)$ of minimizing curves satisfying (13) be denoted normal MPPs. The system (13) has $2(d+d^2)$ degrees of freedom, in contrast to the usual $2d$ degrees of freedom for the classical geodesic equation. Of these, d^2 describes the current frame at time t, while the remaining d^2 allows the curve to "twist" while still being horizontal. We will see this effect visualized in Section 6.

In a local canonical trivialization $z = (u, \xi)$, (13) gives the Hamilton–Jacobi equations

$$\begin{aligned} \dot{u} &= \partial_\xi H(u,\xi) = g_{FM}(u,\xi) = h_u\left(u(\,\xi(H_1(u)),\ldots,\xi(H_d(u))\,)^T\right) \\ \dot{\xi} &= -\partial_u H(u,\xi) = -\partial_u \frac{1}{2}\|\xi \circ h_u \circ u\|^2_{(\mathbb{R}^d)^*} = -\partial_u \frac{1}{2}\sum_{i=1}^{d} \xi(H_i(u))^2. \end{aligned} \tag{14}$$

Using (3), we have for the second equation

$$\begin{aligned} \dot{\xi} &= -\sum_{i=1}^{d} \xi(H_i(u))\xi(\partial_u h_u(ue_i)) \\ &= -\sum_{i=1}^{d} \xi(H_i(u))\xi\left(\psi(u, R_u(ue_i, \pi_*(\partial_u))) + \partial_{h_u(ue_i)}\psi(u, \psi^{-1}(\mathcal{C}(\partial_u))) + \partial_{h_u(ue_i)}h_u(\pi_*(\partial_u))\right) \\ &= -\xi\left(\psi(u, R_u(\pi_*(\dot{u}), \pi_*(\partial_u))) + \partial_{\dot{u}}\psi(u, \psi^{-1}(\mathcal{C}(\partial_u))) + \partial_{\dot{u}}h_u(\pi_*(\partial_u))\right). \end{aligned} \tag{15}$$

Here $\partial_{\dot{u}}$ denotes u-derivative in the direction $\dot{u}$, equivalently $\partial_{\dot{u}}h_u(v) = \partial_t(h_u)(v)$. While the first equation of (14) involves only the horizontal part of ξ, the second equation couples the vertical part of ξ through the evaluation of ξ on the term $\psi(u, R_u(\pi_*(\dot{u}), \pi_*(\partial_u)))$. If the connection is curvature-free, which in non-flat cases implies that it carries torsion, this vertical term vanishes. Conversely, when M is Riemannian, $\mathcal{C}$ the g_R Levi–Civita connection, and u_0 is g_R orthonormal, $g_{FM}(h_u(v), h_u(w)) = g_R(v,w)$ for all $v, w \in T_{\pi(u_t)}M$. In this case, a normal MPP $\pi(u_t)$ will be a Riemannian g_R geodesic.

4.2. Evolution in Coordinates

In coordinates $u = (x^i, u^i_\alpha, \xi_i, \xi_{i_\alpha})$ for T^*FM, we can equivalently write

$$\begin{aligned} \dot{x}^i &= g^{ij}\xi_j + g^{ij_\beta}\xi_{j_\beta} = W^{ij}\xi_j - W^{ih}\Gamma^{j_\beta}_h \xi_{j_\beta} \\ \dot{X}^i_\alpha &= g^{i_\alpha j}\xi_j + g^{i_\alpha j_\beta}\xi_{j_\beta} = -\Gamma^{i_\alpha}_h W^{hj}\xi_j + \Gamma^{i_\alpha}_k W^{kh}\Gamma^{j_\beta}_h \xi_{j_\beta} \\ \dot{\xi}_i &= -\frac{1}{2}\left(\partial_{y^i} g_y^{hk}\xi_h\xi_k + \partial_{y^i} g_y^{hk_\delta}\xi_h\xi_{k_\delta} + \partial_{y^i} g_y^{h_\gamma k}\xi_{h_\gamma}\xi_k + \partial_{y^i} g_y^{h_\gamma k_\delta}\xi_{h_\gamma}\xi_{k_\delta}\right) \\ \dot{\xi}_{i_\alpha} &= -\frac{1}{2}\left(\partial_{y^{i_\alpha}} g_y^{hk}\xi_h\xi_k + \partial_{y^{i_\alpha}} g_y^{hk_\delta}\xi_h\xi_{k_\delta} + \partial_{y^{i_\alpha}} g_y^{h_\gamma k}\xi_{h_\gamma}\xi_k + \partial_{y^{i_\alpha}} g_y^{h_\gamma k_\delta}\xi_{h_\gamma}\xi_{k_\delta}\right) \end{aligned}$$

with $\Gamma^{h_\gamma}_{k,i}$ for $\partial_{y^i}\Gamma^{h_\gamma}_k$, and where

$$\partial_{y^l} g^{ij} = 0\,, \quad \partial_{y^l} g^{ij_\beta} = -W^{ih}\Gamma^{j_\beta}_{h,l}\,, \quad \partial_{y^l} g^{i_\alpha j} = -\Gamma^{i_\alpha}_{h,l} W^{hj}\,, \quad \partial_{y^l} g^{i_\alpha j_\beta} = \Gamma^{i_\alpha}_{k,l} W^{kh}\Gamma^{j_\beta}_h + \Gamma^{i_\alpha}_k W^{kh}\Gamma^{j_\beta}_{h,l}\,,$$

$$
\begin{aligned}
&\partial_{y^{l_\zeta}} g^{ij} = W^{ij}{}_{,l_\zeta}\,, \quad \partial_{y^{l_\zeta}} g^{ij_\beta} = -W^{ih}{}_{,l_\zeta}\Gamma^{j_\beta}_{h} - W^{ih}\Gamma^{j_\beta}_{h,l_\zeta}\,, \quad \partial_{y^{l_\zeta}} g^{i_\alpha j} = -\Gamma^{i_\alpha}_{h,l_\zeta}W^{hj} - \Gamma^{i_\alpha}_{h}W^{hj}{}_{,l_\zeta}\,,\\
&\partial_{y^{l_\zeta}} g^{i_\alpha j_\beta} = \Gamma^{i_\alpha}_{k,l_\zeta}W^{kh}\Gamma^{j_\beta}_{h} + \Gamma^{i_\alpha}_{k}W^{kh}{}_{,l_\zeta}\Gamma^{j_\beta}_{h} + \Gamma^{i_\alpha}_{k}W^{kh}\Gamma^{j_\beta}_{h,l_\zeta}\,,\\
&\Gamma^{i_\alpha}_{h,l_\zeta} = \partial_{y^{l_\zeta}}\left(\Gamma^{i}{}_{hk}u^k_\alpha\right) = \delta^{\zeta\alpha}\Gamma^{i}{}_{hl}\,, \quad W^{ij}{}_{,l_\zeta} = \delta^{il}u^j_\zeta + \delta^{jl}u^i_\zeta\,.
\end{aligned}
$$

Combining these expressions, we obtain

$$
\begin{aligned}
&\dot{x}^i = W^{ij}\xi_j - W^{ih}\Gamma^{j_\beta}_{h}\xi_{j_\beta} \quad , \quad \dot{X}^i_\alpha = -\Gamma^{i_\alpha}_{h}W^{hj}\xi_j + \Gamma^{i_\alpha}_{k}W^{kh}\Gamma^{j_\beta}_{h}\xi_{j_\beta}\\
&\dot{\xi}_i = W^{hl}\Gamma^{k_\delta}_{l,i}\xi_h\xi_{k_\delta} - \frac{1}{2}\left(\Gamma^{h_\gamma}_{k,i}W^{kh}\Gamma^{k_\delta}_{h} + \Gamma^{h_\gamma}_{k}W^{kh}\Gamma^{k_\delta}_{h,i}\right)\xi_{h_\gamma}\xi_{k_\delta}\\
&\dot{\xi}_{i_\alpha} = \Gamma^{h_\delta}_{k,i_\alpha}W^{kh}\Gamma^{k_\delta}_{h}\xi_{h_\gamma}\xi_{k_\delta} - \left(W^{hl}{}_{,i_\alpha}\Gamma^{k_\delta}_{l} + W^{hl}\Gamma^{k_\delta}_{l,i_\alpha}\right)\xi_h\xi_{k_\delta} - \frac{1}{2}\left(W^{hk}{}_{,i_\alpha}\xi_h\xi_k + \Gamma^{h_\delta}_{k}W^{kh}{}_{,i_\alpha}\Gamma^{k_\delta}_{h}\xi_{h_\gamma}\xi_{k_\delta}\right)\,.
\end{aligned}
$$

4.3. Acceleration and Polynomials for C

We can identify the covariant acceleration $\nabla_{\dot{x}_t}\dot{x}_t$ of curves satisfying the MPP equations, and hence normal MPPs through their frame coordinates. Let (u_t, ξ_t) satisfy (13). Then, u_t is a horizontal lift of $x_t = \pi(u_t)$ and hence by (1), (3), (10), and (15),

$$
\begin{aligned}
u_t^{-1}\nabla_{\dot{x}_t}\dot{x}_t &= \frac{d}{dt}\begin{pmatrix}\xi(h_{u_t}(u_te_1))\\ \vdots\\ \xi(h_{u_t}(u_te_d))\end{pmatrix} = \begin{pmatrix}\dot{\xi}(h_{u_t}(u_te_1))\\ \vdots\\ \dot{\xi}(h_{u_t}(u_te_d))\end{pmatrix} + \begin{pmatrix}\xi(\partial_t h_{u_t}(u_te_1))\\ \vdots\\ \xi(\partial_t h_{u_t}(u_te_d))\end{pmatrix}\\
&= -\begin{pmatrix}\xi(\partial_{h_{u_t}(u_te_1)}h_{u_t}(\pi_*(\dot{u}_t)))\\ \vdots\\ \xi(\partial_{h_{u_t}(u_te_d)}h_{u_t}(\pi_*(\dot{u}_t)))\end{pmatrix} + \begin{pmatrix}\xi(\partial_{h_{u_t}(\pi_*(\dot{u}_t))}h_{u_t}(u_te_1))\\ \vdots\\ \xi(\partial_{h_{u_t}(\pi_*(\dot{u}_t))}h_{u_t}(u_te_d))\end{pmatrix}\\
&= \begin{pmatrix}\xi(\psi(u_t, R_{u_t}(u_te_1, \pi_*(\dot{u}_t))))\\ \vdots\\ \xi(\psi(u_t, R_{u_t}(u_te_d, \pi_*(\dot{u}_t))))\end{pmatrix}.
\end{aligned}
\tag{16}
$$

The fact that the covariant derivative vanishes for classical geodesic leads to a definition of higher-order polynomials through the covariant derivative by requiring $(\nabla_{\dot{x}_t})^k\dot{x}_t = 0$ for a kth order polynomial (e.g., [26,27]). As discussed above, compared to classical geodesics, curves satisfying the MPP equations have extra d^2 degrees of freedom, allowing the curves to twist and deviate from being geodesic with respect to C while still satisfying the horizontality constraint on FM. This makes it natural to ask if normal MPPs relate to polynomials defined using C. For curves satisfying the MPP equations, using (16) and (15), we have

$$
u_t^{-1}(\nabla_{\dot{x}_t})^2\dot{x}_t = \frac{d}{dt}\begin{pmatrix}\xi(\psi(u_t, R_{u_t}(u_te_1, \pi_*(\dot{u}_t))))\\ \vdots\\ \xi(\psi(u_t, R_{u_t}(u_te_d, \pi_*(\dot{u}_t))))\end{pmatrix} = \begin{pmatrix}\xi(\psi(u_t, \frac{d}{dt}R_{u_t}(u_te_1, \pi_*(\dot{u}_t))))\\ \vdots\\ \xi(\psi(u_t, \frac{d}{dt}R_{u_t}(u_te_d, \pi_*(\dot{u}_t))))\end{pmatrix}.
$$

Thus, in general, normal MPPs are not second order polynomials in the sense $(\nabla_{\dot{x}_t})^2\dot{x}_t = 0$ unless the curvature $R_{u_t}(u_te_i, \pi_*(\dot{u}_t))$ is constant in t.

For comparison, in the Riemannian case, a variational formulation placing a cost on covariant acceleration [28,29] leads to cubic splines

$$
(\nabla_{\dot{x}_t})^2\dot{x}_t = -R(\nabla_{\dot{x}_t}\dot{x}_t, x_t,)\dot{x}_t\,.
$$

In (16), the curvature terms appear in the covariant acceleration for normal MPPs, while cubic splines leads to the curvature term appearing in the third order derivative.

5. Cometric Formulation and Low-Rank Generator

We now investigate a cometric $g_{F^kM} + \lambda g_R$, where g_R is Riemannian, g_{F^kM} is a rank k positive semi-definite inner product arising from k linearly independent tangent vectors, and $\lambda > 0$ a weight. We assume that g_{F^kM} is chosen so that $g_{F^kM} + \lambda g_R$ is invertible, even though g_{F^kM} is rank-deficient. The situation corresponds to extracting the first k eigenvectors in Euclidean space PCA. If the eigenvectors are estimated statistically from observed data, this allows the estimation to be restricted to only the first k eigenvectors. In addition, an important practical implication of the construction is that a numerical implementation need not transport a full $d \times d$ matrix for the frame, but a potentially much lower dimensional $d \times k$ matrix. This point is essential when dealing with high-dimensional data, examples of which are landmark manifolds as discussed in Section 6.

When using the frame bundle to model covariances, the sum formulation is natural to express as a cometric compared to a metric because, with the cometric formulation, $g_{F^kM} + \lambda g_R$ represents a sum of covariance matrices instead of a sum of precision matrices. Thus, $g_{F^kM} + \lambda g_R$ can be intuitively thought of as adding isotropic noise of variance λ to the covariance represented by g_{F^kM}.

To pursue this, let F^kM denote the bundle of rank k linear maps $\mathbb{R}^k \to T_xM$. We define a cometric by

$$\langle \xi, \tilde{\xi} \rangle = \delta^{\alpha\beta}(\xi | h_u(u_\alpha))(\tilde{\xi} | h_u(u_\beta)) + \lambda \langle \xi, \tilde{\xi} \rangle_{g_R}$$

for $\xi, \tilde{\xi} \in T_u^*F^kM$. The sum over α, β is for $\alpha, \beta = 1, \ldots, k$. The first term is equivalent to the lift (9) of the cometric $\langle \xi, \tilde{\xi} \rangle = (\xi | g_u(\tilde{\xi}))$ given $u : \mathbb{R}^k \to T_xM$. Note that in the definition (6) of g_u, the map u is not inverted; thus, the definition of the metric immediately carries over to the rank-deficient case.

Let (x^i, u^i_α), $\alpha = 1, \ldots, k$ be a coordinate system on F^kM. The vertical distribution is in this case spanned by the dk vector fields $D_{j_\beta} = \partial_{u^j_\beta}$. Except for index sums being over k instead of d terms, the situation is thus similar to the full-rank case. Note that $(\xi | \pi_*^{-1} w) = (\xi | w^j D_j) = w^i \xi_i$. The cometric in coordinates is

$$\langle \xi, \tilde{\xi} \rangle = \delta^{\alpha\beta} u^i_\alpha \xi_i u^j_\beta \tilde{\xi}_j + \lambda g_R^{ij} \xi_i \tilde{\xi}_j = \xi_i \left(\delta^{\alpha\beta} u^i_\alpha u^j_\beta + \lambda g_R^{ij} \right) \tilde{\xi}_j = \xi_i W^{ij} \tilde{\xi}_j$$

with $W^{ij} = \delta^{\alpha\beta} u^i_\alpha u^j_\beta + \lambda g_R^{ij}$. We can then write the corresponding sub-Riemannian metric g_{F^kM} in terms of the adapted frame D

$$g_{F^kM}(\xi_h D^h + \xi_{h_\gamma} D^{h_\gamma}) = W^{ih} \xi_h D_i \tag{17}$$

because $(\xi | g_{F^kM}(\tilde{\xi})) = \langle \xi, \tilde{\xi} \rangle = \xi_i W^{ij} \tilde{\xi}_j$. That is, the situation is analogous to (11), except the term λg_R^{ij} is added to W^{ij}.

The geodesic system is again given by the Hamilton–Jacobi equations. As in the full-rank case, the system is specified by the derivatives of g_{F^kM}:

$$\begin{aligned}
&\partial_{y^l} g^{ij}_{F^kM} = W^{ij}{}_{,l}\,, \quad \partial_{y^l} g^{ij_\beta}_{F^kM} = -W^{ih}{}_{,l} \Gamma^{j_\beta}_h - W^{ih} \Gamma^{j_\beta}_{h,l}\,, \quad \partial_{y^l} g^{i_\alpha j}_{F^kM} = -\Gamma^{i_\alpha}_{h,l} W^{hj} - \Gamma^{i_\alpha}_h W^{hj}{}_{,l}\,, \\
&\partial_{y^l} g^{i_\alpha j_\beta}_{F^kM} = \Gamma^{i_\alpha}_{k,l} W^{kh} \Gamma^{j_\beta}_h + \Gamma^{i_\alpha}_k W^{kh}{}_{,l} \Gamma^{j_\beta}_h + \Gamma^{i_\alpha}_k W^{kh} \Gamma^{j_\beta}_{h,l}\,, \\
&\partial_{y^{l_\zeta}} g^{ij}_{F^kM} = W^{ij}{}_{,l_\zeta}\,, \quad \partial_{y^{l_\zeta}} g^{ij_\beta}_{F^kM} = -W^{ih}{}_{,l_\zeta} \Gamma^{j_\beta}_h - W^{ih} \Gamma^{j_\beta}_{h,l_\zeta}\,, \quad \partial_{y^{l_\zeta}} g^{i_\alpha j}_{F^kM} = -\Gamma^{i_\alpha}_h W^{hj}{}_{,l_\zeta} - \Gamma^{i_\alpha}_{h,l_\zeta} W^{hj}\,, \\
&\partial_{y^{l_\zeta}} g^{i_\alpha j_\beta}_{F^kM} = \Gamma^{i_\alpha}_{k,l_\zeta} W^{kh} \Gamma^{j_\beta}_h + \Gamma^{i_\alpha}_k W^{kh}{}_{,l_\zeta} \Gamma^{j_\beta}_h + \Gamma^{i_\alpha}_k W^{kh} \Gamma^{j_\beta}_{h,l_\zeta}\,, \\
&\Gamma^{i_\alpha}_{h,l_\zeta} = \partial_{y^{l_\zeta}} \left(\Gamma^i{}_{hk} u^k_\alpha \right) = \delta^{\zeta\alpha} \Gamma^i{}_{hl}\,, \quad W^{ij}{}_{,l} = \lambda g_R{}^{ij}{}_{,l}\,, \quad W^{ij}{}_{,l_\zeta} = \delta^{il} u^j_\zeta + \delta^{jl} u^i_\zeta\,.
\end{aligned}$$

Note that the introduction of the Riemannian metric g_R implies that W^{ij} are now dependent on the manifold coordinates x^i.

6. Numerical Experiments

We aim at visualizing most probable paths for the driving process and projections of curves satisfying the MPP Equation (13) in two cases: On 2D surfaces embedded in $\mathbb{R}^3$ and on finite dimensional landmark manifolds that arise from equipping a subset of the diffeomorphism group with a right-invariant metric and letting the action descend to the landmarks by a left action. The surface examples are implemented in Python using the Theano [30] framework for symbolic operations, automatic differentiation, and numerical evaluation. The landmark equations are detailed below and implemented in Numpy using Numpy's standard ODE integrators. The code for running the experiments is available at http://bitbucket.com/stefansommer/mpps/.

6.1. Embedded Surfaces

We visualize normal MPPs and projections of curves satisfying the MPP Equation (13) on surfaces embedded in $\mathbb{R}^3$ in three cases: The sphere $\mathbb{S}^2$, on an ellipsoid, and on a hyperbolic surface. The surfaces are chosen in order to have both positive and negative curvature, and to have varying degree of symmetry. In all cases, an open subset of the surfaces are represented in a single chart by a map $F : \mathbb{R}^2 \to \mathbb{R}^3$. For the sphere and ellipsoid, this gives a representation of the surface, except for the south pole. The metric and Christoffel symbols are calculated using the symbolic differentiation features of Theano. The integration are performed by a simple Euler integrator.

Figures 4–6 show families of curves satisfying the MPP equations in three cases: (1) With fixed starting point $x_0 \in M$ and initial velocity $\dot{x}_0 \in TM$ but varying anisotropy represented by changing frame u in the fiber above x_0; (2) minimizing normal MPPs with fixed starting point and endpoint $x_0, x_1 \in M$ but changing frame u above x_0; (3) fixed starting point $x_0 \in M$ and frame u but varying V^*FM vertical part of the initial momentum $\xi_0 \in T^*FM$. The first and second cases thus show the effect of varying anisotropy, while the third case illustrates the effect of the "twist" that the d^2 degrees in the vertical momentum allows. Note the displayed anti-developed curves in $\mathbb{R}^2$ that for classical C geodesics would always be straight lines.

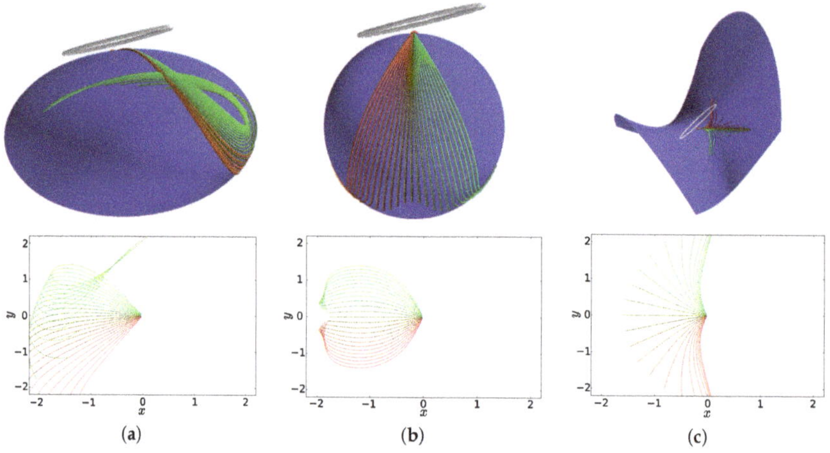

Figure 4. Curves satisfying the MPP equations (top row) and corresponding anti-development (bottom row) on three surfaces embedded in $\mathbb{R}^3$: (**a**) An ellipsoid; (**b**) a sphere; (**c**) a hyperbolic surface. The family of curves is generated by rotating by $\pi/2$ radians the anisotropic covariance represented in the initial frame u_0 and displayed in the gray ellipse.

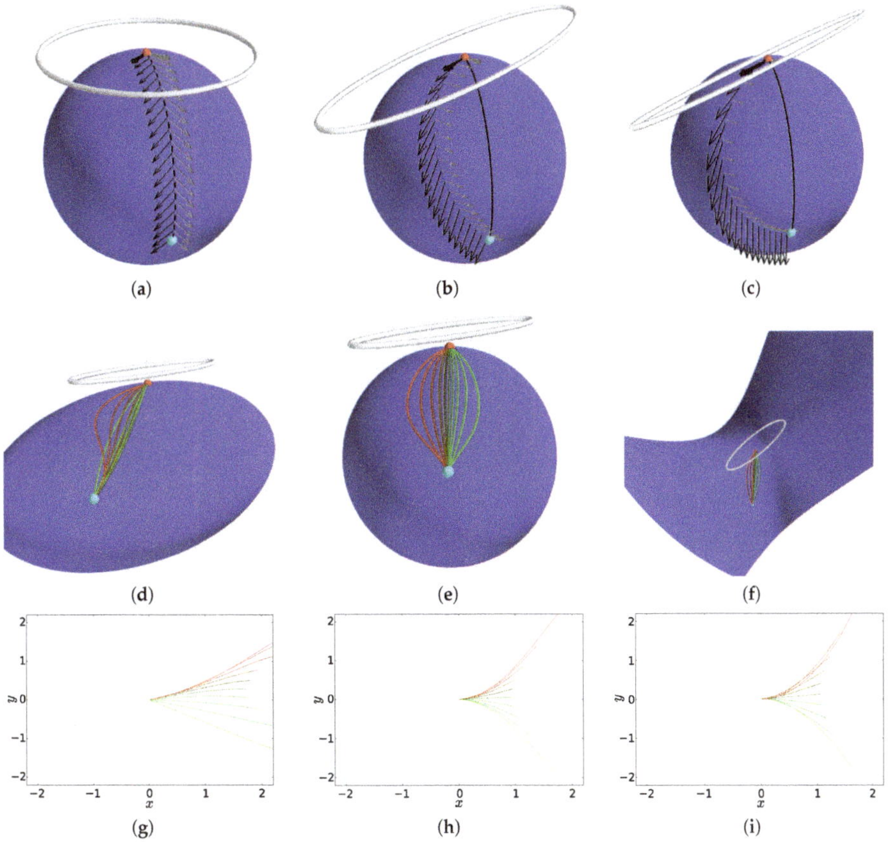

Figure 5. Minimizing normal MPPs between two fixed points (red/cyan). From isotropic covariance (top row, (**a**)) to anisotropic (top row, (**c**)) on $\mathbb{S}^2$. Compare with minimizing Riemannian geodesic (black curve). The MPP travels longer in the directions of high variance. Families of curves (middle row, (**d–f**)) and corresponding anti-development (bottom row, (**g–i**)) on the three surfaces in Figure 4. The family of curves is generated by rotating the covariance matrix as in Figure 4. Notice how the varying anisotropy affects the resulting minimizing curves, and how the anti-developed curves end at different points in $\mathbb{R}^2$.

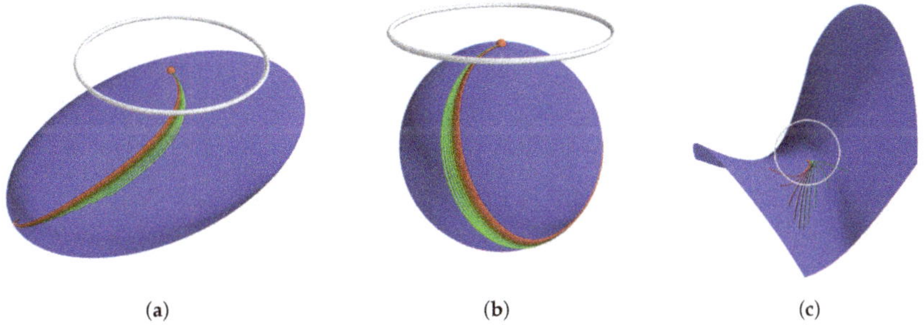

Figure 6. *Cont.*

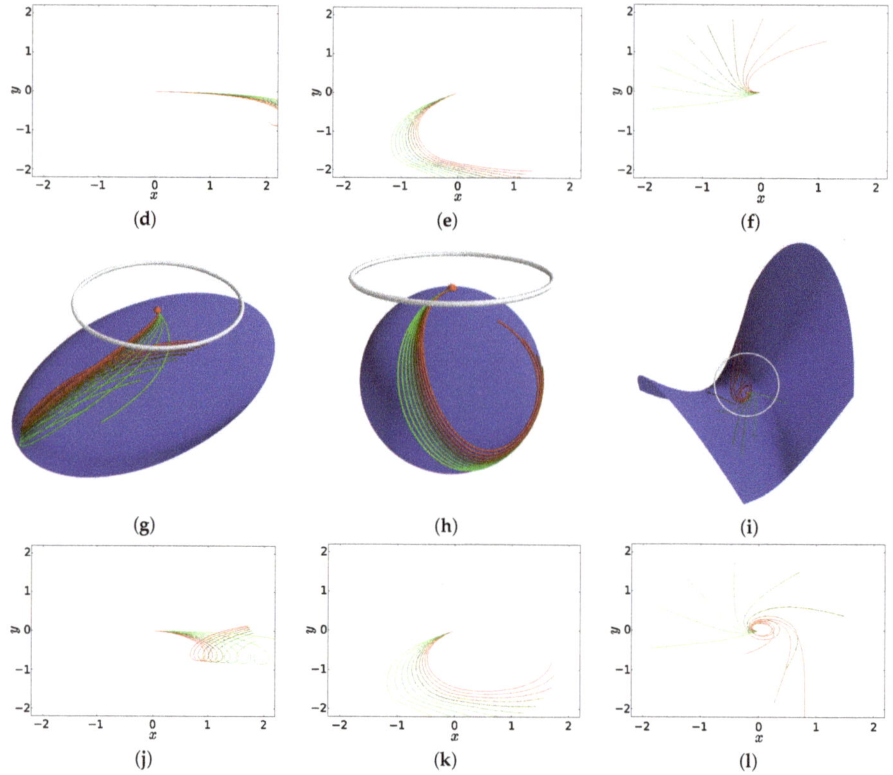

Figure 6. **(a–l)** With the setup of Figures 4 and 5, generated families of curves by varying the vertical V^*FM part of the initial momentum $\xi_0 \in T^*FM$ but keeping the base point and frame u_0 fixed. The vertical part allows varying degree of "twisting" of the curve.

6.2. LDDMM Landmark Equations

We here give a example of the MPP equations using the finite dimensional landmark manifolds that arise from right invariant metrics on subsets of the diffeomorphism group in the Large Deformation Diffeomorphic Metric Mapping (LDDMM) framework [8]. The LDDMM metric can be conveniently expressed as a cometric, and, using a rank-deficient inner product g_{F^kM} as discussed in Section 5, we can obtain a reduction of the system of equations to $2(2N + 2Nk)$ compared to $2(2N + (2N)^2)$ with N landmarks in $\mathbb{R}^2$.

Let $\{p_1, \ldots, p_N\}$ be landmarks in a subset $\Omega \subset \mathbb{R}^d$. The diffeomorphism group $\mathrm{Diff}(\Omega)$ acts on the left on landmarks with the action $\varphi.\{p_1, \ldots, p_N\} = \{\varphi(p_1), \ldots, \varphi(p_N)\}$. In LDDMM, a Hilbert space structure is imposed on a linear subspace V of $L^2(\Omega, \mathbb{R}^d)$ using a self-adjoint operator $L : V \to V^* \subset L^2(\Omega, \mathbb{R}^d)$ and defining the inner product $\langle \cdot, \cdot \rangle_V$ by

$$\langle v, w \rangle_V = \langle Lv, w \rangle_{L^2} .$$

Under sufficient conditions on L, V is reproducing and admits a kernel K inverse to L. K is a Green's kernel when L is a differential operator, or K can be a Gaussian kernel. The Hilbert structure on V gives a Riemannian metric on a subset $G_V \subset \mathrm{Diff}(\Omega)$ by setting $\|v\|^2_\varphi = \|v \circ \varphi^{-1}\|^2_V$; i.e., regarding $\langle \cdot, \cdot \rangle_V$ an inner product on $T_{\mathrm{Id}} G_V$ and extending the metric to G_V by right-invariance. This Riemannian metric descends to a Riemannian metric on the landmark space.

Let M be the manifold $M = \{(p_1^1, \ldots, p_1^d, \ldots, p_N^1, \ldots, p_N^d) | (p_i^1, \ldots, p_i^d) \in \mathbb{R}^d\}$. The LDDMM metric on the landmark manifold M is directly related to the kernel K when written as a cometric $g_p(\xi, \eta) = \sum_{i,j=1}^N \xi^i K(p_i, p_j)\eta^j$. Letting i^k denote the index of the kth component of the ith landmark, the cometric is in coordinates $g_p^{i^k j^l} = K(p_i, p_j)_k^l$. The Christoffel symbols can be written in terms of derivatives of the cometric g^{ij} [31] (recall that $\delta_j^i = g^{ik}g_{kj} = g_{jk}g^{ki}$)

$$\Gamma^k{}_{ij} = \frac{1}{2} g_{ir}\left(g^{kl}g^{rs}{}_{,l} - g^{sl}g^{rk}{}_{,l} - g^{rl}g^{ks}{}_{,l}\right) g_{sj} \, . \tag{18}$$

This relation comes from the fact that $g_{jm,k} = -g_{jr}g^{rs}{}_{,k}g_{sm}$ gives the derivative of the metric. The derivatives of the cometric is simply $g^{i^k j^l}{}_{,r^q} = (\delta_r^i + \delta_r^j)\partial_{p_r^q} K(p_i, p_j)_k^l$. Using (18), derivatives of the Christoffel symbols can be computed

$$\begin{aligned}\Gamma^k{}_{ij,\xi} &= \frac{1}{2} g_{ir,\xi}\left(g^{kl}g^{rs}{}_{,l} - g^{sl}g^{rk}{}_{,l} - g^{rl}g^{ks}{}_{,l}\right) g_{sj} + \frac{1}{2} g_{ir}\left(g^{kl}g^{rs}{}_{,l} - g^{sl}g^{rk}{}_{,l} - g^{rl}g^{ks}{}_{,l}\right) g_{sj,\xi} \\ &+ \frac{1}{2} g_{ir}\left(g^{kl}{}_{,\xi}g^{rs}{}_{,l} + g^{kl}g^{rs}{}_{,l\xi} - g^{sl}{}_{,\xi}g^{rk}{}_{,l} - g^{sl}g^{rk}{}_{,l\xi} - g^{rl}{}_{,\xi}g^{ks}{}_{,l} - g^{rl}g^{ks}{}_{,l\xi}\right) g_{sj} \, .\end{aligned}$$

This provides the full data for numerical integration of the evolution equations on F^kM.

In Figure 7 (top row), we plot minimizing normal MPPs on the landmark manifold with two landmarks and varying covariance in the $\mathbb{R}^2$ horizontal and vertical direction. The plot shows the landmark equivalent of the experiment in Figure 5. Note how adding covariance in the horizontal and vertical direction, respectively, allows the minimizing normal MPP to vary more in these directions because the anisotropically-weighted metric penalizes high-covariance directions less.

Figure 7 (bottom row) shows five curves satisfying the MPP equations with varying vertical V^*FM initial momentum similarly to the plots in Figure 6. Again, we see how the extra degrees of freedom allows the paths to twist, generating a higher-dimensional family than classical geodesics with respect to $\mathcal{C}$.

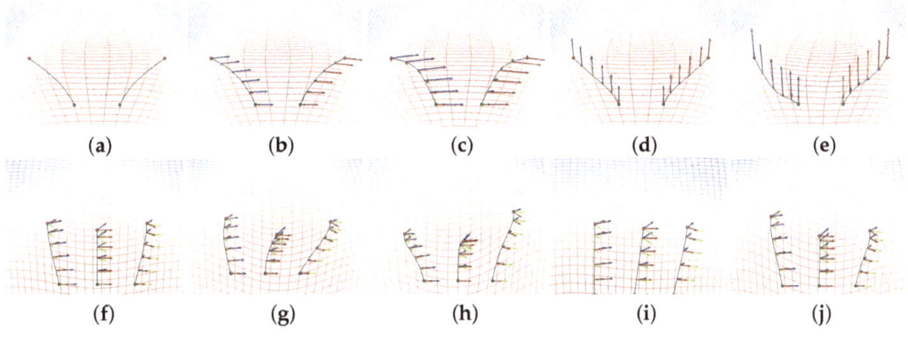

Figure 7. (Top row) Matching of two landmarks (green) to two landmarks (red) by (**a**) computing a minimizing Riemannian geodesic on the landmark manifold, and (**b–e**) minimizing MPPs with added covariance (arrows) in $\mathbb{R}^2$ horizontal direction (**b**,**c**) and vertical (**d**,**e**). The action of the corresponding diffeomorphisms on a regular grid is visualized by the deformed grid which is colored by the warp strain. The added covariance allows the paths to have more movement in the horizontal and vertical direction, respectively, because the anisotropically weighted metric penalizes high-covariance directions less. (bottom row, (**f–j**)) Five landmark trajectories with fixed initial velocity and anisotropic covariance but varying V^*FM vertical initial momentum ξ_0. Changing the vertical momentum "twists" the paths.

7. Discussion and Concluding Remarks

Incorporating anisotropy in models for data in non-linear spaces via the frame bundle as pursued in this paper leads to a sub-Riemannian structure and metric. A direct implication is that most probable paths to observed data in the sense of sequences of stochastic steps of a driving semi-martingale are not related to geodesics in the classical sense. Instead, a best estimate of the sequence of steps $w_t \in \mathbb{R}^d$ that leads to an observation $x = \varphi_u(w_t)|_{t=1}$ is an MPP in the sense of Definition 1. As shown in the paper, these paths are generally not geodesics or polynomials with respect to the connection on the manifold. In particular, if M has a Riemannian structure, the MPPs are generally neither Riemannian geodesics nor Riemannian polynomials. Below, we discuss the statistical implications of this result.

7.1. Statistical Estimators

Metric distances and Riemannian geodesics have been the traditional vehicle for representing observed data in non-linear spaces. Most fundamentally, the sample Frechét mean

$$\hat{x} = \text{argmin}_{x \in M} \sum_{i=1}^{N} d_{g_R}(x, x_i)^2 \tag{19}$$

of observed data $x_1, \ldots, x_N \in M$ relies crucially on the Riemannian distance d_{g_R} connected to the metric g_R. Many PCA constructs (e.g., Principal Geodesics Analysis [6]) use the Riemannian Exp. and Log maps to map between linear tangent spaces and the manifold. These maps are defined from the Riemannian metric and Riemannian geodesics. Distributions modelled as in the random orbit model [32] or Bayesian models [15,33] again rely on geodesics with random initial conditions.

Using the frame bundle sub-Riemannian metric g_{FM}, we can define an estimator analogous to the Riemannian Frechét mean estimator. Assuming the covariance is a priori known, the estimator

$$\hat{x} = \text{argmin}_{u \in s(M)} \sum_{i=1}^{N} d_{FM}\left(u, \pi^{-1}(x_i)\right)^2 \tag{20}$$

acts correspondingly to the Frechét mean estimator (19). Here $s \in \Gamma(FM)$ is a (local) section of FM that to $x \in M$ connects the known covariance represented by $s(x) \in FM$. The distances $d_{FM}\left(u, \pi^{-1}(x_i)\right)$, $u = s(x)$ are realized by MPPs from the mean candidate x to the fibers $\pi^{-1}(x_i)$. The Frechét mean problem is thus lifted to the frame bundle with the anisotropic weighting incorporated in the metric g_{FM}. This metric is not related to g_R, except for its dependence on the connection C that can be defined as the Levi–Civita connection of g_R. The fundamental role of the distance d_{g_R} and g_R geodesics in (19) is thus removed.

Because covariance is an integral part of the model, sample covariance can also be estimated directly along with the sample mean. In [3], the estimator

$$\hat{u} = \text{argmin}_{u \in FM} \sum_{i=1}^{N} d_{FM}\left(u, \pi^{-1}(x_i)\right)^2 - N \log(\det_{g_R} u) \tag{21}$$

is suggested. The normalizing term $-N \log(\det_{g_R} u)$ is derived such that the estimator exactly corresponds to the maximum likelihood estimator of mean and covariance for Euclidean Gaussian distributions. The determinant is defined via g_R, and the term acts to prevent the covariance from approaching infinity. Maximum likelihood estimators of mean and covariance for normally distributed Euclidean data have unique solutions in the sample mean and sample covariance matrix, respectively. Uniqueness of the Frechét mean (19) is only ensured for sufficiently concentrated data. For the estimator (21), existence and uniqueness properties are not immediate, and more work is needed in order to find necessary and sufficient conditions.

7.2. Priors and Low-Rank Estimation

The low-rank cometric formulation pursued in Section 5 gives a natural restriction of (21) to $u \in F^kM$, $1 \leq k \leq d$. As for Euclidean PCA, most variance is often captured in the span of the first k eigenvectors with $k \ll d$. Estimates of the remaining eigenvectors are generally ignored, as the variance of the eigenvector estimates increases as the noise captured in the span of the last eigenvectors becomes increasingly uniform. The low-rank cometric restricts the estimation to only the first k eigenvectors, and thus builds the construction directly into the model. In addition, it makes numerical implementation feasible, because a numerical representation need only store and evolve $d \times k$ matrices. As a different approach for regularizing the estimator (21), the normalizing term $-N \log(\det_{g_R} u)$ can be extended with other priors (e.g., an L^1-type penalizing term). Such priors can potentially partly remove existence and uniqueness issues, and result in additional sparsity properties that can benefit numerical implementations. The effects of such priors have yet to be investigated.

In the $k = d$ case, the number of degrees of freedom for the MPPs grows quadratically in the dimension d. This naturally increases the variance of any MPP estimate given only one sample from its trajectory. The low-rank cometric formulation reduces the growth to linear in d. The number of degrees of freedom is however still k times larger than for Riemannian geodesics. With longitudinal data, more samples per trajectory can be obtained, reducing the variance and allowing a better estimate of the MPP. However, for the estimators (20) and (21) above, estimates of the actual optimal MPPs are not needed—only their squared length. It can be hypothesized that the variance of the length estimates is lower than the variance of the estimates of the corresponding MPPs. Further investigation regarding this will be the subject of future work.

7.3. Conclusions

The underlying model of anisotropy used in this paper originates from the anisotropic normal distributions formulated in [2] and the diffusion PCA framework [1]. Because many statistical models are defined using normal distributions, this approach to incorporating anisotropy extends to models such as linear regression. We expect that finding most probable paths in other statistical models such as regressions models can be carried out with a program similar to the program presented in this paper.

The difference between MPPs and geodesics shows that the geometric and metric properties of geodesics, zero acceleration, and local distance minimization are not directly related to statistical properties such as maximizing path probability. Whereas the concrete application and model determines if metric or statistical properties are fundamental, most statistical models are formulated without referring to metric properties of the underlying space. It can therefore be argued that the direct incorporation of anisotropy and the resulting MPPs are natural in the context of many models of data variation in non-liner spaces.

Acknowledgments: The author wishes to thank Peter W. Michor and Sarang Joshi for suggestions for the geometric interpretation of the sub-Riemannian metric on FM and discussions on diffusion processes on manifolds. The work was supported by the Danish Council for Independent Research, the CSGB Centre for Stochastic Geometry and Advanced Bioimaging funded by a grant from the Villum foundation, and the Erwin Schrödinger Institute in Vienna.

Conflicts of Interest: The author declares no conflict of interest.

References

1. Sommer, S. Diffusion Processes and PCA on Manifolds. Available online: https://www.mfo.de/document/1440a/OWR_2014_44.pdf (accessed on 24 November 2016).
2. Sommer, S. Anisotropic distributions on manifolds: Template estimation and most probable paths. In *Information Processing in Medical Imaging*; Lecture Notes in Computer Science; Springer: Berlin/Heidelberg, Germany, 2015; Volume 9123, pp. 193–204.
3. Sommer, S.; Svane, A.M. Modelling anisotropic covariance using stochastic development and sub-riemannian frame bundle geometry. *J. Geom. Mech.* **2016**, in press.

4. Hsu, E.P. *Stochastic Analysis on Manifolds*; American Mathematical Society: Providence, RI, USA, 2002.
5. Fréchet, M. Les éléments aléatoires de nature quelconque dans un espace distancie. *Annales de l'Institut Henri Poincaré* **1948**, *10*, 215–310.
6. Fletcher, P.; Lu, C.; Pizer, S.; Joshi, S. Principal geodesic analysis for the study of nonlinear statistics of shape. *IEEE Trans. Med. Imaging* **2004**, *23*, 995–1005.
7. Vaillant, M.; Miller, M.; Younes, L.; Trouvé, A. Statistics on diffeomorphisms via tangent space representations. *NeuroImage* **2004**, *23*, S161–S169.
8. Younes, L. *Shapes and Diffeomorphisms*; Springer: Berlin/Heidelberg, Germany, 2010.
9. Pennec, X. Intrinsic statistics on riemannian manifolds: Basic tools for geometric measurements. *J. Math. Imaging Vis.* **2006**, *25*, 127–154.
10. Karcher, H. Riemannian center of mass and mollifier smoothing. *Commun. Pure Appl. Math.* **1977**, *30*, 509–541.
11. Huckemann, S.; Hotz, T.; Munk, A. Intrinsic shape analysis: Geodesic PCA for Riemannian manifolds modulo isometric lie group actions. *Stat. Sin.* **2010**, *20*, 1–100.
12. Jung, S.; Dryden, I.L.; Marron, J.S. Analysis of principal nested spheres. *Biometrika* **2012**, *99*, 551–568.
13. Pennec, X. Barycentric subspaces and affine spans in manifolds. In Proceedings of the Second International Conference on Geometric Science of Information, Paris, France, 28–30 October 2015; Nielsen, F., Barbaresco, F., Eds.; Lecture Notes in Computer Science; pp. 12–21.
14. Sommer, S. Horizontal dimensionality reduction and iterated frame bundle development. In *Geometric Science of Information*; Springer: Berlin/Heidelberg, Germany, 2013; pp. 76–83.
15. Zhang, M.; Fletcher, P. Probabilistic principal geodesic analysis. In Proceedings of the 26th International Conference on Neural Information Processing Systems, Lake Tahoe, Nevada, 5–10 December 2013; pp. 1178–1186.
16. Tipping, M.E.; Bishop, C.M. Probabilistic principal component analysis. *J. R. Stat. Soc. Ser. B* **1999**, *61*, 611–622.
17. Elworthy, D. Geometric aspects of diffusions on manifolds. In *École d'Été de Probabilités de Saint-Flour XV–XVII, 1985–1987*; Hennequin, P.L., Ed.; Number 1362 in Lecture Notes in Mathematics; Springer: Berlin/Heidelberg, Germany, 1988; pp. 277–425.
18. Mok, K.P. On the differential geometry of frame bundles of Riemannian manifolds. *J. Reine Angew. Math.* **1978**, *1978*, 16–31.
19. Taubes, C.H. *Differential Geometry: Bundles, Connections, Metrics and Curvature*, 1st ed.; Oxford University Press: Oxford, UK; New York, NY, USA, 2011.
20. Kolář, I.; Slovák, J.; Michor, P.W. *Natural Operations in Differential Geometry*; Springer: Berlin/Heidelberg, Germany, 1993.
21. Andersson, L.; Driver, B.K. Finite dimensional approximations to wiener measure and path integral formulas on manifolds. *J. Funct. Anal.* **1999**, *165*, 430–498.
22. Fujita, T.; Kotani, S.i. The Onsager-Machlup function for diffusion processes. *J. Math. Kyoto Univ.* **1982**, *22*, 115–130.
23. Strichartz, R.S. Sub-Riemannian geometry. *J. Differ. Geom.* **1986**, *24*, 221–263.
24. Bloch, A.M. Nonholonomic mechanics and control. In *Interdisciplinary Applied Mathematics*; Springer: New York, NY, USA, 2003; Volume 24,
25. Marsden, J.E.; Ratiu, T.S. Introduction to mechanics and symmetry. In *Texts in Applied Mathematics*; Springer: New York, NY, USA, 1999; Volume 17,
26. Leite, F.S.; Krakowski, K.A. *Covariant Differentiation under Rolling Maps*; Centro de Matemática da Universidade de Coimbra: Coimbra, Portugal, 2008.
27. Hinkle, J.; Fletcher, P.T.; Joshi, S. Intrinsic polynomials for regression on riemannian manifolds. *J. Math. Imaging Vis.* **2014**, *50*, 32–52.
28. Noakes, L.; Heinzinger, G.; Paden, B. Cubic splines on curved spaces. *IMA J. Math. Control Inf.* **1989**, *6*, 465–473.
29. Camarinha, M.; Silva Leite, F.; Crouch, P. On the geometry of Riemannian cubic polynomials. *Differ. Geom. Appl.* **2001**, *15*, 107–135.
30. Team, T.T.D.; Al-Rfou, R.; Alain, G.; Almahairi, A.; Angermueller, C.; Bahdanau, D.; Ballas, N.; Bastien, F.; Bayer, J.; Belikov, A.; et al. Theano: A Python framework for fast computation of mathematical expressions. *arXiv* **2016**, arXiv:1605.02688.

31. Micheli, M. The Differential Geometry of Landmark Shape Manifolds: Metrics, Geodesics, and Curvature. Ph.D. Thesis, Brown University, Providence, RI, USA, 2008.
32. Miller, M.; Banerjee, A.; Christensen, G.; Joshi, S.; Khaneja, N.; Grenander, U.; Matejic, L. Statistical methods in computational anatomy. *Stat. Methods Med. Res.* **1997**, *6*, 267–299.
33. Zhang, M.; Singh, N.; Fletcher, P.T. Bayesian estimation of regularization and atlas building in diffeomorphic image registration. In *Information Processing for Medical* Imaging (IPMI); Lecture Notes in Computer Science; Springer: Berlin/Heidelberg, Germany, 2013; pp. 37–48.

entropy

MDPI

Article

Non-Asymptotic Confidence Sets for Circular Means †

Thomas Hotz *,‡, Florian Kelma ‡ and Johannes Wieditz ‡

Institut für Mathematik, Technische Universität Ilmenau, 98684 Ilmenau, Germany; florian.kelma@tu-ilmenau.de (F.K.); johannes.wieditz@tu-ilmenau.de (J.W.)

* Correspondence: thomas.hotz@tu-ilmenau.de; Tel.: +49-3677-69-3627

† This paper is an extended version of our paper published in Proceedings of the 2nd International Conference on Geometric Science of Information, Palaiseau, France, 28–30 October 2015; Nielsen, F., Barbaresco, F., Eds.; Lecture Notes in Computer Science, Volume 9389; Springer International Publishing: Cham, Switzerland, 2015; pp. 635–642.

‡ These authors contributed equally to this work.

Academic Editors: Frédéric Barbaresco and Frank Nielsen

Received: 15 July 2016; Accepted: 13 October 2016; Published: 20 October 2016

Abstract: The mean of data on the unit circle is defined as the minimizer of the average squared Euclidean distance to the data. Based on Hoeffding's mass concentration inequalities, non-asymptotic confidence sets for circular means are constructed which are universal in the sense that they require no distributional assumptions. These are then compared with asymptotic confidence sets in simulations and for a real data set.

Keywords: directional data; circular mean; universal confidence sets; non-asymptotic confidence sets; mass concentration inequalities; Hoeffding's inequality

MSC: 62H11; 62G15

---

1. Introduction

In applications, data assuming values on the circle, i.e., *circular data*, arise frequently, examples being measurements of wind directions, or time of the day that patients are admitted to a hospital unit. We refer to the literature, e.g., [1–5], for an overview of statistical methods for circular data, in particular the ones described in this section.

Here, we will concern ourselves with the arguably simplest statistic, the *mean*. However, given that a circle does not carry a vector space structure, i.e., there is neither a natural addition of points on the circle nor can one divide them by a natural number, what should the meaning of "mean" be?

In order to simplify the exposition, we specifically consider the unit circle in the complex plane, $S^1 = \{z \in \mathbf{C} : |z| = 1\}$, and we assume the data can be modelled as independent random variables $Z_1, \ldots, Z_n$ which are identically distributed as the random variable Z taking values in S^1. In the literature, however, the circle is often taken to lie in the real plane $\mathbf{R}^2$, i.e., while we denote the point on the circle corresponding to an angle $\theta \in (-\pi, \pi]$ by $\exp(i\theta) = \cos(\theta) + i\sin(\theta) \in \mathbf{C}$ one may take it to be $(\cos\theta, \sin\theta) \in \mathbf{R}^2$.

Of course, $\mathbf{C}$ is a real vector space, so the *Euclidean sample mean* $\bar{Z}_n = \frac{1}{n}\sum_{k=1}^{n} Z_k \in \mathbf{C}$ is well-defined. However, unless all Z_k take identical values, it will (by the strict convexity of the closed unit disc) lie inside the circle, i.e., its *modulus* $|\bar{Z}_n|$ will be less than 1. Though $\bar{Z}_n$ cannot be taken as a mean on the circle, if $\bar{Z}_n \neq 0$, one might say that it specifies a direction; this leads to the idea of calling $\bar{Z}_n / |\bar{Z}_n|$ the circular sample mean of the data.

Observing that the Euclidean sample mean is the minimiser of the sum of squared distances, this can be put in the more general framework of *Fréchet means* [6]: define the *set of circular sample means* to be

$$\hat{\mu}_n = \operatorname*{argmin}_{\zeta \in \mathbf{S}^1} \sum_{k=1}^{n} |Z_k - \zeta|^2, \tag{1}$$

and analoguously define the *set of circular population means* of the random variable Z to be

$$\mu = \operatorname*{argmin}_{\zeta \in \mathbf{S}^1} \mathbf{E}\,|Z - \zeta|^2. \tag{2}$$

Then, as usual, the circular sample means are the circular population means with respect to the empirical distribution of $Z_1, \ldots, Z_n$.

The circular population mean can be related to the Euclidean population mean $\mathbf{E}\,Z$ by noting that $\mathbf{E}\,|Z-\zeta|^2 = \mathbf{E}\,|Z - \mathbf{E}\,Z|^2 + |\,\mathbf{E}\,Z - \zeta|^2$ (in statistics, this is called the *bias-variance decomposition*), so that

$$\mu = \operatorname*{argmin}_{\zeta \in \mathbf{S}^1} |\,\mathbf{E}\,Z - \zeta|^2 \tag{3}$$

is the set of points on the circle closest to $\mathbf{E}\,Z$. It follows that μ is unique if and only if $\mathbf{E}\,Z \neq 0$ in which case it is given by $\mu = \mathbf{E}\,Z/|\,\mathbf{E}\,Z|$, the *orthogonal projection* of $\mathbf{E}\,Z$ onto the circle; otherwise, i.e., if $\mathbf{E}\,Z = 0$, the set of circular population means is all of S^1. We consider the information of whether the circular population mean is not unique, e.g., but not exclusively because Z is uniformly distributed over the circle, to be relevant; it thus should be inferred from the data as well. Analogously, $\hat{\mu}_n$ is either all of S^1 or uniquely given by $\bar{Z}_n/|\bar{Z}_n|$ according to whether $\bar{Z}_n$ is 0 or not. Note that $\bar{Z}_n \neq 0$ a.s. if Z is continuously distributed on the circle, even if $\mathbf{E}\,Z = 0$. $\bar{Z}_n$ is what is known as the *vector resultant*, while $\bar{Z}_n/|\bar{Z}_n|$ is sometimes referred to as the *mean direction*.

The expected squared distances minimised in Equation (2) are given by the metric inherited from the ambient space $\mathbf{C}$; therefore, μ is also called the set of *extrinsic* population means. If we measured distances intrinsically along the circle, i.e., using arc-length instead of chordal distance, we would obtain what is called the set of *intrinsic* population means. We will not consider the latter in the following, see e.g., [7] for a comparison and [8,9] for generalizations of these concepts.

Our aim is to construct *confidence sets* for the circular population mean μ that form a superset of μ with a certain (so-called) *coverage probability* that is required to be not less than some pre-specified significance level $1 - \alpha$ for $\alpha \in (0, 1)$.

The classical approach is to construct an *asymptotic confidence interval* where the coverage probability converges to $1 - \alpha$ when n tends to infinity. This can be done as follows: since Z is a bounded random variable, $\sqrt{n}(\bar{Z}_n - \mathbf{E}\,Z)$ converges to a bivariate normal distribution when identifying $\mathbf{C}$ with $\mathbf{R}^2$. Now, assume $\mathbf{E}\,Z \neq 0$ so μ is unique. Then, the orthogonal projection is differentiable in a neighbourhood of $\mathbf{E}\,Z$, so the δ-method (see e.g., [1] (p. 111) or [4] (Lemma 3.1)) can be applied and one easily obtains

$$\sqrt{n}\,\mathbf{Arg}(\mu^{-1}\hat{\mu}_n) \xrightarrow{\mathcal{D}} \mathcal{N}\left(0, \frac{\mathbf{E}(\mathbf{Im}(\mu^{-1}Z))^2}{|\,\mathbf{E}\,Z|^2}\right), \tag{4}$$

where $\mathbf{Arg} : \mathbf{C} \setminus \{0\} \to (-\pi, \pi] \subset \mathbf{R}$ denotes the *argument* of a complex number (it is defined arbitrarily at $0 \in \mathbf{C}$), while multiplying with μ^{-1} rotates such that $\mathbf{E}\,Z = \mu$ is mapped to $0 \in (-\pi, \pi]$, see e.g., [4] (Proposition 3.1) or [7] (Theorem 5). Estimating the asymptotic variance and applying Slutsky's lemma, one arrives at the asymptotic confidence set $C_A = \{\zeta \in S^1 : |\,\mathbf{Arg}(\zeta^{-1}\hat{\mu}_n)| < \delta_A\}$ provided $\hat{\mu}_n$ is unique, where the angle determining the interval is given by

$$\delta_A = \frac{q_{1-\frac{\alpha}{2}}}{n|\bar{Z}_n|}\sqrt{\sum_{k=1}^{n}\left(\mathbf{Im}(\hat{\mu}_n^{-1}Z_k)\right)^2}, \tag{5}$$

with $q_{1-\frac{\alpha}{2}}$ denoting the $(1-\frac{\alpha}{2})$-quantile of the standard normal distribution $\mathcal{N}(0,1)$.

There are two major drawbacks to the use of asymptotic confidence intervals: firstly, by definition, they do not guarantee a coverage probability of at least $1-\alpha$ for finite n, so the coverage probability for a fixed distribution and sample size may be much smaller. Indeed, Simulation 2 in Section 4 demonstrates that, even for $n = 100$, the coverage probability may be as low as 64% when constructing the asymptotic confidence set for $1-\alpha = 90\%$. Secondly, they *assume* that $\mathbf{E}\,Z \neq 0$, so they are not applicable to all distributions on the circle. Since in practice it is unknown whether this assumption hold, one would have to test the hypothesis $\mathbf{E}\,Z = 0$, possibly again by an asymptotic test, and construct the confidence set conditioned on this hypothesis having been rejected, setting $C_A = S^1$ otherwise. However, this sequential procedure would require some adaptation taking the pre-test into account (cf. e.g., [10])—we come back to this point in Section 5—and it is not commonly implemented in practice.

We therefore aim to construct *non-asymptotic* confidence sets for μ, guaranteeing coverage with at least the desired probability for any sample size n, which in addition are *universal* in the sense that they do not make any distributional assumptions about the circular data besides them being independent and identically distributed. It has been shown in [7] that this is possible; however, the confidence sets that were constructed there were far too large to be of use in practice. Nonetheless, we start by varying that construction in Section 2 but using Hoeffding's inequality instead of Chebyshev's as in [7]. Considerable improvements are possible if one takes the variance $\mathbf{E}(\mathbf{Im}(\mu^{-1}Z))^2$ "perpendicular to $\mathbf{E}\,Z$" into account; this is achieved by a second construction in Section 3. Of course, the latter confidence sets will still be conservative but Proposition 2(iv) shows that they are (for $1-\alpha = 95\%$) only a factor $\sim \frac{3}{2}$ longer than the asymptotic ones when the sample size n is large. We further illustrate and compare those confidence sets in simulations and for an application to real data in Section 4, discussing the results obtained in Section 5.

2. Construction Using Hoeffding's Inequality

We will construct a confidence set as the acceptance region of a series of tests. This idea has been used before for the construction of confidence sets for the circular population mean [7] (Section 6); however, we will modify that construction by replacing Chebyshev's inequality—which is too conservative here—by three applications of Hoeffding's inequality [11] (Theorem 1): if $U_1, \ldots, U_n$ are independent random variables taking values in the bounded interval $[a,b]$ with $-\infty < a < b < \infty$. Then, $\bar{U}_n = \frac{1}{n}\sum_{k=1}^{n} U_k$ with $\mathbf{E}\,\bar{U}_n = \nu$ fulfills

$$\mathbf{P}(\bar{U}_n - \nu \geq t) \leq \left[\left(\frac{\nu-a}{\nu-a+t}\right)^{\nu-a+t}\left(\frac{b-\nu}{b-\nu-t}\right)^{b-\nu-t}\right]^{\frac{n}{b-a}} \tag{6}$$

for any $t \in (0, b-\nu)$. The bound on the right-hand side—denoted $\beta(t)$—is continuous and strictly decreasing in t (as expected; see Appendix A) with $\beta(0) = 1$ and $\lim_{t\to b-\nu}\beta(t) = \left(\frac{\nu-a}{b-a}\right)^n$ whence a unique solution $t = t(\gamma,\nu,a,b)$ to the equation $\beta(t) = \gamma$ exists for any $\gamma \in \left(\left(\frac{\nu-a}{b-a}\right)^n, 1\right)$. Equivalently, $t(\gamma,\nu,a,b)$ is strictly decreasing in γ. Furthermore, $\nu + t(\gamma,\nu,a,b)$ is strictly increasing in ν (see Appendix A again), which is also to be expected. While there is no closed form expression for $t(\gamma,\nu,a,b)$, it can without difficulty be determined numerically.

Note that the estimate

$$\beta(t) \leq \exp(-2nt^2/(b-a)^2) \tag{7}$$

is often used and called Hoeffding's inequality [11]. While this would allow to solve explicitly for t, we prefer to work with β as it is sharper, especially for ν close to b as well as for large t. Nonetheless, it shows that the tail bound $\beta(t)$ tends to zero as fast as if using the central limit theorem which is why it is widely applied for bounded variables, see e.g., [12].

Now, for any $\zeta \in \mathrm{S}^1$, we will test the hypothesis that ζ is a circular population mean. This hypothesis is equivalent to saying that there is some $\lambda \in [0,1]$ such that $\mathbf{E}\,Z = \lambda\zeta$. Multiplication by ζ^{-1} then rotates $\mathbf{E}\,Z$ onto the non-negative real axis: $\mathbf{E}\,\zeta^{-1}Z = \lambda \geq 0$.

Now, fix ζ and consider $X_k = \mathbf{Re}(\zeta^{-1}Z_k)$, $Y_k = \mathbf{Im}(\zeta^{-1}Z_k)$ for $k = 1,\ldots,n$ which may be viewed as the projection of $Z_1,\ldots,Z_k$ onto the line in the direction of ζ and onto the line perpendicular to it. Both are sequences of independent random variables taking values in $[-1,1]$ with $\mathbf{E}\,X_k = \lambda$ and $\mathbf{E}\,Y_k = 0$ under the hypothesis. They thus fulfill the conditions for Hoeffding's inequality with $a = -1$, $b = 1$ and $\nu = \lambda$ or 0, respectively.

We will first consider the case of non-uniqueness of the circular mean, i.e., $\mu = \mathrm{S}^1$, or equivalently $\lambda = 0$. Then, the critical value $s_0 = t(\frac{\alpha}{4}, 0, -1, 1)$ is well-defined for any $\frac{\alpha}{4} > 2^{-n}$, and we get $\mathbf{P}(\bar{X}_n \geq s_0) \leq \frac{\alpha}{4}$, and also, by considering $-X_1,\ldots,-X_n$, that $\mathbf{P}(-\bar{X}_n \geq s_0) \leq \frac{\alpha}{4}$. Analogously, $\mathbf{P}(|\bar{Y}_n| \geq s_0) \leq 2\frac{\alpha}{4} = \frac{\alpha}{2}$. We conclude that

$$\mathbf{P}\big(|\bar{Z}_n| \geq \sqrt{2}s_0\big) = \mathbf{P}\big(|\bar{X}_n|^2 + |\bar{Y}_n|^2 \geq 2s_0^2\big) \leq \mathbf{P}\big(|\bar{X}_n|^2 \geq s_0^2\big) + \mathbf{P}\big(|\bar{Y}_n|^2 \geq s_0^2\big) \leq \alpha.$$

Rejecting the hypothesis $\mu = \mathrm{S}^1$, i.e., $\mathbf{E}\,Z = 0$, if $|\bar{Z}_n| \geq \sqrt{2}s_0$ thus leads to a test whose probability of false rejection is at most α (see Figure 1). Of course, one may work with $|\bar{X}_n|^2 \geq s_0^2$ and $|\bar{Y}_n|^2 \geq s_0^2$ as criterions for rejection; however, we prefer working with $|\bar{Z}_n| \geq \sqrt{2}s_0$ since it is independent of the chosen ζ.

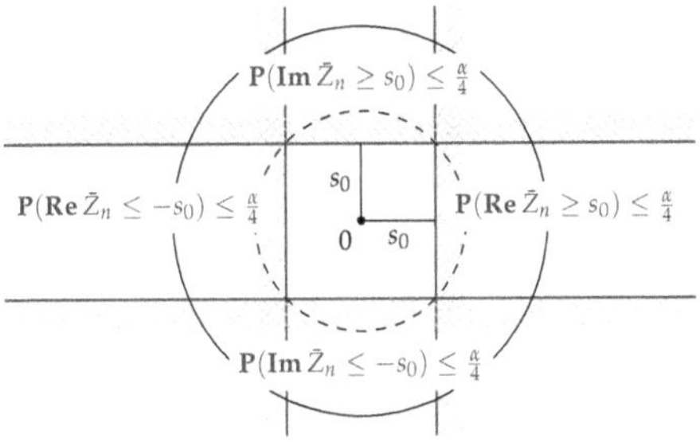

Figure 1. The construction for the test of the hypothesis $\mu = \mathrm{S}^1$, or equivalently $\mathbf{E}\,Z = 0$.

In the case of uniqueness of the circular mean, i.e., for the hypothesis $\lambda > 0$, we use the monotonicity of $\nu + t(\gamma,\nu,a,b)$ in ν and obtain

$$\mathbf{P}\big(\bar{X}_n \leq -s_0\big) = \mathbf{P}\big(-\bar{X}_n \geq t(\tfrac{\alpha}{4},0,-1,1)\big) \leq \mathbf{P}\big(-\bar{X}_n \geq -\lambda + t(\tfrac{\alpha}{4},-\lambda,-1,1)\big) \leq \frac{\alpha}{4}$$

as well. For the direction perpendicular to the direction of ζ (see Figure 2), however, we may now work with $\frac{3}{8}\alpha$, so for $s_p = t(\frac{3}{8}\alpha, 0, -1, 1)$—which is well-defined whenever s_0 is since $\frac{3}{8}\alpha > \frac{\alpha}{4} > 2^{-n}$—we obtain

$$\mathbf{P}\big(\bar{Y}_n \geq s_p\big) + \mathbf{P}\big(\bar{Y}_n \leq -s_p\big) \leq 2 \cdot \tfrac{3}{8}\alpha.$$

Rejecting if $\bar{X}_n \leq -s_0$ or $|\bar{Y}_n| \geq s_p$, then, will happen with probability at most $\frac{\alpha}{4} + 2 \cdot \frac{3}{8}\alpha = \alpha$ under the hypothesis $\mu = \zeta$. In case that we already rejected the hypothesis $\mu = \mathrm{S}^1$, i.e., if $|\bar{Z}_n| \geq \sqrt{2}s_0$, ζ will not be rejected if and only if $\bar{X}_n > s_0 > 0$ and $|\bar{Y}_n| < s_p < s_0$ which is then equivalent to $|\mathbf{Arg}(\zeta^{-1}\bar{Z}_n)| = \arcsin(|\bar{Y}_n|/|\bar{Z}_n|) < \arcsin(s_p/|\bar{Z}_n|) = \delta_H$ (see Figure 3).

Define C_H as all ζ which we could not reject, i.e.,

$$C_H = \begin{cases} \mathrm{S}^1, & \text{if } \alpha \leq 2^{-n+2} \text{ or } |\bar{Z}_n| \leq \sqrt{2}s_0, \\ \{\zeta \in \mathrm{S}^1 : |\mathbf{Arg}(\zeta^{-1}\hat{\mu}_n)| < \delta_H\} & \text{otherwise.} \end{cases} \tag{8}$$

Then, we obtain the following result:

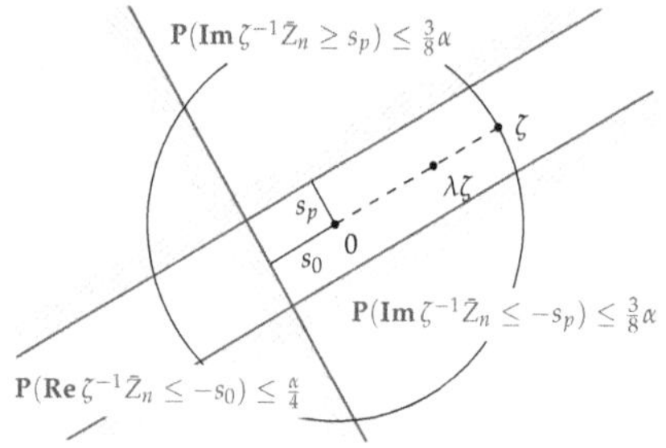

Figure 2. The construction for the test of the hypothesis $\mathbf{E}\,Z = \lambda\zeta$ with $\lambda > 0$.

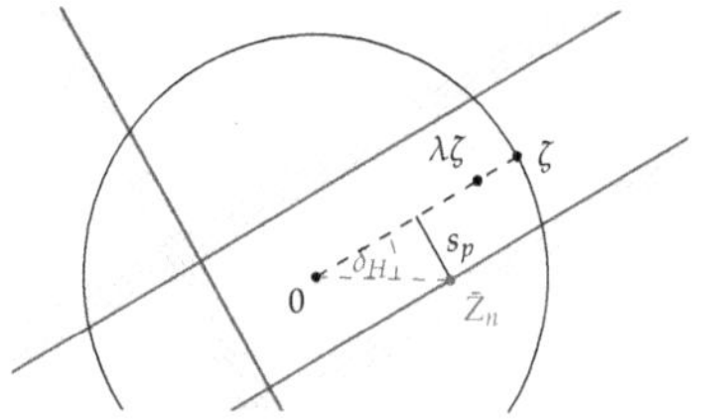

Figure 3. The critical $\bar{Z}_n$ regarding the rejection of ζ. δ_H bounds the angle between $\hat{\mu}_n$ and any accepted ζ.

Proposition 1. *Let $Z_1, \ldots, Z_n$ be random variables taking values on the unit circle S^1, $\alpha \in (0,1)$, and let C_H be defined as in Equation* (8).

(i) C_H *is a* $(1-\alpha)$*-confidence set for the circular population mean set. In particular, if* $\mathbf{E}\,Z = 0$*, i.e., the circular population mean set equals* S^1*, then* $|\bar{Z}_n| > \sqrt{2}s_0$ *with probability at most* α*, so indeed* $C_H = S^1$ *with probability at least* $1-\alpha$.

(ii) s_0 *and* s_p *are of order* $n^{-\frac{1}{2}}$.

(iii) *If* $\mathbf{E}\,Z \neq 0$*, then* $\sqrt{n}\delta_H \to 0$ *in probability and the probability of obtaining the trivial confidence set, i.e.,* $\mathbf{P}(C_H = S^1) = \mathbf{P}(|\bar{Z}_n| \leq \sqrt{2}s_0)$*, goes to 0 exponentially fast.*

Proof. (i) holds by construction.

For (ii), recall Equation (7), from which we obtain the estimates $\frac{\alpha}{4} \leq \exp(-ns_0^2/2)$ resp. $\frac{3}{8}\alpha \leq \exp(-ns_p^2/2)$, implying that s_0 and s_p are of order $n^{-\frac{1}{2}}$; the same holds stochastically for δ_H since $\bar{Z}_n \to \mathbf{E}\,Z$ a.s. Regarding the second statement of (iii), if μ is unique, consider $\zeta = -\mu$; then, $\tau = \mathbf{E}\,\bar{X}_n < 0$ and $-\sqrt{2}s_0$ is eventually less than $\frac{\tau}{2}$ and also $\alpha > 2^{-n+2}$ eventually. Hence, the probability of obtaining the trivial confidence set $C_H = S^1$ is eventually bounded by $\mathbf{P}(\zeta \in C_H) \leq \mathbf{P}(\bar{X}_n > -s_0) \leq \mathbf{P}(\bar{X}_n > \frac{\tau}{2}) = \mathbf{P}(\bar{X}_n - \mathbf{E}\,\bar{X}_n > -\frac{\tau}{2}) \leq \exp(-n\tau^2/8)$, and thus will go to zero exponentially fast as n tends to infinity. □

3. Estimating the Variance

From the central limit theorem for $\hat{\mu}_n$ in case of unique μ, cf. Equation (4), we see that the aymptotic variance of $\hat{\mu}_n$ gets small if $|\,\mathbf{E}\,Z|$ is close to 1 (then $\mathbf{E}\,Z$ is close to the boundary S^1 of the unit disc, which is possible only if the distribution is very concentrated) or if the variance $\mathbf{E}(\mathbf{Im}(\mu^{-1}Z))^2$ in the direction perpendicular to μ is small (if the distribution were concentrated on $\pm\mu$, this variance would be zero and $\hat{\mu}_n$ would equal μ with large probability). While δ_H ($|\bar{Z}_n|$ being the denominator of its sine) takes the former into account, the latter has not been exploited yet. To do so, we need to estimate $\mathbf{E}(\mathbf{Im}(\mu^{-1}Z))^2$.

Consider $V_n = \frac{1}{n}\sum_{k=1}^{n} Y_k^2$ that is under the hypothesis that the corresponding ζ is the unique circular population mean has expectation $\sigma^2 = \mathbf{Var}(Y_k) = \mathbf{E}(\mathbf{Im}(\zeta^{-1}Z))^2$. Now, $1 - V_n = \frac{1}{n}\sum_{k=1}^{n}(1 - Y_k^2)$ is the mean of n independent random variables taking values in $[0,1]$ and having expectation $1-\sigma^2$. By another application of Equation (6), we obtain $\mathbf{P}(\sigma^2 \geq V_n + t) = \mathbf{P}(1 - V_n \geq 1 - \sigma^2 + t) \leq \frac{\alpha}{4}$ for $t = t(\frac{\alpha}{4}, 1-\sigma^2, 0, 1)$, the latter existing if $\frac{\alpha}{4} > (1-\sigma^2)^n$. Since $1 - \sigma^2 + t(\frac{\alpha}{4}, 1-\sigma^2, 0, 1)$ increases with $1-\sigma^2$, there is a minimal σ^2 for which $1 - V_n \geq 1 - \sigma^2 + t(\frac{\alpha}{4}, 1-\sigma^2, 0, 1)$ holds and becomes an equality; we denote it by $\widehat{\sigma^2} = V_n + t(\frac{\alpha}{4}, 1-\widehat{\sigma^2}, 0, 1)$. Inserting into Equation (6), it by construction fulfills

$$\frac{\alpha}{4} = \left[\left(\frac{1-\widehat{\sigma^2}}{1-V_n}\right)^{1-V_n}\left(\frac{\widehat{\sigma^2}}{V_n}\right)^{V_n}\right]^n. \tag{9}$$

It is easy to see that the right-hand side depends continuously on and is strictly decreasing in $\widehat{\sigma^2} \in [V_n, 1]$ (see Appendix A), thereby traversing the interval $[0,1]$ so that one can again solve the equation numerically. We then may, with an error probability of at most $\frac{\alpha}{4}$, use $\widehat{\sigma^2}$ as an upper bound for σ^2. Note that $\widehat{\sigma^2} > V_n$ exists if $\frac{\alpha}{4} > (1-\widehat{\sigma^2})^n$. The latter is fulfilled for any $V_n < 1$ since Equation (9) is equivalent to

$$\frac{\alpha}{4} = (1-\widehat{\sigma^2})^n\left[\underbrace{\left(\frac{1}{1-V_n}\right)}_{>1}\underbrace{\left(\frac{1-\widehat{\sigma^2}}{1-V_n}\right)^{-V_n}}_{>1}\underbrace{\left(\frac{\widehat{\sigma^2}}{V_n}\right)^{V_n}}_{>1}\right]^n.$$

For $V_n = 1$, let $\widehat{\sigma^2} = 1$ be the trivial bound.

With such an upper bound on its variance, we now can get a better estimate for $\mathbf{P}(\bar{Y}_n > t)$. Indeed, one may use another inequality by Hoeffding [11] (Theorem 3): the mean $\bar{W}_n = \frac{1}{n}\sum_{k=1}^{n} W_k$ of a sequence $W_1, \ldots, W_n$ of independent random variables taking values in $(-\infty, 1]$, each having zero expectation as well as variance ρ^2 fulfills

$$\mathbf{P}(\bar{W}_n \geq w) \leq \left[\left(1+\frac{w}{\rho^2}\right)^{-\rho^2-w}\left(1-w\right)^{w-1}\right]^{\frac{n}{1+\rho^2}}, \tag{10}$$

$$\leq \exp(-nt[(1+\tfrac{\rho^2}{t})\ln(1+\tfrac{t}{\rho^2})-1]). \tag{11}$$

for any $w \in (0,1)$. Again, an elementary calculation (analogous to Lemma A1) shows that the right-hand side of Equation (10) is strictly decreasing in w, continuously ranging between 1 and $\left(\frac{\rho^2}{1+\rho^2}\right)^n$ as w varies in $(0,1)$, so that there exists a unique $w = w(\gamma, \rho^2)$ for which the right-hand side equals γ, provided $\gamma \in \left(\left(\frac{\rho^2}{1+\rho^2}\right)^n, 1\right)$. Moreover, the right-hand side increases with ρ^2 (as expected), so that $w(\gamma, \rho^2)$ is increasing in ρ^2, too (cf. Appendix A).

Therefore, under the hypothesis that the corresponding ζ is the unique circular population mean, $\mathbf{P}(|\bar{Y}_n| \geq w(\frac{\alpha}{4}, \sigma^2)) \leq 2\frac{\alpha}{4} = \frac{\alpha}{2}$. Now, since $\mathbf{P}(w(\frac{\alpha}{4}, \sigma^2) \geq w(\frac{\alpha}{4}, \widehat{\sigma^2})) = \mathbf{P}(\sigma^2 \geq \widehat{\sigma^2}) \leq \frac{\alpha}{4}$, setting $s_V = w(\frac{\alpha}{4}, \widehat{\sigma^2})$ we get $\mathbf{P}(|\bar{Y}_n| \geq s_V) \leq \frac{3}{4}\alpha$. Note that $\frac{\rho^2}{1+\rho^2}$ increases with ρ^2, so in case s_0 exists, $\widehat{\sigma^2} \leq 1$ implies $\frac{\alpha}{4} > 2^{-n} \geq \left(\frac{\widehat{\sigma^2}}{1+\widehat{\sigma^2}}\right)^n$, i.e., the existence of s_V.

Following the construction for C_H from Section 2, we can again obtain a confidence set for μ with coverage probability at least $1-\alpha$ as shown in our previous article [13]. In practice however, this confidence set is hard to calculate since $\widehat{\sigma^2} = \widehat{\sigma^2}(\zeta)$ has to be calculated for every $\zeta \in S^1$. Though these confidence sets can be approximated by using a grid as in [13], we suggest using a simultaneous upper bound for the variance of $\mathbf{Im}\,\zeta^{-1}Z_k$.

We obtain a (conservative) connected, symmetric confidence set $C_V \subseteq C_H$ by testing $\zeta \in C_H$ with $\widehat{\sigma^2_{\max}} = \sup_{\zeta \in C_H} \widehat{\sigma^2}$ as a common upper bound for the variance perpendicular to any $\zeta \in C_H$. Note that $\widehat{\sigma^2_{\max}}$ can be obtained as the solution of Equation (9) with

$$\tilde{V}_n = \sup_{\zeta \in C_H} \tfrac{1}{n} \sum_{k=1}^{n} \left(\mathbf{Im}\, \zeta^{-1} Z_k\right)^2.$$

Furthermore, we can shorten C_V by iteratively redefining $\tilde{V}_n = \sup_{\zeta \in C_V} \frac{1}{n} \sum_{k=1}^{n} \left(\mathbf{Im}\, \zeta^{-1} Z_k\right)^2$ and recalculating C_V (see Algorithm 1). The resulting opening angle will be denoted by $\delta_V = \arcsin \frac{s_V}{|\bar{Z}_n|}$.

Algorithm 1: Algorithm for computation of C_V.

Data: observations $Z_1, \ldots, Z_n \in \mathbb{S}^1$; significance level α; stop criterion ε
Result: a non-asymptotic confidence set C_V for the circular population mean

compute the confidence set C_H;
if $C_H = \mathbb{S}^1$ **then**
  $C_V \leftarrow \mathbb{S}^1$
else
  $C_V \leftarrow C_H$; $\widehat{\sigma^2_{\max}} \leftarrow 1$;
  while $\sup_{\zeta \in C_V} \widehat{\sigma^2} < \widehat{\sigma^2_{max}} - \varepsilon$ **do**
    $\widehat{\sigma^2_{\max}} \leftarrow \sup_{\zeta \in C_V} \widehat{\sigma^2}$;
    $s_V \leftarrow w(\frac{\alpha}{4}, \widehat{\sigma^2})$;
    $C_V \leftarrow \{\zeta \in \mathbb{S}^1 : |\mathbf{Arg}(\zeta^{-1} \hat{\mu}_n)| < \arcsin \frac{s_V}{|\bar{Z}_n|}\}$
  end
end

Proposition 2. *Let $Z_1, \ldots, Z_n$ be random variables taking values on the unit circle $\mathbb{S}^1$, and let $\alpha \in (0,1)$.*

(i) The set C_V resulting from Algorithm 1 is a $(1-\alpha)$-confidence set for the circular population mean set. In particular, if $\mathbf{E}\, Z = 0$, i.e., the circular population mean set equals $\mathbb{S}^1$, then $|\bar{Z}_n| > \sqrt{2}s_0$ with probability at most α, so indeed $C_V = \mathbb{S}^1$ with probability of at least $1-\alpha$.
(ii) s_V is of order $n^{-\frac{1}{2}}$.
(iii) If $\mathbf{E}\, Z \neq 0$, i.e., if the circular population mean is unique, then $\sqrt{n}\delta_V \to 0$ in probability, and the probability of obtaining a trivial confidence set, i.e., $\mathbf{P}(C_H = \mathbb{S}^1) = \mathbf{P}(|\bar{Z}_n| \leq \sqrt{2}s_0)$, goes to 0 exponentially fast.
(iv) If $\mathbf{E}\, Z \neq 0$, then

$$\limsup_{n \to \infty} \frac{\delta_V}{\delta_A} \leq \frac{\sqrt{-2 \ln \frac{\alpha}{4}}}{q_{1-\frac{\alpha}{2}}} \quad a.s.$$

with $q_{1-\frac{\alpha}{2}}$ denoting the $(1-\frac{\alpha}{2})$-quantile of the standard normal distribution $\mathcal{N}(0,1)$.

Proof. Again, (i) follows by construction, while (iii) is shown as in Proposition 1.

For (ii), note that $s_V \leq s_0$ since the bound in Equation (10) for $\rho^2 = 1$ agrees with the bound in Equation (6) for $a = -1$, $b = 1$ and $v = 0$, thus s_V and δ_V are at least of the order $n^{-\frac{1}{2}}$.

For (iv), we will use the estimate in Equation (11). Recall that $\ln(1+x) = x - \frac{x^2}{2} + o(x^2)$; therefore, for large n and hence small s_V a.s.

$$\begin{aligned}\frac{\alpha}{4} &\leq \exp\left(-ns_V\left[\left(1+\frac{\widehat{\sigma^2_{\max}}}{s_V}\right)\left(\frac{s_V}{\widehat{\sigma^2_{\max}}} - \frac{s_V^2}{2(\widehat{\sigma^2_{\max}})^2} + o(s_V^2)\right) - 1\right]\right)\\ &= \exp(-ns_V^2/2\widehat{\sigma^2_{\max}} + o(s_V^2)),\end{aligned}$$

thus $s_V \leq \sqrt{-2\widehat{\sigma^2_{\max}}\ln(\frac{\alpha}{4})/n} + o(n^{-\frac{1}{2}})$. Additionally, $\arcsin x = x + o(x)$ for x close to 0 which gives $\delta_V = s_V/|\bar{Z}_n| + o(s_V) \leq \sqrt{-2\widehat{\sigma^2_{\max}}\ln\frac{\alpha}{4}}/(\sqrt{n}|\bar{Z}_n|) + o(n^{-\frac{1}{2}})$ a.s.

Furthermore, $\widehat{\sigma^2_{\max}} \to \sigma^2$ a.s. for $n \to \infty$, and we obtain

$$\limsup_{n\to\infty} \frac{\delta_V}{\delta_A} \leq \frac{\sqrt{-2\ln\frac{\alpha}{4}}}{q_{1-\frac{\alpha}{2}}} \quad \text{a.s.}$$

since

$$\delta_A = \frac{q_{1-\frac{\alpha}{2}}}{\sqrt{n}|\bar{Z}_n|}\underbrace{\sqrt{\frac{1}{n}\sum_{k=1}^{n}\left(\mathbf{Im}(\hat{\mu}_n^{-1}Z_k)\right)^2}}_{\to\sqrt{\sigma^2}}$$

(see Equation (5)). □

4. Simulation and Application to Real Data

We will compare the asymptotic confidence set C_A, the confidence set C_H constructed directly using Hoeffding's inequality in Section 2, and the confidence set C_V resulting from Algorithm 1 by reporting their corresponding opening angles δ_A, δ_H, and δ_V in degrees (°) as well as their coverage frequencies in simulations.

All computations have been performed using our own code based on the software package R (version 2.15.3) [14] .

4.1. Simulation 1: Two Points of Equal Mass at ±10°

First, we consider a rather favourable situation: $n = 400$ independent draws from the distribution with $\mathbf{P}(Z = \exp(10\pi i/180)) = \mathbf{P}(Z = \exp(-10\pi i/180)) = \frac{1}{2}$. Then, we have $|\mathbf{E}\,Z| = \mathbf{E}\,Z = \cos(10\pi i/180) \approx 0.985$, implying that the data are highly concentrated, $\mu = 1$ is unique, and the variance of Z in the direction of μ is 0; there is only variation perpendicular to μ, i.e., in the direction of the imaginary axis (see Figure 4).

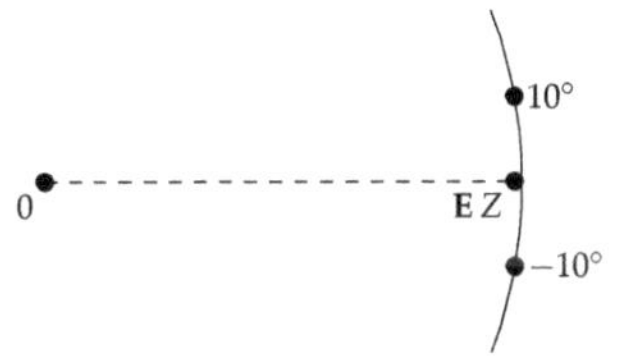

Figure 4. Two points of equal mass at ±10° and their Euclidean mean.

Table 1. Results for simulation 1 (two points of equal mass at $\pm 10°$) based on 10,000 repetitions with $n = 400$ observations each: average observed δ_H, δ_V, and δ_A (with corresponding standard deviation), as well as frequency (with corresponding standard error) with which $\mu = 1$ was covered by C_H, C_V, and C_A, respectively; the nominal coverage probability was $1 - \alpha = 95\%$.

Confidence Set	Mean δ ($\pm$s.d.)	Coverage Frequency ($\pm$s.e.)
C_H	8.2° (±0.0005°)	100.0% (±0.0%)
C_V	2.4° (±0.0025°)	100.0% (±0.0%)
C_A	1.0° (±0.0019°)	94.8% (±0.2%)

Table 1 shows the results based on 10,000 repetitions for a nominal coverage probability of $1 - \alpha = 95\%$: the average δ_H is about 3.5 times larger than δ_V, which is about twice as large as δ_A. As expected, the asymptotics are rather precise in this situation: C_A did cover the true mean in about 95% of the cases, which implies that the other confidence sets are quite conservative; indeed C_H and C_V covered the true mean in all repetitions. One may also note that the angles varied only a little between repetitions.

4.2. Simulation 2: Three Points Placed Asymmetrically

Secondly, we consider a situation which has been designed to show that even a considerably large sample size ($n = 100$) does not guarantee approximate coverage for the asymptotic confidence set C_A: the distribution of Z is concentrated on three points, $\xi_j = \exp(\theta_j \pi i/180)$, $j = 1,2,3$ with weights $\omega_j = \mathbf{P}(Z = \xi_j)$ chosen such that $\mathbf{E}\,Z = |\,\mathbf{E}\,Z| = 0.9$ (implying a small variance and $\mu = 1$), $\omega_1 = 1\%$ and $\mathbf{Arg}\,\xi_1 > 0$, while $\mathbf{Arg}\,\xi_2, \mathbf{Arg}\,\xi_3 < 0$. In numbers, $\theta_1 \approx 25.8$, $\theta_2 \approx -0.3$, and $\theta_3 \approx -179.7$ (in °) while $\omega_2 \approx 94\%$, and $\omega_3 \approx 5\%$ (see Figure 5).

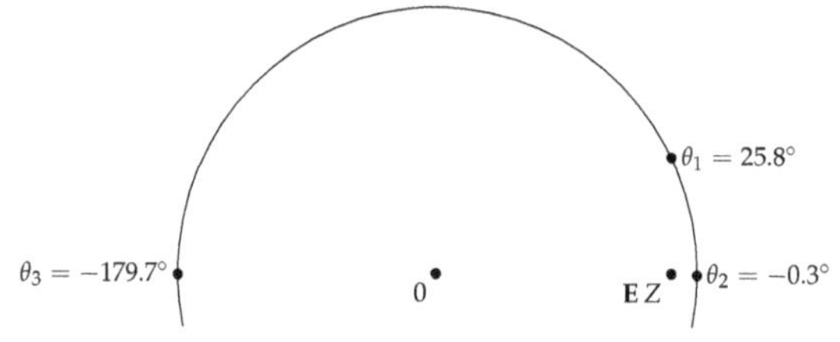

Figure 5. Three points placed asymmetrically with different masses and their Euclidean mean.

The results based on 10,000 repetitions are shown in Table 2 where a nominal coverage probability of $1 - \alpha = 90\%$ was prescribed. Clearly, C_A with its coverage probability of less than 64% performs quite poorly while the others are conservative; $\delta_V \approx 5°$ still appears small enough to be useful in practice, though.

Table 2. Results for simulation 2 (three points placed asymmetrically) based on 10,000 repetitions with $n = 100$ observations each: average observed δ_H, δ_V, and δ_A (with corresponding standard deviation), as well as frequency (with corresponding standard error) with which $\mu = 1$ was covered by C_H, C_V, and C_A, respectively; the nominal coverage probability was $1 - \alpha = 90\%$.

Confidence Set	Mean δ ($\pm$s.d.)	Coverage Frequency ($\pm$s.e.)
C_H	16.5° (±0.85°)	100.0% (±0.0%)
C_V	5.0° (±0.38°)	100.0% (±0.0%)
C_A	0.4° (±0.28°)	62.8% (±0.5%)

4.3. Real Data: Movements of Ants

Fisher [3] (Example 4.4) describes a data set of the directions 100 ants took in response to an illuminated target placed at 180° for which it may be of interest to know whether the ants indeed (on average) move towards that target (see [15] for the original publication). The data set is available as `Ants_radians` within the `R` package `CircNNTSR` [16].

The circular sample mean for this data set is about $-176.9°$; for a nominal coverage probability of $1-\alpha = 95\%$, one gets $\delta_H \approx 27.3°$, $\delta_V \approx 20.5°$, and $\delta_A \approx 9.6°$ so that all confidence sets contain $\pm 180°$ (see Figure 6). The data set's concentration is not very high, however, so the circular population mean could—according to C_V—also be $-156.4°$ or $162.6°$.

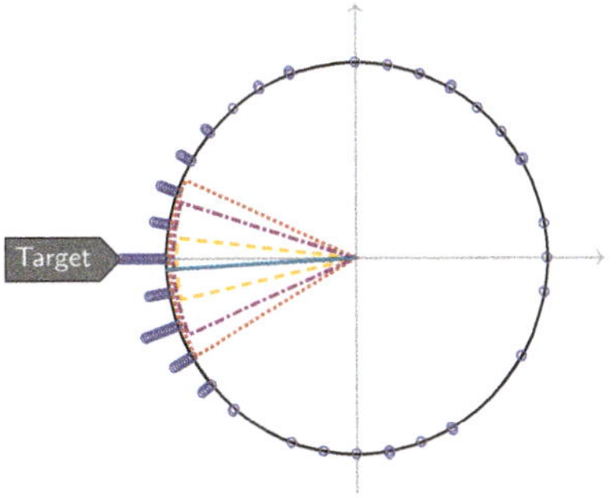

Figure 6. Ant data (○) placed at increasing radii to visually resolve ties; in addition, the circular mean direction (—) as well as confidence sets C_H (·····), C_V (-·-), and C_A (- -) are shown.

5. Discussion

We have derived two confidence sets, C_H and C_V, for the set of circular sample means. Both guarantee coverage for any finite sample size without making any assumptions on the distribution of the data (besides that they are independent and identically distributed) at the cost of potentially being quite conservative; they are non-asymptotic and universal in this sense. Judging from the simulations and the real data set, C_V—which estimates the variance perpendicular to the mean direction—appears to be preferable over C_H (as expected) and small enough to be useful in practice.

While the asymptotic confidence set's opening angle is less than half (asymptotically about 2/3 for $\alpha = 5\%$) of the one for C_V in our simulations and application, it has the drawback that even for a sample size of $n = 100$, it may fail to give a coverage probability close to the nominal one; in addition, one has to assume that the circular population mean is unique. Of course, one could also devise an asymptotically justified test for the latter but this would entail a correction for multiple testing (for example working with $\frac{\alpha}{2}$ each time), which would also render the asymptotic confidence set conservative.

Further improvements would require sharper "universal" mass concentration inequalities taking the first or the first two moments into account; however, this is beyond the scope of this article.

Acknowledgments: T. Hotz wishes to thank Stephan Huckemann from the Georgia Augusta University of Göttingen for fruitful discussions concerning the first construction of confidence regions described in Section 2. We acknowledge support for the Article Processing Charge by the German Research Foundation and the Open Access Publication Fund of the Technische Universität Ilmenau. F. Kelma acknowledges support by the Klaus Tschira Stiftung, gemeinnützige Gesellschaft, Projekt 03.126.2016.

Author Contributions: All authors contributed to the theoretical and numerical results as well as to the writing of the article. All authors have read and approved the final manuscript.

Conflicts of Interest: The authors declare no conflict of interest.

Appendix A. Proofs of Monotonicity

Lemma A1. $\beta(t) = \left[\left(\frac{v-a}{v-a+t}\right)^{v-a+t}\left(\frac{b-v}{b-v-t}\right)^{b-v-t}\right]^{\frac{n}{b-a}}$ *is strictly decreasing in t.*

Proof. We show the equivalent statement that $\tilde{\beta}(t) = \ln\left[\left(\frac{v-a}{v-a+t}\right)^{v-a+t}\left(\frac{b-v}{b-v-t}\right)^{b-v-t}\right]$ is strictly decreasing in t:

$$\begin{aligned}\frac{d}{dt}\tilde{\beta}(t) &= \frac{d}{dt}\Big((\ln(v-a)-\ln(v-a+t))(v-a+t)+(\ln(b-v)-\ln(b-v-t))(b-v-t)\Big)\\ &= \ln(v-a)-\ln(v-a+t)-\tfrac{1}{v-a+t}(v-a+t)-\ln(b-v)+\ln(b-v-t)+\tfrac{1}{b-v-t}(b-v-t)\\ &= \ln\Bigg(\underbrace{\frac{b-v-t}{b-v}}_{<1}\cdot\underbrace{\frac{v-a}{v-a+t}}_{<1}\Bigg) < 0.\end{aligned}$$

Hence, $\tilde{\beta}(t)$ and thus $\beta(t)$ are strictly decreasing in t. □

Lemma A2. *Let $t = t(\gamma, v, a, b)$ be the solution to the equation $\beta(t) = \gamma$. Then, $v + t$ is strictly increasing in v.*

Proof. t is the solution of the equation

$$(v-a+t)\ln\left(\frac{v-a}{v-a+t}\right)+(b-v-t)\ln\left(\frac{b-v}{b-v-t}\right) = \frac{b-a}{n}\ln\gamma. \tag{A1}$$

The derivatives of the left-hand side of Equation (A1) w.r.t. v and t exist and are continuous. Furthermore, the derivative w.r.t. t does not vanish for any $t \in (0, b-v)$, cf. the proof of Lemma A1, whence the derivative $t' = \frac{dt}{dv}$ exists by the implicit function theorem. When differentiating Equation (A1) with respect to v, one obtains

$$\begin{aligned}(1+t')\ln\left(\frac{v-a}{v-a+t}\right)+(v-a+t)\left(\frac{1}{v-a}-\frac{1+t'}{v-a+t}\right)&\\ -(1+t')\ln\left(\frac{b-v}{b-v-t}\right)+(b-v-t)\left(-\frac{1}{b-v}+\frac{1+t'}{b-v-t}\right) &= 0,\end{aligned}$$

or equivalently

$$(1+t')\Bigg[\underbrace{\ln\left(\frac{v-a}{v-a+t}\right)-\ln\left(\frac{b-v}{b-v-t}\right)}_{<0\quad\quad >0}\Bigg] = \frac{t(a-b)}{(v-a)(b-v)} < 0,$$

whence $1 + t' = \frac{d}{dv}(v+t) > 0$ finishes the proof. □

Lemma A3. *The function*

$$\xi(\widehat{\sigma^2}) = \left[\left(\frac{1-\widehat{\sigma^2}}{1-V_n}\right)^{1-V_n}\left(\frac{\widehat{\sigma^2}}{V_n}\right)^{V_n}\right]^n$$

is strictly decreasing in $\widehat{\sigma^2} \in [V_n, 1]$.

Proof. We show the equivalent statement that $n^{-1}\ln\xi(\widehat{\sigma^2})$ is strictly decreasing in $\widehat{\sigma^2}$:

$$\begin{aligned}\frac{d}{d\widehat{\sigma^2}}\left[n^{-1}\ln\xi(\widehat{\sigma^2})\right] &= \frac{d}{d\widehat{\sigma^2}}\left[(1-V_n)\left(\ln(1-\widehat{\sigma^2})-\ln(1-V_n)\right)+V_n\left(\ln(\widehat{\sigma^2})-\ln(V_n)\right)\right] \\ &= -\underbrace{\frac{1-V_n}{1-\widehat{\sigma^2}}}_{>1}+\underbrace{\frac{V_n}{\widehat{\sigma^2}}}_{<1}<0.\end{aligned}$$

□

Lemma A4. *Let $w = w(\gamma, \rho^2)$ be the solution of the equation*

$$\left[\left(1+\frac{w}{\rho^2}\right)^{-\rho^2-w}\left(1-w\right)^{w-1}\right]^{\frac{n}{1+\rho^2}}=\gamma.$$

Then, w is increasing in ρ^2.

Proof. w is the solution of the equation

$$\frac{\rho^2+w}{1+\rho^2}\ln\left(1+\frac{w}{\rho^2}\right)+\frac{1-w}{1+\rho^2}\ln(1-w)=-\frac{\ln\gamma}{n}. \tag{A2}$$

The derivatives of the left-hand side of Equation (A2) w.r.t. ρ^2 and w exist and are continuous. Furthermore, the derivative w.r.t. w does not vanish for any $w \in (0,1)$: this derivative is

$$\frac{1}{1+\rho^2}\left[\ln\left(1+\frac{w}{\rho^2}\right)+\frac{\rho^2+w}{\rho^2(1+\frac{w}{\rho^2})}-\ln(1-w)-1\right]=\frac{1}{1+\rho^2}\left[\ln\left(1+\frac{w}{\rho^2}\right)-\ln(1-w)\right],$$

vanishing if and only if $1+\frac{w}{\rho^2}=1-w$, i.e., if and only if $w(1+\frac{1}{\rho^2})=0$, which does not happen for $w, \rho^2 > 0$. Now, the derivative $w' = \frac{dw}{d\rho^2}$ exists by the implicit function theorem. When differentiating Equation (A2) with respect to ρ^2, one obtains

$$\begin{aligned}&\frac{(1+w')(1+\rho^2)-(\rho^2+w)}{(1+\rho^2)^2}\ln\left(1+\frac{w}{\rho^2}\right) \\ &\qquad+\underbrace{\frac{\rho^2+w}{1+\rho^2}\cdot\frac{\frac{w'}{\rho^2}-\frac{w}{\rho^4}}{1+\frac{w}{\rho^2}}}_{\frac{w'\rho^2-w}{\rho^2(1+\rho^2)}}-\frac{w'(1+\rho^2)+(1-w)}{(1+\rho^2)^2}\ln(1-w)-\frac{w'}{1+\rho^2}=0,\end{aligned}$$

or equivalently

$$w'\underbrace{\left[\ln\left(1+\frac{w}{\rho^2}\right)-\ln(1-w)\right]}_{>0}=\frac{w}{\rho^2}-\frac{1-w}{1+\rho^2}\ln\left(\frac{\rho^2+w}{\rho^2(1-w)}\right).$$

Hence, $w' \geq 0$ if and only if $\frac{w}{\rho^2} \geq \frac{1-w}{1+\rho^2}\ln\left(\frac{\rho^2+w}{\rho^2(1-w)}\right)$, which holds since $\ln\left(\frac{\rho^2+w}{\rho^2(1-w)}\right)=\ln\left(1+\frac{w(1+\rho^2)}{\rho^2(1-w)}\right)\leq \frac{w}{\rho^2}\frac{1+\rho^2}{1-w}$, finishing the proof. □

References

1. Mardia, K.V. *Directional Statistics*; Academic Press: London, UK, 1972.

2. Watson, G.S. *Statistics on Spheres*; Wiley: New York, NY, USA, 1983.
3. Fisher, N.I. *Statistical Analysis of Circular Data*; Cambridge University Press: Cambridge, UK, 1993.
4. Jammalamadaka, S.R.; SenGupta, A. *Topics in Circular Statistics*; Series on Multivariate Analysis; World Scientific: Singapore, 2001.
5. Mardia, K.V.; Jupp, P.E. *Directional Statistics*; Wiley: New York, NY, USA, 2000.
6. Fréchet, M. Les éléments aléatoires de nature quelconque dans un espace distancié. *Annales de l'Institut Henri Poincaré* **1948**, *10*, 215–310. (In French)
7. Hotz, T. Extrinsic vs. Intrinsic Means on the Circle. In Proceedings of the 1st Conference on Geometric Science of Information, Paris, France, 28–30 October 2013; Lecture Notes in Computer Science, Volume 8085; Springer-Verlag: Heidelberg, Germany, 2013; pp. 433–440.
8. Afsari, B. Riemannian L^p center of mass: Existence, uniqueness, and convexity. *Proc. Am. Math. Soc.* **2011**, *139*, 655–673.
9. Arnaudon, M.; Miclo, L. A stochastic algorithm finding p-means on the circle. *Bernoulli* **2016**, *22*, 2237–2300.
10. Leeb, H.; Pötscher, B.M. Model selection and inference: Facts and fiction. *Econ. Theory* **2005**, *21*, 21–59.
11. Hoeffding, W. Probability Inequalities for Sums of Bounded Random Variables. *J. Am. Stat. Assoc.* **1963**, *58*, 13–30.
12. Boucheron, S.; Lugosi, G.; Massart, P. *Concentration Inequalities : A Nonasymptotic Theory of Independence*; Oxford University Press: Oxford, UK, 2013.
13. Hotz, T.; Kelma, F.; Wieditz, J. Universal, Non-asymptotic Confidence Sets for Circular Means. In Proceedings of the 2nd International Conference on Geometric Science of Information, Palaiseau, France, 28–30 October 2015; Nielsen, F., Barbaresco, F., Eds.; Lecture Notes in Computer Science, Volume 9389; Springer International Publishing: Cham, Switzerland, 2015; pp. 635–642.
14. R Core Team. *R: A Language and Environment for Statistical Computing*; version 2.15.3; R Foundation for Statistical Computing: Vienna, Austria, 2013.
15. Jander, R. Die optische Richtungsorientierung der Roten Waldameise (Formica Rufa L.). *Zeitschrift für Vergleichende Physiologie* **1957**, *40*, 162–238. (In German)
16. Fernandez-Duran, J.J.; Gregorio-Dominguez, M.M. *CircNNTSR: An R Package for the Statistical Analysis of Circular Data Using Nonnegative Trigonometric Sums (NNTS) Models*, version 2.1; 2013.

Chapter 6:
Entropy and Complexity in Linguistic

entropy

MDPI

Article

Syntactic Parameters and a Coding Theory Perspective on Entropy and Complexity of Language Families

Matilde Marcolli

Department of Mathematics, California Institute of Technology, Pasadena, CA 91125, USA; matilde@caltech.edu; Tel.: +1-626-395-4326

Academic Editors: Frédéric Barbaresco, Frank Nielsen and Kevin H. Knuth
Received: 14 January 2016; Accepted: 18 March 2016; Published: 7 April 2016

Abstract: We present a simple computational approach to assigning a measure of complexity and information/entropy to families of natural languages, based on syntactic parameters and the theory of error correcting codes. We associate to each language a binary string of syntactic parameters and to a language family a binary code, with code words the binary string associated to each language. We then evaluate the code parameters (rate and relative minimum distance) and the position of the parameters with respect to the asymptotic bound of error correcting codes and the Gilbert–Varshamov bound. These bounds are, respectively, related to the Kolmogorov complexity and the Shannon entropy of the code and this gives us a computationally simple way to obtain estimates on the complexity and information, not of individual languages but of language families. This notion of complexity is related, from the linguistic point of view to the degree of variability of syntactic parameter across languages belonging to the same (historical) family.

Keywords: syntax; principles and parameters; error-correcting codes; asymptotic bound; Kolmogorov complexity; Gilbert–Varshamov bound; Shannon entropy

1. Introduction

We propose an approach, based on Longobardi's parametric comparison method (PCM) and the theory of error-correcting codes, to a quantitative evaluation of the "complexity" of a language family. One associates to a collection of languages to be analyzed with the PCM a binary (or ternary) code with one code word for each language in the family and each word consisting of the binary values of the syntactic parameters of that language. The ternary case allows for an additional parameter state that takes into account certain phenomena of entailment of parameters. We then consider a different kind of parameters: the code parameters of the resulting code, which in coding theory account for the efficiency of the coding and decoding procedures. These can be compared with some classical bounds of coding theory: the asymptotic bound, the Gilbert–Varshamov (GV) bound, *etc.* The position of the code parameters with respect to some of these bounds provides quantitative information on the variability of syntactic parameters within and across historical-linguistic families. While computations carried out for languages belonging to the same historical family yield codes below the GV curve, comparisons across different historical families can give examples of isolated codes lying above the asymptotic bound.

1.1. Principles and Parameters

The generative approach to linguistics relies on the notion of a Universal Grammar (UG) and a related universal list of syntactic parameters. In the Principles and Parameters model, developed since [1], these are thought of as binary valued parameters or "switches" that set the grammatical structure of a given language. Their universality makes it possible to obtain comparisons, at the syntactic level, between arbitrary pairs of natural languages.

A PCM was introduced in [2] as a quantitative method in historical linguistics, for comparison of languages within and across historical families at the syntactic instead of the lexical level. Evidence was given in [3,4] that the PCM gives reliable information on the phylogenetic tree of the family of Indo-European languages.

The PCM relies essentially on constructing a metric on a family of languages based on the relative Hamming distance between the sets of parameters as a measure of relatedness. The phylogenetic tree is then constructed on the basis of this datum of relative distances, see [3].

More work on syntactic phylogenetic reconstructions, involving a larger set of languages and parameters is ongoing, [5]. Syntactic parameters of world languages have also been used recently for investigations on the topology and geometry of syntactic structures and for statistical physics models of language evolution, [6–8].

Publicly available data of syntactic parameters of world languages can be obtained from databases such as Syntactic Structures of World Languages (SSWL) [9] or TerraLing [10] or World Atlas of Language Structures (WALS) [11]. The data of syntactic parameters used in the present paper are taken from Table A of [3].

1.2. Syntactic Parameters, Codes and Code Parameters

Our purpose in this paper is to connect the PCM approach to the mathematical theory of error-correcting codes. We associate a code to any group of languages one wishes to analyze via the PCM, which has one code word for each language. If one uses a number n of syntactic parameters, then the code C sits in the space $\mathbb{F}_2^n$, where the elements of $\mathbb{F}_2 = \{0,1\}$ correspond to the two $\mp$ possible values of each parameter, and the code word of a language is the string of values of its n parameters. We also consider a version with codes on an alphabet $\mathbb{F}_3$ of three letters which allows for the possibility that some of the parameters may be made irrelevant by entailment from other parameters. In this case we use the letter $0 \in \mathbb{F}_3$ for the irrelevant parameters and the nonzero values ± 1 for the parameters that are set in the language.

In the theory of error-correcting codes, see [12], one assigns to a code $C \subset \mathbb{F}_q^n$ two code parameters: $R = \log_q(\#C)/n$, the transmission rate of the code, and $\delta = d/n$ the relative minimum distance of the code, where d is the miminum Hamming distance between pairs of distinct code words. It is well known in coding theory that "good codes" are those that maximize both parameters, compatibly with several constraints relating R and δ. Consider the function $f : \mathcal{C}_q \to [0,1]^2$ from the space $\mathcal{C}_q$ of q-ary codes to the unit square, that assigns to a code C its code parameters, $f(C) = (\delta(C), R(C))$. A point (δ, R) in the range of f has finite (respectively, infinite) multiplicity if the preimage $f^{-1}(\delta, R)$ is a finite set (respectively, an infinite set). It was proved in [13] that there is a curve $R = \alpha_q(\delta)$ in the space of code parameters, the asymptotic bound, that separates code points that fill a dense region and that have infinite multiplicity from isolated code points that only have finite multiplicity. These better but more elusive codes are typically obtained through algebro-geometric constructions, see [13–15]. The asymptotic bound was related to Kolmogorov complexity in [16].

1.3. Position with Respect to the Asymptotic Bound

Given a collection of languages one wants to compare through their syntactic parameters, one can ask natural questions about the position of the resulting code in the space of code parameters and with respect to the asymptotic bound. The theory of error correcting codes tells us that codes above the asymptotic bound are very rare. Indeed, we considered various sets of languages, and for each choice of a set of languages we considered an associated code, with a code word for each language in the set, given by its list of syntactic parameters. When computing the code parameters of the resulting code, one finds that, in a range of cases we looked at, when the languages in the chosen set belong to the same historical-linguistic family the resulting code lies below the asymptotic bound (and in fact below the Gilbert–Varshamov curve). This provides a precise quantitative bound to the possible spread of syntactic parameters compared to the size of the family, in terms of the number of different languages belonging to the same historico-linguistic group.

However, we also show that, if one considers sets of languages that do not belong to the same historical-linguistic family, then one can obtain codes that lie above the asymptotic bound, a fact that reflects, in code theoretic terms, the much greater variability of syntactic parameters. The result is in itself not surprising, but the point we wish to make is that the theory of error-correcting codes provides a natural setting where quantitative statements of this sort can be made using methods already developed for the different purposes of coding theory. We conclude by listing some new linguistic questions that arise by considering the parametric comparison method under this coding theory perspective.

1.4. Complexity of Languages and Language Families

The study of natural languages from the point of view of complexity theory has been of significant interest to linguists in recent years. The approaches typically followed focus on assigning a reasonable measure of complexity to individual languages and comparing complexities across different languages. For example, a notion of morphological complexity was studied in [17]. An approach to defining Kolmogorov complexity of languages on the basis of syntactic parameters was developed in [18]. A notion of language complexity based on the production rules of a generative grammar was considered in [19], in the setting of (finite) formal languages. For a more general computational perspective on the complexity of natural languages, see [20]. The idea of distinguishing languages by complexity is not without controversy in Linguistics. A very interesting general discussion of the problem and its evolution in the field can be found in [21].

In the present paper, we argue in favor of a somewhat different perspective, where we assign an estimate of complexity not to individual languages but to groups of languages, and in particular (historical) language families. Our version of complexity is measuring how "spread out" the syntactic parameters can be, across the languages that belong to the same family. As we outlined in the previous subsections, this is measured by assigning to the language family a code, whose code words record the syntactic parameters of the individual languages in the family, then computing its code parameters and evaluating the position of the resulting code points with respect to curves like the asymptotic bound or the Gilbert–Varshamov line. The reason why this position carries complexity information lies in the subtle relation between the asymptotic bound and Kolmogorov complexity, recently derived by Manin and the author in [16], which we will review briefly in this paper.

2. Language Families as Codes

The Principles and Parameters model of Linguistics assigns to every natural language L a set of binary values parameters that describe properties of the syntactic structure of the language.

Let F be a *language family*, by which we mean a finite collection $F = \{L_1, \dots, L_m\}$ of languages. This may coincide with a family in the historical sense, such as the Indo-European family, or a smaller subset of languages related by historic origin and development (e.g., the Indo-Iranian, or Balto–Svalic languages), or simply any collection of languages one is interested in comparing at the parametric level, even if they are spread across different historical families.

We denote by n be the number of parameters used in the parametric comparison method. We do not fix, a priori, a value for n, and we consider it a variable of the model. We will discuss below how one views, in our perspective, the issue of the independence of parameters.

After fixing an enumeration of the parameters, that is, a bijection between the set of parameters and the set $\{1, \dots, n\}$, we associate to a language family F a code $C = C(F)$ in $\mathbb{F}_2^n$, with one code word for each language $L \in F$, with the code word $w = w(L)$ given by the list of parameters $w = (x_1, \dots, x_n)$, $x_i \in \mathbb{F}_2$ of the language. For simplicity of notation, we just write L for the word $w(L)$ in the following.

In this model, we only consider binary parameters with values ± 1 (here identified with letters 0 or 1 in $\mathbb{F}_2$) and we ignore parameters in a neutralized state following implications across parameters, as in the datasets of [3,4]. The entailment of parameters, that is, the phenomenon by which a particular value of one parameter (but not the complementary value) renders another parameter irrelevant, was addressed in greater detail in [22]. We first discuss a version of our coding theory model that does not incorporate entailment. We will then comment in Section 2.7 below on how the model can be modified to incorporate this phenomenon.

The idea that natural languages can be described, at the level of their core grammatical structures, in terms of a string of binary characters (code words) was already used extensively in [23].

2.1. Code Parameters

In the theory of error-correcting codes, one assigns two main parameters to a code C, the *transmission rate* and the *relative minimum distance*. More precisely, a binary code $C \subset \mathbb{F}_2^n$ is an $[n,k,d]_2$-code if the number of code words is $\#C = 2^k$, that is,

$$k = \log_2 \#C, \tag{1}$$

where k need not be an integer, and the minimal Hamming distance between code words is

$$d = \min_{L_1 \neq L_2 \in C} d_H(L_1, L_2), \tag{2}$$

where the Hamming distance is given by

$$d_H(L_1, L_2) = \sum_{i=1}^{n} |x_i - y_i|,$$

for $L_1 = (x_i)_{i=1}^n$ and $L_2 = (y_i)_{i=1}^n$ in C. The transmission rate of the code C is given by

$$R = \frac{k}{n}. \tag{3}$$

One denotes by $\delta_H(L_1, L_2)$ the relative Hamming distance

$$\delta_H(L_1, L_2) = \frac{1}{n} \sum_{i=1}^{n} |x_i - y_i|,$$

and one defines the relative minimum distance of the code C as

$$\delta = \frac{d}{n} = \min_{L_1 \neq L_2 \in C} \delta_H(L_1, L_2). \tag{4}$$

In coding theory, one would like to construct codes that simultaneously optimize both parameters (δ, R): a larger value of R represents a faster transmission rate (better encoding), and a larger value of δ represents the fact that code words are sufficiently sparse in the ambient space $\mathbb{F}_2^n$ (better decoding, with better error-correcting capability). Constraints on this optimization problem are expressed in the form of bounds in the space of (δ, R) parameters, see [12,13].

In our setting, the R parameter measures the ratio between the logarithmic size of the number of languages encompassing the given family and the total number of parameters, or equivalently how densely the given language family is in the ambient configuration space $\mathbb{F}_2^n$ of parameter possibilities. The parameter δ is the minimum, over all pairs of languages in the given family, of the relative Hamming distance used in the PCM method of [3,4].

2.2. Parameter Spoiling

In the theory of error-correcting codes, one considers *spoiling operations* on the code parameters. Applied to an $[n,k,d]_2$-code C, these produce, respectively, new codes with the following description (see Section 1.1.1 of [24]):

- A code $C_1 = C \star_i f$ in $\mathbb{F}_2^{n+1}$, for a map $f : C \to \mathbb{F}_2$, whose code words are of the form $(x_1,\dots,x_{i-1},f(x_1,\dots,x_n),x_i,\dots,x_n)$ for $w = (x_1,\dots,x_n) \in C$. If f is a constant function, C_1 is an $[n+1,k,d]_2$-code. If all pairs $w,w' \in C$ with $d_H(w,w') = d$ have $f(w) \neq f(w')$, then C_1 is an $[n+1,k,d+1]_2$-code.
- A code $C_2 = C\star_i$ in $\mathbb{F}_2^{n-1}$, whose code words are given by the projections

$$(x_1,\dots,x_{i-1},x_{i+1},\dots,x_n)$$

  of code words $(x_1,\dots,x_{i-1},x_i,x_{i+1},\dots,x_n)$ in C. This is an $[n-1,k,d-1]_2$-code, except when all pairs $w,w' \in C$ with $d_H(w,w') = d$ have the same letter x_i, in which case it is an $[n-1,k,d]_2$-code.
- A code $C_3 = C(a,i) \subset C \subset \mathbb{F}_2^n$, given by the level set $C(a,i) = \{w = (x_k)_{k=1}^n \in C \mid x_i = a\}$. Taking $C(a,i)\star_i$ gives an $[n-1,k',d']_2$-code with $k-1 \leq k' < k$, and $d' \geq d$.

The same spoiling operations hold for q-ary codes $C \subset \mathbb{F}_q^n$, for any fixed q.

In our setting, where C is the code obtained from a family of languages, according to the procedure described above, the first spoiling operation can be seen as the effect of considering one more syntactic parameter, which is dependent on the other parameters, hence describing a function $F : \mathbb{F}_2^n \to \mathbb{F}_2$, whose restriction to C gives the function $f : C \to \mathbb{F}_2$. In particular, the case where f is constant on C represents the situation in which the new parameter adds no useful comparison information for the selected family of languages. The second spoiling operation consists in forgetting one of the parameters, and the third corresponds to forming subfamilies of the given family of languages, by grouping together those languages with a set value of one of the syntactic parameters. Thus, all these spoiling operations have a clear meaning from the point of view of the linguistic PCM.

2.3. Examples

We consider the same list of 63 parameters used in [3] (see Section 5.3.1 and Table A). This choice of parameters follows the *modularized global parameterization* method of [2], for the Determiner Phrase module. They encompass parameters dealing with person, number, and gender (1–6 on their list), parameters of definiteness (7–16 in their list), of countability (17–24), genitive structure (25–31), adjectival and relative modification (32–14), position and movement of the head noun (42–50), demonstratives and other determiners (51–50 and 60–63), possessive pronouns (56–59); see Section 5.3.1 and Section 5.3.2 of [3] for more details.

Our very simple examples here are just meant to clarify our notation: they consist of some collections of languages selected from the list of 28, mostly Indo-European, languages considered in [3]. In each group we consider we eliminate the parameters that are entailed from others, and we focus on a shorter list, among the remaining parameters, that will suffice to illustrate our viewpoint.

Example 1. Consider a code C formed out of the languages ℓ_1 = Italian, ℓ_2 = Spanish, and ℓ_3 = French, and let us consider only the first six syntactic parameters of Table A of [3], so that $C \subset \mathbb{F}_2^n$ with $n = 6$. The code words for the three languages are

ℓ_1	1	1	1	0	1	1
ℓ_2	1	1	1	1	1	1
ℓ_3	1	1	1	0	1	0

This has code parameters $(R = \log_2(3)/6 = 0.2642, \delta = 1/6)$, which satisfy $R < 1 - H_2(\delta)$, hence they lie below the GV curve (see Equation (8) below). We use this code to illustrate the three spoiling operations mentioned above.

- Throughout the entire set of 28 languages considered in [3], the first two parameters are set to the same value 1, hence for the purpose of comparative analysis within this family, we can regard a code like the above as a twice spoiled code $C = C' \star_1 f_1 = (C'' \star_2 f_2) \star_1 f_1$ where both f_1 and f_2 are constant equal to 1 and $C'' \subset \mathbb{F}_2^4$ is the code obtained from the above by canceling the first two letters in each code word.
- Conversely, we have $C'' = C' \star_2$ and $C' = C \star_1$, in terms of the second spoiling operation described above.
- To illustrate the third spoiling operation, one can see, for instance, that $C(0,4) = \{\ell_1, \ell_3\}$, while $C(1,6) = \{\ell_2, \ell_3\}$.

2.4. The Asymptotic Bound

The spoiling operations on codes were used in [13] to prove the existence of an *asymptotic bound* in the space of code parameters (δ, R), see also [16,24,25] for more detailed properties of the asymptotic bound.

Let $\mathcal{V}_q \subset [0,1]^2 \cap \mathbb{Q}^2$ denote the space of code parameters (δ, R) of codes $C \subset \mathbb{F}_q^n$ and let $\mathcal{U}_q$ be the set of all limit points of $\mathcal{V}_q$. The set $\mathcal{U}_q$ is characterized in [13] as

$$\mathcal{U}_q = \{(\delta, R) \in [0,1]^2 \mid R \leq \alpha_q(\delta)\}$$

for a continuous, monotonically decreasing function $\alpha_q(\delta)$ (the asymptotic bound). Moreover, code parameters lying in $\mathcal{U}_q$ are realized with infinite multiplicity, while code points in $\mathcal{V}_q \setminus (\mathcal{V}_q \cap \mathcal{U}_q)$ have finite multiplicity and correspond to the *isolated codes*, see [13,16].

Codes lying above the asymptotic bound are codes which have extremely good transmission rate and relative minimum distance, hence very desirable from the coding theory perspective. The fact that the corresponding code parameters are not limit points of other code parameters and only have finite multiplicity reflect the fact that such codes are very difficult to reach or approximate. Isolated codes are known to arise from algebro-geometric constructions, [14,15].

Relatively little is known about the asymptotic bound: the question of the computability of the function $\alpha_q(\delta)$ was recently addressed in [25] and the relation to Kolmogorov complexity was investigated in [16]. There are explicit upper and lower bounds for the function $\alpha_q(\delta)$, see [12], including the Plotkin bound

$$\alpha_q(\delta) = 0, \quad \text{for } \delta \geq \frac{q-1}{q}; \tag{5}$$

the singleton bound, which implies that $R = \alpha_q(\delta)$ lies below the line $R + \delta = 1$; the Hamming bound

$$\alpha_q(\delta) \leq 1 - H_q(\frac{\delta}{2}), \tag{6}$$

where $H_q(x)$ is the q-ary Shannon entropy

$$x \log_q(q-1) - x\log_q(x) - (1-x)\log_q(1-x)$$

which is the usual Shannon entropy for $q = 2$,

$$H_2(x) = -x\log_2(x) - (1-x)\log_2(1-x). \tag{7}$$

One also has a lower bound given by the Gilbert–Varshamov bound

$$\alpha_q(\delta) \geq 1 - H_q(\delta) \tag{8}$$

The Gilbert–Varshamov curve can be characterized in terms of the behavior of sufficiently random codes, in the sense of the Shannon Random Code Ensemble, see [26,27], while the asymptotic bound can be characterized in terms of Kolmogorov complexity, see [16].

2.5. Code Parameters of Language Families

From the coding theory viewpoint, it is natural to ask whether there are codes C, formed out of a choice of a collection of natural languages and their syntactic parameters, whose code parameters lie above the asymptotic bound curve $R = \alpha_2(\delta)$.

For instance, a code C whose code parameters violate the Plotkin bound (5) must be an isolated code above the asymptotic bound. This means constructing a code C with $\delta \geq 1/2$, that is, such that any pair of code words $w \neq w' \in C$ differ by at least half of the parameters. A direct examination of the list of parameters in Table A of [3] and Figure 7 of [4] shows that it is very difficult to find, within the same historical linguistic family (e.g., the Indo-European family) pairs of languages L_1, L_2 with $\delta_H(L_1, L_2) \geq 1/2$. For example, among the syntactic relative distances listed in Figure 7 of [4] one finds only the pair (Farsi, Romanian) with a relative distance of 0.5. Other pairs come close to this value, for example Farsi and French have a relative distance of 0.483, but French and Romanian only differ by 0.162.

One has better chances of obtaining codes above the asymptotic bound if one compares languages that are not so closely related at the historical level.

Example 2. Consider the set $C = \{L_1, L_2, L_3\}$ with languages L_1 = Arabic, L_2 = Wolof, and L_3 = Basque. We exclude from the list of Table A of [3] all those parameters that are entailed and made irrelevant by some other parameter in at least one of these three chosen languages. This gives us a list of 25 remaining parameters, which are those numbered as 1–5, 7, 10, 20–21, 25, 27–29, 31–32, 34, 37, 42, 50–53, 55–57 in [3], and the following three code words:

L_1	1	1	1	1	1	1	0	1	0	1	0	1	0	1	1	1	1	1	1	0	1	0	0	0	0
L_2	1	1	1	0	0	1	1	0	1	0	1	0	0	1	0	1	1	0	0	1	1	1	1	1	1
L_3	1	1	0	1	0	0	1	0	0	0	1	1	1	0	1	1	0	1	1	1	1	1	1	0	0

This example, although very simple and quite artificial in the choice of languages, already suffices to produce a code C that lies above the asymptotic bound. In fact, we have $d_H(L_1, L_2) = 16$, $d_H(L_2, L_3) = 13$ and $d_H(L_1, L_3) = 13$, so that $\delta = 0.52$. Since $R > 0$, the code point (δ, R) violates the Plotkin bound, hence it lies above the asymptotic bound.

It would be more interesting to find a code C consisting of languages belonging to the same historical-linguistic family (outside of the Indo-European group), that lies above the asymptotic bound. Such examples would correspond to linguistic families that exhibit a very strong variability of the syntactic parameters, in a way that is quantifiable through the properties of C as a code.

If one considers the 22 Indo-European languages in [3] with their parameters, one obtains a code C that is below the Gilbert–Varshamov line, hence below the asymptotic bound by Equation (8). A few other examples, taken from other non Indo-European historical-linguistic families, computed using those parameters reported in the SSWL database (for example the set of Malayo–Polynesian languages currently recorded in SSWL) also give codes whose code parameters lie below the Gilbert–Varshamov curve. One can conjecture that any code C constructed out of natural languages belonging to the same historical-linguistic family will be below the asymptotic bound (or perhaps below the GV bound), which would provide a quantitative bound on the possible spread of syntactic parameters within a historical family, given the size of the family. Examples like the simple one constructed above, using languages not belonging to the same historical family show that, to the contrary, across different historical families one encounters a greater variability of syntactic parameters. To our knowledge, no systematic study of parameter variability from this coding theory perspective has been implemented so far.

Ongoing work of the author is considering a systematic analysis of language families, based on the SSWL database of syntactic parameters, using this coding theory technique. This will include an analysis of how much the conclusions about the spreading of syntactic parameters across language families obtained with this technique depends on data pre-processing like the removal of spoiling features and what can be retained as an objective property of a set of languages. Moreover, a further purpose of this ongoing study is to combine the coding theory approach and the measures of complexity for groups of languages described in the present paper with the spin glass dynamical models of language change considered in [8], which was aimed at studying dynamically the spreading of syntactic parameters across groups of languages. The aim is to introduce complexity measures based on coding theory as part of the energy landscape of the spin glass model, following the suggestion of [28], on analogies between the roles of complexity in the theory of computation and energy in physical theories. These results, along with a more detailed analysis of the codes and code parameters of various language families, will appear in forthcoming work.

2.6. Comparison with Other Bounds

Another possible question one can consider in this setting is how the codes obtained from syntactic parameters of a given set of natural languages compare with other known families of error correcting codes and with other bounds in the space of code parameters.

For instance, it is known that an important improvement over the behavior of typical random codes can be obtained by considering codes determined by algebro-geometric curves defined over a finite field $\mathbb{F}_q$. Let $N_q(X) = \#X(\mathbb{F}_q)$ be the number of points over $\mathbb{F}_q$ of the curve X, and let $N_q(g) = \max N_q(X)$, with the maximum taken over all genus g curves X over $\mathbb{F}_q$. As shown in Theorem 2.3.22 of [12], asymptotically the $N_q(g)$ satisfy the Drinfeld–Vladut bound

$$A(q) := \limsup_{q\to\infty} \frac{N_q(g)}{g} \leq \sqrt{q} - 1,$$

and as shown in Section 3.4.1 of [12], this determines an algebro-geometric bound

$$\alpha_q(\delta) \geq R_{AG}(\delta) = 1 - \frac{1}{A(q)} - \delta$$

and the asymptotic Tsfasman–Vladut–Zink bound

$$\alpha_q(\delta) \geq R_{TVZ}(\delta) = 1 - (\sqrt{q} - 1)^{-1} - \delta.$$

The Tsfasman–Vladut–Zink line $R_{TVZ}(\delta) = 1 - (\sqrt{q} - 1)^{-1} - \delta$ lies entirely below the GV line for $q < 49$ (Theorem 3.4.4 of [12]).

A probabilistic argument given in Section 3.4.2 of [12] shows that highly non-random codes coming from algebraic curves can be asymptotically better than random codes (for sufficiently large q) as they cluster around the TVZ line. However, for $q = 2$ or $q = 3$, as in the case of codes from syntactic parameters of groups of languages that we consider here, the TVZ line lies below the GV line, hence any example that lies above the GV bound also behaves better than the the algebro-geometric bound. Such examples, like the one given above, for the three languages Arabic, Wolof, Basque, are very rare among codes obtained from syntactic parameters of languages, as they require the choice of a group of languages that are all very far from each other syntactically, with very large relative Hamming distances between syntactic parameters.

On the other hand, even for cases of groups of languages for which the resulting code parameters are below the GV line, it is still possible to get some additional information by comparing the position of the code parameters to other curves obtained from other bounds, such as the Blokh–Zyablow bound or the Katsman–Tsfasman–Vladut bound, see Appendix A.2.1 of [12] for a summary of all these different bounds.

For example, the first example given above, with the three languages Italian, Spanish, French and a string of six syntactic parameters, gives a code with code parameters that are below the GV line, but above both the Blokh–Zyablow and the Katsman–Tsfasman–Vladut, according to the table of asymptotic bounds given in Appendix A.2.4 of [12].

2.7. Entailment and Dependency of Parameters

In the discussion above we did not incorporate in our model the fact that certain syntactic parameters can entail other parameters in such a way that one particular value of one of the parameters renders another parameter irrelevant or not defined, see the discussion in Section 5.3.2 of [3].

One possible way to alter the previous construction to account for these phenomena is to consider the codes C associated to families of languages as codes in $\mathbb{F}_3^n$, where n is the number of parameters, as before, and the set of values is now given by $\{-1, 0, +1\} = \mathbb{F}_3$, with ± 1 corresponding to the binary values of the parameters that are set for a given language and value 0 assigned to those parameters that are made irrelevant for the given language, by entailment from other parameters, or are not defined. This allows us to consider the full range of parameters used in [3,4]. We revisit Example 2 considered above.

Example 3. Let $C = \{L_1, L_2, L_3\}$ be the code obtained from the languages L_1 = Arabic, L_2 = Wolof, and L_3 = Basque, as a code in $\mathbb{F}_3^n$ with $n = 63$, using the entire list of parameters in [3]. The code parameters $(R = 0.0252, \delta = 0.4643)$ of this code no longer violate the Plotkin bound. In fact, the parameters satisfy $R < 1 - H_3(\delta)$ so the code C now also lies below the GV bound.

Thus, the effect of including the entailed syntactic parameters in the comparison spoils the code parameters enough that they fall in the area below the GV bound.

Notice that what we propose here is different from the counting used in [3], where the relative distances $\delta_H(L_1, L_2)$ are normalized with respect to the number of non-zero parameters (which therefore varies with the choice of the pair (L_1, L_2)) rather than the total number n of parameters. While this has the desired effect of getting rid of insignificant parameters that spoil the code, it has the undesirable property of producing codes with code words of varying lengths, while counting only those parameters that have no zero-values over the entire family of languages, as in Example 2 avoids this problem. Adapting the coding theory results about the asymptotic bound to codes with words of variable length may be desirable for other reasons as well, but it will require an investigation beyond the scope of the present paper.

More generally, there are various kinds of dependencies among syntactic parameters. Some sets of hierarchical relations are discussed, for instance, in [29].

By the spoiling operations $C \star_i f$ of codes described above, we know that if some of the syntactic parameters considered are functions of other parameters, the resulting code parameters of $C \star_i f$ are worse than the parameters of the code C where only independent parameters were considered.

Part of the reason why code parameters of groups of languages in the family analyzed in [3] end up in the region below the asymptotic and the GV bound may be an artifact of the presence of dependences among the chosen 63 syntactic parameters. From the coding theory perspective, the parametric comparison method works best on a smaller set of independent parameters than on a larger set that includes several dependencies.

Entailment relations between syntactic parameters play an important role in the dynamical models of language evolutions constructed in [8], based on spin glass models in statistical physics.

Notice that the type of entailment relations we consider here are only of a rather special form, where a parameter is made undefined by effect of the value of another parameter (hence the use of the value 0 for the undetermined parameter). There are more general forms of entailment that we do not consider here, but which will be discussed in more detail in upcoming work. For example, one can have a situation with two languages in which a parameter is entailed by the values of two other parameters, but entailed to two different values in the two languages. In this case, the proposal above need to be modified, because this entailed parameter should contribute to the Hamming distance between the two languages. In such a situation the entailed parameter should increase, rather than spoil, the efficiency of the code. Keeping entailed parameters can be used for error-correcting purposes, as contributing to error detection. The role of entailment of parameters was considered in [8], in the use of spin glass models for language change, where the entailment relations appear as couplings at the vertices (interaction terms) between different Ising/Potts models on the same underlying graph of language interactions. In upcoming work, now in preparation, we will discuss how treating different forms of entailment of parameters in the coding theory setting described here related to the treatment of entailment relations in the spin glass model of [8].

3. Entropy and Complexity for Language Families

3.1. *Why the Asymptotic Bound?*

In the examples discussed above we compared the position of the code point associated to a given set of languages to certain curves in the space of code parameters. In particular, we focused on the asymptotic bound curve and the Gilbert–Varshamov curve. It should be pointed out that these two curves have a very different nature.

The asymptotic bound is the only curve that separates regions in the space of parameters that correspond to code points with entirely different behavior. As shown in [13,24], code points in the area below the asymptotic bound are realized with infinite multiplicity and fill densely the region, while code points that lie above the asymptotic bound are isolated and realized with finite multiplicity.

The Gilbert–Varshamov curve, by contrast, is related to the statistical behavior of sufficiently random codes (as we recall in Section 3.2 below), but does not separate two regions with significantly different behavior in the space of code points. Thus, in this respect, the asymptotic bound is a more natural curve to consider than the Gilbert–Varshamov curve.

Thus, a heuristic interpretation of the position of codes obtained from groups of languages, with respect to the asymptotic bound can be understood as follows. The position of a code point above or below the asymptotic bound reflects a very different behavior of the corresponding code with respect to how easily "deformable" it is. The sporadic codes that lie above the asymptotic bound are rigid objects, in contrast to the deformable objects below the asymptotic bound. In terms of properties of the distribution of syntactic parameters within a set of languages, this different nature of the associated code can be seen as a measure of the degree of "deformability" of the parameter distribution: in languages that belong to the same historical linguistic families, the parameter distribution has evolved historically along with the development of the family's phylogenetic tree, and one expects that correspondingly the code parameters will indicate a higher degree of "deformability" of the corresponding code. If a group of languages is chosen that belong to very different historical families, on the contrary, one expects that the distribution of syntactic parameters will not necessarily lead any longer to a code that has the same kind of deformability property: code points above the asymptotic bound may be realizable by this type of language groups.

There is no similar interpretation for the position of the code point with respect to the Gilbert–Varshamov line. An interpretation of that position can be sought in terms of Shannon entropy, as we discuss below. Summarizing: the main conceptual distinction between the Gilbert–Varshamov line and the asymptotic bound is that the GV line represents only a statistical phenomenon, as we review below, while the asymptotic bound represents a true separation between two classes of structurally different codes, in the sense explained above.

3.2. Entropy and Statistics of the Gilbert–Varshamov Line

The Gilbert–Varshamov line $R = 1 - H_q(\delta)$ can be characterized statisticallly. Such a statistical description can be obtained by considering the Shannon Random Code Ensemble (SRCE). These are random codes obtained by choosing code words as independent random variables with respect to a uniform Bernoulli measure, so that a code is described by a randomly chosen set of different words of length n occurring with probability q^{-n}, see [26,27]. There is no a priori reason why the type of codes we consider here, with code words formed using the syntactic parameters of natural languages, would be linear. Thus, we consider the general setting of unstructured codes, as in Section V of [27].

The Hamming volume $Vol_q(n, d = n\delta)$, that is, the number of words of length n at Hamming distance at most d from a given one, can be estimated in terms of the q-ary Shannon entropy

$$H_q(\delta) = \delta \log_q(q-1) - \delta \log_q \delta - (1-\delta) \log_q(1-\delta)$$

in the form

$$q^{(H_q(\delta)-o(1))n} \leq Vol_q(n, d = n\delta) = \sum_{j=0}^{d} \binom{n}{j} (q-1)^j \leq q^{H_q(\delta)n}.$$

The expectation value for the random variable counting the number of unordered pairs of distinct code words with Hamming distance at most d is then estimated as

$$\mathbb{E} \sim \binom{q^k}{2} Vol_q(n,d) q^{-n} \sim q^{n(H_q(\delta)-1+2R)+o(n)}.$$

This estimate is then used (see [26,27]) to show that the probability to have codes in the SRCE with $H_q(\delta) \geq \max\{1-2R, 0\} + \epsilon$ is bounded by $q^{-\epsilon n(1+o(1))}$. By a similar argument (see Section V of [27] and Proposition 2.2 of [16]) it is shown that the probability that $H_q(\delta) \geq 1 - R + \epsilon$ is bounded by $q^{-n\epsilon(1+o(1))}$.

While, by this type of argument, one can see the Gilbert–Varshamov line as representing the typical behavior of sufficiently random codes, the asymptotic bound does not have a similar statistical interpretation. It does have, however, a relation to Kolmogorov complexity, which is relevant to the point of view discussed in the present paper. The relation between asymptotic bound of error correcting codes and Kolmogorov complexity was described in [16]. We recall it in the rest of this section, along with its implications for the linguistic applications we are considering.

3.3. Kolmogorov Complexity

We refer the reader to [30] for an extensive treatment of Kolmogorov complexity and its properties. We recall here some basic facts we need in the following.

Let $T_\mathcal{U}$ be a universal Turing machine, that is, a Turing machine that can simulate any other arbitrary Turing machine, by reading on tape both the input and the description of the Turing machine it should simulate. A prefix Turing machine is a Turing machine with unidirectional input and output tapes and bidirectional work tapes. The set of programs P on which a prefix Turing machine halts forms a prefix code.

Given a string w in an alphabet $\mathfrak{A}$, the prefix Kolmogorov complexity is given by minimal length of a program for which the universal prefix Turing machine $T_\mathcal{U}$ outputs w,

$$\mathcal{K}_{T_\mathcal{U}}(w) = \min_{P: T_\mathcal{U}(P) = w} \ell(P).$$

There is a universality property. Namely, given any other prefix Turing machine T, one has

$$\mathcal{K}_T(w) \leq \mathcal{K}_{T_\mathcal{U}}(w) + c_T,$$

where the shift is by a bounded constant, independent of w. The constant c_T is the Kolmogorov complexity of the program needed to describe T so that $T_{\mathcal{U}}$ can simulate it.

A variant of the notion of Kolmogorov complexity described above is given by conditional Kolmogorov complexity,

$$\mathcal{K}_{T_{\mathcal{U}}}(w \mid \ell(w)) = \min_{P:T_{\mathcal{U}}(P,\ell(w))=w} \ell(P),$$

where the length $\ell(w)$ is given, and made available to the machine $T_{\mathcal{U}}$. One then has

$$\mathcal{K}(w \mid \ell(w)) \leq \ell(w) + c,$$

because if $\ell(w)$ is known, then a possible program is just to write out w. This means that then $\ell(w) + c$ is just number of bits needed for the transmission of w plus the print instructions.

An upper bound is given by

$$\mathcal{K}_{T_{\mathcal{U}}}(w) \leq \mathcal{K}_{T_{\mathcal{U}}}(w \mid \ell(w)) + 2\log \ell(w) + c.$$

If one does not know a priori $\ell(w)$, one needs to signal the end of the description of w. For this it suffices to have a "punctuation method", and one can see that this has the effect of adds the term $2\log \ell(w)$ in the above estimate. In particular, any program that produces a description of w is an upper bound on Kolmogorov complexity $\mathcal{K}_{T_{\mathcal{U}}}(w)$.

One can think of Kolmogorov complexity in terms of data compression: the shortest description of w is also its most compressed form. Upper bounds for Kolmogorov complexity are therefore provided easily by data compression algorithms. However, while providing upper bounds for complexity is straightforward, the situation with lower bounds is entirely different: constructing a lower bound runs into a fundamental obstacle caused by the fact that the halting problem is unsolvable. As a consequence, Kolmogorov complexity is not a computable function. Indeed, suppose one would list programs P_k (with increasing lengths) and run them through the machine $T_{\mathcal{U}}$. If the machine halts on P_k with output w, then $\ell(P_k)$ is an approximation to $\mathcal{K}_{T_{\mathcal{U}}}(w)$. However, there may be an earlier P_j in the list such that $T_{\mathcal{U}}$ has not yet halted on P_j. If $T_{\mathcal{U}}$ eventually halts also on P_j and outputs w, then $\ell(P_j)$ will be a better approximation to $\mathcal{K}_{T_{\mathcal{U}}}(w)$. So one would be able to compute $\mathcal{K}_{T_{\mathcal{U}}}(w)$ if one could tell exactly on which programs P_k the machine $T_{\mathcal{U}}$ halts, but that is indeed the unsolvable halting problem.

Kolmogorov complexity and Shannon entropy are related: one can view Shannon entropy as an averaged version of Kolmogorov complexity in the following sense (see Section 2.3 of [31]). Suppose given independent random variables X_k, distributed according to Bernoulli measure $\mathbb{P} = \{p_a\}_{a\in\mathfrak{A}}$ with $p_a = \mathbb{P}(X = a)$. The Shannon entropy is given by

$$S(X) = -\sum_{a\in\mathfrak{A}} \mathbb{P}(X = a) \log \mathbb{P}(X = a).$$

There exists a $c > 0$, such that, for all $n \in \mathbb{N}$,

$$S(X) \leq \frac{1}{n} \sum_{w\in\mathcal{W}^n} \mathbb{P}(w)\, \mathcal{K}(w \mid \ell(w)) \leq S(X) + \frac{\#\mathfrak{A} \log n}{n} + \frac{c}{n}.$$

The expectation value

$$\lim_{n\to\infty} \mathbb{E}(\frac{1}{n}\mathcal{K}(X_1 \cdots X_n \mid n)) = S(X)$$

shows that the average expected Kolmogorov complexity for length n descriptions approaches the Shannon entropy in the limit when $n \to \infty$.

3.4. Kolmogorov Complexity and the Asymptotic Bound

We recall here briefly the result of [16] linking the asymptotic bound of error correcting codes to Kolmogorov complexity.

As we discussed above, only the asymptotic bound marks a significant change of behavior of codes across the curve (isolated code points with finite multiplicity versus accumulation points with infinite multiplicity). In this sense this curve is very different from all the other bounds in the space of code parameters. However, there is no explicit expression for the curve $R = \alpha_q(\delta)$ that gives the asymptotic bound. Indeed, even the question of the computability of the function $R = \alpha_q(\delta)$ is a priori unclear. This question was formulated explicitly in [25].

It is proved in [16] that the asymptotic bound $R = \alpha_q(\delta)$ becomes computable given an oracle that can list codes by increasing Kolmogorov complexity. Given such an oracle, one can provide an explicit iterative (algorithmic) procedure for constructing the asymptotic bound. This implies that the asymptotic bound is "at worst as non-computable as Kolmogorov complexity".

Consider the set $X = \mathcal{C}_q$ of (unstructured) q-ary codes and the set $Y \subset [0,1]^2$ of code points and the computable function $f : X \to Y$ that assigns to a code $C \in X$ its code parameters $(R(C), \delta(C)) \in Y$. Let Y_{fin} and Y_∞ be, respectively, the subsets of the space of code points that correspond to code points realized with finite and with infinite multiplicity. The algorithm iteratively produces two sets A_m and B_m that approximate, respectively, Y_∞ and Y_{fin} by $Y_{fin} = \cup_{m\geq 1} B_m$ and $Y_\infty = \cup_{m\geq 1}(\cap_{n\geq 0} A_{m+n})$. The inductive construction starts by choosing an increasing sequence of positive integers N_m and setting $B_1 = \emptyset$ and taking A_1 to be the set of code points y with $\nu_Y^{-1}(y) \leq N_1$, where $\nu_Y : \mathbb{N} \to Y$ is a fixed enumeration of the set of rational points $[0,1]^2 \cap \mathbb{Q}^2$ where code points belong.

General estimates on the behavior of (exponential) Kolmogorov complexity under composition of total recursive functions (see [30], Section VI.9 of [32]) show that, for a composition $F = f_0(f_1(t_1,\dots,t_m),\cdots,f_n(t_1,\dots,t_m),t_{m+1},\dots,t_\ell)$ of recursive functions the Kolmogorov complexity satisfies

$$\mathcal{K}(F) \leq c \cdot \prod_{i=1}^{n} \mathcal{K}(f_i) \cdot \left(\log \prod_{i=1}^{n} \mathcal{K}(f_i)\right)^{n-1},$$

for a fixed f_0 and varying f_i, $i \geq 1$.

Consider the total recursive function $F(x) = (f(x), n(x))$ with

$$n(x) = \#\{x' \mid \nu_X^{-1}(x') \leq \nu_X^{-1}(x),\ f(x') = f(x)\}$$

where $\nu_X : \mathbb{N} \to X$ is an enumeration of the space of codes. Consider the enumerable sets $X_m := \{x \in X \mid n(x) = m\}$ and $Y_m := f(X_m) \subset Y$, with $Y_\infty = \cap_m f(X_m)$ and $Y_{fin} = f(X) \smallsetminus Y_\infty$. For $\varphi : f(X) \to X_1$, defined as f^{-1} on $f(X_1) = f(X)$, applying the composition rule for exponential Kolmogorov complexity, it is shown in Proposition 3.1 of [16] that, for $x \in X_1$ and $y = f(x)$, one has $\mathcal{K}(x) = \mathcal{K}(\varphi(y)) \leq c_\varphi \cdot \mathcal{K}(y) \leq c\nu_Y^{-1}(y)$, hence

$$\mathcal{K}_{T_\mathcal{U}}(x) \leq c \cdot \nu_Y^{-1}(y).$$

Using the same type of estimate of Kolmogorov complexity for composition of recursive functions, it is then shown in Proposition 3.2 [16] that, for $y \in Y_\infty$ and $m \geq 1$, and for a unique $x_m \in X$, with $y = f(x_m)$, $n(x_m) = m$ and $c = c(f, u, v, \nu_X, \nu_Y) > 0$, one finds

$$\mathcal{K}_{T_\mathcal{U}}(x_m) \leq c \cdot \nu_Y^{-1}(y)\, m\, \log(\nu_Y^{-1}(y) m).$$

To construct inductively A_{m+1} and B_{m+1}, given A_m and B_m, one takes A_{m+1} to consist of the elements in the list

$$\mathcal{L}_{m+1} = \{y \in f(X) : \nu_Y^{-1}(y) \leq N_{m+1}, \ \exists x \in X, \text{ with } y = f(x) \text{ and } n(x) = m+1\}.$$

Here one invokes the oracle, which ensures that, if such x exists, then it must be contained in a finite list of points $x \in X$ with bounded complexity

$$\mathcal{K}_{T_{\mathcal{U}}}(x_m) \leq c \cdot \nu_Y^{-1}(y)\, m \, \log(\nu_Y^{-1}(y) m).$$

One then takes B_{m+1} to consist of the remaining elements in the list $\mathcal{L}_{m+1}$. We refer the reader to [16] for a more detailed formulation.

More generally, the argument of [16] recalled above shows that, for a recursive function $f : \mathbb{Z}_+ \to \mathbb{Q}$, determining which values have infinite multiplicities is computable given an oracle that enumerate integers in order of Kolmogorov complexity.

As discussed in [16,24], the asymptotic bound can also be seen as the phase transition curve for a quantum statistical mechanical system constructed out of the space of codes, where the partition function of the system weights codes according to their Kolmogorov complexity. This is as close to a "statistical description" of the asymptotic bound that one can achieve.

In comparison with the behavior of random codes (codes whose complexity is comparable to their size), which concentrate in the region bounded by the Gilbert–Varshamov line, when ordering codes by complexity, non-random codes of lower complexity populate the region above, with code points accumulating in the intermediate region bounded by the asymptotic bound. That intermediate region thus, in a sense, reflects the difference between Shannon entropy and complexity.

3.5. Entropy and Complexity Estimates for Language Families

On the basis of the considerations of the previous sections and of the results of [16,24] recalled above, we propose a way to assign a quantitative estimate of entropy and complexity to a given set of natural languages.

As before let $C = \{L_1, \ldots, L_k\}$ be a binary (or ternary) code where the code words L_i are the binary (ternary) strings of syntactic parameters of a set of languages L_i. We define the *entropy* of the language family $\{L_1, \ldots, L_k\}$ as the q-ary Shannon entropy $H_q(\delta(C))$, where q is either 2 or 3 for binary or ternary codes, and $\delta(C)$ is the relative minimum distance parameter of the code C. We also define the *entropy gap* of the language family $\{L_1, \ldots, L_k\}$ as the value of $H_q(\delta(C)) - 1 + R(C)$, which measures the distance of the code point $(R(C), \delta(C))$ from the Gilbert–Varshamov line, that is, from the behavior of a typical random code.

As a source of estimates of complexity of a language family $\{L_1, \ldots, L_k\}$ one can consider any upper bound on Kolmogorov complexity of the code C. A possible approach, which contains more linguistic input, would be to provide estimates of complexity for each individual language in the family and then compare these. Estimates of complexity for individual languages have been considered in the literature, some of them based on the description of languages in terms of their syntactic parameters. For instance, following [18], for a syntactic parameter Π with possible values $v \in \{\pm 1\}$, the Kolmogorov complexity of Π set to value v is given by

$$\mathcal{K}(\Pi = v) = \min_{\tau \text{ expressing } \Pi} \mathcal{K}_{T_{\mathcal{U}}}(\tau),$$

with the minimum taken over the complexities of all the parse trees that express the syntactic parameter Π and require $\Pi = v$ to be grammatical in the language. Notice that, in this approach, the syntactic parameters are not just regarded as binary or ternary values, but one needs to consider actual parse trees of sentences in the language that express the parameter. Thus, such an approach to complexity has the advantage that it is very rich in linguistic information. However, it is at the same time computationally very difficult to realize.

What we are proposing here is a much simpler way to obtain an estimate of complexity for a language family $\{L_1, \ldots, L_k\}$, which is not based on estimating complexity of the individual languages in the family, but which is aimed at detecting how spread out and diversified the syntactic parameters are across the family, by estimating the position of the code point $(R(C), \delta(C))$ of the associated code C with respect to the asymptotic bound $R = \alpha_q(\delta)$. This can be estimated in terms of the distance to other curves in the space of code parameters (R, δ) that constrain the asymptotic bound from above and below, such as the Plotkin bound, Hamming bound, and Gilbert–Varshamov bound, as in the examples discussed in the previous sections.

4. Conclusions

We proposed an approach to estimating entropy and complexity of groups of natural languages (language families), based on the linguistic parametric comparison method (PCM) of [2,22] via the mathematical theory of error-correcting codes, by assigning a code to a family of languages to be analyzed with the PCM, and investigating its position in the space of code parameters, with respect to the asymptotic bound and the GV bound. We have shown that there are examples of languages not belonging to the same historical-linguistic family that yield isolated codes above the asymptotic bound, while languages belonging to the same historical-linguistic family appear to give rise to codes below the bound, though a more systematic analysis would be needed to map code parameters of different language groups. We have also shown that, from these coding theory perspective, it is preferable to exclude from the PCM all those parameters that are entailed and made irrelevant by other parameters, as those spoil the properties of the resulting code and produce code parameters that are artificially low with respect to the asymptotic bound and the GV bound.

The approach proposed here, based on the PCM and the theory of error-correcting codes, suggests a few new linguistic questions that may be suitable for treatment with coding theory methods:

1. Do languages belonging to the same historical-linguistic family always yield codes below the asymptotic bound or the GV bound? How often does the same happen across different linguistic families? How much can code parameters be improved by eliminating spoiling effects caused by dependencies and entailment of syntactic parameters?
2. Codes near the GV curve are typically coming from the Shannon Random Code Ensemble, where code words and letters of code words behave like independent random variables, see [26,27]. Are there families of languages whose associated codes are located near the GV bound? Do their syntactic parameters mimic the uniform Poisson distribution of random codes?
3. The asymptotic bound for error-correcting codes was related in [16] to Kolmogorov complexity, and the measure of complexity for language families that we proposed here is estimated in terms of the position of the code point with respect to the asymptotic bound. There are other notions of complexity, most notably the type of organized complexities discussed in [33–35]. Can these be related to loci in the space of code parameters? What do these represent when applied to codes obtained from syntactic parameters of a set of natural languages?
4. Is there a more direct linguistic complexity measure associated to a family of natural languages that would relate to the position of the resulting code above or below the asymptotic bound?
5. Codes and the asymptotic bound in the space of code parameters were recently studied using methods from quantum statistical mechanics, operator algebra and fractal geometry, [24,36]. Can some of these mathematical methods be employed in the linguistic parametric comparison method?

The observational results reported here are still preliminary. The following topics should be consolidated:

- How much the conclusions obtained for a given family of languages will depend on data pre-processing (removal of "spoiling" features, *etc.*)
- To what extent the proposed criterion (above or below the asymptotic bound) is really an objective property of a set of languages.

This will be addressed more thoroughly in future work. The concern about the effect of data pre-processing in paticular requires more analysis, that will be developed in further ongoing work, as outlined at the end of Section 2.5.

Acknowledgments: The author's research is supported by NSF grants DMS-1201512 and PHY-1205440, and by the Perimeter Institute for Theoretical Physics. The author thanks the referees for their useful comments.

Conflicts of Interest: The author declares no conflict of interest.

References

1. Chomsky, N. *Lectures on Government and Binding*; Foris: Dordrecht, The Netherlands, 1981.
2. Longobardi, G. Methods in parametric linguistics and cognitive history. *Linguist. Var. Yearb.* **2003**, *3*, 101–138.
3. Longobardi, G.; Guardiano, C. Evidence for syntax as a signal of historical relatedness. *Lingua* **2009**, *119*, 1679–1706.
4. Longobardi, G.; Guardiano, C.; Silvestri, G.; Boattini, A.; Ceolin, A. Toward a syntactic phylogeny of modern Indo-European languages. *J. Hist. Linguist.* **2013**, *3*, 122–152.
5. Aziz, S.; Huynh, V.L.; Warrick, D.; Marcolli, M. Syntactic Phylogenetic Trees. 2016, In Preparation.
6. Park, J.J.; Boettcher, R.; Zhao, A.; Mun, A.; Yuh, K.; Kumar, V.; Marcolli, M. Prevalence and recoverability of syntactic parameters in sparse distributed memories. 2015, arXiv:1510.06342.
7. Port, A.; Gheorghita, I.; Guth, D.; Clark, J.M.; Liang, C.; Dasu, S.; Marcolli, M. Persistent Topology of Syntax. 2015, arXiv:1507.05134.
8. Siva, K.; Tao, J.; Marcolli, M. Spin Glass Models of Syntax and Language Evolution. 2015, arXiv:1508.00504.
9. Syntactic Structures of the World's Languages (SSWL) Database of Syntactic Parameters. Available online: http://sswl.railsplayground.net (accessed on 18 March 2016).
10. TerraLing. Available online: http://www.terraling.com (accessed on 18 March 2016).
11. Haspelmath, M.; Dryer, M.S.; Gil, D.; Comrie, B. *The World Atlas of Language Structures*; Oxford University Press: Oxford, UK, 2005.
12. Tsfasman, M.A.; Vladut, S.G. Algebraic-Geometric Codes. In *Mathematics and Its Applications (Soviet Series)*; Springer: Amsterdam, the Netherlands, 1991; Volume 58.
13. Manin, Y.I. What is the maximum number of points on a curve over $\mathbb{F}_2$? *J. Fac. Sci. Univ. Tokyo Sect. 1A Math.* **1982**, *28*, 715–720.
14. Tsfasman, M.A.; Vladut, S.G.; Zink, T. Modular curves, Shimura curves, and Goppa codes, better than Varshamov–Gilbert bound. *Math. Nachr.* **1982**, *109*, 21–28.
15. Vladut, S.G.; Drinfel'd, V.G. Number of points of an algebraic curve. *Funct. Anal. Appl.* **1983**, *17*, 68–69.
16. Manin, Y.I.; Marcolli, M. Kolmogorov complexity and the asymptotic bound for error-correcting codes. *J. Differ. Geom.* **2014**, *97*, 91–108.
17. Bane, M. Quantifying and measuring morphological complexity. In Proceedings of the 26th West Coast Conference on Formal Linguistics, Berkeley, CA, USA, 27–29 April 2007.
18. Clark, R. *Kolmogorov Complexity and the Information Content of Parameters*; Institute for Research in Cognitive Science: Philadelphia, PA, USA, 1994.
19. Tuza, Z. On the context-free production complexity of finite languages. *Discret. Appl. Math.* **1987**, *18*, 293–304.
20. Barton, G.E.; Berwick, R.C.; Ristad, E.S. *Computational Complexity and Natural Language*; MIT Press: Cambrige, MA, USA, 1987.
21. Sampson, G.; Gil, D.; Trudgill, P. (Eds.) *Language Complexity as an Evolving Variable*; Oxford University Press: Oxford, UK, 2009.
22. Longobardi, G. A minimalist program for parametric linguistics? In *Organizing Grammar: Linguistic Studies in Honor of Henk van Riemsdijk*; Broekhuis, H.; Corver, N.; Huybregts, M.; Kleinhenz, U.; Koster, J., Eds.; Mouton de Gruyter: Berlin, Germany, 2005; pp. 407–414.

23. Clark, R.; Roberts, I. A computational model of language learnability and language change. *Linguist. Inq.* **1993**, *24*, 299–345.
24. Manin, Y.I.; Marcolli, M. Error-correcting codes and phase transitions. *Math. Comput. Sci.* **2001**, *5*, 133–170.
25. Manin, Y.I. A computability challenge: Asymptotic bounds and isolated error-correcting codes. 2011, arXiv:1107.4246.
26. Barg, A.; Forney, G.D. Random codes: minimum distances and error exponents. *IEEE Trans. Inf. Theory* **2002**, *48*, 2568–2573.
27. Coffey, J.T.; Goodman, R.M. Any code of which we cannot think is good. *IEEE Trans. Inf. Theory* **1990**, *36*, 1453–1461.
28. Manin, Y.I. Complexity vs Energy: Theory of Computation and Theoretical Physics. 2014, arXiv:1302.6695.
29. Baker, M.C. *The Atoms of Language: The Mind's Hidden Rules of Grammar*; Basic Books: New York, NY, USA, 2001.
30. Li, M.; Vitányi, P. *An Introduction to Kolmogorov Complexity and Its Applications*; Springer: New York, NY, USA, 2008.
31. Grünwald, P.; Vitányi, P. Shannon Information and Kolmogorov Complexity. 2004, arXiv:cs/0410002.
32. Manin, Y.I. *A Course in Mathematical Logic for Mathematicians*, 2nd ed; Springer: New York, NY, USA, 2010.
33. Bennett, C.; Gacs, P.; Li, M.; Vitanyi, P.; Zurek, W. Information distance. *IEEE Trans. Inf. Theory* **1998**, *44*, 1407–1423.
34. Delahaye, J.P. *Complexité Aléatoire et Complexité Organisée*; Éditions Quæ: Paris, France, 2009. (In French)
35. Gell-Mann, M.; Lloyd, S. Information measures, effective complexity, and total information. *Complexity* **1996**, *2*, 44–52.
36. Marcolli, M.; Perez, C. Codes as fractals and noncommutative spaces. *Math. Comput. Sci.* **2012**, *6*, 199–215.

MDPI AG
St. Alban-Anlage 66
4052 Basel, Switzerland
Tel. +41 61 683 77 34
Fax +41 61 302 89 18
http://www.mdpi.com

Entropy Editorial Office
E-mail: entropy@mdpi.com
http://www.mdpi.com/journal/entropy

www.ingramcontent.com/pod-product-compliance
Lightning Source LLC
LaVergne TN
LVHW071356070326
832902LV00028B/4625
9783038424246